ADVANCES IN IMAGING AND ELECTRON PHYSICS

THE GROWTH OF ELECTRON MICROSCOPY

VOLUME 96

EDITOR-IN-CHIEF
PETER W. HAWKES
*CEMES/Laboratoire d'Optique Electronique
du Centre National de la Recherche Scientifique
Toulouse, France*

ASSOCIATE EDITORS
BENJAMIN KAZAN
*Xerox Corporation
Palo Alto Research Center
Palo Alto, California*

TOM MULVEY
*Department of Electronic Engineering and Applied Physics
Aston University
Birmingham, United Kingdom*

Advances in
Imaging and
Electron Physics

The Growth of
Electron Microscopy

EDITED BY
TOM MULVEY

*Department of Electronic Engineering
and Applied Physics
Aston University
Birmingham, United Kingdom*

VOLUME 96

ACADEMIC PRESS
San Diego   New York   Boston
London   Sydney   Tokyo   Toronto

This book is printed on acid-free paper. ∞

Copyright © 1996 by ACADEMIC PRESS, INC.

All Rights Reserved.
No part of this publication may be reproduced or transmitted in any form or by any means, electronic or mechanical, including photocopy, recording, or any information storage and retrieval system, without permission in writing from the publisher.

Academic Press, Inc.
A Division of Harcourt Brace & Company
525 B Street, Suite 1900, San Diego, California 92101-4495

United Kingdom Edition published by
Academic Press Limited
24-28 Oval Road, London NW1 7DX

International Standard Serial Number: 1076-5670

International Standard Book Number: 0-12-014738-6

PRINTED IN THE UNITED STATES OF AMERICA
96 97 98 99 00 01 BB 9 8 7 6 5 4 3 2 1

CONTENTS

CONTRIBUTORS	ix
PREFACE	xiii
PROLOGUE	xix

PART I IFSEM

1.1	Early History of the International Federation of Societies for Electron Microscopy	3
	VERNON E. COSSLETT	
1.2	IFSEM 1995: Objectives, Organization, and Functions . .	21
	ARVID B. MAUNSBACH AND GARETH THOMAS	

PART II SOME INDIVIDUAL SOCIETIES

2.1	Electron Microscopy in Australia	39
	PETER S. TURNER AND MAREK VESK	
2.2A	The Austrian Society for Electron Microscopy	55
	ERWIN M. HÖRL	
2.2B	Reminiscences of Walter Glaser	59
	HANS GRÜMM AND PETER SCHISKE	
2.3	Electron Microscopy in Belgium	67
	DIRK VAN DYCK	
2.4	The History of Electron Microscopy in Canada	79
	GÉRARD T. SIMON AND FRANCES W. DOANE	
2.5	Electron Microscopy in France	93
2.5A	Early Findings in the Life Sciences	93
	FRANÇOISE HAGUENAU	
2.5B	The Development for Physics and Materials Sciences . . .	101
	BERNARD JOUFFREY	
2.6A	The Early History of Electron Microscopy in Germany . .	131
	HEINZ NIEDRIG	
2.6B	The History of the German Society for Electron Microscopy	149
	GERHARD SCHIMMEL	

2.6C	Electron Microscopy in the Former German Democratic Republic	171
	JOHANNES HEYDENREICH, HANS LUPPA, ALFRED RECKNAGEL, AND DANKWART STILLER	
2.7	The Hungarian Group for Electron Microscopy	181
	SZABOLCS VIRÁGH AND ÁGNES CSANÁDY	
2.8	Electron Microscopy in Italy	193
	UGO VALDRÈ	
2.9	The Growth of Electron Microscopy in Japan	217
2.9A	Introduction	217
	KEIJI YADA	
2.9B	The 37th Subcommittee of the Japanese Society for the Promotion of Science, 1939–1947	227
	BUNYA TADANO	
2.9C	History of Electron Microscopes at Tohoku University	245
	KEIJI YADA	
2.9D	Development of Electron Microscopes at Tokyo Imperial University	251
	AKIRA FUKAMI, KOICHI ADACHI, AND KENTARO ASAKURA	
2.9E	Development of Electron Microscopes at the Electrotechnical Laboratory	257
	KOICHI KANAYA	
2.9F	Early Electron Microscopes at Osaka University, 1934–1945	263
	KATSUMI URA	
2.10	Electron Microscopy in the Netherlands	271
2.10A	Earliest Developments	271
	WOUTERA VAN ITERSON	
2.10B	Developments since the 1950s	287
	PIETER KRUIT, FRITS W. SCHAPINK, JOHN W. GEUS, ARIE J. VERKLEIJ, AND CAESAR E. HULSTAERT	
2.11	The Development of Electron Microscopy in Scandinavia	301
	ARVID B. MAUNSBACH AND BJÖRN A. AFZELIUS	
2.12	Electron Microscopy in Southern Africa	323
	MICHAEL J. WITCOMB AND SONIA A. WOLFE-COOTE	
2.13	Highlights in the Development of Electron Microscopy in the United States: A Bibliography and Commentary of Published Accounts and EMSA Records	347
	ROBERT M. FISHER	

Part III HIGHLIGHTS OF THE IFSEM CONGRESSES

3.1 Biology 385
     BJÖRN A. AFZELIUS
3.2 Materials Science. 393
     JOHN L. HUTCHISON
3.3 Electron Optics. 405
     PETER W. HAWKES

Part IV INSTRUMENTAL DEVELOPMENTS

4.1 The Story of European Commercial Electron
     Microscopes 415
     ALAN W. AGAR
4.2 My Early Work on Convergent-Beam Electron
     Diffraction. 585
     GOTTFRIED MÖLLENSTEDT
4.3 Atom images and IFSEM Affairs in Kyoto, Osaka, and
     Okayama 597
     HATSUJIRO HASHIMOTO
4.4 Reminiscences on the Origins of the Scanning Electron
     Microscope and the Electron Microprobe. 635
     MANFRED VON ARDENNE
4.5 Electron Microscopes and Microscopy in Japan 653
4.5A Electron Microscope Development at Hitachi in the
     1940s 653
     TSUTOMU KOMODA
4.5B Development of the Electron Microscope at JEOL . . . 659
     KAZUO ITO
4.5C Development and Application of Electron Microscopes,
     Model SM-1 Series, at Shimadzu Corporation 665
     SHIN-ICHI SHIMADZU
4.5D Electron Microscope Research at the Toshiba
     Corporation 673
     HIROSHI KAMOGAWA
4.5E Development of the Electron Microscope at Kyoto
     Imperial University Faculty of Medicine 679
     YUTAKA TASHIRO AND AKIO OYAMA
4.5F Instrumentation 685
     TSUTOMU KOMODA

4.5G	Application of Electron Microscopy to Biological Science . . . . ATSUSHI ICHIKAWA AND YONOSUKE WATANABE	723
4.5H	Application of Electron Microscopy to Biological Science (Microbiology) . . . . YASUHIRO HOSAKA AND TADASHI HIRANO	735
4.5I	Applications to Materials Science . . . . HIROSHI FUJITA	749
4.5J	Specimen Preparation Techniques . . . . KEIJI YADA	773
4.6	Towards Atomic Resolution . . . . FRIEDRICH LENZ	791
4.7	The Construction of Commercial Electron Microscopes in China . . . . LAN YOU HUANG	805
APPENDIX	Conference Proceedings and Conference Abstracts . . PETER W. HAWKES	849
INDEX		875

CONTRIBUTORS

Numbers in parentheses indicate the pages on which the authors' contributions begin.

KOICHI ADACHI (251), Engineering Research Institute, University of Tokyo, Yayoi-cho, Bunkyo-ku, Tokyo 113, Japan

BJÖRN A. AFZELIUS (301, 385), Department of Ultrastructure Research, Stockholm University, S-10691 Stockholm, Sweden

ALAN W. AGAR (415), Agar Scientific, Ltd., Stanstead, United Kingdom

KENTARO ASAKURA (251), Engineering Research Institute, University of Tokyo, Yayoi-cho, Bunkyo-ku, Tokyo 113, Japan

VERNON E. COSSLETT[1] (3), Cavendish Laboratory, University of Cambridge, Cambridge CB3 0HE, United Kingdom

ÁGNES CSANÁDY (181), Hungalu Engineering and Development Centre, Budapest 1389, Hungary

FRANCES W. DOANE (79), Department of Microbiology, University of Toronto, Toronto, M5S 1A8 Canada

ROBERT M. FISHER (347), Department of Materials Science and Engineering, University of Washington, Seattle, Washington 98195, USA

HIROSHI FUJITA (749), Osaka University, Suita, Osaka 565, Japan

AKIRA FUKAMI (251), Department of Physics, College of Humanities and Science, Nihon University, Sakurajousui, Setagaya-ku, Tokyo 156, Japan

JOHN W. GEUS (287), Department of Inorganic Chemistry, University of Utrecht, 3508 TC Utrecht, The Netherlands

HANS GRÜMM (59), University of Vienna, Vienna, Austria

FRANÇOISE HAGUENAU (93), College de France, Paris, France

HATSUJIRO HASHIMOTO (597), Okayama University of Science, Okayama 700, Japan

PETER W. HAWKES (xiii, 405, 849), CEMES/Laboratoire d'Optique Electronique du Centre National de la Recherche Scientifique, 31055 Toulouse, France

---

[1] Deceased.

JOHANNES HEYDENREICH (171), Akademie der Wissenschaften der DDR, Institut für Festköperphysik und Elektronenmikroskopie, DDR-4010 Halle/Saale, Germany

TADASHI HIRANO (735), Jikei University School of Medicine, Minato-ku, Tokyo 105, Japan

ERWIN M. HÖRL (55), Austrian Research Center, Seibersdorf, Germany

YASUHIRO HOSAKA (735), Department of Virology and Immunology, Osaka University of Pharmaceutical Sciences, Kawai Matsubara, Osaka 580, Japan

LAN YOU HUANG (805), KYKY Scientific Instrument Research and Development Center, Chinese Academy of Sciences, Beijing, China

CAESAR E. HULSTAERT (287), Laboratory for Cell Biology and Electron Microscopy, University of Groningen, 9700 AB Groningen, The Netherlands

JOHN L. HUTCHISON (393), Department of Materials, University of Oxford, Oxford OX1 3PH, United Kingdom

ATSUSHI ICHIKAWA (723), Yokohama City University, Minamiku, Yokohama 232, Japan

KAZUO ITO (659), JEOL, Ltd., Nakagami, Akishima, Tokyo 196, Japan

BERNARD JOUFFREY (101), École Centrale Paris, LMSS-Mat, Grande Voie des Vignes, 92295 Châtenay-Malabry, France

HIROSHI KAMOGAWA (673), Toshiba Research and Development Center, Komukai Toshiba-cho, Saiwai-ku, Kawasaki 210, Japan

KOICHI KANAYA (257), Kogakuin University, Nishishinjuku, Shinjuku-ku, Tokyo 160, Japan

TSUTOMU KOMODA (653, 685), Central Research Laboratory, Hitachi, Ltd., Kokubunji, Tokyo 185, Japan

PIETER KRUIT (287), Department of Applied Physics, Delft University of Technology, 2600 AA Delft, The Netherlands

K. H. KUO (xxviii), China University of Science and Technology, 23006 Hefei, China

FRIEDRICH LENZ (791), Institute of Applied Physics, University of Tübingen, D-72076 Tübingen, Germany

HANS LUPPA (171), Karl-Marx-Universität Leipzig, Sektion Biowissenschaften, DDR-7010 Leipzig, Germany

VIRUL MANGCLAVIRAJ (xxiii), Institute Building 2, Chulalongkorn Soi 62, Bangkok 10500, Thailand

ARVID B. MAUNSBACH (21, 301), Department of Cell Biology, Institute of Anatomy, University of Aarhus, DK-8000 Aarhus, Denmark

GOTTFRIED MÖLLENSTEDT (585), Institute of Applied Physics, University of Tübingen, D-72076 Tübingen, Germany

TOM MULVEY (xix), Department of Electronic Engineering and Applied Physics, Aston University, Birmingham B4 7ET, United Kingdom

HEINZ NIEDRIG (131), Optisches Institut, Technische Universität Berlin, 10623 Berlin, Germany

AKIO OYAMA (679), Department of Microbiology, Kansai Medical University, Moriguchi-shi, Osaka 570, Japan

L. Z. QIAN (xxviii), Chinese Electron Microscopy Society, 100080 Beijing, China

ALFRED RECKNAGEL[2] (171), Technische Hochschule, DDR-8053 Dresden, Germany

FRITS W. SCHAPINK (287), Laboratory of Materials Science, Delft University of Technology, 2600 AA Delft, The Netherlands

GERHARD SCHIMMEL (149), Battelle Institute, Frankfurt, Germany

PETER SCHISKE (59), Fritz-Haber-Institut der Max-Planck-Gesellschaft, Berlin, Germany

SHIN-ICHI SHIMADZU[3] (665), Shimadzu Corporation, Kawaramachi-nijo, Nakagyo-ku, Kyoto 604, Japan

GÉRARD T. SIMON (79), Faculty of Health Sciences, McMaster University, Hamilton, L8N 3Z5 Canada

DANKWART STILLER (171), Martin-Luther-Universität Halle-Wittenberg, Pathologisches Institut, DDR-4020 Halle/Saale, Germany

BUNYA TADANO (227), Hitachi, Ltd., Tokyo, Japan

YUTAKA TASHIRO (679), Department of Physiology, Kansai Medical University, Moriguchi-shi, Osaka 570, Japan

GARETH THOMAS (21), Department of Materials Science and Mineral Engineering, University of California, Berkeley, California 94720, USA

[2] Deceased.
[3] Deceased.

PETER S. TURNER (39), Division of Science and Technology, Griffith University, Brisbane, Queensland 4001, Australia

KATSUMI URA (263), Department of Electric and Electronic Engineering, Faculty of Engineering, Osaka University, Sangyo, Nakagaito, Osaka 574, Japan

UGO VALDRÈ (193), Centre for Electron Microscopy, Department of Physics and INFM, GNSM–CNR, University of Bologna, I-40126 Bologna, Italy

DIRK VAN DYCK (67), University of Antwerp (RUCA), B-2020 Antwerpen, Belgium

WOUTERA VAN ITERSON (271), Section of Molecular Cytology, Institute for Molecular Cell Biology, BioCentrum Amsterdam, University of Amsterdam, The Netherlands

ARIE J. VERKLEIJ (287), Department of Molecular Cell Biology, University of Utrecht, 3508 TC Utrecht, The Netherlands

MAREK VESK (39), Electron Microscope Unit, University of Sydney, Sydney, New South Wales 2006, Australia

SZABOLCS VIRÁGH (181), Postgraduate Medical School, Budapest 1389, Hungary

MANFRED VON ARDENNE (635), von Ardenne Institute for Applied Medical Research, 01324 Dresden, Germany

J. B. VUKOVIĆ (xxv), Institute of Biophysics, Faculty of Medicine, 11000 Beograd, Yugoslavia

YONOSUKE WATANABE (723), Department of Pathology, Keio University Medical School, Shinjuku-ku, Tokyo 160, Japan

MICHAEL J. WITCOMB (323), Electron Microscope Unit, University of the Witwatersrand, WITS 2050, South Africa

SONIA A. WOLFE-COOTE (323), Experimental Biology Programme, Medical Research Council, Tygerberg 7505, South Africa

KEIJI YADA (217, 245, 773), Tohoku University, Katahira, Sendai 980, Japan

PREFACE

The present volume is in a sense a successor to *The Beginnings of Electron Microscopy,* which appeared as Supplement 16 to this series in 1985. My associate editor Tom Mulvey, guest-editor for this special volume, explains in his prologue how *The Growth of Electron Microscopy* came into being at the wish and under the aegis of the International Federation of Societies of Electron Microscopy (IFSEM), and I am delighted that the Federation chose to publish this collection of essays in these *Advances.* It is divided into four large sections, which contain material of very different kinds. Part I retraces the history of IFSEM itself, with an essay by the late Vernon E. Cosslett, who was active in the creation of the Federation in the 1950s, and another by Arvid B. Maunsbach and Gareth Thomas, present and past presidents of IFSEM.

Part II contains the material submitted by a number of the member societies of the Federation. All were invited to contribute and many provided the suitable texts published here; a few sent only short notes, and these have been incorporated in the prologue. The various societies have interpreted the editorial guidelines in very different ways, with the result that these chapters are a rich source of formal and informal history of the instruments and techniques of electron microscopy with many personal anecdotes about the microscopists who developed and used them. I have no doubt that they will awaken many memories among readers who lived through the years thus recorded.

The principal activity of the IFSEM over the years has been to encourage and guide organizers of the international conferences on electron microscopy, which have been held every four years since 1950; there was one earlier meeting, in Delft in 1949. The role played by these congresses is recalled in three short essays on highlights in biology, chosen by Björn A. Afzelius; in materials science, by John L. Hutchison; and in electron optics, by me. These highlights form Part III.

The last part contains a series of essays on particular aspects of instrumental development in electron microscopy. It opens with a remarkable account by Alan W. Agar of the development of the electron microscope in Europe; an enormous amount of half-forgotten information has been assembled here from publications, from manufacturers' brochures and, in many cases, from personal communications. Next come chapters by Gottfried Möllenstedt recalling his early work on convergent-beam electron diffraction and by Hatsujiro Hashimoto on atom images and IFSEM affairs in three major Japanese centers of electron microscopy. The latter is of unusual interest in that Professor Hashimoto has been president of IFSEM and of CAPSEM, the Committee of Asian–Pacific Societies of Electron Microscopy, which gives him a different perspective on the subject than the more familiar views from Europe and America. This is followed by an essay on the beginnings of scanning electron microscopy, by Manfred von Ardenne. In the next long section, we return to Japan; the development of the various instruments is traced in detail in a series of short chapters by many members of the Japanese electron microscope community. Although much of this material was made available to participants at the 1986 International Congress on Electron Microscopy in Kyoto, it is not widely known and it seemed appropriate to publish this slightly revised version here. The next contribution comes from Friedrich Lenz, who traces the approach to atomic resolution. Part IV concludes with a most unusual chapter by Lan you Huang on the construction of commercial microscopes in China. This not only is full of detail of a technical nature, very difficult for Western readers to access, but also contains a vivid evocation of changing conditions in China, which deserves to be put on record.

The proceedings of the conferences organized under the aegis of IFSEM, the international conferences, and the regional conferences, contain an enormous amount of information and it therefore seemed useful to provide full bibliographical details of these volumes. I have prepared such a list, to which the proceedings of a number of related meetings have been added. This list forms an appendix to the book.

As usual, I conclude by thanking all the contributors to this volume, especially Tom Mulvey, who has taken enormous pains with the manuscripts submitted. A list of articles to appear in future volumes, two of which will follow close on the heels of Volume 96, is given below.

Peter W. Hawkes

FORTHCOMING ARTICLES

Nanofabrication	H. Ahmed and W. Chen
Finite-element methods for eddy-current problems	R. Albanese and G. Rubinacci
Use of the hypermatrix	D. Antzoulatos
Image processing with signal-dependent noise	H. H. Arsenault
The Wigner distribution	M. J. Bastiaans
Edge-preserving image reconstruction	L. Bedini, E. Salerno, and A. Tonazzini (Vol. 97)

Hexagon-based image processing	S. B. M. Bell
Microscopic imaging with mass-selected secondary ions	M. T. Bernius
Modern map methods for particle optics	M. Berz and colleagues
Cadmium selenide field-effect transistors and display	T. P. Brody, A. van Calster, and J. F. Farrell
ODE methods	J. C. Butcher
Electron microscopy in mineralogy and geology	P. E. Champness
Electron-beam deflection in color cathode-ray tubes	B. Dasgupta
Fuzzy morphology	E. R. Dougherty and D. Sinha
The study of dynamic phenomena in solids using field emission	M. Drechsler
Gabor filters and texture analysis	J. M. H. Du Buf
Miniaturization in electron optics	A. Feinerman
Liquid metal ion sources	R. G. Forbes
The critical-voltage effect	A. Fox
Stack filtering	M. Gabbouj
Median filters	N. C. Gallagher and E. Coyle
RF tubes in space	A. S. Gilmour
Quantitative particle modeling	D. Greenspan (Vol. 98)
Structural analysis of quasicrystals	K. Hiraga
Formal polynomials for image processing	A. Imiya
Contrast transfer and crystal images	K. Ishizuka
Seismic and electrical tomographic imaging	P. D. Jackson, D. M. McCann, and S. L. Shedlock
Morphological scale-spaces	P. Jackway (Vol. 99)
Quantum theory of the optics of charged particles	R. Jagannathan and S. Khan (Vol. 97)
Optical interconnects	M. A. Karim and K. M. Iftekharuddin

Surface relief	J. J. Koenderink and A. J. van Doorn
Spin-polarized SEM	K. Koike
Sideband imaging	W. Krakow
The recursive dyadic Green's function for ferrite circulators	C. M. Krowne (Vol. 98)
Near-field optical imaging	A. Lewis
Vector transformation	W. Li
SAGCM InP/InGaAs avalanche photodiodes for optical fiber communications	C. L. F. Ma, M. J. Deen, and L. E. Tarof (Vol. 99)
SEM image processing	N. C. MacDonald
Electron holography and Lorentz microscopy of magnetic materials	M. Mankos, M. R. Scheinfein, and J. M. Cowley (Vol. 98)
Electron holography of electrostatic fields	G. Matteucci, G. F. Missiroli, and G. Pozzi
Electronic tools in parapsychology	R. L. Morris
Image representation with Gaussian wavelets	R. Navarro, A. Taberno, and G. Cristobal (Vol. 97)
Phase-space treatment of photon beams	G. Nemes
The imaging plate and its applications	T. Oikawa and N. Mori (Vol. 99)
Representation of image operators	B. Olstad
Z-contrast in materials science	S. J. Pennycook
HDTV	E. Petajan
The wave-particle dualism	H. Rauch
Scientific work of Reinhold Rüdenberg	H. G. Rudenberg
Electron holography	D. Saldin
Space-variant image restoration	A. de Santis
X-ray microscopy	G. Schmahl
Accelerator mass spectroscopy	J. P. F. Sellschop
Applications of mathematical morphology	J. Serra
Set-theoretic methods in image processing	M. I. Sczan
Wavelets and vector quantization	E. A. B. da Silva and D. G. Sampson (Vol. 97)

Focus-deflection systems and their applications	T. Soma
Mosaic color filters for imaging devices	T. Sugiura, K. Masui, K. Yamamoto, and M. Tni
New developments in ferroelectrics	J. Toulouse
Electron gun optics	Y. Uchikawa
Very high resolution electron microscopy	D. van Dyck
Morphology on graphs	L. Vincent
Canonical aberration calculation and magnetic lenses	J. Ximen (Vol. 97)

PROLOGUE

TOM MULVEY

The International Federation of Societies for Electron Microscopy (IFSEM) was founded at the International Meeting on Electron Microscopy held in London in 1954. Vernon E. Cosslett was elected Secretary and Bodo von Borries was elected Chairman. At the London Conference, it was generally recognized that European and international collaboration between electron microscopists was essential for rapid progress in the field, but it was not clear what form this collaboration should take or even whether it was feasible to hold conferences in which all microscopists, including physicists, materials scientists, life scientists, and instrument designers, could participate together. In Europe, English, French, and German were adopted as the conference languages and no translation facilities were provided. Few electron microscopists were fluent in all three languages, however, and eventually "Conference English" was adopted as a standard language. Nevertheless, from the beginning there was a strong feeling of solidarity among all participants and a desire to help and to be helped. It was readily agreed to hold the subsequent international meeting in Berlin in 1958. Bodo von Borries, a strong supporter of international EM collaboration, was to have been the convener of the Berlin meeting. This did not come to pass, however, since he died unexpectedly of a stroke in 1957. Somewhat reluctantly, since he never got around to learning English, and also because he had little time to devote to administrative affairs, Ernst Ruska took on the role of leadership. In 1958 Berlin still bore the scars of battle and was still occupied by the Allied forces, but it was, in fact, easily accessible by electron microscopists from both Eastern and Western countries. Ruska's opening speech to the participants, delivered and published in German, was a masterpiece of his own personal convictions, which coincided strongly with the ideas developed at the London Conference. The English translation follows:

THE IVTH INTERNATIONAL CONGRESS ON ELECTRON MICROSCOPY,
BERLIN 1958
OPENING ADDRESS BY ERNST RUSKA, IFSEM PRESIDENT

Ladies and Gentlemen,

It is an especial honor for me on this festive occasion to be privileged to welcome you, the participants at the Fourth International Congress on Electron Microscopy in Berlin. As President of IFSEM and at the same time the leader of the Berlin Congress organization, I bid you all a hearty welcome as our guests and thank you for following up our invitation to this city. Many of you have come from distant continents and several of you have not been deterred even by a journey over the Arctic region. I greet with especial warmth the two guests of honor at our congress, Dr. Francis O. Schmitt from Cambridge, USA, and Professor Max von Laue from Berlin. I thank you both for agreeing so readily to open and enrich our Congress with your lectures. I also greet the representatives of both the city and the state authorities here present and thank them for their keen interest in our activities and for their sympathetic promotion of our work.

At the center of our multidisciplinary community stands the electron microscope, in its various forms. We are concerned to make it ever more powerful and strive to use and exploit its advantages in all scientific areas in which the resolving power of our eyes is no longer adequate, even when assisted by the light microscope. The annually increasing number of talks and participants at our conferences demonstrate the vast extent of the new subject areas that are being enriched by electron microscopical investigations. We have to ascribe this happy situation to the untiring and ingenious development of new techniques in specimen preparation, and so I want at this point to thank, in the name of the physicists and instrument designers, all those who overcame the difficulties of specimen preparation, giving our instrument an ever-growing esteem and reputation. Every scientific task performed with dedication gives pleasure and satisfaction to all others seeking after knowledge. Our work, it seems to me, can bring us a special bonus of pleasure and satisfaction, through its very nature of close contact with colleagues of different disciplines and our ability to peep into the inexhaustible world of the smallest forms of existence. No one, even those far removed from science, who ever looked at an exhibition of electron microscope pictures can escape from the feelings of reverent astonishment at the intangible beauty and harmony of this world of shapes; we, as enquiring scientists, may also get some premonition about many laws of physics whose discovery still lies in the lap of the future. Such thoughts recognize no national or ideological frontiers. We rejoice in the fact that scientific collaboration still allows person-to-person contact, which, in our time, has been made so difficult by many political developments. Therefore we particularly welcome every international meeting of scientists and we want at this congress, too, to learn a bit more about the secrets shrouded within the universal laws and, armed with this knowledge, to serve the good of mankind. In this spirit I open and launch the Fourth International Congress on Electron microscopy.[1]

The subsequent IFSEM conferences are listed at the end of this book and form the subject of many of the individual contributions to the volume.

---

[1] Translated from: Ruska, E. (1960). Eröffingnungs Ansprache. In Proceedings of the Fourth International Conference on Electron Microscopy. (p. 1).

A pivotal conference for IFSEM was the 1984 European Congress at Budapest. This conference marked the 30th anniversary of IFSEM. According to the rules, however, it could not be an International conference. Nevertheless, the strategic geographical position of Budapest enabled visitors from the Eastern bloc and from the West to attend without much difficulty. For all intents and purposes it was an International Congress. It became clear at this Congress that IFSEM had come to stay and through its loosely, nonpolitically-organized structure, IFSEM was accepted by the government of any country whose subjects wished to participate in electron microscopy.

It was at Budapest that the idea of writing a history of IFSEM, already on the agenda at the 1982 Hamburg Meeting, began to crystallize. The publication by Ernst Ruska of *Die fruehe Entwicklung der Elektronenlinsen und der Elektronenmikroskopie* (Leopoldina) was a historical landmark in electron microscopy. The treatise described the early beginnings of electron optics and concluded with a description of the production of the world's first serially-produced transmission electron microscope. The English translation, *The Early Development of Electron Lenses and Electron Microscopy* (Hirzel Verlag), appeared in 1980; this was widely read and it stimulated an interest among electron microscopists to produce further historical records. In particular, in 1981 Peter W. Hawkes was inspired to plan a sequel to be entitled *The Beginnings of Electron Microscopy* and he began to solicit articles towards the end of that year for publication as a supplement to *Advances in Electronics and Electron Physics*. This volume was not to appear until 1985, but in his foreword he wrote: "it was the publication of Professor Ruska's meticulous volume that spurred me into action; that and the macabre reality that time was running out." It was therefore not surprising to find that on the Agenda for the IFSEM Committee Meeting at Hamburg 1982, item 16 read: "History of 50 Years of Electron Microscopy—an IFSEM Project?" Peter W. Hawkes, who attended the meeting as a member of the French delegation, informed the committee of his own plans and hospitably offered the *Advances in Electronics and Electron Physics* for publication of the IFSEM volume. This offer was turned down in favor of setting up an IFSEM Editorial Board, under the leadership of J. B. LePoole (Netherlands), B. Jouffrey (France), and H. Hashimoto (Japan). The three were to manage the coordination, financing, publishing, and distribution of the volume from within IFSEM. In view of the poor state of world communications at that time and the widespread political distrust between West and East, it is not surprising that progress was exceptionally slow and tedious. By the time *The Beginnings . . .* had appeared in 1985, no progress could be reported on the IFSEM volume, although it was discussed afresh at the 1984 Budapest Conference thanks to the enthusiasm of H. Hashimoto. Even ten years later, with vastly improved communications, things do not happen quickly in IFSEM with its tolerant, democratic structure; many members and the officers of societies, who are not professional administrators, have to be consulted before a consensus emerges and a financial plan is agreed upon. By late 1988, however, an outline plan was in place and I agreed to act as Editor. After investigating several possible publishing routes, it was eventually arranged with Academic Press that a sequel to *The Beginnings of Electron Microscopy* should be produced in the same series of *Advances.* The title *The Growth of Electron Microscopy* was deemed appropriate. The aim was to solicit original contributions from all over the world, primarily within the framework of IFSEM, but also from known pioneers in different countries. Great efforts were made to have an input from China and from Russia, following a plea made by the editor Peter W. Hawkes, in the Preface to *The Beginnings,* asked for further information from these countries. Readers will discover, from the articles by Alan Agar and by Lan you Huang in the present volume, the enormous efforts put in by both China and Russia, which have not been previously reported in the western literature. The extensive contribution from Japan, now the biggest supplier of electron microscopes, contains material that has been published previously, sometimes in Japanese or in piecemeal form. All this material has been considerably revised, rearranged, and reedited by Keiji Yada and it has been incorporated with new material to form an impressive whole. It is interesting to realize, as described in the Japanese account here, that all this scientific activity was triggered in 1942 when a German submarine succeeded in bringing a copy of von Ardenne's seminal book on electron microscopy to Japan.

Inevitably there are some gaps in what this volume covers. Several National Societies (that of Switzerland [1], for example) have already published comprehensive historical accounts of their activities. Other contributors, even after several years of effort, were unable in the end to meet what turned out to be a four-year deadline. This often happened for understandable reasons. Nevertheless, this volume gives a flavor of IFSEM and the infectious enthusiasm it engenders on an international scale and on a voluntary basis.

The Editor would like to thank sincerely all contributors for their efforts and for the patience of those contributors who submitted their work early and had to wait for their articles to appear in print.

Three short accounts from National Societies are appended to this prologue. The first is from Thailand. It was too brief to constitute a separate contribution, but it is included here since it illustrates how IFSEM, as formed by its President, H. Hashimoto, and its Secretary, G. Thomas, can act as a catalyst in the critical formation period of a new electron microscope society. The second account is from Yugoslavia, as it was known when this report was received. Again, the presence of Vernon E. Cosslett and W. Bernhard at the First Balkan Conference in Sarajevo was clearly appreciated by the Organizing Committee and the Conference participants. In view of the present tragic situation in that country, it was felt important to include this report in the Volume to express our sympathy with our colleagues there.

The third report concerns the Chinese Electron Microscope Society. The revised manuscript by Kuo and Qian arrived just in time to be incorporated in these pages. IFSEM "invaded" China as early as 1957, when our second President, Ernst Ruska, visited China and was warmly received by the small group of Chinese enthusiasts for electron microscopy who faced many difficulties in building their own electron microscopes and using them in research and development. Ruska's visit was to be followed later by another IFSEM President, Hatsujiro Hashimoto, who collaborated closely with Professor Kuo, Chairman of the Chinese Electron Microscopy Society, in creating scientific and technical links in electron microscopy between Japan and China.

Reference

1. *History of Electron Microscopy in Switzerland*, (J. R. Günther, Ed.), Basel: Birkhäuser Verlag, 1990.

Electron Microscopy in Thailand

VIRUL MANGCLAVIRAJ[2]

Institute Building 2
Chulalongkorn Soi 62
Bangkok 10500, Thailand

The historical development of electron microscopy in Thailand dates from 1952, when the first transmission electron microscope was installed at the Science Division of the Thai Red Cross Society. Since then, the use of electron microscopes has gradually increased for studying biology and medicine; it has taken almost thirty years for electron microscopy to be applied significantly in the other fields of natural science, engineering, and industry.

[2] General Secretary of EMST (1985–1989).

Visits from world-famous electron microscopists produced a strong impetus to the development of electron microscopy in Thailand. In December 1982, on his way to Singapore to attend the CAPSEM meeting, Professor Hatsujiro Hashimoto, President of the International Federation of Societies for Electron Microscopy (IFSEM), visited Chulalongkorn University. Inspired by his visit, a seminar and workshop on electron microscopy were organized on 20 April 1983 at the Scientific and Technological Research Equipment Centre (STREC) at Chulalongkorn University. The prominent guest speaker was Professor G. Thomas, General Secretary of IFSEM, whose lecture was entitled, "Latest Trends of High Voltage Electron Microscopy in Materials Science." Discussions on other topics, including scanning and transmission electron microscopy, were presented by specialists from JEOL Ltd.

During the seminar, a discussion about forming a Thai Society for Electron Microscopy occurred. On 3 July 1983, ten people who were interested in the establishment of a Society for Electron Microscopy met at STREC and, after the meeting, circulars were distributed among scientists urging them to join the Electron Microscopy Society of Thailand (EMST).

On 8 February 1984, the first general meeting of the EMST took place. The meeting, the first of its kind in the history of electron microscopy in the country, was highlighted by the presence of Professor Hatsujiro Hashimoto who lectured on "New Trends and Future Prospects of Electron Microscopy" and "Atom Resolution Electron Microscopy." In addition, Dr. Y. Noguchi, General Manager and Director of the Overseas Operation Division, JEOL Ltd., presented "Recent Advancements in the Application of the Scanning and X-Ray Microprobe."

The Executive Committee of the EMST met several times in 1984 and 1985 to consider application for membership of IFSEM and the readiness to host the Fourth Asia-Pacific Conference and Workshop on Electron Microscopy (4th APEM) in 1988. A working committee composed of electron microscopists from various institutes and universities was set up to organize the Fourth APEM; the workshop of the APEM was held in Bangkok at Chulalongkorn University, 26–30 July, and the conference at the Bangkok Convention Centre, Central Plaza Hotel, 31 July–4 August.

Since its inception, several seminars and workshops were held on behalf of the EMST and other institutions, producing sustained activity in the work of electron microscopy. There has been in the 1990s a sharp increase in the use of electron microscopy in teaching, research, and industry, both in biology and materials science.

Although the EMST was formed in 1984, it was not until August 1985 that the Thai Government officially approved the establishment of the Electron Microscopy Society of Thailand with Professor Dr. Thavorn Vajrabhaya, Dr. Aphirat Arunin, and Associate Professor Virul Mangclaviraj as the founders of the Society. Professor Thavorn Vajrabhaya was elected as the first President of the Society at the Annual Meeting, 14 March 1986, and he remained as president until June 1989; by that time the number of members had increased to 169.

Twenty Years of Electron Microscopy in Yugoslavia, 1969-1989

J. B. VUKOVIĆ

Institute of Biophysics, Faculty of Medicine, 11000 Beograd, Yugoslavia

The first Yugoslav Symposium on Electron Microscopy [1] was held in Ljubljana on 20-21 November 1969. Dr. V. Marinković, Prof. Dr. N. Pipan, and Prof. Dr. Aleš Strojnik sat on its organizing committee. The symposium was hosted by the Jožef Štefan Institute of Nuclear Sciences. The invited speakers were A. Strojnik, V. Pantić, Z. Devidé, and V. Marinković. Unfortunately, no written record exists of their talks. Twenty-one papers and five reviews were presented, the abstracts of which may be found in the meeting proceedings.

The first YUSEM, in 1969, brought together almost the whole Yugoslav electron microscopy community. The papers and results discussed reflect the maturity of the work then underway. It was Yugoslavia's first organized world debut. 1969 can be considered the year of the founding of the Yugoslav Electron Microscopy Society, although Yugoslav research workers had already participated in the proceedings of the IV European Congress in Rome in 1968. International Congresses have been organized since the first one in 1949 in Delft and European Congresses have been organized since 1956, with the first being held in Stockholm. The former are organized by the International Federation of Societies for Electron Microscopy (IFSEM) and the latter by the Committee of European Societies for Electron Microscopy (CESEM). Our Society has been a member of CESEM since 1977 and IFSEM since 1990. Professor N. Pipan is still a CESEM Executive Board Member.

With the increased availability of commercial electron microscopes, their use increased nationwide, although the first electron microscope in Yugoslavia, at the Rudjer Bošković Institute of Nuclear Sciences, started operating in 1953. This Trüb-Täuber microscope had electromagnetic and electrostatic lenses and an electron energy of 50 keV. Around it, under the direction of Prof. Z. Devidé and Prof. M. Wrischer, the Zagreb School of Electron Microscopy came into being. The school proved to be a major force in the development of electron microscopy in our country since 1954.

The first electrostatic electron microscope (made by C. Zeiss, Oberkochen, West Germany) started operating in 1954 at the Jožef Štefan Institute in Ljubljana. Dr. V. Marinković headed the group. I was initiated into the field at his laboratory for which I am truly grateful.

The work in Beograd started in 1955 with the founding of a University Laboratory. The activities of the lab were entrusted to a board consisting of representatives of the various departments. The lab was headed by Prof. S. Harisijades. An electrostatic ELMI D-2 C. Zeiss (Jena, East Germany) microscope was bought and the work started in 1956 after its arrival.

Nedeljko Košut designed the first Yugoslav laboratory electron microscope in 1954. However, this work did not lead to a working instrument. The Yugoslav electron microscope designer emeritus is Dr. Aleš Strojnik, Professor in the Electrical Engineering Department of Ljubljana University, and the designer of a whole range of microscopes. The unique electromagnetic 50 kV microscope he described at the Berlin Congress in 1958 [2] marks the emergence of the Ljubljana Design School that was instrumental in the development of the electron microscopy field in Yugoslavia. The designs were designated LEM for Ljubljana Electron Microscope, that is, LEM-1, LEM-2, LEM-3, LEM-4, LEM-4B (1965), and LEM-5 (1969). Iskra Electronics manufactured and marketed these instruments (LEM-4C). The first order was placed by the Yugoslav Army for its Military Medical Academy, getting the endeavor off the ground. The medical schools at Novi Sad and Rijeka followed suit. A microscope was delivered to a buyer in London, for instance. A number of instruments were sold in Ljubljana. The 100-kV, 1-nm advanced LEM-3 operated for years at the Jožef Štefan Institute. However, in the course of time, the number of manufacturers worldwide dwindled to a few multinationals (Philips, JEOL, Zeiss, Tesla), with many of the firms making electron microscopes going out of business (Trüb-Täuber, RCA, Siemens, Zeiss–Jena). Thus the fact that electron microscopes are no longer made in Yugoslavia may be viewed as a natural development. Delving into the history of the electron microscopy field in Yugoslavia has merit on its own as well as being instructive. It also parallels the activities initiated within CESEM regarding the history of European electron microscopy.

The period preceding the First Symposium was crammed with activity—a veritable golden age. *The Basic Electron Microscope Handbook* dates from this period and is a product of the first meeting of electron microscopy workers in Ljubljana in 1959. Later the first postgraduate textbook in the field, *Basic Electron Microscopy,* edited by Prof. V. R. Pantić (Beograd, 1962) appeared as a joint Ljubljana–Beograd effort with A. Strojnik, V. Marinković, J. Vuković, B. Navinšek, V. Pantić, and N. Pipan [3] as the contributors. No follow-up to this fine work was made. At the time of the First Symposium, however, a 400-page book [4] entitled *The Electron Microscope,* by Prof. A. Strojnik, was available. This book dealt with the physics and design parameters of TEM in a straightforward and detailed manner (Ljubljana, 1968).

The First Symposium was so successful that the Second Symposium was scheduled at the Veterinary Department at Beograd University, organized by V. R. Pantić in 1970. Unfortunately the proceedings were not published.

International meetings ensued. The organization of the First Balkan Congress in Sarajevo, 22–25 May 1974, was a pioneering endeavor in this direction. The Proceedings, edited by L. Jerkovic, contain the 41 papers presented. Luminaries such as Vernon E. Cosslett and W. Bernhard attended.

Yugoslav cooperation intensified in the meantime. Yugoslav workers attended the Second Balkan Symposium in Istanbul in 1977. The realization that national symposia should be continued took hold and resulted in the founding of Republic societies. In actual fact, these societies were already in existence (except in Montenegro, which did not have any electron microscopes). The first Society for Electron Microscopy to come into being was that of Serbia in 1979 with Prof. V. Pantić as chairman. This society created the impetus for the organization of the Yugoslav Symposium. The members of the first board of the Yugoslav Society were Prof. V. Pantić (President), J. Vuković (Secretary), N. Pipan, V. Marinković, Z. Devidé, R. Milin, L. Jerković, L. Grozdev, and M. Japundžić.

The Third Yugoslav Symposium on Electron Microscopy (YUSEM '80) was held at the Beograd Medical School, 16–18 May 1980, and was the most widely attended to date. There were seven invited speakers and 85 papers and posters by about 150 authors, as evident in the *Proceedings* [5]. Scanning electron microscopes, TEMs, and analytical EMs were in operation. Electron microanalysis progresses.

YUSEM '83 (The Fourth Yugoslav Symposium on Electron Microscopy, 26–28 May 1983) was organized by the Ljubljana Society, presided over by Prof. N. Pipan. The venue was the beautiful Kranjska Gora. Eighty papers were devoted to biomedicine and 35 to materials science [6].

The last symposium was unforgettable, because both the setting, Plitvice, and the time, 27–30 May 1986, were ideal. Academician Zvonimir

Devidé presided as the Chairman of the Section for Electron Microscopy of the Croation Natural Sciences Society. The *Proceedings* contain 106 papers [7].

The First Scientific Meeting of the Serbian Electron Microscopy Society was held in 1982 [8]. The second, in 1986, was organized jointly with the Serbian Academy of Science as "30 Years of Electron Microscopy in Serbia" [9]. Two international electron microanalysis symposia (1978 and 1984) have been held in Beograd as well (organizer, M. Pavićević). The Bosnian EM Society (B. Plavšić, Chairman) held its first symposium [10] in 1987 jointly with the Bosnian Academy of Science.

Two decades after the first Yugoslav Symposium, the III Balkan Congress [11] shows how high the standard of electron microscopy has become in the Balkans. The Union of Yugoslav Societies has played an important role in this. This text was originally written for the 20th Anniversary of YUSEM and for the VI YUSEM (YUSEM 1989 [12]) and the III Balkan Congress [11], but was not published until now.

REFERENCES

1. *First Yugoslav Symposium on Electron Microscopy* (V. Marinković, Ed.), Ljubljana, 1969.
2. A. Strojnik, *Ein vereninfachtes 50 kV Elektronenmikroskop,* IV Int. Kongress für EM, Berlin: pp. 182, 184, 1958.
3. *Basic Electron Microscopy* (in Serbo-Croat) (V. R. Pantić, Ed.), Naučna Knjiga, Beograd, 1962.
4. A. Strojnik, *The Electron Microscope,* The University in Ljubljana, Electrical Engineering Department, 1968.
5. *Proceedings of the Third Yugloslav Symposium on Electron Microscopy* (J. B. Vuković, T. M. Nenadović, Eds.), Beograd, 1980.
6. *Proceedings of the IV Yugoslav Symposium on Electron Microscopy* (M. Pšeničnik, B. Drinovec, Eds.), Kranjska Gora, 1983.
7. *Proceedings of the V Yugoslav Symposium on Electron Microscopy* (M. Wrischer, A. Hioušek-Radojčić, N. Ljubešić, Eds.), Plitvice Lakes, p. 304, 1986.
8. *Modern Electron Microscopy and Microanalysis in Biomedical Research* (J. B. Vuković, Ed.), Beograd, 1982.
9. *30 Years of Electron Microscopy in Serbia* (J. B. Vuković, Ed.), Beograd, 1986.
10. *Proceedings of the First Symposium of the Bosnian Society for Electron Microscopy* (B. Plavšić, Ed.), Sarajevo, 1987.
11. *Proceedings of the III Balkan Congress on Electron Microscopy* (L. Margaritis, Ed.), Athens, p. 492, 1989
12. *Proceedings of the VI Yugoslav Symposium on Electron Microscopy* (B. Plavšić, Ed.), Sarajevo-Igman, p. 294, 1989

The Chinese Electron Microscopy Society

L. Z. QIAN* AND K. H. KUO[†]

*China University of Science and Technology, 23006 Hefei, China and [†]Chinese Electron Microscopy Society, 100080 Beijing, China

The Chinese Electron Microscopy Society (CEMS) was founded 1 November 1980 with 286 members at an inaugural conference on electron microscopy. An election was held at this first meeting; Dr. L. Z. Qian was elected President, Drs. T. Ko, K. H. Kuo, Z. B. Wu, and L. Y. Huang were elected Vice-Presidents, and Mr. Xian Penju was elected General Secretary. Dr. Hatsujiro Hashimoto and Dr. G. Thomas were present at the ceremony, so the CEMS had good international connections from the start. Since 1980, Chinese electron microscopists have worked hard to keep up with the rapid development of science and technology. As a result, the Society experienced a steady expansion both in the quality of microscopy and in the numbers of members. Thus when the Society celebrated its Tenth Anniversary in 1990, the membership had grown to 2461. This increase prompted us to start gathering information on the historical background of electron microscopy in China. For economic and other reasons it was necessary for us to begin manufacturing our own electron microscopes, both transmission and scanning types, as well as electron probe X-ray microanalyzers. The enormous and difficult task of accomplishing this goal is described in some detail by Dr. Lan you Huang elsewhere in this volume.

For various reasons, China made a late start in electron microscopy. Little was done in this area before 1949, the year in which the Chinese Peoples' Republic was founded. In 1951, however, an important and unexpected event took place. In this year, some scientists at the Institute of Physics of the Chinese Academy of Sciences were asked to inspect some "scientific equipment" that had been imported before 1949 by the Bureau of Broadcasting. This equipment had been lying in a warehouse in Chongqing, formerly the "War Capital" during the Second World War. To their utter astonishment, the equipment turned out to be an EM3 transmission electron microscope (TEM), made by the Metropolitan Vickers Electrical Company of Trafford Park, Manchester, United Kingdom. How it got to Zhongjing is still an unsolved mystery. The discoverers were, of course, delighted and the microscope was immediately transported to Beijing and installed in the Institute of Physics. A group of young scientists led by Dr. L. Z. Chien (later spelled Qian Linzhao [L. Z. Qian] when the spelling reform occurred later). Dr. Qian had worked previously with Dr. E. N. da Andrade at London University in the 1930s, researching slip-phenomena in body-centered cubic metals. He now started to study slip traces in aluminum single-crystals. In the EM3 electron microscope, the oxide replicas showed clearly the localized slip in the early stages of formation and the cross-slip in the later stages. This was the first piece of research work with the electron microscope to be performed in China. These results were reported at the First Asian–Pacific Electron Microscopy Conference, originally known as the Pacific and Oceana Conference, held in Tokyo in 1956 and referred to by Professor Hashimoto elsewhere in this volume. In the same year, a group of physicists came together in Beijing and they discussed the possibility of constructing electron microscopes in China. This idea received instant and enthusiastic support from Dr. Ernst Ruska when he visited China in the winter of 1957. He was impressed by the machine tool factories that he saw in Shenyang and he expressed his confidence in the capability of Chinese physicists and engineers to produce electron microscopes entirely from their own resources.[3]

The "Great Leap Forward" started in 1958. In spite of its many negative features, illustrated graphically in the article by L. Y. Huang referred to above, the Great Leap Forward hastened the production of the first electron microscope constructed in China.

At the suggestion of Dr. Daheng Wang, director of the Institute of Optics and Fine Mechanics (IOFM), a plan was prepared in 1958 within the framework of the Great Leap Forward to design and construct a transmission electron microscope by the end of 1960. Dr. L. Y. Huang had received part of his education in the United States and also in Germany where he had completed a Ph.D. under the supervision of Prof. G. Möllenstedt. Dr. Huang was persuaded to return to China and his timely arrival greatly hastened the design and manufacturing process. While in Tübingen, he had designed a magnetic lens and had constructed electron optical columns. In addition, he had met many of the leading designers and users of electron microscopes. As related in more detail elsewhere in this volume, the work was begun on 13 June 1958 at the IOFM and was completed three months

---

[3] Ernst Ruska, upon his return from China, made it clear to his friends and colleagues that he had been impressed by the high scientific and technological level of the specialists he had met in China. He believed that they possessed the same sort of motivation that he had experienced in wartime Berlin, namely to press ahead, regardless of the difficulties, with the development and production of the electron microscope. But he also sensed the deep level of civilization and courtesy that he believed was innate in even the poorest Chinese citizen.

later, in September, with a resolution of 10 nm; within a year this resolution was improved to 5 nm. Considering the complexity of the instrument, this was a considerable feat of organized teamwork. A huge number of young scientists and engineers worked around the clock to get this prototype microscope ready in record time. The many technical difficulties encountered included that of acquiring suitable materials in a developing country. For example, iron–cobalt alloys for the lens polepieces were specially made in the Institute of Metal Research for this sole purpose. The work under the technical supervision of Dr. Huang continued with the construction of a new model and, in 1959, resolutions of 2.5 nm were obtained. One of the fruits of all this activity was that a considerable number of young engineers, such as Junen Yao and Zufu Zhu, emerged from the task as experts in electron optics, precision engineering, and electronics. They all subsequently played important roles in the development of electron optical instruments in China.

Later this kind of development work was transferred to the Scientific Instrument Factory (KYKY) in Beijing, which belongs to the Chinese Academy of Sciences. The Shanghai Electron Optical Factory and the Nanjing Jiangnan Optical Factory also became involved. As a result of these efforts, the resolution of the commercially manufactured TEM was improved to 0.5 nm by 1965 and to 0.2 nm by 1980. Among the scientists responsible for these developments were Lan you Huang, Junen Yao, and Hongyi Wang.

FIGURE 1. Visitors to the 1986 Beijing Symposium on Electron Microscopy. Symposium excursion to the Great Wall. In the foreground (left to right): D. Dorignac (France), Tom Mulvey (United Kingdom), and Peter W. Hawkes (France).

In the late 1950s and early 1960s, diffraction contrast imaging was developing rapidly worldwide, but this kind of work did not commence in China until 1964, when a JEM 150 was installed in the Institute of Metal Research at the Chinese Academy of Sciences. Unfortunately, this work was interrupted at an early stage during the so-called "Cultural Revolution" of 1966, in which electron microscopy in China was brought to a virtual standstill. When it became possible in 1980 to resume work in electron microscopy, high resolution microscopy was the main interest. Drs. K. H. Kuo and F. H. Li, respectively, took the lead in this development and many graduate students were trained in the techniques of high resolution microscopy. In the field of life sciences, Drs. Q. F. Pang and Kun Li, among others, were pioneers in applying electron microscopy to the study of biomedical problems, such as the investigation of viruses peculiar to China. Cytochemistry was also high on the list of suitable topics.

Dr. T. Ko took a special interest in training young electron microscopists in applying their skills to materials research. It was Dr. Ko who instituted the first graduate program in transmission electron microscopy at the Beijing University for Iron and Steel Technology. As a result, a large cohort of highly qualified young electron microscopists are now scattered not only through China itself, but also worldwide, especially in Europe, the United States, and Japan. It may be mentioned that the above-mentioned scientists were the main promoters in 1980 of the founding of the Chinese Society of Electron Microscopy. Enthusiasm for electron microscopy is now widespread throughout China. The International Symposium on Electron Microscopy (1986) held in Beijing, for example, was a stimulating specialist meeting and attracted experts from all over the world. Some of the IFSEM visitors to the Symposium may be seen in Fig. 1, taken at the Great Wall of China. Likewise, the Fifth Asia–Pacific EM Conference (1992) was held in Beijing (see the conference list at the end of this volume). These are just two of many meetings on electron microscopy that are being staged in China for the benefit of electron microscopy worldwide. In addition, the China–Japan Seminar on Electron Microscopy, coordinated by Drs. Hashimoto and Kuo, has been held several times since 1981 and this greatly enhanced the development of electron microscopy in China.

Part I

IFSEM

1.1
Early History of the International Federation of Societies for Electron Microscopy

VERNON E. COSSLETT

Cavendish Laboratory
University of Cambridge
Cambridge CB3 0HE, United Kingdom

The International Federation of Societies for Electron Microscopy grew out of the common interests of the various national societies for electron microscopy (EM), but not without a few false starts. An attempt was made in 1953–1954 to set up some sort of joint organization under the umbrella of the International Council of Scientific Unions (ICSU). Owing to bureaucratic opposition, that attempt failed and it was replaced in 1954/1995 by an independent federation. Since then the IFSEM has gradually developed in numbers and in activities. A tenuous connection with the ICSU has now been established.

I. Origins

It is not clear who, in particular, first had the idea of some sort of federation of the early existing electron microscope societies. Probably the general intention developed from the example of the X-ray crystallographers, who during World War II had adopted the habit of holding informal meetings, largely on the initiative of Sir Lawrence Bragg. Most of the early relevant correspondence has been lost, and the memories of the initial participants have faded. The earliest documented reference to the need for some sort of continuing international organization occurs in the minutes of the British Electron Microscopy Group of the Institute of Physics. Those for the committee meeting held on 20 May 1949 include a section on the forthcoming Delft EM conference: "Before any steps were taken towards the formation of any more permanent international body, it was agreed that the advice of Professor Stratton, Secretary of the International Union of Scientific Conferences [sic] should be sought."

Thus it is clear that the idea of an international body linking the various EM societies was already in the air before the first International Conference took place at Delft in July 1949. A number of national organizations had been set up during and after World War II. A large amount of correspondence had grown up among them, largely fueled by personal visits of electron microscopists among the different countries, especially the Netherlands, Britain, and United States.

The Electron Microscope (later Microscopy) Society of American (EMSA) was founded formally in January 1944 (Reisner, 1982), after some previous informal meetings of interested individuals, especially those connected with the Radio Corporation of America (RCA), which was actively manufacturing electron microscopes. In Britain a series of semiformal gatherings of users of the RCA microscopes imported under the Lend-Lease scheme were held on the initiative of the head of the National Physical Laboratory (NPL), Sir Charles Darwin. These meetings began in November 1943 and culminated in the formation of the EM Group of the Institute of Physics in September 1946 (with this author as secretary) (Cosslett, 1971). Close contact had already been established with Netherlands scientists, as Dr. Le Poole attended the EM Group meeting in Manchester in January 1946, and Dr. van Dorsten with others from the Philips Laboratories came to the founding meeting in Oxford that September.

Out of these contacts, and others formed at this first "official" conference held by the EM Group, the plan for an international conference developed. It took place at Delft in July 1949 (Houwink *et al.*, 1950), organized by the Netherlands EM Society with the active collaboration of the British. As the minutes of the Institute of Physics Group make clear, several of its members took a prominent part in the preparation and running of the conference. Professor Astbury, chairman of the Group, Dr. Cosslett, and Dr. MacFarlane were members of the Advisory Committee.

Doubtless, further private discussions about international connections went on at Delft. Microscopists from 13 countries were present, including Britain and the United States, but not Germany. Contacts certainly continued during the rather larger international conference held in Paris in September 1950, attended by some 600 participants from 17 countries, now including Germany (Locquin, 1953). As a result, more positive steps were taken toward setting up some type of international organization. The minutes of the committee meeting of the Electron Microscopy Group of the Institute of Physics dated 2 February 1951 state: "The Honorary Secretary (Mr. F. W. Cuckow) had suggested there that an international body was needed with powers to call such meetings and to ensure the co-operation of all interested parties in the selection of dates and meeting places. It was agreed to attempt the formation of such a body within the framework of the International Council of Scientific Unions." I was deputed to seek the advice of its General Secretary, Professor Stratton. My own notes mention the possibility of a "mixed Commission or Association," i.e., mixed as between the Scientific Unions of different subjects.

The next committee meeting, on 23 April 1951, heard a report from V. E. Cosslett and F. W. Cuckow on the progress of these talks. It was agreed to circulate among members of the committee a draft proposal to form a Joint Commission on Electron Microscopy, associated with the Unions of Pure and Applied Physics (IUPAP), Chemistry, Biology and Crystallography. This proposal was reaffirmed at a subsequent meeting of the Committee on 20 June 1951, and it was decided "to send a copy to Professor Fleury [General Secretary of IUPAP] at once and to seek supporting signatures from other countries at the same time." The ICSU discussed the proposal at its meeting in Washington later that year, with a positive result, so that the secretary of the Group (F. W. Cuckow) then "sent a wire to [Professor Robley] Williams saying that the Joint Commission is now an established fact."

The establishment of the Joint Commission was reported in the next circular of the EM group to members, dated 29 January 1952. It will have "3 members from the International Union of Pure and Applied Physics, 3 from the I.U. of Pure and Applied Chemistry, 3 from the I.U. of Biological Sciences and 1 from the I.U. of Crystallography. Your present Honorary Secretary [i.e., F. W. Cuckow) will become Provisional Secretary of the Joint Commission." The "Parent Union" was to be the I.U. of Pure and Applied Physics, in spite of some objections that this was too narrow a basis for such an interdisciplinary subject as electron microscopy. As Cuckow said in his letter to this author of 25 September 1951: "I know only too well that to secure this cooperation we must have an organization in which physicists do not even appear to play the predominant role."

II. THE RISE AND FALL OF THE JOINT COMMISSION

Delay occurred because of the need for the separate international unions to elect their representatives. At the same time, differences of opinion arose about the method of election. Many electron microscopists felt that the members of the Joint Commission should be selected and nominated by the various national societies, even if they were to be formally elected as representatives of the individual international unions. Others thought that the national EM socieities ought in any case to have a predominant say in the activities of the Joint Commission.

These doubts and difficulties about matters of organization and function were brewing and began to be voiced during the preparations for the next

International EM Conference, to be held in London in July 1954 (Ross, 1956). They were not mentioned, however, by R. W. G. Wyckoff in his opening address to the conference as president of the Joint Commission. He played a leading part in all these discussions, being a representative on the governing body of ICSU for the I.U. Crystallography.

Evidently, dissatisfaction was taking practical form, and an alternative system was being put forward. Under the heading, "Formation of International Representative Body," the minutes of the Annual General Meeting of the British EM Group (17 July 1954) report: "The Chairman, Dr. V. E. Cosslett, who had represented the group at a meeting of the Joint Commission and National Representatives, outlined the proposed scheme for the formation of an International Assembly for Electron Microscopy. Under this scheme the British group would have three delegates, who would attend the provisional meeting of the Assembly on Tuesday, 20th July. At this meeting the International Committee for Electron Microscopy would be elected from among the members of the Assembly. This committee would arrange future International Conferences and carry on other activities under the authority of the Joint Commission."

The International Assembly for Electron Microscopy (and its Committee) was duly set up. The record of its first meeting states that the then-Secretary of the Joint Commission (J.-B. Le Poole) had suggested its formation. "As a result the respective national societies invited by him nominated representatives with full power of decision to attend a joint meeting with the members of the Joint Commission." That meeting took place on 15 July 1954, with Dr. Wyckoff of the Joint Commission in the chair, followed by the inaugural meeting of the International Assembly on 20 July. From it an executive committee was elected consisting of T. F. Anderson, B. von Borries, V. E. Cosslett, S. Sjöstrand, and M. Terada, plus two representatives appointed by the Joint Commission (R. W. G. Wyckoff and J. B. Le Poole).

At the subsequent meeting of the Committee on 21 July 1954, with Prof. Wyckoff again in the chair, Prof. B. von Borries (see Fig. 1) was elected as its chairman and Dr. V. E. Cosslett its secretary (see Fig. 2). The future program of international conferences was discussed (see Appendix A). Dr. Wyckoff reported that the formal constitution of the International Committee was to be considered by the Executive Committee of the ICSU in October 1954.

Not surprising, the ICSU was not willing to countenance such a hybrid organization. The standard rules for joint commissions laid down that the executive committee must consist of representatives of the cooperating international unions only. The then-Secretary General of the International Union of Pure and Applied Physics (Prof. Fleury) adhered rigidly to the

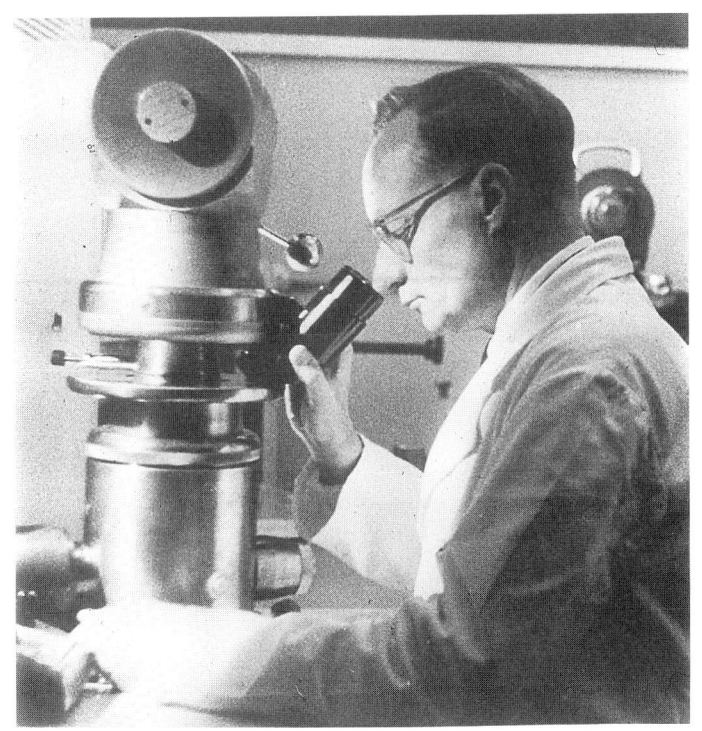

FIGURE 1. Bodo von Borries (1905–1956). Founding Chairman of the IFSEM (1954–1958).

rules, and so the proposed scheme fell to the ground. Without attempting to give a blow-by-blow account of later developments, it is enough to say that the International Federation was born out of the ruins of the Joint Commission plan.

III. THE FORMATION OF THE INTERNATIONAL FEDERATION

The last action of the International Committee was to inform all member societies of the failure of the Joint Commission plan and the proposal to replace it by the establishment of an international federation (Appendix B). Replies were requested to three short questions about the new constitution and the election of members of the Committee of the Federation. The setting of this new body was approved by all the member societies, and the IFEMS came into being in October 1955. The Joint Commission had been formally dissolved at the meaning of the Executive Committee of the ICSU in August 1955, with the consequence that Prof. Wyckoff and Dr.

FIGURE 2. Left to right: Gaston DuPouy (France); Ellis Cosslett (UK), Founding Secretary of the IFSEM; Robert M. Fisher (USA); Hatsujiro Hashimoto (Japan); Albert Szirmae (USA). Venue: First International HVEM Conference, Monroeville, PA, USA, June 1969.

Le Poole were no longer members of the International Committee. They were replaced, by postal ballot, by Dr. Locquin and Dr. Le Poole.

Otherwise, the International Federation inherited the clothes of the International Committee, including its constitution (the "Main Rules"). A few alterations were made in the latter in 1955, particularly by a clause increasing the size of the Executive Committee to 7, including the president and secretary, from the previous total of 5 (Appendix C). Professor von Borries continued as president and this author as Secretary. At that time the assembly was made up of 9 national societies. The word "Microscope" in the title was later replaced by "Microscopy," so that from the meeting of the Assembly at Berlin in 1958 we became the Federation of Societies for Electron Microscopy, IFSEM instead of IFEMS.

This author's "Report for the Period of 1954–58," prepared for the 1958 Assembly, stated that "the establishment of the new body, and the revised Constitution" was approved by all the member socities and the IFEMS came into existence in October 1955. . . . The Committee of IFEMS addressed a letter of thanks to Dr. R. W. G. Wyckoff on 24 January 1956 in appreciation of his part in these complicated negotiations with ICSU." The foundation members of the International Comittee were Belgium, France, Germany, Great Britain, Japan, the Netherlands, Scandinavia, Switzerland, and the United States. To these 9 societies were added at the 1958 Assembly those of Czechoslovakia, Hungary, Italy, and Spain.

Thus, although the International Federation came into existence *de jure* only in October 1955, it should *de facto* date from July 1954 as the logical and constitutional successor to the International committee for Electron Microscopy. On this basis it was 30 years old in August 1984.

IV. The Development of the International Federation

Almost as soon as it was formed, the IFEMS lost its first president, Prof. Bodo von Borries. He was taken ill and died of a brain tumor in July 1956, at the early age of 51. He had been the prime mover in the creation of an international organization, and his death was a great blow to the infant federation. This was especially so on the eve of the first conferences to be mounted under its sponsorship, the First European EM Conference, to be held in Stockholm in September 1956 (Sjöstrand and Rhodin, 1957), and that for the Far East and Oceania, in Tokyo in October 1956 (Tani *et al.*, 1957). I believe that the idea of holding regional meetings between the four-yearly world conference was his in the first place. On the proposal of Dr. J. Hillier, the Stockholm conference was dedicated to his memory.

Before that meeting the committee met and decided to nominate E. Ruska, von Borries' long-term collaborator, as president in his place. This proposal was unanimously approved by the member societies. At the next meeting of the Assembly at Berlin in 1958, he was succeeded by T. F. Anderson (USA). V. E. Cosslett was replaced by J. B. Le Poole (The Netherlands) at the 1962 meeting in Philadelphia. A list of presidents and secretaries is given in Appendix D.

The number of foundation national societies of the Federation was added to rapidly. By 1962 it had already grown to 15, by 1972 it was 20, and at present (1984) it totals 34. Each society was allocated by mutual agreement to one of the groups I to IV, which determined its voting power in the Assembly and the amount of its annual subscription, in accordance with the constitution.

Initially the constitution of the International Committee, and following it of the Federation, was a simple document of a single page in length (Appendix C). At the 1958 Assembly meeting it was reported that "the Committee was engaged in a complete overhaul of the Constitution. . . .

A change of name would also be proposed to 'International Federation of Societies for Electron Microscopy.' " This new constitution was adopted, after much revision, at the Philadelphia meeting in 1962; it now ran to 20 pages (triple spaced).

Our constitution remains as that of an independent federation, but relations with the ICSU have been under discussion from time to time. After some changes in the constitution and activities of ICSU in 1957, the IFSEM made an approach about the possibility of becoming an Affiliated Commission, but nothing came of it. The committee concluded "that as yet no suitable place for IFEMS [as it then was] exists in the structure of unions and commissions. The situation will be kept under review, in case some change is made in the ICSU regulations."

Such a change took place in 1975, when the category of Scientific Associate was created by ICSU. An application for membership under this new rule was made and was approved in October 1976. Since that time IFSEM has been an Associate Member of ICSU.

V. Concluding Remarks

Electron microscopy has developed in the past 30 years to become an essential tool for morphological research, and increasingly for analysis also. Along with its associated techniques it represents a whole battery of approaches to any major problem in materials science as well as in biology. Inspection of the program of any recent EM congress will reveal what a range of methodologies is now covered in it. That of the most recent World Congress (Hamburg, 1982) include symposia on electron holography and lithography, image processing and energy loss spectroscopy, as well as on the more obvious subjects of electron optics, electron diffraction, and instrumentation.

The International Federation has worked hard to keep pace with this rapid expansion of subject material, which now includes such recondite techniques as photoelectron and electron-acoustic microscopy. This proliferation of topics, as well as the growth in the number of participants, has required a corresponding adjustment in the format of the International meetings. They now consist mainly of morning symposia, followed by afternoon poster sessions.

Apart from its major functions of coordinating the activities of the various national EM societies and arranging for the regular international and regional conferences, it sponsors summer schools, workshops, and seminars. The detailed arrangements are left in the hands of one of the national societies, for whom it is usually helpful to be operating "under the auspices" of the IFSEM.

So much for the first 30 years of our International Federation. In the light of its growth and present diversity, it would be difficult to forecast its development in the next two decades, up to its golden jubilee in 2004.

Vernon E. Cosslett
August 1984

Editorial postscript: The first-ever transmission electron micrograph (TEM) was taken at a magnification of 17.4 times, on April 7, 1931, by Ernst Ruska, operating the two-stage TEM designed and built by himself and Max Knoll at the Technische Hochschule Berlin. To celebrate the 50th anniversary of this event, the Imperial Science Museum at South Kensington, London, put on, in April 1981, a historical exhibition of electron microscopes including a replica of this instrument. Figure 3, which Cosslett

FIGURE 3. V. E. Cosslett (left) congratulating Ernst Ruska at the Imperial Museum of Science and Technology on April 7, 1981, the 50th anniversary of his taking the first-ever electron micrograph in the newly designed and constructed Knoll-Ruska TEM.

had originally intended to include in this article, shows him at the museum congratulating Ernst Ruska, IFSEM Chairman 1956–1958, on his outstanding achievement with Max Knoll. The electron gun of their replica instrument is just visible in the background. Max Knoll had, in fact died in 1969. Ruska died in 1988, soon after receiving the Nobel Prize in Physics for, among other things, the design of this and later electron microscopes. Cosslett died in 1990. Both Ruska and Cosslett made immense contributions to the design and construction of electron microscopes and to their widespread use worldwide. The remarkable international collaboration that now exists, fostered to a large extent by the IFSEM's activities, has ensured its rapid development. Few in 1931 could have predicted the enomous benefits that the electron microscope would rapidly bring to science, technology, and medicine.

REFERENCES

Reisner J. H. (1982). *Bull. Electron Microsc. Soc. Am.* **12,** 11.
Cosslett, V. E. (1971). *Phys. Bull.* (London) **22,** 339.
Houwink, A. K., Le Poole, J. B., and Le Rütte, W. A. (Eds.) (1950). *Proc. Conf. Electron Microscopy,* Delft, July 1949.
Locquin, M. (Ed.) (1953). *Proc. Int. Conf. Electron Microscopy,* Paris, 1950. Rev. d'Optique, Paris.
Ross, R. (Ed.) (1956). *Proc. Third Int. Conf. on Electron Microscopy,* London, 1954. Royal Microscopical Society, London.
Sjöstrand, F. S., and Rhodin, J. (Eds.) (1957). *Electron Microscopy, Proc. Stockholm Conf.,* September 1956. Almqvist and Wiksell, Stockholm.
Tani, Y., et al (eds.) (1957). *Electron Microscopy: Proc. First Regional Conf. Asia and Oceania,* Tokyo, 1956. Electrotechnical Laboratory, Tokyo.

APPENDIX A

International Committee for Electron Microscopy

Record of the meetings establishing the Committee (Sent to Assembly members 3 October 1955)

1. The Joint Commission on Electron Microscopy of the International Council of Scientific Unions had raised the question, in two circular letters from its Secretary Dr. J. B. Le Poole in April and May 1954, of the foundation of an International Committee for Electron Microscopy. As a result the respective national societies invited by him nominated representatives

with full power of decision to attend a joint meeting with the members of the Joint Commission, called by its circular letter of 7 July 1954.

2. The joint meeting of national representatives with the Joint Commission was held at the Science Office of the American Embassy in London on 15 July 1954. There were present:

From the Joint Commission: Dr. Wyckoff (Chairman)
Dr. Le Poole (Secretary)
Prof. Bugnard
Prof. Dupouy
Dr. McFarlane
Prof. F. O. Schmitt

From the national societies: Prof. von Borries (Germany)
Dr. Cosslett (Great Britain)
Dr. van Dorsten (Holland)
Dr. Itoh (Japan)
Dr. Kellenberger (Switzerland)
Dr. Locquin (France)
Dr. Picard (U.S.A.)
Dr. Vandermeerssche (Belgium)

On behalf of the Joint Commission the plan was put forward that it would in future concern itself with the wider field of Micro-Structure and would in due course be transformed into a "Joint Commission on Micro-Structure." Complete responsibility for matters concerned with electron microscopy would then be transferred to a committee of this Joint Commission, to be known as the International Committee for Electron Microscopy. After thorough discussion the following resolution was adopted unanimously:

> This meeting agrees that steps shall be taken to set up an International Committee of the Joint Commission for Electron Microscopy to further international collaboration.

The draft "Main Rules" for the new Committee were then discussed. The principles of each section were agreed and a Drafting Committee was set up, composed of Prof. v. Borries, Dr. Le Poole and Dr. Locquin, to draw up the definitive form of the Rules. It was agreed that the Assembly of National Societies, provided for in the Rules, should hold its inaugural meeting during the London Conference. At this meeting the voting power of the several countries represented was agreed to be as follows:

United States	4
Japan	4
Germany	3
Great Britain	3
France	2
Australia	1
Belgium	1
Holland	1
Scandinavia	1
Switzerland	1

3. The Drafting Committee carried out its appointed task in a meeting held in University College Hall, London, on the afternoon of July 17, 1954. Invitations were then issued to the representatives of national societies and groups to attend the inaugural meeting.

4. The Inaugural Meeting of the International Assembly was held at 6.30 pm on July 20, 1954, in the London School of Hygiene and Tropical Medicine. Dr. Wyckoff presided and the following representatives were present:

United States	Dr. Anderson, Miss Cooper, Dr. Picard and Dr. Porter
Japan	Prof. Terada, (Profs. Higashi, Tani & Yamashita absent)
Germany	Prof. v. Borries, Dr. Mahl and Dr. Wolpers
Great Britain	Dr. Cosslett, Dr. Dawson and Mr. Haine
France	Prof. Lepine and Dr. Locquin
Australia	Dr. Mercer
Belgium	Dr. Vandermeerssche
Holland	Dr. van Dorsten
Scandinavia	Prof. Sjöstrand
Switzerland	Dr. Kellenberger

In addition Dr. Guba (Hungary) was present as an observer; Dr. Hercik (Czechoslovakia), announced as an observer, was absent. Dr. Fraser was present as representative of ICSU.

  a. The Main Rules (attached) were adopted, as formulated by the Drafting Committee with the addition of an additional paragraph as proposed by the Joint Commission:

  11). Alterations in these rules can be proposed by the International Committee. They are subject to the approval of the Joint Commission and must be accepted by a majority vote of the Assembly.

b. The members of the International Committee were then elected by the Assembly, according to a voting system proposed by the Drafting Committee. In the first stage of the procedure three Committee members were to be elected and two more in the second stage. However in the first vote only two candidates received an absolute majority, so that three remained to be elected in the second stage. As only two candidates obtained the required absolute majority in the second vote also, a third vote was held to elect the remaining member. No candidate obtained an absolute majority in the third vote, and it was agreed to appoint the candidate who received the highest vote. The full details of the voting are given in the following table:

	Votes		
Candidate	I	II	III
Picard	2	4	1
von Borries	$\underline{13}$		
Anderson	8	$\underline{14}$	
Haine	1	1	
van Dorsten	5	7	3
Sjöstrand	5	$\underline{12}$	
Cosslett	$\underline{11}$		
van Iterson	1	1	
Mahl	1	1	1
Vandermeerssche	1	6	3
Terada	5	8	$\underline{8}$
Locquin	4	6	3
Wolpers	2		
McFarlane	1	1	
Kellenberger		2	1
Invalid	3		1
Totals	63	63	21

The Committee accordingly consists of the following five elected members:

    Dr. T. F. Anderson
    Prof. B. von Borries
    Dr. V. E. Cosslett
    Prof. S. Sjöstrand
    Prof. M. Terada

The Joint Commission appointed Dr. R. W. G. Wyckoff and Dr. J. B. Le Poole as its representatives on the Committee.

5. The first *Meeting of the Committee* was held in the Common Room of the London School of Hygiene and Tropical Medicine on 21 July 1954. All members of the Committee were present, except for Prof. Terada, who had not been reached with the notice of the meeting. With Dr. Wyckoff as Chairman of the meeting, the work of the Committee was discussed in general and its officers were elected. On the proposal of Prof. von Borries, Dr. Cosslett was elected Secretary and on the proposal of Prof. Sjöstrand, Prof. von Borries was elected Chairman.

In accordance with paragraph 9 of the Main Rules, the amount of the annual contribution from national groups was discussed and was finally fixed at 10 dollars per vote.

The Committee decided, in accordance with the general opinion, that the interval between meetings should be four years, that the next International Conference should be held in 1958. It agreed to enquire of the Deutsche Gesellschaft für Elektronenmikroskopie whether it would be willing to issue an invitation and undertake the preparation for that meeting. Prof. von Borries undertook to bring this question before the next meeting of that society in Münster in March 1955.

After further discussion, the following schemes for future congresses was agreed upon:

1955	National meetings
1956	Regional meetings
1957	National meetings
1958	International Congress

As regards the formal constitution of the International Committee, Dr. Wyckoff reported that it would be considered by the Executive Committee of ICSU at its meeting in October 1954, which he hoped to attend. As soon as the decision of that body was known it should be reported to all the member national societies. It seemed desirable that the establishment and constitution of the International Committee should be ratified by the full assembly of members of each national body, rather than by their respective committees.

APPENDIX B

International Committee for Electron Microscopy

Cavendish Laboratory,
Free School Lane,
Cambridge, United Kingdom
10 March 1955

Dear _____

Circular to Assembly Members

We enclose for your information a Record of the meetings held in July 1954, during the London Conference, at which the International Committee for Electron Microscopy was established. The plans then made for it to work in parallel with the Joint Commission have had to be abandoned and we now wish to propose an alternative course of action. The objections raised by the International Council of Scientific Unions to the original proposals have been explained to you in a letter from Professor von Borries dated 22 December 1954 (enclosed is a copy of a letter from Dr. Wyckoff). The Joint Commission was unable to accept the ruling of the ICSU and has requested that the ICSU should now dissolve it.

After consulting with Dr. Le Poole and Dr. Wyckoff, it seems to us that the best course is to proceed with the establishment of an international body of electron microscopists, without further reference to the ICSU. To indicate the independence of this new body, we suggest it be called "Federation" instead of "Committee" or "Assembly." As a full title, we suggest *International Federation of Electron Microscope Societies,* which describes exactly the nature of its organization. The alternative, "International Federation of Electron Microscopists" is shorter, but is open to the objection that it implies that individual members can join.

A copy of the "Main Rules," as adopted in London, is attached to this letter together with a record of the setting up of the abortive International Committee. On a separate sheet we propose a new form of constitution, modeled on the Main Rules, but with certain changes necessitated by its reconstitution as an independent federation. The term Constitution would appear to be preferable to that of Main Rules.

The proposed changes in the original Main Rules are automatic consequences of the new situation. As the Committee will no longer contain two representatives of the Joint Commission, there is no reason to leave to the Committee the election of the president and secretary. This privilege should revert to the Assembly, as expressed in the new form of paragraph 4, in accordance with the normal practice in scientific bodies. It appears desirable to maintain the size of the Committee at 7 members, all 7 now being elected by the Assembly.

In order to proceed as rapidly as possible with setting up of a functioning body, it seems best to seek approval of these changes from the delegates who met on July 20, 1954, and approved the original constitution. Therefore we wish to ask you, as one of those delegates, to vote on the following questions by letter to the Secretary:

1. Do you agree to the new constitution?
2. Do you agree that the election of Committee members and officers held under the old Main Rules shall be valid under the new constitution?
3. For which two candidates do you vote as sixth and seventh members of the Committee?

In respect of question 3 we draw your attention to paragraph 4b of the record, showing that France, Holland, and Belgium are not so far represented on the Committee. On the other hand, bearing in mind the great activity in America in the field of electron microscopy, the election of a second member from the United States should be considered.

If and when a majority vote in favor of these proposals is received, the International Federation will be formally set up and can become active. We shall then ask all the adhering bodies for ratification of the constitution as soon as possible. There can be no objection, of course, to your requesting ratification before that date by the Executive Committee or the general meeting of your Society, if it should happen that it will be meeting in the near future. We hope that the national EM societies will be able to ratify during the next 6 to 9 months at latest, so that planning for the International Conference in 1958 can go ahead on a solid basis and in good time.

Yours sincerely,

B. von Borries (Chairman)
V. E. Cosslett (Secretary)

APPENDIX C

International Federation of Electron Microscope Societies Constitution

1. The International Federation of Electron Microscope Societies shall consist of adhering bodies, which shall be organized national groups. The

representative organs of the Federation shall be the Assembly of delegates and its Executive Committee. The delegates are appointed by the adhering bodies.

2. Voting at Assemblies shall be by countries, which shall have votes in accordance with the group in which they adhere as follows:

Group	I	II	III	IV
Number of votes	1	2	3	4

(See also paragraph 7)

The adhering body of each country shall make known to the Secretary of the Committee before the opening of each Assembly the names of the delegates (and of their substitutes) appointed the vote on behalf of that country. Each of the delegates present at an Assembly shall normally have one vote only. Any adhering body unable to be fully represented can make known to the Secretary of the Committee which other member of the Assembly has power to vote instead of an absent person.

3. Should any question arise between the time of Assemblies, on which a conclusive decision is needed, the Committee shall have power to ask the adhering bodies to send their votes by letter.

4. The Assembly shall elect a president, a secretary, and 5 Committee members. These 7 persons together form the Executive Committee, serving through the next Assembly.

5. The Assembly shall normally meet and elect the officers and members of the Committee on the occasion of quadrennial scientific congresses which the Committee will organize.

6. The Committee shall have the responsibility of arranging for the holding of periodic Congresses and for such other forms of international cooperation between electron microscopists as it considers desirable. It shall have the responsibility for administering the funds it collects.

7. Each adhering body shall pay an annual subscription according to the group in which it adheres. The number of unit contributions shall be equal to the number of the group (see paragraph 2). The amount of the unit contribution shall be set by the Committee every four years to apply to the four years that follow.

8. The group in which any national body adheres shall reflect the international contribution of this country to electron microscopy subject to the approval of the Committee.

9. Alterations in this Constitution can be proposed by the Committee or by any adhering body. In order to be included in the agenda of an Assembly, notice of any such proposal must be received by the Secretary at

least one month before the date of meeting of the Assembly. An alteration requires a majority vote of the Assembly for its acceptance.

July 1955

APPENDIX D

PRESIDENTS AND SECRETARIES OF THE INTERNATIONAL FEDERATION

Year elected	President	Secretary
1954	Prof. B. von Borries	Dr. V. E. Cosslett
1956	Prof. E. Ruska	Dr. V. E. Cosslett
1958	Dr. T. F. Anderson	Dr. V. E. Cosslett
1962	Prof. N. Higashi	Prof. J. B. Le Poole
1966	Prof. G. Dupouy	Prof. J. B. Le Poole
1970	Dr. V. E. Cosslett	Dr. W. Bernhard (resigned December 1974)
1974	Prof. D. W. Fawcett	Prof. G. Thomas
1978	Prof. J. B. Le Poole	Prof. G. Thomas
1982	Prof. H. Hashimoto	Prof. G. Thomas

1.2
IFSEM 1995: Objectives, Organization, and Functions

ARVID B. MAUNSBACH

Department of Cell Biology, Institute of Anatomy, University of Aarhus
DK-8000 Aarhus C, Denmark

GARETH THOMAS

Department of Materials Science and Mineral Engineering, University of California,
Berkeley, California 94720, USA

I. Introduction

The creation in 1954 of the International Federation of Societies for Electron Microscopy, IFSEM, as described by Ellis Cosslett, was a complex interaction among the international pioneers in electron microscopy. The resulting small organization of 10 electron microscopy societies has grown steadily to its present 39 members. This chapter—written 40 years after the foundation of IFSEM—describes the past two decades of the growth of IFSEM and outlines its organization and cooperative roles in the field of electron microscopy today.

II. Objectives of the Federation

The objectives of the International Federation of Societies for Electron Microscopy, as stated in its present constitution, are "to promote international cooperation in electron microscopy; and to contribute to the advancement of electron microscopy in all its aspects." For this purpose, the Federation initiates different types of activities, of which the most important is the IFSEM International Congress on Electron Microscopy every 4 years. Furthermore, it sponsors regional meetings and conferences, and initiates and promotes electron microscopy research in several other ways. Most of these activities are carried out in collaboration with the national societies. IFSEM therefore serves as a coordinating organization for electron microscopy societies worldwide.

III. IFSEM General Assembly and Executive Committee

The highest ruling body in the IFSEM is the General Assembly. It consists of delegates from all member societies and meets once every 4 years in connection with an International Congress for Electron Microscopy (ICEM). The General Assembly is definitive in all principal IFSEM matters, notably the acceptance of new member societies, the acceptance of venue for the next ICEM, confirmation of regional congresses, and the financial contributions to IFSEM from the member societies.

The General Assembly invariably has many important items on the agenda and, with the large number of delegates business must be conducted efficiently. Some delegates' inclinations to discuss at length are quite impressive, but the fact that recent General Assemblies have been followed by dinner sponsored by the Congress organizers has automatically introduced a certain time discipline! Earlier, when a good dinner preceeded the General Assembly, some meetings lasted well past midnight and more than one distinguished scientist has been observed asleep on such occasions.

Despite its somewhat cumbersome format, the General Assembly has, from an overall point of view, provided significant leadership in the field of electron microscopy. The most important results are that it has succeded in choosing good venues and practices for the ICEMs and secured, through its elected officers, that the Congresses have had a high scientific standard. Furthermore, it has stimulated the formation of national societies in many countries. The influence of the General Assembly on IFSEM has clearly been enhanced by the fact that the member societies generally choose some of their most experienced scientists as delegates.

The General Assembly elects the IFSEM Executive Committee, which deals with urgent matters in between the ICEMs. It consists of 9 members: the President, Vice President, General Secretary/Treasurer, and 6 ordinary committee members (See Figs. 1 and 2). Candidates for the Committee should be active in research and be recognized internationally for their contributions to electron microscopy and its applications. Other important aspects when electing committee members are their fields of research—for example, biology or physics/materials science—as well as their geographic location, and a reasonable balance among all of these parameters has usually been aimed at. So far, committee members have come from all parts of the world, except that the African continent has not yet been fully represented.

The President and the Vice President serve for 4 years, while the General Secretary and the ordinary committee members serve for 8 years. IFSEM Presidents and General Secretaries from 1954 to the present are listed in Table 1. Presidents and General Secretaries now assume office at the

FIGURE 1. Hatsujiro Hashimoto, IFSEM President, and Gareth Thomas, IFSEM General Secretary, conducting the IFSEM General Assembly during ICEM 11 in Kyoto, 1986.

beginning of the calendar year following that in which they were elected, but before 1978 the President, according to the earlier rules, assumed office "12 months after the close of the calendar year in which he was elected." Figures 3–11 show a few of the well-known scientists from different fields of electron microscopy who have served as committee members during the the last two decades.

The Executive Committee meets separately from the General Assembly at the ICEMs and at least once in between, normally during a regional congress. In recent years the Presidents and other members of the Committee have been very active in participating in national or regional congresses to lecture and to visit different countries to stimulate new membership (Fig. 9).

In the daily running of the IFSEM, most of the activities are carried out by the General Secretary: correspondence with member societies, interactions with other international scientific organizations, distribution of information regarding EM meetings in different parts of the world, evaluation of applications for international fellowhsips for ICEMs, and handling of the IFSEM

FIGURE 2. Meeting of the IFSEM Executive Committee during ICEM 12 in Seattle, 1990, including newly elected members and guest. Standing (from left): David J. H. Cockayne, Australia (General Secretary from 1995; committee member 1987–1994), Kazuo Ogawa, Japan (committee member 1987–1994), Ernie F. J. van Bruggen, The Netherlands (committee member 1983–1990), K. H. Kuo, Beijing, China (guest, Chairman of Organizing committee for 5th Asia-Pacific Electron Microscopy Conference, 1992), Ágnes Csanády, Hungary (committee member from 1991). Sitting (from left): Robert M. Fisher, United States (committee member from 1991), Arvid B. Maunsbach, Denmark (President from 1995; General Secretary 1987–1994; committee member 1983–1986), Gareth Thomas, United States (President 1987–1990; Vice President 1991–1994; General Secretary 1975–1986), Hatsujiro Hashimoto, Japan (President 1983–1986; Vice President 1987–1990), and Elmar Zeitler, Germany (President 1991–1994; Vice President from 1995; committee member 1983–1990). Not in photo: Archie Howie, U.K. (committee member 1987–1994), Armin Delong, Czechoslovakia (committee member 1987–1990), and Hans Geuze, The Netherlands (committee member from 1991).

TABLE 1

Presidents and General Secretaries of the
IFSEM: 1954 to Present

Year elected	President	General Secretary
1954	B. von Borries	V. E. Cosslett
	E. Ruska [a]	
1958	T. F. Anderson	V. E. Cosslett
1962	N. Higashi	J. B. Le Poole
1966	G. Dupouy	J. B. Le Poole
1970	V. E. Cosslett	W. Bernhard
1974	D. W. Fawcett	G. Thomas
1978	J. B. Le Poole	G. Thomas
1982	H. Hashimoto	G. Thomas
1986	G. Thomas	A. B. Maunsbach
1990	E. Zeitler	A. B. Maunsbach
1994	A. B. Maunsbach	D. J. H. Cockayne

[a] Appointed 1956 due to the death of Professor von Borries.

FIGURE 3. Wilhelm Bernhard, France (General Secretary 1971–1974).

FIGURE 4. Gérard T. Simon, Canada (committee member 1978–1986). Chairman of Organizing Committee of ICEM 11 in Toronto, 1978.

economy. Other important tasks are to invite and distribute proposals for ICEM venues and Committee candidates and to prepare the meetings of the IFSEM General Assembly. The General Secretary has many duties and certainly gets to know the member societies with respect to the strengths and weaknesses of organization, quality of research, and ability to communicate.

IV. IFSEM Member Societies

In 1995 the IFSEM has 31 member societies and 8 associated member societies (Table 2). The largest societies are the Microscopy Society of America, with approximately 4600 individual members, and the Chinese Electron Microscopy Society, with 3500 members; the majority of the member societies range between 200 and 600 members. The associate member societies are smaller, typically with 30–100 individual members, and were

FIGURE 5. Ewald R. Weibel, Switzerland (committee member 1975–1978).

FIGURE 6. Kazuo Ogawa, Japan (committee member 1987–1994). Chairman for Organizing Committee of ICEM 11 in Kyoto, 1986.

FIGURE 7. Robert M. Fisher, United States (committee member 1991–present). Co-Chairman of ICEM 12 in Seattle, 1990.

FIGURE 8. Bernard Jouffrey, France (committe member 1980–1986). President of ICEM 13 in Paris, 1994.

formed during the last few years. The total number of individual scientists belonging to the 39 IFSEM societies is about 20,000.

The following chapters in this book illustrate that the member societies each have their specific histories of formation, development, and function. However, they have in common that their individual members are microscopists within different fields of research, including biology, medicine, physics, and materials science. Most member societies have members from only one country, but there are a few exceptions. Thus, the Scandinavian Society has members from Denmark, Finland, Norway, and Sweden, and the Czechoslovak Society represents both the Czech and the Slovak Republic.

The number of votes allocated to each member society at the IFSEM General Assembly ranges from 1 to 4 (Table 2) and is related both to the number of individual scientists who are members of the society as well as to the general impact of the society in the field of electron microscopy. Associate member societies have no voting rights and do not play annual dues to IFSEM, but are in all other respects equal to Member Societies.

FIGURE 9. Elmar Zeitler (left) and Arvid Maunsbach (right) at the 5th Asia-Pacific Conference on Electron Microscopy in Bangkok in 1988. Center, conference participant Dr. Noriko Suzuki, Japan.

FIGURE 10. Ellis Cosslett presenting the new European Committee of European Societies for Electron Microscopy (CESEM) at the 6th European Congress on Electron Microscopy in Jerusalem, 1976.

V. IFSEM REGIONAL COMMITTEES

In addition to member societies, the Federation has two regional committees: the Committee of European Societies of Electron Microscopy, CESEM, and the Committee of Asia-Pacific Societies for Electron Microscopy, CAPSEM.

Towards the mid-1970s some scientists in Europe expressed the view that the European Societies for Electron Microscopy should form a regional subgroup of IFSEM specifically to enhance the coordination of electron microscopy within Europe. In May 1975 a meeting was therefore held at Schipol International Airport, Amsterdam, following a Belgian/Dutch initiative and with representatives from Belgium, France, the German Federal Republic, the Netherlands, Scandinavia, and the United Kingdom. It was agreed to form a Preparatory Committee to draw up a constitution, which would then be put before all European Societies. The representatives elected to this Committee were V. E. Cosslett (chairman), A. B. Maunsbach (secretary), K. von Bassewitz, M. Bouteille, L. Simar, and E. Wisse. A constitution, aimed at simplicity, was drafted over the following months,

FIGURE 11. Tom Mulvey (second from left), IFSEM committee member 1975–1982, discussing with participants at the inaugural meeting of CESEM in Jerusalem, 1976; Antonio Coimbra, Portugal (left), Gerhard Schimmel, West Germany (third from left), and newly appointed CESEM Secretary Arvid Maunsbach (right).

and the proposal to form a Committee of European Societies for Electron Microscopy, CESEM, was formalized 19 September 1976, during the European Congress on Electron Microscopy in Jerusalem. Ellis Cosslett (Fig. 10) was elected Chairman and Arvid Maunsbach Secretary/Treasurer, and initially there were 10 European member societies (Fig. 11).

The European Committee works under the auspices of the IFSEM and was officially recognized by the IFSEM as a Regional Organization of the Federation in 1978. The Committee has gradually increased in membership and had 23 European member societies in 1994. It has contributed to the organization of subsequent European Congresses by appointing scientific advisory boards. To stimulate contacts among the European societies, Arvid Maunsbach, while he was CESEM Secretary, initiated a European EM Newsletter with information about coming events, particularly coming meetings related to electron microscopy.

Suggestions within CESEM to arrange full-scale European Congresses on Electron Microscopy every second year have been rejected, mainly on the grounds that there are already too many electron microscopy congresses. However, joint meetings between neighboring European societies have

TABLE 2

IFSEM MEMBERSHIP, 1995

	Member since	Individual Members	Votes in EFSEM
Member societies			
Microscopy Society of America	1954	4600	4
Australian National Committee for Electron Microscopy	1962	280	2
Austrian Society for Electron Microscopy	1974	280	1
Belgian Society for Electron Microscopy	1954	220	2
Brazilian Society for Electron Microscopy	1992	400	1
British Joint Committee for Electron Microscopy	1954	1000	3
Microscopical Society of Canada	1974	620	2
Chinese Electron Microscopy Society	1986	3500	1
Electron Microscopy Society, Taipei, China	1982	120	1
Czechoslovak Society for Electron Microscopy	1958	130	1
Deutsche Gesellschaft für Elektronenmikroskopie	1954	880	4
Dutch Society for Electron Microscopy	1954	525	2
French Society for Electron Microscopy	1954	1180	4
Hungarian Society for Electron Microscopy	1958	180	1
Electron Microscope Society of India	1962		1
Microscopial Society of Ireland	1982	70	1
Israel Society for Microscopy	1966	300	2
Italian Society for Electron Microscopy	1958	630	2
Japanese Society for Electron Microscopy	1954	2800	4
Korean Society for Electron Microscopy	1978	300	1
Mexican Society for Electron Microscopy	1992	150	1
New Zealand Society for Electron Microscopy	1982	140	1
Polish Commission for Electron Microscopy	1974	150	1
Portuguese Society for Electron Microscopy and Cell Biology	1970	200	1
Russian Society for Electron Microscopy	1970		2
Scandinavian Society for Electron Microscopy	1954	750	3
Electron Microscopy Society of Southern Africa	1966	300	1
Spanish Society for Electron Microscopy	1958	300	1
Swiss Society for Electron Microscopy	1954	450	2
Electron Microscopy Society of Thailand	1986		1
Venezuelan Society for Electron Microscopy	1992	150	1
Associate member societies			
Armenian Electron Microscopy Society	1994	33	0
Bulgarian Society for Electron Microscopy	1994		0
Croatian Society for Electron Microscopy	1994	42	0
Latvian Society for Electron Microscopy	1994	34	0
The Electron Microscopy Section, Moldova	1994	60	0
Rumanian Society for Electron Microscopy	1994		0
Microscopy Society (Singapore)	1994	160	0
Slovenian Society for Electron Microscopy	1994	120	0

been held several times recently, for example, the Austrian-Hungarian Joint Conference on Electron Microscopy in 1987, the Dreiländertagung among Austria, Germany, and Switzerland in 1993, and the Multinational Congress on Electron Microscopy arranged jointly by the Italian, Czechoslovak, Hungarian, and Slovenian Societies for Electron Microscopy in 1993.

Although there was considerable activity in Japan and also some informal Asian regional meetings in IFSEM's early years, it was not until 1980 that a similar initiative was launched for the Asia-Pacific region. Vigorous activity followed the admission of South Korea, Taiwan, Thailand, and later China to IFSEM membership, especially as the result of efforts by Gareth Thomas and Hatsujiro Hashimoto to stimulate electron microscopy activities in Southeast Asia and their attendances at many national conferences. The Committee of Asia-Pacific Societies for Electron Microscopy, CAPSEM, was founded in 1984 at the Asia-Pacific Conference on Electron Microscopy in Singapore under the presidency of Hatsujiro Hashimoto and with David J. F. Cockayne as secretary. The next APEM Conferences were held in Bangkok in 1988 (Fig. 12) and in Beijing in 1992. The 1996 conference will be held in

FIGURE 12. Meeting of the Committee for Asia-Pacific Societies for Electron Microscopy (CAPSEM) in Bangkok, 1988. Left: CAPSEM President Hatsujiro Hashimoto; right: CAPSEM General Secretary David J. F. Cockayne.

Hong Kong. Thus, as with CESEM, meetings are held every 4 years interlaced between the ICEMs and likewise the constitution of CAPSEM occupies only a single page. CAPSEM has incorporated several groups of electron microscopists from countries that are not yet members of IFSEM, e.g., Burma, Hong Kong, and Pakistan.

Due to communication problems, the Latin American Society was formally disbanded at the 1990 IFSEM Congress in Seattle. Following this, several informal meetings were held at regional or national congresses, first in Maracaibo, Venezuela, and then in Cancun, Mexico, in 1992. As a result, a new organization, the Inter-American (or Latin American) Committee for Electron Microscopy was formed. The Committee has not yet asked for official recognition as a regional organization of the IFSEM, and it remains to be determined which are the member societies of the Committee. IFSEM expects the situation to be clarified before ICEM XIV, to be held in Cancun, Mexico, in 1998.

VI. IFSEM Constitution

The current constitution of the IFSEM was ratified at the International Congress on Electron Microscopy in Kyoto on 2 September 1985, after thorough work by Gerard Simon, Canada, and with important contributions by Gareth Thomas and other committee members. The principle of the new constitution is precision and simplicity. The old constitution of IFSEM in the 1970s became increasingly cumbersome to cope with, due to the complicated voting rules.

The constitution of IFSEM has functioned smoothly since 1986, but a few modifications have been necessary. Thus, in 1992 it was decided to create associate memberships of the Federation, following a proposal by Gareth Thomas at the European Congress in Granada. As a result of the new developments, particularly in Eastern Europe, a number of new societies were formed or emerged as independent organizations, but were comparatively small. Such societies are now eligible for associated membership of the Federation, but are without voting rights and without financial contribution, albeit in other respects equal to regular members. Another addition to the constitution was introduced in 1994 when it was decided that nominees for the Executive Committee must confirm their willingness to accept election.

VII. International Congresses

The International Congresses for Electron Microscopy represent major events in the field of electron microscopy. The number of participants has ranged in the last 20 years between 1400 in Hamburg 1982 and 4500 in Seattle 1990. An International Advisory Committee has been appointed by the IFSEM for each ICEM. While the Organizing Committee has had the final word regarding the congress program, the International Advisory Committee has always been available to facilitate the planning.

The General Assembly of the IFSEM decides the venue for the International Congresses. The choice is based on proposals from member societies, which must contain information on congress facilities, exhibition possibilities, and program structure. Up to now there have always been several good proposals to choose among for each International Congress. In fact, the competition among different proposals at the General Assemblies has been increasingly fierce over the last two decades. The principle has been to choose the best proposal regardless of geographic location. It has sometimes been argued that a certain sequence should be established, e.g., among the continents, but as late as in 1994 this was rejected by the IFSEM General Assembly.

The scientific content of the international congresses has gradually changed over the last 20 years. Emphasis now is on instruments, methods, and analytical procedures, rather than specific information about, e.g., ultrastructural characteristics of cells or nanostructures in materials science. While there seems to be a general consensus about this development, it has, at the same time, the consequence that many scientists using the electron microscope as a routine tool often prefer to attend meetings within their own specialized fields, rather than EM meetings. It is therefore crucial that future organizers emphasize the common ground for microscopists, i.e., instruments, preparation procedures, analystical methods, interpretation, image processing, etc. Thus, in our opinion, plenary lectures and symposia should focus on recent developments in these areas. Specialized applications should also be accepted, since it is often a requirement from granting agencies that a communication is actually presented by the participant, but such contributions should in most cases be presented in poster sessions. This format will be adhered to at APEM 1996 and ICEM 1998.

Undoubtedly, the International Congresses have stimulated personal contacts among scientists from different parts of the world and also enabled participants to become acquainted with the cities and countries where the congresses have been held. A highlight of the 1986 congress in Kyoto was the attendance of Crown Prince (now Emperor) Akihito of Japan, himself a marine biologist and a user of the scanning electron microscope.

VIII. Relations with Industry

Over the last 5–10 years, the IFSEM has developed increasing contact with the manufacturing industry, particularly in connection with the instrument exhibitions held at the international or regional congresses. It is the undisputed opinion of the IFSEM that cooperation with industry is indispensable in the field of electron microscopy.

During the last few years discussions with the manufacturers have centered around financial problems. While industrial representatives have argued that exhibition fees are too high, congress organizers advocate the necessity to charge for exhibition space. To facilitate negotiations, IFSEM has invited industrial representatives for discussions of practical and financial questions.

There is no doubt that the instrument exhibition forms a central part of an international congress. During recent ICEMS, the instrument exhibitions have been large and very well attended and, most important, contained many new technical developments. Undoubtedly, the exhibitions for many participants represent one of the major reasons for attending ICEM and it is essential that in the future they also continue to be well planned.

IX. The Expansion of Microscopies

Over the last decade, a number of new microscopic methods have appeared, notably scanning probe microscopes and confocal microscopes. The research aims of these new microscopies coincide to some extent with these of electron microscopy, and several national EM societies have incorporated the new microscopies in their programs. As a consequence, a few member societies of IFSEM have changed their names accordingly. Thus, for many years the Canadian Society has been the Microscopy Society of Canada, the Electron Microscopy Society of America, EMSA, is now the Microscopy Society of America, MSA, and the Israel Society recently became the Israel Society for Microscopy. While some societies, such as the Scandinavian Society, have delayed decisions in this matter, this change of outlook is likely to appear in many EM societies. The IFSEM General Assembly has not discussed this matter yet. One possibility is to change the name of the IFSEM in the same way, but this metamorphosis probably lies a few years in the future. Furthermore, this almost certainly will require negotiations with the International Council of Scientific Unions, of which the IFSEM is an associate member.

X. Conclusions

The International Federation of Societies for Electron Microscopy is an active organization promoting electron microscopy worldwide. It has continued to grow over the last two decades and now has 39 member societies. The main contributions of the Federation are to bring together scientists within the field of electron microscopy for international meetings and consequent interactions and in doing so to secure a high scientific standard in the various fields of electron microscopy.

Part II

SOME INDIVIDUAL SOCIETIES

2.1
Electron Microscopy in Australia

PETER S. TURNER[1]

Division of Science and Technology, Griffith University
Brisbane, Queensland 4001, Australia

MAREK VESK

Electron Microscope Unit, University of Sydney
Sydney, New South Wales 2006, Australia

I. Introduction

Australia is large in area, with a relatively small population but a significant reputation for creative scientific work. Dependent on primary produce from pastoral, agricultural, and mining sectors for export income, Australia has a notably weak involvement of business and industry in support of scientific research. The federal government has been, and continues to be, the major source of research funding, through the Commonwealth Scientific and Industrial Research Organisation (CSIRO) originally established as the Council for Scientific and Industrial Research, and through support for universities, especially since the mid-1950s. University research has been supported both directly through operating grants, and through funding agencies such as the National Health and Medical Research Council and the Australian Research Council (ARC).

It is not surprising, then, that the history of electron microscopy (EM) in Australia is dominated by the achievements of CSIRO scientists and their colleagues in the universities. No electron microscopes have been manufactured, nor, with one exception, has Australia's small scientific instrument industry contributed to the production of electron microscope accessories to any significant degree. Nevertheless, electron microscopists worldwide depend on several major developments by Australian scientists.

For over 30 years the country's microscopists have been affiliated with the IFSEM through the Australian Academy of Science, rather than through a society for electron microscopy. The Academy also sponsored the successful

---

[1] Present address: CSIRO Division of Wool Technology, Geelong Laboratory, Belmont, Victoria 3216, Australia.

series of biennial EM conferences from 1968 to 1986, as well as the Eighth International Congress held in Canberra in 1974. The Australian Society for Electron Microscopy (ASEM) was formed quite recently, to take over the organization of the conferences, but the Academy of Science, through its National Committee for Electron Microscopy, remains the corresponding member of the IFSEM.

In this brief article, we review the early history of electron microscopy in this country, and the contributions of the first microscopists to the development of the technique and the training of other scientists; the activities of the Academy's National Committee for Electron Microscopy, and the recent formation of the ASEM; as well as the establishment of major electron microscope laboratories across the country. We also outline briefly several topics central to contemporary electron microscopy worldwide, which are based on Australian research.

II. Early History

In Australia, interest in electron diffraction and its application in studies of surfaces and gases appears to have developed at the universities of Melbourne and Adelaide in the late 1930s. In Melbourne, in 1937, A. L. G. Rees presented a colloquium on "Electron Diffraction of Gaseous Molecules" to the Chemistry Department, as part of his M.Sc. program. His doctoral studies were later conducted at Imperial College, London, from 1939 to 1941, where both electron diffraction and microscopy were developing rapidly within groups led by G. P. Thomson, G. I. Finch, and L. C. Martin. Dr. Rees then worked at the Philips Electrical Industries Laboratory in England, on a surface emission electron microscope. In 1943 he was appointed head of the new Chemical Physics Section within the CSIR's Division of Industrial Chemistry, to introduce electron microscopy, X-ray and electron diffraction, and optical spectroscopy to the work of the Division.

The earliest major electron-optical instrument in Australia was a Finch-type electron diffraction camera built by the Edwards Company installed in the Physics Department at the University of Adelaide for surface physics research by Dr. R. S. Burdon. This was used for RHEED studies by J. M. Cowley during his M.Sc. studies (1943–1945), and subsequently by J. V. Sanders.

The first electron microscope was an RCA EMU, delivered in 1945 to CSIR's Chemical Physics Section. John M. Cowley and John L. Farrant were appointed by Rees to the staff of the section, and the stage was set for the major contributions to the fields of electron diffraction and microscopy from what became the CSIRO Division of Chemical Physics, over a period of some 40 years.

To many of today's younger microscopists, the names of John Cowley and Alex Moodie are probably those most closely associated with Australian contributions to electron microscopy. However, that was a later development, considered below. Cowley was appointed to CSIR to develop electron diffraction as a research technique, and after some early work with the RCA microscope, he and Rees designed an electron diffraction camera for studies of the fine structure in dynamical scattering from small crystals. A. F. Moodie joined the Division in 1948. Together, based on their experimental diffraction studies, Cowley and Moodie developed the optical approach to the dynamical theory of electron diffraction, now known as the multi-slice method. Figure 1 shows Alex Moodie (left) and John Cowley operating the first electron diffraction camera built at CSIRO. The impact of their work on electron microscopy began in the 1960s with Cowley's move to the University of Melbourne, and with the installation of an Hitachi HU-125 at Chemical Physics.

In electron microscopy itself, the significant early Australian contributions were made by John Farrant and his collaborators, through their work with the RCA EMU, and through the development of improvements to that instrument, such as the addition of an externally adjustable objective aperture, auto-bias electron gun, and objective lens stigmators. The ultimate resolution achieved with the instrument was the then excellent level of 0.7 nm; by 1950, high-resolution images of ferritin were recorded at magnifications of 320,000 × at 50 kV, showing detail of the iron core (see Goodchild and Dowell, 1986). A considerable variety of materials was studied over the period 1945–1950, including wool fibers (Mercer and Rees, 1946a, 1946b); minerals (Bayliss et al., 1948); virus particles (Farrant and O'Connor, 1949); minerals in skeletal muscle (Draper and Hodge, 1949), and wood fibers (Hodge and Wardrop, 1950). Thus the new techniques of microscopy were introduced to a number of Australian scientists through use of the RCA EMU at Chemical Physics in their work across a wide range of disciplines.

A second RCA EMU was purchased for the Defence Research Laboratories and installed in 1948 under the supervision of Dr. R. I. Garrod. In addition to particle size analysis, replica techniques were used in studies of the deformation of high-purity aluminum, in collaboration with the University of Melbourne (DRL Annual Reports, 1949, 1950; Garrod et al., 1952). Two of the smaller RCA EMU console-model instruments were installed in 1947, at the Sydney Technical College and in the Postmaster General's Research Laboratory. A Philips EM 100 was installed at the University of Adelaide in 1952, for biophysical studies of cell membrane structures by Dr. S. G. Tomlin. The fourth large electron microscope in Australia was a Siemens UM100, installed at the Walter and Eliza Hall Institute of Medical Research in Melbourne in 1953, for work on viruses. Further details of these early developments have been described by Goodchild and Dowell (1986).

III. THE ORGANIZATION OF ELECTRON MICROSCOPY IN AUSTRALIA

A. *The National Committee for Electron Microscopy*

Two Australians, J. L. Farrant and J. V. Sanders, attended the First International Conference on Electron Microscopy in Delft in 1949. Following the 1950 International Conference in Paris, and the proposal to form a Joint Commission on Electron Microscopy under the International Council of Scientific Unions (ICSU), support was requested from Australian microscopists; in 1951, this resulted in the formation of the National Committee for Electron Microscopy (NCEM), as a committee of the Australian National Research Council. J. L. Farrant was appointed convenor, with R. Mitchell (Walter and Eliza Hall Institute), S. G. Tomlin (University of Adelaide), and A. L. G. Rees.

Initially, the committee acted merely as a mechanism for communication between Australian electron microscope laboratories and overseas societies. E. H. Mercer was appointed to represent the country at the Joint Commission on EM held in London in July 1954, when Australia was accepted as a member of Group I of the adhering bodies. In 1954, the role of the Australian National Research Council was taken over by the new Australian Academy of Science (AAS), which became the corresponding body for the ICSU. Since the recently formed International Federation of Electron Microscope Societies had been established independently of the ICSU, the Academy decided that it could not take over the NCEM. So at the 4th International Congress in Berlin in 1958, J. L. Farrant attended the IFEMS general assembly as an "unofficial" observer.

These difficulties were resolved by the AAS Council in 1961, following an approach by A. L. G. Rees. The NCEM was reestablished as a committee of the Academy, with J. L. Farrant as convenor. The new committee paid the IFEMS subscription to establish membership and was represented by M. F. C. Day and D. G. Drummond in 1962 at the General Assembly held during the 5th International Congress.

The most significant work of the National Committee was initiated in 1966 and 1967 at the ANZAAS (Australia and New Zealand Association for the Advancement of Science) meetings, with the decision to hold a

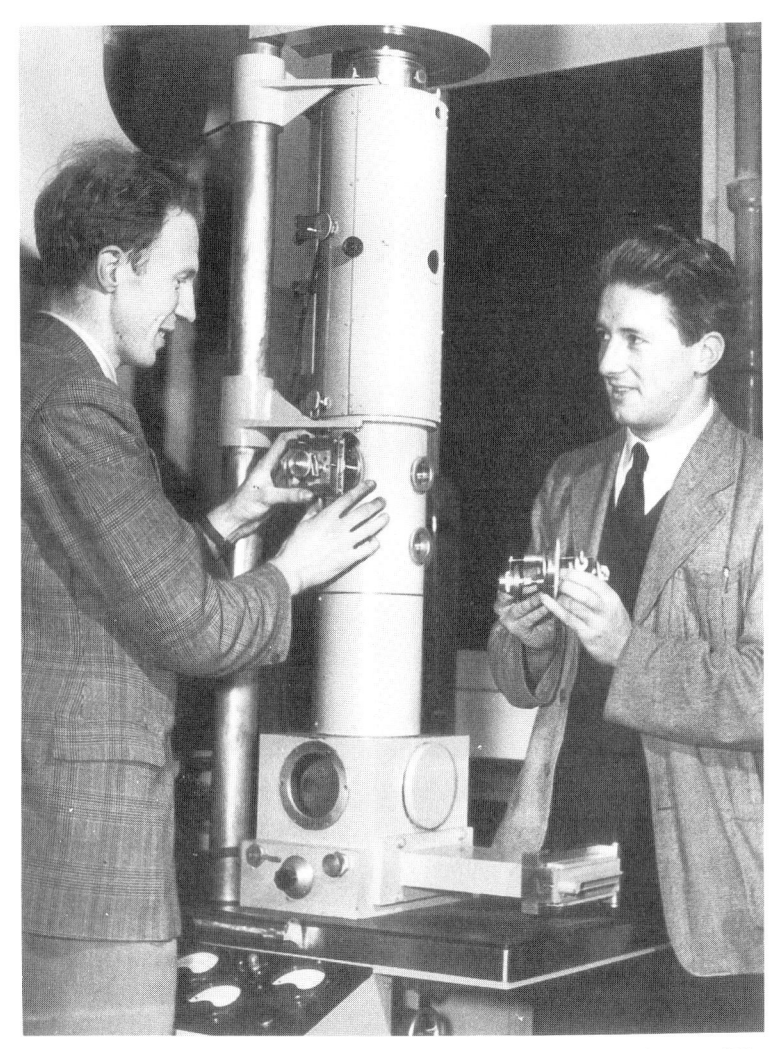

FIGURE 1. Alex Moodie (left) and John Cowley operating the first electron diffraction camera built at CSIRO (Courtesy of Professor J. M. Cowley).

national conference in February 1968, in Canberra. Although there was at that time considerable support for the formation of a society for EM, the final decision was that the conference should be held under the auspices of the National Committee, with support from the Academy of Science. Also discussed at the 1967 ANZAAS meeting was the possibility of holding the 1974 International Congress in Australia.

A further factor in the initiation of the Australian series of EM conferences may have been the highly successful international conference on "Electron Diffraction and the Nature of Defects in Crystals," held at the University of Melbourne during August 1965, which attracted many scientists from overseas. Electron microscope image contrast, and TEM applications in the study of crystal defects, featured strongly in the program. In addition, an instrument exhibition included several electron microscopes.

Following this conference, a workshop was held at the Warburton Chalet near Melbourne. Lectures were presented on X-ray and electron diffraction, both kinetic and dynamic, and on diffraction contrast, by authorities from around the world and from Australia. This meeting came at a crucial time in the development of the physical basis of electron microscopy, and it had a major impact on many of the students and younger Australian scientists who attended. Figure 2 shows the group photograph at Warbuton. It contains many well-known faces, including that of Ewald, and many of John Cowley's students or younger staff, who have now become the "older generation": Cockayne, Spargo, Johnson, Turner, and others. It also marks the coming together of the Oxbridge (Hirsch, Howie, Whelan, *et al.*) and the Melbourne approaches to n-beams and image contrast.

The 1968 First Australian Conference on EM, organized by a committee led by David Goodchild and Barry Filshie, was most successful in bringing together many of Australia's users of EM, including students and technical staff, in a format which resulted in the attendance of a number of invited speakers from overseas. The conference started with plenary sessions attended by both biological and physical microscopists, followed by parallel physical and biological sessions. This format has been continued for the subsequent series of biennial meetings. The major equipment exhibitions associated with these meetings have proved particularly valuable for both manufacturers and microscopists, because of the isolation of most users from examples of the latest instrumentation and accessories.

The success of the 1968 conference resulted in the National Committee making a bid to host the 8th International Congress in Canberra in August 1974, which was accepted by the IFSEM General Assembly at the 7th Congress in Grenoble. The 8th Congress was attended by 1000 delegates, 600 of whom were from outside Australia. Australian contributions to the international organization of EM have continued, through scientific delegations at subsequent conferences, representation at IFSEM assemblies, and through membership on the IFSEM executive (J. L. Farrant, 1970–1978; D. J. H. Cockayne, 1986–1994).

In addition to the Australian conferences and to its international role, the NCEM has promoted electron microscopy within the country in representations to government from time to time, and initiated the *Directory of*

FIGURE 2. Group photograph of the workshop on electron and X-ray diffraction held at the Warburton Chalet, near Melbourne, in August 1965.

Electron Microscope Units and Staff. The first edition, produced by D. J. Goodchild after the 1968 meeting, lists 124 microscopes and 400 users. The directory includes brief descriptions of equipment and areas of interest for each laboratory, and has proved valuable for people seeking information about techniques from colleagues. It has been revised every two years since 1968 for the biennial conferences.

A further initiative of the NCEM has been the Australian *EM Newsletter.* In 1983 the Committee persuaded Dr. G. C. Cox to take on the task of producing the newsletter, and the first issue was published in time for the 8th ACEM in Brisbane. During his period as editor, Guy Cox built up an excellent publication, with a balance between general articles of interest to microscopists, news items and reports from special-interest groups, and strong support from the EM suppliers. The newsletter has proved to be most valuable in strengthening communication among EM users in this country. In 1984, the Microbeam Analysis Society of Australia adopted the *EM Newsletter* as its official organ, followed by the Microscopical Society of Australia and the newly formed ASEM in 1986. The newsletter has thus become a valuable link between the country's electron and optical microscopists and analysts. Since 1988, the editor has been John Terlet.

B. *The Australian Society for Electron Microscopy*

The Australian Society for Electron Microscopy (ASEM) is a child of necessity. It was formed in 1986 to take over the responsibility for the biennial Australian EM conferences, after the Academy of Science decided to cease its sponsorship of such meetings. The Society, which has a membership of over 400, an elected executive committee, and a formal constitution, was seen as providing the coordinating framework within which the conference series could continue to be organized. At the same time it was realized that the Society could also have a function in coordinating other EM activities within the country. For example, from 1990 the Society, in conjunction with the (optical) Microscopy Society of Australia, has taken over responsibility for publication of the successful *EM Newsletter.* It administers three scientific awards (see below), and has provided the focus for a number of special-interest groups based in the major Australian cities. However, the Academy of Science, through the NCEM, remains the official Australian representative to the IFSEM, while collaborating with the Society in determining issues of significant policy.

C. *Awards*

The first Australian award for work in the field of EM resulted from an initiative of Philips Scientific and Industrial (Australia). The Cowley-

Moodie Award commemorates the significance of the work of J. M. Cowley and A. F. Moodie by providing a young electron microscopist with the opportunity to visit international laboratories. Initially, the award was made annually, and was open to both physical and biological scientists; the first Cowley-Moodie award was made to Dr. T. J. White. From 1988, the C-M Award has been biennial, restricted to the physical and chemical sciences.

Subsequently, two further awards have been established. The David Goodchild Award, which recognizes the contributions to Australian electron microscopy of the late Dr. David Goodchild, is also a travel grant, offered biennially for a young microscopist in the biological and medical sciences. The Sanders Medal, in memory of Dr. John V. Sanders, is awarded biennially to an Australian scientist of international distinction for excellence in developing or applying EM techniques in the physical and chemical sciences over the previous five years.

All three awards are now established under the auspices of the ASEM. The inaugural Goodchild and Sanders presentations were made at the 11th ACEM, held in Melbourne in 1990, to Dr. Michelle Williams and to Dr. Peter Tulloch, respectively.

IV. Major Australian Centers for Electron Microscopy

The crucial role of the CSIRO Division of Chemical Physics as the major early center for electron microscopy in this country has been described earlier. A list from 1951 lists seven microscopes installed or ordered, and 11 staff, 4 of these at Chemical Physics. By 1968, the first *Directory of EM Units and Staff* included references to 124 microscopes (including 1 Cambridge Stereoscan), and about 400 users (staff and postgraduate students). These instruments and people were distributed through all the universities and many CSIRO divisions and other government research institutions, hospitals, and a few industrial laboratories. Central EM units had been established at the Universities of Sydney, Queensland, and Western Australia.

Until about 1960, most Australian electron microscopists received their research training in the field overseas, usually in the UK. Since then, two centers have had a major impact on research training, in both the biological and physical sciences—the EM unit at the University of Sydney, under Dr. D. G. Drummond, and the Diffraction Group at the University of Melbourne's Physics Department, established by Professor J. M. Cowley in 1963.

Within most universities, microscopes were (and still are) found in individual departments, resulting in large variations in the extent and quality of usage and (especially) of training. At the University of Sydney the EM Unit began operating in 1958 with the appointment of Dr. D. G. Drummond and the installation of a Siemens Elmiskop I and ancilliary preparative facilities. From the beginning it was envisaged that this "new type of service unit. . . [would be] independent of any department or faculty,. . . [with] facilities available for any research project in the University which demanded them" (Drummond, 1960). The success of this (for the time) unusually far-sighted policy has been reflected in high instrument usage, professional training of users in both the scientific and technical aspects of electron microscopy, and the availability to all users of very highly qualified advice from research scientists active in the development of EM applications. The University of Sydney remains well ahead of most institutions, both nationally and internationally, in the provision of centralized EM facilities and services. The principles of central units (Drummond, 1971) are even more relevant now, given the costs of both instruments and expert staff, and have been adopted in the reorganization of several units over the past few years. In addition, a number of groups of institutions have collaborated to form interinstitutional structures for the purchase and operation of major new microscope facilities.

In addition to serving the local needs, Dr. Drummond established a series of Vacation Schools in Electron Microscopy, which have provided expert training of scientists and research students from all parts of the country, thus continuing the role which originated at Chemical Physics. Figure 3 shows Dr. and Mrs. Drummond at the 1978 Christmas party of the University of Sydney Electron Microscope Unit, five years after Dr. Drummond's retirement. The schools have continued under the present director, Dr. D. J. H. Cockayne, and have expanded into the Asian-Pacific region through Australia's close involvement with the Committee for Asian-Pacific Societies for EM (CAPSEM), of which Dr. Cockayne has been General Secretary since 1986. CAPSEM, which became a member of the IFSEM in 1990, has a responsibility for fostering EM in Asia, especially through sponsorship of the four-yearly Asia-Pacific conferences. CAPSEM has invited the EM Unit at the University of Sydney to organize training workshops immediately prior to the conferences, and these attracted approximately 100 participants at both Singapore (1984) and Thailand (1988).

The growth of the Diffraction Group at Melbourne University under Professor John Cowley was rapid from 1963, with a small group of staff, including A. E. C. Spargo, and a large number of graduate students. The influence of the Melbourne group is well recognized, and has continued under Spargo and L. A. Bursill since Cowley's move to Arizona in 1969, especially in the area of crystal structure determination by high-resolution imaging and image matching.

FIGURE 3. Dr. and Mrs. D. G. Drummond at the 1978 Christmas Party of the University of Sydney Electron Microscope Unit, five years after Dr. Drummond's retirement.

The close interaction between Cowley's group and CSIRO researchers—Moodie, Goodman, and Johnson at Chemical Physics, and Sanders, Head, and Loretto at Tribophysics—was of great benefit to a number of students. The separation between the Australian "multislice" groups in Melbourne and the "Bloch wave" groups of Hirsch and Whelan in Oxford and Howie in Cambridge was bridged during the late 1960s and early 1970s through postgraduate students such as D. J. H. Cockayne, and postdoctoral research workers (P. S. Turner, D. J. Smith, J. C. Barry) and other visitors.

While these two centers were probably the strongest, a substantial number of smaller groups, in universities and CSIRO, contributed greatly to the overall development of electron microscopy during the 1960s and 1970s. There is insufficient space to list them here, but an indication of their importance, and of the growth of the discipline, can be seen in the 1988 edition of the *Directory,* which lists over 1100 users of some 220 transmission microscopes, 180 SEMs, and 23 microprobes, installed within 240 laboratories, including about 25 EM units or centers (Beaton, 1988).

V. Australian Contributions to Electron Microscopy

A. Instrumentation

In addition to the significant improvements made to the performance of the RCA EMU-1 by J. L. Farrant and his colleagues during the late 1940s and early 1950s, the group at CSIRO Chemical Physics was very concerned with the problems of specimen preparation and staining. Farrant and A. J. Hodge investigated microtome designs for thin sectioning; Hodge's collaboration with H. E. Huxley resulted in the construction at CSIRO of a prototype instrument, and the subsequent production by the Fairey Aviation Company in Australia of 50 SIROFLEX microtomes (Farrant and Powell, 1956).

Very advanced levels of scientific instrument design and construction were evident at CSIRO in the work of J. C. Mills, who was largely responsible, in collaboration with J. M. Cowley and A. F. Moodie, for a series of electron diffraction instruments, culminating in a convergent-beam camera based on an inverted HU-11 objective lens—a geometry seen later in dedicated STEM instruments, but used here to achieve the small crossovers essential for the new CBED technique (see later) (Cockayne *et al.,* 1967). The Mills sample holder, with tilting and heating facilities, was also fitted to the HU-125. A further development of this type of top-entry tilt holder and stage was made at the Melbourne University Physics Department, and subsequently released in a commercial version by JEOL (Bursill *et al.,* 1977, 1979).

The development in the United States of the field emission scanning TEM by A. V. Crewe stimulated worldwide interest. At the University of Melbourne, A. Strojnik and J. M. Cowley designed and built a 600-keV STEM, with EELS detector, and foreshadowed applications to the imaging of samples in air (Cowley and Strojnik, 1968; Cowley *et al.,* 1970).

The best-known Australian-designed and -made accessory for electron microscopy is the Robinson detector for back-scattered electrons in the SEM. Despite great improvements to solid-state BSE detectors, this development by V. N. E. Robinson provides very high signal–noise performance at TV scanning rates for compositional imaging (Robinson, 1974a). Robinson applied this detector to the problem of imaging samples under conditions of poor vacuum and high water vapor pressure (Robinson, 1974b). In a subsequent development, G. D. Danilatos realized a new type of detector, involving the ionization of gases in the sample chamber, resulting in the Electroscan "environmental" SEM recently released commercially—but not by an Australian manufacturer (Danilatos, 1981, 1988).

B. Theory and Techniques

The development by J. M. Cowley and A. F. Moodie of the multislice formulation of the n-beam dynamical electron diffraction problem has been the major contribution by Australian scientists to the theoretical framework of electron microscopy. The history of this work has been described by Cowley (1981) and by Sanders and Goodman (1981), and need not be repeated here. The multislice theory resulted in the multislice computational method (Goodman and Moodie, 1974), which in turn proved critical in the development of structure imaging of crystal lattices (Allpress and Sanders, 1973); new insights into phase-contrast microscopy (Moodie and Warble, 1971); accurate methods for determination of crystal potentials, and the determination of crystal symmetry from convergent beam diffraction patterns (Gjønnes and Moodie, 1965; Goodman, 1975).

The significance of the multislice approach to the determination of crystal structure and in particular to the analysis of defective superlattice stuctures in nonstoichiometric compounds is widely recognized. A range of sophisticated image computation packages are in use worldwide, permitting the interpretation of the very-high-resolution lattice images now obtainable; many of these have been introduced by young Australian scientists trained in the Melbourne University Diffraction Group founded by John Cowley. The initially surprising (to the proponents of n-beam dynamical diffraction) close correlation between optimally focused images and the projected structure of thin large-cell crystals was understood in terms of the phase object approximation to the multislice theory, although detailed calculations including lens aberrations are required in general (see Cowley, 1981).

Major progress was made in determining the parameters of crystalline defects when A. K. Head, of the CSIRO Division of Tribophysics, introduced a mathematical technique for rapidly simulating images of defects

(Head, 1967), based upon the Howie-Whelan two-beam equations. Head and his colleagues included the effects of elastic anisotropy, and developed the technique to such a degree (Head *et al.*, 1973) that it is now a standard tool in all groups involved with defect analysis. To display the computed images, Head introduced the use of computer line printer output of simulated micrographs, a technique subsequently adapted for multislice image simulations.

The revival by Goodman and Lehmpfuhl (1965) of the Kossel-Möllenstedt technique for convergent beam electron diffraction, combined with multislice calculations of n-beam dynamical rocking curves, introduced new levels of precision to the determination of electron structure factors. Of much wider significance was the subsequent development of CBED analysis of crystal symmetry, based on the dynamical extinctions resulting from symmetrical multiple scattering. The theoretical analysis was made by Gjønnes and Moodie (1965); precision CBED patterns were obtained using the CSIRO CBED instrument (Cockayne *et al.*, 1967), and led to methods for determination of absolute symmetry and handedness (Goodman, 1975). The use of CBED for symmetry determination has been extended greatly by Professor Steeds' group at Bristol and by Professor Tanaka in Japan, and is now commonly utilized in conjunction with structure imaging in the high-resolution TEM.

VI. Conclusions

Australian scientists were quick to adopt the new science of electron microscopy as it emerged during the 1940s, and many have contributed substantially to its development over the past 40 years or so. In this brief account it is not possible to include due recognition to all of those who have made such contributions, particularly in the biological sciences. No one biological electron microscopist or group in Australia stands out well above the large number who have made improvements to EM techniques in biology since about 1960.

From the early days, Australia has been involved with the international organization of electron microscopy, with strong support over many years from the Academy of Science. An active society of electron microscopists can look forward to continued growth of the field in this country, and further strengthening of links with the international centers of electron microscopy.

Acknowledgments

A considerable body of material for this article was prepared by the late Dr. David J. Goodchild, whose contributions to the organization of Australian electron microscopy over the past 25 years have been very substantial. Dr. Lloyd Rees also provided notes on the early activities in the field, before his death in 1989.

We have also depended heavily on articles by Goodchild and Dowell (1986), Cowley (1981), and Sanders and Goodman (1981) for details of the early history of Australian electron microscopy. Other people who have provided material include D. J. H. Cockayne, J. M. Cowley, and C. E. Nockolds.

REFERENCES

Allpress, J. G., and Sanders, J. V. (1973) *J. Appl. Cryst.* **6,** 165.
Bayliss, N. S., Cowley, J. M., Farrant, J. L., and Miles, G. L. (1948). *Aust. J. Sci. Res. Ser. A* **1,** 343.
Beaton, C. D. (1988). "Australian Directory of Electron Microscope Units and Staff." CSIRO Division of Entomology, Canberra.
Bursill, L. A., Spargo, A. E. C., Stone, G., Wilson, A. R., and Wentworth, D. (1977). *Jeol News* **15,** 5.
Bursill, L. A., Spargo, A. E. C., Wentworth, D., and Wood, G. J. (1979). *J. Appl. Cryst.* **12,** 279.
Cockayne, D. J. H., Goodman, P., Mills, J. C., and Moodie, A. F. (1967). *Rev. Sci. Inst.* **38,** 1093.
Cowley, J. M. (1981). *In* "Fifty Years of Electron Diffraction" (P. Goodman, ed.), p 271. D. Reidel, Dordrecht.
Cowley, J. M., Smith, D. J., Strojnik, A., and Sussex, G. A. (1970). *Grenoble,* Vol. 1, p. 117.
Cowley, J. M., and Strojnik, A. (1968). *Rome,* Vol. 1, p. 71.
Danilatos, G. D. (1981). *Scanning* **4,** 9.
Danilatos, G. D. (1988). *In* "Advances in Electronics and Electron Physics," Vol. 71, p. 109. Academic Press, Boston.
Draper, M. H., and Hodge, A. J. (1949). *Nature* (London) **163,** 576.
DRL Annual Reports (1949, 1950). Defense Research Laboratory, Melbourne.
Drummond, D. G. (1960). The University of Sydney. *Gazette,* April 1960, p. 276.
Drummond, D. G. (1971). The University of Sydney. *Gazette,* September 1971, p. 21.
Farrant, J. L., and O'Connor, J. L. (1949). *Nature* (London) **163,** 260.
Farrant J. L., and Powell, S. E. (1956). International Federation of Electron Microscope Societies: *Tokyo 1956,* p. 260.
Garrod, R. I., Suiter, J. W., and Wood, W. A. (1952). *Phil. Mag.* **43,** 677.
Gjønnes, J. K., and Moodie, A. F. (1965). *Acta Cryst.* **19,** 65.
Goodchild, D. J., and Dowell, W. C. T. (1986). *Micron and Microscopica Acta* **17,** 101.
Goodman, P. (1975). *Acta Cryst.* **A31,** 804.
Goodman, P., and Lehmpfuhl, G. (1965). *Z. Naturforsch.* **20a,** 110.
Goodman, P., and Moodie, A. F. (1974). *Acta. Cryst.* **A30,** 280.
Head, A. K. (1976). *Aust. J. Phys.* **20,** 557.
Head, A. K., Humble, P., Clarebrough, L. M., Morton, A. J., and Forwood, C. T. (1973). "Computed Electron Micrographs and Defect Identification." North Holland, Amsterdam.
Hodge, A. J., and Wardrop, A. B. (1950). *Nature* (London) **165,** 272.
Mercer, E. H., and Rees, A. L. G. (1946a). *Nature* (London) **157,** 589.
Mercer, E. H., and Rees, A. L. G. (1946b). *Aust. J. Exp. Biol. Med. Sci.* **24,** 147.
Moodie, A. F., and Warble, C. E. (1971). *J. Cryst. Growth* **10,** 26.
Robinson, V. N. E. (1974a). *J. Phys. E: Sci. Instrum.* **7,** 650.
Robinson, V. N. E. (1974b). *J. Microsc.* **103,** 71.
Sanders, J. V., and Goodman, P. (1981). *In* "Fifty Years of Electron Diffraction," (P. Goodman, ed.), p. 281, D. Reidel, Dordrecht.

2.2A
The Austrian Society for Electron Microscopy

ERWIN M. HÖRL[1]

Austrian Research Center, Seibersdorf, Germany

What is now known as the Austrian Society for Electron Microscopy was first founded in 1965 as the Austrian Working Group for Ultrastructure. Even before that time, numerous Austrian scientists with different specializations had become members of the German Society for Electron Microscopy and played a full part in their activities. However, for the purpose of improved and more efficient cooperation at the national level, as well as the need to organize both national and international events, the foundation of a distictive Austrian Society became essential.

Pischinger was elected Chairman of the new Society; not only did he head the Institute for Histology and Embryology at the University of Vienna, he also made the facilities of this Institute available to the Secretariat of the new Society. The members of the Governing Board were selected on the basis of their different specializations, as well as taking into consideration the need to have the Austrian universities well represented. Grasenick, an acknowledged Austrian pioneer of electron microscopy and Head of the Center for Electron Microscopy at Graz, was elected Deputy Chairman.

In the years that followed, numerous small-scale meetings covering a great variety of topics took place in Vienna, Graz, and Salzburg.

On 22–25 September 1969, the Austrian Working Group for Ultrastructure Research held its first joint Meeting with the German Society for Electron Microscopy. The chosen city was Vienna. This meeting was attended by more than 400 participants and proved an outstanding success from both a scientific and a social point of view. In May 1970, Pichinger, who had been the Chairman up to then, resigned from this post upon his retirement from the University. L. Stockinger, who had been the Society's Secretary until then, was elected as his successor, and E. M. Hörl from the Austrian Research Center, Seibersdorf, was elected Deputy Chairman. Pischinger and Grasenick continued as honorary members of the Board. In 1974, the Annual General Meeting decided to change the name of the Society to the Austrian Society of Electron Microscopy. In the same year

---

[1] Present address: Institute of Applied and Technical Physics, Technical University of Vienna, 1040 Vienna, Austria.

the Austrian Society became a member of the IFSEM and has been represented on this body with one vote ever since.

In addition to organizing numerous smaller events, the Society has always considered the organization of specialist symposia, in collaboration with other specialist societies, as one of its most important tasks. Among these may be listed the following symposia:

- 1973 Neurotransmitters, Vienna
- 1975 Peroxisomes (with Nobel Prize winner de Duve), Vienna
- 1976 Morphometry of Biological Objects with the Aid of Electron Microscopy, Innsbruck
- 1977 Structure and Pathology of the Skin, Vienna
- 1988 Histology as the Basis of Modern Cytogenetics and Cell Biology, Vienna
- 1988 Andrology Today, Vienna

The close co-operation between our Society and those of neighboring countries led to the joint organization of further Regional Annual Meetings:

- 1969 Austro-German Meeting on Electron Microscopy, Vienna
- 1981 Austro-German Meeting on Electron Microscopy, Innsbruck
- 1985 German-Swiss-Austrian Meeting on Electron Microscopy, Constance
- 1989 Austro-German-Swiss Meeting on Electron Microscopy, Salzburg
- 1993 Austro-German-Swiss Meeting on Electron Microscopy, Zurich

Given the pivotal location of Austria's geographical position, the society regards establishing contacts with its Eastern neighbors as one of its principal tasks. Relevant events in this field include the following:

- 1985 Hungarian-Austrian Meeting on Electron Microscopy, Balatonaliga
- 1987 Austro-Hungarian Meeting on Electron Microscopy, Seggau, Styria

Plans for further joint meetings with the Hungarian, Yugoslav, and Italian societies for electron microscopy are under discussion.

Numerous members of the Austrian Society regularly take part in European and international meetings on electron microscopy. The Society also provides financial support for research students and other young electron microscopists to take part in the IFSEM and other meetings on electron microscopy.

By its publication of complete lists of all electron microscopes and ancillary equipment and its collection of other useful data for electron microscopists, contacts among the different laboratories have been greatly facilitated. The lists are updated at regular intervals.

At present, the society has 250 members. In this account, we have not singled out for attention the work of any particular Austrian electron microscopists, but an exception must be made for Walter Glaser, one of the most influential researchers in electron optics and electron microscopy. The following accounts, by Hans Grümm and by Peter Schiske, give great personal insights into Glaser's theoretical work, preserved in the famous *Grundlagen der Elektronenoptik,* on the guiding and focusing of charged particles in magnetic and electrical fields.

2.2B
Reminiscences of Walter Glaser

HANS GRÜMM[1]

University of Vienna, Vienna, Austria

PETER SCHISKE[2]

Fritz-Haber-Institut der Max-Planck-Gesellschaft, Berlin, Germany

I. Walter Glaser (1906–1960) and the Grundlagen der Elektronenoptik[3]

Walter Glaser was the most outstanding pioneer of electron optics, working in Austria after World War II. As an expert of international reputation, Glaser decisively influenced, among other things, the understanding and protecting of the magnetic electron microscope with his famous and exhaustive treatise, *Grundlagen der Electronenoptik,* on the guiding and focusing of charged particles in electric and magnetic fields. I first met Glaser in 1949 when I was looking for a subject for my doctoral dissertation. At that time, it was difficult to find a supervisor for a doctoral thesis. H. Thirring referred me to a colleague, W. Glaser, who worked as a kind of Senior Assistant in the Institute. At that time Glaser was having a difficult time, struggling to gain a foothold somewhere. He had, in fact, been called as a full professor to Breslau (now Wroclaw), but could not take up his professorship because of the war situation. I was told that Glaser had a long list of subjects for dissertations, and indeed he showed me a list of 30 project titles. I soon realized that I could not make out very much from the titles alone, so I asked for more information. He gave me a manuscript of some hundred pages; this was, in fact, the first draft of the *Grundlagen*. After plowing through this manuscript, I chose the subject of the caustic surface of electron lenses, and found it a very fruitful area of research. This meant that I was able to finish my dissertation in record time.

Glaser, who in the meantime had been called to a full professorship at the University of Technology of Vienna, invited me to help him, as his

---

[1] Present address: Joanelligasse 7/9, A-1060 Vienna, Austria.
[2] Present address: Jägergrund 9, 2304 Orth a.d. Donau, Austria.
[3] Authored by Hans Grümm.

FIGURE 1. Walter Glaser (1906–1960).

assistent, in the compilation and publishing of the *Grundlagen.* I vividly remember the many months during which my home was essentially in the basement of the university. Although Glaser was allowed a part-time secretary, she was, in fact, hijacked by his colleague for his work, when she should really have been working for Glaser. So it fell to me to type the complete manuscript, all 800 pages of it, using the well-known "two-finger" method of typing. Each morning, whenever this was feasible, Glaser stood at the blackboard, sketching in the text in rough outline. In the afternoon I typed it out and the next day we did the final editing. Glaser was extremely creative and original. On several occasions I pointed out to him open questions whose solution, I thought, would take considerable time. In fact, he would usually accept the challenge instantly and often solved the problem in a few hours.

When we were working on the *Grundlagen,* I noticed that Glaser often came quite close to solving problems that later proved to be of considerable importance and contributed greatly to the international reputation of those physicists who solved them completely later on. An early example was the strong focusing synchrotron. Glaser did not restrict himself to the electron microscope; he also wanted to include in the book such applications as mass spectrometers and particle accelerators. He also dealt with the steering of bunches of charged particles traveling in circular orbits, naturally assuming cylindrical symmetry of the field around the circle. I happened to remark that this meant, in effect, imposing a restriction on a general problem. In a special case, after all, the field could be developed in a Fourier series, but then I excused myself for my cheekiness by remarking that this would be a purely abstract exercise. Glaser, as an experienced theoretician, was not of this opinion. According to him, such an approach would lead to a theory concerning the correction of real fields deviating from ideal ones. Unfortunately, at that time, he did not proceed along this line of thought, because the deadline for finishing his book was approaching. Naturally he would immediately have recognized that a sequence of converging and diverging fields can result, in total, in focusing.[4]

Later on, I went to other parts of Europe, particularly Germany, where I worked in the field of reactor theory. When I returned to Vienna, Glaser was working on the translation of the standard text, *The Elements of Nuclear Reactor Theory*, by Glasstone-Edlund. I commented that any student intending to work in the field of nuclear reactors would have to know English anyway. Glaser did not share this opinion, but thought that life had to be made easier for students. However, owing to his serious illness and early death, he never completed his translation. I took it upon myself, as an expression of my gratitude to him, to complete the unfinished work.

It is very difficult today to visualize the working conditions that an expert in the field of theoretical electron optics had to cope with in Glaser's time. As long as general analytical formulas were being developed, there were no problems. The troubles began when all but the simplest numerical computation was needed. There were no digital computers, so all calculations had to be done by hand. It was a matter for rejoicing if, during this tedious work, one chanced on functions that were tabulated in "Jahnke and Emde." Glaser himself discovered and applied the famous "bell-shaped field" $1/(1 + z^2)$ field. Schiske, in a moment of inspiration, discovered an electrostatic lens field leading to elliptical functions. I myself succeeded in hitting on a generalization that resulted in about a dozen such fields. Today, surrounded by computers, one no longer faces such problems, although the excessive use of computers can, indeed, lead to the loss of physical perspective and intuitive "feeling" for the problem.

When I look back on Glaser, I do so with feelings of gratitude and admiration to an exemplary man and a scientist who contributed greatly

---

[4] Peter Schiske adds: "I believe that Glaser's ideas on the properties of a series of alternate converging and diverging fields greatly stimulated the discussion and later development of the correction of aberrations in electron optics by means of nonrotationally symmetric fields, based on the idea of imaging by a series of quadrupole lenses."

to my career and with whom I managed to develop a close rapport. It was tragic that he left us so early, depriving us of many further contributions that we had come to expect of him.

II. Further Reminiscences of Walter Glaser[5]

My introduction to Walter Glaser came about from attending his seminar. Later, in 1948, I joined his group and was in contact with him until his death in 1960. Like Gruemm, I also started by asking for a suitable theme for a doctoral dissertation and was likewise presented with a large manuscript comprising the draft of the *Grundlagen*. In my case, I settled for a proposal to treat the electron/lens field interaction in the same way as the electron/atom interaction, using standard quantum mechanical scattering methods.

At that time, I could think only of Born's approximation. The first difficulty was the necessity of introducing some lateral limitation, since an infinite plane wave would spread beyond the region where the series expansion of the field was valid. Glaser suggested multiplication by a discontinuous factor, represented by an integral. Browsing through *Atomic Collisions* by Mott and Massey evoked the idea of Gaussian wave packets. The formulation of the Hamilton-Jacobi equation for paraxial rays is characterized by the omission of terms containing the square of the longitudinal momentum. By making the same omission in the Schrödinger equation, one obtains a very manageable differential equation, containing only first-order derivatives with respect to the longitudinal coordinates. Transverse momenta and coordinates are, however, included up to the second degree. A paper by Kennard on the movement of wave packets in systems with a second-order Hamiltonian proved to be very helpful. It had been written in the first years of quantum mechanics and is cited in Heisenberg's *Physikalische Prinzipien der Quantum-Mechanik*.

The name "paraxiale Schrödingergleichung" (paraxial Schrödinger equation) was not coined by me! It was included in the *Grundlagen* under that name and also in a separate joint publication. At any rate, it constituted a framework in which Gaussian wave packets as well as Fresnel and Fraunhofer diffraction were included without further approximation. Disappointment came when I could not find a manageable quantum mechanical perturbation expansion that would include geometric third-order aberrations, and perhaps, in addition, effects such as electron spin. I still believe that an expansion of this kind is feasible, and that only under special assumptions will it coincide with the well-known WKB approximation. Here I am think-

---

[5] Authored by Peter Schiske.

ing of the summation methods used today in quantum field theory and many-body physics. At that time, however, WKB seemed to be the only way forward, and a second part of the construction of Green's function was written, drawing heavily on an idea taken from Glaser and Sommerfeld's textbook, *Theoretical Physics: Optics*. The finally published text on the "paraxiale Schrödingergleichung" differs greatly from that of my dissertation. On the other hand, the wording of the subsection, "Konstruktion der Greensche Funktion," follows closely that of a draft written by me.

The electrostatic "single lens" (Einzellinse) permitting rigorous calculation, as mentioned by Grümm, can be regarded as the field of two point charges situated at complex locations. The problem of two fixed charges had been discussed in Glaser's group. Glaser's idea was to use this system as an ideal fictitious test specimen for a theoretical test of resolution. To my knowledge, this project was never carried out. My colleague F. Katscher, who tried his hand at it, had to give it up. He kindly presented me with a pile of notes on the relevant literature, since I had shown some interest in the subject. Of course, this was about two centers in real locations. In his lectures on optics, the experimentalist Przibram had repeatedly told us: "In physics, you should not be afraid of complex quantities." For the description of Gaussian wave packets, complex wave packets had been very useful.

The work done by H. Grümm for Glaser and the *Grundlagen* can hardly be overestimated. Glaser, as a thorough-going scientist, wished to include all the latest developments made by himself and other people, whereas Grümm stressed the time that would be needed to do this and the need for a deadline. I cannot prove it, but I have the feeling that Grümm's authority can be sensed even in the names given to various subsections. To this, one has to add that Grümm was responsible for all the diagrams in the book.

Later, other colleagues acted as "unofficial assistants" to Glaser. For example, O. Bergmann worked on the concept of "osculating cardinal elements" for arbitrary lens fields.

At that time, many researchers were looking for methods that could appropriately be called the diffraction theory of aberrations. In a certain sense, that goal is attacked by the evaluation of two-dimensional integrals by the method of stationary phase. In Glaser's group, G. Braun successfully improved and applied this method. Later, he went to the Philips Laboratories in the Netherlands and eventually became a Professor of Applied Mathematics there. Unfortunately, he was killed in a traffic accident, as I learned from Professor Grümm in the 1970s.

While at Prague, Glaser could delegate numerical work to people paid specifically for the task. Because of that, we probably overestimated the work involved in the evaluation of the diffraction integrals. Numerical work by E. Gütter on Fresnel diffraction by a circular aperture, for example, was reported in the *Grundlagen.*

When I was in the fortunate position of "unofficial assistant," my main task consisted of helping, in collaboration with F. Putz, in developments for Glaser's contribution to Flügge's *Handbook.* My work was quite restricted and in no way comparable with that of Grümm's on the *Grundlagen.* In addition, I participated in a paper on axial astigmatism. On that occasion I made an error in a figure caption, wrongly ascribing the two sheets of the caustic surface. This error was never corrected, since the editor of *Zeitschrift für Angewandte Physik* flatly refused to correct it! Glaser was very tolerant and understanding about this inadvertent error of mine. At about the same time, F. Putz and I assisted in the preparation of a highly speculative paper on the foundations of wave mechanics. Playing around with general formulas was great fun, but afterward we both felt a bit uneasy about it. Glaser was very much concerned about the fundamental and the didactic problems of transferring from classical mechanics to wave mechanics. This can be seen immediately from the amount of space devoted to these questions in the *Grundlagen.* In addition, there were many special publications. His point of view in the philosophy of science reflected, in an undogmatic way, the Prague tradition of Ernst Mach and, to a lesser degree, that of his teacher Philipp Frank. The papers of D. Bohm and L. de Broglie, around 1952, on hidden variables, fitted in well with Glaser's development of wave mechanics, and he followed them assiduously. The first discussion of what was eventually to be called the Aharanov-Bohm effect was in a paper by Ehrenberg and Siday, which was concerned mainly with Glaser's introduction of the electron optical index of refraction. The bibliographic details of the ensuing discussion with Glaser can be found in the "Anmerkungen" of the *Grundlagen.* In one of my last visits to Glaser, the Aharanov-Bohm effect came up in conversation, and he indicated that he really doubted the existence of such a seemingly paradoxical phenomenon. It is very probable that Glaser, who died on 3 February 1960, never actually saw the seminal paper of Aharanov and Bohm of 1959.

In his last years, Glaser was, as mentioned by Professor Grümm, occupied with reactor physics. During my short visits at the time, I never realized that he was having severe health problems. Nevertheless, I am tempted to compare his scepticism in that area and also his initial scepticism about holography, not so much with failing health but with the general attitude of Ernst Mach, in his last years, toward the physics of his time. For example, Glaser could not accept that Gabor's proposed method of holography would be able to record both amplitude and phase of an electron wave leaving an electron microscope specimen.[6]

Glaser's treatment of the theory of relativity also had some original traits. At a public lecture given by Glaser and organized by the Chemisch-Physikalische Gesellschaft, he sketched a model of special relativity kinematics, possibly including dynamics. Each particle was associated with a Zitterbewegung of instantaneous velocity equal to that of light, but steadily changing its direction. In this model, he attempted to give a causal explanation of time dilation and the Lorentz contraction. I cannot remember if mass variation with (mean) velocity was included.

On the electron microscope development front itself, Glaser had been retained as a consultant by Siemens und Halske, right from the start of their development of the TEM in Berlin under von Borries and Ruska in 1936. His relations with Ernst Ruska were very cordial, but as Glaser recalled, their completely different attitudes toward physics and electron microscopy certainly made for difficulties. It was quite hard for Glaser to convey to Ruska his theoretical point of view and to translate it into practical engineering terms. Only after a considerable incubation period would Ernst Ruska have grasped the full significance of the point in question. On the other hand, once the point had sunk in, Ruska incorporated it deeply into his thinking. Glaser also mentioned that when J. Dosse, an engineer with strong theoretical interests, joined the Siemens group, communication with Ernst Ruska was greatly facilitated.

Concerning Glaser's Breslau appointment, I have in fact only just learned about it from Grümm's report. From my own conversations with Glaser, I gathered that during the war years in Praque, he had experienced some difficulties with the Party and with the State authorities, but that in 1945 the new regime would have tolerated his remaining in Prague. There was, incidentally, a close connection between Glaser and the mathematician Paul Funk, who was also initially in Prague and then in Vienna. Funk was interned in a concentration camp during the war.

If asked to give a list of the most important contributions of Glaser to electron optics, I would suggest the following:

1. The introduction of the refractive index for electrostatic-magnetic fields.

---

[6] *Editorial footnote:* This is understandable. Gabor's concept of holography was based on a sudden, unexpected intuition, which he could not explain satisfactorily, even to himself. It was not based on any prior theory or experiment; these came much later. Gabor and Glaser met in 1951 at a symposium on electron physics at the National Bureau of Standards, Washington, DC. By then, the theory of electron holography and its experimental verification had been firmly established. Glaser was converted.

2. A systematic calculation of all third-order aberrations by the powerful eikonal method and hence the discovery of anisotropic aberrations. (The best explanation of the eikonal method is to be found in Glaser's *Handbook* article.)
3. The bell-shaped magnetic field model and the evaluation of its aberrations. In particular, his suggestion of placing the specimen at the center of the field distribution (condensor-objective mode) has proved remarkably fruitful in all modern electron microscopes, whether TEM, STEM, or SEM. For a practical appraisal of that suggestion from someone not belonging directly to the Siemens group, see Manfred von Ardenne's autobiography, *Ein glueckliches Leben fuer Technik und Forschung* (ISBN 3 463 005220, Copyright 1972, Kindler Verlag, Zurich and Munich, p. 144).
4. The establishment and evaluation of the combined diffraction-spherical abberration limit of the electron microscope.
5. In the 1940s, the performance of electron microscopes was limited by the deviation of the optical system from axial symmetry. Glaser's work on axial astigmatism helped to analyze and hence, eventually, to overcome this problem.
6. Generalization of the work of Finsterwalder on caustic figures by third-order aberrations as applied to electron optics. This was, incidentally, carried out in two places, due to postwar turmoil, by Grümm and Hofmann.

It has no doubt to be conceded that items 1–5, and to a considerable extent 6, belong to Glaser's Prague period. Of course, the *Grundlagen* itself was actually produced in Vienna, and through fortunate circumstances, it could be followed by Glaser's contribution to the Flügge *Handbook,* which many people preferred because of its condensed style. The success of the *Grundlagen* was soon demonstrated by the fact that, in addition to its citation by electron microscopists, even the authors of light optical papers cited it as a reference for statements on instrumental light optics. Glaser's aim, in fact, was not just the clever derivation and manipulation of formulas, but the introduction of new ways of thinking about the basic physics of electron optics and electron microscopy. He will be remembered with admiration and affection.

2.3
Electron Microscopy in Belgium

DIRK VAN DYCK

University of Antwerp (RUCA), B-2020 Antwerpen, Belgium

I. Introduction

Despite its small size, Belgium has played an important role, not only in the development of electron microscopy as a technique, but also in the dissemination of the use of electron microscopy in materials science as well as in biomedicine. The number of electron microscopes (and hence users) per capita still ranks among the highest in the world.

II. The Pioneer

As early as 1932, immediately after the first results were obtained by Knoll and Ruska, Ladislaus ("Bill") Laszlo Marton realized the potentialities of this new technique. Marton was born in Budapest (August 15, 1901). He obtained his Ph.D. in Zurich on X-ray spectrometry and in 1928 he began working in the small group of Prof. Henriot at the Faculty of Science of the Free University of Brussels (ULB). Professor Henriot (1855–1961) stimulated him to devote his research to the construction of an electron microscope, although the financial means were very limited. His only source of funds was a modest grant from the Institut National de Physique Solvay, whose scientific secretary was Prof. Henriot himself and who provided Marton with a fellowship for several years.

His first prototype with a horizontal column (Fig. 1) was ready by the end of 1932. His first paper about this subject was published in Flemish in 1933 together with Maurice Nuyens in the journal *Wis- en Natuurkundig Tijdschrift* under the title "Meetkundige Optiek der Elektronen" (Marton and Nuyens, 1933). It had only one electromagnetic lens, outside the vacuum. Originally Marton wanted to use the instrument to investigate photoelectricity, but in view of the fact that the AEG laboratory in Berlin had already started similar research with ample financial means, he abandoned this idea (Süsskind, 1986).

FIGURE 1. Photograph taken in 1935 of the two electron microscopes built by L. Marton and used for the earliest biological work. The horizontal instrument (right, in foreground) was built in 1932, the vertical one (left, in background) in 1934.

With this instrument, Marton was able to image the spiral cathode. The image was not homogeneous, which Marton attributed to different work functions of the different crystalline phases. This result encouraged him to construct a second instrument in 1933 with a vertical column consisting of a condenser and two lenses (Fig. 1). The instrument had an electron optical magnification of 1000 × in two stages over a total imaging length of 72 cm. Only the projector lens had polepieces. The objective focal length was estimated to be about 10 mm and the projector focal length 12.5 mm. The goal of this project was twofold: to explore the possible application of electron microscopy, and to understand the mechanism of image formation. Before starting the investigation of biological samples in April 1934, Marton consulted his colleagues in the life sciences at the ULB, the zoologist Paul Brien and the botanist Marcel Homès (later president of the Royal Academical Society of Belgium). Homès provided him with microtome sections impregnated with osmium tetroxide. The thickness of the slices was still several micrometers. The object was then mounted on a copper grid. In this way Marton obtained the first micrographs of a sundew plant (Fig. 2). Although the organic material was destroyed by electron bombardment, the osmium skeleton remained unchanged to reveal the structure and provided sufficient contrast to obtain an interpretable image. Marton immediately realized that by staining biological objects with heavy metals, the structure can remain visible; and even without staining, biological specimens can be resistant to electron irridiation for a sufficient time to record the photographs, so as to reveal details beyond the resolution of the optical microscope. This opened enormous perspectives for the investigation of biological samples.

He presented his results at the meeting of the Belgian Academy on May 8, 1934, and published a short note in *Nature* (Marton, 1934a).

Marton then visited Ruska, Knoll, and Brüche in Berlin, who did not believe his results. In Belgium too, he was criticized, even by the eminent

FIGURE 2. (left and right) First biological electron micrograph, obtained with the horizontal microscope; the specimen is a 15-μm-thick microtome section of *Drosera intermedia*, stained with osmium tetroxide (April 1934).

bacteriologist Jules Bordet (Nobel Prize for the discovery of immuno factors in blood serum), who, after Marton's lecture, said: "Oh no, not an electron microscope, we have enough trouble trying to interpret the images we get with a light microscope!"

In the autumn of 1934, Marton started to build a third instrument, again vertical, at the cost of enormous effort. The university funding was now nearly nil. Furthermore, he could not persuade the biomedical sector to invest in this project. Marton had to buy parts in the flea market, usually with his own money. The instrument was equipped with airlocks for the specimen and the photographic plates and a beam shutter. The instrument operated at an accelerating voltage of 45–50 keV, and the exposure times were of the order of 0.1 s. The total imaging length was almost 1400 mm. Both objective and projector lenses had polepieces. In principle, a magnification of about 5000 could be reached, but he used a magnification of 700. In order to reduce the irradiation of the specimen, Marton used a very convenient method of focusing first on a radiation-resistant test specimen and then bringing the biological specimen to the same axial position by means of the specimen airlock. In order to focus the image, which was not visible on the screen, he calibrated the lens currents.

To improve the technique, he cooled the sample by supporting it on a very thin (0.5-μm) cooled supporting grid of aluminum. In this way he was able to improve the resolution to about 1 μm. He was able to visualize the cell wall and the nucleus of an orchid. Marton even succeeded in observing samples without fixation, by mounting them on thin (20-nm) supporting foils of collodion (nitrocellulose) (Fig. 3). He also contributed to the understanding of the image formation. He realized that the image contrast was not caused by absorption (as in optical microscopy) but by scattering and used the Bethe formula to calculate and tabulate mean scattering angles for multiple scattering as a function of thickness and electron velocity for different materials.

Marton also studied the resolving power of the electron microscope (Marton, 1936) and in particular he showed how the contrast decreases with increasing numerical aperture, so that the upper bounds for the resolution are limited by the spherical aberration. He also introduced the concept of depth resolving power.

Marton devoted most of his efforts toward improving the electron microscope for biomedical research and particularly to reducing the necessary electron dose. He introduced a number of improvements that are now in common use, such as:

- Airlocks for the specimen and photographic plate
- Movable photographic plates for multiple exposure

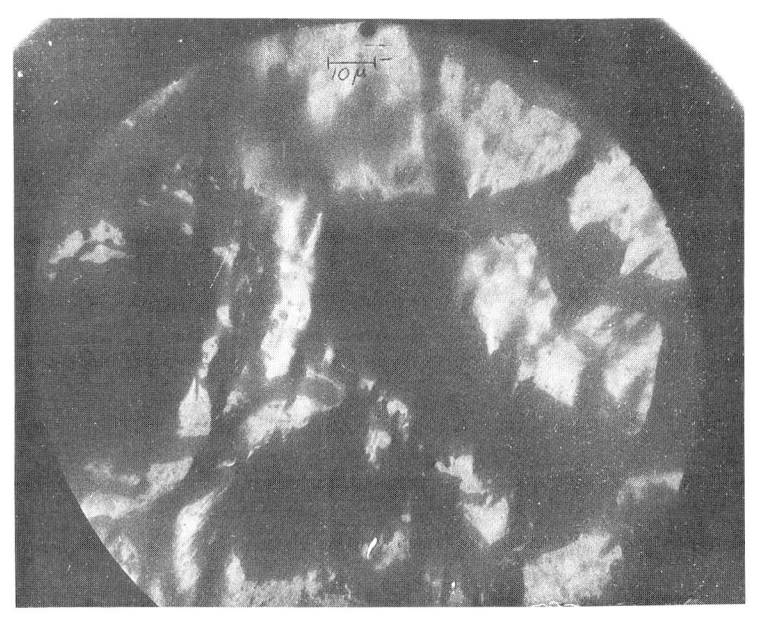

FIGURE 3. Root of *Neottia nidus avis* on collodion films (January 1936); micrograph obtained with the vertical microscope.

- Specimen stage with mechanical $x-y$ control
- Specimen carousel for multiple specimens
- Electronic exposure control
- Beam blanking, beam shift
- Better phosphor for increased contrast

Marton also succeeded in interesting André Callier, the director and founder of a Belgian manufacturing company (Société Belge d'Optique, Ghent) in the construction of an electron microscope according to his design, but unfortunately Callier died before the construction was started.

Probably owing to lack of funding, but also in view of the political situation in Europe, the electron microscopy project in Brussels was abandoned before World War II and Marton left Brussels for the United States. He later introduced techniques for determining the thickness of the object, stroboscopy for time-varying objects, stereoscopy (Marton, 1944), electron optical shadowing, electron interferometry, X-ray shadow microscopy, and electron energy loss studies. In 1945 he obtained the medal of the ULB, and in 1955 he became a member of the Royal Academy of Belgium.

III. THE WAR PERIOD

Immediately after Marton left Belgium, activity in the field became nearly nonexistent. Since commercial instruments were not available, activities were started at the universities of Brussels and Ghent to construct homemade diffraction instruments. At the University of Brussels, Dr. O. Goche developed an electron diffraction apparatus, based on the design introduced by Finch. This instrument was then used for the study of thin layers. At the University of Ghent, in 1943, A. Lagasse, in the framework of his Ph.D. research, constructed an electron diffraction instrument. The instrument is shown in Fig. 4. The quality of the diffraction patterns obtained was excellent, as can be judged from the example in Fig. 5, obtained from graphite. Lagasse used his instrument to investigate thin metal layers and the structural relation between deposits and the crystal substrate.

Immediately after the war, a "real" electron microscope facility, one of the first in the world, was set up at the University of Ghent; this provided services for biologists as well as for physicists. The first director was Van der Meersch, who was after a short period succeeded by Lagasse.

The first commercial TEM was installed in 1949 at the University of Liège. It was an RCA instrument (type EMU2) operating at 50 keV and used for materials science as well as biology. The first results in metallurgy using replica techniques were obtained by P. Cohen and L. Habraken and were presented at the Delft Conference (July 1949). The microscope remained in operation until 1970.

The first person to introduce electron microscopy in biomedical research was M. H. A. De Groodt-Lasseel, a former student of B. von Borries, who also purchased the first microtome and started a very productive scientific career. Another pioneer of the use of electron microscopy in biomedicine was Albert Claude (Bordet institut) (Porter *et al.*, 1945), the later Nobel Prize winner, who first observed the structure of tissue cells in the electron microscope, which, fortunately, were sufficiently thin and did not need microtomy.

In the early 1950s, other instruments were installed in Leuven (RCA), Ghent (Siemens), and ULB (RCA). The work in Leuven was carried out by Van Itterbeek and L. De Greve, who studied the structure of epitaxial films in relation to electron conductivity. The first JEOL microscope (JEM 100) was installed at the University of Namur.

In the 1950s and 1960s several electron microscopic groups were established by different scientists, all prominent in their research fields: in biomedicine, A. Claude (Brussels), Ch. De Duve (Leuven, Brussels), J. Drochmans and P. Dustin (Brussels), Firket (Liège), M. De Groodt

FIGURE 4. Electron diffraction instrument built by P. Lagasse. It was constructed and located at the Faculty of Science of the University of Ghent and operated at 70 kV. The accelerating voltage was generated with a Siemens & Halske 220-V/80-kV transformer. The vacuum was obtained by means of a Holweck pump and an oil diffusion pump.

and D. Scheuermann (Antwerp), pioneering freeze-fracturing and X-ray microscopy; in materials science, Amelinckx (Antwerp), Habraken, Greday (Liège), De Berghezan, Deruyttere, and Delaey (Leuven).

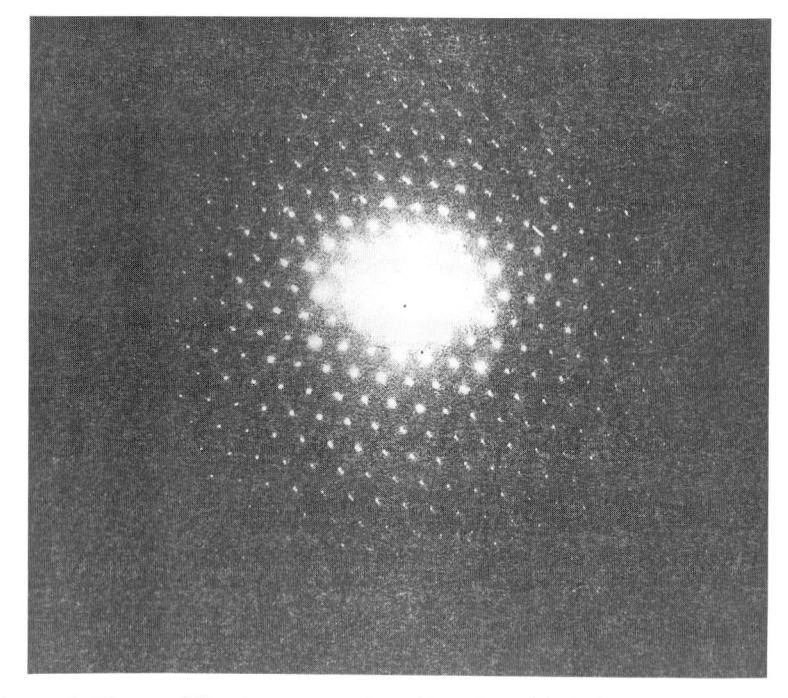

FIGURE 5. Electron diffraction pattern of graphite taken with the instrument of Lagasse.

In the 1980s more than 100 electron microscopes (TEM and SEM) were in operation in Belgium.

IV. Society for Electron Microscopy

A number of Belgian microscopists were present at the international conference in Delft (1949): A Claude (Bordet), P. Cohen, Ch. Gregoire, L. Habraken (Liège), L. De Greve (Leuven), A. Lagasse, J. Roose, and J. Voets (Ghent). At the international conference in Paris (September 1950), presentations were made in materials science (Habraken, Cohen) and biology (Gregoire and Florkin).

In the early 1950s, a first attempt had been made to found a Belgian Society for Electron Microscopy, with L. Habraken as President and Van der Meersch as Secretary. This society did not survive. In 1957, under the initiative of L. Habraken, A. De Gueldre, M. Desirant, A. Lagasse, M. De Groodt, J. Fripiat, and A. Bruaux, the Comité Belge de Microscopie Electronique—Belgische Comiteit voor Elektronenmicroscopie was founded and officially recognized by the IFSEM in 1957–1958 at the international conference in Berlin (and the official fees were paid!). At the moment the committee consisted of 40 members, 8 from the biomedical sciences, 27 from physics, chemistry, and metallurgy, and 5 others. The President was A. Lagasse, and the Secretary/Treasurer was T. Greday. This committee remained in office until 1966. The present Society was established in 1966 with President A. Claude, Van Itterbeek (Vice President), Lagasse (Administrator), Firket (Treasurer), Baudhoin (Assistant Secretary), and Greday (Secretary). Greday retained this function until 1990. The society now has about 200 members.

V. INTERNATIONAL MEETINGS AND CONFERENCES

Liège (May 1956)	Journées internationales de microscopie electronique appliquée à l'industrie
Brussels (May 1967)	Joint Meeting of the BVEM/SBME with the SFME (French society)
Liège (May 1973)	Joint Meeting of the BVEM/SBME with the DGEM (German society) and NVEM (Dutch society)
Rotterdam (May 1974)	Joint Meeting of the BVEM/SBME and the NVEM
Antwerp (September 1970)	International Congress on High Voltage Electron Microscopy
Liège (May 1983)	Joint Meeting of the BVEM/SBME with the SFME
Antwerp (September 1983)	Joint Meeting of the BVEM/SBME with the DGEM (German society for electron microscopy)
Lille (May 1984)	Joint Meeting of the BVEM/SBME with the SFME
Antwerp (September 1985)	Joint Meeting of the BVEM/SBME and the Deutsche Anatomengesellschaft
Wageningen (December 1990)	Joint Meeting of the BVEM/SBME with the NVEM
Antwerp (December 1992)	Joint Meeting of the BVEM/SBME with the NVEM
Papendal (December 1994)	Joint Meeting of the BVEM/SBME with the NVEM

VI. Nobel Prizes

Albert Claude, who worked at the Rockefeller Institute (New York) and afterwards at the Bordet Institute (Brussels), was the first to visualize tissue culture cells in the electron microscope. His paper together with K. Porter and E. Fulham was a real milestone in the use of the electron microscope in cell research. Fortunately, the cells he investigated were so thin in natural form that they did not need to be sectioned and could reveal the intrinsic potentialities of the technique. Albert Claude continued his research on the structural and functional organization of the cell mainly using electron microscopy. He improved the technique almost without help, and also developed the differential centrifugal method for fractionating cells. In this way he paved the way for morphologists as well as for the biochemists to enter and discover the unexplored domain of the interior of the cell, and he witnessed the success that resulted from this collaboration. His work lies directly at the origin of the discovery and study of the subcellular organelles. In 1974 he received the Nobel Prize for his work together with

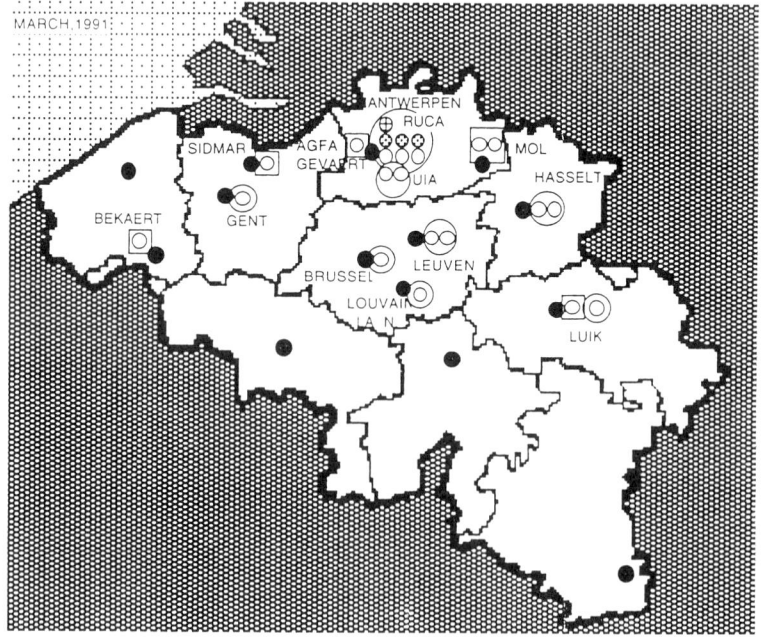

FIGURE 6. Distribution of transmission electron microscopes for materials research in Belgium. Instruments dedicated to: ⊕, high voltage; ⊛, high resolution; ○, analytical applications or general purposes.

Ch. De Duve (Univ. Leuven and Brussels) and George Palade. Albert Claude died in 1983.

VII. ANTWERP AND THE BRITE-EURAM PROJECT

The center of mass of electron microscopy in materials research in Belgium is the University of Antwerp. Due to the pioneering work of S. Amelinckx, first in Ghent (1950–), then in Mol (1960–), and afterwards in Antwerp (1965–), electron microscopy became recognized in Belgium as a major research technique and Antwerp became the center of mass of electron microscopy in materials science in Belgium (Fig. 6).

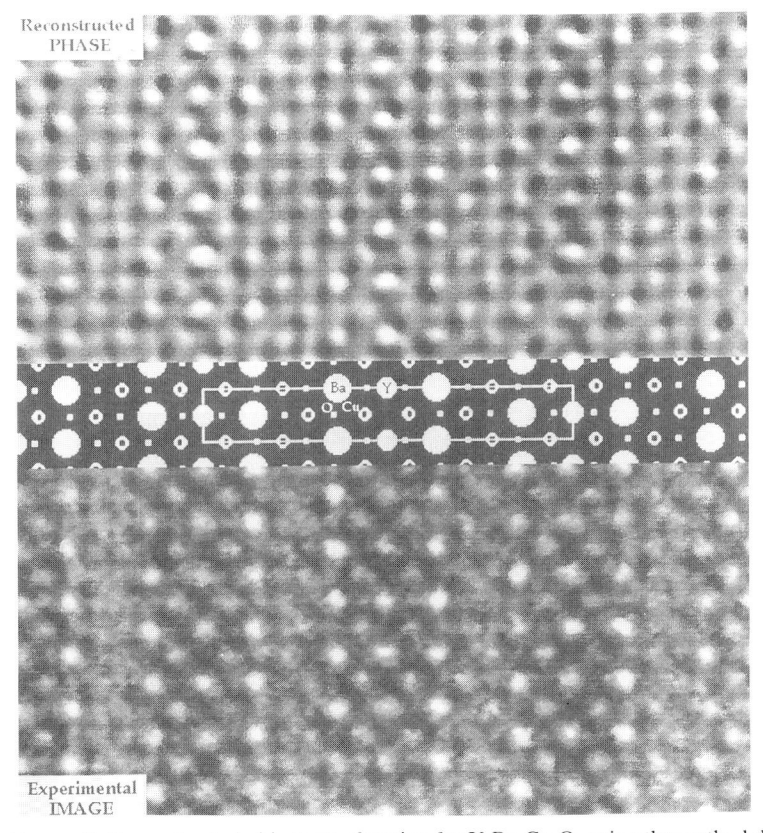

FIGURE 7. Reconstructed object wavefunction for $Y_1Ba_2Cu_4O_8$ using the method developed in Antwerp. Below: normal high-resolution image with a resolution of about 0.2 nm taken with a Philips CM20 ST FEG. Middle: structure model. Above: Reconstructed object phase, showing all atom columns with a resolution better than 0.15 nm.

Under the stimulus of Amelinckx and the guidance of Van Landuyt, Van Tendeloo, and Van Dyck, the EMAT group (Electron Microscopy in Materials) in Antwerp has become one of the leading centers in the world, with 7 electron microscopes and a team of 20 researchers. Over a period of about 25 years, it has welcomed more than 200 visiting scientists and published more than 1000 papers. The group is particularly known for its work in defects, alloys, semiconductors, superconductors, and fullerenes.

A culminating point is the construction of a new electron microscope, the working principle of which is based on ideas developed in Antwerp. It has been shown that by combining images taken at different focus values it is in principle possible, using image processing techniques, to reconstruct the structure of the object without prior knowledge about this structure and with a resolution well beyond the point resolution of the microscope (Fig. 7). In the framework of a Brite-Euram project, this instrument (a 300-keV FEG microscope) is now being constructed by Philips in collaboration with the Universities of Antwerp, Delft, and Tübingen. The microscope is a 300-keV instrument, equipped with a field emission gun, a $(1024)^2$ slow-scan CCD camera, a powerful image processing system (100 Mflops) and complete computer control and alignment of the instrument. The first prototype has a resolution of 0.1 nm and will be installed at the EMAT laboratory in Antwerp, where it will be evaluated on a variety of materials.

Acknowledgments

The author wishes to thank S. Amelinckx, M. H. A. De Groodt-Lasseel, T. Greday, and P. Lagasse for providing valuable information.

References

Marton, L., and Nuyens, M. (1933). Meetkundige optiek der elektronen. *Wis- en Natuurkundig Tijdschrift* **6,** 159.
Marton, L. (1934a). Electron microscopy of biological objects. *Nature* **133,** 911.
Marton, L. (1934b). Electron microscopy of biological objects. *Phys. Rev.* **46,** 527.
Marton, L. (1934–1935). La microscopie electronique des objects biologiques. *Bull. Acad. Royale de Belgique* Part I, **20,** 439 (1934), Part II **21,** 553 (1935), Part III **21,** 606 (1935).
Marton, L. (1936). Quelques considérations concernant le pourvoir séparateur en microscopie électronique. *Physica* **9,** 959.
Marton, L. (1944). Steroscopy with the electron microscope. *J. Appl. Phys.* **15,** 726.
Marton, L. (1965). Historical background of image formation. *Lab. Invest.* **14**(6), 2.
Porter, K. R. Claude, A., and Fullam, E. F. (1945). *J. Exp. Med.* **81,** 233.
Süsskind, Ch. (1986). L. Marton, 1901–1979. *Adv. Electronics and Electron Physics* **Suppl. 16,** 501.

2.4
The History of Electron Microscopy in Canada

GÉRARD T. SIMON[1]

Faculty of Health Sciences, McMaster University, Hamilton, L8N 3Z5 Canada

FRANCES W. DOANE

Department of Microbiology, University of Toronto, Toronto, M5S 1A8 Canada

I. Toronto in the 1930s

In 1930 Canada had been a confederated nation for only 63 years, and the total population in this the second largest country (geographically) in the world was less than 11 million. Like other countries, Canada had been plunged into an economic depression in 1929, but by the early 1930s many parts of the country had begun to recover. One such area was Toronto, capital of the province of Ontario. Expansion was particularly evident at the University of Toronto, which had been formed in the 1880s through the federation of several smaller universities, and which had achieved international fame in 1921–1922 with the discovery of insulin by Frederick Banting and Charles Best.

The University of Toronto had established a reputation for its strong Department of Physics under the chairmanship of Sir John Cunningham McLennan, who was the second in the world to prepare liquid helium, and who established at Toronto one of only four cryogenic laboratories in the world capable of producing helium (Franklin *et al.*, 1978). In 1932 Eli Franklin Burton succeeded McLennan as Head of the Physics Department. He had graduated from the University of Toronto in 1901 with honors in Mathematics and Physics, and had worked for two years as a demonstrator under McLennan before being awarded a scholarship which took him to the Cavendish Laboratory, Cambridge University. In Cambridge, Burton worked under the direction of J. J. Thomson, the discoverer of the high charge-to-mass ratio of electrons, and earned a B.A. for his work on colloidal solutions. He returned to Toronto in 1906 to become a Senior Demonstrator and after completing his Ph.D. in 1910 he was promoted to Associate Professor.

---

[1] Present address: P.O. Box 2000, Station A, Hamilton, L8N 3Z5 Canada.

Under Burton's direction, the Department of Physics broadened its research interests to include applied physics as well as fields that today would be classed as biophysics. A diabetic, Burton was particularly interested in solving problems in which both medical research and physics could be applied. One example of this interest was his collaboration with some Toronto physicians to study the possible use of colloidal arsenic solutions in the treatment of cancer.

Since his years at Cambridge, Burton had followed closely the major developments in electron theory and the feasibility of the construction of an electron microscope. He was aware of Louis de Broglie's hypothesis of 1924 that electrons had wave properties, and of Hans Busch's subsequent calculation that magnetic or electric fields having axial symmetry can act as lenses for an electron beam. Although surrounded by sceptical colleagues, Burton could see that electron optics, and electron microscopy in particular, could lead not only to commercial television, but also had the potential for revealing structural detail beyond that attainable by light microscopy. In 1926, Eli Burton was invited by the Mayo Foundation to participate in a series of lectures on "Biological Aspects of Colloidal Chemistry." It is noteworthy that one of his presentations was entitled "Physics of the Ultramicroscope," reflecting his enthusiasm for the development of such an instrument, a project that culminated in 1938 in the construction of the first electron microscope in North America.

An important collaborator on this project was Walter H. Kohl, who had arrived in Toronto in 1930 after completing a doctorate in Engineering Physics from the Technical University in Dresden. He had obtained a position as a development engineer with a Canadian company, Rogers Radio Tubes Limited, which was involved in pioneer work in television. Kohl's work with Rogers dealt with the deflection of electron beams by means of magnetic and electrostatic fields, and the development of cathode-ray tubes and luminescent screens. Burton, whose knowledge of German was limited, was to rely on Walter Kohl not only for his technical expertise in electronics, but also for translating scientific publications, all in German, on the newest electron optics developments in Germany.

From 1932 on, at Burton's invitation, Kohl gave regular seminars and lectures, pioneering the modern concept of a journal club. Among his lectures, a few titles are highly significant. In 1935 one finds seminars on "The Fundamental Principles of Electron Optics" and "The Electron Microscope"; in 1936 his seminars included "Electrostatic Lenses," "Electromagnetic Lenses," "The Electron Microscope," and even "Applications of the Electron Microscope." He repeated much of the experimental work of von Ardenne, Knoll, Borries, Ruska, Scherzer, and Johannson, and in April 1934 he demonstrated one of his specially constructed tubes and the electron optic image of an oxide cathode produced by means of a Johannson immersion objective. This is believed to be the first demonstration of its kind in North America.

Kohl's lectures and demonstrations encouraged Burton to initiate the construction of an electron microscope in the Department of Physics. Burton's high regard for Kohl is evident in the fact that he included Kohl as a co-author in his book entitled *The Electron Microscope* (Burton and Kohl, 1942). Burton acknowledged several times that Kohl had participated actively in the construction of the 1938 electron microscope. Many years later, however, Kohl modestly noted that although this was a generous gesture on Burton's part, it was hardly justified since he took no part whatsoever in the design and construction of the instrument.

II. C. E. Hall's Microscope

In the summer of 1935, Burton attended a meeting in Berlin on "possible areas for the application of the electron microscope." When he returned to Toronto, he assigned his graduate student Cecil Hall the M.A. project of constructing a simple electrostatic transmission electron microscope. Hall successfully completed the project, and in the University of Toronto President's Report of 1935–1936, the submission from the Physics Department notes that "C. E. Hall, B. Sc., Alberta (holder of the Alumni Federation Fellowship), has been working in the new field of electron optics and has completed, almost entirely by his own efforts, and electron microscope of the electrostatic type. The work is so promising that the National Research Council has given the department a research grant for the continuation of this work during the session of 1936–1937."

Canada had been hard hit by the Depression, and financial support for research was reduced to a minimum. Nevertheless, the excellent results obtained by Hall permitted Burton to obtain a grant of $800 from the National Research Council to cover both equipment and a stipend. Thus Hall was able to extend his project and construct an electromagnetic lens which alowed him to obtain images of the cathode at a magnification of about 3000 times. In May 1937, Burton returned to the National Research Council with a Progress Report and asked for further funds. "The next step in this research," he explained in his written application, "is to attempt to take electron pictures of sections of some substance placed in the electron stream." Burton asked for $724.50, of which $250 was for the condenser and the balance was for payment of the investigator's salary at the rate of $62.50 a month. On the grounds that the work could be assigned to a scholarship holder, the National Research Council refused the request. Unfortunately, this meant that Hall had to leave Toronto, and he joined the Research Laboratory of the Eastman Kodak Company in Rochester, New York, where he developed the first electron microscope constructed in the United States, in 1939. Although his graduate work formed the basis for the 1938 electron microscope built in Toronto, his results were never published in a scientific journal, and the only official record is contained in his M.A. thesis of 1936.

III. The 1938 Prototype TEM

After Hall's departure, Burton asked James Hillier, a Mathematics and Physics graduate from the University of Toronto, and Albert Prebus, who had just received an M.A. from the University of Alberta, to undertake the construction of a high-voltage, magnetic compound electron microscope with the aim of applying it to the investigation of biological specimens. They started to work together at the beginning of the winter of 1937, designed the microscope over the Christmas holidays, and constructed the instrument in the astonishingly short period of four months in the beginning of 1938 (Hillier, 1939; Prebus and Hillier, 1939). To accomplish this feat, they had at their disposal the thesis of Hall and the publications of Knoll, Ruska, and Marton. They "borrowed" two high-tension condensers from the University of Alberta and obtained fluorescent screens from Kohl. The rest of the instrument was manufactured by themselves with the help of the Physics Department workshop technicians. "Our greatest mechanical challenge was the design and construction of the components of the instrument" stated Hillier (Franklin et al., 1978). Prebus recalls that "the shopwork was done on a two-shift basis; the professional machinists worked the day shift. Without their unreserved approval, Hillier and I worked the night shift, often until 4 a.m., and occasionally until the day shift was about to start" (Franklin et al., 1978). Many parts of the design were innovative, and much of the machining was of a very high quality.

The transmission electron microscope completed in 1938 by Hillier and Prebus (Fig. 1)—the first of its kind in North America—is now displayed at the Ontario Science Centre in Toronto. It was constructed in six sections, which were sealed together by means of plane-lapped grease joints. The first section contained the electron gun and the condenser lens. For an electron source, Hillier and Prebus chose a hard-vacuum thermionic cathode, although they had made many attempts to use a cold cathode gas discharge tube. A maximum of 45,000 V could be applied to the gun, and

ELECTRON MICROSCOPY IN CANADA 83

FIGURE 1. The 1939 microscope, installed in the Department of Physics at the University of Toronto. Left to right: Professor E. F. Burton, A. F. Prebus, W. A. Ladd, and J. A. Hillier (inset). (Photo courtesy of W. A. Ladd and *The New York Times.*)

the filament current was supplied by two 2-V batteries. The filament was made from 6-mil (150-μm) tungsten wire, and the radius of the curvature of the tip had to be less than 0.5 mil (25 μm) to ensure that the emission approximated that of a point source. The filament could be raised or lowered with respect to the rest of the cathode by a screw and bellows arrangement. The whole cathode assembly was sealed by wax to the upper end of a glass tube. The wax seal had to be water-cooled during operation, creating obvious practical difficulties. To replace the filament, the seal had to be broken, this seal was eventually replaced by a Neoprene joint so that the filament could be changed in a few minutes. The condenser was of conventional design.

The second section of the microscope contained the object chamber. Specimens were inserted through a simple conical greased joint opening and rested on a hollow cartridge which fitted into a collar, about 5 mm above the center of the gap between the upper and lower polepieces of the objective lens. The specimens were suspended over a small hole in the center of a circular platinum diaphram which was clamped at the lower end of the specimen holder.

The objective lens, comprising the third section of the column, was designed so that sufficient latitude was given to allow changes to be made in polepiece geometry. The first polepiece design to be made had matching upper and lower polepieces separated by a brass spacer, which fixed the length of the gap between the polepieces and aligned their axes of symmetry. A micrograph taken with this polepiece design showed a resolving power of 140 Å. Unfortunately, the inevitable misalignment of the upper and lower polepieces was one of the limitations of the design. Therefore, the polepiece was redesigned so that the upper and lower polepieces and spacer could be made from a single piece of metal. With this new design a resolution of better than 60 Å was obtained, as reported by Hillier in the *Canadian Journal of Research* in April 1939. The objective lens was capable of magnifying 100 to 125 times, and the first-stage image could be viewed on an intermediate fluorescent screen at the lower end of a brass tube which constituted the fourth section of the microscope. Inside this tube was a soft iron cylinder to shield the electron beam from magnetic fields.

The fifth section of the column consisted of the projector lens. The initial image could be further magnified up to 330 times by this lens, giving a total magnification of approximately 40,000 times. In the sixth section were located the camera and final image screen, which could be lifted by means of a conical grease joint fixed in the wall of the brass tube. A chamber was later added so that a series of images could be recorded on photographic plates without letting air into the main body of the microscope.

The high-tension system consisted of a step-up transformer, hard-vacuum half-wave rectifier, and a ripple-voltage filter system. All of the high-tension apparatus and connecting leads were carefully shielded. The fluctuation in accelerating potential was less than 1 in 50,000 V.

The success of this venture, and the subsequent improvement in available financial support, are referred to in the President's Report for 1938–1939: "Mr. James Hillier, assistant demonstrator, and Mr. Albert Prebus, holder of a studentship from the National Research Council, have continued the work of perfecting the electron microscope, and have succeeded in taking many photographs of sub-microscopic structures up to a primary magnification of 30,000 times. This is equivalent to being able to separate two points on an object at a distance of .0000004 in., or .00000100 cm., or 100 Å apart. In addition to the studentship held by Mr. Prebus, the National Research Council gave a small grant during the present year to enable these two workers to continue the work during the summer vacation of 1938. The electron microscope is so promising that assistance has been offered by the National Research Council and the Banting Institute to keep these two workers employed for the next calendar year, beginning July 1, and we are hoping for some outstanding results."

The optimism expressed in this report was well founded. From the beginning, the resolving power of the microscope was very good, and the pictures produced were of excellent clarity. In 1938, in an article in the *Zeitschrift für Technische Physik,* Ruska and von Borries had reported a resolving power of approximately 100 Å for their new Siemens Elmiskop. Having achieved a resolving power of 60 Å with their instrument, the Toronto group were understandably proud. In his Ph.D. thesis in 1940, Prebus states:[2] "Comparison of the reproduction of better photographs of the German workers with prints of the best photographs obtained with the apparatus described herein, indicates that higher resolving power has been obtained with the Toronto apparatus."

In 1939 the graduate students W. A. Ladd and J. H. L. Watson joined Burton's group (Watson, 1974). With Prebus and Hillier, Ladd collaborated on the design and construction of a second microscope (Fig. 2), underwritten by and intended for the Columbian Carbon Company of New York. Watson worked on basic electron-optical problems, introduced improvements in the design of the second microscope, and investigated a wide range of specimens from biological and materials sciences (Fig. 3).

The development of electron microscopes in the Physics Department of the University of Toronto ended with the construction of a third microscope, completed in 1944, involving Watson, S. Glenn Ellis, G. David Scott, and Beatrice R. Dean. In due course, most of the personnel directly involved in electron microscope development at the University of Toronto moved to the United States. Among the reasons for this "emigration" must be included the limited funding available in Canada for such enterprises and the surge of interest exhibited at this time by several American companies.

---

[2] *Editor's note:* It should be remembered, however, that von Borries and Ruska had made enormous progress from 1938 to 1940 with their commercially produced Siemens Übermikroskop, so that by 1940 they had also achieved a similar resolution on a day-to-day basis. It is likely that, due to wartime postal restrictions, these results were not available to Prebus at the time.

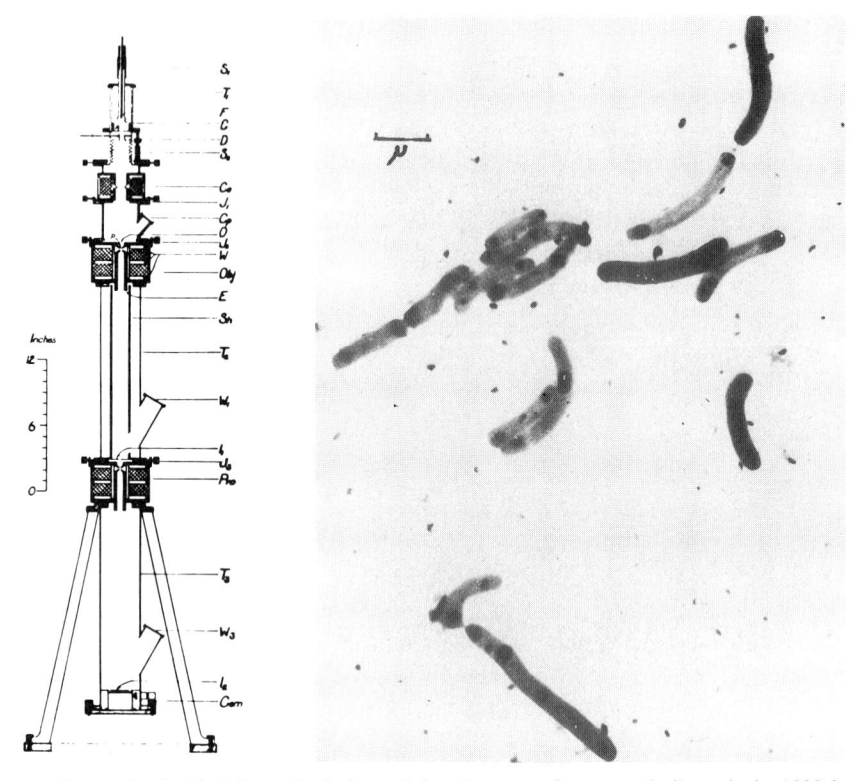

FIGURE 2. (Left) Schematic design of the electron microscope built early in 1938 by Prebus and Hillier. (From Prebus and Hillier, 1939.)

FIGURE 3. (Right) Electron micrograph of vole bacilli taken on the 1939 microscope. (From Ph.D. thesis of J. H. L. Watson, 1943.)

James Hillier (who graduated from Toronto with an M.A. in 1938 and a Ph.D. in 1941) joined the Radio Corporation of America in Camden, New Jersey. With V. K. Zworykin and A. W. Vance he designed the first electron microscope to be made commercially in the United States (Zworykin et al., 1941). Albert Prebus accepted a post at Ohio State University, Bill Ladd moved to the Columbian Carbon Company in charge of the microscope he had helped to build in Toronto in 1939, and John Watson eventually settled in Detroit. With their departure, the development of transmission electron microscopes shifted away from Toronto to the United States, but the technology pioneered at the Department of Physics formed the foundation for the massive production of electron microscopes by RCA that was subsequently to follow. Many of early commercial instruments were purchased by Canadian institutes across the country, as scientists in both the materials and the biological sciences found increasing applications for the high-resolution instruments.

An article by A. D. G. Stewart in 1985, entitled "The Origins and Developments of Scanning Electron Microscopy," notes that in 1958 a scanning electron microscope built by K. C. A. Smith of the Cambridge University Engineering Department was marketed via the Metropolitan Vickers Electrical Company, Manchester, U.K., and delivered to the Pulp and Paper Research Institute in Shawinigan, Quebec. This was evidently the first scanning electron microscope to be made commercially.

IV. The Burton Society of Electron Microscopy

The first attempt to form an organization of Canadian electron microscopists was made in October in 1958, at the newly opened Ontario Cancer Institute in Toronto. Present at this meeting were 20 representatives from laboratories engaged in electron microscopy in London, Guelph, Kingston, Ottawa, and Toronto. Naming themselves "The Ontario Group of Electron Microscopists," they decided to meet three or four times a year, the meetings to take the form of "one or two formally presented papers, informal discussion and the viewing of the equipment in the laboratories visited" (Howatson, 1982). By 1960 the mailing list of the group had increased to over 50, including several members from outside the province of Ontario. In May 1960 the members voted on and accepted a constitution whereby the group became a society, to be known as the Burton Society of Electron Microscopists, and Dr. A. F. Howatson of the Ontario Cancer Institute in Toronto was elected President. The inaugral meeting of the new Society was held in Toronto 28–29 October, and featured Dr. James Hillier, Vice President of RCA Laboratories, who spoke on his association with Professor Burton and the early work on electron microscopy in Toronto.

In the following years the Burton Society continued to expand and flourish, and by 1965 the number of members approached 100. That same year, the officers of the Society planned an innovation: a meeting to be held in Ottawa following the Annual Meeting in June of the Canadian Federation of Biological Societies. However, the outcome was not what the organizers anticipated. As recounted by Allan Howatson several years later: "The combination of mental fatigue after attending four days of Federation meetings and an exceptionally fine sunny day in June was too much for all but the most dedicated members, and the turnout at the final session on Saturday afternoon was dismally poor. The session ended without the usual business meeting being convened and the audience left to enjoy what was left of the sunshine. This meeting marked the end of the Burton Society" (Howatson, 1982).

V. The Microscopical Society of Canada/Société de Microscopie du Canada

The dream of grouping together the Canadian microscopists did not, however, die on this sunny Saturday afternoon in 1965. Five years later, in Grenoble, at the 7th International Congress on Electron Microscopy, Allan Howatson from Toronto and Huntington Sheldon from Montreal sat as observers at the General Assembly of the IFSEM, but Canadian electron microscopists could not be directly represented without a national organization. It was during this international meeting that the Canadian scientists in attendance at Grenoble seriously contemplated the creation of a Canadian Society. In the spring of 1972, largely on the initiative of Frances Doane and Felix de la Iglesia, a small group of microscopists in the Toronto area met to create a society which would be nationwide in scope and would encompass all branches of microscopy in physics and biology. In June 1972, a meeting of those interested in the proposal was held at Laval University in Quebec, at the time of the Annual Meeting of the Canadian Federation of Biological Societies, and the decision was taken to launch the new society. Letters Patent were applied for and obtained on 16 October 1972, and the Microscopical Society of Canada/Société de Microscopie du Canada came officially into existence (Fig. 4).

A quarterly *Bulletin* was initiated in 1973, and the first issue—published in February of that year—records the names of 169 members. The first Annual Meeting was held in 1974 at the Ontario Science Centre, where the 1938 Toronto electron microscope had been installed. More than 150 people attended the first meeting, and the guest speakers included Keith Porter and Charles Leblond. Since that founding scientific meeting, the MSC has shown sustained growth and interest in light and electron microscopy and their application to a broad range of scientific disciplines, through the *Bulletin of the Microscopical Society of Canada/Société de Microscopic du Canada*, edited by Frances Doane, and its annual meetings.

In August 1974, at the 8th International Congress on Electron Microscopy in Canberra, the Microscopical Society of Canada/Société de Microscopic du Canada was officially granted membership in the International Federation of Societies for Electron Microscopy. It was during the Canberra meeting that the new Canadian Society, represented by Jennifer Sturgess and Gérard Simon, presented the bid for Toronto to host the 9th Interna-

FIGURE 4. Founding Members and first Council of the Microscopical Society of Canada/Société de Microscopie du Canada, 1972.

tional Congress in 1978. Canada was awarded the honor of organizing the 9th Congress, to be sponsored jointly by the Microscopical Society of Canada, the Electron Microscopy Society of America, and the University of Toronto.

The 9th International Congress, with approximately 3500 registrants, was a major achievement for the young Canadian Society. Under the chairmanship of Gérard Simon, the local organizing committee gave special emphasis in the program to symposia in both biology and physics, many of them with an interdisciplinary flavor. A special volume of the *Proceedings* contained the contributions of the invited speakers, and three volumes of papers were published and available at the Congress (Sturgess, 1978).

An important innovation was the collaboration with other national Societies to sponsor young scientists. By providing free registration, loading, and meals, more than 120 young scientists from many parts of the world were able to attend the Congress, and to meet socially with senior microscopists in Toronto.

The 1978 Congress coincided with the 40th anniversary of the construction of the first Toronto electron microscope, and provided an excellent opportunity for microscopists to pay tribute to five pioneers in electron microscopy. Through the cooperation of the University of Toronto, honorary degrees were awarded during the Congress to Ernst Ruska, and to four native-born Canadians—Cecil Hall, James Hillier, Albert Prebus, and Keith Porter.

Ten years later, in 1988, the Canada Post Corporation issued a commemorative stamp to mark the 50th anniversary of the construction of the Toronto electron microscope by Hillier and Prebus, under the supervision of Dr. Eli Burton.

The Microscopical Society of Canada/Société de Microscopie du Canada, with a membership of more than 650 in 1995, has brought together electron microscopists from all parts of the country, and provided them with a forum for scientific discussion and exchange of information. There are now five geographic Sections within the Society, and Annual Meetings have been held in many different Canadian centers, from Newfoundland to British Columbia. On three occasions the Annual Meeting has been held jointly with EMSA and MAS—in Detroit in 1984, in Milwaukee in 1988, and in Boston in 1992. Among the projects undertaken by the MSC/SMC are the compilation of a national directory of electron microscopy facilities in Canada, the introduction of Presidential Awards for young scientists, which since 1995 bear the name of Gérard Simon, and the publication of selected topics in microscopy (Klosevych, 1989). Electron microscopes are firmly entrenched in many fields of Canadian science, and have become essential tools in research and industry. The joy of discovery still remains, however, for those who use these instruments, and the observation made in 1974 by MSC/SMC member John H. L. Watson continue to hold true in 1989: "Electron microscopists are still privileged travellers in the largely unknown world of inner space."

APPENDIX I: FOUNDING MEMBERS AND FIRST COUNCIL OF THE MICROSCOPICAL SOCIETY OF CANADA/SOCIÉTÉ DE MICROSCOPIE DU CANADA

President — Allan F. Howatson
First Vice President — Harry Pullan
Second Vice President — Gérard T. Simon
Secretary — Felix de la Iglesia
Treasurer — Frances W. Doane
Councillor — Eric J. Chatfield
Councillor — Ivan Grinyer

REFERENCES

Burton, E. F., and Kohl, W. H. (1942). "The Electron Microscope." Reinhold, New York.
Franklin, U. M., Weatherly, G. C., and Simon G. T. (1978). A history of the first North American electron microscope (1938). Toronto, Vol. III.
Hillier, J. (1939). The effect of chromatic error on electron microscope images. Can. J. Res. A17, 64.
Howatson, A. F. (1982). The Burton Society of electron microscopists. Bull. Microsc. Soc. Can. 10, 9.
Klosevych, S. (1989). "Principles and Practice of Microscopy and Scientific Photography." Microsc. Soc. Can.
Prebus, A., and Hillier, J. (1939). The construction of a magnetic electron microscope of high resolving power. Can. J. Res. A17, 49.
Stewart, A. D. G. (1985). The origins and development of scanning electron microscopy. J. Microsc. 139, 121.
Sturgess, J. M. (1978). "Electron Microscopy 1978." "Proc. 9th Int. Cong. Elect. Microsc. Toronto. Vols I–III.
Watson, J. H. L. (1974). Toronto "daze." Bull. Microsc. Soc. Can. 2, 10.
Zworykin, V. K., Hillier, J., and Vance, A. W. (1941). An electron microscope for practical laboratory service. Elec. Eng. 60, 157.

2.5
Electron Microscopy in France
2.5A Early Findings in the Life Sciences[1,2]

FRANÇOISE HAGUENAU

College de France, Paris, France[3]

When the first electron microscopes became available to biologists, there already existed in France a long tradition in the domain of cytology. The event presented a unique opportunity to end many discussions, often fiery, which divided scientists on the structure and significance of certain cell organelles. Thus, to take some examples of important questions raised, the specific structure of what was long known in France as ergastoplasm (discovered by Charles Garnier in 1897; see Garnier, 1899) was revealed, the reality of the Golgi apparatus was ascertained, and the noncontinuity of the contact between neurons or between neurons and muscle fibers (synapses) was shown. For the first time also, and this represented a major contribution, the structure of many viruses was illustrated and a new classification of viruses based on their ultrastructural morphology was presented, which served as the foundation for the present international nomenclature.

In the beginning, only a few laboratories were equipped. Among those centers which played a pioneering role at almost the same time, one can single out Pierre Lépine's laboratory with Odile Croissant at the Institut Pasteur, the group of Marcel Bessis with Janine Breton-Gorius and Jean-Paul Thiery at the Centre National de Transfusion Sanguine, Pierre Paul, Grassé's group with Nina Carasso and Pierre Favard at the Faculté des Sciences, where also was located Marcel Buvat's laboratory for plant biology (Ecole Normale Supérieure), and Charles Oberling's team with Wilhelm Bernhard, Alain Gautier, and myself in Villejuif. Many of the contributions from these laboratories represent first discoveries or descriptions, and they are important landmarks in the history of cell ultrastructure. We report here some of the early findings.

The first electron microscopy laboratory set up in France by a biologist was that of Pierre Lépine at the Institut Pasteur in Paris. He was at the time the head of the Service des Virus, and as early as 1945 was enthusiastic

---

[1] See also the more complete monograph: *Biology of the Cell* **80** nos. 2–3, 1994.
[2] References, added in proofs for Chapter 2.5A, are found on pages 129 and 130.
[3] Present address: 138, Boulevard Haussman, 75008 Paris, France.

about the advent of electron microscopy. From the start he had established close collaboration with the phycisists involved in the construction of the instrument. P. Grivet, H. Brück, and J. Guintini. The former two had constructed the CSF (French Compagnie Générale de Télégraphie sans Fil) electrostatic electron microscope, while J. Guintini was responsible for one of the prototypes of the Siemens electron microscopes equipped with magnetic lenses supplied by batteries.

From Pr. Lépine's laboratory stem some of the very first ultrastructural studies in France on bacteriophages and viruses. Lepine established close links with Dr. R. W. G. Wyckoff, who developed the remarkable techniques of direct observation and shadowing; with Odile Croissant and Joseph Guintini he published a series of pioneer papers on animal viruses—rabies virus, pox virus, herpes virus, polio virus—plant and insect viruses—tobacco mosaic virus, polyedrosis virus (with C. Vago)—to cite some examples only (see the chapter, "Ultrastructural Morphology of Viruses, French Studies" in *Biology of the Cell* **80,** nos. 2–3, 1994.)

Among other meaningful early work accomplished in that department one should point to the studies on rabies virus: P. Atanasiu succeeded in obtaining the *in vitro* multiplication of two strains, a "street" rabies virus strain and the Pasteur strain. He showed that viral particles were formed by budding at the level of the host cell membranes and demonstrated the specificity of the observed viral particles by the binding of ferritin-conjugated antirabies immunoglobulins.

One should also recall that the purity of the vaccines at the time developed at the Pasteur Institute were controlled there by electron microscopy of viral proteins (rabies) or viral envelopes (hepatitis B).

In that laboratory, the role played by Odile Croissant should be stressed. Because of her competence both in virology and as an expert in the use of the instrument, she formed a link among the different workers using the electron microscope on the campus as well as in numerous other institutes. The strong association she established with Gérard Orth, who worked first in Villejuif but soon came to the Institut Pasteur, was never loosened. It led to the early outstanding work on human papillomaviruses (HPV), in which the role of electron microscopy should be especially stressed. Indeed, because of the lack of an *in vitro* system, knowledge of the biological properties of these viruses could be approached only by ultrastructural studies of papillomavirus-associated lesions. For the first time in the annals of electron microscopy, Gérard Orth and Odile Croissant had recourse to, besides hybridization *in situ* with the light microscope, ultastructural *in situ* molecular hybridization techniques and thus studied the cytopathogenic effect of papillomavirus infection in rabbit and human warts. The remarkable plurality of HPVs was disclosed and illustrated by heteroduplex analysis—to mention electron microscopy work only. Conserved nucleotidic sequences between the genomes of bovine and one human (HPV1) papillomavirus were first detected by this technique prior to sequencing methods, suggesting the possible evolution of PVs from a common ancestor.

A captivating electron microscope finding by Luc Montagnier and co-authors should also be mentioned here, although it dates from more recent times: the discovery and first description of a new retrovirus in cell cultures from a lymph node biopsy of a patient presenting a lymphoadenopathy (Barré-Sinoussi *et al.*, 1983). Later the virus was recognized as responsible for AIDS (acquired immunodeficiency syndrome); it is now called HIV and classified as a Lentivirus. The tremendous impact of that discovery need not be stressed, and everyone knows of the battle it engendered.

Another pioneer in France, Marcel Bessis, was one of the very first biologists to acquire an electron microscope (1949). He was a hematologist who published a well-known treatise, *Cellules du sang normal et pathologique,* in 1972. He had been interested before 1949 by the Rh factor and was responsible for the first ex-sanguino transfusion practiced in a newborn with hemolytic anemia due to this factor. He then turned to the study of cell structure and organization, a subject to which he decided to devote himself entirely. He was working at the time with Prof. A. Tzanck, who had created the Centre de Transfusion Sanguine, which was a private foundation located at the Hopital St. Antoine in Paris (Fondation Deutsch de la Meurthe). He had heard that P. Grivet had constructed an electrostatic lens electron microscope and that one of the two existing instruments was for sale by the CSF Company. He convinced Prof. Tzanck, who managed to obtain the necessary funds, and the microscope was set up in the basement of his department in that hospital. Not being himself a physicist, he recruited as engineer M. Bricka, who left the CSF Company for the purpose. He also enlisted a technician, Janine Gorius (later Mrs. Breton-Gorius, Ph.D). Together with Jean-Paul Thiery, who joined them later (1954), they published some of the very first and important papers on white blood cells and platelets.

The personality, the "style," of Marcel Bessis accounts for the striking quality of the documents published; some of his pictures could in truth be compared to works of art in a gallery as attested by the book on red cell shapes by scanning electron microscopy, published in 1976. Indeed, Marcel Bessis's concern for fine arts and his admiration for craftwork are manifest all the time. When one penetrated into his laboratory at the Centre National de Transfusion Sanguine, one was struck by the importance given to light microscopy, and immediately became aware of the care with which, in their glistening brass the earlier types of microscope were maintained as well as the latest phase-contrast, fluorescent and interference microscopes and also how clever were the devices which enabled him, one of the first, to use either UV or a laser beam to destroy targeted cell organelles.

Among the early findings, one can point to the work on sickle cell anemia, in which, with the replica method, was shown the organization of the polymerized S hemoglobin molecule responsible for the change in the shapes of the red cell (1954). One may also quote the original description (1958) of the "erythroblastic island" (the association, in the bone marrow, between a central macrophage surrounded by maturing erythroblasts), which constitutes the first example of the creation of a microenvironment which influences erythroblastic differentiation, the role of ferritin molecules in normal and pathological erythroblasts (1957) and in myeloblastic leukemia cells, the description of filaments which much later were to be known as vimentins (1955). One should also mention how, later, when scanning electron microscopes became available, various pathological shapes of red cells were strikingly illustrated.

Needless to say, the Center created in 1966 by M. Bessis on the premises of the Hopital Bicêtre had a worldwide reputation. The scientists and artists who gathered there, either to work, to learn, or to teach, are important. Among them were A. Policard, who had supported him from the start and who played a definite role in the launching of electron microscopy in France, J. L. Binet, P. Tambourin, and the Americans G. Brecker, E. Conkrite, H. Mel, M. M. Chediak, F. Stohlman, E. Ponder, W. Jensen, L. L. Lessin, and M. Murphy, and the Australian B. Morris, to mention only the acknowledged international hematologists of the time.

During the same period, P. P. Grassé, who was the Director of the Laboratoire d'Evolution des Etres Organisés at the Faculté des Sciences in Paris, created there with the support of the Central National de la Recherche Scientifique (CNRS) an electron microscopy department which was first sponsored by the university. That laboratory played an important role as a welcome center where many scientists from different horizons gathered. Both Prof. Grassé and his collaborators, Nina Carasso and Pierre Favard, had a great talent for teaching. The latter lectured at the Faculté des Sciences of the Université of Pierre et Marie Curie. He played an important part in the proliferation of electron microscopy in France by establishing the headquarters of the Société Française de Microscopie Electronique (SFME) in the Centre de Biologie Cellulaire of the CNRS in Ivry. It is he who created the *Journal de Microscopie*, edited by the SFME, and he was also responsible for the editing of excellent and very successful books on cell biology for students. That group brought forth some of the basic information on protozoans first with E. Holland, an eminent specialist of protists and afterwards with Prof. Grassé, Nina Carasso, and Pierre Favard. They contributed to important knowledge not only in the realm of protozoans, but in that of molluscs and worms. One should mention here studies in collaboration with E. Fauré-Fremiet, from the Collège de

France, on endocytosis in ciliates and on the contractile system of vorticella, where they were able to localize the Ca^{2+} in the endoplasmic reticulum, thus establishing a comparison with the sarcoplasmic reticulum of muscles. As far as metazoans are concerned, one should cite A. Berkaloff, in particular his studies on malpighian tubules in insects. It is in that laboratory also that R. Couteaux and J. Taxi developed their research on the vertebrate neuronal system and striated muscles. In fact, R. Couteaux was to succeed to P. P. Grassé as head of that electron microscopy laboratory.

Prof. Grassé's school is notable for having contributed to establishing in 1955 the reality of the Golgi apparatus which was disputed by the school of M. Parat. The latter had indeed considered the Golgi apparatus as an artefact formed by silver deposits in between a system of vacuoles characterized by their affinity to vital stains. That battle had also been won for mammalian cells by the same date by the Villejuif team.

Finally, also at the Ecole Normale Supérieure, M. Buvat was the main leader of research in plant cell biology, an active discipline in this country. Investigations were undertaken in his own laboratory in Paris, and in many other centers in France, including Lyon, Bordeaux, and Marseille. Marcel Buvat was the first to describe the ultrastructure of the Golgi apparatus (in the root cells of allium cepa) in 1957, and that of the ergastoplasm with Nina Carasso at the same date. He has made himself well known for his analysis of the "vacuolar compartment" and its relation to lysosomes, also first described in plant cells by him and his collaborators. Indeed, his collaborators and disciples were many and stand on their own. More detailed reports of their work will be found in the chapter of the monograph, "French Contribution to the Studies of Plant Cytoplasm by Electron Microscopy" (*op. cit.*).

The role played by the group in Villejuif should now come to the fore in this account of early times. The great achievement here was the launching of the viral theory of cancer through the detection, in cells and in tumors, of a series of oncogenic viruses which were described and classified for the first time. Charles Oberling was its champion and herald, at a time when the notion of viruses as a possible cause of cancer was disputed and considered by many as a heresy. It is he who created the center of research at the Institut de Recherches sur le Cancer which was sponsored by the CNRS.

He had just arrived from Strasbourg, invited to Villejuif by Gustave-Roussy, the Dean of the Faculty of Medicine, who had asked him to direct the Pathology Department of the Institute and he was actively searching for young students to come and work for him.

I will always remember his visit at the Hospital Broussais in the department of Louis Pasteur's grandson, René Pasteur Valéry-Radot, who accompanied him to the laboratory of pathology while Wilhelm Bernhard, a postdoc from Bern University (Switzerland), and myself were at the bench reading pathology slides. Oberling immediately noticed him, stopped to talk to him for a few minutes, then went on. Almost the next day he called and asked him if he would agree to head the electron microscopy department that he wished to set up in Villejuif. When a year later Bernhard asked me to join them, the laboratory already possessed a Trüb-Täuber microscope, equipped with electrostatic lenses.

The ambience was unique and was due no doubt to Bernhard's and Oberling's personalities—incredibly different yet alike in many respects. They were profoundly linked by their eagerness and their passion for science. They also shared a great and humanistic culture heightened by their perfect knowledge of French, English, and German literature (to say nothing of music). They were quite the opposite, however, in character. Oberling was heavily built, Bernhard was lean. Oberling, warm and congenial, could be rough, even violent sometimes; Bernhard was extremely courteous, amiable, and controlled.

It was no surprise, then, that both graveness and mirth reigned in that laboratory, which played a leading role in revealing the fine structure of cells and viruses. It played a major role also in the formation of electron microscopists and in the spreading of the discipline in France and abroad. It is impossible to give an idea of the extraordinary attractive power of the Villejuif center. Students and post-docs as well as internationally known scientists haunted the place to acquire the new art or to keep in touch with new developments. The list is impressive, and personalities came from everywhere. There was J. Beard, A. Claude, L. Dmochowski, A Graffi, L. Gross, D. H. Moore, and L. A. Zilber, to mention some of the noted viral oncologists; famous pathologists such as H. Braunsteiner, P. Dustin, H. Hamperl, and H. U. Zollinger; and considerable numbers of future outstanding electron microscopists who themselves later created important research centers: J. André (Orsay), Y. Clermont (Montréal), D. Ferreira (Lisbonne), N. Hinglais-Guillaud (Paris), J. Izard (Toulouse), A. Porte (Strasbourg), C. Rouillier (Genève), B. Stevens (Toulouse), G. de Thé (Paris), and Y. Yotsuyanagi (Paris) (the list is far from exhaustive). Since the program was oriented toward the search for viruses in cells, it necessarily concerned the description of normal and pathologic cells and of course, to start with, improvements in techniques to reach these goals.

In this respect, practically no problem of the time was left unapproached—fixatives, embedding media (an ultramicrotome had been designed by A. Gautier in 1951), spreading of cells, culture devices—all were explored. The challenge of cutting appropriate ultrathin sections so as to secure sectioning transparency to electrons was taken up with amazing gusto, and those who came there to learn will always remember the sessions in the cold room where they remained for hours dressed like astronauts, shivering, to obtain sections of tissues embedded in the first premethacrylate waxes which were being tested.

W. Bernhard's constant concern with advances in methods is, in particular, well illustrated by his and his collaborator's exploratory approach to ultrastructural cytochemistry. The search for water-miscible embedding resins (durcupan and glycol methacrylate, with W. Staubli and with E. Leduc), the introduction later of the so-called regressive staining technique, as well as the early work on ultrastructural immunocytochemistry with peroxydase (S. Avrameas, M. Bouteille, and E. Leduc), the labeling of cell membranes by concanavalin A (C. Huet), and the adjustment of specific stains for DNA (A. Gautier, G. Moyne) should also be emphasized. In conjunction were also developed at an early date the techniques of high-resolution autoradiography with P. Granboulan and the first attempts at cryo-ultramicrotomy with E. Leduc.

Equipped with such tools, basic findings on the ultrastructure of normal and cancerous cells in the mammal were published. Thus were described for the first time the nucleolus (W. Bernhard *et al.*, 1952a), of which the strandlike structure was first shown in the electron microscope; the ergastoplasm, which the Rockefeller group, unaware at the time of the early finding of Charles Garnier, termed endoplasmic reticulum (W. Bernhard *et al.*, 1952b); the Golgi apparatus (F. Haguenau and W. Bernhard, 1955a), the structure of which was revealed; the nuclear membrane with its pores (F. Haguenau and W. Bernhard, 1955b); and the centriole, of which the remarkable cylinder structure made of a wall with probably nine tubules was first reported (E. de Harven and W. Bernhard, 1956). The elucidation of the structure of the interphase nucleus later became one of the important goals of the laboratory. Interchromatin granules, perichromatin fibrils, and nuclear bodies were in turn described (A. Monneron, M. Bouteille, V. Marinozzi, etc.) and characterized biochemically. Nucleolar and extranucleolar transcription were further studied with E. Puvion and F. Puvion-Dutilleuil, and the demonstration of the existence of "Christmas tree" configurations in mammalian cells illustrates well the turn taken by that laboratory toward molecular biology.

One may conclude this report on the Villejuif activities with what may stand out as perhaps the most important contribution of the team: the discovery and description of oncogenic viruses.

Except for Rous sarcoma virus (RSV), which was earlier seen by A. Claude, K. R. Porter, and E. G. Pickels in 1947, almost all other types of oncogenic viruses were first described here. Between 1953 and 1960, W. Bernhard and his collaborators reported on the structure and mode of formation in the cell of RSV (W. Bernhard *et al.*, 1953), Shope fibroma virus (W. Bernhard *et al.*, 1954), molluscum contagiosum (R. Dourmashkin and H. Febvre, 1958), mouse mammary tumor virus (W. Bernhard and A. Bauer, 1955), Murrary Begg avian endothelioma virus (F. Haguenau *et al.*, 1955), the viruses of avian erythroblastosis and myeloblastosis (W. Bernhard *et al.*, 1958; L. Benedetti and W. Bernhard, 1958), the murine leukemias and sarcoma viruses (W. Bernhard and L. Gross, 1959), and polyoma (W. Bernhard *et al.*, 1959). After that date many other viruses were described, in particular, SV40 (N. Granboulan and P. Tournier) and adenovirus (N. Granboulan and A. Martinez Palomo). For more detailed developments, see the chapter on "Ultrastructural Morphology of Viruses" in the monograph *Biology of the Cell* **80,** nos. 2–3, 1994.

A summary of these findings on the ultrastructure of viruses was presented by W. Bernhard in Columbus (Ohio) at the XVIIth meeting of the EMSA (1959).

His masterly description of the various fowl and rodent tumors of the well-known A, B and C particles (to which D particles were later to be added) played a determining role in the elaboration of new rules for the classification of viruses (Bernhard, 1960). Indeed, the images observed in the electron microscope [together with results obtained by X rays and superbly enunciated by D. L. D. Caspar and A. Klug (1960, 1962)] had such an impact that it led the virology community to found modern nomenclature not only on the viral genome nucleic acid constituents but on ultrastructural morphology as well (Propositions of the Provisional Committee for Nomenclature of Viruses PCNV, 1965).

These directions hold true to this day; they constitute the basis of a single, universal taxonomic scheme promoted by the International Committee on Taxonomy of Viruses (ICTV).

These advances, the detailed morphological and beautiful images of viruses, that of the adoption by the international community of the ultrastructural description and denomination proposed by Bernhard and the Villejuif group for the now-called retroviruses, set France in the forefront of the domains of electron microscopy and viruses.

REFERENCES

References, added in proofs for Chapter 2.5A, are found on pages 129 and 130.

2.5
Electron Microscopy in France
2.5B The Development for Physics and Materials Sciences

BERNARD JOUFFREY

École Centrale Paris, LMSS-Mat, Grande Voie des Vignes, 92295 Châtenay-Malabry, France

I. THE PIONEERS

The early developments of electron microscopy in France are due essentially to G. Dupouy and P. Grivet, who independently developed a transmission microscope at the beginning of World War II. G. Dupouy had been interested in electron microscopy since 1937, when he decided to develop electron microscopy in Toulouse. He chose this area almost certainly due to his previous experience in magnetic fields. On the other hand, P. Grivet used electrostatic lenses to make his electron microscope (1942), which was commercialized by the CSF (Compagnie Générale de Télégraphie sans Fil).

There were, therefore, from the start two groups working on electron optics in France, one in Toulouse, the other in Paris. Another group was working in Besançon on electron diffraction. It was under the guidance of J. J. Trillat.

It seems that the first mention in France of electron microscopy was in 1932, when P. Grivet heard from a Mlle. Meyer about the invention of "electron microscopy." At this time the *thèse de Doctorat d'État* included two dissertations, one describing the work of the thesis itself, and what was called the second subject, which was a bibliographic study proposed by the *faculté*, in fact by the president of the jury. The idea of this second thesis was to gain an intimate knowledge of a given aspect of a subject, different from that of the first thesis. The second thesis of Mlle. Meyer was on electron microscopy. Finding the subject rather tricky she asked P. Grivet, a young *étudiant normalien* that she had heard about as a clever student, to help her to understand the scientific publications. This he did. This encounter had two consequences; the first was the strong interest of P. Grivet toward electron microscopy, and the second was that Mlle. Meyer became Mrs. P. Grivet in 1934.

Before this time, in 1929, M. Ponte, following C. J. Davisson and L. Germer (1927) and then G. P. Thomson (1928–1929), had confirmed that it is possible to obtain an electron diffraction pattern as with X rays. This point is important, because Ponte was very helpful to P. Grivet, in particular during the war, in facilitating the continuation of his work at the CSF.

In the 1930s it was known that L. Marton (1932) had constructed an electron microscope and developed a method of preparing biological samples to be observed in an electron microscope, by staining the sample with osmium. The samples so treated were not destroyed by the electron beam. It is also quite well known that he came to France to the Laboratoire du Grand Aimant de Bellevue, around 1936 to give a talk on electron microscopy. It is difficult to be sure that G. Dupouy was present at this conference, but it seems probable that he was, because he was at this time joint director of this laboratory, and also *maître de conférence* at Rennes (1935–1937). On the occasion of some conversation that A. Guinier and I had with him, we wanted to confirm the matter, and put the question to him. The answer was not completely clear. He told us that he had heard about electron microscopy around 1934. Whatever the truth, it is obvious that the work of Knoll and Ruska gave rise to a large amount of interest in some researcher's mind. However, we did not find a written reference of these first approaches in electron microscopy in France.

In particular, it is quite astonishing that no paper has been found on the resolving power of the electron microscope. The first mention of this problem is due to R. Fritz, an assistant of J. J. Trillat, in 1936. The thesis of L. de Broglie had appeared about 10 years before. This paper, "La microscopie électronique," was published in *Revue générale scientifique*. Therefore, it seems he was probably the first in France to consider this important question of resolving power in electron microscopy. Remember that J. J. Trillat was working by this time in the field of electron diffraction. It is interesting to notice here that P. M. Duffieux, who became famous as the first person to introduce the use of Fourier transform (1946) into light optics, also worked in Besançon.

The next indication of the interest in electron microscopy is found in the first application for support; this came from G. Dupouy, upon his promotion to professor of physics at the University Paul Sabatier of Toulouse. This request, dated September 1937, is reproduced here (Fig. 1). G. Dupouy was originally a specialist in the measurement of magnetic fields (see his thesis of 1930, in which he studied the magnetic properties of crystals an during which he constructed a direct-reading gaussmeter).

The first instruments became operational during World War II, in 1942. One was in Toulouse (a microscope with magnetic lenses); the other one, constructed by P. Grivet (electrostatic lenses), was in Paris.

The first experimental results, and the improvements, which were also obtained in Germany and the United States, in particular, stimulated a big interest in this new technique. One consequence of this popularity was the

TOULOUSE, le 15 Septembre 1937.

Monsieur Gaston DUPOUY
Professor à la Faculté des Sciences
Laboratoire de Physique
TOULOUSE

à Monsieur le Ministre
de l'Education Nationale,
(Service Central de la Recherche Scientifique)

Monsieur le Ministre,

J'ai l'honneur de vous adresser sous ce pli une demande de crédits concernant:

1°) L'Achat d'appareils destinés aux recherches que je compte entreprendre au laboratoire de Physique de la Faculté des Sciences de Toulouse.

2°) L'Achat d'appareils et de matériel destinés aux travaux pratiqués et aux expériences de Cours, ainsi qu'à l'équipement général du Laboratoire.

3°) L'Achat de machines-outils en vue de l'installation d'un atelier dans ce laboratoire.

Permettez-moi d'attirer respectueusement votre attention sur le fait que ce Laboratoire est complétement démuni des appareils dont je désire faire l'acquisition, et qu'il est indispensable, pour que je puisse y créer un centre actif de travail, de m'accorder l'appui financier que je sollicite.

C'est pourquoi je vous serais très reconnaissant de vouloir bien examiner avec bienveillance la demande ci-jointe.

Veuillez agréer, Monsieur le Ministre, l'expression de mes sentiments très respectueux et dévoués.

FIGURE 1. These two pages are a copy of the first request for financial help to make an electron microscope in France. It was prepared by G. Dupouy when he arrived in Toulouse in September 1937. (Courtesy of *J. Microsc. Spectrosc. Electron.* (1987) **12**(6), 508–509.) (*Continues on page 104.*)

DEMANDE DE CREDITS POUR LES RECHERCHES
DE MONSIEUR G. DUPOUY, PROFESSEUR A LA FACULTE DES SCIENCES
DE TOULOUSE
-:-:-:-:-:-:-:-:-:-:-:-:-:-:-:-:-:-:-:-

I   RECHERCHES CONCERNANT LES PROPRIETES MAGNETIQUES ET
    MAGNETO-OPTIQUES DES SELS DE TERRES RARES.

Electro-Aimant (Beaudoin)	51.000 Frs
Batterie d'accumulateurs (120 v- 500 A.H.)	47.405 Frs
Un ensemble d'appareils pour la mesure des champs magnétiques comprenant: gaussmètre, fluxmètre, galvanomètre balistique, potentiomètre, inductance mutuelle, résistances étalonnées etc.	34.250 Frs
Un monochromateur double (Gobin)	35.500 Frs
Un héliostat	4.230
Un régulateur de température au 1/50 ième de degré avec groupe-moteur pompe pour circulation d'huile	10.475 Frs
Une balance de précision	5.150 Frs
Un arc à avance automatique avec rhéostat et condenseur	7.450 Frs
2 arcs à mercure (Bruhat) et une lampe à vapeur de mercure avec transformateur	4.750 Frs
1 boîte de contrôle (A.O.I.P.) et deux appareils à cadre à échelle dilatée	4.350 Frs
1 spectrographe pour le visible	10.200 Frs
1 Balance de Curie-Chéneveau	4.230 Frs
Divers supports; universels, à crémaillère et échelles de galvanomètres	5.925 Frs
Quelques boîtes d'accumulateurs portatifs	2.800 Frs
TOTAL	227.715 "

II—RECHERCHES CONCERNANT LA MICROSCOPIE ELECTRONIQUE

Générateur à Haute-tension constante	114.680 Frs
Microscope électronique (par transmission et par réflexion)	46.100 Frs
Groupe moteur-pompe (vide préliminaire)	8.500 Frs
Pompe Holweck	11.000 Frs
Pompe à condensation 5 (Mercure)	5.850 Frs
Pompe à condensation (Huile)	5.280 Frs
Appareillage pour le mesure du vide (jauge à ionisation et divers)	7.350 Frs
Accessoires	950 Frs
	199.710 Frs
A reporter:::	227.715 "
TOTAL	427.425

FIGURE 1. (*continued*)

organization of the *conférences du lundi*, periodic meetings held under the patronage of L. de Broglie. At least some of them were on electron optics. In the *Collection des réunions et mises au point*, an issue called *L'optique électronique* was published in 1945 by the *Revue d'optique théorique et instrumentale.* It gathered the proceedings of meetings to survey the state of the art in this field. L. de Broglie was by this time *secrétaire perpétuel* of the French Academy of Sciences, and a professor at the Sorbonne. In this publication, he regrets that electron optics was not developed first in France. He mentioned that in 1928–1929, "I pointed out to one of the first among my students that it would be interesting to work in the field of geometrical electron optics. Unfortunately, he did not carry on this work, and I myself was too absorbed by a more general research on wave mechanics and I did not go deeply into the matter." However, it appeared that five or six French researchers worked in the field from the end of the 1930s. It is clear that L. de Broglie was interested in the field, since he wrote a book on electron optics (*Optique électronique et corpusculaire*) in 1950. The electrostatic electron microscope designed by P. Grivet, which was operational in 1942, was mentioned above. It was built in a difficult situation, in a laboratory of the CSF. It seems that another one was built by M. Magnan in the Collège de France. On the other hand, the magnetic electron microscope (*l'ancêtre*) of G. Dupouy was built independently in Toulouse and was operational in 1942. It can still be seen in Toulouse in the entrance of the Laboratoire d'Optique Electronique, which recently slightly altered its name.

It is also interesting to notice, in a field which is not so different, that M. Lallemand carried out his work at the laboratory of l'Observatoire de Paris on an electronic telescope (the Lallemand camera), which he had previously started at the observatory of Strasbourg. There also L. Cartan worked on electron optics, due to its possible interest for nuclear physics. It is also good to recall that, by 1938, M. Cotte had finished his thesis at Institut Henri Poincaré. The title of it was *Recherches sur l'optique électronique.* This work was the starting point of the research of Castaing and co-workers in the 1960s on energy filtering and on secondary ion spectroscopy. As early as 1937, M. Cotte published in *Compte-rendus de l'Académie des Sciences* his results on the equations of charged particles trajectories, in particular in the presence of a magnetic field. He developed the equations concerning the orthogonal systems which were used, for instance, by J. F. Hennequin, L. Henry, G. Slodzian, and others. It is strange, because later on M. Cotte became a professor at the Sorbonne and I went to his lectures on thermodynamics. I did not know at the time that he had worked on a subject which would interest me later. I was very surprised to hear of him from R. Castaing, and to meet him at the presentation of the *diplome d'études supérieures* of J. F. Hennequin in 1961 (*Étude d'un dispositif de filtrage magnétique d'images électroniques*). He mentioned then that he thought that his work would have no interest for researchers.

Concerning the early work due to P. Grivet; he did not forget to mention, when talking with him about this period, that it is indispensable to acknowledge the work of H. Brück, who shared the responsibility for this venture, E. Regenstreif, who set up and maintained the apparatus, J. Vastel, who improved the fabrication, and the designer H. Blattmann. He also mentioned D. Charles and J. Laplume, who helped him with some theoretical problems. By this time, a committee on electron optics had been created at Centre National de la Recherche Scientifique (CNRS), which had been founded in 1939.

It is also at this time that J. Léauté emphasized how much metallurgists waited impatiently for improvements in electron microscopy. "The purpose is to establish relations between the microscopic behaviour and the macroscopic properties of the material." He was helped toward this goal by a scanning microscope, inspired by the work of von Ardenne.

II. The Interest of Metallurgists and the Improvement of Electron Microscopes

Many people working in fields other than electron optics itself were present at these conferences. In particular, A. Guinier appeared in the discussions early on, as being interested by the potentiality of electron microscopy. In 1949, R. Castaing was the first to show how useful microscopy could be to metallurgists. In a *Note aux Compes rendus de l'Académie des Sciences* (April 1949), he published the results of work on precipitation in aluminium alloys (Al-4%wt Cu, annealed for 36 h at 200°C). He used the oxide replica technique. Two other *Notes* followed in the same year, published with A. Guinier, in which the distribution of the θ' phase was shown (the thickness was below 5 nm). Castaing mentioned that the irregular structure due to the annealing was closely related to the lattice defects and that the precipitates are associated with the glide planes after deformation caused by quenching. He claimed that the precipitates in the polygonization structure revealed individual dislocations. The last paper of that year was on the annealing of Al-Mg-Si alloys, which exhibit needlelike structures. An excellent review of all these findings was published by Castaing in *La Recherche aéronautique* (No. 13) in 1950. Thus Castaing played a very important role in pioneering the use of electron microscopy for shedding light on metallurgical problems. M. Dargent, an assistant of G. Dupouy, later worked on replicas. He finished his thesis in 1957 on "Film Support and Replica Techniques in Electron Microscopy." This same year, two other relevant theses were examined, which were very close to electron microscopy. The first was the well-known work of P. Durandeau, "Studies on Magnetic Electron Lenses," which was useful for many years to people who wanted to build lenses. The second was the thesis of P. Gautier, who later became Professor of Physics at the Paul Sabatier University in Toulouse (Paul Sabatier won the Nobel Prize in 1912), and a few years ago Director of the Solid State Physics Laboratory at the same university, before he recently retired. These years were very productive in Toulouse, since we also notice the thesis of P. Pilod (1958) on "Fabrication, Critical Study and Applications of an Analogic Rheographic Installation" and the one of R. Saporte on "Studies on Electron Scattering Aspects by a Bulk Metal in Reflection Electron Microscopy." One year later, in 1959, F. Pradal finished his thesis on "Fabrication of a Magnetic Spectrometer and Its Application to the Study of Characteristic Energy Losses in Solids." Also M. Simon completed his work, "Emission Electron Microscopy: Application to the Study of Secondary Electron Emission."

Obviously, around this time, many meetings on electron microscopy were organized. It was in 1950 that G. Locquin organized, at the Museum d'Histoire Naturelle in Paris, from 14 to 22 September, the first International Congress on Electron Microscopy. It is quite instructive to read the different contributions and in particular, since we are discussing electron microscopy in France, the French contribution.

The opening paper was given by L. de Broglie. Reading at random the various contributions, we notice the presentation of R. Bernard and E. Pernoux (Electron Optics Laboratory in Lyon). It concerns the relation between some fringes (bend contours) and Bragg reflections. M. Françon, a well-known personality in the field of light optics, was by the time *Chef de travaux à la Sorbonne* and Professor at the *Institut d'Optique,* and he presented a contribution on this theme. Other well-known people in the same field, A. Maréchal and G. Pieuchard, wrote a paper on "The Influence of Aberrations on the Image of a Point, a Line, or a Plane." Therefore, specialists of light optics were at that time also interested in electron optics.

It is also extremely instructive to look at the section devoted to electron optics; this filled 268 pages, compared with the 768 pages of the proceedings in total. Electron microscopy was in an active instrumental state, and the microscopists were more opticians than metallurgists or materials specialists, generally speaking.

It was during this meeting that R. Castaing and A. Guinier, following the initial paper in Delft, one year before (1949), presented their paper, "On the Exploration and Elemental Analysis of a Sample by Means of an Electron Probe." Elemental analysis by means of X-ray emission under the impact of incident electrons was born. It was a revolutionary method. In this paper, the method was used to study the evolution of a silver–zinc compound which was obtained by sticking the two metals face to face, which were then annealed for a few hours at 400°C. On the diagram which shows the composition as a function to the distance from the interface, a gross change of concentration at the crossing through the interface is clearly visible, passing from one phase to the other formed by diffusion from this interface, and a continuous variation inside the same phase. The capability of the method was proved. At the same congress, we notice also a contribution of G. Möllenstedt on the chromatic losses of electrons passing through a thin foil of material. He used, for this purpose, his electrostatic analyzer. It appears obvious that, by the time of the 1950s, areas such as electron optics, electron microscopy, and all the related instrumental problems were being actively pursued in France as they were throughout other countries.

In 1955, 4–8 April, an international CNRS meeting, "Recent Techniques in Electron Microscopy," was organized, in Toulouse, by C. Fert (Fig. 2). At that time, G. Dupouy was the General Director of CNRS, and he gave the opening talk. The French participation includes a contribution from C. Fert, B. Marty, and R. Saporte on the continuation of their work on reflection electron microscopy under large angles, transmission electron microscopy, and microdiffraction. The micrographs of a surface of white iron and electrolytic copper are very attractive. The first results of the Toulouse research on reflection electron microscopy had been presented one year before at the European meeting in Gand (1954), and at the London International Congress, in the same year, in a paper by G. Dupouy and C. Fert. At

FIGURE 2. Electron microscopy meeting at Toulouse (4–7 April 1955). In the first row from left to right: V. E. Cosslett, L. Marton, E. Ruska, G. Dupouy, J. Le Poole, and Ch. Fert. We recognize also, among others, J. J. Trillat on the left, W. Bernhard, F. Perrier (third row toward the right), A. Septier, and R. Castaing at the center of the top fourth and fifth rows. (Courtesy of G. Möllenstedt.)

the same congress, A. Septier and M. Gauzit (Laboratoire d'Electronique et de Radioélectricité de l'Université de Paris) gave a presentation of "The Study and the Fabrication of an Electrostatic Energy Analyser." The spectra which were shown were acquired using 30-keV incident electrons. The range of losses is up to about 60 eV with a resolution of 1.5 eV. At the same congress, L. Marton, from the National Bureau of Standards in Washington, DC, produced a paper with L. B. Leder, H. Mendlowicz, J. A. Simpson, and C. Marton on the influence of the collecting angle, thickness, and the invariability of the position of the peaks of energy losses, whatever the incident energy is. The interpretation of these periodic peaks in the low-loss part of the spectra was, some time before, proposed not only by D. Bohm and D. Pines but also by P. Nozières as being due to the oscillations of free electrons (plasmons). R. Castaing was present at this meeting; he was at that time *Maître de conférences* at the University of Toulouse, and he gave a talk on "The State of the Art of Electron Metallography by Direct Transmission." He presented his method of ion milling, which he had used to prepare samples of Al–4% wt Cu annealed for 24 h at 200°C, 240 h at 150°C, and 24h at 100°C. One year before, at the London congress, he had proposed a slightly different method. The sample was first electropolished following the technique that had been proposed in 1949 by R. D. Heidenreich, and completed by ion milling (3000-eV Ar ions) on the two sample faces. In order to improve the understanding and the calibration of the ion dose, R. Castaing had set up a cold-cathode G. Induni gun inside the microscope itself. This system facilitated the bombardment of the sample on its top side. This technique also had the merit of avoiding the buildup of contamination. A little before, N. Takahashi and J. J. Trillat used evaporation to prepare thin films. J. J. Trillat was at Besançon. He was more interested by electron diffraction than by electron microscopy, as he admitted to A. Guinier and myself during a talk we had in his home at Le Chesnais in 1988.

It was at the London congress also, in 1954, that P. Gauzit (Laboratoire de Radioélectricité de La Sorbonne), gave a paper on ion microscopy with lithium ions. There is also a contribution of C. Magnan and P. Chanson on the question of a possible image contrast in proton microscopy (Laboratoire de Physique Atomique et Moléculaire du Collège de France). We now know that this domain has not been very conclusive. However, it was about this time that R. Castaing conceived the idea of using secondary ions under the impact of primary ions to develop a new method of microanalysis. We notice also, in the same meeting, a contribution of M. Y. Bernard and P. Ehringer on the three-electrode-thick electrostatic lenses. This work was of great interest at that time and was linked to the work in electron optics of P. Grivet, E. Regenstreif, and F. Bertein. The same year, A. Septier, who was in the same laboratory of radioelectricity of La Faculté des Sciences de Paris, worked on electrostatic lenses and more precisely on the optics of immersion lenses used as an objective. He speculated that the resolving power could be of the order of 10 nm. Due to this part of electron optics, a problem appeared related to the microdischarges inside electrostatic lenses. This question was studied by R. Arnal. In Toulouse, E. Durand, who became Dean of the University of Toulouse, and his student, M. Laudet, developed research on the numerical calculation of the fields of magnetic lenses. In particular, they attacked the important problem of using a grid in order to limit the errors in solving the Laplace equation. Clearly, the essential part of the French contribution was still in electron optics. Nevertheless, C. Fert and R. Dargent, for instance, were interested in the alumina replicas obtained by evaporation of aluminium. The original idea was put forward by G. Hass and M. E. MacFarland.

From that time on, the proliferation of electron microscopy was very rapid in France, with two distinct tendencies. One was obviously based on electron optics, but this diminished continuously from 1968 even if elegant experiments have indeed been made since this time. The other one was centered on defects in materials. R. Castaing finished his thesis in 1951, and this was the starting signal for intense activity in microanalysis (see Sec. V). In addition, the Castaing microprobe conferred, in France, the mark of credibility on electron microscopy, which was seen to be indispensable as a technique. More generally speaking, it was an important milestone for the credibility of instrumentation.

III. Thin Films

In 1956, R. Castaing, now in Toulouse, was still interested in electron metallography. In particular, he published a paper in *La Revue universelle des mines,* which was published in Liège. In the same years, J. Plateau, G. Henry, and C. Crussard used a replica technique based on evaporated carbon films previously proposed by J. Nutting for the study of fracture surfaces. R. Castaing tried the method of the oxide replica, which enabled one to study directly precipitates extracted from alloys (a technique first proposed by H. Mahl). As mentioned above, following the remark of R. D. Heidenreich, in 1949, who thought it would be fascinating to study the interior of materials themselves, R. Castaing tried with P. Laborie and G. Lenoir to observe thin films of an aluminium–copper alloy with different annealings. In 1953, R. Castaing and P. Laborie again used this method, once more finishing the preparation by ion thinning. Micrographs clearly show how interesting it is to examine the interior of alloys. He studied Guinier-Preston zones and also used this method for α-brass or refractory alloys such as Nimonic. Unfortunately, he did not observe dislocations. It is important to mention here how much the subject of electropolishing is indebted to P. A. Jacquet. He proposed it for the observation of metals and alloys in light microscopy to prepare a nice reflecting surface. The sample was placed in an electrolytic cell with the sample as anode. Based on this invention of P. A. Jacquet in 1933, R. D. Heidenreich made his proposal in 1949 to polish a material on both faces instead of one to obtain a thin film. It was also in the same years that C. Fert and R. Dargent made replicas of surfaces either by oxidation of metallic surfaces or by spreading out a solution of plastic material or by electrolytic deposit. A. Saulnier also was interested in replicas. He used anodic oxidation for light alloys. The metal is dissolved in a saturated solution of mercuric chloride or the thin layer is pulled off by electrolytic action (P. Lacombe and L. Beaujard, 1943). The extraction technique of precipitates was adapted to stainless steels by G. Henry, J. Plateau, and J. Philibert (1958) using bromine. It was from 1956 that C. Crussard, J. Plateau and colleagues from IRSID (Institut de Recherche de la SIDérurgie) at Saint Germain en Laye, close to Paris, made a famous catalog of fracture micrographs of stainless steels using carbon replicas. The metal was then dissolved by means of an alcoholic solution of bromine. It was possible to distinguish the brittle or ductile-type fractures, and a classification of the different types of fractures was established.

During the years 1950–1960, many papers were published in this field. It is quite impossible to detail the different aspects that appeared. The paper of R. Castaing (1956) in *La Revue universelle des mînes* has already been mentioned to show the level of research in the application of electron microscopy to metallurgy at that time. We can also notice a little later, on 15 February 1962, the colloquium which was held in Saint-Cloud near Paris on "the contribution of electron microscopy to modern metallurgy." There also exists a report published by A. Saulnier, Chief Engineer at the Metallurgy Research Centre of Péchiney Company in Chambéry, and J. J. Trillat, who was acting head of the section for structural physics on the "Direction of Studies and Fabrication of Arming." It shows some very instructive micrographs, in particular a replica of a car piston, and the features of a fracture of a soft iron due to C. Crussard and J. Plateau. The surface of a single crystal of gold cleaned by ion bombardment and many other micrographs of different materials are shown as well. At this time, the role of industrial research laboratories was rather important in the development of electron micrographs for the application of materials studies, but essentially in metallurgy.

Later on, much research was done in scanning electron microscopy. Some of the first work in this field was done in Lyon around 1958 by R. Bernard and F. J. Davoine. They also studied voltage contrast. The same group developed a method of imaging surfaces by means of negative secondary ions (R. Bernard and R. Goutte, 1958), contrary to Castaing, who advocated the use of positive ions (1958). We should notice here the development of the elegant backscattering imaging method that was carried out in Lyon later on by the group of G. Fontaine, E. Viccario, M. Pitaval (thesis in 1977), and P. Morin (thesis in 1981). This method has been used successfully for the observation of defects in semiconductors. This group had previously worked, at the beginning of the 1970s, on pseudo-Kikuchi patterns obtained by scanning electron microscopy. In this domain of semiconductors, more recent work has been done and published on electrical characterization (see the work of A. Rocher and his group and J. F. Bresse, 1977), for the measurement of lifetimes of minority carriers, or the diffusion length. The mixing of EBIC with cathodoluminescence was studied by P. Hénoc (who died in 1993) and continues to interest his group at CNET-Bagneux. We may also mention a portable scanning electron microscope, developed during 1981–1983 and built by J. L. Franceschi, J. Trinquier, and myself. It was presented in a talk that I gave at the 10th ICXOM, organized together with F. Maurice, R. Tixier, and J. Sévely, in Toulouse (1983).

Another interesting result, due to J. Beauvillain (3rd cycle thesis in 1969 and state doctorate in 1977), was among the first experiments, as far as I know, on convergent beam diffraction patterns in transmission electron microscopy. This technique was finally used in 1976 to determine the Burgers vector of a dislocation.

IV. High-Voltage Electron Microscopy and Subsequent Evolution of Electron Microscopy

Toward the end of the 1950s, the first high-voltage electron microscope in the world was built in Toulouse. G. Dupouy, F. Perrier, and co-workers obtained the first image at 1000 kV in 1960. The microscope was located in the new and beautiful laboratory which had been built during the end of the period during which G. Dupouy was Director-General of the CNRS (1950–1958) in Paris. The Electron Optics Laboratory was located on a 5-ha campus, 1 km away from the University Paul Sabatier. At that time, the price of land was cheap, of the order of 3 FF per square meter, since the Rangueil area was outside the City. This was no longer the case 20 years later! The situation of this laboratory on the Canal du Midi side was nice and quiet. This microscope, which worked up to 1.2 MeV, was situated in *La Boule* (Fig. 3), which became famous in Toulouse and among the worldwide community of electron microscopists. Every taxi driver knew where *La Boule* was. I remember when I became Director of this laboratory about 10 years later that *La Boule* was a myth in Toulouse. Some inhabitants of the area were fascinated by it. One of them wrote to tell me that he was disturbed because I could read his mind with this ball! This new laboratory was visited by many dignitaries and, in particular, on 14 February 1959, by Charles de Gaulle (Fig. 4).

Everyone knows the superb micrographs that were obtained by G. Dupouy and co-workers. A few years later, such was the success of the first microscope that he decided to built a new one working at 3 MeV. The first micrographs were obtained in May 1969 at 2 MeV. The generator, a Cockcroft Walton type, was capable of going up to 3.5 MeV. It had been constructed by la Compagnie Générale de Systèmes et Projets Avancés (GESPA), a subsidiary of the Compagnie Générale d'Électricité. In the making of this generator and the accelerator, J. Huret, the Director of GESPA, and A. Séguéla, CNRS engineer at Electron Optics Laboratory, played a major role. When G. Dupouy retired in 1970, F. Perrier replaced him as Director for two years. I became Joint Director and J. Trinquier Assistant Director. F. Perrier retired in July 1972. He had played an important role in the making of the electronics, in particular of the 1-MeV and 3-MeV electron microscopes. My arrival, as an ex-student of R. Castaing,

FIGURE 3. *La Boule* (1.2-MeV microscope) and *le Briquet* [cigarette lighter] (3-MeV microscope) at Toulouse.

FIGURE 4. Visit of Charles de Gaulle when Président de la République, 14 February 1959.

oriented the work of the laboratory toward energy losses, in particular in the domain of high-energy electrons and condensed-matter applications. A 90° magnetic spectrometer was rapidly constructed and located under the 1.2-MeV microscope. Experiments between 0.3 and 1.2 MeV worked perfectly. The first results showed how much high-voltage electron microscopy was favorable for the observation, in particular, of large losses corresponding to inner-shell ionization edges. These first results, obtained with J. Sévely and J. Ph. Pérez, were presented at the Oxford Congress on HVEM in 1973. They also showed the relativistic behavior of the plasmon excitation mean free path and the loss profile for thick samples. These experiments and their interpretation are to be found in the thesis of J. Ph. Pérez in 1975. Encouraged by these results, it was decided to adapt the same type of magnetic spectrometer for placing under the 3.5-MeV microscope. Later experiments were made up to 2.5 MeV. With these results in mind, the first symmetric omega filter was constructed by J. Sévely, G. Zanchi, and myself. I had proposed in 1972 to G. Zanchi to prepare his thesis (examined in 1978) on this subject. I obviously knew the work of S. Sénoussi, who had calculated in 1969 at Orsay, under the guidance of R. Castaing and L. Henry, a pure magnetic filter for 100 keV. The project, which had been calculated and set up at that time, was not concerned with a symmetric solution. The experimental difficulty had been investigated by C. Colliex (who was preparing his thesis under my guidance) and myself in 1970. To test this pure magnetic semiomega system, we used a Hitachi HU IIB which we had transformed to accept different types of filters in the column. J. M. Rouberol, from Caméca, was testing a filter for ions. He observed experimentally at Caméca that the symmetric solution, which he chose mainly by intuition was much better. This solution was, I remember, understood by R. Castaing, who had come back after a weekend and was quite excited about this solution, which is described in the 3rd Cycle thesis of S. Sénoussi (1971), but without any experimental tests. It was therefore tempting to use such a symmetric filter on my arrival at Toulouse, since I was in a High Voltage Electron Microscope Centre and the Castaing-Henry system could not be used under these conditions. I remember discussing all this before leaving Orsay with L. Henry and R. Castaing. I proposed the subject to J. Sévely and very soon after to J. P. Pérez and later on to G. Zanchi. The filter (slightly modified) was made in 1975–1977, and the first experimental results were published in 1977 by G. Zanchi, J. Sévely, and myself. The same year, the detailed calculations, including aberrations, were published; an earlier paper by J. Sévely, G. Zanchi, and J. Ph. Pérez had appeared in *Optik* in 1975. The fabrication was made in particular with the Toulouse workshop personnel, and the practical studies involved R. Sirvin and J. C. Lacaze, engineers of the laboratory. This filter worked well up to 1.2 MeV, right up to the time the microscope was decommissioned in 1992.

The 3.5-MeV microscope produced images at its highest voltage (3140 kV) in 1974, during the work of A. Rocher on critical voltages in copper (a study of the disappearance of the second-order reflection 622 at 2740 kV). A. Rocher prepared his thesis with me on the penetration of high-energy electrons in materials. He showed quite clearly that the gain in penetration is better for light materials, as was also found by G. Thomas and J. C. Lacaze in the same period. These experiments were carried out with the help of C. Jourret, I. Crouigneau, and J. Bilotte. R. Ayroles and A. Mazel, who were in the laboratory for several years, had worked also in the field with N. Uyeda.

In 1975, I organized the Fourth Conference on High Voltage Electron Microscopy in Toulouse. On this occasion many famous people came. The atmosphere was convivial, as can be seen in two pictures from the banquet (Fig. 5 and Fig. 6). At this time, P. Favard and N. Carasso often came to use the 3-MeV microscope. A. Rambourg also worked also on Golgi apparatus, and M. Maurette worked with A. Jourret on samples from the moon. Many other celebrities visited the laboratory. A typical group is shown in Fig. 7.

The tradition of electron optics of magnetic lenses was maintained in the Toulouse laboratory by J. Trinquier and his student J. L. Balladore,

FIGURE 5. (Top) Back view of R. Castaing and H. Hashimoto.
FIGURE 6. (Bottom) At the banquet of the IVth Congress on HVEM in 1975 at Toulouse. From left to right: G. Dupouy, B. Jouffrey, V. E. Cosslett, J. Racadot, P. B. Hirsch, and R. Fisher.

FIGURE 7. A visit to Toulouse in 1971. From left to right: F. Perrier, R. Chabbal, B. Jouffrey, unidentified person, J. Trinquier, H. Curien (who later became Minister of Research and Technology), G. Dupouy.

and by J. Sévely and others who worked on spectrometers and field emission guns. The laboratory received notable support when P. W. Hawkes, a theoretician, joined us in 1975, with his wife, a cell biologist. He arrived with a reputation of being an excellent specialist in electron optics and image processing. His contribution was not only scientific, but the level of English in our papers suddenly improved. Everyone knows the important role of P. W. Hawkes in scientific publishing.

Around 1969–1970, it became clear that high-voltage electron microscopy could give important information if it were possible to tackle the plastic deformation of samples *in situ*. This was developed during the preparation of the thesis of L. P. Kubin, who first worked in my group in Orsay in the Solid State Physics Laboratory. His subject was the plastic deformation of niobium, a cubic-centered material, at low temperature. In 1971, I went to Toulouse with the majority of my group. L. P. Kubin was finishing his thesis on plastic deformation at low temperature (1972). A. Rocher was also preparing a thesis on the problems of penetration (1975). M. O. Ruault worked on radiation effects (1976). C. Colliex did not come to Toulouse. The first specimen holders for *in situ* experiments were fabricated by R. Valle and J. L. Martin as well as by B. Genty and A. Marraud. B. Genty was a good friend of P. Selme, who was the talented engineer in charge of the development of the OPL microscope. The specimen holder was made partly for creep experiments and worked from room temperature up to about 1000°C. It was made for the 1.2-MeV GESPA microscope which was located at ONERA and began work in 1969. The Grenoble GESPA 1.2-MeV microscope was working in 1970. These microscopes were "offsprings" of the Toulouse microscope, but modified to be used more easily by specialists in condensed matter. A. Marraud became Head of the High Voltage CNRS ONERA service in 1978, until 1985 when it was decided by CNRS to decommission the microscope. The same specimen holder was built in Toulouse when J. L. Martin came to Toulouse in 1973. In the meantime, L. P. Kubin constructed with F. Louchet a low-temperature tensile specimen holder, which was used for elegant experiments on BCC materials such as niobium. The different domains of behavior of screw dislocations (straight or curved as a function of temperature) as well as their creation were very pleasingly observed.

Several teams in France worked for a long time on electron guns. Most experiments since 1969 concerned field emission. M. Troyon worked in Reims, G. Fontaine in Lyon, and M. Denizart in Toulouse. Also, A. Séguéla constructed a gun for the 1.6-MeV scanning transmission microscope. The gun worked well, but was never tested on the accelerating tube.

As the diameter of the lenses was an important parameter, in order to obtain a sufficient and stable magnetic field for focusing high-energy electrons, several researchers decided to use superconducting lenses. P. Bonjour, a student of A. Septier, developed a lens with holmium pole pieces (thesis 1973). On the other hand, J. L. Balladore made a lens and adapted in on a 300-keV microscope in Toulouse under the guidance of J. Trinquier. However, the most advanced work in the field was carried out by A. Laberrigue and his group, first at the Collège de France in Paris, and then in Reims. They succeeded in making a complete microscope working at 400 keV. They obtained 5 Å resolution. It would have been able to work up to 1 MeV. It was very successful, but due to lack of funds, they had, unfortunately, to stop the work.

The study of defects was quite popular in France in the 1970s. In particular, the CENG (Centre d'Études Nucléaires de Grenoble) played an important role in the application of electron microscopy to the understanding of radiation effects. D. Dautreppe asked A. Bourret in 1965 to investigate the effect of electron and neutron irradiation of nickel and in particular to see what the origin of swelling could be. When Bourret completed his thesis, it was clearly demonstrated that electron microscopy is indispensible for understanding the mechanisms of irradiation, the kinetics of the growing of loops due to point defects precipitation and their nature as interstitial or vacancy type. If *in situ* experiments were made several years before with ions at room temperature, the Grenoble team succeeded in a very nice experiment. A. Pluchery constructed a system which enabled the introduction in the microscope of a sample irradiated and maintained at low temperature. The essential result of this work was the demonstration, from the beginning of the neutron irradiation, of the presence of small interstitial defects. Following this incisive result, it was decided to try to use high resolution to determine, at the atomic level, the structure of defects. The first attempt was made with a 1.2-MeV GESPA-CNRS high-voltage electron microscope installed in Grenoble (CENG). The results, which were presented in 1970 at the Grenoble International meeting (ICEM 7), showed 0.2-nm resolution in the 200 plane. The microscope, fabricated by the GESPA company, was installed in 1969, following, by a few months, the CNRS-ONERA microscope. Tests were made over a year but, in spite of these efforts, the results were not totally convincing, and A. Bourret decided to buy a special low-voltage microscope for this purpose. The 1.2-MeV microscope, essentially under the responsibility of J. Pelissier, was then used for *in situ* experiments. In 1972, J. Thibault started her thesis, which was defended in 1977. It is really during this time that high resolution was developed in Grenoble. In parallel, crucial experiments and interpretation were carried out. At the beginning, the calculations were performed essentially by hand. An important step forward was the observation of a dislocation core in germanium. The result was presented in 1978 in a paper in *Nature*. Afterwards (1979), grain boundaries were studied in detail successively by J. Thibault and J. M. Pénisson from the group of A. Bourret in Ge, Al, and Si. Another question, which was a continuation of this work, was the study of the interaction of dislocations with grain boundaries following plastic deformation. It is set out in the thesis of J. L. Putaux (1989). We can notice here, in the field of semiconductors, the work of F. Glas at CNET-Bagneux (thesis in 1985).

From 1980 on, high-voltage electron microscopy was proving to be expensive. A meeting was organized in Toulouse (Fig. 8) to discuss the capabilities of high-voltage operation and the domains to be developed in the study of materials. In spite of an interesting and realistic report, CNRS, in close relation with ONERA, soon decided to close the CNRS-ONERA microscope (1985).

On the hand, people from chemistry were interested by electron microscopy in the 1970s. In 1972, D. Ballivet (who became Mrs. Tkatchenko), from l'Institut de Catalyse de Lyon, published in *Compte-Rendus de l'Académie des Sciences* a paper showing a high-resolution micrograph of L-type zeolith [$K_6Na_2(AlO_2)_9(SiO_2)_{27}$]. Also, in the Laboratoire de Métallurgie de Vitry, directed by M. Fayard, some works used largely electron microscopy in particular for the defects in oxides. We may mention here, in

FIGURE 8. Participants in an RCP (Recherche Coopérative sur Programme, part of the CNRS funding program) meeting on HVEM in Toulouse in 1986.

particular, the thesis of R. Portier in 1976 ("Étude microstructurale d'oxydes mixtes ordonnés dérivant de structures spinelles ou perovskite"), and two years later the work of D. Gratias on "Cristallographie des interfaces dans les cristaux homogènes." This last work was more toward theory, and we know the role played by D. Gratias in the discovery of quasicrystals which have also been studied by P. Guyot by means of electron microscopy.

People from light optics were also interested in the structure of thin films. P. Croce developed a group involved with electron microscopy. I remember the thesis of M. Gandais on "l'Étude de contours d'extinction dans l'image électronique de cristaux d'or." She made interesting contrast calculations in order to understand the aspect of zone axis contrasts (flowers) in gold. D. Renard, a little later, worked on the same domain. A. Marraud worked in the same group at the beginning of the 1960s. M. Gandais and P. Croce were among the first, following M. Tournarie (1960), P. Haymann, D. Taupin, A. Rocher, and myself, to start work on contrast techniques in the field before 1970. I remember the paper we wrote with D. Taupin on the noncolumn approximation (1967).

V. MICROANALYSIS

Electron probe microanalysis was first developed as an analytical tool by R. Castaing. His original work is described comprehensively in his thesis (1951). The subject of X-ray elemental and local analysis was proposed to him by A. Guinier, who became his supervisor in 1948. The work was carried out at ONERA (Châtillon-sous-Bagneux), where he equipped a

CSF microscope, designed some six years earlier by P. Grivet and sold commercially by the CSF, with a probe-forming lens, an optical device for localizing the point being analyzed on the sample, and an X-ray spectrometer. The diameter of the probe was about 1 μm. Detailed optical studies (astimatism detection, etc.) and the principles of the quantitative analysis are to be found in Castaing's thesis, submitted in 1951. Some early results were presented at the Delft Congress in 1949. The ONERA subsequently built two microanalysers, on which the famous Castaing microprobe, marketed by Caméca, was based. Applications were very soon pursued at ONERA by people such as J. Descamps. The ONERA, a government research center, played an important role in the career of R. Castaing: He became Scientific Director of the OM Division (Materials Office), later becoming Director of the ONERA for five years (1965–1970).

Many applications have been carried out following the original work of R. Castaing, and there is now a strong tradition in France in microanalysis. There is a microanalysis group linked to ANRT, meeting regularly. Its central activity is concerned with training in the field of microanalysis. More recently, at the retirement of R. Castaing (Fig. 9), a microanalysis group at SFME, developed as a club, has been founded to discuss fundamental and quantitative questions in microanalysis. The photograph of Fig. 10 was taken in Lille in 1985 on the occasion of a meeting organized by SFME in honor of R. Castaing for his retirement. Four generation of researchers are present.

FIGURE 9. R. Castaing (on the left) and T. Mulvey at the retirement ceremony for R. Castaing in 1987.

After the work of Castaing, innumerable people have used his method. We can mention the work of J. Philibert, who soon after the work of R. Castaing, began, toward the end of his thesis on oxidation problems, another subject under the direction of C. Crussard, who suggested that he should work on the microprobe that had just arrived at IRSID.

Following this work and that of Y. Adda, a number of studies were made on diffusion. The second domain of applications concerned stainless steels. They were initiated at IRSID by J. Philibert and then by R. Tixier (thesis in 1972 on microanalysis by electron microprobe of thin samples). G. Pomey was involved also in this work at IRSID. A third domain was centered on radioactive materials in collaboration with CEA and Caméca.

In the domain of diffusion there are many papers from 1965, by J. Philibert, J. Manenc, Mlle. F. Maurice, J. Levasseur, Y. Adda, P. Bastien, P. Guiraldenq, J. Descamps, and G. Lenoir on diffusion coefficient in particular, which were published essentially in *Compte-Rendus de L'Académie des Sciences, Métaux-corrosion-industrie, Journal de microscopie et spectroscopie electroniques,* and CEA reports.

As far as equilibrium diagram determination is concerned, there are also many investigations which have been performed on ternary alloys based on uranium.

On stainless steel, the detection of carbon has been a kind of work developed by J. Ruste, L. Merry, and co-workers. Many works on other

FIGURE 10. Four generations of researchers in Lille (1988) at the meeting in honor of R. Castaing. From left to right: M. Bernheim, R. Quettier, A. Guinier, G. Blaise, R. Castaing, P. Joyes, Mrs. Castaing, G. Slodzian, L. Henry, and B. Jouffrey.

elements, phase changes, and inclusions have been published in *Les Mémoires scientifiques de la revue de métallurgie, Métaux corrosion, Documentation sidérurgique,* and *Compte-Rendus de L'Académie des Sciences.* In the earlier works we recall the names of B. Hocheid, J. Manenc, J. Philibert, J. Plateau, G. Henry, J. Hénoc, P. Hénoc, C. Wache, and E. Weinryb. Other works have been done around the problems of treatments (R. Tixier and co-workers), on the influence of impurities on the mechanical properties, and the question of segregations in "*les fontes.*"

In mineralogy, the role of microanalysis is also important. It is an essential tool to describe the different mineral species. In 1960, only 10% had been described by an elemental analysis. In 1980, by contrast, 80% had been described by elemental analysis.

Many other domains, such as petrology, environment, cosmochemistry, cathodoluminescence in mineralogy (G. Renard and co-workers), glasses, ceramics, solding glasses-refractory materials, concretes (M. Jeanne, 1968, *Revue des matériaux de construction*), archeological materials (1969, *Bull. Soc. préhistorique Française*), catalysis (J. M. Leroy and co-workers in the 1970s), textile fibers (P. Kassenbeck, 1967, *C.R. Acad. Sci.*), and composite materials (5th Int. Symp. on Boron Fiber-Titanium Composite Materials) were studied with the help of the Castaing microprobe.

R. Castaing came to Paris as a professor in the beginning of 1958. G. Slodzian, one of Castaing's students at Toulouse University, came with him to Fontenay-aux-Roses to prepare his thesis on the possible use of characteristic secondary ion emission as a microanalysis tool. Before coming to Paris, G. Slodzian had prepared in Toulouse, with R. Castaing, a *Diplôme d'Études Supérieures,* which was a one-year research program to be prepared after the *licence* and was compulsory for candidates for the *Agrégation.* The subject was electron secondary emission under the impact of primary electrons. At the time of his coming to Paris, it was not clear if secondary ion emission could be used for this purpose. I entered the Fontenay-aux-Roses Laboratory in September 1958. The laboratory of R. Castaing, only one large room, was situated in the laboratory of P. Grivet in LCIE, Laboratoire Central des Industries Électriques. I developed an ion–electron converter to enhance the brightness of ion images. The idea of this converter, based on the use of an immersion lens, had been put forward by G. Möllenstedt and W. Hubig (1958). The idea was good, and the converter worked at the beginning of 1960. With R. Castaing and G. Slodzian, I used this converter to try the first experiment in secondary ion microanalysis. The apparatus for this experiment (*Diplôme d'Études Supérieures,* B. Jouffrey, 1961, and also described in the thesis of G. Slodzian, 1963) consisted of two immersion lenses. The ion beam provided by an ion gun (a high-frequency plasma system, ~ 1 MHz) was deflected to impinge on the cathode (two blocks of Cu and Ta) of the first immersion lens. Characteristic ions were focused and imaged onto the cathode of the second immersion lens, where the conversion characteristic ion–electron was used by means of the appropriate voltage. The introduction of a very crude magnet succeeded in separating the image due to Cu and that due to Ta. This result was published in *Compte-Rendus de l'Académie des Sciences* in 1961. It was clear that secondary ions could be used for microanalysis. The credit for studying in depth and, soon after, performing the first secondary-ion microanalysis is due to G. Slodzian, who calculated, assembled, and adjusted the apparatus. It worked in 1962; the details can be found in his thesis (1963), with the first applications.

By that time, and with the purpose of studying the properties of spectrometers with the final idea of using them for ions, R. Castaing began to study the optical properties of magnetic sectors. To this purpose, he started from the work of Cotte, who, in 1938, published his thesis, prepared in Paris, on "Recherches sur l'optique électronique," as I already mentioned.

J. F. Hennequin, in 1960, prepared his *Diplôme d'Études Supérieures* on a 90° magnetic spectrometer to test experimentally the validity of the work of M. Cotte. It was successful with electrons. The Cotte theory was correct, and so was the delicate experimental work. In particular, the correction of chromatic aberration was not easy. But the fundamental idea of R. Castaing was to use this kind of magnetic sector with ions. The problem at the time was to filter the image, which was not easy with a simple magnetic sector placed under a microscope (it was a CSF electrostatic microscope). R. Castaing thought of two ways to filter the image, by using either a Wien filter or a symmetric system composed of a triangular magnetic sector associated with an electrostatic mirror. L. Henry entered the laboratory in 1961, to study the image filter. To help him in the first stage of this project, Miss N. Paras came to the laboratory; she studied the characteristics of both systems theoretically. It was soon decided by Castaing to make the Castaing-Henry filter. It was assembled in 1961–1962 on an RCA microscope. The assembly and experiments were performed by L. Henry. The practical centring was extremely difficult, and he came very close to giving up. Finally, one night in 1962, he succeeded in getting a crossover and a filtered image. It was a tremendous success. The first results were published in a *Compte-Rendus de l'Académie des Sciences*. After this original experiment, many applications followed. The first of them was the observation of Fresnel fringes in the inelastic image. El Hili, in his thesis (1967), illustrated a variety of applications. He used the low losses (close to the plasmon losses) to perform a kind of microanalysis for light alloys by means of plasmon losses. He also studied the preservation of contrasts (bend contours, stacking faults fringes, etc.) in filtered images. I have to mention here that all this assembling of instruments in the laboratory of R. Castaing was greatly facilitated by the presence of a very good workshop man, R. Quettier. The laboratory of R. Castaing gained an enviable reputation, and many visitors came to Orsay (Fig. 11).

I had submitted my thesis in 1964. It was on "Étude de perturbations cristallines produites par des ions de moyenne énergie." I had studied the effect of ion bombardment (4 keV), as we used for ion milling of samples. I observed the formation of defects by *in situ* experiments and the implantation of ions giving the formation of bubbles in epitaxial thin films of gold at 200°C. I had the benefit of a jury of famous names: A. Guinier, J. Friedel, and R. Castaing. Following this work, I spent nearly a year in Cambridge in the Metal Physics Group of P. B. Hirsch in the Cavendish Laboratory and worked with A. Howie.

In 1966, it was decided to construct a new Castaing-Henry filter on a modern microscope. I had a new student, C. Colliex. By that time the question of the possibility of observing vortex lines either by phase contrast (A. Bourret and G. Renaud) through the Bohm-Aharonov effect, or by deviation of the electron beam (B. Jouffrey, C. Colliex, and M. Kleman) was exciting everybody. The idea was first to try the observation of vortex lines using a filter to improve the contrast, which we calculated would be poor. A liquid-helium specimen holder with tilting capabilities was constructed. V_3Si was chosen as superconducting material. No definitive results were observed, in spite of strange contrast effects. But this contrast bore

FIGURE 11. A visit of an important Japanese delegation to Orsay in 1971.

no resemblance to those that could be expected. Recently, A. Tonomura showed, following in part the first ideas of G. Möllenstedt, that both methods were able to detect the presence of vortex lines. He found similar contrast effects to those we had calculated at that time. The filter had new facilities, such as adding the energy loss under study to the accelerating voltage of the gun in order the keep the energy of the electrons in the column and hence the optical state of the column constant; it was also possible to remove the filter easily so as to work as with a conventional microscope, and it was possible to insert other filters. This possibility was very useful a few years later to test the Senoussi filter. The observation of superconducting materials being unsuccessful, it was decided to move to another subject more closely related to energy losses. In 1968 it was decided to work on characteristic energy losses. I and Colliex were oriented toward this domain because during the tests of the new version of the filter, we observed energy loss structures beyond the plasmon losses in aluminum which had been studied by L. Henry, A. El Hili, and P. Hénoc. We remembered the excellent work of L. Henry and P. Hénoc on characteristic plasmons of bubbles in aluminum. At the same time, L. Henry, P. Duval, and Hoan presented filtered diffraction patterns. They had been obtained with an OPL (Optique de Précision de Levalois) microscope equipped with the Castaing-Henry filter and a new lens permitting one to place the diffraction pattern at the object crossover for the filter. The result was quite fascinating; most of the background disappeared, especially at small angles.

OPL constructed electron microscopes for many years. This microscope formed part of a collaboration between the Society, Optique et Précision de Levallois, and the Toulouse Electron Optics Laboratory. Those involved with its construction were, in particular, C. Fert, P. Selme (see the paper of Ch. Fert and P. Selme on "Le Microscope electronique OPL," which appeared in 1956), and le Comte A. de Gramont, in charge of this company, who was convinced by electron microscopy. In 1943 they fabricated an electron lens; in 1949–1950 came the first microscope with two stages of magnification; in 1953–1954, a microscope with a microdiffraction capability; in 1955 the "universal electron microscope" (MEU I A); and one year later the MEU I B. The last microscope was working at 75 keV (Micro 75) and was equipped with a Castaing-Henry filter. This microscope was in commercial production until 1975. The Mega 125 had a point-to-point resolving power better than 6 A.U. The last commercial production was carried out under the SOPELEM trade mark.

Miss A. Daniel, who became later Mrs. Tiberghien, spent two years (1964–1966) in my group studying thin films, which were obtained directly by ultrarapid quenching from the liquid. A little later, P. Petroff (now in Santa Barbara) spent his time of military service in my group in the Laboratoire de physique des solides at Orsay (1968). A paper published in 1970 by P. Petroff, C. Colliex, A. Rocher, and myself described the results obtained with lithium, which presented characteristic losses quite well separated with fine structures. The same year, C. Colliex and myself presented the first images taken with electrons corresponding to a K loss in graphite and showing that it was possible (it was shown, in these experiments, in the case of a Si–Cu system), to detect the different components. C. Colliex submitted his thesis in 1970 on the "Contribution à l'étude des excitations électroniques créées dans une couche mince par un faisceau d'électrons de moyenne énergie." He went on using energy losses, first with C. Mory and P. Trebbia, who prepared his thesis under the common guidance of R. Castaing and C. Colliex. The title was "Contribution au développement de la microscopie électronique analytique par utilisation quantitative de la spectroscopie des pertes d'énergies" (1979). C. Mory had earlier prepared her 3rd Cycle thesis on diffraction problems in high voltage, with A. Rocher and me (1972). From this time she prepared a thesis with C. Colliex on STEM, which is a basic and useful study in the field. The title is "Étude théorique et expérimentale de la formation de l'image en microscopie à balayage en transmission" (1985). Other theses in the field were prepared in Orsay under the guidance of C. Colliex, such as the one of Manoubi (1989), who was first a student of A. El Hili. Following my nomination at Toulouse as Joint Director (1971) and then Head of the Laboratoire d'Optique Électronique (1972), two laboratories worked on questions related to the analysis of materials by means of energy losses. They were the group under the guidance of C. Colliex which inherited the one I developed in Orsay in the laboratory of R. Castaing, and the other one in Toulouse, where I proposed to Sévely that he should also work in this domain.

It soon became apparent, following the development of the Castaing microprobe, that microanalysis could be very useful in biology and medical studies. In this last area, P. Galle showed in his medical thesis in 1965, on the "Local Chemical Analysis of Intracellular Inclusions," its importance for diagnosis. This work was developed with the Castaing microprobe. Galle worked in collaboration with R. Castaing and G. Slodzian in the 1970s to show the value of the secondary ion microprobe in the field. He carried out exploratory work comparing the advantages of the two methods. Most of this work was carried out with J. P. Berry and his group. They also used the method to detect impurities, such as aluminum, in biological tissues, by energy loss methods and in particular went to Toulouse to use the 1.2-MeV microscope equipped with the omega filter. This work using the characteristic losses is analogous to that of F. Hawkes, which was also performed with the same microscope from 1975 onwards.

VI. Société Française de Microscopie Électronique (SFME)

The French Society for Electron Microscopy (Société Française de Microscopie Électronique, SFME) was founded 6 June 1959. The president was R. Castaing. The SFME had two Honorary Presidents, L. de Broglie and A. de Gramont. The Vice President was E. Fauré-Fremiet. The two General Secretaries for Biology and Physics, respectively, were R. Buvat and Ch. Fert; the two Secretaries were J. Faget and P. Favard. The Treasurer was N. Carasso. The Committee included W. Bernhard, M. Bessis, R. Couteaux, G. Dupouy, S. Goldstaub, P. Lépine, C. Magnan, A. Oberlin, and Ch. Oberling (see, for instance, the paper of R. Couteaux in the bulletin of the SFME in 1989).

The role of the SFME has been important in the development of electron microscopy in France. Every year since 1959, this society has maintained the organization of its annual meeting in different cities in France following a delightful *tour de France;* meetings have also been held together with neighboring countries, notably Belgium (Liège), Switzerland (Lausanne), and Spain (Barcelona).

VII. Some Publications and Schools

We mention here that several schools have been organized, some of them in connection with the SFME. The first one was in Perros-Guirec, in Brittany (1969). I chose this area because my wife comes from this part of France. In 1978, another school (with Association Nationale pour la Recherche Technique), specially on microanalysis, was held in Saint-Martin-d'Hères. It was organized by R. Tixier, L. Mény, and J. L. Maurice. Another one on electron microscopy in 1977 at Villars-de-Lans was under the responsibility of G. Fontaine and E. Vicario. I organized the last one in 1981 in Bombannes with A. Bourret and C. Colliex. Each of these schools resulted in the publication of a book, with the exception of the 1977 school.

Other works on electron optics and electron microscopy have been published. Books by G. Dupouy (*Élements d'optique électronique,* Armand Colin, Paris, 1952) and by C. Magnan (*Traité de microscopie électronique,* Hermann, Paris, 1961), give interesting information about the history of electron microscopy; a two-volume treatise by P. Grivet (*Optique electronique,* Bordas, Paris, 1955, 1958) was revised by A. Septier (1965) and translated into English by P. W. Hawkes (*Electron Optics,* Pergamon, Oxford, 1965); A. Septier edited a two-volume collection entitled *Focusing of*

Charged Particles (Academic Press, New York, 1967), and a more ambitious three-volume collection, *Applied Charged-Particle Optics,* in 1980 and 1983.

Acknowledgments

It would take an entire book to write a complete history of electron microscopy in France. I hope that colleagues and friends who have not had their work mentioned in these few pages will forgive me. Most of the facts in this document arose from discussions or talks I had with many microscopists or more simply from events I was lucky to be mixed up with. I would like in particular to thank A. Guinier. I have had many useful talks with the different pioneers I have mentioned above. I was also lucky to make my first steps in research with R. Castaing, who was my *Maître*. I also thank all the people who worked with me, and also Dr. Maurice, who gave me a lot of information on various aspects of microanalysis. I would like to thank Miss D. Lotthé for refreshing my memory about dates and chronology.

References[1]

Barré-Sinoussi, F., Chermann, J. C., Rey, F., Nugeyre, M. T., Chamaret, S., Gruest, J., Dauguet, C., Axper-Blin, C., Vézinet-Brun, F., Rouzioux, C., Rozenbaum, W., and Montagnier, L. (1983). Isolation of a T lymphotropic retrovirus from a patient at risk for acquired immune deficiency syndrome (AIDS). *Science* **220**, 868–871.

Benedetti, E. L., and Bernhard, W. (1958). Recherches ultrastructurales sur les virus de la leucémie erythroblastique du poulet. *J. Ultrastruct. Res.* **1**, 309–336.

Bernhard, W., Haguenau, F., and Oberling, C. (1952a). L'ultrastructure du nucléole de quelques cellules animales revélée par le microscope électronique. *Experientia* **8**, 58–59.

Bernhard, W., Haguenau, F., Gautier, A., and Oberling, C. (1952b). La structure microscopique des éléments basophiles cytoplasmiques dans le foie, le pancréas et les glandes salivaires. *Z. Zellforsche. Mikroskop. Anat.* **37**, 281–300.

Bernhard, W., Dontcheff, A., Oberling, C., and Vigier, P. (1953). Corpuscules d'aspect virusal dans les cellules du sarcoma de Rous. *Bull. Cancer* **40**, 311–321.

Bernhard, W., Bauer, A., Harel, J., and Oberling, C. (1954). Les formes intracytoplasmiques du virus fibromateux de Shope. Etude de coupes ultrafines au microscope électronique. *Bull. Cancer* **41**, 423–444.

Bernhard, W., and Bauer, A. (1955). Mise en évidence de corpuscules d'aspect virusal dans des tumeurs mammaires de la souris. *C. R. Acad. Sci. Paris* **240**, 1380–1382.

Bernhard, W., Bonar, R. A., Beard, D., and Beard, J. W. (1958). Ultrastructure of viruses of myeloblastosis and erythroblastosis isolated from plasma of leukemic chickens. *Proc. Soc. Exp. Biol. Med.* **97**, 48–52.

---

[1] References, added in proofs here, are from Chapter 2.5A.

Bernhard, W., and Gross, L. (1959). Présence de particules d'aspect virusal dans les tissus tumoraux de souris atteintes de leucémie induite. *C. R. Acad. Sci. Paris* **248,** 160–163.

Bernhard, W., Fébvre, H. L., and Cramer, R. (1959). Mise en évidence au microscope électronique d'un virus dans des cellules infectées in vitro par l'agent du polyome. *C. R. Acad. Sci. Paris* **249,** 483–485.

Bernhard, W. (1960). The detection and study of tumor viruses with the electron microscope. *Cancer Res.* **20,** 712–727.

Bessis, M. (1972). *Cellules du Sang Normal, Pathologique.* Paris: Masson.

deHarven, E., and Bernhard, W. (1956). Etude au microscope électronique de l'ultrastructure du centriole chez les vertébrés. *Z. Zellforsch. Mikroskop. Anat.* **45,** 378–398.

Dourmashkin, R., and Fèbvre, H. (1958). Culture in vitro sur des cellules de la souche de HeLa et identification au microscope électronique du virus du molluscum contagiosum. *C. R. Acad. Sci. Paris* **246,** 2308–2310.

Garnier, C. (1899). Contribution à l'etude de la structure et du fonctionnement des cellules glandulaires séreuses. Du rôle de l'ergastoplasme dans la sécrétion. *Thèse,* Nancy, France.

Haguenau, F., Rouillier, C., and Lacour, F. (1955). Corpuscules d'aspect virusal dans l'endothélioma de Murray-Begg. Etude au microscope électronique. *Bull. Cancer* **42,** 350–357.

Haguenau, F., and Bernhard, W. (1955a). L'appareil de Golgi dans les cellules normales et cancéreuses des vertébrés. Rappel historique et étude au microscope électronique. *Arch. Anat. Microscop. Morphol. Comp.* **44,** 27–55.

Haguenau, F., and Bernhard, W. (1955b). Particularités structurales de la membrane nucléaire. Etude au microscope électronique de cellules normales et cancéreuses. *Bull. Cancer* **42,** 537–544.

2.6A
The Early History of Electron Microscopy in Germany[1]

HEINZ NIEDRIG[2]

Optisches Institut, Technische Universität Berlin, 10623 Berlin, Germany

I. Introduction

The bestowal in 1986 of the Nobel Prize for Physics on Ernst Ruska for the construction of the first electron microscope has aroused interest in the early history of the electron microscope in the 1930s. At that time, there appeared in Berlin practically all the development routes of electron microscopy that are relevant today: the electron microscope with magnetic lenses, associated with Ernst Ruska and Max Knoll and also Bodo von Borries; the electron microscope with electrostatic lenses, associated with Ernst Brüche, Helmut Johannson, Hans Boersch, and Hans Mahl; the scanning electron microscope associated with Max Knoll and Manfred von Ardenne; and the electron field emission microscope of Erwin Müller. A comprehensive account can be found in Ernst Ruska's book, *The Early Development of Electron Lenses and Electron Microscopy* [1].

The invention of the electron microscope is closely connected with the development of the cathode-ray oscillograph. At the end of the 1920s, both at the Technische Hochschule in Berlin-Charlottenburg (TH Berlin) and at the AEG Research Institute in Berlin-Reinickendorf, work was being carried out to improve the recording of fast electrical transients. In particular, the recording of the so-called traveling waves following lightning strikes and switching surges in high-voltage overhead lines called for an extremely high writing speed. For this, the beam intensity on the fluorescent screen or photographic plate was normally increased by concentrating the beam by means of a so-called concentrating coil. Denis Gabor, who was later to receive the Nobel Prize for the invention of holography, had built such an oscillograph as part of his doctoral thesis (1924/1926) at the Electrotechnical Institute of the TH Berlin, with a concentrating coil of novel form [2]. Gabor, from purely practical considerations, replaced the conventional long solenoid by a short solenoid contained within an outer iron cylinder,

---
[1] Translation of an article in *Physik und Didaktik* **15** (1987) with slight modifications.
[2] Translated, with additions, by T. Mulvey.

so as to reduce the external stray field. This was the first primitive iron-shrouded magnetic electron lens, but Gabor did not realize this at the time, nor could he explain satisfactorily, as he admitted in his dissertation, how it worked. A long solenoid does not act as a lens, since there is no radial field component. This only became clear a year later, in 1926, when Hans Busch at the Physics Institute of the University of Jena delivered the theoretical tool for the newly emerging electron optics; he calculated the trajectories of cathode rays in short, rotationally symmetrical electrical and magnetic fields and found that they behave toward electron beams as lenses do in light optics, with the same formula (lens equation) known from optics [3]. However, experiments that he had made previously with electrons agreed only qualitatively with his calculations. Perhaps this is the reason why he did not draw any conclusions about the possibility of a science of "electron optics."

II. The "Magnetic" Electron Microscope

In 1928, A. Matthias, Professor of high-voltage technology and electrical installations at the TH Berlin, had set up a working group for the technical development of the cathode-ray oscillograph under the leadership of *Assistent* (one who helps the Professor) M. Knoll. Co-workers were Ernst Ruska and Dr. B. von Borries. During his *Studienarbeit* (student project) of 1928/1929 and *Diplomarbeit* (final-year project) of 1930, Ruska investigated the properties of concentrating coils and of the electrostatic spherical condenser for cathode-ray beams; he achieved one-stage imaging of the cathode at a magnification of 3.5:1 and an image of a T-shaped aperture at 10:1. Thus, plane objects were imaged electron optically for the first time, whereas with the more well known "converging" coils, it was only possible to achieve a more or less pointlike spot of higher beam density on the fluorescent screen of the cathode-ray oscillograph. Because of the poor economic situation at the time, Ruska could not find a job after his Diploma examination, but he was allowed to continue his electron optical experiments in the TH Berlin without financial support. He immediately constructed an improved experimental apparatus with two coils placed in line along the ray path (Fig. 1) and was able, on 7 April 1931, to achieve the first-ever two-stage electron optical imaging of a grid in transmission, at a magnification of 16×. Thus these investigations of the optimization of the cathode-ray oscillograph led to the design and first realization of an electron microscope [1]. At this stage, Ruska and Knoll discussed whether by electron optical magnification a better resolution than that of the electron microscope might

HISTORY OF ELECTRON MICROSCOPY IN GERMANY   133

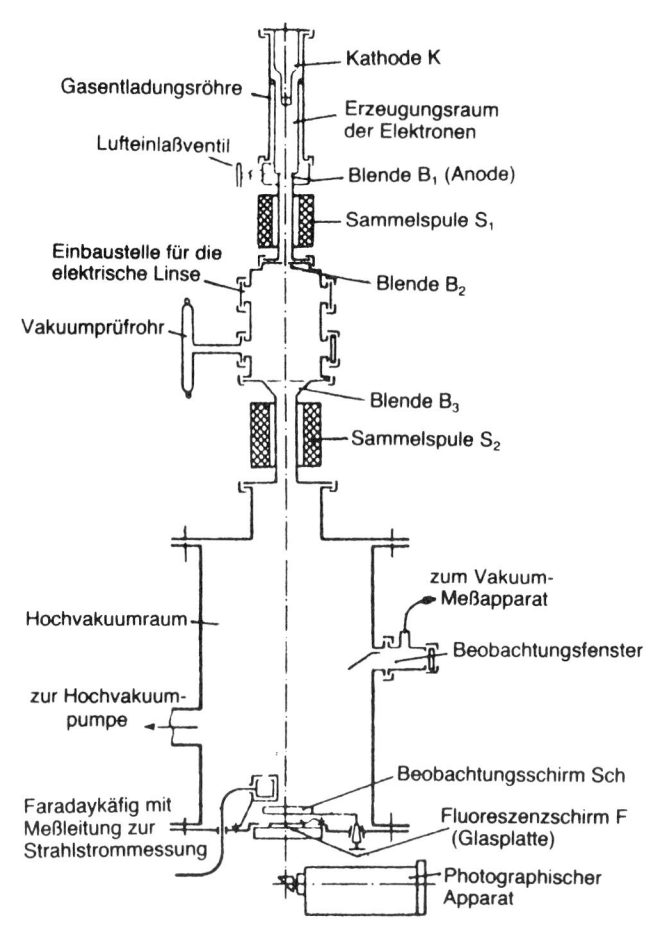

FIGURE 1. First two-stage electron microscope with magnetic lenses (1931), designed and realized by M. Knoll and E. Ruska [6]. Key: Gasentladungsröhre = gas discharge tube; Erzeugungsraum der Elektronen = electron production space; Lufteinlassventil = air inlet valve; Blende B1 (Anode) = diaphragm B1; Sammelspule S1 = converging coil S1; Blende B2 = diaphragm B2; Einbaustelle für die elektrische Linse = place for an electrostatic lens; Sammelspule S2 = converging coil S2; Hochvakuumraum = high-vacuum region; Zum Vakuum-Messapparat = for vacuum measuring equipment; Beobachtungsfenster = observation window; Beobachtungsschirm Sch = observation screen Sch; Fluoreszenzschirm F (Glasplatte) = fluorescent screen (glass plate); Photographischer Apparat = photographic apparatus.

be possible, since the limitation set by the wavelength of light was now removed. In this they were thinking of the extremely small size of the corpuscular "electron." The possibility of a resolution limit set by the associated *material wavelength* that de Broglie had introduced in 1924/1925 [4] was only raised later on [5]. The foregoing results on electron optical imaging were publicly disclosed in a lecture given by Knoll on 4 June 1931 in the open Cranz colloquium of the TH Berlin. The expression "electron microscope" was first used in a comprehensive paper [6] submitted on 10 September 1931 by Knoll and Ruska. The first patent application on the electron microscope, however, was not made by Knoll and Ruska, but by Professor Reinhold Rüdenberg, Chief of the Scientific Department of the Siemens Schuckertwerke, on Saturday, 30 May 1931. We now know that Rüdenberg had been informed by his assistant, Dr. Max Steenbeck who had made a visit to the laboratory of Knoll and Ruska before the date of the patent submission, about the experimental work there [1,7].

The patents were not granted in Germany, for reasons that have never been adequately explained. Knoll and Ruska, for example, knew nothing about these patents at the time. The patent was, however, granted in France on 10 October 1932 (see Ruska's book). More significantly, the patents were granted in the United States in October 1936. The patents had been assigned originally to Siemens Schuckertwerke, but after Rüdenberg became an American citizen during World War II, he brought a legal action in 1947 to have the patents assigned to himself by an American court (Custodian of Alien Property). The action was successful, and so Rüdenberg (now written as Rudenberg) became the inventor of the electron microscope in patent law. In 1953 and 1954, the German Patent Office belatedly granted the three patents applied for in 1931 on "magnifying electron optical imaging" (DBP 895635, 889660, 906737) (see T. Mulvey, *Phys. Bull.*, March 1973, pp. 147–154.) No participation by Rüdenberg in the construction and development of the electron microscope, however, has been established by scientific publications.

In order to achieve high resolution, the focal length of the lenses must be kept small. A decisive step in the development of magnetic lenses was taken by Ruska in his dissertation work in the years 1932 and 1933, by the introduction of extensive iron cladding of the lens enclosing an air gap formed between a pair of funnel-shaped iron polepieces [8]. Already on 17 March 1932, von Borries and Ruska, having become cautious in view of their previous experience, applied for, among other things, a patent on the iron-encapsulated magnetic polepiece lens (DRP 680284, 679857). This was granted in 1939.

In 1933 a two-stage electron microscope was built to Ruska's design, with magnetic lenses (objective and projector) as well as a condenser lens, and at the end of November it was installed at the *Arbeitsgruppe* (working group) of the High-Voltage Laboratory of the TH Berlin, which had been set up in the meantime at Neubabelsberg. In December 1933, success was obtained with the new instrument in taking the first images of cotton fibers and aluminum foils at a magnification of 12,000 and at a resolution limit just beyond the limit of the light microscope.

In the meantime, the *Arbeitsgruppe* collapsed. Max Knoll went to Telefunken Berlin on 4 April 1932, to undertake work on television projects. Bodo von Borries had become *Assistent* to Professor Matthias after completing his dissertation on 1 April 1932, but by the end of April he had taken up a post at the Rheinisch-Westfälische Elektrizitätswerke in Essen. Ernst Ruska himself joined the Fernseh GmbH Berlin on 1 December 1933, where he developed image receiver and transmitter tubes, as he wrote: "there is not a hope that a university institute or an industry has the means at its disposal for a real step forward in the work on the electron microscope [1].

Ruska, however, remained in contact with research students at TH Berlin who were investigating the applications of the instrument, mainly for biological specimens, and improving it technically. The electrotechnic students E. Driest and H. O. Müller, for example, fitted out the microscope with an internal camera and obtained images in 1934 of the hairy wings and legs of the housefly, with a resolution somewhat better than that of the light microscope [9]. In 1936, the medical student F. Krause obtained images of diatoms with a grating structure of 260 nm [10], and shortly afterward, of 130 nm [11]. This was indubitably beyond the resolution limit of the light microscope, an achievement that Ernst Ruska held to be of great importance. There were at the time considerable misgivings about this scientific goal of wanting to image sublight microscopic detail, especially for biological specimens. Moreover, there existed among some biologists the fear that such submicroscopic structures would be changing too quickly for them to be followed by eye because of the persistence of vision, so that the relation between structure and function in muscle, for example, would not be accurately interpretable. From the physics point of view, there was the fear of strong structural variations in the specimen, on the one hand from dehydration because of the vacuum that was needed in the microscope column, and on the other hand from specimen heating arising from the absorption of energetic electrons, which was assumed at that time to be necessary for the production of image contrast. Concerning this last objection, it fortunately became apparent by the end of 1933 that image contrast arose mainly from the scattering of the beam electrons, i.e., from small angular deflections at the specimen atoms. Thus only a small amount of energy need be deposited in the specimen [12]. By making the specimen very thin, the heating up could be kept within the damage-free regime. However, "ultramicrotomes," which were able to cut sections down to 10 nm, were not developed until 1950 [1,13].

In the meantime, Ladislaus Marton in Brussels, stimulated by the early work of Knoll and Ruska, had also begun to work in electron microscopy and had concerned himself with developing suitable specimen preparation techniques for biological specimens. In order to improve the scattering contrast as well as the heat transfer properties, he impregnated his samples with heavy metals, in particular, osmium salts [14]. These preparation methods, in various forms, are extremely important for biological specimens. As early as 1937, F. Krause devised a method that would today be called "negative staining" [15].

In 1934 von Borries came back to Berlin and took up a position with the Siemens-Schuckertwerke. In this and subsequent years, Bodo von Borries and Ernst Ruska, supported by Ernst's younger brother, the medical doctor Helmut Ruska, busied themselves with numerous discussions with the representatives of Siemens, Zeiss, and other firms as well with Professor Max Planck of the Kaiser-Wilhelm-Gesellschaft, to gain support for the further development of the electron microscope [1]. In this connection, it was of decisive help that Professor Richard Siebeck, Director of the first University Clinic of the Charité in Berlin, and incidentally Helmut Ruska's clinical tutor, issued a Referee's Report in October 1936, following a discussion with Bodo von Borries, Ernst Ruska, and Helmut Ruska, on the significance of a high-resolution electron microscope for medicine and biology. This report set out the possible advantages for medical science and practice that could come about from the electron microscope. It pointed out the significant foreseeable benefits for the investigation of normal and pathological cell structures and especially of infectious agents of sub-light microscopical size. Happily, the negotiations of von Borries and Ruska with Siemens and Halske and with Zeiss eventually led to an offer by both firms to develop, in house, an electron microscope with adequate financing. Von Borries and Ruska decided to go for Siemens and Halske as an electrotechnical firm, in view of the need to develop power sources of high constancy for the lens currents and the high accelerating voltage of the electron microscope, and also because the patent situation was better at Siemens. Bodo von Borries and Ernst Ruska were thus able, at the beginning of 1937, to set up the laboratory of Siemens and Halske in Spandau and engage W. Glaser, H. O. Müller, and H. Ruska as collaborators. By 1938 two improved electron microscopes had been produced as prototypes for a serially produced instrument; all the components were made of iron, to act as screening against external stray magnetic fields. Airlocks were provided for the specimen and the photographic plates. However, the gas discharge cathode was retained. With this instrument a maximum magnification of 30,000× was attainable; the resolution was 13 nm. Under the guidance of Helmut Ruska, electron microscopists of different disciplines worked there, including F. Schmieder from IG Farbenwerk Hoechst; he was able to show that the electron microscope is also suitable for the investigation of synthetic fibers and dyestuffs. As a result, at the end of 1938, four such instruments were ordered by IG Farben. That, in turn, stimulated Siemens to begin serial production with a batch of 12 instruments fitted with thermionic cathodes and Wehnelt cylinders. The resolving power was also upgraded to 7 nm. The first of these Siemens, "Ruska–von Borries" microscopes [16] was delivered at the end of 1939 to the Physics Laboratory of IG-Farben at Hoechst.

In 1940, a "Siemens and Halske AG. Laboratory for Electron Microscopy" was set up as a guest laboratory for visiting scientists under H. Ruska (Charité) and G. A. Kausche (Biologische Reichsanstalt), which operated until it was destroyed in 1944 during an air raid. Nevertheless, the development of magnetic electron microscopes proceeded successfully with the Elmiskop 1, which was produced and sold in large numbers.

In the 1970s, the development and production of electron microscopes was stopped, a decision that seems puzzling today in view of the demand for high-performance equipment in the field of electron beam lithography and electron beam measuring techniques for the production and testing of highly integrated solid-state circuitry.

III. THE ELECTROSTATIC ELECTRON MICROSCOPE

Max Knoll had already patented in 1929 an arrangement of two hollow cylinders held at the same potential with a plane-apertured electrode sandwiched between them and held at either a positive or a negative potential. This constituted, in principle, an electrostatic Einzel lens as a "device for concentrating the beam of a cathode-ray oscillograph," although its action was not yet fully understood [1]. This patent was granted in 1940 (DRP 690809). In addition, Ernst Ruska, as mentioned above, had investigated in his Diploma work the concentrating power of a bored-out spherical condenser arrangement as an electrostatic beam-concentrating element. However, the development of electron microscopes with electrostatic lenses was not carried out at the TH Berlin, but in another Berlin institute: the AEG Research Institute in Berlin-Reinickendorf, under the direction of Professor Carl Ramsauer. Here, in 1930, at the instigation of Ernst Brüche, the leader of the Physics Laboratory of the institute, the study of electron optics was taken up on a broad basis. AEG thus became the first industrial establishment to undertake the systematic study of geometric electron optics, long before any industrial company in Germany had turned its attention to experimental or development work in electron microscopy [17]. Their first concern was the further development of the Braun cathode-ray tube and, among other things, the emission from the cathode. As soon as Brüche was informed in 1931 about the work of Knoll and Ruska at the TH Berlin, he let it be known in a "short communication" [18] that similar work was being pursued at the AEG Research Institute, and he also showed one of the first results obtained by H. Johannson, an electron optical image at a magnification of 60×, of the electron distribution of the emitting surface of the central region of a plane cathode. A more detailed paper by Brüche and Johannson appeared in 1932 and dealt with the single-stage imaging of the cathode surface onto the fluorescent screen by two apertured electrodes held at different potentials and placed in front of the cathode [19], an arrangement that was later to be known as an "immersion objective" or "cathode lens." This was more closely investigated by Johannson in his 1933 doctoral dissertation [20]. Without being aware of Knoll's proposal of 1929, Brüche and Johannson also used electrostatic Einzel lenses consisting of three electrodes with central apertures (two external electrodes with one charged electrode sandwiched between them [21,22], whose mathematical treatment was subsequently given by H. Johannson and O. Scherzer [23]. As early as 1934, the first monograph on electron optics was produced by this research group; Ernst Brüche and Otto Scherzer [24] were the authors. In succeeding years studies were made of the electron emissions resulting from thermal, photoelectric, and ion bombardment and high applied electrical field strengths. Immersion objectives and iron-shrouded lenses were investigated for this purpose.

The applications of emission microscopy yielded new results even in the resolution range of the light microscope. The electron emission superscope was first achieved in 1941 [17], after A. Recknagel had concluded that there were no reasons in principle to prevent the emission microscope from attaining the resolution of the light microscope [25]. At the AEG Institute the immersion objective was also used for imaging transmission films with fast electrons (R. Behne [26]).

In 1936 Hans Boersch, after his doctorate in Vienna dealing with electron interference, returned to his home city of Berlin and joined Brüche's research group. There he began to investigate electron optical image formation from a wave mechanics point of view, transferring Abbe's theory of the light microscope to the electron microscope. With the aid of a two-stage transmission electron microscope with a thermionic hairpin cathode and a pair of iron-encapsulated lenses, he showed that in the back focal plane of the objective, a diffraction pattern (Abbe's primary image) of a selected area of a polycrystalline gold foil is formed. The electron waves propagating from the primary image then interfere in the image plane to form a "secondary image," i.e., the actual image of the object. By inserting small apertures in the diffraction image plane, he showed that the image contrast can be altered in a systematic way: A reduction of the aperture in the back focal plane in bright-field imaging results in an increase in scattering contrast and a reduction of the lack of sharpness due to inelastically scattered electrons. Moreover, the orientation contrast of crystallites in dark-field imaging from particular diffraction reflections can be verified [27]. Furthermore, he found a new method of selected-area diffraction in two-stage imaging, by aperturing the intermediate image [28]. That was the basis of the "Boersch ray path" method later provided in various commercial transmission electron microscopes.

In 1938, Boersch made the first attempt at what is called "two-wavelength microscopy," the principle of which had been proposed by M. Wolfke in 1920 [29]. According to this method, the imaging process is split up into two stages in the way described in Abbe's theory. In step 1, the diffraction pattern (primary image) of the object is produced and photographed with X rays or electrons. From this pattern, the image (secondary image) is then obtained with light optics. Without any knowledge of Wolfke's proposal, Boersch formed the diffraction pattern of lattice-type objects, formed from arrays of holes in a plate. By illuminating these with coherent light, he obtained "images" of (pseudo) lattice structures. Boersch also was aware of the necessity of controlling the phase in his artificial "diffraction patterns [30]."

Later, D. Gabor solved the problem of recording simultaneously both amplitude and phase by the invention of holography [31; see also Ref. 32, and footnote 3 by T. Mulvey].

The electrostatic lens used in 1932 by Brüche and Johannson was further developed by Mahl and Boersch as a high-voltage (50-kV) Einzel lens [17]. The special advantage of the electrostatic Einzel lens over the magnetic lens is that its focal length does not depend on the lens potential, as long as this is kept at a constant fraction of the cathode potential [23]. This was an important property in view of the poor stability of the high-voltage power supplies at that time. On the other hand, there were also problems: voltage breakdown of the electrodes, the sensitivity of the electron beam to stray charges caused by impurities on the electrodes, especially in the vicinity of the retarding field produced by the central electrode. The aberrations were also higher than those of magnetic lenses. With such electrostatic lenses, the breakthrough to supermicroscopic resolution occurred in 1939, with two instruments of different design, both originating at the AEG Research Institute. Hans Mahl developed a two-stage 50-kV transmission microscope that showed a resolution better than 15 nm [33], and soon afterward 8–10 nm. [34]. In the next two years, Mahl discovered an important method of making visible, electron optically, the surface of massive objects that could not be viewed in transmission. By covering the surface to be viewed with a thin layer of lacquer, which could be removed on hardening, one obtained a thin-film replica that could be viewed in a transmission electron microscope. Later, this was shadowed by a thin metal layer to increase the contrast [17,35,36]. Hans Boersch, on the other hand, developed an electron shadow (projection) microscope, in which he demagnified the electron source, using two electrostatic lenses, to a lateral extent of less than 10 nm. Close behind this small spot he placed a transmission specimen. The magnification of the resulting shadow projection of this fundamentally simple microscope is given, according to the ray diagram, by the ratio of the distance from the source to the projection screen to that between the source and the specimen. A resolution better than 50 nm was attained [17,37,38]. With the projection microscope Boersch discovered, among other things, Fresnel diffraction fringes of electrons at macroscopic edges, for example, those of zinc oxide needles.[3]

---

[3] *Translator's comment.* It is interesting that in the same year James Hillier in the United States observed Fresnel fringes in the transmission electron microscope [J. Hillier, *Phys. Rev.* **58,** 842 (1940)]. Zworykin sent Gabor an original optically enlarged print of the micrograph. Gabor refused to believe that the fringes were Fresnel fringes! In his book, *The electron microscope* (Hulton Press, London, 1946), with a Preface dated 22 July 1944, Gabor reproduced this micrograph as Fig. 15 on page 34; the Fresnel fringes are clearly visible, but Gabor supplied a provocative legend: "Extremely thin fibres of a synthetic rubber 'Koroseal' photographed with RCA electron microscope. The dark centres are presumably molecules. The fringes surrounding the fibres are 'chromatic' fringes corresponding to the characteristic energy loss of electrons in carbon (multiples of 24 eV). Magnification about 100,000." Acknowledgment was made to RCA. Gabor held similar views about Boersch's fringes. In a footnote on page 35 of the same book, we read: "These contours were first observed in 1940 by J. Hillier, who instead of connecting them with his observations of 1939, tried to explain them tentatively as Fresnel diffraction fringes. . . . Hillier has not attempted a quantitative explanation which would have shown that Fresnel fringes cannot account for his observations. . . ."

What neither Gabor, Hillier, or Boersch realized at the time was that these micrographs were indeed the first Gabor holograms. Boersch had inadvertently produced a way of determining the amplitude and the phase of the electron wave from the object, but he just couldn't recognize it when he found it! Gabor was also looking for a way to correct spherical aberration in electron microscopes, but he thought that the fringes were not interference fringes at all and hence not capable of reproducing the wavefront behind the specimen. However, by the time Gabor wrote the Preface to the second edition (6 May 1947), Hillier had convinced him that the fringes obtained by Boersch and himself were indeed genuine Fresnel fringes. What Gabor did not reveal in this book was that, just before he had sent the second edition off to the press, he had "had a vision during the Easter holiday of 1947, while sitting on a bench with his wife, waiting their turn for a game of tennis." In this vision, which probably took place on Easter Monday, 15 April 1947, "effortlessly and with no action on my part, the whole concept of holography was revealed. "Take a photograph of the object with coherent illumination, recording both amplitude and phase, then illuminate the negative with a coherent source to reconstruct the wave and hence view the object" (extract from T. E. Allibone's Royal Society Memoir of Gabor).

Gabor must have realized at this point that Boersch's projection microscope would allow him to produce such a negative, a Fresnel interferogram which he promptly named a "hologram," since it contained simultaneously both amplitude and phase information.

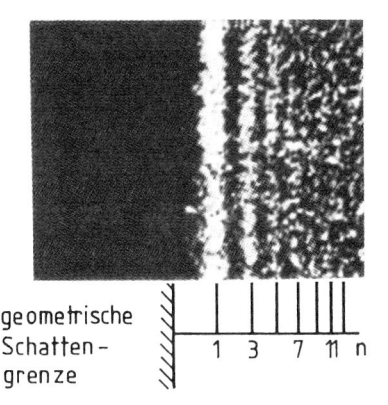

FIGURE 2. Fresnel diffraction of electrons at an edge (1940). Micrograph taken by Boersch [39] in his specially designed "shadow" electron microscope. Key: Geometrische Schattengrenze = limit of geometric shadow.

The special significance of such fringes at the time (see Fig. 2) is that the separation of the interference fringes does not depend on the geometric dimensions of the scattering body; in particular, atomic spacings play no role (contrary to the case of crystalline diffraction). This provided a definitive proof of the wave nature of electrons [39]. The number of such Fresnel fringes at a known image defocusing distance serves today as a measure of the quality of the coherent illumination.

In 1941, Boersch left the Berlin AEG Research Institute and went to Vienna, from where, after some intermediate positions in Württemberg and Braunschweig, he returned in 1954 as full Professor of the 1st Physikalisches Institut at the recently renamed Technische Universität (TU Berlin). In the 1950s, G. Möllenstedt in Tübingen finally developed the electrostatic Fresnel biprism, thereby making electron interferometry and off-axis holography possible [40].[4]

Electrostatic transmission electron microscopes (see, for example, Ref. 17) were serially produced for about 15 years, first by AEG and later by Carl Zeiss Oberkochen and by VEB Carl Zeiss Jena, and also by several firms abroad. Because of the problems, previously mentioned, with voltages higher than 50 kV and the higher aberrations compared with those of magnetic lenses, these firms also went over to the construction of magnetic electron microscopes.

---

[4] *Translator's comment.* It seems that Möllenstedt did not realize at that time that these electron interferograms were in fact off-axis holograms. Such holograms do not suffer from the grave defects of "in-line" holograms. Off-axis electron holograms were investigated in detail during the 1970s and 1980s in Germany by H. Wahl, K.-J. Hanszen, and H. Lichte.

IV. The Scanning Electron Microscope

Max Knoll, who since 1932 had undertaken television development projects at Telefunken Berlin, devised in 1935 a new method of investigation, based on television technology, for displaying the localized distribution of secondary electrons from inhomogeneous bodies [41]. With his electron "scanner," as he called it [42], an electron beam was scanned, in television fashion, by suitable deflection coils (Fig. 3). The secondary signal from the specimen was used, after amplification, to modulate the intensity of a cathode-ray tube, whose writing beam scanned in synchronism with that of the microscope. In this way one obtained a "secondary electron image" of the scanned specimen. This principle gave the possibility of imaging massive opaque objects. Knoll did not develop this method further for high resolution; this took place, as far as secondary electron emission is concerned, elsewhere. He was, however, the first to observe "voltage contrast" [42], which is of considerable importance today in electron beam measuring techniques for highly integrated microcircuits.

In 1938, Manfred von Ardenne advanced to submicroscopic resolution in the scanning electron microscope in his laboratory in Berlin-Lichterfelde

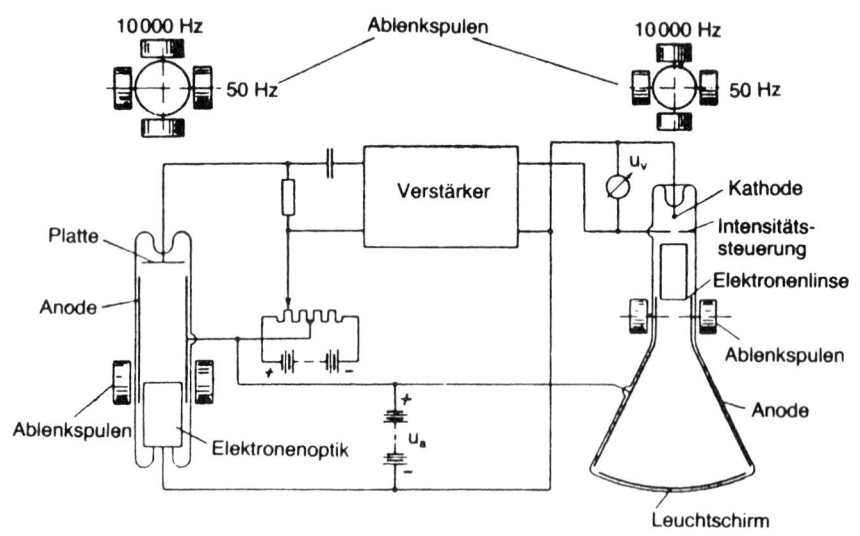

Figure 3. Schematic arrangement of the imaging of secondary emission from a metal plate irradiated by a focused scanning beam of electrons: the first scanning electron microscope (1935). Design and construction realized by M. Knoll [41]. Key: Ablenkspulen = deflection coils; Platte = plate; Anode = anode; Elektronenoptik = electron optics; Verstärker = amplifier; Kathode = cathode; Intensitätssteuerung = intensity control; Elektronenlinse = electron lens; Leuchtschirm = fluorescent screen.

[43], when he demagnified the electron source in two stages to about 10 nm, thereby creating the preconditions for such a resolution. Von Ardenne called his instrument, working on the electron transmission principle, an "electron scanning microscope" rather than the usual "scanning electron microscope." Ardenne also recognized that a particular advantage of the scanning electron microscope compared with the transmission electron microscope is that the chromatic aberration of the electron lenses, strongly evident in thick specimens in the TEM, plays no part in the scanning method, since the electron lenses that focus the beam onto the specimen operate with a uniform energy, apart from the normal thermal energy spread in the beam. The scanning electron microscope was first developed commercially in the 1950s and 1960s, in England, the United States, and Japan, to the present manifold areas of applications, once the technology of processing the sequential image signals had reached an appropriate development stage.

As a further development of fine electron probes, von Ardenne published, in 1939, the principle of the X-ray projection microscope [44]. In this, a stationary electron probe is focused onto an anticathode, creating a very fine source of X rays. In this way, an X-ray projection microscope can be built, analogous to the projection electron microscope of Boersch [38]. Because of the weak interaction of X rays with matter, one may magnify, in point projection, even thick specimens without undue loss of resolution.

V. The Field Emission Microscope

In Berlin there also arose a fourth direction of development in electron microscopy: field emission microscopy. One may well treat this as a special variant of projection microscopy, in which the distance between the specimen and illuminating source has been reduced to the smallest possible size. The specimen is located on the surface of the illuminating source, or indeed, the surface itself may constitute the specimen to be imaged. It was with this form of microscopy that atoms were first made visible. In 1936, at Research Laboratory II of the Siemens Works in Berlin-Siemensstadt, under the leadership of Gustav Hertz, E. W. Müller, in his dissertation, investigated electron emission at high field strengths [45]; this emission could be explained, according to Fowler and Nordheim [46], in terms of quantum mechanical tunneling effects. In a subsequent publication [47], Müller described, for the first time, the principle of the field emission microscope; it consists of a fine tungsten tip, of about 1 μm in radius, as a cathode at the center of a glass vessel whose inner surface is coated with a conducting fluorescent screen and serving as an anode. At a voltage of about 10 keV, the cathode emits electrons; these are accelerated almost radially toward the fluorescent screen, creating a central projection of the tip surface with a lateral magnification of some 10,000–100,000×. On the fluorescent screen, one then sees cathode regions of high emission, as a consequence of local low work function. In 1937, E. W. Müller published field emission patterns from cathodes of different materials and with different orientations [48]. He was able, shortly afterward, working meanwhile at Stabilovolt GmbH in Berlin, to make visible singly evaporated barium atoms as a consequence of the locally reduced work function, at a magnification of around 1 million and a resolution of 2 nm [49]. Müller achieved atomic resolution in 1951, while working at the former Kaiser Wilhelm Institute for Physical Chemistry and Electrochemistry (at that time the Deutsche Forschungshochschule, at present the Fritz Haber Institute of the Max Planck Gesellschaft in Berlin-Dahlem). Here he operated the field emission microscope as a field ion microscope, with the aid of an auxiliary gas, helium or hydrogen, forming positive ions by field ionization at the positively charged tip surface [50]. The ionizing probability is especially large at the edges and at corner atoms of the tip surface. These details are thus imaged preferentially; the resolution compared with electron imaging then lies in the region 0.1–0.2 nm, because of the shorter wavelength of the ions. Field emission microscopy, however, is limited to imaging the tip surface.

VI. Conclusion

Since the pioneering era described above, electron microscopy has made great progress. Its further development after World War II was pushed forward with great investment by America, Europe, and Japan. By removing several disturbing factors, the resolution of TEM and STEM has been pushed to the imaging of single atoms. Ernst Ruska, when he was Director of the Institute for Electron Microscopy at the Fritz Haber Institute in the 1950s and 1960s, devoted himself particularly to these tasks and contributed significantly to the very high resolution of today. The present range of accessories of electron microscopy in many research fields in medicine, biology, physics, materials research, etc., can hardly be encompassed. In 1941, the Prussian Academy bestowed the Leibniz Medal on seven experimenters for their services in the development of electron microscopy: Manfred von Ardenne, Hans Boersch, Bodo von Borries, Ernst Brüche, Max Knoll, Hans Mahl, and Ernst Ruska [51]. To this list of pioneers in electron microscopy one should perhaps add the theoreticians Otto Scherzer, Alfred Recknagel, and Walter Glaser.

The Nobel Committee itself honored Ernst Ruska with the highest scientific recognition, the Nobel Prize. By his development of the first two-stage transmission electron microscope for the imaging of plane specimens and the invention of the magnetic polepiece lens, as well as his unremitting efforts in carrying out this development in the face of original scepticism and in overcoming industrial difficulties, Ernst Ruska undoubtedly made a major contribution to this important instrument.

REFERENCES

1. Ruska, E. (1979). Die frühe Entwicklung der Elektronenlinsen und der Elektronenmikroskopie. "Acta Historica Leopoldina," Heft 12. Barth, Leipzig; English translation Hirzel-Verlag, Stuttgart.
2. Gabor, D. (1926). Oszillographieren von Wanderwellen. *Arch. Elektrotech.* **16,** 296–302.
3. Busch, H. (1926). Berechnung der Bahn von Kathodenstrahlen im axialsymmetrischen elektromagnetischen Felde. *Ann. Phys. 4. Folge* **81,** 974–993. Busch, H. (1927). Über die Wirkungsweise der Konzentrierungsspule bei der Braunschen Röhre. *Arch. Elektrotech.* **18,** 583–594.
4. De Broglie, L. (1924). Recherches sur la theorie des quanta. Thèse. Masson, Paris. De Broglie, L. (1925). *Ann. de Physique* **3,** 22–128.
5. Knoll, M., and Ruska, E. (1932). Das Elektronenmikroskop. *Z. Phys.* **78,** 318–339.
6. Knoll, M., and Ruska, E. (1932). Beitrag zur geometrischen Elektronenoptik I und II. *Ann. Phys.* **12,** 607–640, 641–661.
7. Ruska, E. (1984). Die Entstehung des Elektronenmikroskops. *Arch. Gesch. Naturwiss.*, 525–551.
8. Ruska, E. (1933). Über ein magnetisches Objektiv für das Elektronenmikroskop. Diss., TH Berlin. Ruska, E. (1934). *Z. Phys.* **89,** 90–128.
9. Driest, E., and Müller, H. O. (1935). Elektronenmikroskopische Aufnahmen (Elektronenmikrogramme) von Chitinobjekten. *Z. wiss. Mikroskopie* **52,** 53–57.
10. Krause, F. (1936). Elektronenoptische Aufnahmen von Diatomeen mit dem magnetischen Elektronenmikroskop. *Z. Phys.* **102,** 417–422.
11. Krause, F. (1937). Neuere Untersuchungen mit dem magnetischen Elektronenmikroskop. *In* "Beiträge zur Elektronenoptik" (H. Busch and E. Brüche, Eds.). Barth, Leipzig.
12. Ruska, E. (1934). Über Fortschritte im Bau und in der Leistung des magnetischen Elektronenmikroskops. *Z. Phys.* **87,** 580–602.
13. Lickfeld, K. G. (1979). "Elektronenmikroskopie." Ulmer, Stuttgart.
14. Marton, L. (1934). Electron microscopy of biological objects. *Nature* **133,** 911 and *Phys. Rev.* **46,** 527–528.
15. Krause, F. (1937). Das magnetische Elektronenmikroskop und seine Anwendung in der Biologie. *Naturwiss.* **25,** 817–825.
16. "Das Übermikroskop als Forschungsmittel" (1941). Walter de Gruyter, Berlin.
17. Ramsauer, C. (Ed.) (1942). Elektronenmikroskopie. Bericht über Arbeiten des AEG Forschungs-Instituts 1930–1941. Springer, Berlin.
18. Brüche, E. (1932). Elektronenmikroskop. *Naturwiss.* **20,** 49.

19. Brüche, E., and Johannson, H. (1932). Elektronenoptik und Elektronenmikroskop. *Naturwiss.* **20,** 353–358.
20. Johannson, H. (1933, 1934). Über das Immersionsobjektiv der geometrischen Elektronenoptik. I, *Ann. Phys. 5. Folge* **18** (1933), 385–413; II, *Ann. Phys. 5. Folge* **21** (1934), 274–284.
21. Brüche, E. (1932). Die Geometrie des Beschleunigungsfeldes in ihrer Bedeutung für den gaskonzentrierten Elektronenstrahl. *Z. Phys.* **78,** 26–42.
22. Brüche, E., and Johannson, H. (1932). Einige neue Kathodenuntersuchungen mit dem elektrischen Elektronenmikroskop. *Phys. Z.* **33,** 898–899.
23. Johannson, H., and Scherzer, O. (1933). Über die elektrische Elektronensammellinse. *Z. Phys.* **80,** 183–192.
24. Brüche, E., and Scherzer, O. (1934). "Geometrische Elektronenoptik." Springer, Berlin.
25. Recknagel, A. (1941). Theorie des elektrischen Elektronenmikroskops für Selbststrahler. *Z. Phys.* **117,** 689–708.
26. Behne, R. (1936). Die Eigenschaften des Immersionsobjektives für die Abbildung mit schnellen Elektronen. Folienabbildung mit dem Immersionsobjektiv. *Ann. Phys.* **26,** 372–384, 385–397.
27. Boersch, H. (1936). Über das primäre und sekundäre Bild im Elektronenmikroskop I. Eingriffe in das Beugungsbild und ihr Einfluβ auf die Abbildung. *Ann. Phys. 5. Folge* **26,** 631–644.
28. Boersch, H. (1936). Über das primäre und sekundäre Bild im Elektronenmikroskop II. Strukturuntersuchung mittels Elektronenbeugung. *Ann. Phys. 5. Folge* **27,** 75–80.
29. Wolfke, M. (1920). Über die Möglichkeit der optischen Abbildung von Molekulargittern. *Phys. Z.* **21,** 495–497.
30. Boersch, H. (1938). Zur Bilderzeugung im Mikroskop. *Z. Tech. Phys.* **19,** 337–338 and Plate III.
31. Gabor, D. (1948). A new microscopic principle. *Nature* **161,** 777–778.
32. Boersch, H. (1967). Holographie und Elektronenoptik. *Phys. Blätter* **9,** 393–404.
33. Mahl, H. (1939). Über das elektrostatische Elektronenmikroskop hoher Auflösung. *Z. Tech. Phys.* **20,** 316–317.
34. Mahl, H. (1940). Über das elektrostatische Elektronen-Übermikroskop und einige Anwendungen in der Kolloidchemie. *Kolloid-Z.* **91,** 105–117.
35. Mahl, H. (1940). Metallkundliche Untersuchungen mit dem elektrostatischen Übermikroskop. *Z. Tech. Phys.* **21,** 17–18 and Plate I.
36. Mahl, H., (1941). Plastisches Abdruckverfahren bei Oberflächen. *Z. Tech. Phys.* **22,** 33.
37. Boersch, H. (1939). Das Schatten-Mikroskop, ein neues Elektronen-Übermikroskop. *Naturwiss.* **27,** 418.
38. Boersch, H. (1939). Das Elektronen-Schattenmikroskop I. Geometrisch-optische Versuche. *Z. Tech. Phys.* **20,** 346–350.
39. Boersch, H. (1940). Fresnelsche Elektronenbeugung. *Naturwiss.* **28,** 709–711. Boersch, H. (1940). Fresnelsche Beugungserscheinung im Übermikroskop. *Naturwiss.* **28,** 711–712.
40. Möllenstedt, G., and Düker, H. (1956). Beobachtungen und Messungen an Biprisma-Interferenzen mit Elektronenwellen. *Z. Phys.* **145,** 377–397.
41. Knoll, M. (1935). Aufladepotential und Sekundäremission elektronenbestrahlter Körper. *Z. Tech. Phys.* **16,** 467–475.
42. Knoll, M. (1941). Steuerwirkung eines geladenen Teilchens im Feld einer Sekundäremissionkathode. *Naturwiss.* **29,** 335–336.
43. Ardenne, M. von (1938). Das Elektronen-Rastermikroskop. Theoretische Grundlagen. *Z. Phys.* **109,** 553–572. Ardenne, M. von (1938). Praktische Ausführung. *Z. Tech. Phys.* **19,** 407–416.

44. Ardenne, M. von (1939). Zur Leistungsfähigkeit des Elektronen-Schattenmikroskopes und über ein Röntgenstrahlen-Schattenmikroskop. *Naturwiss.* **27,** 485–486.
45. Müller, E. W. (1936). Die Abhängigkeit der Feldelektronenemission von der Austrittsarbeit. Diss. TH Berlin and *Z. Phys.* **102,** 734–761.
46. Fowler, R. H., and Nordheim, L. (1928). Electron emission in intense electric fields. *Proc. R. Soc. A* **119,** 173–181.
47. Müller, E. W. (1936). Versuche zur Theorie der Elektronenemission unter Einwirkung hoher Feldstärken. *Phys. Z.* **37,** 838–842.
48. Müller, E. W. (1937). Elektronenmikroskopische Beobachtung von Feldkathoden. *Z. Phys.* **106,** 541–550.
49. Müller, E. W. (1938). Weitere Beobachtungen mit dem Feldelektronenmikroskop. *Z. Phys.* **108,** 668–680.
50. Müller, E. W. (1951). Das Feldionenmikroskop. *Z. Phys.* **131,** 136–142.
51. Brüche, E., and Scherzer, O. (1947). Stand und Zukunft des Elektronenmikroskops. *Phys. Blätter* **3,** 263–266.

2.6B
The History of the German Society for Electron Microscopy

GERHARD SCHIMMEL[1,2]

Battelle Institute, Frankfurt, Germany

I. INTRODUCTION

The history of the German Society for Electron Microscopy is inextricably bound up with the existence of the electron microscope. The same pioneers who had taken part in the development of the electron microscope were also prominent in the founding of the German Society. For this reason it is difficult to state a point in time at which the history of the society can be said to have begun. The actual founding day is certainly later, since many important preliminary efforts had already been taken. The early development of electron lenses and electron microscopes in Germany have been competently described in several publications, so here one is referred to the relevant literature.

One of the most important development phases of the electron microscope was attained in 1939. In this year, the firm of Siemens and Halske unveiled the first model of a serially produced set of electron microscopes, designed by Bodo von Borries and Ernst Ruska. This instrument was sold to the firm Farbwerke Hoechst and placed in the care of Dr. Helmut Kehler, a founding member of the German Society. It worked there successfully until 1961.

In the meantime, the working group under Prof. Ernst Brüche had made notable progress. With the aid of the electrostatic high-voltage lens of Boersch and Mahl, it was possible to build a particularly simple electron microscope that was ready for production around 1940. A small series of these AEG electrostatic instruments was delivered up to the end of the war.

In 1941, the Prussian Academy bestowed the Silver Leibniz Medal on the following scientists: M. von Ardenne, H. Boersch, B. von Borries, E. Brüche, M. Knoll, H. Mahl, and E. Ruska. By this high citation not only were the pioneers themselves honored, but the scientific progress associated with the electron microscope was also recognized officially. The bestowal

---

[1] Translated by T. Mulvey.
[2] Present address: Am Sonnenberg 17A, D-64385 Reichelscheim, Germany.

of the Silver Leibniz Medal on the above scientists may also be regarded as the endpoint of the early development history of the electron microscope.

Two further events fostered considerably the spreading of electron microscopy in these years. First, in 1940 the Siemens firm instituted a guest laboratory. Here, guest scientists could become acquainted with the operation and servicing of the Siemens electron microscope and immerse themselves in the mysteries of electron microscope specimen preparation.

The other significant event in this year was the development of the replica technique by Hans Mahl at AEG. This technique enables the surfaces of nontransparent objects to be observed in the transmission electron microscope. This method, which attracted widespread interest in its own right, extended enormously the range of application of the electron microscope. With the availability of serially produced electron microscopes and the possibility of learning how to use them in the guest laboratory, and with the development of improved and more versatile preparation methods, the requirements for an expansion of electron microscopy in science and technology were met, so that very soon a group of electron microscopists was formed, who kept in touch with each other and could form the seed for a later society. The timing of this development, however, was not favorable.

The end of World War II and the early postwar years marked a sharp cutoff point. At the end of the war, three electron microscope centers were formed, in which development was once more taken up and pushed ahead. In Berlin, at the firm of Siemens, the interrupted work on the electron microscope was resumed under the leadership of Prof. E. Ruska, and the ÜM 100 was produced. In Mosbach (Baden), Prof. E. Brüche succeeded in gathering round him a gifted circle of electron microscope scientists and, in the Süddeutsche Laboratorien, a daughter member of AEG, the production of electron microscopes was resumed in 1945. It was here that the instruments EM 7, EM 8, and EM 8/2 came into being. In Düsseldorf, through the initiative of Bodo von Borries, the Society for Supermicroscopy e.V. (Gesellschaft für Übermikroskopie e.V.) at Düsseldorf was brought into being, whose purpose was to be the promotion of high-resolution electron microscopy on a basis exclusively and directly of common interest. To this end, the Association set up and operated the Rhine-Westphalian Institute for Supermicroscopy (Übermikroskopie).

II. The Founding of the Society

Professor Brüche sent out invitations to an electron microscope conference to be held in Mosbach in April 1949. It can be assumed that on this occasion, he hoped that an impulse would be given to the founding of a society for electron microscopy. Before this conference, in February 1949, Bodo von Borries had organized a foundation meeting for a German Society for Electron Microscopy in Düsseldorf. In a footnote about the foundation meeting it was said under point 2: "after extensive discussion it was unanimously resolved to found the Association 'Deutsche Gesellschaft für Elektronenmikroskopie' ('German Society for Electron Microscopy')." To the board of the society were elected:

1. Chairman: Ernst Ruska
2. Chairman: Hans Mahl
1. Committee member: Fritz Jung
2. Committee member: Walter Kikuth
3. Committee member: Otto Scherzer
   Secretary: Bodo von Borries

On this board, the Berlin Siemens group was represented by Ernst Ruska, the Mosbach group of Ernst Brüche by Hans Mahl and Otto Scherzer, the Rhine-Westphalian Institute for Supermicroscopy by Bodo von Borries, and the University by Walter Kikuth, Professor of medicine in Düsseldorf. The composition of the first Board of the German Society must be seen as being extremely well balanced. Helmut Kehler was chosen as auditor of the society's accounts.

At the foundation meeting it was further decided that for the year 1950 Ernst Brüche would be First Chairman. Bodo von Borries announced that he was empowered by Prof. Ruska to explain that he [Ruska] accepted the vote to become the First Chairman.

The minutes of this historic meeting are reproduced in Appendix I.

Hans Mahl suggested that the meeting in Mosbach, planned by Ernst Brüche in February 1949, should be carried out as a conference of the newly founded society. He would ask Ernst Brüche to put back the date to the end of April. The meeting unanimously approved this proposal. The minutes of the foundation meeting concluded with the insertion of a handwritten conclusion: "The assembled members unanimously elected Hans Busch in Darmstadt as an Honorary Member as a tribute to his services to electron optics."

On 23 and 24 April 1949 the first conference of the German Society for Electron Microscopy took place in Mosbach (Baden) in the Council Chamber of the Town Hall. The full title was "Meeting for Electron Microscopy, United with the First Conference of the German Society for Electron Microscopy." The conference was introduced by the presentation of the Certificate of Honorary Membership to Hans Busch. The topics of the scientific program of this first German national conference are listed on page 152.

1. E. Ruska (Berlin). Present status of the development of the electron microscope.
2. H. König (Göttingen). New results in electron microscopy.
3. V. E. Coslett (Cambridge). Work in the Cavendish Laboratory.
4. F. Jung (Würzburg). Structure problems with red blood corpuscles.
5. W. Lindemann (Hamburg). Carbodian ringbodies of erythrocytes.
6. C. Wolpers (Lübeck). Cross-striped phase.
7. H. König (Göttingen). Formation of cell extensions in the human tooth.
8. A. Jakob (Nürnberg). Particular development stages of leptospire culture.
9. G. A. Kausche (Heidelberg). New results in the area of virus research.
10. W. Schäfer (Tübingen). Change of shape of the virus of atypical fowl pest.
10a. K. Liebermeister (Frankfurt/Main). Pleuropneumonie.
11. B. von Borries (Düsseldorf): The origin of contrast in the electron microscope image by elastic and inelastic scattering.
12. G. Ruthemann (Schwäbisch-Gmünd). On the temperature rise of specimens in the electron microscope.
13. H. Haardick (Bonn). On the interaction of evaporated layers and microscope specimens.
14. W. Kossel (Tübingen). On the imaging of crystalline materials.
15. H. W. Kohlschütter (Darmstadt). Hydrolysis of silicon compounds.
16. K. Bayersdorfer (Mosbach). Imaging of glass surfaces.
17. H. Boersch (Braunschweig). Velocity filters.
18. G. Möllenstedt and F. Heise (Mosbach). The electron lens as high-resolution velocity analyzer (presented by G. Möllenstedt).
19. R. Rühle (Stuttgart). A special arrangement of illuminating and specimen system of an electron microscope.
20. K.-H. Steigerwald (Mosbach). An electron source for the production of a high current density with low beam current.
21. G. Möllenstedt and F. Heise (Mosbach). A new method of obtaining stereoimages in the electron microscope (presented by F. Heise).
22. H. König (Göttingen). Schlieren optics in the electron microscope.
23. O. Rang (Mosbach). An electron optical stigmator.
24. R. Seeliger (Mosbach). An attempt to correct spherical aberration with nonrotationally symmetric lenses.

The themes of the lectures clearly reflect the contemporary themes of electron microscopy, both in applications and in instrument development.

Several of the themes held their immediacy over several years, even to the present day. Here may be mentioned the problem of temperature rise of the specimen, or the interaction between evaporated layers and microscopic objects—and especially the possibility of correcting spherical aberration with nonrotationally symmetric lenses. On the eve of the conference, i.e., on 22 April, the first meeting of the organizing committee of the newly founded society took place. According to the minutes of the meeting, the following were present: Ernst Ruska (chairman), Bodo von Borries (secretary), Hans Mahl, Walter Kikuth, Fritz Jung, and Otto Scherzer. Figure 1 shows a group photograph of the participants.

The following items were discussed:

1. Admission of members
2. Decision on circular letters and membership cards
3. Applications for amending the constitution
4. Communication with the press, specialist journals, and with other societies active in the same field
5. Report by the Secretary on the financial situation
6. Place of next meeting
7. Discussion of the activity of the society

In a preliminary discussion, in which Ernst Brüche took part, items 3 and 7 were considered in detail. The first meeting of the members of the society, at that time still called a business meeting, took place on 23 April.

The second meeting of the DGE (Deutsche Gesellschaft für Elektronenmikroskopie) took place in Bad Soden on 14–26 April 1950, at the invitation of Dr. Kehler in Bad Soden im Taunus.

At the foundation meeting and the first board meeting of the newly founded society, the Constitution of the Society was also formulated (see Appendix II). The structure of the constitution remained the same over the years, despite various adjustments that were necessary to meet changed circumstances. According to the original constitution, the executive organs of the Association were:

1. The chairman and his deputies
2. The board
3. The Members' Assembly

The executive members of the board consisted of the two chairmen, three committee members, and the secretary.

Particularly important was Sec. 6 of the constitution, whereby the executive in the sense of Sec. 26 BGB (Bürgerliches Gesetzbuch) consisted of the secretary alone. From the very beginning, the office of secretary was held by Bodo von Borries. Appendix III shows the names of all executive

FIGURE 1. Group participants at the Mosbach Meeting in February 1949. Among the participants were: 1, Kehler; 2, Mahl; 3, Seeliger; 4, van Murwyck; 5, Kinder; 6, Recknagel; 7, Bernard; 8, Herr and Frau Brüche; 9, Kossel; 10, Pfefferkorn; 11, Ruska; 12, Möllenstedt; 13, von Borries; 14, Sommerfeld; 15, Rang.

members since the foundation of the society. It soon became clear that the main tasks of the executive lay with the secretary. In a letter written in 1954, Chairman Hans Mahl and Secretary Bodo von Borries stated that the actual position of the secretary within the society was not accurately described from an outsider's point of view by the designation "Secretary." In view of this, they suggested the designation "Geschäftsführender Vorsitzender" ("Business Chairman"). At the Member's Assembly at the conference at Münster, this proposal was accepted and the constitution was amended correspondingly. The organs of the association were still:

1. The two chairmen and their deputies
2. The board
3. The Assembly of Members

Concerning the chairmen and their deputies, "The two members, of whom one carries out the business, will be elected by the Assembly of Members with a simple majority and be the business conducting chairman for 4 years, the other is chairman for 2 years, after which he is the deputy chairman for a further 2 years."

In the following paragraphs it was again stated: "The 'Vorstand' (Executive) in the sense of Sec. 26 of the BGB is formed by the business conducting chairman alone." On this occasion it was also decided to expand the Executive Committee by the office of Treasurer. Bodo von Borries was elected as the first business chairman.

At the International Conference for Electron Microscopy in Paris in 1950, talks began among the various national societies for electron microscopy concerning the founding of an International Association for Electron Microscopy. On 12 July 1951, Dr. S. W. Cuckov, Secretary of the Electron Microscopy Group of the Institute of Physics, circulated a proposal, headed "Proposal for the Establishment of a Joint Commission on Electron Microscopy" and addressed to the International Council of Scientific Unions. This paper concluded with the following proposal: "We, the undersigned, being representatives of the various electron microscopy societies or otherwise interested in this work, therefore wish to propose that the Joint-Commission on Electron Microscopy be set up within the framework of the International Council of Scientific Unions."

The document was signed on behalf of the German Society by Chairman Helmut Ruska and Secretary Bodo von Borries. Bodo von Borries in the following period actively engaged himself in writing many letters for the international association. At the International Conference in London in 1954 the formation of an International Committee of Electron Microscopy was accepted by the participating societies. On 15 July 1954, the first joint meeting of the national representatives took place in London. The German

Society was represented by Professor von Borries. On this occasion Dr. Cosslett was elected Secretary and Prof. von Borries Chairman.

At the same time it was arranged that international conferences should be held at four-year intervals. Agreement was reached that the German Society for Electron Microscopy should be asked whether it would organize the next international conference in Berlin in 1958. In July 1955, Dr. Cosslett, as Secretary, made it known that the International Association would now adopt the name "International Federation of Electron Microscope Societies." This name was later modified to "International Federation of Societies for Electron Microscopy (IFSEM)." By March 1955, the Founding Committee definitely decided that international congresses should be held every four years, with various regional congresses in the middle of this period and congresses of the national societies in the other years.

This rule also affected the series of national conferences of the German society. It had been originally planned to hold yearly conferences, but the new rules required two years between the national conferences. On 17 July 1956, Bodo von Borries, the Business Chairman of the German society and at the same time director of the Rhine-Westphalian Institute of Supermicroscopy, died. His death was a great loss for the IFSEM and the German Society for Supermicroscopy, as well as putting into question the further existence of the Rhine-Westphalian Institute. Dr. Kehler, as the then First Chairman of the German Society, saw it as his task to leave no avenue unexplored in order to safeguard the continuation of the Rhine-Westphalian Institute. In this connection many letters were sent to many places. From this correspondence, particular attention should be drawn to the letter of 31 October 1956 of the German society, which is reproduced in Appendix IV. In this letter, which was addressed primarily to the Executive of the Society for Supermicroscopy at Düsseldorf and the Ministry of Culture of the Länder of the Bundesrepublik, it was proposed that the new head of the independent institute at Düsseldorf should hold simultaneously a chair in the Medical Academy. Further, Prof. Helmut Ruska was named as a possible director. This intensive concern on the part of the German Society met with success, and Prof. Helmut Ruska was called to Düsseldorf as the successor to Prof. von Borries at the restructured Institute for Electron Microscopy at the Medical Academy, Düsseldorf.

At the International Congress in Berlin 1958, Dr. Helmut Kehler was elected as Business Chairman of the German Society, as successor to Bodo von Borries. He held this office up to 1969.

The 4th International Conference in Berlin 1958 was a high point in the history of the German Society. It was the first conference held under the umbrella of the International Federation of Societies for Electron Microscopy (IFSEM). A second international conference was organized by the

German Society in 1982 in Hamburg. There it was possible, in the Congress Centre of the city, to hold lectures, poster displays, and the exhibition of instruments under one roof. The development of membership numbers of the German Society was steady. The only setback to be noted was in the years 1969/1970, when all members who were domiciled in the former German Democratic Republic, on the orders of the GDR government, were compelled to resign from the society in writing, and to send a duplicate copy of their letter of resignation to the German Academy of Science in Berlin. The only citizen of the GDR who was allowed to keep his membership was Prof. Manfred von Ardenne. At the founding meeting of the German Society on 16 February 1949, Prof. Busch had been elected as the first Honorary Member of the German Society, thereby honoring his contribution to electron optics. The German Society also concerned itself subsequently with nominating other members for Honorary Membership for their services to electron microscopy. (Appendix V contains a list of Honorary Members elected up to January 1992.) It is perhaps worth pointing out that in 1975, Prof. Dr. med. Helmut Ruska, Prof. Dr. Walter Glaser, and Prof. Dr.-Ing. Bodo von Borries were named postmortem as Honorary Members. The German Society wished to honor and acknowledge, even after their deaths, the outstanding service of these scientists to the development of electron microscopy.

The broad area of work covered by the members of the society led to the formation and encouragement of working groups for specialized fields of activity. The first of these, the working group Mikrosonde, was brought into being in 1967. At the Members Meeting in 1969 it was decided to form a working group in reflection microscopy. Professor Pfefferkorn was asked to lead this group. The working group Mikrosonde later merged with this group, which then assumed the name "Working Group for the Electron Optical Direct Imaging of Surfaces (EDO)." This working group was subsequently very active and held their own colloquia in the years when no national meetings were scheduled. The scientific contributions presented at these colloquia were published in the series *Contributions to Electron Optical Direct Imaging of Surfaces* (BEDO).

In 1987 working groups for cryoelectron microscopy and in 1989 scanning tunnel microscopy were formed.

From Prof. Dr. Bethge of the working group for electron microscopy of the German Academy of Sciences at Berlin in Halle/Saale, there came the suggestion of publishing a compendium of techniques in electron microscopy, to be sponsored by the German Society for Electron Microscopy. The realization of this plan foundered initially on organizational problems. Thanks to the previous Business Chairman, Dr. Kehler, and Hans Rotta from the Wissenschaftliche Verlagsgesellschaft mbH (Scientific Publishing

Company), following the Annual Conference of the German Society in Marburg in 1967, the first steps for carrying out the plan could be taken. Professor W. Vogell and Dr. G. Schimmel had already declared their willingness to undertake the task of publication. Thus in 1970 the first edition of *Methodensammlung der Elektronenmikroskopie* (*Compendium of Techniques in Electron Microscopy*) appeared. At first, this was thought of largely as a collection of recipes for specimen preparation, but it soon became apparent that this concept was too narrow. Instrumental accessories and evaluation procedures were also incorporated. With the aid of decimal classification, the whole area was subdivided according to specialization, so that the possibility of ongoing extension remained open.

In 1978, the firms of Philips, Siemens, and Carl Zeiss jointly founded the Ernst Ruska Prize Foundation. The aim of this syndicate was the promotion of research and development in the area of electron microscopy. This goal was realized by, among other things, the presentation of the Ernst Ruska Prize to young scientists for notable prowess in the area of electron microscopy and its application to the promotion of scientific progress. The prize was first presented in 1980. After Siemens gave up the production of electron microscopes, it withdrew from the Ernst Ruska Prize Foundation. A modification of the constitution enabled the German Society to join the foundation as a new member. According to the constitution, the various prize winners were determined by a selection committee. The chairman of this committee was Prof. Dr. Ernst Ruska until 1986, when he asked to be relieved of this activity and responsibility. From then on, Prof. Dr. K.-H. Herrmann served as Chairman of the Selection Committee. To the Selection Committee belonged one designated representative of the members of the association, together with three internationally recognized specialists in the field of electron microscopy, who were designated and called in by the chairman. The Ernst Ruska Prize is a cash prize, whose magnitude is determined by the board; it must be at least DM 7.500. The prize is formally bestowed within the framework of the German Society conferences.

The financial position of the society in the early years was determined by the relatively low member's subscription of 5 DM for individual members, 50 DM for legal persons, and 100 DM for universities and institutes. The resulting low cash input limited the activities of the society. So, in a minute from the third board meeting in Hamburg on 17 May 1951, it was announced that the influence of the society in inaugurating training courses was seen as desirable, but that practical possibilities of becoming active in this direction did not exist at present. Since members' subscriptions had been fixed in Sec. 3 of the constitution, any increase could be made only by amending the constitution, which was approved in 1970 by the Members' Assembly. The corresponding section of the constitution provided henceforth that the annual subscription could be proposed by the Executive and approved by the Members' Assembly. Thus an adjustment was made to the membership subscription for the general development of the society.

Increasingly, the German Society switched over to the financial and administrative sponsoring of conferences. If the society in the early years limited itself mainly to the scientific preparation of conferences and left the organizational details and financial accounting largely to the inviting institute, so from now on the society began to take over the organization and financial accounting of conferences. In particular, the society organized instrument exhibitions in connection with conferences. This increased financial room for maneuvering enabled the society to give financial support to EM laboratory courses and symposia, as well as enabling more generous contributions to be made toward traveling expenses.

On 14 October 1986, the Royal Swedish Academy of Science bestowed one half of the 1986 Nobel Prize for Physics on Prof. Dr. Ernst Ruska and the other half on Dr. Gert Binnig and Dr. Heinrich Rohrer. Ernst Ruska obtained the prize for his fundamental electron optical work and the construction of the first electron microscope. This brought to fullfilment the secretly nurtured wish of many members of the German Society, that the efforts of Prof. Ruska in the development of the electron microscope should be recognized in this form.

Appendix I

*The Foundation Meeting
of the German Society for Electron Microscopy*

Düsseldorf 16 February 1949
v.Bs/Ro

Minutes of the Foundation Meeting of the
German Society for Electron Microscopy,
held in Düsseldorf, Eisenhüttenhaus, Breitestrasse 27,
16 February 1949

Present: The Society for Supermicroscopy (Gesellschaft für Übermikroskopie e.V.) at Düsseldorf, including Messrs: Mahl, Friedrich-Freska, Peters, Gönnert, Gottsacker, Kircher, Wolpers, Ruthemann, Jung, Fresen, Grün, Kikuth, Schonhofer, Schrader, Wiester, Seifert, van Marvyck, Rühle, Pfefferkorn, Haussmann, Koch, Kehler, Radczewski, Langenwalter, v.Borries, Haardick, Flesh, H. Ruska (part of the time).

Herr von Borries opened the meeting at 1740 hours. The Agenda had already been set out at the Welcome Evening on 15 February and communicated at the Colloquium on the morning of 16 February.

Item 1: Herr von Borries and Herr Jung reported on the lengthy discussions.

Item 2: After a verbal exchange in some depth, it was decided unanimously to found the

"German Society for Electron Microscopy"

The undersigned above joined as members.

Item 3: The Statutes attached to these minutes were accepted unanimously.

Item 4: The following were elected as Officers (of the Board):
> First Chairman: Ernst Ruska
> Second Chairman: Hans Mahl
> Committee Member: Fritz Jung
> Committee Member: Walter Kikuth
> Committee Member: Otto Scherzer
> Secretary: Bodo von Borries

It was further resolved that for the year 1950, Ernst Brüche be elected First Chairman. Herr von Borries announced that he had been authorized by Herr Ernst Ruska to explain that he [Ruska] accepts this choice of First Chairman. Herr von Borries then handed over the chair to Herr Mahl.

Herr Mahl proposed that the meeting planned by Herr Brüche for the Spring of 1949 in Mosbach should be re-arranged as a conference of the newly formed Society. He would ask Herr Brüche to place it slightly later, at the end of April. The meeting unanimously approved this proposal.

The Members' Assembly voted unanimously that Hans Busch of Darmstadt should be elected Honorary Member, in view of his outstanding contribution to the development of electron optics.

Herr Mahl closed the meeting at 1915 hours.

Düsseldorf, 16 February 1949

Below: Official Certificate of Registration of the German Society for Electron Microscopy as No. 1428 in the Register of Societies held by the Local Authority in Düsseldorf. Registration date: 17 March 1949.

Es wird hiermit bescheinigt, dass umstehend aufgeführter Verein heute unter der Nr. 1 4 2 8 in das Vereinsregister des Amtsgerichts in Düsseldorf eingetragen worden ist.

Düsseldorf, den 17. März 1949

Justizangestellter
als Urkundsbeamter der Geschäftsstelle des Amtsgerichts.

Appendix II

Constitution of the Association
"German Society for Electron Microscopy e.V."

Name, Seat, and Objects of the Association

§1

The Association has the name: "German Society for Electron Microscopy e.V." The seat of the Association is Düsseldorf.

§2

The Association pursues exclusively and directly objectives useful for the community in the meaning of the section "Tax Exemption Purposes" of the German Tax Regulations. It is non-profit making; it does not pursue, in the first instance, its own business ends.

The Association has the task of furthering electron microscopy in research, techniques, and industry, the setting up of conferences, and the fostering of international relations.

The financial resources of the Association must be used solely for purposes according to the Constitution. The members obtain no financial benefits from the financial resources of the Association and no person must be favored by payments that are foreign to the purposes of the Association or through excessive compensation.

Membership and Members Subscription

§3

Members of the Association can be individual persons, legal persons, and groups of persons. The acceptance of the Association is effected through the Executive (Vorstand).

The annual subscriptions are proposed by the Executive and approved by the Members' Meeting. The amount is due at the beginning of the calender year, in January.

Membership is terminated by a written declaration of departure, however, only at the end of a calender year. Membership is also terminated in the case of a member being two years in arrears with payment in spite of two reminders. It is terminated as well with immediate effect when the Executive for a serious reason decides on expulsion. Such an expulsion can take place only if not more than one vote is against it, including the written votes of any absent members of the Executive.

The Members' Assembly can, on a proposal by the Executive, proclaim persons as Honorary Members, who have rendered outstanding services to electron microscopy. The Honorary Members have all the rights of a member, but pay, however, no subscription.

Membership and Administration

§4

The organs of the Association are:

a) The two Chairmen and their deputies
b) The Board
c) The Members' Assembly

§5

The two chairmen, of whom one carries out the business, are voted with a simple majority by the Members' Assembly, and where the First Chairman is elected for two years, after which he becomes the Deputy Chairman. The Business Managing Chairman is elected for four years. A reelection is permitted. The two chairmen lead the Association in accordance with the Constitution. They announce the meetings of the Board and the Members' Assembly and lead them. If the Chairman is prevented from attending, his rights are transferred to the Deputy Chairman. The Chairmen can entrust the exercise of certain rights and tasks to another member of the Board. The division of tasks between the two chairmen can be regulated by a business procedure laid down by the Board.

§6

The Board consists of the two chairmen, the Deputy Chairman, the Treasurer, and five ordinary Board members. The Treasurer and the five ordinary Board members are elected by the Members' Assembly by a simple majority: the Treasurer for four years, the ordinary members for two years. A reelection is allowed.

The legal representation of the Association against a third party is carried out under all circumstances by the Business Conducting Chairman after consultation with the First Chairman or the Deputy Chairman, in cases of assets or property additionally with the Treasurer.

The Board in the sense of § 26 BGB (Bürgerliches Gesetzbuch) may be formed alone by the Business Chairmen.

Should within the terms provided by the Statutes a Members' meeting not be held and thus the new election of the chairmen cannot be carried out, the old Members of the Board remain in place until the election of a new Board into office.

The Board carries out the business in accordance with the Constitution and with the decisions of the Members' Assembly.

The Board has the right, for consultative cooperation, from time to time, to call on particular persons, whose collaboration in such cases appears expeditious.

The Board is empowered to act when, inclusive of the two chairmen, at least half the members are present. In the absence of the Business Chairman, motions passed need his subsequent agreement. The Board decides by a simple majority. Each member has one vote. Equal number of votes counts as nonapproval. The Board can handle matters by postal ballots.

In important cases, where a decision of the Members' Assembly in this connection has to be made, the Board is empowered to decide the matter when the decisions cannot wait until a Members' Assembly can be called. For such a decision, approval of the next Members' Assembly must be obtained.

A meeting of the board must be called if at least one-third of the members of the Board wish it to be called.

§7

An ordinary Members' Assembly is called normally once a year. Extraordinary Members Assemblies are called by the Board when required or when requested by at least one-fifth of the membership. The appointed date should be made known at least eight weeks in advance. Items for the Agenda must be sent out by the Business Chairman six weeks in advance. The final invitation is carried out with an enclosed Agenda at the latest two weeks before the Members' Assembly. These requirements are reduced to a half in the case of Extraordinary Meetings. Emergency items can be allowed at the meeting only with the approval of two-thirds of the votes; they must not be related to change of the Statutes.

The Members' Assembly is empowered to decide if at least a tenth of all members are represented. If this is not the case, the Members' Assembly is called for a second time within four weeks and is then empowered to decide regardless of the number of representative members or votes.

The Agenda of an ordinary Members' Assembly contains, *inter alia*, the following points:

a) Report of the Business Chairman on the foregoing financial year.
b) Presentation of the Annual Accounts by the Treasurer and agreement to the discharge (formal approval) according to the Report of the Auditor of Accounts.
c) Presentation of the Budget Plan and approval of current and one-off payments.
d) Election of the Board and two Auditors from the circle of members.
e) Various decisions in the affairs of the Association.

Proposals for the election of the Board together with a declaration by the candidate of willingness to serve must be sent to the Business Chairman at least six weeks before the Members' Assembly. The Board can extend the list of proposals, especially when insufficient nominations are received. In this case the multidisciplinary character must be taken into consideration.

In voting and in election, with the exception of cases dealt with in § 10, a simple majority of the members in good standing present is definitive. In the case of equality of votes, the motion is not accepted; in elections the casting of lots is definitive.

A report is to be prepared on every Members' Assembly that must be signed by the First Chairman and the Business Chairman.

Accounting Procedure and Property of the Association
§8

The cash in hand and property of the Association are managed by the Business Chairman and the Treasurer. Financial accounting takes place annually and is audited by two chosen members of the Association and presented to the Members' Assembly for their formal approval.

The financial year is the calendar year.

Publications
§9

The Board can grant or withold permission to journals to represent themselves as "Organ der Deutschen Gesellschaft für Elektronenmikroskopie e.V." (Organ of the German Society for Electron Microscopy e.V.).

Amendments to the Constitution and Termination of the Association
§10

Amendments to the Constitution as well as to the termination of the Association can be passed by the Members' Assembly, but a majority of at least two-thirds of the votes cast is required.

The following clause is excluded from the possibility of amendment.

If the Association is terminated or if its aims and ends as defined in § 2 are changed, its property must be made available to ends which are recognized, exclusively and directly, to be useful to the community and serve the promotion of science. Any such decisions may be implemented only after the Inland Revenue Service has given its approval.

Interim Regulations

§11

The Members' Assembly transfer to the Board the right to make amendments to the Constitution if such amendments are required by the legally constituted legal authorities.

Passed: Innsbruck 26 August 1981

APPENDIX III: BOARD MEMBERS OF THE GERMAN SOCIETY FOR ELECTRON MICROSCOPY

Year Period of office Conference venue	Chairmen Business chairman Secretary treasurer	Committee members	Auditor
1949 Mosbach	E. Ruska (Phys.) Mahl von Borries	Jung Kikuth Scherzer	
1950 Bad Soden	Brüche (Phys.) E. Ruska von Borries	Boersch Jung Kikuth	Kircher Kehler
1951 Hamburg	H. Ruska (Med.) Brüche von Borries	Radczewski Boersch Jung	Kircher Kehler
1952 Tübingen	Hofmann (Chem.) H. Ruska von Borries	Möllenstedt Radczewski Boersch	Kircher Kehler
1953 Innsbruck	Glaser (Phys.) Hofmann von Borries	Schramm Möllenstedt Radczewski	Kircher Kehler
1954/1955 Münster	Mahl (Phys.) Glaser von Borries	Wolpers Schramm Möllenstedt	Kircher Kehler
1956/1957 Darmstadt	Kehler (Phys.) von Borries + 17.7.56 Mahl Fr. Beckers	Peters Pfefferkorn Wolpers	Kircher Thal
1958 Berlin	Möllenstedt (Phys.) Kehler Bargmann Lippert	Bethge Peters Wolpers	Kircher Thal
1959 Freiburg	Bargmann (Med.) Möllenstedt		

(*continues*)

APPENDIX III (*Continued*)

Year Period of office Conference venue	Chairmen Business chairman Secretary treasurer	Committee members	Auditor
1960/1961 Kiel	Peters (Med.) Kehler Bargmann Möllenstedt Lippert	Bethge Büchner H. Sitte Wegmann Wohlfahrt-B.	Kircher Thal
1962/1963 Zürich	Ruska (Phys.) Kehler Peters Lippert	Bethge Miller H. Sitte Wegmann Wohlfahrt-B.	Kircher Thal
1964/1965 Aachen	Boersch (Phys.) Kehler H. Ruska Lippert	Bethge van Dorsten Miller H. Sitte Vogell	Kämpf Thal
1966/1967 Marburg	Wohlfahrt-B. (Biol.) Kehler Boersch Lippert	van Dorsten Geiger Miller P. Sitte Vogell	Kämpf Thal
1968/1969 Wien	Wegmann (Phys.) Kehler Wohlfahrt-B. Lippert	van Dorsten Geiger Giesbrecht Peters Vogell	Kämpf Thal
1970/1971 Karlsruhe	Giesbrecht (Biol.) Schimmel Wegmann Lippert	Kehler (E) van Dorsten Geiger Peters Reimer Ibemann	Kämpf Weitsch
1972/1973 Lüttich	Reimer (Phys.) Schimmel Giesbrecht von Bassewitz	Kehler (E) Geiger Herrmann Kölbel Themann Wegmann	Kämpf Weitsch
1974/1975 Lüttich	P. Sitte (Biol.) Schimmel Reimer von Bassewitz	Kehler (E) Blaschke Herrmann Kölbel Themann Wegmann	Kämpf Weitsch

(*continues*)

Appendix III (*Continued*)

Year Period of office Conference venue	Chairmen Business chairman Secretary treasurer	Committee members	Auditor
1976/1977 Münster/W.	G. Pfefferkorn (Phys.) Schimmel P. Sitte von Bassewitz	Kehler (E) Blaschke Herrmann Falk Haanstra Kölbel	Kämpf Weitsch
1978/1979 Tübingen	H. Themann (Med.) Schimmel G. Pfefferkorn von Bassewitz	Kehler (E) Engel Falk Haanstra Nickel Weichan	Herrmann Weitsch
1980/1981 Innsbruck	F. Lenz (Phys.) G. Schimmel H. Themann von Bassewitz	Kehler (E) Engel Haanstra Weichan Ehrenwerth Kubalek	Herrmann Weitsch + 22.3.80
1982/1983	E. Zeitler (Phys.) G. Schimmel F. Lenz von Bassewitz	Kehler (E) Weichan Ehrenwerth Fromme de Groodt-Lasseel Plattner	Herrmann Jönsson
1984/1985 Antwerpen	H. Plattner (Biol.) G. Schimmel E. Zeitler H. Rotta	Kehler (E) Lenz Weichan von Bassewitz W. Baumeister B. Tesche	Herrmann Jönsson
1986/1987 Konstanz	K.-H. Herrmann (Phys.) B. Tesche H. Plattner H. Rotta	Baumeister Lenz Mannweiler Niedrig Partsch Schimmel (E)	Jönsson Schwarz
1988/1989 Bremen	W. Baumeister (Biol.) B. Tesche Herrmann H. Rotta	Schimmel (E) Lenz Lichte Mannweiler Urban Zierold	Jönsson Schwarz

(*continues*)

APPENDIX III (*Continued*)

Year Period of office Conference venue	Chairmen Business chairman Secretary treasurer	Committee members	Auditor
1990/1991 Salzburg	H. Niedrig (Phys.) B. Tesche W. Baumeister H. Rotta	Schimmel (E) Herrmann Junger K. Mannweiler Rose Urban	Jönsson Schwarz
1992/1993 Darmstadt	K. Zierold (Phys.) B. Tesche H. Niedrig H. Rotta	Schimmel (E) Herrmann Heydenreich Rose Rühle Schaefer	Lichte Schwarz

Note: (Phys.), physics; (Med.), medicine; (Biol.), biology.

APPENDIX IV

From the Deutsche Gesellschaft, . . .
Letter sent by Dr. Kehler concerning the
Rhine-Westphalian Institute for
Supermicroscopy (RWI)

On 17 September 1956, the German Society for Electron Microscopy addressed itself to the executive and the financial sponsors of the RWI to interested civic and communal bodies and to the directors of institutes interested in EM of the German universities and technological universities with several considerations regarding the continuance of the RWI. At the Stockholm European Conference for Electron Microscopy, the undersigned chairmen of the Society for Electron Microscopy and the other members of the committee had the opportunity to hear the opinion of recognized scientists of this country and abroad on the problem of dissolving the Rhine-Westphalian Institute. Moreover, a stream of letters has been sent to the German society, all underlining the importance of preserving this institute, while some additionally make suggestions for the continuation of the institute. Some relevant extracts are quoted in an appendix to this letter.

Moreover, the Committee of the International Federation of Electron Microscope Societies, whose members, with the exception of the Japanese delegaties, were present at Stockholm, has written to the Executive of the Society for Supermicroscopy at Düsseldorf favoring the preservation of the institute. On the basis of the letters that have been sent to us and the verbal discussion, we feel ourselves duty-bound, herewith, once more to address the board of the Society for Supermicroscopy in Düsseldorf and all relevant circles in Germany, from whom support and suggestions is to be expected.

In the RWI, Germany possesses a unique facility in the area of the further development and coordination of specimen techniques, which are of extreme importance for practical electron microscopy. This institute in the few years of its existence has found worldwide recognition and has produced extraordinarily useful work. We are now convinced that the continuation of the institute under a first-class specialist is of great general importance. Though it cannot be expected that the few firms who together with the representative bodies of the Land Nordrhein-Westfalen are making high contributions will in the long run continue such payment, then, in any case, the closure of the institute must be prevented until all ways of negotiation are exhausted. Everything must be explored, for example, the community powers of the Minister of Culture of the Länder of the Bundesrepublik or from the funds of the Max-Planck-Gesellschaft, possibly with a certain backing from industry, to retain an independent institute in Düsseldorf, whose director holds at the same time a chair at the Medical Academy. Just one of those two positions, i.e., the leadership of an institute without the possibility of a teaching activity or of an Extraordinary Professorship in the framework of an existing Medical Academy alone, would be for a qualified person as, for example, Herr Professor Helmut Ruska, unattractive, and only under the leadership of such a man is progress and a further enhancement of the institute to be expected.

A splitting up of the institute, except that already proposed in our letter of 17 September, namely to bring to an end instrumental development, should in any case be avoided.

Appendix V: Honorary Members of the German Society for Electron Microscopy

Title	Date of birth	Deceased	Date of honorary membership
1. Prof. Dr. phil. Hans Busch	27.02.1884	16.12.1973	16.02.1949
2. Prof. Dr. Max von Laue	09.10.1879	24.04.1960	16.09.1953
3. Prof. Dr.-Ing. Ernst Ruska	25.12.1906	25.05.1988	25.09.1961
4. Prof. Dr.-Ing. Ernst Brüche	28.03.1900	08.02.1985	27.09.1965
5. Prof. Dr.-Ing. Max Knoll	17.07.1897	06.11.1969	29.09.1965
6. Dr.-Ing. Hans Mahl	17.07.1909	25.11.1988	22.09.1975
7. Prof. Dr. phil. Otto Scherzer	09.03.1909	15.11.1982	22.09.1975
8. Prof. Dr. phil. Hans Boersch	01.06.1909	09.06.1986	22.09.1975
9. Prof. Dr. med. Helmut Ruska[a]	07.06.1908	30.08.1973	22.09.1975
10. Prof. Dr. Walter Glaser[a]	31.07.1906	03.02.1960	22.09.1975
11. Prof. Dr.-Ing. Bodo von Borries[a]	22.05.1905	17.07.1956	22.09.1975
12. Prof. Dr.-Ing. Gottfried Möllenstedt	14.09.1912		10.09.1979
13. Prof. Dr. Gerhard Pfefferkorn	31.03.1913	29.06.1989	17.09.1985
14. Dr.-Ing. Helmut Kehler	07.04.1908		14.09.1987
15. Prof. Dr. Heinz Bethge	15.11.1919		11.09.1989
16. Prof. Dr. Friedrich Lenz	21.03.1922		02.09.1991

[a] Awarded postmortem.

2.6C
Electron Microscopy in the Former German Democratic Republic

JOHANNES HEYDENREICH[1]

Akademie der Wissenschaften der DDR
Institut für Festkörperphysik und Elektronenmikroskopie
DDR-4010 Halle/Saale, Germany

HANS LUPPA[2]

Karl-Marx-Universität Leipzig, Sektion Biowissenschaften
DDR-7010 Leipzig, Germany

ALFRED RECKNAGEL

Technische Hochschule, DDR-8053 Dresden, Germany

DANKWART STILLER[3]

Martin-Luther-Universität Halle-Wittenberg
Pathologisches Institut
DDR-4020 Halle/Saale, Germany

Based on the worldwide pioneering work done in electron microscopy and electron optics in Germany between 1930 and the end of the 1940s by E. Ruska, B. von Borries, E. Brüche, O. Scherzer, M. von Ardenne, H. Mahl, and H. Boersch, many activities related to electron beam imaging techniques were explored in the German Democratic Republic (GDR) as early as the 1950s. At that time A. Recknagel was engaged in emission microscopy, J. Picht in electron optics, M. von Ardenne and his colleagues in surface electron microscopy, H. Bethge and his group in the application of electron microscopy to crystal physics and surface physics (including thin films), and D. Schulze *et al.* in the microstructure of metals and alloys. In the field of the application of electron microscopy to biology and medicine, the activities in the GDR were initiated by F. Jung, who worked

---

[1] Present address: Max-Planck-Institut für Mikrostrukturphysik, Weinberg 2, D-06120 Halle/Saale, Germany.
[2] Present address: Paul-Michael-Strasse 11, 04179 Leipzig, Germany.
[3] Fraunhofer-Strasse 2a, 06118 Halle, Germany.

together with H. Ruska as early as the end of the 1930s, followed by H. David, M. Girbardt, and K. Zapf in the 1950s.

Up to the end of the 1960s, commercial electron microscopes were produced in the GDR, viz., in Jena (VEB Carl Zeiss: A. Recknagel, E. Guyenot, E. Hahn) and in Berlin-Oberschöneweide (VEB Werk für Fernsehelektronik: F. Eckart, B. Schramm). In Jena, initially electrostatic transmission electron microscopes (e.g., ELMI-D2, Fig. 1a) were produced, followed by more universal electron-optical instruments (Elektronenoptische Anlage EF, including attachment for emission microscopy). The electron microscope production in Berlin-Oberschöneweide was concentrated on conventional transmission electron microscopes (TEMs) working with magnetic lenses (Fib. 1b) and on versatile tabletop TEMs.

In the past decades, the Ardenne Institute (Dresden) focused their main interest in electron optics on the development of instruments for electron beam processing of materials. Thus, now effective electron beam micromachining automatons are available (Fig. 1c), which are particularly suitable for the formation and structuring of thin layers on surfaces.

For the last two decades, the VEB Carl Zeiss Jena plants have produced electron-optical instruments rather than electron microscopes to be used in microelectronics, viz., an electron-beam exposure system (ZBA 21) for microlithography, and an electron-beam inspection system (ZRM 20) to measure structures in integrated circuits.

The work of GDR research groups in the field of electron microscopy is characterized by the fact that, on the one hand, specialized electron-optical instruments were constructed, and, on the other hand, based on the above-mentioned types of commercial electron microscopes from the GDR and also on commercial ones from abroad, during the past three decades electron microscopy has been used in the wide fields of solid-state physics, solid-state chemistry, materials science, biology, and medicine.

Typical of the electron microscopical instrumentation in research institutes of the GDR (Recknagel and his colleagues at the Technical University of Dresden, von Ardenne's group in Dresden, Bethge and his co-workers at the Academy of Sciences of the GDR in Halle) is the construction of home-made surface electron microscopes, e.g., emission electron microscopes [Recknagel (Fig. 2a), Heisig, Bethge] and mirror electron microscopes (Bethge and Heydenreich, Schwartze, Kuhlmann). In further developing emission electron microscopes, Bethge succeeded in constructing an ultrahigh-vacuum photoelectron emission equipment (Fig. 2b). Scanning electron microscopes were completed, e.g., for *in situ* investigations (deformation stages) and laser irradiation of specimens under observation (LASEM: combination of laser source and SEM, K. Wetzig, Fig. 2c), and

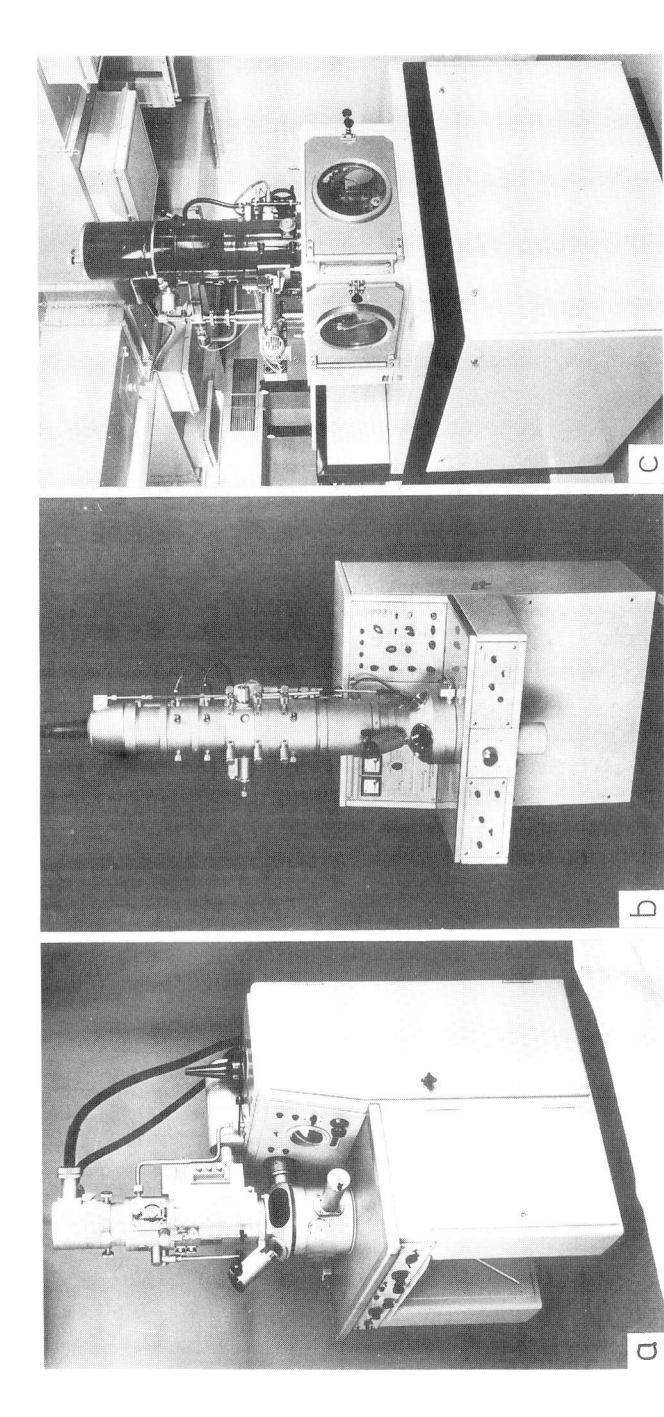

FIGURE 1. Commercial electron-optical instruments: (a) electrostatic transmission EM, ELMI-D2 (VEB Carl Zeiss Jena); (b) magnetic transmission EM, SEM 3-2 (VEB Werk für Fernsehelektronik Berlin); (c) electron beam micro-machining automaton (Institut Manfred von Ardenne Dresden).

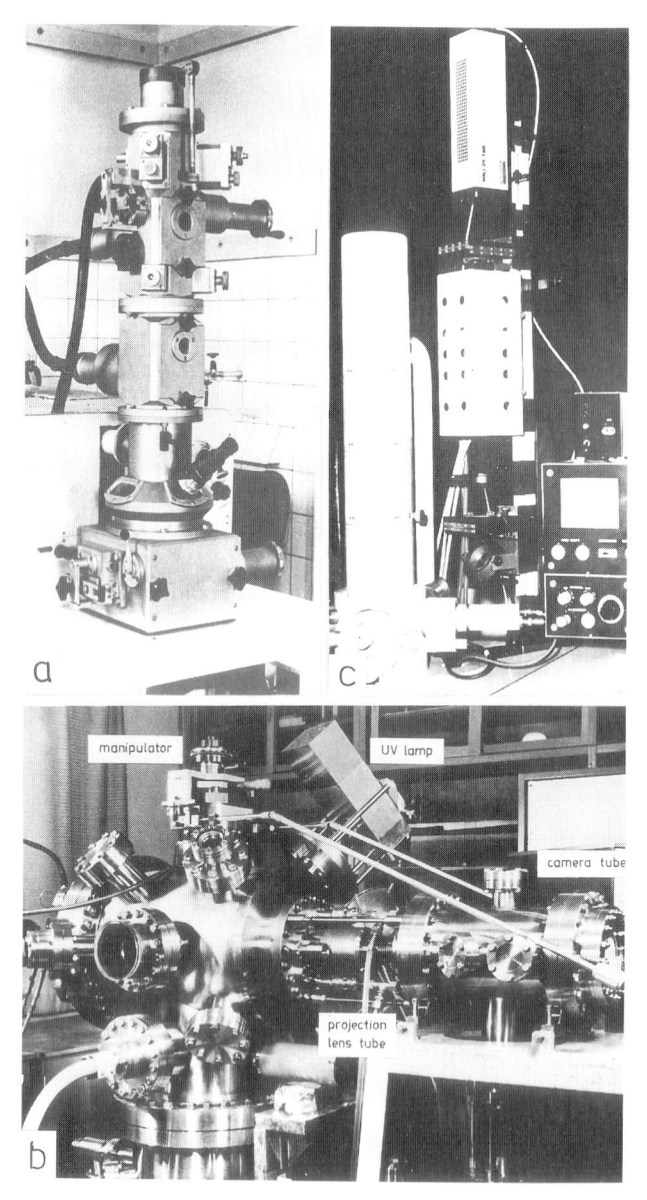

FIGURE 2. Home-made surface electron microscopes: (a) emission electron microscope (A. Recknagel/Dresden); (b) UHV photoemission electron microscope (H. Bethge/Halle); (c) completion of a commercial SEM with a laser source for specimen treatment (Wetzig/Dresden).

SDLTS work (scanning deep-level transient spectroscopy) carried ahead by Breitenstein.

In the GDR, during the past four decades the application of electron microscopy to biomedical research comprised investigations in the fields of cell biology and molecular biology. As an example, F. Jung's research work on the morphology of erythrocytes should be mentioned. Later, the immunoelectron microscopical detection of bioactive substances and the practical field of medical diagnostics gained increasing importance. These fundamental topics were treated mainly in electron microscopical centers established at institutes of the Academy of Sciences of the GDR in Berlin-Buch and in Jena and at universities in Berlin (Humboldt University), in Jena (Friedrich-Schiller University), in Leipzig (Karl Marx University), and in Halle (Martin Luther University). Figure 3 presents some selected examples of the GDR work done in biology and medicine. The process of virus-induced cancerogenesis is demonstrated in Fig. 3a for a virus of the myeloic leukemia of the mouse (Graffi, Bierwolf/Berlin). Figure 3b shows a micrograph of the glomerulus of the human kidney (David/Berlin). As an example of the ultrahistochemical detection of thiamine pyrophosphatase, Fig. 3c shows the Golgi apparatus with positive reaction products in the mouse nucleus supraopticus (J. Weiss *et al.*/Leipzig). Figure 3d is the SEM image of a spore cluster of yellow rust on barley (H. B. Schmidt/Aschersleben). Furthermore, regarding biomedical electron microscopy in the GDR, investigations were carried out on the cytoskeleton (M. Girbardt, E. Unger/Jena), on the nervous system of invertebrates (J. Ude, H. Penzlin/Jena) and vertebrates (H. Müller, G. Sterba/Leipzig), on the ultrastructural localization of enzymes of the second messenger system (W. Schulze, H. Wollenberger/Berlin, G. Poeggel, H. Luppa/Leipzig), and in diagnostic pathology (D. Stiller/Halle). The wide application of the freeze-etching technique to cell biological research (W. Meyer/Jena) should also be mentioned.

With respect to the application of electron microscopy to nonbiological fields, first should be cited Bethge's studies of the real structure of crystalline materials and of characterizing thin films carried out as early as the late 1950s (using home-made transmission electron microscopes at that time), by applying the surface decoration technique (Fig. 4a). At about the same time, D. Schulze in Dresden started microstructural investigations of metals and alloys (powder metallurgical products, magnetic and electrotechnical materials).

Electron microscope investigations in the field of materials research were carried out in many laboratories in the GDR, studying metals, alloys, semiconductors, oxides (ceramics), minerals, polymers, and glasses, always related to bulk materials as well as to surfaces, thin layers, and interfaces.

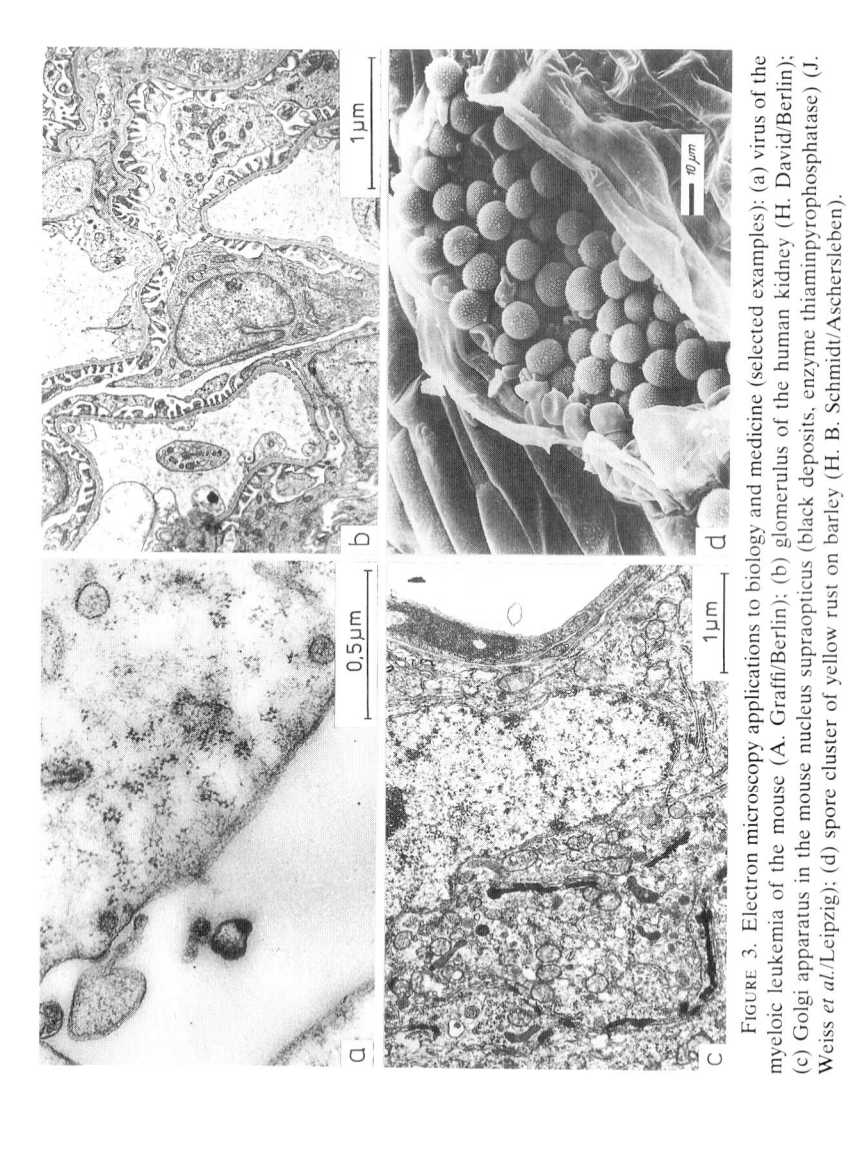

FIGURE 3. Electron microscopy applications to biology and medicine (selected examples): (a) virus of the myeloic leukemia of the mouse (A. Graffi/Berlin); (b) glomerulus of the human kidney (H. David/Berlin); (c) Golgi apparatus in the mouse nucleus supraopticus (black deposits, enzyme thiaminpyrophosphatase) (J. Weiss et al./Leipzig); (d) spore cluster of yellow rust on barley (H. B. Schmidt/Aschersleben).

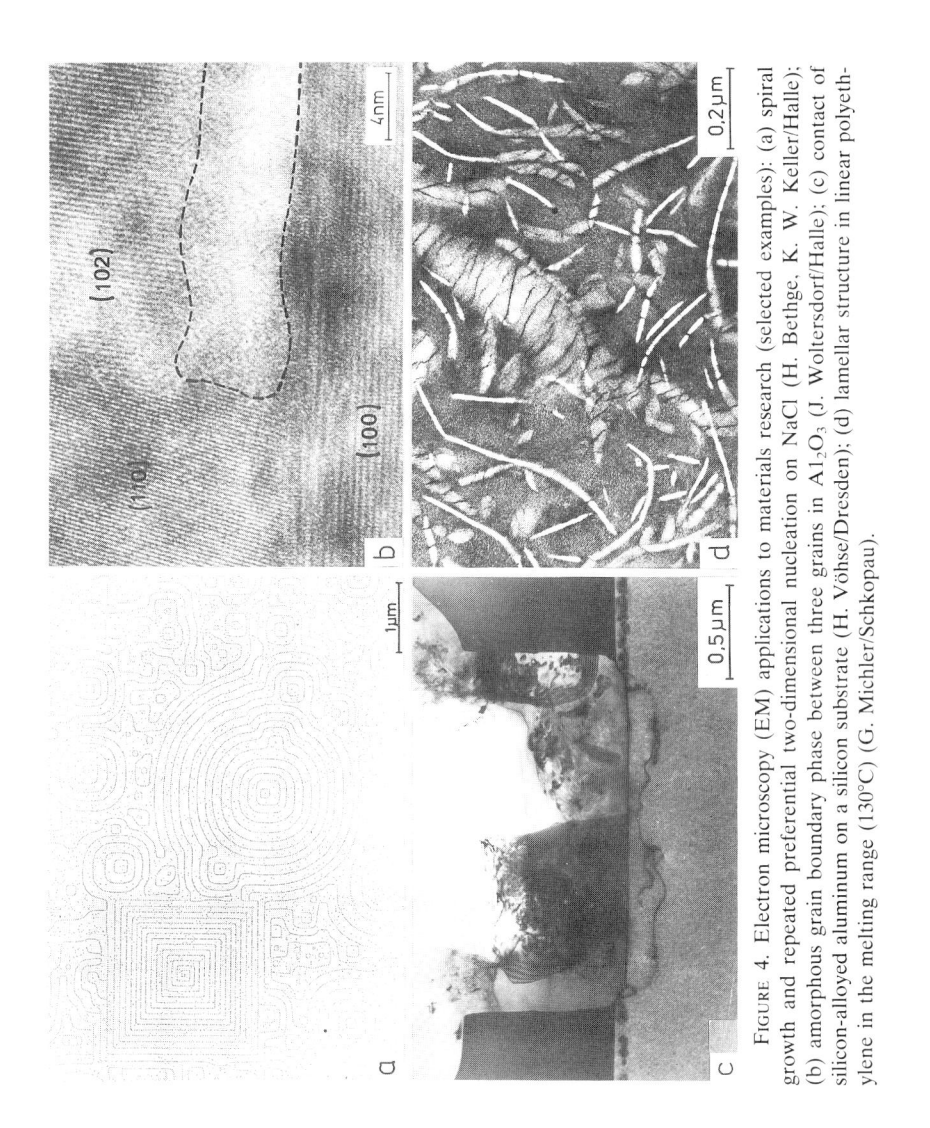

FIGURE 4. Electron microscopy (EM) applications to materials research (selected examples): (a) spiral growth and repeated preferential two-dimensional nucleation on NaCl (H. Bethge, K. W. Keller/Halle); (b) amorphous grain boundary phase between three grains in Al$_2$O$_3$ (J. Woltersdorf/Halle); (c) contact of silicon-alloyed aluminum on a silicon substrate (H. Vöhse/Dresden); (d) lamellar structure in linear polyethylene in the melting range (130°C) (G. Michler/Schkopau).

While three decades ago the replica technique was commonly applied, later diffraction contrast electron microscopy was increasingly used, including high-voltage electron microscopy (in Halle), especially for *in situ* work. Since the beginning of the 1970s, high-resolution electron microscopy and analytical electron microscopy have also been used. Examples of the application of electron microscopy to materials research in the GDR are given for the field of ceramics (Fig. 4b: Al_2O_3, Woltersdorf/Halle), of semiconductors (Fig. 4c: contact of silicon-alloyed aluminum on silicon substrate, H. Vöhse/Dresden), and of polymers (Fig. 3d: linear polyethylene, G. Michler/Schkopau).

As to the literature published, fundamental books on electron optics and electron microscopy, by Picht (*Einführung in die Theorie der Elektronenoptik,* 1939), von Ardenne (*Elektronen-Übermikroskopie,* 1940), and Brüche and Recknagel (*Elektronenstrahlgeräte,* 1941) are today still highly esteemed in the former GDR. Table I gives a survey of books published in the GDR in the past three decades that are related to electron microscopy in general, and to its applications to materials science, biology, and medicine.

Besides the research work carried out in different university laboratories of the GDR, at the Academy of Sciences of the GDR, in industry and other research centers, scientific activities have been organized by the Section of Electron Microscopy of the Physical Society (PS) of the GDR for nonbiological fields, and (since 1970) by the Society of Topochemistry and Electron Microscopy (GTE) for biological fields. Both societies organize regular national conferences on electron microscopy, related to progress in the methods (electron optics, instrumentation, image formation, image processing, preparation technique) and to applications of electron microscopy to materials science, biology, and medicine. According to Table II, since the first meeting in 1959, twelve national conferences on electron microscopy have been held in the GDR, always with invited electron microscopists from abroad taking part.

Within the Electron Microscopy Section of the Physical Society of the GDR there are divisions for scanning electron microscopy and transmission electron microscopy. In addition, there is a large group engaged in microprobe techniques that have organized microprobe conferences since 1971; these were always held after the regular conferences on electron microscopy. Thus good contacts were provided between electron microscopists and specialists in the field of microprobe techniques (not only related to electron microprobes). Before the foundation of the Society of Topochemistry and Electron Microscopy, the electron microscope research in biology and medicine was coordinated by an interdisciplinary working group for morphology, and partly by the Physical Society of the GDR.

TABLE I

MONOGRAPHS OF GDR AUTHORS

Authors	Title	Published
M. von Ardenne	Tabellen zur Elektronenphysik, Ionenphysik und Übermikroskopie	Berlin, 1956
H. Bartsch et al.	Elektronenmikroskopische Querschnittsabbildung von Interfaces und Heterostrukturen in Halbleitern	Berlin, 1987[a]
H.-D. Bauer	Analytische Transmissions-Elektronenmikroskopie	Berlin, 1986[a]
H. Bethge J. Heydenreich (Eds.)	Elektronenmikroskopie in der Festkörperphysik	Berlin, 1982
L. Cossel	Die menschliche Leber im Elektronenmikroskop	Jena, 1964
L. Cossel H. Bräuer	Excerpta Hepatologica—Bildatlas zur Physiologie und Pathophysiologie der Leber, Bd. 1	München, 1980
H. David	Submikroskopische Ortho- und Pathomorphologie der Leber (Textbuch und Atlas)	Berlin, 1964
H. David	Elektronenmikroskopische Organpathologie	Berlin, 1967
G. Geyer	Ultrahistochemie: Histochemische Arbeitsvorschriften für die Elektronenmikroskopie	Jena, 1969
R. Hillebrand et al.	Bildinterpretation in der Hochauflösungs Elektronenmikroskopie	Berlin, 1984[a]
H. Müller	Präparation von physikalisch-technischen Objekten für die elektronenmikroskopische Untersuchung	Leipzig, 1962
K.-G. Nestler	Einführung in die Elektronenmetallographie: Eisen und Stahl	Leipzig, 1961
J. Picht	Einführung in die Theorie der Elektronenoptik	Leipzig, 1939, 1957, and 1963
J. Picht J. Heydenreich	Einführung in die Elektronenmikroskopie	Berlin, 1966
H. G. Schneider J. Woltersdorf (Eds.)	Strukturen kristalliner Phasengrenzen—Elektronenmikroskopischer Bildkontrast	Leipzig, 1977
J. Ude M. Koch	Die Zelle—Atlas der Ultrastruktur	Jena, 1982
K. Zapf J. Ludvik	Einführung in die elektronenmikroskopische Präparationstechnik der Mikrobiologie	Jena, 1961

[a] Series: Beiträge zur Forschungstechnologie (D. Schulze et al., Eds.).

International symposia on "In Situ High-Voltage Electron Microscopy—Application to Plasticity and Further Topics of Materials Research" and on "Electron Microscopy in Plasticity and Fracture Research of Materials" were held in Halle (1979) and in Holzhau (1989), respectively, both orga-

TABLE II

NATIONAL CONFERENCES ON ELECTRON MICROSCOPY IN THE GDR

No.	Year	Place	Organizer[a]
1	1959	Halle	PS
2	1962	Dresden	PS
3	1964	Jena	PS
4	1966	Erfurt	PS
5	1969	Dresden	PS
6	1971	Berlin	PS
7	1973	Berlin	PS, GTE
8	1975	Berlin	GTE, PS
9	1978	Dresden	PS, GTE
10	1981	Leipzig	GTE, PS
11	1984	Dresden	PS, GTE
12	1988	Dresden	GTE, PS

[a] PS, Physical Society; GTE, Society for Topochemistry and Electron Microscopy.

nized by the Institute of Solid State Physics and Electron Microscopy (Halle/Saale) of the Academy of Sciences of the GDR. As early as in 1975, this institute also established an International Center of Electron Microscopy of the Academies of Sciences of the Socialist Countries. The program of its main activities comprises the organization of courses (twice a year) on different topics of EM research (spring schools, autumn schools) and the realization of individual training courses for different methods and topics in the field of electron microscopy.

The National Commission of Electron Microscopy of the GDR, made up of physicists as well as biologists, has been a member of the IFSEM since 1968 and of the CESEM since 1982. H. Bethge was a member of the Executive Committee of the IFSEM from 1978 to 1986.

2.7
The Hungarian Group for Electron Microscopy

SZABOLCS VIRÁGH

*Postgraduate Medical School
Budapest 1389, Hungary*

ÁGNES CSANÁDY

*Hungalu Engineering and Development Centre
Budapest 1389, Hungary*

The Hungarian Group for Electron Microscopy was founded in 1955 and has over 200 members. The nucleus of this group came into being in 1950, when the Hungarian Academy of Sciences purchased Hungary's first electron microscope, a Trueb-Taueber instrument manufactured in Switzerland. The director of the laboratory was M. Gerendás. Soon after this, the laboratory acquired two additional instruments, and during the 1950s the Faculty of Technology, as well as faculties of medicine in Hungarian Universities, were able to set up centralized facilities for electron microscopy. Although the number of instruments grew only slowly in the 1950s and 1960s, Hungarian scientists tried to ensure that they were represented at international meetings in electron microscopy. They were certainly present at the foundation conference of the International Federation of Societies for Electron Microscopy (IFSEM) in 1954. The official Hungarian representatives, Irén Sugár and Ferenc Guba, can be picked out in Fig. 1 at the coordinates N-12.5 and N-14.5, respectively. Figure 1 is a group photograph of the well-remembered Conference Banquet in Soho Square, London. This was a truly international conference, in which the feeling of international cooperation in scientific work completely overshadowed the understandable negative feelings of international animosity engendered by the horrors of World War II.

I. HUNGARIAN SCIENTISTS AND ELECTRON MICROSCOPY

International acceptance of Hungarian scientists in the postwar years was also greatly facilitated by the significant number of Hungarian scientists working outside the country. Two names should perhaps be singled out in connection with electron microscopy. The first is Dennis Gabor, whose

FIGURE 1. Group photographs of the Banquet in Soho Square, at the International Conference on Electron Microscopy in London (1954). Poster-size enlargements were prominently displayed at the 30th Anniversary IFSEM Meeting (1984) in Budapest.

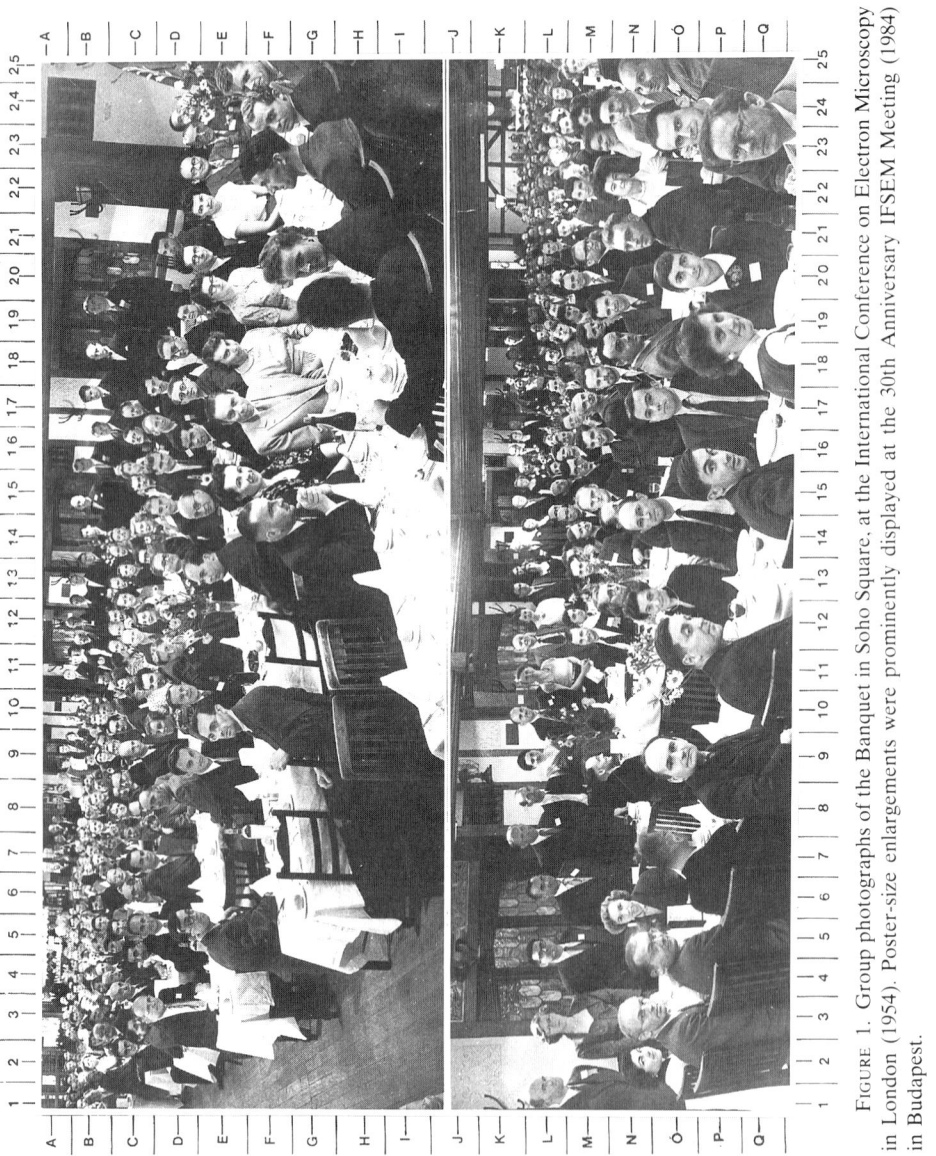

contribution of stimulating ideas and inventions in electron optics and electron microscopy kept electron microscope designers constantly on their toes for many years. His "crazy" idea of electron beam holography, which came to him suddenly and unexpectedly while working in the United Kingdom, was put forward by him as a means of correcting spherical aberration in the transmission electron microscope. However, it involved electron beam technology that would take a further 40 years or more to develop commercially. Unfortunately, he did not live to see his ideas brought to a successful conclusion in the transmission electron microscope (TEM), which has only recently achieved the goal of aberration-free atomic imaging in practice.

Nevertheless, the unexpected (by Gabor himself) applications in light optics earned him a Nobel Prize. The second ex-patriate to notice is perhaps Ladislas Marton, known outside Hungary as "Bill" Marton, whose pioneer work in the early days, first in Belgium and later in the United States, should be mentioned here. "Bill" was a congenital optimist, never put off by the enormous practical difficulty of preparing biological specimens for the TEM in the absence of an ultramicrotome. He made full use of his infectious Hungarian sense of humor to encourage his colleagues when faced with any difficulty, financial or political. On the physics side of TEM, he developed the theory of electron scattering and its role in image formation and contrast in the electron microscope. In the area of biological applications, he developed some of the early innovative preparation techniques for electron microscopy, notably the use of osmium tetroxide for specimen fixation and enhancement of contrast. He also pioneered the use of camera airlocks and internal photography in the TEM, so as to reduce irradiation damage to the specimen. His were probably the first electron micrographs that had genuine biological significance. Marton emigrated from Belgium to the United States at the outset of World War II and worked on electron microscopy for the Radio Corporation of America (RCA). He designed and built the RCA model A TEM, a two-stage instrument. At about the same time, J. Hillier built his first TEM in Canada. RCA had to choose between the two designs and in the end chose the Hillier design as being easier for mass production. Marton then moved on to Stanford University, where he designed an innovative multilens TEM which was probably too far ahead of its time for commercial production.

There are many other Hungarian scientists who have contributed to the field, mainly working in other countries. One could mention the name Szent-Györgyi (Albert and Andrew), along with Andrew Somlyó and John Gergely. Their distinguished contributions have facilitated Hungary's scientific communication on the international scene.

II. The Development of Electron Microscopy in Hungary

Due to postwar political, organizational, and especially financial difficulties, the development of electron microscopy in Hungary in materials science was slower than in the biological and medical fields. Nevertheless, during the early 1970s, these applications became more adequately represented in the overall scientific effort, thus allowing Hungarian scientists to contribute internationally in a broad range of applications of electron microscopy.

F. Guba served as the first president of the Hungarian Society of Electron Microscopy (1955–1979), followed by Sz. Virágh (see Fig. 3, below). Over this period, the secretaries were Magda Mészáros (1969–1969), G. Groma (1970–1979), and Ágnes Csanády (see Fig. 3.). The Hungarian Society organizes two-yearly conferences, the first of which was in 1959 and the fourteenth in 1987. As noted above, Hungarian scientists have always been anxious to develop and maintain contacts with their colleagues in other countries and have tried to involve as many of them as possible in their conferences. Initially, these contacts came from neighboring countries, but this gradually expanded geographically. In 1975, the international aspect assumed great importance, so it was decided that the use of English as the conference language would be both practical and beneficial to all concerned. Ten years later, in 1985, Hungarian and Austrian scientists joined forces to organize the first Hungarian-Austrian conference in Hungary; the second such conference took place in Austria in 1987. Scientists in other neighboring countries have recognized the value of these conferences and have expressed the wish that their scope be extended to include other countries of the Danube Basin.

III. The 8th European Congress on Electron Microscopy in Budapest (EUREM 1984)

The Hungarian Society for Electron Microscopy welcomed with enthusiasm the chance to host this congress. The cover photograph of the Congress Booklet, reproduced in Fig. 2, shows the Danube Bridge spanning the East and West Banks, a symbol of the corresponding scientific links being forged at the congress. Figure 3 shows the scene at the Opening Ceremony. Professor J. Szentágothai, President of the Hungarian Academy of Sciences, is welcoming the participants. Seated along the platform are (left) A. Csanády, Congress Secretary, Sz. Virágh, Congress President, and on the right, H. Hashimoto, IFSEM Chairman. Although this congress was planned as a European congress, participants turned up, in fact, from all over the world,

FIGURE 2. Cover of the EUREM 8 Congress booklet. The Danube Bridge spans East and West.

East and West, and the atmosphere was more like that of an international congress. Many participants from the Eastern Bloc were able to get travel permits, and several came at great personal expense and difficulty.

H. Hashimoto, IFSEM President, is shown in Fig. 4, addressing the participants at the Opening Ceremony, reminding them that IFSEM was now well established and that the year 1984 was in fact its thirtieth anniversary. Poster-sized enlargements had been made by the Hungarian Society of the group photographs taken in 1954 at Soho Square, London (see Fig. 1) and were prominently displayed at the congress. The scientific success of this meeting owed much to the Hungarian Academy of Sciences, in particular to its President, Prof. Szentágothai, who took part personally in all the congress activities. A special feature of the congress was that several distinguished pioneers had been invited to give summarizing talks of a historical nature illustrating the development of the IFSEM and its various

FIGURE 3. Prof. J. Szentágothai, President of the Hungarian Academy of Sciences, greets participants at the Opening Ceremony of EUREM 8. Left to right, Á. Csanády, Congress Secretary, Sz. Virágh, Conference Chairman, H. Hashimoto, IFSEM President.

subject areas. Among these speakers were M. von Ardenne, shown in Fig. 5, at the lectern, talking about the early beginnings of SEM and STEM. In Fig. 6, V. E. Cosslett is shown reviewing the steady progress of the IFSEM since its beginnings in 1954. In Fig. 7, G. Möllenstedt is seen in a characteristic pose, making a point about the early days of convergent beam diffraction and electron microscopy. Finally, H. Hashimoto, already shown in Fig. 4, reviewed the difficult immediate postwar period in Japan, which had already been inspired by the early pioneer work on electron diffraction associated with the names of Kikuchi and others. He recounted the enormous scientific impact in Japan in 1942, when von Ardenne's book on electron microscopy was brought to Japan by a wartime German submarine. It was immediately translated into Japanese, and this provided the impulse that launched Japan headlong into the world of electron microscopy and ensured its eventual world preeminence as a supplier of electron microscopes. These fascinating talks were not in fact published in the Congress Proceedings, which were already overloaded with the contributions from congress participants. However, updated versions of the talks given by von Ardenne and by Hashimoto are to be found elsewhere in this volume. The posthumously published article written by V. E. Cosslett and placed at the beginning of this volume also follows closely his talk given at this congress.

FIGURE 4. H. Hashimoto speaking on behalf of the IFSEM at the Opening Ceremony.

FIGURE 5. M. von Ardenne giving his invited talk on the early days of SEM and TEM.

FIGURE 6. V. E. Cosslett reviewing 30 years of development of the IFSEM.

FIGURE 7. G. Möllenstedt recalling the early development of convergent beam electron diffraction and electron microscopes.

THE HUNGARIAN GROUP FOR ELECTRON MICROSCOPY 189

FIGURE 8. Congress President Sz. Virágh (left) thanks retiring President E. Kellenberger and welcomes new President A. Robards (right).

EUREM8 was of course a European congress, and the photograph shown in Fig. 8, taken at the EUREM Cultural Event, shows Congress President Sz. Virágh (left) expressing his thanks to the retiring European President E. Kellenberger (center). He then turned to welcome the new president, A. Robards (right), who instantly invited all participants to take part in the 9th EUREM Congress to be held in York in 1988.

In retrospect, the Budapest meeting was a most friendly and enjoyable occasion, with plenty of room in the building for informal discussion between electron microscopists who, in many cases, had not met in person for some years. There was also an excellent social program prepared by the very active Organizing Committee. The Conference Proceedings themselves contained 952 papers, totaling 2400 pages spread over three volumes. At the plenary session, the co-chairmen were Gareth Thomas (Fig. 9), currently Secretary of the IFSEM, and Arvid Maunsbach (Fig. 10), currently Secretary of the CESEM.

IV. FUTURE DEVELOPMENTS

The Hungarian Society for Electron Microscopy continues to promote the materials science and the life science aspects of electron microscopy. To

FIGURE 9. Gareth Thomas, IFSEM Secretary, co-chairs the Plenary Session of EUREM 8.

FIGURE 10. Arvid Maunsbach, CSEM Secretary, co-chairs the Plenary Session of EUREM 8.

this end, many courses and workshops have been organized over the years in both fields. These have included topics such as X-ray microanalysis, scanning electron microscopy, image analysis, etc. In 1987 a one-year training course for the training of laboratory electron microscope technicians was inaugurated, and a relevant training manual was published.

2.8
Electron Microscopy in Italy[1]

UGO VALDRÈ

Centre for Electron Microscopy
Department of Physics and INFM, GNSM–CNR, University of Bologna
I-40126 Bologna, Italy

I. Introduction

Electron Microscopy (EM) in Italy originated both in Rome at the Istituto Superiore di Sanità (ISS) under the Ministry of Health and in Bologna at the EM Center of the Physics Department under the Ministry of Public Education (now MURST, Ministry for the Universities and for Scientific and Technical Research). Some of the early users of EM worked in industrial laboratories but did not spread the EM gospel or take part in development of EM outside their own in-house applications.

Hence, for the very early days of EM in Italy, one has to look at the activity of the ISS in Rome for (mainly) biomedical applications and to the EM Center of Bologna for (mainly) physics/metallurgy work. The history presented here refers essentially to the activity of the first University Center of Electron Microscopy (EM Center), which was created in the Department (formerly Institute) of Physics of Bologna.

A. The EM Laboratory of the Istituto Superiore di Sanità in Rome

The very first Italian Laboratory which acquired an electron microscope was the High Institute of Health, Rome. A Siemens & Halske Übermikroskop (ÜM) was installed in the spring of 1943 (the order had been placed in 1939), during World War II, at the Physics Laboratory of ISS [1–4]. The Head of the Laboratory was Prof. Giulio Cesare Trabacchi (the same person who had previously built in his laboratory the ion accelerator designed by Enrico Fermi for nuclear physics research). On 8 October 1943, before the Germans retreated from Rome, the column of the microscope was taken back to the Siemens factory in Berlin by the German troops, since its value

---

[1] With particular reference to the Electron Microscopy Laboratory of the Department of Physics in the University of Bologna, Italy.

was estimated as equivalent to the second installment payment, which had not yet been paid by ISS to Siemens (the selling agreement envisaged a trial period before full payment would be made, which had not yet expired). It is believed that the instrument would have been taken back to Germany regardless of the administration reason given above, since at that time electron microscopes were considered strategic instruments. The microscope was supposed to be removed immediately after the opening of the sealed letter handed to Trabacchi, containing the order of "confiscation in order to be put in a safe place" [2]. However, Trabacchi managed to convince the Germans to postpone the removal until the next day; during that night, drawings were made of the most important parts of the microscope. These drawings were to be the basis for the design and construction of an entirely new and improved version of the Siemens ÜM.

The construction took place at the Physics Laboratory of the ISS under the supervision of Trabacchi in a difficult period of power cuts and of material shortages (the photographic chamber was made of a melted-down bronze bust of the defeated dictator Mussolini). Half of the instrument was ready by June 1944, but owing to the random electric power cuts, the complete microscope was tested only at the end of 1946 [2]. Figure 1 shows the ISS electron microscope; its resolution was about 5 nm.

B. The EM Laboratory of the Istituto di Fisica at the University of Bologna

The role of the EM Center was initially that of a regional (Emilia-Romagna Region) facility for the entire scientific community, with the statutory duty of training electron microscopists and of disseminating basic EM techniques.

From 1959 on, the EM Center became a Laboratory with its own research program; nevertheless, even then, it continued its original role as a reference point in Italy for the development of, and training in, new techniques and methodologies, for the dissemination of EM culture, and for services in several fields of science. Consequently, apart from the very early stages, the ongoing activity in the biomedical field is not covered in this account.

Internationally, the EM Center has been engaged in collaborations with various universities and industries (the latter particularly in England and West Germany), and the outcomes of its activities have been beneficial to various fields of research and applications (from solid-state physics to biology, from instrumentation to metallurgy and the science of materials).

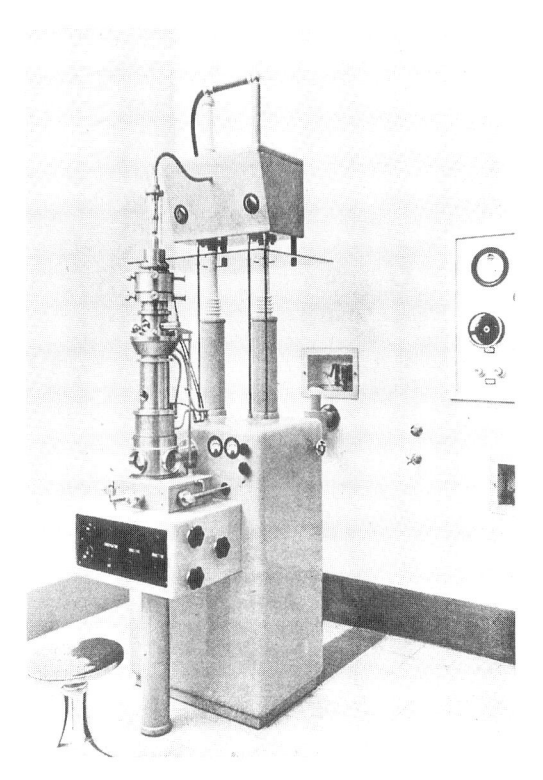

FIGURE 1. The transmission electron microscope built at the Istituto Superiore di Sanità, Rome, during the period 1944–1946.

II. Early History (1947–1959)

A. The Founding of the EM Center and the Acquisition of the First Electron Microscope

The University of Bologna, which is the oldest Western university (Alma Mater Studiorum), was the first Italian university to organize an electron microscopy laboratory. This took place in 1947 at the Istituto di Fisica (now the Department of Physics). The laboratory was the result of the efforts of the newly appointed Professor of Physics and Head of the Institute, Giorgio Valle. His research interest was actually in the field of electrical discharges in gases but, having completed his doctorate at the Vienna Polytechnic (he became an Italian citizen after the World War I), he had strong connections with the German scientific community. In particular, he was acquainted with the work of Ernst Ruska. Evidence for this is given in a review article on electron lenses and on the new electron microscope that Valle wrote as early as 1933 [5].

The laboratory was even initially an ambitious project, which brought about an extension of the institute building with the construction of a second floor over the rear of the central block at via Irnerio 46, to provide an area of about 400 m^2. The work was supported financially by the Italian Ministry of Public Education (MPI). Funds for acquiring the microscope and a few accessories were collected from public and private institutions (universities, ministries, the Emilia-Romagna County Council, various town councils, industry, the Lions Club, etc.). For budgetary reasons, and also because of a certain eagerness to install an instrument as soon as possible, an electrostatic-type microscope was ordered (CSF model M IV, manufactured by the French firm Compagnie Générale de Telegraphie sans Fils). It was not a good choice, and the limitations of the microscope (low accelerating voltage, no condenser lenses, unreliable alignment, poor resolution, inadequate X-ray screening, etc.) were immediately apparent; this greatly hampered its practical use. It is interesting to note that Trabacchi had advised against this choice, suggested other machines, and even contacted an Italian firm which gave a quotation for an instrument to be built on the lines of the ISS microscope. Italy might thereby have missed the opportunity to have its own electron optics industry!

Although the EM Laboratory was intended to be a regional facility, it served in practice, in its first years at least, a much larger geographic area, covering most of Italy. It essentially attracted medical doctors and biologists.

After the great excitement of creating the EM Laboratory in Bologna, the subsequent poor performance of the microscope and the lack of funds placed the future of the microscope in doubt. With the death of Valle (9 December 1953), a heavy smoker, by cancer of the throat, a young nuclear physicist, Giampietro Puppi, became Director of the Department. He decided that the EM Laboratory activity should continue. There was a transition period during which a procession of various people were put in charge of the EM Laboratory, with part-time help from a technical assistant. However they all became, in turn, frustrated by the unreliability of the CSF microscope. At the end of December 1953, the person in charge gave Puppi advance notice that she had decided to leave. Puppi then asked Ugo Valdrè if he were willing to become responsible for the maintenance of the CSF and for the service to external users. The part-time help of the technical assistant would be maintained.

Valdrè was at that time a research student, just finishing writing his thesis on extensive air showers; he accepted, and from January 1954 was paid as a "freelance" for a couple of years. Incidentally, the study of cosmic rays was the traditional field of research in high-energy physics in Italy before

CERN (Centre Europeanne de Rechèrches Nucleaire) in Geneva (CH) was established, since the primary source of high-energy particles was available at zero cost! Puppi's decision and the ensuing lecturership of Valdrè created the basis for the continuation, development, and consolidation of the electron microscopy activity within the Department of Physics up to the present day.

By the second half of 1955, as a result of Puppi's own research interest, funding from the National Institute of Nuclear Physics and MPI, and expected funding from the Bologna Town Council (see Sec. II. B), the work in nuclear and particles physics in the department had grown so much that the newly built EM area was greatly coveted by the high-energy particle physicists as a possible extension of their empire. As a result the EM Laboratory was transferred to the ground floor of the building, its present location, but with, what was at that time a large loss of floor space. The move took place over a period of about nine months.

B. *The Ultramicrotomy Service (1956–1962) and the Construction of a Field Emission Electron Microscope*

Under the stimulus of a politically active member of the Istituto di Fisica of Bologna, Dr. Protogene Veronesi, a member of the Communist Party, later an Italian MP and then a European MP, the Bologna Town Council committed on 28 February 1956 to sponsoring research in nuclear physics at the Istituto di Fisica with an annual budget of some 50 million lire for 10 years [6]. It was a very large sum at that time, much greater than any other Ministerial contribution. Puppi must have felt a moral duty to help less well funded laboratories and/or less fortunate university colleagues.

In that spirit, in order to help biologists and physicians, an ultramicrotome (LKB, Sjoestrand model), in addition to other beneficial initiatives, was ordered in 1955 and installed in the EM Laboratory in June 1956 for the benefit of external users. This happened just a couple of years after the Sjoestrand model was first produced commercially, in the period 1953–1954. Two technical assistants were recruited to service the CSF instrument and, more specifically, to section embedded specimens with the ultramicrotome; they were paid from Town Council funds. Service activities on metal shadowing and SiO and carbon coating were also performed and, sometimes, embedding service was provided. The actual observation of the specimens was usually done elsewhere. Because time and people were now available, the EM Laboratory started its own research activity. Up to that time, in fact, no independent research activity had been undertaken, and the laboratory worked only in a service capacity.

The first two research projects were (1), a systematic investigation into the preparation and performance of glass knives used for cutting thin sections of biological material in the ultramicrotome [7]; and (2) the construction of a Mueller-type field emission electron microscope (Fig. 2).

C. The Founding of the Italian Society of Electron Microscopy

The Società Italiana di Microscopia Elettronica (SIME) was founded in 1956 by 11 Italian biologists/physicians who attended the 1st European Congress of Electron Microscopy, held in Stockholm in the same year. G. C. Trabacchi, Director of the Physics Laboratory of the ISS, was elected President and remained in this post until his death in 1959. The first meeting of SIME took place at the ISS in Rome on 7 May 1957; it was a one-day meeting. The Bologna EM Laboratory presented two papers based on the work mentioned above; these were the only nonbiological papers presented at the meeting.

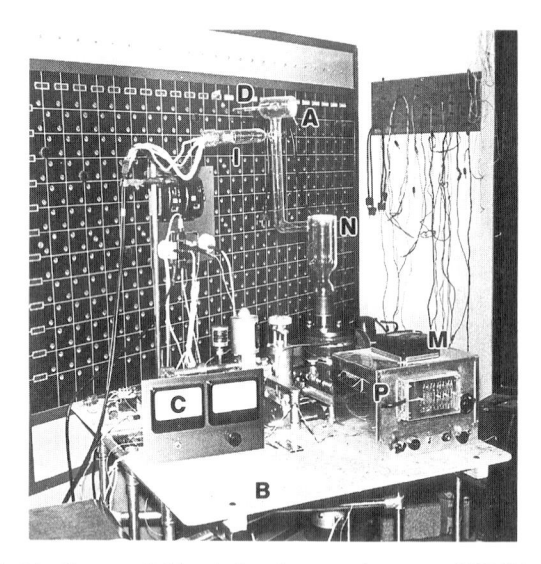

FIGURE 2. The Mueller-type field emission electron microscope (FEM) built at the Center for Electron Microscopy of Bologna in 1956. The all-glass construction is typical of the ultrahigh vacuum equipment of that time. The instrument was located in the DC voltage supply room on the ground floor of the Istituto di Fisica. The DC distribution panel at the back of the FEM has nothing to do with the FEM. Underneath the bench (B) on which the FEM is mounted is the pumping system. On the bench is the high-voltage power supply, P (adapted from a TV set), pressure gauges (home-made thermocouples, C), an ion gauge (I), a liquid nitrogen trap (N), and the field emission current meter (M). The FEM itself is at the top and contains the demountable emitting filament tip (D), and the combined fluorescent screen and anode (A).

Originally SIME had 68 members, of which about 44 came from Milan and Rome. SIME currently has around 700 members, 70% working in biomedical fields. The members are distributed among universities (58%), public research laboratories (21%), industry (18%), and hospitals (3%). The actual number of users of electron microscopes is estimated at 3000. In Italy there are now about 1000 electron microscopes, are almost equally divided between TEMs and SEMs [8]. Currently, sales favor SEM over TEM in the ratio 5:1, most of the SEMs being equipped with attachments.

D. The Ispra Center

In the late 1950s the Italian Atomic Energy Authority (Comitato Nazionale per l'Energia Nucleare, CNEN), now Ente Nazionale Energie Alternative (ENEA), i.e., National Authority for Alternative Energies, created a research center at the village of Ispra, in the province of Varese, North Italy, near lake Maggiore. The center was devoted to the study and exploitation of nuclear energy and included an Electron Microscopy Laboratory in its Department of Materials. The laboratory was equipped with one electron microprobe, one ion microprobe (Cameca), and one TEM; the microprobes were both industrial prototypes based on Castaing's design. When the EEC treaty was signed (1960), the CNEN Center of Ispra became one of the (Euratom) research centers of the European Community. The EM Laboratory was run by Dr. E. Ruedl, an Italian citizen, who had graduated in Chemistry at the University of Innsbruck, and had spent a couple of years in Belgium for training in the electron microscopy of metals in the group of Professor S. Amelinckx. Until the early 1980s, when nuclear power came under strong criticism, the Ispra Center enjoyed good funding, as demonstrated by the replacement of transmission electron microscopes every 6–7 years. High-quality work was performed at the Ispra EM Laboratory, although restricted to the study of materials for use in nuclear reactors. Particular reference should be made to the results obtained on sintered aluminum powder (SAP) as fuel cladding material, and on void and bubble formation in steel under irradiation at various temperatures by several kinds of elementary particles; notable were investigations on chemical and precipitation reactions and on corrosion studies in various types of steels which are believed to be best suited for the walls of future fusion reactors [cf.9 for references].

E. The Formation of the Solid-State Physics Group in Bologna: The Siemens Elmiskop I

As part of Professor Puppi's program to open, in Bologna, research activity in solid-state physics (SSP) [6,10], a new microscope, the Siemens Elmiskop

I, serial number 277, was ordered in 1958; it was commissioned in July 1959 at the EM Laboratory. The Structure of Matter Group was formed in the same year, stemming from and with the support of the Electron Microscopy Laboratory. The Elmiskop I model proved very reliable, robust, and successful. This microscope is still in use at the present time (January 1996), together with two other Elmiskop I's, which were acquired second-hand in the late 1970s. Their serial numbers were 142 and 474. Serial number 142 must have been one of the first few made, i.e., in the mid-1950s. Number 142 was extensively modified by us for use in low-temperature work; the model 474, donated to the EM Laboratory by the EEC Establishment of Ispra, is used mainly for teaching purposes. Several other Elmiskops, given to the EM Laboratory by various Institutions (Cavendish Laboratory, Cambridge, Metallurgy at Oxford University, and ISS) are now used for spare parts.

The SSP group planned, among other activities, to apply transmission electron microscopy to problems of solid-state physics. To that end it was necessary to become acquainted with the thinning of metal foils and the interpretation of the micrographs, as well as in performing *in situ* experiments. A cooling and straining stage with (single-axis) tilting capabilities was then required. Very few types of stages were offered at that time by the manufacturers, and each stage could perform only one function (e.g., either stereo-tilting, or cooling, or straining). The cooling and straining stage that was required was not available commercially. There were small firms offering services and accessories for biologists, but they did not offer the kind of stages that were required. Valdrè's initial training as a mechanical engineer led in the early 1960s to a decision to design a stage capable of performing the three combined functions of cooling, tilting, and straining the specimen. This stage, manufactured "in house," was constructed by using the main workshop facilities of the institute. Cooling was achieved by the use of liquid nitrogen, and straining and stereo-tilt were both obtained by means of the tilt controller. This successful project had important consequences in the later activity of the EM Laboratory, as described in Sec. III. A.

F. An Electron Microscopy Support Service for External Users

The electron microscopy support services of the EM Laboratory (already in place for specimen embedding, sectioning, coating, and shadowing, as reported in Sec. II. B) increased greatly with the arrival of the Siemens microscope; many users were attracted, mainly from the medical and biological institutions of the University of Bologna, for the observation of their own specimens. Customers from other universities and sometimes from industries of the north and northeast of Italy (e.g., Modena, Parma, Piacenza, Padua, and Ferrara) were, however, not uncommon. This led to several theses being written for university degrees and for further qualifications in medicine.

One of the most frequent external users of the time of the Elmiskop I was Renzo Laschi, a medical doctor in the Department of Human Pathology. He devoted himself entirely to electron microscopy. A few years later, in 1962, he founded the Bologna Center for the Development of Electron Microscopy, providing a service facility in Biomedicine [11]; the center had its own personnel and instrumentation. Later, Laschi succeeded in creating the Institute of Clinical Electron Microscopy (formally from 1 November 1972), the only university institution with this name, devoted entirely to diagnosis by electron microscopy of disease in patients [11]. It is believed to be one of the few such centers worldwide. The institute is located in the Sant'Orsola (University) Hospital and is one of its clinical laboratories; its personnel can provide electron micrographs, implemented normally with an interpretation of the sectioned material, within 24 hours from the extraction of a biopsy. Laschi died of cancer of the brain in 1989; his institute is continuing these services and research.

Another successful biomedical group which had its initial training at the EM Center operates in Modena at the Institute of General Pathology; the senior staff is formed by Profs. Paolo Buffa, Umberto Muscatello, and Ivonne Pasquali Ronchetti. The group still collaborates with the EM Center.

Attempts to run intensive collaborative projects with the biomedical faculties include a few cases that are worth mentioning.

1. A joint supervision between the Departments of Physics and Zoology led to results of some interest related to the structure on the Bidder organ of a particular frog (*Bufo bufo*) [12].

2. The development of the cryoelectron microscopy technique in collaboration with Dr. Michael Sjoestroem of the University of Umea (Sweden) [13]. M. Sjoestroem specialized in cryoultramicrotomy at Prof. W. Bernhard's laboratory in Paris; with his expertise, cryosections were cut at the Institute of General Pathology of the University of Modena and transferred to Bologna inside a liquid nitrogen flask . At Bologna a low-temperature manipulator and transfer device, and a cold specimen stage had been built, thanks to the experience gained in the study of specimens in the superconducting state (see Sec. III. C). Low-temperature work on biological specimens dates from as early as 1972 [14] and the project to observe frozen hydrated sections in an electron microscope was probably one of the first attempts in this area.

3. Another collaboration was set up with the John Innes Institute at Norwich (UK) for the study of viruses by low-temperature microscopy [14]. The local partner was Prof. Robert W. Horne.

4. Other collaborative work took place on the development of specimen stages for tomography with the Wadsworth Research Center of the New York State Department of Health in Albany, United States (local partner Dr. James N. Turner) [15], and work is currently being carried out with the Department of Biomedical Sciences of the University of Modena [12] (partner Prof. Umberto Muscatello) on the study of phospholipids (cardiolipin) by electron diffraction and microscopy and by scanning force microscopy [16].

G. Other EM Centers and Their Activities in Italy

In 1951 three RCA electron microscopes arrived in Italy from the United States as part of the Marshall Plan (ERP, 1948) for the reconstruction of Europe after the destruction of World War II. One of the three microscopes went to Milan Polytechnic, where a good support service tradition has been established in the general applications of electron microscopy. The other two instruments went (however strange it may appear) to chemistry departments, one in Naples and the other in Rome.

The interest of industry in electron microscopy started in northern Italy, in about 1953. Probably the first instrument to be installed was a Swiss model, made by the firm Trüb-Täuber. This microscope was acquired by the firm Montevecchio and went to a factory near Venice that produces zinc alloys; it was an unconventional, hybrid instrument. It had an electron gun of the gas-discharge type, a magnetic objective lens, and an electrostatic projector. The EM Laboratory of Montevecchio was run by Dr. Primo Gondi, who had obtained his degree in Physics in Bologna.

The Istituto Donegani and the Istituto Sperimentale Metalli Leggeri (Experimental Laboratory for the study of Light Metals, ISML), both based in Novara, were among the first industry-related research laboratories (both belonged to the firm Montecatini) to exploit electron microscopy on a regular basis from about 1959.

H. Developments and Achievements in the Biomedical Sciences

The regular use of the electron microscope as a tool to investigate biological and medical problems arose in Italy toward the end of the 1940s at the Istituto Superiore di Sanità in Rome. The very first studies were performed on microorganisms, either untreated or metal-shadowed. These early attempts by microbiologists are easily understandable since ISS, which acts as the Italian Central Health Institution, was at that time engaged largely in the control of infectious diseases.

Very soon, however, the new imaging method gained a much larger following in biology and medicine than in physics and metallurgy: aggregates of actin filaments were studied by Aloisi, Ascenzi, and Bonetti [17,18]; fragments from normal and dystrophic muscle were investigated by Bompiani [19]; the submicroscopic structure of collagen fibers was studied by Bairati [20]; while Archetti and Steve Bocciarelli extended the initial microbiological interest of the ISS to the study of viruses [21].

In these pioneering studies, enthusiasm for the new, beautiful images hindered the understanding of the severe limits imposed by the lack of suitable preparation techniques for biological material and of the inevitable artefacts. Thus the results obtained may now appear to be of rather limited validity. However, these early studies were of unique value in that they revealed the potentialities of the electron microscope as a new tool to develop biological sciences. Furthermore, they stimulated very fruitful contacts between biologists and electron microscopists from both Italian and foreign laboratories; particularly active collaborations were established between U. Valdrè of the Istituto di Fisica in Bologna, A. Claude of the Université Libre, Bruxelles, R. W. Horne of the Cavendish Laboratory in Cambridge and later with the John Innes Institute in Norwich, F. Sjoestrand of the Karolinska Institutet, and A. Afzelius of the Wenner Grens Institutet in Stockholm.

In this way, the Italian biologists interested in electron microscopy were involved from the very beginning in the central problems of that time, i.e., the search for suitable preparative and observation techniques that would allow the preservation of ultrastructural details in biological specimens. A relevant contribution in this field was provided by Millonig with the proposal of a suitable buffer solution to counteract pH variations during fixation and staining [22].

The availability of reasonably good fixation and staining procedures, as well as the introduction of improved embedding media by the Glauerts, opened up the new area of ultrastructural anatomy to detailed studies. In a few years, the number of Italian biologists using the electron microscope as a specific research tool increased tremendously, as can be seen from the contributions to international congresses on electron microscopy.

In these types of studies, the electron microscope was used merely as an instrument capable of higher resolution than that of the optical microscope, whereas the approach, both theoretical and experimental, was that of traditional morphologists, i.e., to describe the structural organization of tissue and cells. Nevertheless, the results obtained were basic for the foundation of modern cell science. In Italy there was also a particular interest in describing the submicroscopic bases of human diseases in view of clinical applications. A well-known center of diagnostic ultrastructural pathology was developed at the Medical School in Bologna by R. Laschi (Sec. II. F).

A less conventional use of the electron microscope for the study of cell physiology, essentially in line with the pioneering work of A. Claude, G. Palade, and Ph. Sickevitz, was developed by U. Muscatello. Basic to this approach is the integration of ultrastructural and biochemical data to gain information about the function of a given cell organel. With this approach, they succeeded in identifying the function of the sarcoplasmic reticulum in muscle and obtaining basic insight into the problem of muscle fiber relaxation [23,24].

The information obtainable with the electron microscope was used by P. Buffa and his colleagues at Modena University in a series of works aimed at revising the concept of the "biochemical lesion." This term was first introduced by Sir Rudolf Peters to signify functional damage of the cell without any structural modification. Buffa collaborated with Sir Rudolf in Oxford in an attempt to give an experimental basis to this concept. According to Buffa, however, first they had to investigate whether the absence of a structural counterpart was an essential part of the definition, or was merely a consequence of the limited resolving power of light microscopy. By studying a number of biochemical lesions relevant to the mechanisms of energy transformation, Buffa and Ronchetti were able to show that the biochemical lesions in mitochondria were closely associated with configurational changes [25], a result that has to be taken into account in a modern redefinition of the biochemical lesion.

Surprisingly enough, comparatively less attention was paid to the new technologies. In fact, even when a suitable observation technique was developed by Horne and colleagues, very few attempts were made by Italian biochemists to use the unique power of the electron microscope to study the structure of macromolecules and of their biological aggregates that could not be resolved by the traditional X-ray diffraction technique.

III. The Elmiskop I Period (1960–1980)

A. The Cambridge Connection

At the beginning of the 1960s, i.e., about one year after the Elmiskop I was commissioned, Valdrè applied for a NATO fellowship to be spent at the University of Cambridge (UK), the birthplace of transmission electron microscopy of thin crystals and the best place for its application. In 1956, in fact, Michael J. Whelan, working at the Cavendish Laboratory under the supervision of Dr. Peter B. Hirsch, demonstrated the great potential of electron microscopy in the study of defects in thin crystals. The purpose of the NATO grant application was to update the EM Laboratory in order to assist members of the Solid-State Group in the solution of their research problems.

During a visit to Cambridge on the occasion of the 1960 International Union of Crystallography Congress, arrangements were made between the Head of the Physics Department, Sir Nevill Mott, P. B. Hirsch, and U. Valdrè. A research program on the rearrangement of dislocations during specimen thinning was agreed, and Valdrè started work in November of the same year. It was a lucky chance that P. B. Hirsch's group, the EM group under V. E. Cosslett, and the EM Laboratory of Bologna were equipped with the same type of electron microscope: the Siemens Elmiskop I. The one-year grant (1 November 1960–31 October 1961) was spent with the Crystallography Group of the Cavendish Laboratory (headed by Dr. W. H. Taylor), of which P. B. Hirsch was a member before becoming the leader of a newly formed group: the Metal Physics (or MP) Group. The agreement between Bologna and Cambridge allowed Valdrè to keep his commitments in Bologna while making use of the NATO grant. This marked the beginning of Valdrè's regular commuting between the two universities, which is still (1996) taking place.

In the late Spring of 1961, Valdrè was informed, during a conversation with A. Howie, of an important standing problem. In fact, it was a technical problem arising from the nonavailablility of a stage capable of tilting the specimen through large angles, in any direction. The lack of such a stage greatly hindered the practical application of the diffraction contrast theory to thin crystalline materials. The early training of Valdrè as a mechanical engineer and the experience subsequently gained in the construction of the combined tilt-cooling-straining stage for the Elmiskop (Sec. II. E) led the EM Laboratory to design, construct, and test a double-tilting specimen holder for the Elmiskop. The stage provided a tilting angle of about 23° around two orthogonal axes. The success of the double-tilting stage was quite unbelievable. The holder was simple and cheap; it had the required tilting angle and could be mounted with little modification on the standard specimen stage of the Elmiskop, the most advanced electron microscope at that time. There is little doubt that the above factors contributed to the success of the tilting stage. However, a psychological factor also helped in boosting the popularity of the Valdrè stage, as it became known: the rapidity (about 10 days) of its realization after waiting for a few years for something similar to be developed. The double-tilting stage enabled physicists, metallurgists, and molecular biologists to start performing long-postponed experiments.

The success of the stage had a decisive influence in the setting up of the future activity of the EM Laboratory of Bologna. The laboratory, in fact, devoted itself mostly to the development of electron microscope accessories, particularly specimen stages, and to the establishment and application of new EM methodologies. Several assistants involved in the development of specimen stages and devices must be named: Antonio Grilli (a chemist by training, but adept in several fields, such as electronics, photography, and fitting, 1957–1974), Primo Ricciotti (draftsman, 1959–1985), the skilled instrument makers Libero Morini (1960–1973), Vittorio Monti (1974–1979), and Alberto Costa (1983–), Andrea Valdrè (electronics, photography, and other skills, 1975–1980), Raffaele Berti (electron microscope service and specimen preparation, 1980–), Attilio (Teo) Ponti-Bartolucci (electrician and troubleshooter, 1961–1969), and Luciano Pizzirani (1959–1967), Head of the main workshop.

The main results of the above activities were:

1. The development of double-tilting cartridges based on different principles and designed for various types of electron microscopes (Fig. 3).

2. The construction of combined cartridges capable of performing several functions, such as double tilting and heating up to 800°C (Fig. 2) (see [26] for reviews about items 1 and 2).

3. The liquid helium cooling stages (Fig. 4) together with the setting up of the basic technique of low-temperature transmission EM. These enabled the observation of a martensitic phase transformation in V_3Si down to the 10-nm level, improvement of the resolution of the observation of the fine lamellae structure of superconducting lead specimens to the 10-μm level,

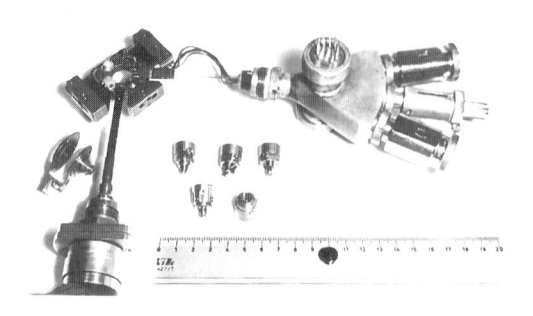

FIGURE 3. Universal specimen stage for double-tilting, rotation, heating, cooling (liquid nitrogen), with various specimen holders and decontamination caps.

FIGURE 4. (a) First prototype of the He cryostat of the reservoir type (which was rather cumbersome to operate). (b) First prototype of the Mark 2 of He circulation type, which produced a wealth of results. (c) Mark 3 with large specimen chamber containing Helmholtz coils for superconducting experiments in an auxiliary magnetic field (note the movie camera for filming dynamic events). (d) Liquid He cold stage mounted in the Cambridge 750-kV electron microscope.

and the first observations of condensed rare gases by electron microscopy (see [27] for a review).

4. The conversion of the specimen chamber of a conventional EM into an ultrahigh-vacuum (10^{-8} torr) deposition chamber for the *in situ* study of nucleation and coalescence phenomena (in collaboration with scientists of the Tube Investment Research Laboratory, Hinxton Hall, Cambridge) [28].

5. The setting up of several phase-contrast techniques and experiments performed by controlled phase shifts, by electron interferometry, and by electron holography [29].

6. The development of the HV STEM/EBIC and Stereo-EBIC (electron beam induced current) techniques [30,31].

Several international patents were obtained, and firms such as Siemens & Halske of Berlin, AEI Ltd (Associated Electrical Industries, at Manchester and Harlow) and AEON Laboratories (Engelfield Green, Surrey) manufactured, under licence, specimen stages designed by U. Valdrè and produced as prototypes in the EM Laboratory. The collaboration with the Cavendish Laboratory has continued uninterrupted up to the present day with the remaining members of the MP Group (about one-third and, in particular, Profs. L. M. Brown and A. Howie), who remained in Cambridge after P. B. Hirsch left in 1966 to become the Isaac Wolfson Professor of Metallurgy at Oxford.

Collaborations also took place with the Electron Microscopy (EM) Group of the Cavendish Laboratory, lead by Dr. V. E. Cosslett—in particular, on the development and application of the HV STEM. The latter work continued after Dr. Cosslett's retirement (1975). The HV(S)TEM (500 kV), the only instrument in the world of that kind at that time, was for a few years funded by a grant of the University of Bologna and run by U. Valdrè, assisted by T. G. Sparrow.

B. *EUREM 1968 in Rome*

During the Prague Conference (1964) SIME applied to the IFSEM to host the 4th European Congress of Electron Microscopy to be held in Rome in 1968; the proposal was accepted. At that time SIME members amounted to about 200, most of them biologists and medical doctors. This situation is common in young EM societies, since biological structures can be readily observed by electron microscopy; in addition, fundamental and applied research in physical metallurgy and in the science of materials are somewhat neglected in countries lacking high-technology industries. It was therefore a hard job for the small number of active members of SIME to organize an international congress. The result was, however, a great success, scientifically, culturally, and socially. It attracted over 2000 participants. The two volumes of the Proceedings comprised some 1200 pages (over 500 papers); an amazingly high number of contributions dealt with innovations in various fields, and several of them are still quoted in the literature today.

Mrs. Daria Steve Bocciarelli of the ISS was the local Chairman and, thanks to her dedication and knowledge of Rome, it was possible to have (as a rare, if not unique, occasion) Castel Sant'Angelo (the early and infamous prison used during the temporal power of the Popes, made famous by Puccini's opera *Tosca*) as the venue for the farewell party. Rome has so much to offer artistically, historically, and religiously that it made the social program very attractive. One may mention visits to Vatican City and to several museums, a Papal audience in Castel Gandolfo, as well as excursions in the surrounding hills and to the catacombs.

At the recent EUREM 92 held in Granada, SIME proposed Florence as the venue for EUREM 96. The proposal was not accepted on the grounds of high living and organization costs. Dublin was chosen as the place for the next meeting; nevertheless, the issuance of the proposal indicates that SIME is a lively society, and present plans are for a new proposal for a future IFSEM conference.

C. The Institute LAMEL-CNR

In 1968 the Italian Research Council (Consiglio Nazionale delle Ricerche, CNR), founded in Bologna a laboratory devoted to the study of materials for electronics (Laboratorio Materiali per l'Elettronica, Istituto LAMEL). The founding was instigated by chemists and engineers of the Bolognese area (some of them belonged to the Solid-State Physics Group) who had good political connections. The EM Laboratory was called on to set up the electron microscopy section within LAMEL. Several members of the EM Laboratory and of the SSP group moved to the new institution, which initially obtained direct support and, later, indirect support from the EM Laboratory. The Electron Microscopy Section of LAMEL has become one of its most important sections and a leading institution in electron microscopy. This was achieved mainly through the work of A. Armigliato on microanalysis, C. Donolato on EBIC contrast theory [32], and P. G. Merli on electron microscopy in general and, in particular, on electron optics. LAMEL has kept close contacts with the EM Laboratory by way of collaborative research and as a source of research students and personnel. Merli and LAMEL have been instrumental in the political and technical organization of a consortium funded with public and private money in South Italy (at Mesagne, near Taranto, in the Puglia Region) devoted to the study of materials. Merli became Director of LAMEL in 1991 and this institution acts as a scientific consultant for the Mesagne Center.

D. Editorial and Teaching Activity

The EM Laboratory at Bologna has had, and continues to have, great importance in the progress and dissemination of electron microscopy through various initiatives, such as sponsoring the publications of both basic and specialized books (in English, in Italian, and, in preparation, in bilingual English/Italian); providing training to newcomers and beginners in the field, and supporting the International School of Electron Microscopy of Erice, Sicily. This school started in 1970 and was the second school to be opened at the "Ettore Majorana" Center of Scientific Culture.

A vital contribution is made to electron microscopy in Italy by the degrees in Physics taken at the EM Laboratory of Bologna. Many research students from the EM Laboratory have continued to apply electron microscopy in solid-state physics and in the science of materials after graduation; several former students have formed EM groups in various parts of Italy. Besides the LAMEL Institute already mentioned (Sec. III. C), another group of growing importance has been established in Rome by Marco Vittori-Antisari at the ENEA, formerly CNEN, the National Committee for Nuclear Energy, (see Sec. II. D). Giulio Pozzi, a member of the EM Laboratory, was appointed Professor at the University of Lecce where the small, existing electron microscopy group was boosted up; from 1993 on he was back in Bologna. In 1991 former students Roberto Galloni and P. G. Merli won a competition for Directors of Research in the CNR, a position equivalent to that of a full Professor. As already reported (Sec. III. C), Merli was instrumental in the founding of the Mesagne Center, where electron microscopy is one of the basic techniques. Both U. Valdrè and P. G. Merli were elected and served as Presidents of SIME.

IV. THE RECENT PERIOD (1980–1994)

A. Power Unbalanced

The life of the EM Laboratory of Bologna has never been easy, owing to the overwelming power in the Physics Department of the high-energy physicists: since the end of World War II they have always been nationally privileged by the Italian government [33,34]. Locally, they benefited from space and permanent research and technical positions granted by the various directors of the Physics Department. This means that particle physicists have always had an absolute majority in all local issues and are very influential nationwide. Two examples will help to illustrate and quantify the situation. Moneywise, the 1992 budget of the Bologna INFN Section, that is, the local section of the National Institute of Nuclear Physics, INFN (excluding personnel and not accounting for the quota paid by the Italian government in support of CERN activity), was much greater than the entire national budget of the groups working in the fields of the physics of matter [34]. The total number of full Professors of the Physics Department of Bologna is 27; they are distributed as follows: 16 in nuclear and high-energy physics, 3 theoreticians in field theory and 2 in mathematical physics, 4 geophysicists, 2 in electron microscopy and solid-state physics (the writer, and G. Pozzi, but only from 1993). Besides the roughly even distribution of a few technical assistants, particle physicists have additional technical staff and researchers in the framework of INFN.

Professor G. Puppi, the founder of the modern Physics Department and its Head from 1954 to 1969, intended to support the various branches of modern physics [10]. In practice, however, his policy turned out to favor high-energy physics and to support, in order of importance, radio astronomy, theoretical physics, geophysics and fluid dynamics, solid-state physics and electron microscopy, and health physics. In the early 1960s, the Institute became a polychair institution; from 1 chair it grew to 12 in 1976, U. Valdrè holding the only chair in the field of solid-state physics and electron microscopy. After student unrest started in 1968, Puppi resigned as Head of the Institute and half a dozen professors followed each other in the Directorship. In 1980 the Istituto di Fisica became the Department of Physics, as a result of a government bill which modernized the statutes of the university.

In a so-called rationalization scheme, the Bologna Senate decided to transform most of the service laboratories into centers of research. In practice, that meant the laboratories continued to provide service without having the direct support of the university in terms of personnel, funds, and administrative autonomy. Some heads of departments, including Physics, liked this policy, for it increased their power. This operation, which was intended to avoid abuses from the laboratories, resulted in a loss of personnel and of regular, basic funding in the case of the EM Laboratory.

B. The University Diploma for Technical Experts in Microscopies

The Italian government has recently modified the university laws in order to harmonize Italian university degrees and curriculae with those of the

European Union. In the past, there were no three-year degrees in Italy (minimum four years), whereas it is now possible for a student to apply for one of the three-year courses approved by MURST; the courses provide successful applicants with a University Diploma (DU).

The EM Center has formally proposed to the Faculty of Sciences, with the support of scientists of other Faculties and Concerns (e.g., Medicine, LAMEL/CNR, industry, and the Assistant Union), the institution of a 3-year University Diploma for Assistants in Laboratories of Analytical Microscopies. This initiative of the EM Center will fill the gaps in the field of the various types of microscopies.

The aim of the Diploma is the training of high-class technicians for the organization, servicing, and improvement of image formation instrumentation. The DU should provide the student with an understanding of the foundations of image contrast formation, regardless of the nature and kind of radiation and probe used to form the images. From the second semester of the second year onward, the course will provide four specializations: biomedical, earth, materials, and miscellaneous sciences; the latter will include fields such as archeometry, art preservation and environmental sciences, and forensic and historical applications. The number of students will be limited to 25 in the first year; the graduates will have a very high chance of immediate employment in Italy. Applications for the DU course from students of EU member states are, of course, accepted and welcome after an exam in proficiency in the Italian language.

The matter has so far obtained the approval of the Faculty of Sciences and of the Senate of the University of Bologna; although the proposal has now to go through two national committees and then more local committees, the major hurdles seem to have been overcome.

If the proposal is approved by MURST, it is believed that it will be the first of this kind not only in Italy but in several other countries.

V. Conclusions

The history of electron microscopy in Italy reported here refers only marginally to biomedical applications. Suffice it to say that the largest Italian centers in these fields are the High Institute of Health (ISS), Rome, and the Institute of Clinic Electron Microscopy (IMEC), Bologna.

The Italian contribution to electron microscopy in the fields of instrumentation, physics, and science of materials may be summarized as follows.

1. Collaboration with various European institutions, universities, and industries.

2. Design and development of specimen stages for *in situ* electron microscopy; the most successful were the double-tilting cartridges, and the combined specimen holders and stages for double tilting (DT) and cooling to He temperatures, DT heating, and DT-UHV for deposition studies (the latter in collaboration with Tube Investment Research Laboratory at Hinxton Hall, Cambridge, UK).

3. Development of the low-temperature (He) and magnetization technique for the study of structural and magnetic properties of superconductors. The application of these techniques to superconducting lead in the mixed state has shown the presence of a very fine lamellae structure (~ 1 μm) and, in the case of V_3Si, the occurrence of a martensitic phase transformation characterized by extremely fine (~ 10-nm) lamellae (in collaboration with the Metal Physics Group of the Cavendish Laboratory of Cambridge University, UK).

4. Study of superconductors and semiconductors, particularly the successful Donolato phenomenological theory of EBIC contrast and, more recently, the setting up of the stereo-EBIC method.

5. Study of radiation damage in organic materials, materials used in fission reactors, and some steels that are suitable candidates for future fusion reactors.

6. Training and dissemination of the EM culture, in particular, the organization of seven courses at an International School of Electron Microscopy (Erice) with total attendance of over 500 scientists, including teachers and participants; the publication of at least 12 books on electron microscopy, three in Italian, seven in English (two with British co-authors), and two with articles in English and in Italian. In addition, the Italian Society of Electron Microscopy regularly organizes short training courses and updating meetings on recent developments in methodologies and/or instrumentation. Courses and meetings run from a few days to two weeks. A magazine is published regularly by SIME twice a year. The most cohesive group of Italian microscopists comes from the Bologna schools. Well worth mentioning is the application presented to the Italian Ministry of Eduction (MURST) to establish a three-year Universiy Diploma for Assistants in Analytical Microscopies.

7. The organization by SIME of the Fourth European Regional Conference of Electron Microscopy (Rome, 1968); the regular biannual meetings, with outstanding invited speakers and the publication of the Proceedings; the several joint meetings with other (usually Eastern) EM societies.

ACKNOWLEDGMENTS

Prof. U. Muscatello of the Department of General Pathology of the University of Modena is thanked for his basic contribution to Sec. II. H.

References

1. Bocciarelli, D., and Trabacchi, G. C. (1946). *Rend. Ist. Sup. Sanità* **IX**(Parte VI), 762–768 (in Italian).
2. Trabacchi, G. C. (1947). Proc. Lecture given 18 June 1947 to the members of Associazione Elettrotecnica Italiana (AEI), Proc. of AEI, 1222–1234 (in Italian).
3. Ageno, M. (1947). *Nuovo Cimento* **3**, 59–68 (in Italian).
4. Bocciarelli, D., and Trabacchi, G. C. (1948). *Rend. Ist. Sup. Sanità* **XI**(III), 794–800 (in Italian).
5. Valle, G. (1933). *L'Elettrotecnica* **XX**(30), 1–11 (in Italian).
6. Il Comune per gli Studi Nucleari di Bologna. Suppl. to *Bologna, Rivista del Comune*, No. 1–3 (Gen.-Febbr.-Mar. 1960) (in Italian).
7. Valdrè, U. (1958). *Boll. Soc. Ital. Biol. Sperim.* **34**, 203; **34**, 207 (in Italian).
8. Donelli, G., Merli, P. G., Pasquali Ronchetti, I., and Valdrè, U. (1982, 1985). *Ann. Ist. Sup. Sanità* **18**, 153–162 (1982) (in Italian); *Proc. Roy. Microsc. Soc.* **20**, 249–253.
9. Ruedl, E, Valdrè, G., Delavignette, P., and Valdrè, U. (1988). *Phys. Stat. Sol. A* **107**, 745–758.
10. Puppi, G. (1962). *Suppl. Nuovo Cimento* **XXV,** Series X, No. 2, 71–76 (in Italian).
11. Favilli, G. (1963). L'attività del Centro per lo Sviluppo della Microscopia Elettronica dopo un anno dalla sua creazione. Internal report, Bologna (in Italian). Laschi, R. (1983). 20 anni di microscopia elettronica applicata alla clinica a Bologna, 1962–1982. Università degli Studi di Bologna (in Italian).
12. Gurrieri, M., Grilli, A., and Valdrè, U. (1964). *Boll. Soc. Ital. Biol. Sperim.* **40,** 764; **40,** 766.
13. Sjoestroem, M., and Valdrè, U. (1979). In "Microbeam Analysis in Biology," Academic Press, New York, pp. 427–444. Valdrè, U., Sjoestroem, M., and Edman, A.-C. (1978), *Proc. 9th Int. Cong. on Electron Microscopy*, Micr. Soc. of Canada, Toronto, Vol. II, p. 92.
14. Valdrè, U., and Horne, R. W. (1972). *Proc. 5th Eur. Cong. on Electron Microscopy*, Manchester, p. 332. Valdrè, U., and Horne, R. W. (1975). *J. Microsc.* **103,** 305–317.
15. Turner, J. N., and Valdrè, U. (1992). Tilting stages for biological applications. In "Electron Tomography" (Joachim Frank, Ed.). Plenum Press, New York, pp. 167–196.
16. Valdrè, G., Muscatello, U., and Valdrè, U. (1994). Imaging of phospholipid phase organization towards atomic resolution. In "Advances in Free Radicals in Disease," CLEUP Publ., Padua, pp. 193–206.
17. Aloisi, M., Ascenzi, A., and Bonetti, E. (1952). *Experientia* **8,** 266.
18. Aloisi, M., Ascenzi, A., and Bonetti, E. (1952). *Rend. Ist. Sup. Sanità* **15,** 430.
19. Bompiani, G. D. (1954). *Rend. Ist. Sup. Sanità* **17,** 1021.
20. Bairati, A. (1956). *Sci. Med. Ital.* **4,** 560.
21. Archetti, I. (1954). *Arch. Virusforschung* **6,** 29.
22. Millonig, G. (1961). *J. Appl. Phys.* **32,** 1837.
23. Muscatello, U., Andersson-Cedergren, E., Azzone, G. F., and von der Decken, A. (1961). *J. Biophys. Biochem. Cytol.* **10,** 201.
24. Muscatello, U., Andersson-Cedergren, E., and Azzone, G. F. (1962). *Biochim. Biophys. Acta* **63,** 55.
25. Buffa, A., Pasquali-Ronchetti, I., Barasa, A., and Godina, G. (1977). *Cell Tissue Res.* **183,** 1.
26. Valdrè, U. (1968). *Nuovo Cimento* **53B,** 157–173.
27. Hawkes, P. W., and Valdrè, U. (1977). *J. Phys. E: Sci. Instrum.* **10,** 309–328.
28. Valdrè, U., Robinson, E. A., Pashley, D. W., Stowell, M. J., and Law, T. J. (1970). *J. Phys. E: Sci. Instrum.* **3,** 501–506.

29. Missiroli, G. F., Pozzi, G., and Valdrè, U. (1981). *J. Phys. E: Sci. Instrum.* **14,** 649–671.
30. Sparrow, T. G., and Valdrè, U. (1977). *Phil. Mag.* **36,** 1517–1528.
31. Valdrè, U., Bergonzoni, A., and Merli, M. (1993). *Ultramicroscopy* **49,** 366–381.
32. Donolato, C. (1989). In "Point and Extended Defects in Semiconductors" (G. Benedek, A. Cavallini, and W. Schroeter, Eds.), NATO ASI Series, Plenum Press, London, pp. 225–241.
33. Boato, G., Careri, G., Chiarotti, G., Fieschi, R., Gondi, P., Gozzini, A., Montalenti, G., and Santangelo, M. *et al.* (1964). Suppl. to *La Ricerca Scientifica, Anno 34,* Vol. 3, No. 4, Series 2, pp. 203–241 (in Italian).
34. Cabibbo, N. (1993). *Physics World* **6,** 40–43; Rizzuto, C. (1993). *Physics World* **6,** 44–47.

2.9
The Growth of Electron Microscopy in Japan
2.9A Introduction

KEIJI YADA[1]

Tohoku University, Katahira, Sendai 980, Japan

The rise of electron optics and its application to electron microscopy in the early 1930s in Germany had been watched with keen interest by foresighted scientists in Japan. The stimulating innovation of the electron microscope, which had exceeded the light microscope in resolution at the hands of Knoll and Ruska (1932), although the report came only later to Japan, inspired us to initiate without delay independent studies of electron microscopy in several places in Japan (Tohoku Imperial University, Tokyo Imperial University, Hitachi Ltd., Electrotechnical Laboratory, Kyoto Imperial University, Osaka Imperial University, etc.). In order to promote studies of electron microscopy and its applications more comprehensively, the need for a research committee was strongly promoted by Prof. S. Seto, Tokyo Imperial University, Dr. K. Kasai, Electrotechnical Laboratory, *et al.* Thus a committee, the 37th Subcommittee of the Japanese Society for the Promotion of Science (JSPS), was started on 6 May 1939 in Tokyo, gathering active research workers. The subcommittee lasted until 1947 with some changes of members throughout World War II, and was re-formed as a collaborative research committee on electron microscopy supported by a grant-in-aid from the Ministry of Education, Science and Culture, Japan, given on the same day of the last meeting of the 37th Subcommittee. Prof. Seto was Chairman, continuously, for both committees.

Electron microscopes, both electromagnetic and electrostatic types, became commercially available from Hitachi, Shimadzu, Toshiba, and JEOL around 1948. Owing to a great increase in numbers of researchers in various fields of instrumentation and application of electron microscopy, the Japanese Society of Electron Microscopy (JSEM) was established in 1949, consisting of about 100 members. Prof. Seto became the first President of the JSEM. The research committee was discontinued at the same time. The JSEM joined the International Federation of Societies for Electron

---

[1] Present address: Aomori Public College, 153-4 Yamazaki, Goshizawa, Aomori 031-01, Japan.

TABLE I

SETO PRIZE WINNERS AND TITLES FROM 1956 TO 1990

1956 1. Koichi Kanaya, Improvement of imaging system of electron microscope; 2. Noboru Higashi, Bacteriological and virological studies by electron microscopy; 3. Tadatosi Hibi, Study of metallic shadowing; 4. Akira Fukami, Studies of replica preparation methods.

1957 5. Bunya Tadano, Study of the reduction of aberrations of electron lenses; 6. Gonpachiro Yasuzumi, Electron microscopic studies on chromosomes and spermatocytes; 7. Eiji Suito, Application of electron diffraction to studies of microcrystals; 8. Kazuo Ito and Goro Honjo, Development of specimen cooling device.

1958 9. Eizi Sugata, Improvement of electron microscope attained by study of electron gun characteristics; 10. Eichi Yamada, Electron microscopic studies on the centrioles and the retinas; 11. Shiro Ogawa, Denjiro Watanabe, Hiroshi Watanabe, and Tsutomu Komoda, Direct observation of the long period structure in ordered alloy lattice; 12. Keinosuke Kobayashi and Tadashi Fujiwara, Theoretical study of ultra thin sectioning.

1959 13. Ryozi Uyeda, Study of image formation of electron microscope; 14. Noboru Takahashi, Electron microscopic direct observation of thin films of alloys.

1960 15. Hiroshi Watanabe, Study of electron energy loss; 16. Konosuke Fukai, Study on the *Influenza* virus by means of electron microscopy; 17. Hatsujiro Hashimoto, Study of crystal growth by means of electron microscope.

1961 18. Yonosuke Watanabe, Electron microscopic studies on the fine structure of cytoplasm; 19. Shinji Sasaki and Ryuzo Uyeda, Electron microscopic study of oxidation and reduction of metals.

1962 20. Yasumasa Tani, Yoneichiro Sasaki, Keinosuke Kobayashi, Shinichi Shimadzu, and Bunya Tadano, Development of high voltage electron microscope; 21. Kenji Takeya and Seijun Koike, Electron microscopic studies on acid-fast bacteria; 22. Zenji Nishiyama and Kenichi Shimizu, Electron microscopic study of martensitic transformation in metallic thin films; 23. Hiroshi Kusuda, Studies on the embedding methods of biological specimens for electron microscopy.

1963 24. Tadatosi Hibi, Shoichi Takahashi, and Susumu Maruse, Development of pointed cathode for electron microscopy; 25. Kiyoshi Hama, Electron microscopic study on the fine structure of the synapses; Sho Yoshida, Electron microscopic study of lattice defects in aluminum; 27. Shigeo Sakata, Development of micro grid preparation technique and its application to resolution estimation.

1964 28. Tsutomu Komoda, Observation of lattice images of (111) plane in gold with 2.35-Å spacing; 29. Kazumasa Kurozumi, Electron microscopic study on the morphology of secretion; 30. Natsu Uyeda: Electron microscopic studies on the thin films of organic semiconductors.

1965 31. Toyotaro Hori, Masaya Iwanaga, Yoshio Sakamoto, Hiroichi Kimura, Shinjiro Katagiri, and Masayuki Nishigaki, Development of 500-kV electron microscope; 32. Toshio Nagano, Electron microscopic study on the testes; 33. Shigeyasu Koda, Keisuke Matsuura, and Minoru Nemoto: Studies on the segregation and the mutual reaction between segregates and dislocation.

1966 34. Kanichi Ashinuma, Masaru Watanabe, Yoshio Ohnuma, and Hiroshi Akahori, Design of high-performance electron microscope and study of easy operation; 35. Takuzo Oda, Contributions to the molecular structures and biochemical function of mitochondrial and intestinal-microvillous membranes; 36. Vinchi Mizuhira, Electron microscopical application of radio autography methods to biological specimens.

1967 37. Takeo Ichinokawa, Invention and development of magnetic type of electron velocity analyzer; 38. Yasuhiro Hosaka, Electron microscopic study on the fine structure of virus particles; 39. Tadami Taoka and Hiroshi Fujita, Applications of high-voltage electron microscope to metallurgy; 40. Keiji Yada, High-resolution lattice imaging by axial illumination.

1968 41. Masaru Watanabe and JEOL 1000-kV HVEM Group, Development of 1000-kV electron microscope; 42. Shinjiro Katagiri and Hitachi 1000-kV HVEM Group, Development of 1000-kV electron microscope; 43. Ryohei Honjin, High-order crystalloid structure of yolk protein molecules.

1969 44. Hisazo Kawakatsu, Studies of electron lenses; 45. Tokio Nei, Electron microscopic observation of the freezing and drying processes involving biological specimens; 46. Shozo Ino, Studies of epitaxial growth mechanism of evaporated metallic particles and their structures; 47. Koichi Kato, Development of fixation and staining methods with OsO_4 for complex polymer materials.

1970 48. Yoshihiro Kamiya, Study of image contrast in electron microscopy; 49. Toku Kanaseki, Study of the coated vesicle; 50. Toru Imura, Dynamic observation of behavior of dislocations by means of television-equipped high-voltage electron microscopy.

1971 51. Shizuo Kimoto, Hiroshi Hashimoto, Masayuki Sato, and Hideo Eguchi, Development of high-performance scanning electron microscope; 52. Taro Takeyama, Electron microscopic study of precipitates by aging in alloys; 53. Yoichi Ishida, Studies of ordered lattice structure and grain boundary dislocation in the metallic grain boundaries.

1972 54. Hatsujiro Hashimoto, Akihiro Kumao, Haruo Yotsumoto, and Akishige Ono, Observations of molecule and single atom in crystalline specimens; 55. Tsuneo Fujita, Hiroshi Sakaguchi, and Junichi Tokunaga, Medical and biological application of scanning electron microscopy; 56. Kazuhiro Mihama, Studies of growth and structure of fine particles of metal and alloy by means of high-resolution electron microscope.

1973 57. Hirotami Koike, Development of high-resolution scanning electron microscope using a highly excited objective lens; 58. Kazuo Ogawa, Contributions to the advancement of electron microscopic cytochemistry; 59. Keiichi Tanaka, Critical point drying method and its application to scanning electron microscopy; 60. Yoshimi Tanabe, Studies on the crystal growth and structure of electroplated films by means of electron microscope.

1974 61. Yuzo Yashiro and Shogo Nakamura, Basic research and application of field-ion microscope; 62. Eishiro Shikata, Electron microscopic studies on plant virus; 63. Kazuo Kimoto and Isao Nishida, Electron microscopic and electron diffraction studies of fine metal particles prepared by evaporation in inactive gas at low pressure.

219

(*continues*)

TABLE I (*Continued*)

1975 64. Hiroshi Shimoyama, Fundamental research of electron gun; 65. Morphological analysis of some biological specimens by means of several modern techniques for electron microscopy: (1) Akira Matsumoto, Fine structure of *Chlamydia* organism, with special reference to its cell envelope; (2) Yoshiaki Nonomura, Reexamination of structure of actin filament. 66. Michio Kiritani and Naoaki Yoshida, Measurement of mobility of point defects by electron microscopy; 67. Hiroyasu Saka, Making the attachments for the dynamic observation and the study of plastic deformation of crystals.

1976 68. Hiroshi Akahori, Electron microscopic study of the aluminum anodic oxide film; 69. Mitsugu Nishiura, Improvement of metal block-type freeze replication apparatus.

1977 70. Jutaro Tawara, Kazunobu Amako, and Hiromi Kumon, The application of the high-resolution scanning electron microscope to microbiological studies; 71. Yoshiharu Shimomura, Studies of point defects in pure metals by electron microscopy; 72. Koichi Adachi, A fundamental study of techniques for specimen preparation in electron microscopy.

1978 73. Atsushi Ichikawa and Misao Ichikawa, Fine structure analysis on the secretory process of the salivary gland.

1979 74. Harunori Ishikawa, Micro filaments in cell motility.

1980 75. Akira Tonomura, Studies on electron beam holography; 76. Tomio Kawata, Electron microscopic studies on surface structures of bacterial cells with special reference to regular arrays on bacterial surfaces; 77. Sumio Iijima, Studies of crystal structure analysis and lattice defects by high-resolution electron microscopy; 78. Teruo Suzuki, Technical devices and applications in electron microscopy of biological specimens.

1981 79. Katsumi Ura and Hiromu Fujioka, Development of stroboscopic scanning electron microscope and its application to semiconductor devices; 80. Hisao Fujita, Electron microscopic studies on the secretory mechanism of thyroid gland; 81. Masako Osumi, Electron microscopy on yeast cells; 82. Katsumichi Yagi, Kunio Takayanagi, and Kunio Kobayashi, Observations on the thin films and surfaces of solids by means of ultrahigh-vacuum electron microscope.

1982 83. Hiromoto Yasuda, Ultrastructural studies on the lung with special reference to the surfactant; 84. Shigeo Horiuchi, Fine structures in inorganic compound crystals studied by high-voltage high-resolution electron microscopy; 85. Makoto Hirabayashi, Kenji Hiraga, and Daisuke Shindo, High-resolution observation of alloy structures by high-voltage electron microscopy and its impact on metals research; 86. Akira Tonosaki and Hiroshi Washioka, Improvement and practice in the complementary freeze-replica methods for the biological membrane research.

1983 87. Michiyoshi Tanaka, Technical development of convergent-beam electron diffraction; 88. Kazushige Hirosawa, Electron microscope autoradiographic study on the Vitamin A-storing cell system; 89. Masakazu Okada, Crystal growth and surface structure of linear chain molecular materials.

1984 90. Ryuichi Shimizu, Computer simulation study on scattering and excitation processes of incident electrons in solids; 91. Yasuo Uehara, three-dimensional structure of peripheral nerve endings: A scanning electron microscope observation as disclosed by removal of extracellular connective tissue components; 92. Denjiro Watanabe, Observation of magnetic domains by high-voltage electron microscopy.

1985 93. Yutaka Shimada, Fine structure of myogenic cells *in vitro*; 94. Yukitomo Komura and Yasuyuki Kitano, Observation on the lattice defects in Laves phase alloys by high-resolution electron microscopy; 95. Nobutaka Hirokawa, Three-dimensional architechture of the plasma membrane and cytoskeletons revealed by the quick-freeze technique.

1986 96. Norio Hibino, Correction of the spherical aberration of magnetic electron lenses; 97. Torao Yamamoto, An electron microscope study on the absorption mechanism in enterocytes; 98. Shigemaro Nagakura, Tempering of carbon steel studied by high-resolution electron microscopy; 99. Kenjiro Yasuda and Keiichi Watanabe, Contributions to advancement in electron microscopic immunohistochemistry techniques.

1987 100. Kazuo Ishizuka, Image formation theory of high-resolution electron microscope; 101. Masuyo Nakai, Ultrastructure and morphogenesis of human-related retroviruses; 102. Makoto Shiojiri, High-resolution electron microscopy studies of *Chalcogenides* grown by solid–solid reactions; 103. Hiroshi Hirano, Lectin histo- and cytochemistry—An electron microscopic study.

1988 104. Kazunobu Hayakawa and Kazuyuki Koike, Spin-polarized scanning electron microscopy; 105. Shoichiro Tsukita and Sachiko Tsukita, Electron microscopic studies on the molecular architecture of cell-attachment apparatuses; 106. Takashi Kobayashi, Lattice defects in organic crystals studied by direct imaging of molecules; 107. Yoshinori Fujiyoshi, Experimental study on electron beam damage.

1989 108. Takashi Nagatani and Shoubu Saito, Development of a high-resolution scanning electron microscope with a field emission gun; 109. Toshi-Yuki Yamamoto, Some new findings on the fine structure of the retina; 110. Hideki Ichinose, Study of grain boundaries by high-resolution electron microscopy; 111. Toshio Sakai, Development and application study of ultra-thin sectioning technique.

1990 112. Takuma Saito, Enzyme histochemistry at the electron microscopic levels—An example application for the retina; 113. Chiken Kinoshita, Study of the radiation-induced phenomena in alloy and ceramics by high-voltage electron microscopy.

TABLE II

Distinguished Paper Prize Winners and Titles from 1986 to 1990

1986 1. Yoshiki Uchikawa, Kazuyuki Ozaki, and Toshimi Ooe. Electron optics of point cathode electron gun [*JEM* **32**(2) (1983)]; 2. Jiro Usukura, Hiroshi Akahori, Hiroshi Takahashi, and Eichi Yamada. An improved device for rapid freezing using liquid helium; 3. Akira Tonomura. Application of electron holography using a field-emission electron microscope [*JEM* **33**(2) (1984)]; 4. Takao Inoue and Hitoshi Nagatake. A freeze-polishing method for observing intra cellular structure by scanning electron microscopy [*JEM* **33**(4) (1984)].

1987 5. Katsuhiko Kuroda, Shigeyuki Hosoki, and Tsutomu Komoda. Observations for crystal surface of W(110) field emitter tip by SEM [*JEM* **34**(3) (1985)]; 6. Jeman Kim. Structural study of the cell fusion with HVJ (Sendai virus) [*Denshikenbikyo* **19**(3) (1985)]; 7. Yoshiji Tomokiyo, Sho Matsumura, and Naomi Toyohara. Strain contrast of coherent precipitation in Cu–Co alloys under excitation of high-order reflections [*JEM* **34**(4) (1985)]; Kurio Fukushima, Akira Ishikawa, and Akira Fukami. Injection of liquid into environmental cell for in situ observations [*JEM* **34**(1) (1985)].

1988 9. Keiji Yada and Hiroshi Shimoyama. Brightness characteristics of carbide emitters for electron microscopy [*JEM* **34**(3) (1985)].

1989 10. Akinori Oshita. Hiroki Teraoka, Masao Mametani, and Hiroshi Tomita. Effect of Fresnel diffraction on measurement of degree of coherent of electron beam with electron biprism [*JEM* **35**(2) (1986)]; 11. Masao Hamazaki and Masahiro Murakami. Three-dimensional profiles of sertoli cell processes and associated appendices of biological materials [*JEM* **36**(4) (1986)]; 12. Tatsuo Arii and Kiyoshi Hama. Method of extracting three-dimensional information from HVTEM stereo images of biological materials [*JEM* **36**(4) (1987)].

1990 13. Ryo Iiyoshi, Hideo Takematsu, and Susumu Maruse. Point cathode electron gun using electron bombardment of cathode tip heating [*JEM* **37**(1) (1988)].

TABLE III

HISTORICAL ACTIVITIES IN ELECTRON MICROSCOPY IN JAPAN

1939	Start of the 37th Subcommittee of the JSPS.
1942	Translation of *Elektronen Übermikroskopie*, by M. von Ardenne, into Japanese.
1947	Start of Research Committee on Electron Microscopy supported by Ministry of Education, Science and Culture
1949	Founding of Japanese Society of Electron Microscopy (JSEM).
1950	First issue of *Denshikenbikyo* (*Journal of Electron Microscopy*), in Japanese.
1953	First issue of *Journal of Electron Microscopy* (annual edition in English).
1954	Affiliation of the JSEM in the IFSEM.
1956	First Regional Conference on Electron Microscopy in Asia and Oceania, Tokyo.
1956	Establishment of the Seto Prize.
1959	Celebration of the tenth anniversary of the JSEM, Tokyo; Publication of *Electron Microscopy* in Japanese by the JSEM, consisting of three volumes: Vol. 1, *Basic Theory and Operation* (Y. Tani, Ed.), Vol. 2, *Applications to Biology and Medicine* (M. Terada, Ed.), Vol. 3, *Applications to Science and Technology* (K. Kubo, Ed.). Maruzen, Tokyo.
1960	*Denshikenbikyo* ceases publication.
1961	Quarterly edition of *Journal of Electron Microscopy*.
1966	6th International Congress on Electron Microscopy, Kyoto.
1968	Publication of *Electron Microscopes in Japan, 1936–1965* (E. Sugata, Ed.), in recognition of the tenth anniversary of the JSEM.
1969	Celebration of the twentieth anniversary of the JSEM, Tokyo; special issue of *Journal of Electron Microscopy*.
1975	Publication of *Denshikenbikyo* resumes.
1977	5th International Conference on High-Voltage Electron Microscopy, Kyoto, organized by the JSEM under the auspices of the IFSEM and MESC, Japan.
1978	Start of technical-grade recognition system for electron microscopy technicians.
1979	Celebration of the thirtieth anniversary of the JSEM, Takarazuka; supplemental issue, "Development of Electron Microscopy and Its Future" (*J. Electron Microsc.* **28**, Suppl.).
1985	Establishment of the Distinguished Paper Prize.
1986	11th International Congress on Electron Microscopy, Kyoto; International Conference on High-Voltage Electron Microscopy, Kyoto.
1987	Bimonthly edition of *J. Electron Microscopy*.
1989	Celebration of the fortieth anniversary of the JSEM, Osaka; comprehensive compilation of all of old documents in Japanese and special issue of *J. Electron Microscopy*.

TABLE V

SUCCESSIVE PRESIDENTS AND HONORARY MEMBERS OF THE JAPANESE
SOCIETY OF ELECTRON MICROSCOPY (1949–1990)

Charter board members at the beginning of the society
President: Shoji Seto
Directors: Noboru Higashi, Hideo Yamashita, Hiroshi Kamogawa,
  Keinosuke Kobayashi, Eizi Sugata, Shigeo Suzuki, Bunya
  Tadano, Yasumasa Tani, Bun-ichi Tamamushi, Teruo Hayashi
Inspectors: Kyugo Sasagawa, Nobuyoshi Kato, Kenji Kazato

Succesive presidents:
Shoji Seto	(May 1949–April 1954)
Hideo Yamashita	(April 1954–May 1955)
Yasumasa Tani	(May 1955–May 1956)
Masanaka Terada	(May 1956–May 1957)
Noboru Higashi	(May 1957–May 1958)
Tadatosi Hibi	(May 1958–May 1959)
Eizi Sugata	(May 1959–May 1960)
Bunya Tadano	(May 1960–May 1961)
Yoneichiro Sakaki	(May 1961–May 1962)
Jyun Hidaka	(May 1962–May 1963)
Gonpachiro Yasuzumi	(May 1963–March 1964)
Fumikazu Takagi	(March 1964–April 1964)
Eiji Suito	(May 1964–May 1965)
Kazuhiko Akashi	(May 1965–April 1966)
Eichi Yamada	(April 1966–May 1967)
Noboru Takahashi	(May 1967–May 1968)
Kenji Kazato	(May 1968–May 1969)
Konosuke Fukai	(May 1969–May 1970)
Koichi Kanaya	(May 1970–May 1971)
Keinosuke Kobayashi	(May 1971–May 1972)
Akira Fukami	(May 1972–May 1973)
Shin-ichi Shimadzu	(May 1973–May 1974)
Yonosuke Watanabe	(May 1974–May 1975)
Ryozi Uyeda	(May 1975–May 1976)
Kenji Takeya	(May 1976–May 1977)
Tokio Nei	(May 1977–June 1978)
Hatsujiro Hashimoto	(June 1978–May 1979)
Tadami Taoka	(May 1979–May 1980)
Atsushi Suganuma	(May 1980–May 1981)
Isamu Kondo	(May 1981–May 1982)
Susumu Maruse	(May 1982–May 1983)
Kiyoshi Hama	(May 1983–June 1984)
Taro Takeyama	(June 1984–June 1985)
Kazuo Ogawa	(June 1985–May 1986)
Atsushi Ichikawa	(May 1986–May 1987)
Keiji Yada	(May 1987–June 1988)
Hiroshi Fujita	(June 1988–May 1989)
Kenjiro Yasuda	(May 1989–May 1990)

(*continues*)

TABLE IV (*Continued*)

Honorary Members:		
Kyugo Sasagawa	Shoji Seto	Yasumasa Tani
Masanaka Terada	Hideo Yamashita	Gaston Dupouy
Ernst Ruska	Fritiof Sjöstrand	H. Stanley Bennett
Noboru Higashi	Jean Jacques Trillat	Keinosuke Kobayashi
Tadatosi Hibi	Eizi Sugata	Bunya Tadano
Sir Peter B. Hirsch	Don W. Fawcett	Yoneichiro Sakaki
Shin-ichi Shimadzu	Gottfried Möllenstedt	Keith R. Porter
Ryozi Uyeda	Kenji Kazato	Eiji Suito
Konosuke Fukai	Koichi Kanaya	

Microscopy (IFSEM) in 1954. Now, 40 years later, the active members of the JSEM total about 2900.

The JSEM organized an editorial committee in 1959 to compile a historical survey of electron microscopes in Japan in honor of the tenth anniversary of the society, and a book, *Electron Microscopes of Japan, 1936–1965,* was published in 1968 by Maruzen Co. Ltd., through a long-lasting effort of Prof. E. Sugata, chairman of the Editorial Committee, and other collaborators (Sugata, 1968).

On the occasion of the thirtieth anniversary of the JESM in 1979, at Takarazuka, activities in electron microscopy in Japan were reviewed in a supplement, "Development of Electron Microscopy and Its Future," which reported that: "By 1979, Japanese companies produced 8260 microscopes (i.e. 4640 microscopes for foreign countries and 3620 microscopes for Japan). The resolving power of electron microscopes reaches atomic size, ~ 3 Å. There are 2338 active members in the JSEM and a total of 7060 registered members" (Hashimoto, 1979).

On the occasion of the X1th International Congress on Electron Microscopy in Kyoto in 1986, the JSEM published a book written in English entitled *History of Electron Microscopes 1986,* (Fujita, 1986) and distributed it to the participants of the Congress. In the book, the activities of these committees and the pioneer works in the active centers are described vividly and in detail.

All the old documents of the activities of these subcommittees were comprehensively compiled in Japanese into five big volumes as a memorial publication for the fortieth anniversary of the Japanese Society for Electron Microscopy in 1989. Another special issue was published as a supplement to the *Journal of Electron Microscopy* (1989) for the fortieth anniversary of the JSEM to survey the progress in the different fields in electron microscopy.

From these publications, we can fully review the early history of electron microscopy in Japan. The contents of *History of Electron Microscopes 1986* seem to meet well one aim of the present volume, the history of the individual member societies of the IFSEM and the development of electron microscopy, so that the activities of the subcommittee as written by Dr. B. Tadano and early local activities as written by the others are reproduced here with only slight modification.

The JSEM founded the Seto Prize in 1956 in memory of Prof. Seto and awarded it annually to workers who made distinguished contributions in the fields of instrumentation (A), biological science (B), materials science (C), and general applications (D) of electron microscopy. At present, the number of Seto Prizes awarded adds up to 113: 28 in field (A), 33 in (B), 34 in (C), and 18 in (D).

Another prize, the Ronbun Prize (Distinguished Paper Prize) of the JSEM for excellent papers published in the *Journal of Electron Microscopy*, was founded in 1985; the research fields concerned are in parallel to those of the Seto Prize. Papers awarded the Ronbun Prize add up to 13. The developments of different fields of research and applications of electron microscopy, another aim of this volume, are mostly involved in the works to which the Seto Prize and the Ronbun Prize were awarded. Many of them are revolutionary developments that were widely influential thereafter in Japan. The titles, years, and winners of these prizes are shown in Table I and Table II, respectively.

The main historical activities on electron microscopy in Japan are listed in Table III. Successive Presidents of the JSEM and Honorary Members of JSEM are shown in Table IV.

Developments of commercial electron microscopes by every maker have been outlined in the *History of Electron Microscopes 1986,* so no detailed description of them is given here. The commercial electron microscopes produced in Japan amounted to about 10,000 and a little more by the end of the 1980s. At present, more than 70% of the electron microscopes worldwide are of Japanese manufacture.

REFERENCES

Fujita, H. (Editor in Chief) (1986). "History of Electron Microscopes 1986." Komiyama.
Hashimoto, H. (1979). Preface to "Development of Electron Microscopy and Its Future," *J. Electron Microsc.* **28,** Suppl.
Japanese Society of Electron Microscopy. (1989). *J. Electron Microsc.* **38,** Suppl.
Sugata, E. (Editor in Chief) (1968). "Electron Microscopes of Japan, 1936–1965: A Historical Survey." Maruzen.

2.9
The Growth of Electron Microscopy in Japan
2.9B The 37th Subcommittee of the Japanese Society for the Promotion of Science, 1939–1947

BUNYA TADANO[1]

Hitachi, Ltd., Tokyo, Japan

I. Introduction

The 37th Subcommittee of the 10th Committee, Japanese Society for the Promotion of Science (JSPS), was established 6 May 1939 for collaborative research on electron microscopy. Although World War II erupted in the same year and various difficulties arose, committee activity was continued without interruption. On 26 November 1947, the 47th and last meeting was held. By then, the electron microscopes constructed in several institutions had attained a resolution of about 3 nm; the first aim of this subcommittee was therefore attained. On the same day, the 37th Subcommittee was reformed into the collaborative Research Committee on electron microscopy, which obtained a grant-in-aid from the Ministry of Education. The Chairman was Dr. S. Seto (Professor, Tokyo Imperial University, at that time) for both committees. He also became the first President of the Japanese Society for Electron Microscopy (JSEM), which was established in 1949.

The 37th Subcommittee gave birth to electron microscopy in Japan; its contribution was very large. Its records on research activities are regarded as first-class documents from the viewpoint of technological history.

Based on these records, research activities by Japanese pioneers in electron microscopy and several episodes are set out in the present article as the history of the development.

II. Establishment of the 37th Subcommittee

A. Background

In Japan, the technical term "electron microscope" first appeared in the *Journal of the Institute of Electrical Engineering* in February 1932, in a

---

[1] Present address: Japan Techno-Economics Society, Masuda Bldg., 2-4-5 Iidabashi, Chiyodaku, Tokyo 102, Japan.

translation by Dr. H. Yamashita. This translated abstract was concerned with the emission electron microscope.

Several papers on the transmission electron microscope were reported by M. Knoll, E. Ruska, B. von Borries (Knoll and Ruska 1931, 1932; von Borries and Ruska, 1933), L. Marton (1934), K. Krause (1936, 1937), and others. In Japan, however, only a few people were interested in these papers; it was never regarded as an important subject.

A paper which was published in *Siemens Werk* in January 1938 showed a comparison of an electron micrograph of Bacterium coli with its optical micrograph (reproduced in Fig. 1) (von Borries and Ruska, 1938; von Borries *et al.*, 1938); it gave a large impulse to the Japanese, who were very interested in a microscope with a very high magnification. Since that time, the importance of the electron microscope has been widely appreciated.

Based on this situation, Prof. S. Seto and Mr. K. Kasai (Engineer, Electrotechnical Laboratory) made a proposal to establish the 37th Subcommittee and to start collaborative research on electron microscopy. In later years, Prof. Seto (1974) recalled the situation as follows: "In 1939, I took the chair of the 10th Committee, JSPS. I was informed that B. von Borries and E. Ruska had succeeded in the construction of an electron microscope with a magnification of about 20,000. At that time, the Japanese scientists supposed that they could also obtain high-resolution micrographs by using electrons. But we did not know how to do it. We also thought that we could do it if the Germans could. Let us do it! Then I made a proposal to Prof. H. Nagaoka, Director, who was very positive about it. A decision was made to establish the 37th Subcommittee and I took the chair. I started to invite the people who were interested in electron microscopy."

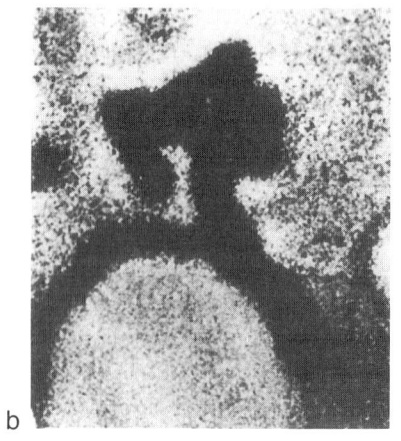

FIGURE 1. Comparison of an electron micrograph (a) with an optical micrograph (b) of bacterium coli (von Borries *et al.*, 1938).

B. The First Meeting

The first meeting was held at the Denki Club, Yurakucho Bld., Tokyo, on 6 May 1939. All committee members attended. Chairman S. Seto explained the process of establishment, research subjects, conditions, and expenses of the 37th Subcommittee. Secretary K. Kasai explained the prospectus, which was written by the chairman. Each committee member explained his situation at that time. Finally, the members commented on research trends.

1. *Committee Members*

Chairman: S. Seto, Prof., Faculty of Engineering, Tokyo Imperial University
Secretary: S. Suzuki, Engineer, Electrotechnical Laboratory
Assistant Secretary: H. Wada, Electrotechnical Laboratory
Members:

- S. Asao, Engineer, Tokyo Electric Corp.
- N. Kato, Professor, Kyoto Imperial University
- K. Kasai, Engineer, Electrotechnical Laboratory
- E. Sugata, Associate Professor, Osaka Imperial University
- K. Tada, Engineer, Yokogawa Electric Works
- Y. Tani, Associate Professor, Tokyo Imperial University
- H. Yamashita, Professor, Tokyo Imperial University
- T. Shimizu, Staff, Institute of Physical and Chemical Research (IPCR)
- J. Ookubo, Professor, Tohoku Imperial University
- H. Toyoda, Engineer, Hitachi Co.
- T. Takeshita, Engineer, Military Science Laboratory
- K. Sasagawa, Professor, Osaka Medical College

2. *Prospectus of the Establishment of the 37th Subcommittee*

"As a result of the recent development of geometric electron optics, the electron microscope has reached a high magnification which a conventional microscope cannot attain. Such a very high magnification microscope would enable us to open up new fields in bacteriology, colloid chemistry, and many other branches of natural science; it would contribute very much to the development of science.

It is most important for the design and construction of a very-high-magnification electron microscope to overcome various technical difficulties, to improve it, and to open up its fields of application. In our country, some researchers have begun to study the electron microscopy. Because of the importance of electron microscope research, this subcommittee should gather more experts, allot them research subjects, synthesize their results, achieve the research work as soon as possible, and then contribute to the development of each branch of natural science.

The first step of this research project should be the basic study of the design of the electron microscope. After this, we should apply it in cooperation with experts in the applied fields and finally synthesize them."

3. *Research Subjects*

The main research subjects on the very-high-magnification electron microscope which were proposed at the first meeting were as follows:

1. Design of high-tension electron guns
2. Stabilization of high tension
3. Design of electron lenses
4. Specimens and specimen-support membranes
5. Electron microscopes for specialized uses (metallurgy, thermionic cathodes, etc.)

4. *Term*

The first term shall be 3 years.

5. *Expenses*

Total: ¥80,000
1939: ¥15,000
1940: ¥25,000
1941: ¥40,000

6. *Some Interesting Comments by Committee Members at the First Meeting*

1. E. Sugata: "I have studied an electron microscope which has a low magnification and an accelerating voltage of 3 kV (maximum 8 kV). I regard the mechanical design as very important. Now I am planning to remove the chromatic aberration. I'd like to take charge of lens design."
2. N. Kato: "T. Inoue of my laboratory has calculated lens aberrations. I'd like to conduct an experiment in cooperation with Prof. Sugata."
3. K. Sasagawa: "The biologist user wishes that the change of biological specimen in vacuum be small. A magnification of 10,000 will be sufficient for biological studies."

4. S. Asao: "We have studied a low-magnification electron microscope to observe the thermionic cathode and the oxide films."
5. K. Kasai and S. Suzuki: "We shall combine our efforts to construct a simple electron microscope which has a high tension and high magnification, and then experiment with it."
6. J. Ookubo: "I have studied thermionic cathode by the Knoll-type emission microscope."
7. Y. Tani: "I have studied a specialized electron microscope and the aberration of the electrostatic lenses."

From the above, one may visualize the research situation of the electron microscope in Japan.[2]

C. Polices of the 37th Subcommittee

The following was determined at the first meeting.

1. Research policy: The research process should be divided into three stages. In the first stage, the basic data should be derived from simple experimental apparatus. Based on this, more elaborate apparatus should be constructed in the second stage. In the third stage, a practical machine should be constructed.
2. Research subjects of each member: Each member was expected to present his research subject to the secretary within two weeks of the first meeting. The secretary would arrange and present them at the second meeting.
3. Acquisition and survey of references: After 1938, the development of electron microscopes by foreign countries became rapid. In 1939, Siemens put the first practical electron microscope on the market (von Borries and Ruska, 1939). In the same year, H. Mahl, AEG, reported an electrostatic electron microscope which had a resolution of 8 nm (1939). At Toronto University, Canada, E. F. Burton together with J. Hillier and A. Prebus constructed a magnetic electron microscope (1939). From such a situation, the members agreed with each other to endeavor to acquire these references.

Chairman Seto conducted the 37th Subcommittee in a different way than other committees. First, the chairman allotted a theme to each member,

---

[2] At the first meeting in 1939, K. Tada, Yokogawa Electric Works, proposed the fundamental idea of an electrostatic or electromagnetic stigmator. He explained the precision limit of mechanical manufacturing and proposed that this should be compensated by using some electric or magnetic field. Unfortunately, this proposal was too early to be accepted.

who was expected to report his result at the committee; it was open to discussion by the other members. Second, each member was expected to construct his own electron microscope and to study his own research subject. He dared to make a contradiction between the construction race and the public offering of results, in order to accelerate the research.

III. Early Japanese Electron Microscopy, 1939-1947

Early Japanese electron microscopy may be divided into the following three periods.

1. Cradle period: From the start of the 37th Subcommittee until the earliest electron micrographs could be taken, though with rather poor resolution.
2. Improvement and reformation period: several electron microscopes attained a resolution of 3–5 nm, and their applications were started. Some practical machines were supplied commercially.
3. Period of application and newcomers in manufacturing: After World War II, the research and development of electron microscope became active again, and applications began to be active with the machines which had not suffered from the war. Some new manufacturing companies joined in, taking account of the future of electron microscopes.

In the following, some details will be described.

IV. Cradle Period

The cradle period includes the time from the first meeting (May 1939) to the 13th meeting (May 1941). (The assistant secretaryship was transferred from K. Kasai to Y. Sakaki.) The activities may be summarized as follows.

A. Aberration Calculations in Geometric Electron Optics

In this period, most of the presented papers were on aberration calculations: N. Kato and T. Inoue presented 13 papers, and Y. Tani calculated aberrations of an electrostatic lens. E. Sugata reported a design of electron lens, and S. Suzuki discussed the form of polepieces and ray tracing.

B. Stabilization of High Tension

H. Yamashita (Tokyo Imperial University) and K. Kasai (Electrotechnical Laboratory, later moved to Hitachi) succeeded in the stabilization of high tension within 0.01% for 5 s.

C. Construction of Electron Microscopes

At the sixth meeting (24 February 1940), E. Sugata, K. Sasagawa, and S. Suzuki presented photographs of their electron microscopes, but no micrograph taken with them was reported.

At the 11th meeting (13 January 1941), K. Sasagawa was accompanied by the young N. Higashi (Kyoto Imperial University). They presented four electron micrographs of polished alumina and cromium oxide, and explained their conditions, but no record remains. It seems that micrographs were passed around. At the twelfth meeting (29 March 1941), K. Sasagawa presented electron micrographs of diphteroid and staphylococcus, and then requested permission to make a presentation at the meeting of bacteriologists. These micrographs, it seems, were those (Fig. 2) included in a reprint which were presented at the 14th meeting (9 July 1941).

The 12th meeting was held at the Faculty of Engineering, Osaka Imperial University; Sugata's laboratory was opened to the committee members. After this meeting, the opening of a member's laboratory became a custom which was one of features of the 37th Subcommittee.

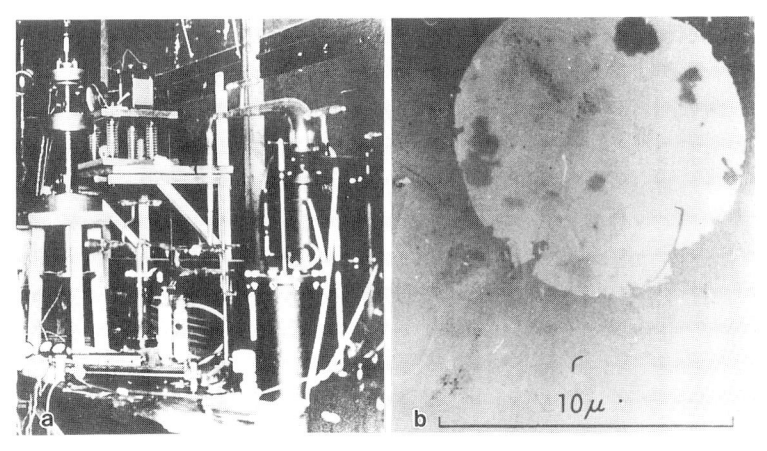

FIGURE 2. (a) Electron microscope of Medical School, Kyoto Imperial University. (b) Electron micrograph of polished alumina which might have been passed around at the 11th meeting (13 January 1941).

E. Sugata was a pioneer who was the first in succeeding in taking low-magnification electron micrographs of a mesh in Japan. His electron micrographs with high magnification are not found in the minutes and proceedings up to the 14th meeting (9 July 1941). At the 15th meeting (20 September 1941), E. Sugata presented a document which had been presented to another journal. At the 17th meeting (17 January 1942), he presented a reprint of a publication which reported the results by using the Mark I machine at another Institute. For details, see Prof. Ura's article in this book, "Early Electron Microscopes at Osaka University, 1934–1945."

S. Suzuki was also one of the pioneers, but he did not present micrographs, just a photograph of the instrument.

Although no proceedings of the 37th Subcommittee remain, research activities of S. Asano and H. Inuzuka (Tokyo Electric) were very high. At the 10th meeting (19 November 1940), S. Asao interpreted electron micrographs which were taken with an electrostatic machine (30 kV) and a magnetic machine. They constructed the above two machines at the same time in 1940–1941 and took electron micrographs. H. Inuzuka reported the construction of the magnetic machine. In his paper, he said: "This structure is exactly the same as the machine which E. Ruska reported and has no new features." This means that the Tokyo Electric Co. had a high level of basic technology at that time. It is very regrettable that this company ceased doing research and development of electron microscopes in later years and then gave up production.

D. Some Episodes

At the 13th meeting (10 May 1941), the chairman said: "Recently, some outsiders have criticized the activities of this subcommittee without understanding our research policy and mission. They demand centralization. They do not understand the collaborative research. I have always refuted it. Gentlemen, please exert yourselves much more so as to achieve high performance!"

K. Sasagawa commented: "I heard that the Military Medical School, Manchuria Empire (now, the North-East region of China), intends to import a Siemens electron microscope." The chairman said: "It is most important to have free use of an excellent imported apparatus to study well the fields concerned." This plan was never realized due to the aggravation of the war.

The Japanese Army and Navy at that time stressed the priority of the intensification of war potential. They intended to prohibit the use of materials and labor for construction of electron microscopes. The chairman always opposed these pressures from outsiders, so that the members could concentrate on their own researches.

V. Improvement and Reformation Period

The improvement and reformation period began with the 14th meeting (July 1941) and ended with the 38th meeting (August 1945). Twenty-five meetings are included. Japan surrendered to the Allied Forces on 15 August 1945. Because of the postwar confusion, the minutes of the 38th meeting were lost: Even the record of its date did not survive.

The features of the 37th Subcommittee in this period were: (1) several electron microscopes attained a resolution of 3–5 nm; (2) applied research was started and both preparation of specimen-support membranes and magnification measurement were studied; (3) bacilli, viruses, carbon black, etc., were observed; (4) each member's laboratory was opened to other members; (5) some makers supplied electron microscopes to university researchers; (6) M. von Ardenne's book was translated into Japanese by the members.

A. Studies on Aberrations

Following the preceding period, the aberration theory was studied by N. Kato, C. Inoue, A. Tani, E. Sugata, and S. Suzuki and consulted in the course of design of the instrument.

B. Stabilization of the High-Tension and High-Frequency Voltage Generator

Experiments on stabilization were reported by H. Yamashita, R. Sato, K. Kasai, B. Tadano, E. Sugata, etc. The high-voltage generator excited by high-frequency current was also reported. K. Tada's circuit was excited at 34 kHz and had a negative-feedback circuit. U. Yoshida (Sumitomo Communication Co). obtained a voltage stability of 0.8×10^{-4} and a voltage ripple of 1.2×10^{-5} for 5 s by using a 20-kHz double-rectifying circuit. He also constructed a magnetic electron microscope.

C. Development Status of Electron Microscopes

After the middle of 1941, the collaborative research of the 37th Subcommittee proved to be effective and results increased rapidly. The electron microscope seemed to be the most complicated and subtle apparatus for Japanese technologies at that time. Many problems had to be solved: on the theoretical side, ray tracings, aberration calculations, estimations of specimen temperature rise, and so on; on the manufacturing side, stabilization of high tension, very-high-precision machining of polepieces, fine adjustment mechanisms for specimen holders, shielding of disturbing magnetic field, antivibration of column, lens alignment in vacuum, exchange of specimens and micrographs in vacuum, boring of small apertures; on the material side, magnetic materials for polepieces, cathodoluminescent materials and their spreading on the screen; on the observation side, preparation of specimen-support membranes and of specimens. These problems appeared as the research went on.

At that time, some references could be acquired. At present, one may conclude that then, the electron microscope could easily be constructed. But, because the Japanese technologies were rather primitive at that time and the members had little experience, the members had to discuss everything, to open their laboratories to each other, to improve their knowledge and technologies, and to solve the problems. During World War II, the research became difficult day by day.

In spite of this situation, the chairman and committee members continued earnestly in their work. I think this was the epoch-making interdisciplinary research among the industry, the government, and the university.

Table I shows the development status of the electron microscopes. As is seen in Fig. 3, several electron microscopes attained a resolution of 3–5 nm. It may be noted that this table was arranged from the proceedings of the 37th Subcommittee.

As is seen in Table I, the electron microscopes were constructed at three universities, one national laboratory, and four manufacturers, that is, a total of eight organizations. Y. Tani constructed an electrostatic machine and the Shiraimatsu Company a magnetic one, though these were not recorded in the proceedings.

D. *Open-Invitation to Members' Laboratories*

As stated above, this was one of the features of the 37th Subcommittee. Each member's laboratory was opened to all the members; there the meeting was held and results were presented and discussed.

1. Sugata's laboratory, Osaka Imperial University, 29 March 1941, the 12th meeting
2. Sasagawa's laboratory, Kyoto Imperial University, 9 July 1941, the 14th meeting
3. Asao's laboratory, Tokyo Shibaura Electric Co., 20 September 1941, the 15th meeting

TABLE I
Trial Construction of Japanese Electron Microscopes Derived from the Proceedings of the 37th Subcommittee

Reporter(s), organization	Document no. Date[a]	Title and features
S. Sasagawa, T. Kimura, H. Higashi, Kyoto Imperial University	14-6 9 July 1941	Bacteriological studies by using a Japanese electron microscope: the first micrographs on biology.
E. Sugata, H. Yokoya, I. Fukui, Osaka University	17-2 17 Jan. 1941	The Mark I electron microscope and its results.
H. Hamada, Hitachi Co.	20-2 11 July 1941	On the image sharpness to mechanical vibration and voltage variation: with an HU-1.
T. Kimura, N. Higashi, Kyoto Imperial University	20-4 11 July 1942	Bacteriological study by using an electron microscope: in cooperation with Sasagawa, IPCR, YEW, and Hitachi.
B. Tadano, Hitachi Co.,	23-1 6 Jan. 1943	The Mark II electron microscope: HU-2.
N. Higashi, Kyoto Imperial University	24-2 2 April 1943	Electron microscopical study on bacilli and viruses: improvement in cooperation with Shimadzu Co. Higher resolution than Fig. 2. The first electron micrographs of a virus by a Japanese microbiologist.

(*continues*)

TABLE I (*Continued*)

H. Hamada, Hitachi Co.	25-2, 3 29 May 1943	Observation of carbon black with an electron microscope HU-2 (Fig. 3).
S. Shimadzu, Shimadzu Co.	25-5 29 May 1943	Mark I electron microscope: micrographs of ZnO, MgO, carbon black.
U. Yoshida, Sumitomo Communication Co.	26-1 10 July 1943	Mark I electron microscope and high-frequency voltage generator: the first HF voltage generator in Japan.
N. Kato, Kyoto Imperial University	27-4 27 Sept. 1943	Mark I magnetic electron microscope.
K. Sakaki, Nagoya Imperial University	27-5 27 Sept. 1943	Some results with an electron microscope: with an HU-2.
S. Suzuki, T. Ochi, Electrotechnical Laboratory	28-2 27 Sept. 1943	On structure of asbestos: with an electron microscope built there.
K. Kobayashi, T. Chikatsuchi, Kyoto Imperial University	29-2 22 Jan. 1944	Study on wood structures with an electron microscope: with a Toshiba machine and resolution of 15 nm.
M. Ohara, T. Hori, Shimadzu Co.	33-1 16 Sept. 1944	Relation between sizes and dispersion process of carbon blacks.

[a] "Document no." means that of the proceedings. "Date" means that of the document. Often the document was the reprint from another journal: in this case, "Date" does not means the real published one.

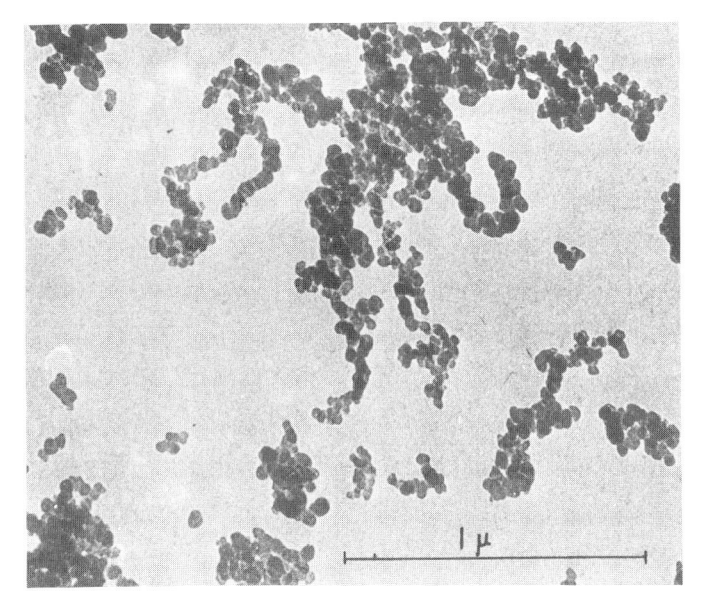

FIGURE 3. Micrograph of carbon black with an HU-2 (Tadano, 1943).

4. Kasai's laboratory, Hitachi Co., 22 November 1941, the 16th meeting
5. S. Suzuki's laboratory, Electrotechnical laboratory, 6 April 1942, the 18th meeting
6. After the 27th meeting, on 27 September 1943, the following were opened:
   Kato's laboratory, Kyoto Imperial University
   Kobayashi's laboratory, Kyoto Imperial University
   Ohara's laboratory, Shimadzu Co., in cooperation with Suzuki
   Sasagawa's laboratory, Kyoto Imperial University
7. Tadano's laboratory, Hitachi Co., 16 February 1946, the 40th meeting

The 31st meeting was held at West Chiba at 3 p.m. on 20 May 1945. Twenty persons attended. At about 6:40 p.m., an air-raid warning was sounded. The meeting was immediately closed without a decision on the date of the next meeting. The chairman said that each member should take care of himself! The members returned home with the horrible feeling that they might not meet again. However, some of them did not go directly home: they drank some Japanese sake and then returned home! This fact was leaked later; the chairman scolded them but laughed at it!

At any rate, these visits undoubtedly contributed to the development of Japanese electron microscopy.

E. Translation of M. von Ardenne's Book into Japanese

At the ninth meeting (28 September 1940), the secretary reported that he had received the book *Elektronen Übermikroskopie,* by M. von Ardenne. Some parts of it were copied and distributed to each member. Since this book contained the latest information, its translation was prescribed. At the 15th meeting (20 September 1941), its planning was decided. It was sponsored by the Science Section, Ministry of Education. The secretary was A. Yamashita and the translators were S. Asao (Tokyo Shibaura Electric Co.), A. Inuzuka (the same), N. Kato (Kyoto Imperial University), C. Inoue (the same), S. Sasagawa (the same), N. Higashi (the same), E. Sugata (Osaka Imperial University), Y. Sakaki (Nagoya Imperial University), S. Suzuki (Electrotechnical Laboratory), S. Nakamura (NHK), R. Fukushima (the same), Y. Tani (Tokyo Imperial University), H. Yamashita (the same), and S. Watanabe (IPCR).

The translated book was published by Maruzen on 25 March 1942. This book was eagerly received by researchers, and it influenced them greatly.

F. Young Power of Researches

The committee members were representatives from each research organization. They stressed the necessity of their research, determined their own research plans, gathered the labor, materials, and funds. Furthermore, they engaged themselves in research and trained young researchers. As a result, the young researchers gained power after the middle of this period. The members of the 37th Subcommittee after 12 May 1945 were as follows: M. Ohara (Shimadzu Co.), K. Kobayashi (Kyoto Imperial University), K. Sakaki (Nagoya Imperial University), B. Tadano (Hitachi Co.), M. Nagayama (Military Weapon Bureau), U. Yoshida (Sumitomo Communication Co), S. Watanabe (IPCR).

While N. Higashi was accompanied by S. Sasagawa (not formally a member), he had engaged in research since the cradle period and they reported their results to the 37th Subcommittee as stated previously. A. Fukami also was accompanied by Y. Tani; the situation was the same as that of N. Higashi.

G. Diverging Trends Between Construction and Application of Electron Microscopes

The research trends of electron microscopy in Japan began to change after 1943. First, the makers had gradually accumulated experience which surpassed that of the universities and the national laboratory. As a result, the research by the latter tended toward basic or application research. Second, researchers who had no electron microscope began to cooperate with makers in applications; they soon began to present various specifications on perfomance, structure, and maintenance of the electron microscope.

Such trends were just the goals at the beginning of the 37th Subcommittee. One may say that one half of its objectives were achieved.

VI. Period of Beginning of Applications and Newcomers in Manufacturers

This period started at the 39th meeting (10 November 1945) and ended at the 47th meeting (26 November 1947), where the 37th subcommittee was disbanded.

The 39th meeting was held after the war, at the Tokyo Imperial University. The chairman asked if each member could take up action as a committee member thereafter. Secretary H. Yamashita reported: "Prof. Yagi, the Director of the JSPS, hopes that the 37th Subcommittee will continue in research activity." The chairman expressed his determination that if the head office of the JSPS agreed the 37th Subcommittee should renew its efforts to apply the electron microscope to scientific problems.

A. Postwar Status of Electron Microscopes

The electron microscopes that were not damaged in the war or were not evacuated to the countryside were only four: at Sasagawa's laboratory, Kyoto Imperial University (with N. Higashi); Sugata's laboratory, Osaka Imperial University; Tadano's laboratory, Hitachi Co.; and Yoshida's laboratory, Sumitomo Communication Co. Research with the other microscopes was delayed because of difficulties of transportation from refuge and from material shortages.

B. Influx of Foreign References

At that time, Japan was occupied by the Allied Forces. At the Hibiya Library, journals of foreign Institutes and PB reports were open to the public. There we could learn about the development of foreign electron microscopy.

It was found that Japanese research on the electron microscope itself was not behind that of foreign countries. It was regrettable that the German researchers, who had been the frontrunners, had not only been in more danger than ourselves during the war but also had to pursue their research with more distress due to the partition of their country.

As to applications, the United States was the most advanced; their surface studies were much advanced compared with ours. At the 47th meeting (26 November 1947), B. Tadano reviewed the foreign and domestic papers that could be acquired at that time.

C. Surface Observation Using Electron Microscopes

The surface observation method which gave an impulse to Japanese researchers was the polystyrene-silica method which had been developed in the United States (Heidenreich and Macheson, 1944; von Borries *et al,* 1945). Inspired by this, B. Tadano (1946) soon reported surface observations as shown in Fig. 4, and the other organizations applied this to metallurgy, dentistry, and so on.

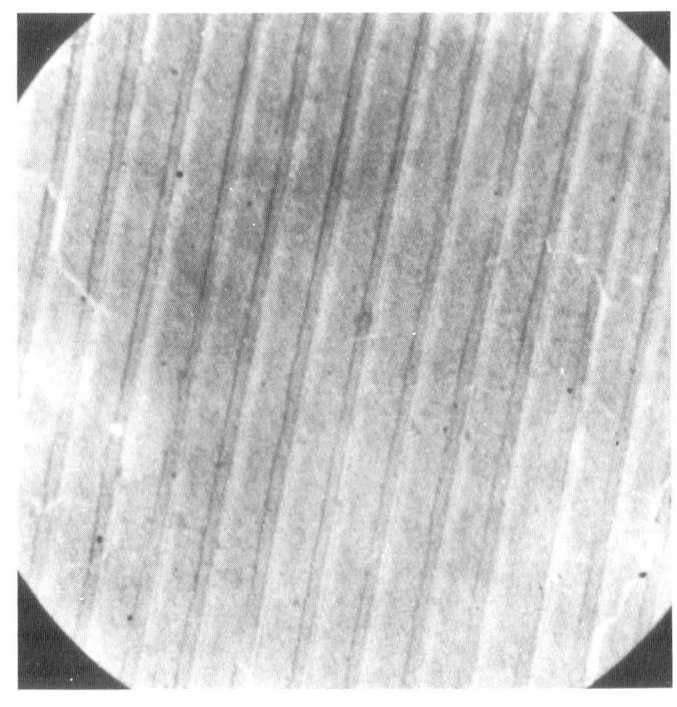

FIGURE 4. Replica (celluloid-aluminum) micrograph of a diffracting grating (Tadano, 1946).

D. Theoretical Study on Aberrations and Resolution

After the war, the research of Suzuki's laborary stressed aberration theory; K. Kanaya and Y. Inoue took part.

E. Newcomers in Manufacturers

At the 40th meeting, which was held at the Central Laborary, Hitachi Co., on 16 February 1946, the chairman was accompanied by K. Kazato, who later founded JEOL; this signaled an epoch in the electron microscopy in the world as well as in Japan.

F. Rapid Increase in Numbers of Researchers

After the war, many researchers and engineers came back to university, national, and industrial laboratories. Every organization was disoriented, while the research group on electron microscopy conducted by Chairman Seto engaged in development; the young researchers gathered to study electron microscopy. As a result, the research population rapidly increased and research was very much activated.

G. Cooperative Research and Development Between Users and Makers

In the 37th Subcommittee, the user who was an application scientist had constructed his own machine, while the researcher or engineer who was an instrumentologist had applied his machine in cooperation with other people and/or by himself. While electron optical theory had been studied primarily at universities and the Electrotechnical Laboratory, the manufacturing companies employed researchers to study aberrations theoretically and experimentally.

The researchers in theory, manufacturing and application began to feel that they should have access to common information. This brought about the organization of the Japan Society of Electron Microscopy.

VII. CONCLUSION

Japanese electron microscopy was certainly founded on the activities of the 37th Subcommittee, JSPS, into which all the committee members put their hearts and souls.

Chairman Seto was honored later by the Order of Cultural Merit from the Japanese Government. At that time, he addressed a message to the members of the Japan Society of Electron Microscopy: "I would like to request one thing to the Members. Although I think Japan is No. 1 all over the world in constructing electron microscopes, we should never be behind the foreign countries in application observing all the microstructures which could infinitely extend mankind's knowledge."

REFERENCES

Heidenreich, R. D., and Macheson, L. A. (1944). *J. Appl. Phys.* **15,** 423.
Knoll, M., and Ruska, E. (1931). *Z. Tech. Phys.* **12,** 389.
Knoll, M., and Ruska, E. (1932). *Ann. Phys.* **12,** 607.
Krause, F. (1936). *Z. Phys.* **102,** 417.
Krause, F. (1937). *Naturwiss.* **25,** 817.
Mahl, H. (1939). *Z. Tech. Phys.* **20,** 316.
Marton, L. (1934). *Bull. de Belg.* 439.
Prebus, A., and Hillier, J. (1939). *Can. J. Res.* **A17,** 49.
Seto, S. (1974). A recollection and impression on early electron microscopy (in Japanese), *J. Electron Microsc.* **9,** 2.
Tadano, B. (1946). Report of the 37th Subcommittee of JSPS, No. 43-3.
von Borries, B., and Ruska, E. (1933). *Z. Phys.* **83,** 187.
von Borries, B., and Ruska, E. (1938). *Siemens Werk* **17-1,** 99.
von Borries, B., Ruska, E., and Ruska, H. (1938). *Siemens Werk* **17-1,** 107.
von Borries, B., and Ruska, E. (1939). **27,** 577.
von Borries, B., Burton, J., and Scott, G. (1945). *J. Appl. Phys.* **16,** 730.

2.9
The Growth of Electron Microscopy in Japan
2.9C History of Electron Microscopes at Tohoku University

KEIJI YADA[1]

Tohoku University
Katahira, Sendai 980, Japan

I. THE MAGNETIC-TYPE EMISSION MICROSCOPE

Professor Kōtaro Honda, a president of Tohoku Imperial University, was interested in the high-temperature microscopic observation of the phase transformation of metals, especially successive stages of the α–γ transformation in iron, and he suggested to Professor Junzo Okubo, the successor of Professor Honda and the teacher of Professor Tadatosi Hibi, the construction of an emission electron microscope because optical microscopes were not able to obtain the necessary images with good quality at high temperature. Therefore, Professor Okubo asked Professor Tadatosi Hibi, who was a research assistant at that time, to start work on the emission microscope in 1936 with his movement to the Department of Physics from the Research Institute for Iron, Steel and Other Metals, Tohoku Imperial University, where he was engaged in the study of magnetism. He attempted to build a magnetic-type emission microscope, which had not been tried in Japan at that time. The original work of Professors E. Brüche and O. Scherzer was known only from one book on electron optics (1934). Therefore, he hesitated before deciding to embark on such a project. However, he started work on the production of a magnetic-type emission microscope after the model by Drs. M. Knoll, F. G. Houtermans, and W. Schulze (1932). The construction (a) and its cross section (b) are shown in Fig. 1. The emission microscope consisted of two magnetic lenses, each having a magnification of 10×. The focal point of each lens was adjusted by moving the lens along the optical bench. The electron image obtained on a fluorescent screen was recorded by a 35-mm camera. The accelerating voltage of 2 kV was supplied from a storage battery, and a vacuum of 10^{-5} mmHg was obtained by means of an oil rotary pump and a mercury diffusion pump with a liquid-air trap. Mechanical construction of this microscope was finished only 1 year later, but a fine electron emission image was not obtained for about 3 years. The main cause of this trouble was the existence of a disturbing magnetic field

---

[1] Present address: Aomori Public College, 153-4 Yamazaki, Goshizawa, Aomori 031-01, Japan.

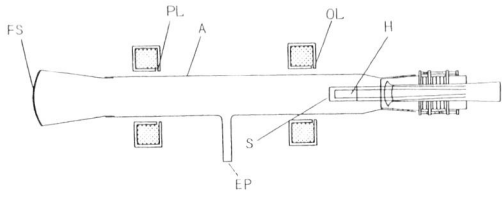

FIGURE 1. (a) The magnetic-type emission microscope built in 1936. (b) Its cross-sectional diagram: A, brass tube; FS, fluorescent screen; H, cathode; S, guard ring; PL, projector lens; OL, objective lens; EP, exit to vacuum pump.

from electric current in the wiring in the room. Finally, simple magnetic shielding was introduced, and consequently, comparatively fine images at about 100× were obtained. Professor Hibi, my teacher, selected the subject of revealing the mechanism of thermionic emission by using this instrument, though it was different from the original aim of Profs. Honda and Okubo. One of the reasons was that he was not in the Research Institute for Iron, Steel and Other Metals nor the Department of Metallurgy but in the Department of Physics. Figure 2a shows the electron image obtained from barium azide coated on a nickel ribbon. It is believed that this is the first electron image obtained in Japan (Hibi, 1942a, 1942b). Figure 2b is the reversed electron image obtained by lowering the temperature of (a), and Fig. 2c is the reproducibly recovered image by raising the temperature. Using the apparatus, several electron optical studies of the emission mechanism of oxide cathodes were made, for example, comparing the thermionic emissions of BaN_6, BaO and SrO (Hibi, 1949), which played an introductive role to further continued studies on the emission mechanism of oxide-

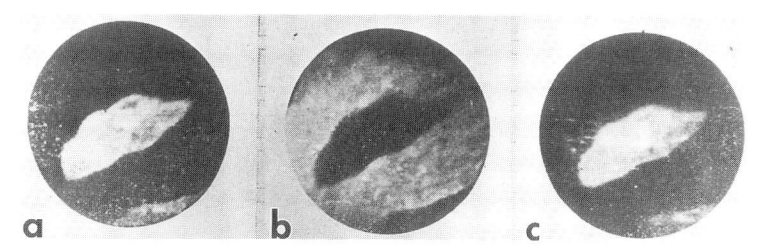

FIGURE 2. Electron images taken by the magnetic-type emission microscope. (a) Electron emission pattern of a nickel cathode coated with barius azide, where the bright part corresponds to the barium azide coated part and the dark part corresponds to the non coated part. (b) The pattern reversed by lowering temperature and (c) the pattern reproducibly recovered by raising the temperature again.

coated cathodes by methods other than electron optical ones (Hibi, 1943, 1950; Hibi and Matsumura, 1951; Hibi and Ishikawa, 1951a, 1951b, 1952).

II. THE ELECTROSTATIC-TYPE ELECTRON MICROSCOPE

An electrostatic-type electron microscope, which Professor Hibi had ordered from the metal workshop of the Department of Physics around 1940, was not completed until 1943, when he was drafted into the army! After the interruption of the study for 2 years due to participation in World War II, construction by the workshop of the Research Institute for Scientific Measurements was re-ordered and completed in 1948. High-magnification electron microscopy had progressed in the magnetic-type microscope rather than the electrostatic-type microscope. Moreover, the latter was technically difficult to manufacture, especially precise machining of the electrostatic lens, so that well-known makers selected the magnetic-type microscope. The electrostatic-type microscope, however, seemed to be suitable to the university laboratory, where research in instrumentation could be conducted steadily, which is the reason why he selected the electrostatic-type microscope. The only Japanese maker who chose the electrostatic-type microscope was Toshiba. Ceramic insulators for separation of the electrodes and high-voltage insulation of his electrostatic lens were supplied by Dr. H. Kamogawa, Toshiba. This electrostatic-type microscope was equipped with a magnetic condenser lens in addition to two electrostatic lenses as shown in Fig. 3. The column and the specimen position were aligned with movable parts on a horizontal optical bench and bellows. It is interesting to note that the electron microscope produced by Dr. Tadano, the magnetic high-magnification type, had exactly the same system as his electrostatic-type. Ideas of many persons are not so different. The image

FIGURE 3. The electrostatic-type electron microscope built in 1948. Only the condenser lens is magnetic.

magnified by the first-stage lens was obtained by this microscope. Before obtaining the highly magnified image by two-stage lenses, Professor Okamura proposed to buy a magnetic-type electron microscope as an instrument for public use in Tohoku University, and a Hitachi HU-6 was installed in 1949, whose resolution was about 30Å at an accelerating voltage of 50 kV. This instrument was actively used for a long time for our research projects.

For instance, the first observation of the dislocation networks in mica and a creative and unique study of the pointed cathode (Hibi, 1954, 1955, 1956) were made with this instrument. By using the pointed cathode and the highly excited objective and projector lenses which were produced in our workshop, the original resolution of 30 Å was greatly improved to about 4 Å later (Hibi et al., 1962; Hibi and Yada, 1964).

REFERENCES

Brüche, E., and Scherzer, O. (1934). "Geometrische Elktronenoptik." Springer-Verlag, Berlin and New York.
Hibi, T. (1942a). Sci. Rep. Tohoku Imp. Univ. 30, 372. (1942b). Ibid. 30, 384.
Hibi, T. (1943). Nihon Sugaku Butsurigaku Kaishi 17, 270 (in Japanese).
Hibi, T. (1949). Sci. Rep. Res. Inst., Tohoku Univ., Ser. A 1, 231. (1950). Ibid. Ser. A 2, 157.
Hibi, T., and Ishikawa, K. (1951a). Phys. Rev. 83, 659. (1951b). Ibid. 84, 1254. (1952). Ibid. 87, 673.
Hibi, T., and Matsumura, T. (1951). Phys. Rev. 81, 884.

Hibi, T. (1954). *Proc. Int. Conf. E. M. London,* p. 636. (1955). *J. Electron Microsc.* **3,** 15. (1956). *Ibid.* **4,** 10.
Hibi, T., Yada, K., and Takahashi, S. (1962). *J. Electron Microsc.* **11,** 244.
Hibi, T., and Yada, K. (1964). *J. Electron Microsc.* **13,** 94.
Knoll, M., Houterman, F. G., and Schulze, W. (1932). *Z. Phys.* **78,** 340.

2.9
The Growth of Electron Microscopy in Japan
2.9D Development of Electron Microscopes at Tokyo Imperial University

AKIRA FUKAMI

Department of Physics
College of Humanities and Science
Nihon University
Sakurajousui, Setagaya-ku
Tokyo 156, Japan

KOICHI ADACHI AND KENTARO ASAKURA

Engineering Research Institute
University of Tokyo
Yayoi-cho, Bunkyo-ku
Tokyo 113, Japan

I. HISTORICAL SURVEY

At the University of Tokyo, Prof. Hideo Yamashita designed a magnetic-type electron microscope and succeeded experimentally in keeping the fluctuation of the direct-current high-voltage power within 0.02%. The fabrication of the electron microscope started in 1943, but was not finished because of World War II. On the other hand, Asst. Prof. Yasumasa Tani had finished designing the lenses of an electrostatic-type electron microscope in the fall of 1939. The Tokyo First Arms Factory was requested to manufacture the device, and an agreement was reached. In 1941, an electron microscope with electrostatic lenses composed of three parallel planelike electrodes was ordered. The accelerating voltage was designed as 50 kV, the magnification 15,000 times, and the resolution 1 nm. The electron microscope was set up in July 1942 and was put in use the next year. However, the image was not very good because of insufficient machining accuracy of the electrodes. At that time, even the supply of film was severely limited because of the war, but the electron microscope was improved continuously until a resolving power better than 10 nm had been obtained.

II. Development of the Magnetic-Type Electron Microscope at the University of Tokyo

Professor Yamashita took charge of the studies: (1) stabilizing the high-voltage DC power supply for the electron microscope, and (2) designing the high-voltage cathode-ray discharge tube. The collaborator was Mr. Ryosaku Sato. In 1942, Prof. Yamashita began the design of a magnetic-type electron microscope, taking care to diminish the aberration resulting from the fluctuation of the high-voltage power supply. In 1943, he had achieved a design with a theoretical power supply fluctuation less than 0.02%, and manufacture was started.

The power supply was designed as follows. To mitigate the dramatic change in AC power supply, a DC shunt motor and an AC generator were used to drive the high-voltage transformer under the control of a mercury rectifier. The motor rotation rate was kept steady by an electronic tube regulator. The rectified DC high voltage was then regulated by the electronic tube on the high-voltage side so that the fluctuation was kept within 0.02%. Though there is a record showing that a magnetic-type electron microscope was ordered and manufacturing had already started in 1943 (Yamashita, 1942), it was interrupted by World War II.

III. Development of the Electrostatic-Type Electron Microscope at the University of Tokyo

Together with the foundation of the 37th Committee of the Japanese Society for the Promotion of Science, Asst. Prof. Tani began to develop an electrostatic-type electron microscope. It is said that Asst. Prof. Tani *et al.* paid attention to this type because no one in Japan had developed it, although in Germany H. Mahl (1939) had made one in 1938.

Assistant Prof. Tani's studies were (1) the focal distance and the aberration of the electrostatic lens and (2) the trial manufacture of an electrostatic-type electron microscope. Mr. Akira Fukami was the collaborator. In 1939, the design of the electrostatic-type electron microscope was finished according to Tani's study, and in the fall, the materials needed were ordered from the Technique Head Office of Arms. The Tokyo First Arms Factory was requested to manufacture the device under the direction of Mr. Mitsuo Nagayama.

The microscope was designed as follows: accelerating voltage, 50 kV; magnification, ×15,000; resolution, 1 nm; and 35-mm film for photography. The electrostatic lenses used for the lens system included the condenser lens, the objector lens, and the projector lens. This microscope was finished in July 1942. Assistant Prof. Tani began the adjustment at once and obtained an enlarged picture the next year. Because it was during the war, there was an extreme shortage of material, so only a little film was supplied. Therefore, experiments could be carried out only on a limited scale. Nevertheless, many papers on electrostatic lenses and the calculation of aberration were presented by him.

Figure 1 shows the microscope (TU 1), which is still preserved as an exhibit in the Engineering Research Institute, University of Tokyo. Figure 2 shows the configuration of the insulator, the central and the external electrodes of the lens. Figure 3 is an image of a razor's edge with a magnification of 15,000 times, taken by this microscope at the first stage. From Fig. 3 it can be seen that the picture was very bad, because of insufficent

FIGURE 1. Electrostatic-type electron microscope (TU 1).

FIGURE 2. Electrostatic lens for TU 1.

machining accuracy of the lens core, the discharging caused by the poor vacuum condition, or the insufficient insulation of the lenses. The focal distance was made longer to improve the image, and therefore the magnification was decreased to 4000 from the designed 15,000.

The microscope developed at the University of Tokyo had the following characteristics:

1. The body was slim. This resulted from the fact that it was difficult to shorten the focal distance of the lens, so the body had to be extended to increase magnification.
2. The oldest rotary cock system was used for the exchange of specimens.
3. The microscope was focused by adjusting of the position of the specimen, which is a primitive method.
4. Because the electron emission part of a CRT was used for the electron gun, it was fixed in special glass to insulate the high voltage. The glass tube was fitted at the top of the body, and could be taken off to change the filament.
5. The special glass tube was used for the insulating of high-voltage parts in the objective and projector lenses.
6. The material used for the body was all mild steel because of the shortage of material during World War II.

During 1945–1946, the picture became much better from the improvement in the machining the lens core, especially the outside one, and the improvement in insulating material. The resolution became better than 10 nm. This microscope was used until 1950 and is the only one of this type still preserved in Japan. Figure 4 shows zinc oxide crystals taken after improvement of this electron microscope.

IV. Change After the Development of the Microscope

As mentioned above, the development study of the electron microscope at the University of Tokyo started in 1939. Though the manufacture of the magnetic-type electron microscope designed by Prof. Yamashita was not finished, the electrostatic-type electron microscope designed by Asst. Prof. Tani was set up in July 1942. Thereafter, the microscope was improved again and again, and after World War II it became possible to take a picture like Fig. 4.

In 1947, the electrostatic-type electron microscope and its supplementary apparatus were contributed to the Engineering Research Institute. This put an end to the practical development and improvement of the body of the microscope, and the emphasis was moved to studying the preparation of specimens for electron microscopes, mainly the study of the replica method which is used to observe the surface of a specimen (Fukami and Tani, 1954).

FIGURE 3. A razor's edge taken before the improvement.

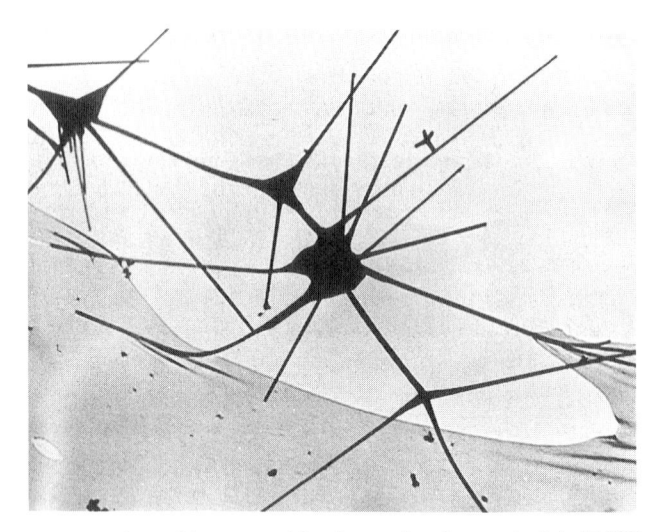

FIGURE 4. Zinc oxide prepared by the combustion method (×15,000).

REFERENCES

Yamashita, H. (1942). *Ann. Rep. Eng. Res. Inst. Faculty Eng., Univ. Tokyo* **1,** 17.
Fukami, A., and Tani, Y. (1954). "Proceedings of the International Conference on Electron Microscopy (Supplement)," London, p. 474.
Mahl, H. (1939). *Z. Tech. Phys.* **20,** 316.

2.9
The Growth of Electron Microscopy in Japan
2.9E Development of Electron Microscopes at the Electrotechnical Laboratory

KOICHI KANAYA

Kogakuin University
Nishishinjuku, Shinjuku-ku
Tokyo 160, Japan

I. INSTRUMENTATION OF ELECTRON MICROSCOPES IN THE ELECTROTECHNICAL LABORATORY

According to a historical survey by L. Marton (1968), the transmission electron microscope was initially constructed by E. Ruska in Prof. M. Knoll's laboratory at the Berlin Technische Hochschule in Germany in June 1931 and was developed into the first Siemens microscope of B. von Borries and E. Ruska (1938) with an estimated resolving power of 100 Å. This transmission electron microscope and its results attracted the attention of several ambitious studies on electron microscopy in Japan, as referred to in the books by Kanaya (1954, 1985).

The instrumentation prototype electron microscope in the Electrotechnical Laboratory (ETL) was developed first by K. Kasai and S. Suzuki, and K. Kanaya joined in 1942. After World War II, the instrumentation and electron optics were pursued by K. Kanaya as the chief of electron microscope section in ETL. Figure 1 shows a preliminary electron microscope and micrographs of butterfly ramentum taken at 50 kV in 1940. This microscope was constructed using a system of three magnetic lenses located outside the column as in a cathode-ray oscillograph. Figure 2 shows the second electron microscope, which was built in 1941. A glass tube was used for the microscope column, so stable operation was difficult.

Figure 3 shows an improved electron microscope (a) and a micrographs (b) of zinc oxide taken at 50 kV (S. Suzuki and K. Kanaya, 1942–1945). The contrast stability was improved by the self-biasing in the electron gun.

In 1941, the Hitachi and Shimadzu companies simultaneously started manufacturing commercial electron microscopes with the aid of the numerous constructional details given in von Ardenne's book (1940) and the

FIGURE 1. Preliminary electron microscope (Suzuki, 1940).

fundamental theories on electron optics developed in Zworykin's book (Zworykin et al., 1945). In 1947, the Japan Electron Optics Laboratory (JEOL) began to manufacture a magnetic electron microscope; Akashi Seisakusho has also been manufacturing such instruments since 1950.

II. Early Work on the "Optimum Lens" and the Temperature Rise of the Specimen Due to Electron Bombardment

One of the earliest contributions to the electron lens was that of Y. Sugiura and S. Suzuki (1943), who introduced the idea that the unipole magnetic lens gives minimum spherical aberration.

The unipole magnetic lens was developed in practice by K. Kanaya as an asymmetrical lens with different bore radii (1949b, 1950) in which the spherical and axial chromatic aberration coefficients C_s and C_c take their minimum values when the front and rear half-width ratios of the magnetic field distribution are $a_1/a_2 = 1/2$ and $1/5$, respectively.

Figure 4a shows the magnetic electron microscope made by Shimadzu Seisakusho Co. Ltd. designed during 1943–1946 under the guidance of the Electrotechnical Laboratory. Figures 4b and 4c show the crystal growth of PbO and zinc taken at 50 kV in 1948; the microscope was designed by S. Suzuki and K. Kanaya; the ripple in the applied voltage is 5×10^{-4}; the resolution, 20 Å. The column is composed of a three-lens system and is a modified version of a von Ardenne design. Early investigations of the temperature of specimens and their structural changes due to electron bombardment in the electron microscope were made by K. Kanaya (1948) using the microscope, as shown in Figs. 4b and 4c.

When using the microscope, it was found that the specimens changed very much due to electron bombardment. Kanaya (1949a) has reported a theoretical formula for calculating the temperature of specimens, consider-

FIGURE 2. Prototype magnetic electron microscope designed by ETL (1941).

FIGURE 3. Electron microscope: (a) column; (b) zinc oxide (Suzuki, 1941).

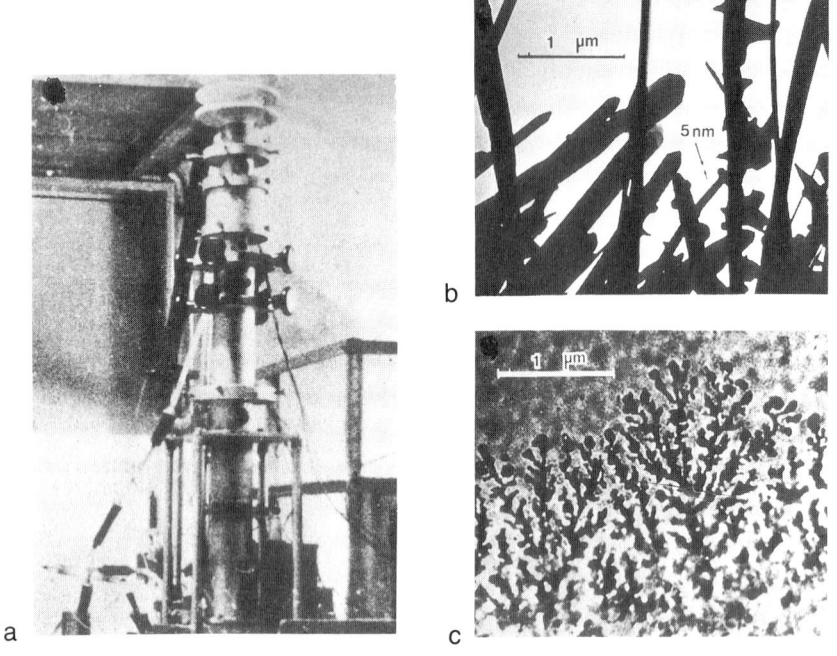

FIGURE 4. Electron microscope: (a) column; (b) litharge crystal grown after electron bombardment; (c) zinc crystal grown (Suzuki and Kanaya, 1947).

ing the energy absorbed by the specimen and that radiated at its surface. The specimen temperature t is given theoretically by

$$t = \left[293^4 + \frac{0.112Fdi_o}{c\Phi A} \sum NZ \left(\log \frac{e\Phi}{IZ} + \frac{1}{2} \right) \right]^{1/4} - 273$$

The summation is taken over all species of atoms involved; it was assumed that the temperature of the instrument is 20°C (corresponding to 293 K) in this calculation; F is the surface area surrounding the specimen, d the thickness, i_o the electron beam current density, c the emissivity, Φ the accelerating voltage, A the cross section bombarded by the beam, N the number of atoms with atomic number Z, e the electronic charge, and I the ionization energy.

Figure 4b shows one of the changed structures of red litharge crystals, where the crystals have grown after heating to 500–890°C in the dense electron beam (Kanaya and Oshina, 1949b, 1949c; Kanaya et al., 1949). Figure 4c also shows the growth of zinc crystals produced by heating the brass mesh in the electron microscope and suddenly introducing air into it (Kanaya and Oshina, 1949a).

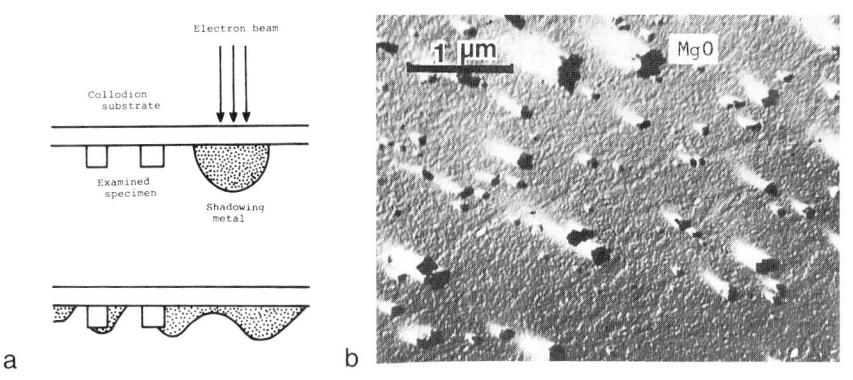

FIGURE 5. Direct shadow-casting principle (a) and (b) a result of MgO crystal shadowed by evaporated Pb, where Pb was bombarded with electron beam $i_o = 0.8$ A/cm^2 for 5 min at 50 kV (Kanaya, 1949a).

From the remarkable facts that metals and metallic oxides melt and evaporate under dense electron bombardment with a current density $i \geqq 0.1$ A/cm^2 at 50 kV, the direct shadow-casting method *in situ* in the electron microscopes has been successfully employed, as shown in Fig. 5. The specimens supported on the collodion film are shadowed very clearly by the evaporated metallic films of Pb and Bi (Kanaya, 1949a). Afterward, following the work of B. von Borries and W. Glaser (1944), the distribution of temperature in thin films supported over a circular opening (Kanaya, 1954), the temperature distribution of specimens on thin substrates (Kanaya, 1955a, 1955b), and the temperature distribution along a rod specimen (Kanaya, 1955c, 1956) were theoretically discussed and formulated taking account of thermal conduction as well as radiation.

It is a remarkable fact that under microbeam illumination, a part of the energy absorbed is used for the conduction of energy, so the temperature at the center of a substrate can be reduced to $(R_2/R_1)^{1/2}T$ in comparison with that of the ordinary illumination, where R_2 and R_1 are the radii of a microbeam and a circular hole, respectively. This was confirmed by the fact that the structure of Pb$_3$O$_4$ is not affected with microbeam illumination produced by a double-condenser system, but melts under ordinary illumination with $R_1 = R_2 = 2$ μm at 50 kV.

REFERENCES

Kanaya, K. (1948). On the temperature rise of specimens in electron microscopes. *Mem. Res. Rep. Electrotech. Lab.*, p. 33.

Kanaya, K. (1949a). Investigation on the temperature of specimens and their structure changes due to electron bombardment in electron microscopes. *Res. Rep. Electrotech. Lab.*, No. 509.
Kanaya, K. (1949b). *Bull. Electrotech. Lab.* **13,** 338.
Kanaya, K. (1950). *J. Electron Microsc.* **1,** 35.
Kanaya, K. (1954). "Electron Microscope—Theory and Practice." Denkishoin, Kyoto.
Kanaya, K. (1955a). Electron optics theory on the design of magnetic lenses for electron microscopes. *Res. Rep. Electrotech. Lab.*, No. 548.
Kanaya, K. (1955b). *Bull. Electrotech. Lab.* **19,** 217.
Kanaya, K. (1955c). *Bull. Electrotech. Lab.* **19,** 680.
Kanaya, K. (1956). *J. Electron Microsc.* **4,** 1.
Kanaya, K. (1985). "Reminiscence of the Development of Electron Optics and Electron Microscope Instrumentation in Japan." In Advances in Electronics and Electron Physics (P. W. Hawkes, Ed.). Academic Press, pp. 317–385.
Kanaya, K., and Ishikawa, A. (1951). *Bull. Electrotech. Lab.* **15,** 264.
Kanaya, K., and Oshina, M. (1949a). *Bull. Electrotech. Lab.* **13,** 104.
Kanaya, K., and Oshina, M. (1949b). *Bull. Electrotech. Lab.* **13,** 112.
Kanaya, K., and Oshina, M. (1949c). *Bull. Electrotech. Lab.* **13,** 284.
Kanaya, K., Oshina, M., and Shibuya, S. (1949). *Bull. Electrotech. Lab.* **13,** 284.
Marton, L. (1968). "Early History of the Electron Microscope." San Francisco Press, San Francisco, CA.
Sugiura, Y., and Suzuki, S. (1943). *Proc. Imp. Acad. Jpn.* **19,** 293.
von Ardenne, M. (1940). "Elektronen-Übermikroskopie." Springer-Verlag, Berlin and New York.
von Borries, B., and Ruska, E. (1938). Wiss. Veroeff. *Siemens-Werken* **17,** 99.
von Borries, B., and Glaser, W. (1944). *Kolloid-Z.* **106,** 123.
Zworykin, V. K., Morton, G. A., Ramberg, E. G., Hillier, J., and Vance, A. W. (1945). "Electron Optics and Electron Microscope." John Wiley, New York.

2.9
The Growth of Electron Microscopy in Japan
2.9F Early Electron Microscopes at Osaka University, 1934–1945

KATSUMI URA

Department of Electric and Electronic Engineering
Faculty of Engineering
Osaka University
Sangyo,
Nakagaito, Osaka 574, Japan

The study of electron microscopes at Osaka University was started in 1934 by Asst. Prof. Eizi Sugata. He was an Emeritus Professor of Osaka University and Chairman of the Board of Directors, Osaka Electro-Communication University, when he died on 13 July 1988. Based on his earlier publications and memoranda, the present article is written by the author, who was one of Prof. Sugata's students. Professor Sugata kindly read the manuscript, corrected some mistakes, and added some recollections.

E. Sugata graduated from the Department of Electrical Engineering, Osaka University, in 1932. His first concern was magnetic materials for engineering use. In 1934 he read the famous paper by M. Knoll and E. Ruska. He was greatly stimulated by this and decided to study the electron microscope, but he had to wait for Prof. Y. Shichiri, who was his research leader and at that time visiting foreign countries. At the end of that year, he started his new study.

E. Sugata had neither high-voltage apparatus nor a vacuum pump. Every thing had to be designed by himself. The emission electron microscope became his first target. The first apparatus was made of hard glass as shown in Fig. 1. The images of the cathode were reported in April 1938 (Sugata, 1938). The Mark II emission microscope was made of metal in 1939. Sugata inserted a mesh after the cathode and studied its electron optical images (Sugata, 1940). These were the first experiments in electron optics in Japan. The Mark III apparatus was completed in 1941 (Sugata, 1941); this remains as it was.

In May 1939, the subcommittee on electron microscopy, the 37th Subcommittee, was established inside the 10th Committee of the Japanese Society for the Promotion of Science (JSPS). E. Sugata was an Associate Professor of Osaka University and was appointed as a committee member at the

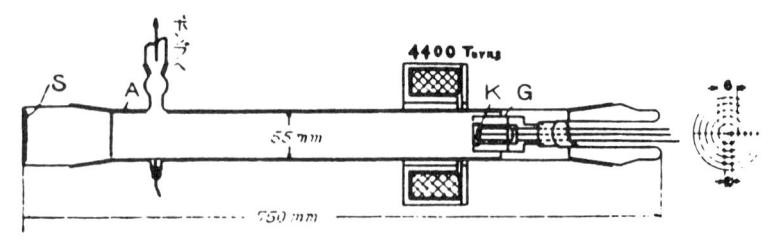

FIGURE 1. Cross section of the Mark I emission electron microscope built in 1936 (Sugata, 1938). The envelope was hard glass, and the accelerating voltage was about 3 kV.

beginning. The financial support from this subcommittee enabled him to construct the Mark I "super" microscope. The magnetic type was chosen. One reason was that it would allow easy adjustment of the lens, while the vacuum discharge in the electrostatic lenses seemed too troublesome to be controlled. Another was his affection for magnetism, which was his former concern. At the end of this year, the column was completed. A photograph of the microscope was presented to the sixth meeting of the 37th Subcommittee, JSPS, on 14 February 1940. *Electronics* magazine (USA) introduced this machine together with young Prof. Sugata in April 1940 in an article entitled "Japan Gets Electron Microscope." Figure 2 shows the Mark I "super" electron microscope.

After its test run, its gun insulator was replaced by porcelain. Its poor mechanical precision forced E. Sugata to devise various mechanisms. The first electron micrograph of a mesh was taken on 18 December 1940, according to the micrograph diary at that time. Figure 3 shows one of the earliest micrographs, which appeared in a popular scientific magazine (Sugata, 1941). On 11 and 12 September of the same year, Prof. E. Sugata took electron micrographs of the suppurative virus of silkworm as shown in Fig. 4, which were passed around at the 15th meeting (20 September 1941) of the 37th Subcommittee, JSPS, according to the minutes and later published (Sugata, 1943). It is noted here that the late Prof. N. Higashi, Kyoto University, previously presented his electron micrographs of a suppurative virus to the 24th meeting of the 37th Subcommittee (2 April 1943).

In 1942, both specimen and camera chambers were air-locked, using a metal bellows. With the updated Mark I machine, many micrographs were taken. The accelerating voltage was 25 kV at the beginning. In 1941, it was 40–50 kV, 40–60 kV in 1942. Some micrographs were taken at 90–100 kV from December 1942 to February 1943, but fine structures were not observed. Figure 5 shows an electron micrograph of cotton fiber which was taken on 16 January 1943. The electron optical magnification ranged from ×3000 to ×13,000. Figure 6 shows the statistics of micrographs per month. At the beginning, one micrograph was taken per day. Later, three micro-

FIGURE 2. General view of the Mark I transmission electron microscope with two-stage magnetic lenses built in 1939. On the right is the young Prof. E. Sugata.

graphs were taken in a day on average. Nine micrographs were recorded on 16 February 1944, where the specimen was megaterium with and without mechanical processing. It took a rather long time to evacuate the column after exchange of specimens, even though the specimen chamber was airlocked. After the middle of 1944, the aggravation of the war made research very difficult; the number of micrographs decreased and finally fell to zero after December 1944.

The specimens were: meshes, MgO, ZnO, iron oxide, tungsten powder, carbon black, collodion membrane, feather of a day-flyer, butterfly ramentum, silk fiber, cotton fiber, diatom, pulp, megaterium, colitis germ, staphylococcus, tubercle bacillus, suppurative virus of silkworm, tobacco virus, and so on. Prof. E. Sugata cooperated with the experts in applications. The fine structure of silk fiber ($\times 24{,}000$) was observed together with a scientist of the Institute of Silk Science (Ozaki et al., 1943). The micro-microbe in bacillus was extracted from a colitis germ by the supersonic vibration and observed with a magnification of 8700 in collaboration with medical scientists (Kasahara et al., 1943).

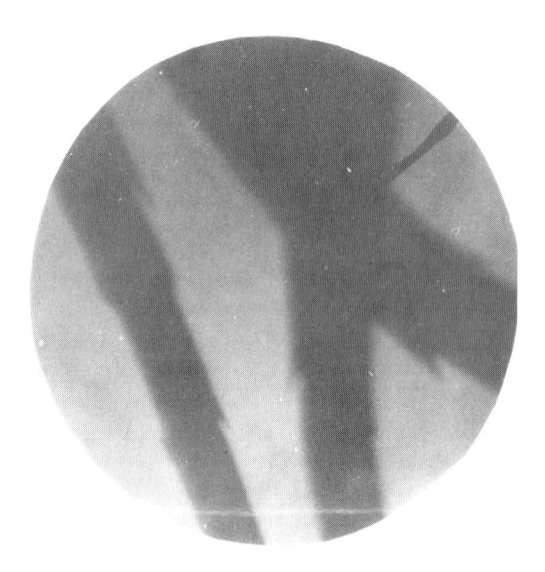

FIGURE 3. Electron micrographs of feather of a day-flyer (Sugata, 1941).

In 1943, the Mark II machine was newly constructed based on acquired experience; it had a new gun alignment mechanism, exchangeable apertures, and a camera box which had a more modern style. Study by this machine and the Mark I machine was interrupted due to the aggravation of World War II. Later, the Mark II was converted to the Mark III machine (1952).

Through a series of trial constructions, Prof. Sugata realized that the electron gun was the fatal component in the electron microscope. He started a systematic study of it. Both radius and angle of the crossover were measured as a function of the Wehnelt voltage, where the parameters were the hole size and thickness of the Wehnelt electrode, the distance from a hairpin

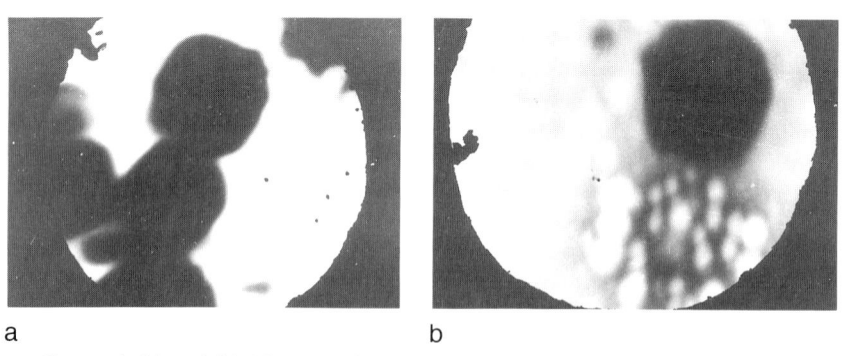

a   b

FIGURE 4. (a) and (b) Electron micrographs of suppurative virus of silkworm (Sugata, 1943). These were passed around at the 15th meeting of the 37th Subcommittee, JSPS (20 September 1941).

ELECTRON MICROSCOPES AT OSAKA UNIVERSITY 267

FIGURE 5. Electron micrograph of cotton fiber (16 January 1943). The acceleration voltage was 61 kV and the electron optical magnification was 12,000 times.

cathode to the Wehnelt, and their mutual alignment. The first report was read at the 26th meeting of the 37th Subcommittee, JSPS, on 10 July 1943. According to the minutes, it was evaluated highly by the other committee members. Unfortunately, this study was also interrupted due to the war and postwar affairs. The second and third reports were presented on 30 August and 26 November 1947 to the 37th Subcommittee; finally it was published in 1950 (Sugata, *et al.*, 1950). Figure 7 shows some results.

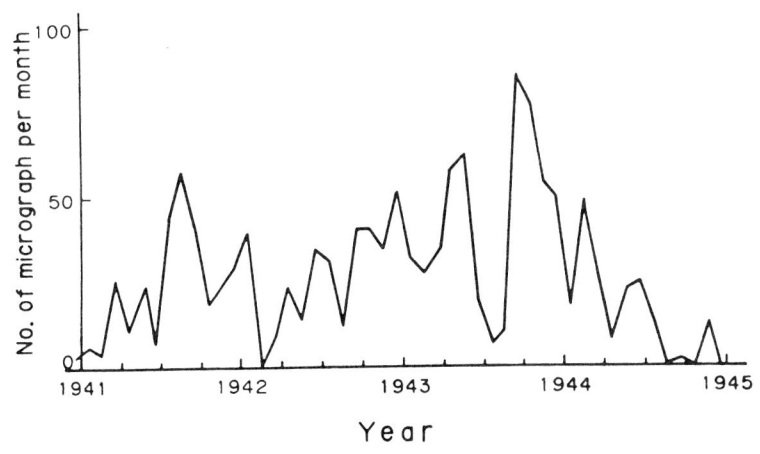

FIGURE 6. Statistics of micrographs per month in Sugata's laboratory, 1940–1945.

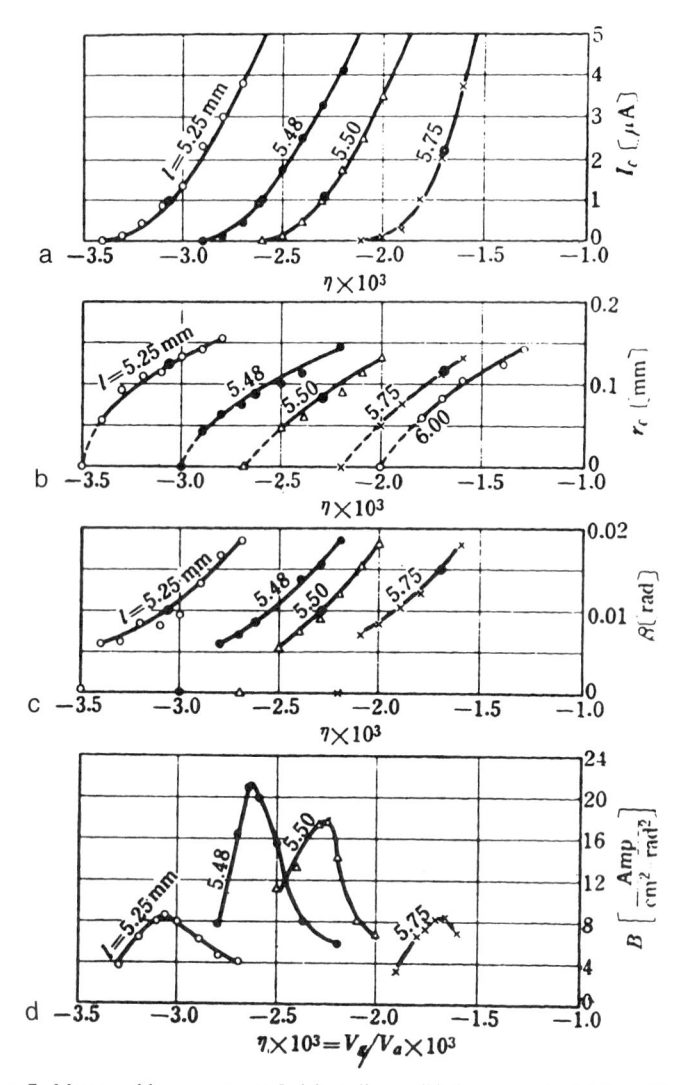

FIGURE 7. Measured beam current I_c (a), radius r_c (b), beam angle β (c), and brightness B (d) as a function of normalized bias voltage $V_g/V_a \times 1000$ (Sugata *et al.*, 1950). The parameter is the distance from the emitter tip to the Wehnelt electrode surface.

At present, not only the revised Mark I machine and Mark III machine, but also a large number of blueprints, micrographic plates and diaries are preserved in a room of Faculty of Engineering, Osaka University.

REFERENCES

Kasahara, M., *et al.* (1943). *J. Inst. Osaka Med. Sci.* **42,** 1538.
Ozaki, S., *et al.* (1943). "Proc. 22nd Meeting of Electrical Engineers," p. 12.
Sugata, E. (1938). "Proc. 13th Meeting of Electrical Engineers," p.13.
Sugata, E. (1940). "Proc. 4th Meeting of Electrical Engineers," p.5.
Sugata, E. (1941). *Kagakujin*, p.137.
Sugata, E., *et al.* (1942). "Proc. 21st Meeting of Electrical Engineers," p.11.
Sugata, E. (1943). *Jpn. Soc. Sci.* **17,** 218.
Sugata, E., *et al.* (1950). *Oyobutsuri*, **19,** 81 (accepted 14 February 1949).

2.10
Electron Microscopy in the Netherlands
2.10A Earliest Developments

WOUTERA VAN ITERSON[1]

Section of Molecular Cytology
Institute for Molecular Cell Biology
BioCentrum Amsterdam
University of Amsterdam, The Netherlands

I. Early Days

In 1939 Delft was only a small town with a famous past. It was here that in 1584 William the Silent, the so-called Father of the Fatherland, was assassinated, and it is in the Nieuwe Kerk in Delft (dating from 1384) that the burial vaults of the members of the Royal House of Orange can be found. It was also in Delft that Antoni van Leeuwenhoek, the founder of microbiology, studied his "little animals" through home-made glass lenses. However, were it not for its Polytechnische School, the Delft University of Technology, and for its innovating industry, the overall impression of Delft shortly before World War II was that of an old-fashioned town. Within the context of this memoir, a particular part of Delft's industry, the Nederlandsche Gist en Spiritus Fabriek, or "Yeast Factory" in short, plays a major role.

For one thing, the Yeast Factory has contributed strongly to the development of the most important tradition of microbiological research in the country in the technological environment of the Delft University of Technology. In 1885 J. C. van Marken, the managing director of the Yeast Factory, invited M. J. Beyerinck to join the factory. Beyerinck was to become Professor of microbiology in 1895 and has been called the father of microbiology. A. J. Kluyver, the father of microbiologists, succeeded Beyerinck in 1921. Kluyver combined his tenure as professor with a consultancy at the Yeast Factory. What is the link between all this and electron microscopy? The yeast cell.

In the summer of 1939, an engineering student at the Delft University of Technology by the name of Jan B. Le Poole (Fig. 1) approached his professor in physics, H. B. Dorgelo, with the somewhat startling request that he should build an electron microscope for his major in engineering. The time was indeed ripe, since by a curious coincidence, Dorgelo, along

---

[1] Present address: De Paasberg, Overzicht 64, 6862 CT Oosterbeek, The Netherlands.

FIGURE 1. Prof. Dr. Ir. J. B. Le Poole, the founding father of electron microscopy in the Netherlands and first President of the Dutch Society for Electron Microscopy.

with F. G. Waller, President of the Yeast Factory, and A. J. Kluyver had just returned from Berlin, where they had been visiting the Siemens Company on 6 July 1939. At the time, Kluyver was familiar with the recently published pictures of microorganisms and other objects at the comparatively high magnification afforded by an electron microscope. The question was whether it would be possible to decide with such an instrument whether a yeast cell is equipped with a true, condensed nucleus with chromosomes or whether it resembles bacteria, where it was not certain that a clear differentiation could be made between nuclear material and cytoplasm. In view of the practical implications of this question, Waller and Kluyver had taken the matter up with Dorgelo, and these three men decided to travel to the Mecca of transmission electron microscopy and its theoretical background: prewar Germany. As early as 1939, Siemens had managed to sell its first commercial electron microscopes, based on a design by von Borries and Ruska. The magnification was up to 40,000 and the resolution was

[2] Waller's report of the visit to Siemens.

much better than that of the light microscope. Its price of approximately Dfl. 80.000, however, was not really in line with the possibilities it offered.[2] Furthermore, in Berlin they did in fact observe a yeast cell at "high" magnification, but this had been no more than an ugly dark spot, whereas with the light microscope they were familiar with a neatly integrated organism, possessing within the cell wall protoplasm, vacuoles, and various other structures; only the nucleus remained dark.

Generally speaking, at that time, the usefulness of such an instrument for biological research was debatable. With the whole of the cell in focus, would one ever be able to distinquish important details? Moreover, electrons had always been regarded as corpuscles, until in 1924, through the work of De Broglie, it was realized that in traveling they also had a wave nature. Still, this did not change the fact that corpuscles would certainly bombard, and thus destroy, organic materials. On top of that, the essence of life is in the high percentage of water in the cell, and were not cells dehydrated under the vacuum conditions of the instrument? When the invention of the electron microscope had become more generally known, in certain biological circles the slogan could be heard: "The electron microscopist is a collector of artifacts." After all, had not Frey Wyssling in Switzerland, and others, analyzed the general structure of the cell sufficiently well by indirect methods? And had not important treatises been written on the structural properties of biological membranes almost to the molecular level? Could the electron microscope really add anything to this important arsenal of knowledge of the 1930s? These objections constituted both the adventurous side and the good fortune of the enterprise for the young scientists-to-be at the Delft University of Technology.

In view of all the uncertainties, it turned out to be fortunate that the young Jan Le Poole desired to be a pioneer. He built a two-stage electron microscope with which the first electron micrographs could be taken in 1941. However, the accelerating voltage of 40 kV proved very restrictive. Thus, it was decided that a 150-kV electron microscope be constructed in collaboration with the Philips Physics Laboratory. At Philips in Eindhoven, A. C. van Dorsten developed a very stable 150-kV unit, while meanwhile Le Poole, by now assisted by H. J. de Heer, was working on the electron optical system in Delft. Here, in the spring of 1944 the new 150-kV electron microscope was ready to take pictures (Le Poole, 1947).

II. Early Organization

It was soon understood that the development of an electron microscope and testing of its applicability in biology and other disciplines would require an organization and funds. In 1941 the TPD was founded as a cooperative

---

[3] Technisch Physische Dienst TNO en TH Delft (henceforth: TPD).

effort between the Organization of Applied Scientific Research (TNO) and the Delft University of Technology.[3] On 1 November 1943, a special Institute for Electron Microscopy was founded as a section of the TPD, although on a separate budget. The institute was subsidized by industries such as the Delft Yeast Factory, Philips, Van Houten (cocoa), Algemene Kunstzijde Unie (AKZO), Heineken Breweries, and by the TPD. Later, Dutch Unilever and Royal Dutch Shell also contributed with at least Dfl. 3000 a year. The institute was supervised by an advisory board, and the technical and daily management was in the hands of Le Poole, while Dorgelo and Kluyver were in charge of the scientific supervision (Spit, 1966).

III. The Delft Microscope

We, from Le Poole's small group, working in isolation before our nation was liberated from the tribulations of war, barely realized that the microscope's design contained a number of exciting innovations. One of these innovations was the introduction of two additional lenses between the objective lens of 40× magnification and the projector lens of 160× magnification. One of the extra lenses had a small bore, which enabled continuous variation of the magnification from 6400× to 80,000×. For magnification up to 6400×, the current was sent through the so-called diffraction lens, with a larger bore. With that diffraction lens, diffraction patterns could be obtained from a selected region of the specimen as small as 3 μm. One could switch back and forth from the electron image to electron diffraction, which in Delft had been found useful for the determination of clay minerals. The principle of selected area diffraction had previously been discovered by H. Boersch, but this was unknown to Le Poole at the time. Another advantage of introducing intermediate lenses was the reduction of the height of the microscope column, the total distance from object to final image amounting to 60 cm. Furthermore, Le Poole introduced a special focusing arrangement providing for accurate focusing, especially at high magnifications, when the intensity on the fluorescent screen is low. By means of a transverse electric field between two sets of parallel plates interposed between the condensor and the object, the incident ray was caused to oscillate with a frequency of 50 Hz. When the objective lens is not exactly focused, this oscillation blurs the image. This facilitated focusing, and substantially improved the quality of the electron micrographs taken at the Delft Institute. A magnetic version of this "wobbler" has been a feature of all Philips transmission microscopes ever since.

The image field in the early microscope was very large (18 cm in diameter) and was projected onto the bottom of an Erlenmeyer flask turned into a

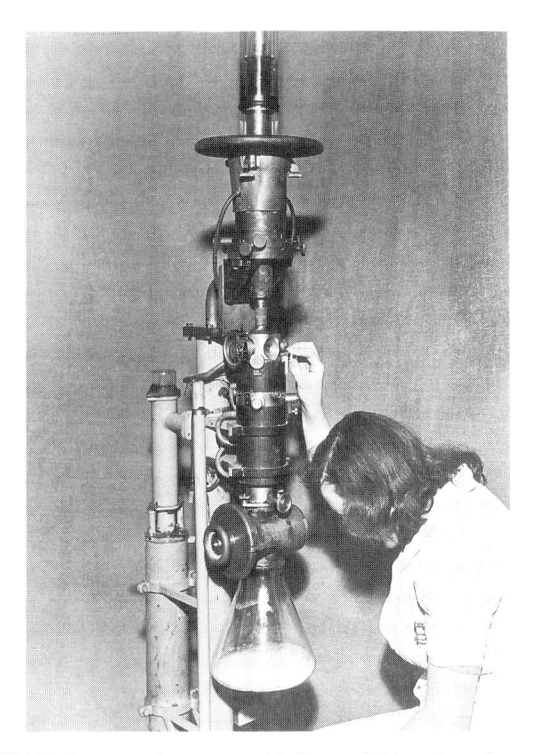

FIGURE 2. 150-kV electron microscope with image field projected onto the fluorescent material deposited on the base of an Erlenmeyer glass flask.

fluorescent screen (Fig. 2). By introducing a 35-mm film above the screen at a location where the cross section of the beam was still sufficiently small, the whole image could be covered in the subsequent photographic enlargements. The emission voltage was variable between 50 and 120 kV, the higher voltages in the case of biological objects often improving penetration by the electron beam.[4]

In Delft an electrostatic electron microscope was constructed as well. This instrument, built by W. A. le Rütte, was ready in 1951 and had a resolution of 8 nm at a fixed magnification. Le Rütte published a thesis on his contribution to the electrostatic electron optics (1952), but this work was discontinued because of the technical advantages of the magnetic version of the electron microscope.

Another interesting development had started in the middle of 1943. Already in 1942, the too-voluminous yeast cell had led Le Poole to propose

---

[4] For further information on the Le Poole electron microscope, consult Le Poole (1947, 1954).

the building of a 1-MeV microscope to improve the penetration of the specimen by electrons. For the construction of such an instrument various problems had to be overcome, and so it was decided to build a 400-kV microscope in the Philips Research Laboratories instead. Le Poole designed the column for this microscope, while at Philips Van Dorsten took care of the high-voltage equipment. Oosterkamp was responsible for the gun, and Verhoeff for the construction (Van Dorsten *et al.*, 1947). In 1947 this apparatus was installed at the Delft Institute.

IV. THE EARLY WORK WITH THE DELFT MICROSCOPE

Not only was research made on the electron microscope but applications with it also started at an early stage. In the years preparing the basic 150-kV electron microscope, the old two-stage microscope had to be used both for testing Le Poole's new ideas and for research. In this work we owe much to Harrie de Heer's introduction of excellent photographic techniques.

A biologist, A. Quispel, started working in October 1942 as a Research Assistant under the supervision of Professor A. J. Kluyver. One of the first things he did was to prepare adequate supporting films of "Geisselthallack" on one-hole specimen holders. Quispel's task was to investigate the usefulness of the instrument for biological research, with special emphasis on the investigation of the alleged chromosomes in the yeast nucleus. In order to do this, Quispel developed a method to "stain" the yeast nucleus, i.e., to raise the contrast in comparison to the rest of the cell. This selective staining required heavy metal, and so, altering the method of Feulgen, he used silver as well as lanthanum salts. However, the yeast did not reveal the secret of its chromosome nucleus, which remained in pitch dark. Quispel then tried to make the cytoplasm more transparent to the electron beam with proteolytic enzymes. When Quispel left Delft in September 1943, the work was transferred to myself, at first with the assistance of Miss J. M. van Brakel. However, it turned out to be premature to work on the much-too-large yeast cell.

In those days, we were much repressed by the war (cf. Le Poole, 1985), but we were young and enthusiastic about the work. Eagerly we studied yeast cells, Cytophaga, tuberculosis bacteria for a sanatorium physician, all sorts of other bacteria, as well as clay minerals in soil samples, pigments, metals, and a variety of other items that were photographed on 35-mm film.

V. The Situation Toward the End of the War

When the 150-kV electron microscope with all its improvements came into operation in 1944, it was only for a few weeks. In the country the situation was becoming very critical with the approach of famine, the "honger winter" of 1944–1945. The Allied troops had liberated the south of the Netherlands, but had been halted near the great rivers. During that winter, our food ration north of the rivers was reduced to 800 calories per week. Naturally, we were preoccupied with food, and our thoughts alternated between a primitive level of wishful thinking and the horrors brought by the, usually illegally, transmitted news or our own personal experiences. We had no electricity, the passenger trains did not run, and we only had bicycles with wooden tires for transportation. Then came a day when it seemed advisable to dismantle the electron microscope to hide the heart of the instrument: the lenses. The cooling oil of the high-voltage generator turned out to be a blessing not intended by Philips: It was distributed for fuel among the workers of the institute. We were glad to use it at home for illumination purposes, where we, being involved in underground activities, kept on trying to resist the dangerously oppressing circumstances, for instance, by hiding under the floorboards when the German occupying forces tried to arrest all male persons between the ages of 18 and 40 for forced labour in Germany.

VI. After the Liberation

After the turmoil of the liberation by the Canadian Armed Forces had died down, and the electron microscope had been reassembled, the main question became whether the activities with the Delft microscope had just kept us busy in our days of isolation, or whether a real contribution had been made to technology and science. The reactions of visitors from the Allied countries gave us the impression that the Le Poole microscope might be something special, but could one rely on that feeling? Anyway, Philips in Eindhoven was not at all prepared to start producing electron microscopes on a commercial basis, since the firm was interested primarily in products that sold at least by the thousands. Was there a means to raise the hopes of my colleagues? Yes, there was. In the first place, I wrote a review on the activities in the field of electron microscopy that had taken place in the United States. This was made possible because shortly after the liberation of the south of the Netherlands, in September 1944, the library of the Dutch State Mines (DSM) had come by specific American scientific journals. Although it was clear that the work of scientists in America had been extensive and impressive, the review convinced the Delft physicists that their achievements had not been in vain. Furthermore, I discussed their concerns with my father, who, being a scientist as well as a member of the board of the Dutch State Mines, was in a position to appreciate both sides: the importance of the new instrument as well as Philips' industrial point of view. The President of Philips, Dr. Anton Philips, had just returned from England, where he had spent the war years. I accompanied my father to Eindhoven, where we had lunch at the president's home. Mr. Philips listened carefully, as he had not yet heard of the construction of the Delft electron microscope, with which his firm was to be so closely involved. In January 1946 Jan Le Poole had the opportunity to visit England and to attend a meeting of the British Electron Microscopy Group. There his last vestiges of doubt disappeared: The Delft microscope really was an innovation (Le Poole, 1985). He met Van Dorsten and in England they discussed the requirements for a commercial Philips electron microscope. In January 1946, the managing board of Philips appeared to have changed its point of view and was now prepared to promote the development of a prototype electron microscope as a basis for commercial production. The instrument could to some extent be developed in the X-ray equipment division, but the prototype (Van Dorsten *et al.*, 1950) was built in the Physics Laboratory of Philips, later known as the Philips Research Laboratories. Later on, a special electron microscope division became part of the group Medical Systems under the main industry group, Science and Industry. In retrospect, this is a true consequence of all the efforts in the early period.

In 1946, the prototype microscope built at Philips was shown at a conference in Oxford, where it stubbornly refused to produce a useful image, but was admired just the same. (After the conference, it was found that an aperture disk had slipped out of place in the column during transport, thereby blocking the electron beam.) Next, the decision was made to build a series of four prototype electron microscopes, of which some parts would be made at the instrument makers' school of the Kamerlingh Onnes Laboratory of Leiden University.

The final design of the Philips EM 100 was completed in 1947. An unusual early feature was that the fluorescent screen was viewed in transmission and was inclined to the horizontal as shown in Fig. 3. In all subsequent Philips microscopes this construction was abandoned, as a vertical column is more stable mechanically than an inclined one.

FIGURE 3. Philips EM 100.

VII. THE POSTWAR PERIOD

The staff at the Delft Institute gradually grew in number: There were four physicists, one biologist, one engineer, two instrument makers and four technicians. From 1946 Le Poole was assisted by J. Kramer, who remained his right-hand man for over 36 years.

In 1946 the priorities of the physicists were the correction of the microscope's astigmatism, the improvement of the high-voltage stability, and the further development of an objective lens of such strength that the chromatic aberration would be sufficiently decreased without the need for further stabilization of the lens current and the high voltage. This work, among other things, provided a background for the design at Philips of a simplified electron microscope (Van Dorsten and Le Poole, 1955).

Besides the development of the electron microscope, working with the instrument became of more and more importance. The latter consisted of research in microbiology and of work done for clients outside the institute. Three electron microscopes would indeed not have been a luxury but there was only one, and that microscope at times had to be taken apart for the development of the instrument.

The quality of an electron microscope is reflected in the quality of the micrographs of well-prepared specimens. The art of making proper preparations was also at a pioneering stage in those days. The specimens would either be too delicate, lacking image contrast, or they would be too thick as in the yeast cell, even for the 90 kV mostly used for biological specimens. The lack of contrast was particularly disturbing when photographing plant juice samples from the Laboratory of Flower Bulb Research at Lisse, in whose samples virus rods had to be discerned (Van Slogteren, 1952). Usually, I photographed these viruses without even being able to observe them. A great improvement was introduced after the visit of Dr. Ralph W. G. Wyckoff from the National Institutes of Health in Bethesda, Maryland, who acquainted us with the shadow-casting technique. This literally added a new dimension to the electron micrographs of bacteria with their long flagellae (Fig. 4) and to numerous other samples.

In 1947, I had the privilege to go and work in the United States on a fellowship from the National Institutes of Health in Bethesda. In December that year, at the EMSA meeting in Philadelphia, I was able to present a paper with the title: "Some Applications of the Delft Electron Microscope in Biology." After the principles of the Delft microscope had been explained, micrographs of various flagellated bacteria were projected (Van Iterson, 1947), followed by micrographs of chloroplasts made for L. Algera (Algera *et al.*, 1947) and by bull sperm micrographs made for L. H. Bretschneider from Utrecht University (Bretschneider and Van Iterson, 1947). One of the sperm cell pictures had the peculiarity that the cell had been fed with an iron sugar complex, an early successful attempt by Bretschneider to raise the contrast of sites where the cell had metabolized most actively.

Owing to my departure for the United States, Dr. A. L. Houwink replaced me in Delft in 1947, and he continued, among other things, the studies of flagella in bacteria and of some Protozoa.

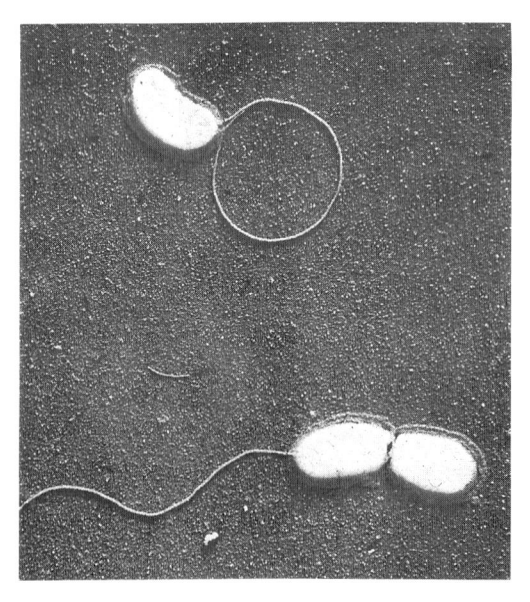

FIGURE 4. *Vibrio metchnikovii.* Field of view 7 μm.

The problems encountered in preparation techniques were considerable in those days. For the Metal Institute of TNO, a replica technique was developed by J. A. Nieuwenhuis in 1944, which was published by Dalitz and Schuchmann (1952) and by Beekhuis and Schuchmann (1952).

When the high-voltage microscope (see Sec. III) was taken from Eindhoven to Delft in 1947, the study of the voluminous yeast cell was still disappointing. At high voltage, unprepared yeast cells, as well as fungus spores, did not reveal important details. Moreover, at the time this prototype high-voltage microscope was ready, the need for such an instrument had faded. The problem of beam penetration had been circumvented by the development of a new strategy: the thin-section technique. Therefore, the development of a high-voltage electron microscope was halted in 1950, but came to life again in a novel design after 1960, when international interest in high-voltage microscopy was revived.

L. H. Bretschneider (1949) experimented with this thin-section technique at the University of Utrecht for his electron microscopical work in Delft. He and his co-worker, P. F. Elbers, clad in heavy overcoats, sectioned embeddings in a mixture of paraffin and hard wax at 4°C with a Cambridge Rocking microtome of 1890 vintage. The technique was further developed in a 1954 study of intestinal cells of Ascaris (Bretschneider, 1954), in which sections in the cold were made on a Cambridge Rocking microtome of 1952 vintage. At the same institute, Elbers constructed a single-pass rotation microtome with a thermal extension device for methacrylate embeddings and also concentrated on the use of electron stains. A little later, H. B. Haanstra (1955) successfully constructed a simple microtome at the Philips Research Laboratories, for which a patent was accorded in 1958.

A great stimulus to all the electron microscopists in the Netherlands was the International Congress on Electron Microscopy held in July 1949 in Delft, where we were given the opportunity to show our best results and to get acquainted with colleagues from abroad (Houwink, 1950).

VIII. More Electron Microscopes in the Country: The Early 1950s

Delft's monopoly on research with the electron microscope came to an end when Philips began to deliver the instruments commercially. The first EM 100, completed in 1949, went abroad to the Statens Serum Institute in Copenhagen for Birch Andersen to experiment with. In the Netherlands, every state university was to get its own microscope, and so were particular institutes such as the Laboratory for Flower Bulb Culture in Lisse, the

Royal Dutch Shell Laboratory, Sikkens (a factory for paints and varnishes), and, of course, the Philips Research Laboratories.

Certainly, it was the work done in Delft that had aroused interest at the universities and institutes. However, there were all sorts of disappointments, as most universities were not yet ready to meet the requirements for well-organized research with the electron microscope, badly underestimating the practical implications.

At the University of Groningen, Prof. E. H. Wiebenga had prepared himself well for research. He had been taught the preparation of protein crystals (edestin and exalsin) by Cecil Hall in the United States; and in England, Wiebenga had acquainted himself with X-ray diffraction of proteins. In November 1950, the first electron micrographs were taken at his university. However, when Wiebenga's work on seed globulins was taken over in October 1951 by a student studying for his doctoral thesis, it was found that the newly installed microscope could not be used. The resolution, approximately 5 nm, offered in the first series of microscopes was insufficient for this kind of work; he had to make do with X-ray diffraction. Electron microscopy in Groningen became more successful around 1952, when G. Boom's study of surface structures of several crystalline materials and E. F. J. van Bruggen's study of denaturation of proteins were supported by a new objective lens in the microscope and by more suitable preparation techniques, such as negative staining. This marked the beginning of fruitful work in structure chemistry of proteins in Groningen.

Thanks to a scheduled visit by Queen Juliana, the Agricultural University in Wageningen was fortunate to be among the first to have an EM 100 installed in 1951. With the Philips technicians still present, the Queen, a very intelligent woman, rightly exclaimed: "I don't see anything at all!" After this initial setback, Christina van der Scheer's work, with the assistence of S. Henstra, mainly concerned the study of virus particles (Rageth *et al.,* 1955).

Few present-day workers are aware of the difficulties experienced at the start. At the University of Amsterdam, the EM 100, delivered in January 1951, was installed in a basement bicycle storage area with a ceiling low enough to bump one's head and without ventilation. Since we were without special funds, the microscope films had to be developed using my kitchenware. Nevertheless, in 1953 I was able to give an invited paper on bacterial flagellation at the 10th Microbiology Congress in Rome (Van Iterson, 1953). In 1959 I obtained a doctorate in science with the monograph, "Gallionella ferruginea in a Different Light" (Van Iterson, 1958).

At Leiden University, one of the four prototype electron microscopes— built in collaboration with the instrument makers' school mentioned in Sec. VI instead of a Philips EM 100—was installed early in 1952 in the Anatomy

Building of the Medical Faculty. The microscope was placed in a kind of Bedouin tent in the middle of a large room. It served as a modest service facility under the care of W. G. Braams. The aim of the EM unit was also to provide a basis for future research in the medical biological field. In 1957, the microscope was replaced by a Philips EM 75 prototype. In December 1958, Braams was succeeded by W. Th. Daems, who had had his microscope training with us in Amsterdam and his training in fixation, embedding, and ultrathin sectioning in Sjöstrand's laboratory in Stockholm. With the replacement of the EM 75 by a Siemens Elmiskop I at the end of 1959, the basis eventually was laid for morphological as well as cell biological studies, among others, with the help of enzyme cytochemistry and autoradiography.

At the University of Utrecht the situation was no easier than in Amsterdam. On 20 March 1952, the official inauguration of the Utrecht EM 100 took place, but the instrument was placed in the physics building at an inconvenient distance from L. H. Bretschneider, who, since 1950, had held a Readership in cytology and electron microscopy, and from P. F. Elbers, his co-worker, who had been trained in Paris by W. Bernhard. It was not until 1954 that the electron microscope became easily accessible to biologists on a daily basis. Bretschneider received a Professorship in 1955 and became head of the Center for Submicroscopic Investigation of Biological Objects, with Elbers in charge of daily affairs.

An EM 100 electron microscope was not installed at the University of Nijmegen until 1957. Here a start was made with one facility for both the medical and science faculties, which was headed by A. Stadhouders, a pupil of Bretschneider and Elbers. The microscope happened to be installed in a building whose foundations were in a gravel bed shared by a railway yard! As a consequence, the microscope could not be used successfully during shunting until this source of disturbance was identified and rectified. The Nijmegen unit has expanded rapidly with important research activities, mainly in the field of human pathology and botany.

Much has been going on in the Philips Research laboratories, where H. B. Haanstra has been in charge of the electron microscopic research for many years. Numerous publications appeared from his hand in the 1950s and 1960s.

The biological work at Le Poole's institute came to an end when, at the University of Technology in Delft, an EM 100 was installed in Prof. A. J. Kluyver's laboratory for microbiology. Here, A. L. Houwink investigated, among other topics, with Professor P. A. Roelofsen, the organization inside plant cell walls, which gave birth to a "multinet growth" theory (1954). With Dr. D. R. Kreger he investigated the cell wall of yeasts (1953). Houwink's finding of a crystalline structure on the wall of a *Spirillum* species

(1953) was later studied extensively by R. G. E. Murray in Canada and became an important topic in molecular biology.

At Le Poole's Institute for Electron Microscopy, interesting developments continued uninterrupted, both in the development of instruments and in electron microscopy for commercial clients. In 1954 Le Poole's doctoral thesis appeared: "Some Designs in Electron and Ion Optics." It contained so many innovations that in his letter of congratulation B. Von Borries wrote: "It could have been three theses." In 1957, Le Poole became Professor at the Delft University of Technology. He continued to investigate the relationship between astigmatism and the lack of circularity of the bore of magnetic lenses and the improvement of resolution by correction of various aberrations. The improvement of image quality continued to hold his attention throughout the years.

IX. In Retrospect

In retrospect, it seems that at the outset, one of the major difficulties in biology was the gap created by what could be seen in the light microscope and what could be seen in the electron microscope.[5] It has taken many years to bridge that gap, and this was only achieved when light microscopists began using preparative procedures developed by electron microscopists. In addition, the electron microscopist long remained a stranger to both his physicist friends and to traditional biologists; what was seen on electron micrographs had to remain purely descriptive morphology for a long time, molecular interpretations being too speculative then. Biochemistry has been one of the major supports in bringing ultrastructure research into the realm of molecular biology.

The first commercially produced microscopes may not have been adequate to meet all the expectations of all the electron microscopists, but has not this been a stimulus for the later production of more and more excellent microscopes?

The considerable efforts in the early days—organizational, technical, as well as purely scientific—laid a basis for fruitful research in the Netherlands.

What has electron microscopy taught me personally? The great idea of Professor Kluyver, who strongly stimulated me in my career, was his concept of unity in biochemistry. What I have been much aware of in my years of active research, as have so many investigators with me, is a unity in diversity that ultrastructural studies have revealed in the basic cell structures. In my

---

[5] Elbers, personal communication.

later years of reflection, the impact of such a unity on the philosophical basis of our lives is slowly beginning to dawn on me.

ACKNOWLEDGMENTS

For this review, use has been made of numerous useful suggestions by the following colleagues: B. J. Spit, P. F. Elbers, H. B. Haanstra, C. E. Hulstaert, E. F. J. van Bruggen, and many others. Apart from these, it is based largely on my own notes and recollections.

REFERENCES

Algera, L., Beyer, J. J., Van Iterson, W., Karstens, W. K. H., and Thung, T. H. (1947). *BBA* **1,** 517.
Beekhuis, D. A., and Schuchmann, J. A. (1952). *Metalen* **7,** 444.
Bretschneider, L. H. (1949). *Proc. Kon. Ned. Acad. Wetensch.* **52,** 301.
Bretschneider, L. H. (1954). *Proc. Kon. Ned. Acad. Wetensch.* **57,** serie C, 524.
Bretschneider, L. H., and Van Iterson, W. (1947). *Proc. Kon. Ned. Acad. Wetensch.* **50,** 88.
Dalitz, V. Ch., and Schuchmann, J. A. (1952). *Metalen* **7,** 153.
Haanstra, H. B. (1955). *Philips Tech. Rev.* **17,** 178.
Houwink, A. L. (Ed.). (1950). *Proc. Conf. on Electron Microscopy,* Delft, 4–8 July 1949, Delft, 1950.
Houwink, A. L.(1953). *BBA* **10,** 60.
Houwink, A. L., and Kreger, D. R. (1953). *Antonie van Leeuwenhoek* **19,** 1.
Houwink, A. L., and Roelofsen, P. A. (1954). *Acta Botan. Neerland.* **3,** 385.
Le Poole, J. B. (1947). *Philips Tech. Rev.* **9,** 33.
Le Poole, J. B. (1954). Some designs in electron and ion optics. Thesis, Delft University of Technology.
Le Poole, J. B. (1985). *In* "Advances in Electronics and Electron Physics" (P. Hawkes, Ed.), Suppl. 16, p. 387. Academic Press, Orlando, FL.
Le Rütte, W. A. (1952). Bijdrage tot de elektrostatische elektronenoptica (Contributions to electrostatic electron optics). Thesis, Delft University of Technology.
Rageth, H. W. J., Van der Scheer, Ch., and Van der Want, J. P. H. (1955). *Tijdschr. Plantenziekten* **61,** 35.
Spit (1966). *In* "Vijf en twintig jaar Technisch Physische Dienst" "Twenty-Five Years of the Technical Physics Service," p. 30.
Van Dorsten, A. C., and Le Poole, J. B. (1955). *Philips Tech. Rev.* **17,** 47.
Van Dorsten, A. C., Nieuwdorp, H., and Verhoeff, A. (1950). *Philips Tech. Rev.* **12,** 35.
Van Dorsten, A. C., Oosterkamp, W. J., and Le Poole, W. B. (1947). *Philips Tech. Rev.* **9,** 195.
Van Iterson, W. (1947). *BBA* **1,** 527.
Van Iterson, W. (1953). *Symp. Citologia Batterica. Roma* **1953,** p. 24.
Van Iterson, W. (1958). *Verh. Kon. Nederl. Ac. Wetensch. 2nd Sec.* **LII-2,** 1.
Van Slogteren, E. (1952). *Philips Tech. Rev.* **14,** 13.

2.10
Electron Microscopy in the Netherlands
2.10B Developments since the 1950s

PIETER KRUIT

Department of Applied Physics
Delft University of Technology
2600 AA Delft, The Netherlands

FRITS W. SCHAPINK

Laboratory of Materials Science
Delft University of Technology
2600 AA Delft, The Netherlands

JOHN W. GEUS

Department of Inorganic Chemistry
University of Utrecht
3508 TC Utrecht, The Netherlands

ARIE J. VERKLEIJ

Department of Molecular Cell Biology
University of Utrecht
3508 TC Utrecht, The Netherlands

CAESAR E. HULSTAERT[1]

Laboratory for Cell Biology and Electron Microscopy
University of Groningen
9700 AB Groningen, The Netherlands

I. INSTRUMENTAL DEVELOPMENTS[2]

After the original design of a 40-kV, a 150-kV, and a 400-kV electron microscope with such inventions as the wobbler and an intermediate lens for continuous magnification and selected area diffraction, Le Poole continued his research as Professor at the Delft University of Technology as a

---

[1] Present address: Faculteitsbureau Geneeskunde, A. Deusinglaan 1, 9713 Av Groningen, The Netherlands.
[2] Authored by Pieter Kruit.

Consultant to Philips and of TPD (Technisch Physische Dienst, TNO-TH). Many prototypes of revolutionary instruments were built in Delft, including an X-ray projection microscope with Ong Sing Poen (ICEM 58), a scanning (mirror) electron microscope using quadrupole lenses with Bok (EUREM 64 and 68), a dedicated microprobe X-ray analyzer using Le Poole's minilenses with Fontijn (Int. Congr. X-ray Opt. 1968) and a very compact 1-MV microscope, folded in the form of a Z, with Van Zuylen, Kramer, and Barth (ICEM 70, EUREM 72).

The instrumental work of Le Poole often concerned the improvement of a section of the microscope or the introduction of a new technique. The minilens (EUREM 64) was used for a while as a second condensor lens. Andersen and Mol studied the energy distribution of electrons emitted from triode guns for optimal operation of such guns. Before energy loss spectroscopy became popular, Andersen and Le Poole developed imaging Wien filters, both in-column and post-column, which were installed on several EM 300s (EUREM 72, ICEM 74). As a high-current alternative for the cold field emission gun, Van der Mast, Barth, Kramer, and Le Poole developed a laser-heated Schottky emission gun, which functioned for some years in the prototype (ICEM 74). There were several experiments on the correction of spherical aberration, using grid lenses, superconducting grids (Dekkers, EUREM 68), or space charge (EUREM 72).

In addition to the pioneering work on magnetic lens design by Van Ments and Le Poole in 1947, the Delft group contributed to the "theory" of instruments by introducing the Fourier analytical method for describing multipole fields (Kramer, 1967) and later by the work on Coulomb interactions in particle beams of Jansen (1990). Recent contributions in computer-aided design of microscopes include advanced finite-element calculations (Lencová and Wisselink, 1990; Barth *et al.*, 1990) and optimization of electrostatic lenses (Van der Steen *et al.*, 1990).

After Le Poole's retirement, the developmental work at the institute in Delft was continued by Van der Mast, with work on automatic alignment and focusing of microscopes (EUREM 84, Koster EUREM 88) and by Kruit with the introduction of Auger and secondary-electron spectroscopy in high-resolution STEMs (ICEM 86, ICEM 90 with Bleeker).

Other Dutch universities also contributed to instrumental developments and to the theory of the instrument. Two groups in Groningen worked in this field. Kamminga, Verster, and Francken were among the first to analyze and design lenses and triode guns with numerical calculations. In the early 1970s, Hoenders, Ferwerda, and Drenth worked on reconstruction problems (Optik 35, 37, 38, 39). Van Heel developed procedures for averaging images of single macromolecules in his Groningen period (EUREM 80, ICEM 82).

Van Dorsten, at the Philips Research Laboratories in Eindhoven, played an important role in the early days, with work on stable high-voltage supplies and on electromagnetic stigmation of electron lenses and later on microscopical methods. He was a pioneer in low-voltage electron microscopy (EUREM 60), and he tried to maximize the image contrast (EUREM 60, EUREM 68). The contribution to differential phase contrast in STEM by Dekkers and De Lang (Optik 41, 1974) originated from the same group.

Probably the best-known Dutch contribution to electron microscopy has come from Philips, with their series EM 75, EM 100, EM 200, EM 300, EM 400, up to the most recent CM series transmission microscopes and XL series scanning microscopes. Of the hundreds of smaller and larger instrumental innovations, we can mention only a few here. The EM 100, of course, contained all Le Poole's early inventions. The EM 200 was equipped with electromagnetic stigmators and a eucentric goniometer stage. The EM 300 had an auxiliary objective lens for low-magnification imaging and a cold finger option. Electromagnetic gun alignment was a novelty in the EM 201. The EM 400 was the first real TEM/STEM with field emission option and the twin-lens configuration for switching between transmission and scanning mode (Van der Mast and Le Poole, EUREM 80). Perhaps the most important instrumental development since the 1950s, but impossible to ascribe to a particular year or place, has been the continuous improvement in stage stability and cleanliness of the vacuum.

References

Barth, J. E., Lencova, B., and Wisselink, G. (1990). *Nucl. Instr. Meth. Phys. Res.* **A298**, 263.
Jansen, G. H. (1990). In "Coulomb Interactions in Particle Beams," Adv. Electron. Electr. Phys. Suppl. 21. Academic Press, San Diego, CA.
Kramer, J. (1967). *Br. J. Appl. Phys.* **18**, 1815.
Lencová, B., and Wisselink, G. (1990). *Nucl. Instr. Meth. Phys. Res.* **A298**, 56.
Van der Steen, H. G. W., Barth, J. E., and Adriaanse, J. P. (1990). *Nucl. Instr. Meth. Phys. Res.* A298, 377.

II. Electron Microscopy in Materials Science

A. *Electron Microscopy of Metallic and Semiconductor Materials*[3]

In a brief historical survey of the application of electron microscopy in materials science in the Netherlands, it is appropriate to start with the

---

[3] Authored by Frits W. Schapink.

work by Burgers and co-workers at the Philips Research Laboratories in Eindhoven around 1935. At that time Burgers worked on the imaging of metal surfaces using an emission electron microscope. Since the emission of electrons from a metal surface is orientation-dependent as well as depending on the crystal structure, images of, for example, the transformation of α- to γ-iron could be made and analyzed. Thus, detailed studies of various surface phenomena on metals were made in those early days.

The major breakthrough in transmission electron microscopy (TEM) of materials after World War II came from the pioneering work of Hirsch and co-workers at the Cavendish Laboratory in Cambridge in 1956, and, independently, by Bollmann at the Battelle Institute in Geneva. They were the first to obtain images of dislocations in thin foils of stainless steel and aluminum, and they were also able to demonstrate the motion of dislocations under mechanical stress. Soon after this discovery the theory of diffraction contrast of lattice defects (dislocations, stacking faults, precipitates) in crystals was developed by Hirsch, Howie, and Whelan and applied to a large number of crystalline substances. At the time, very little work of this nature was being carried out in the Netherlands, in contrast to the situation in Belgium, where the group led by Amelinckx was very active right from the beginning. In the late 1960s, Burgers and co-workers of the Delft Laboratory of Metallurgy started to use TEM for qualitative studies in materials science, for example, the (pre)precipitation phenomena in Al–Cu alloys. Up to then the main emphasis in Burgers' group had been on the application of X-ray diffraction in recrystallization and precipitation.

Around 1970, the investigators in Delft realized that an electron microscope should not be used as a simple magnifying glass. It was thought necessary to study the principles of electron diffraction phenomena of defects in crystals in order to obtain a more detailed understanding of diffraction contrast phenomena. Schapink started a small group and investigated a number of phenomena in a quantitative manner. Among the topics studied were symmetry properties of dislocation images and the interaction of dislocations with thickness contours, especially under weak-beam conditions. In the late 1970s, the group became interested in the structural aspects of grain boundaries using TEM and special electron diffraction effects. A technique was developed which made it possible to obtain thin Au crystals with a grain boundary running parallel to the specimen surface. Many boundaries were studied in this way, especially low-angle twist boundaries and near-coincident twin boundaries in Au crystals. Although the existing TEM techniques provided much information about grain boundary structure, a need was felt for additional methods. Thus, around 1980 the technique of convergent-beam electron diffraction (CBED) was applied to bicrystals. This required extending the existing theory for symmetry determination of single crystals to the case of bicrystals. The symmetry of a bicrystal is related to the state of translation existing at the grain boundary, and this provides a way of distinguishing various rigid-body translations which may exist at a grain boundary in a bicrystal. The technique has become complementary to the existing method of determining the state of translation from α-fringes in TEM.

The study of grain boundaries was continued in the 1980s. The relationship between geometric structure and electrical properties of grain boundaries was investigated in Si, in connection with the application of polycrystalline Si in solar cells. It became clear that the phenomena are very complicated for all types of grain boundaries except for a coherent twin boundary in Si. Apart from Si, III–V semiconducting compounds were investigated, and especially the diffraction effects associated with a superlattice of GaAs/AlAs in CBED. For the first time, extra rings of diffraction, called superlattice higher-order Laue zone (HOLZ) rings, could be observed for superlattices observed in plan view, that is, with the electron beam parallel to the normal of the superlattice (Fig. 1). Lastly, the structure of grain boundaries, especially twin boundaries and low-angle grain boundaries, was studied in ordered alloys. It could be shown that the ordering in the matrix introduces a number of possible boundary structures, depending on the type of ordering.

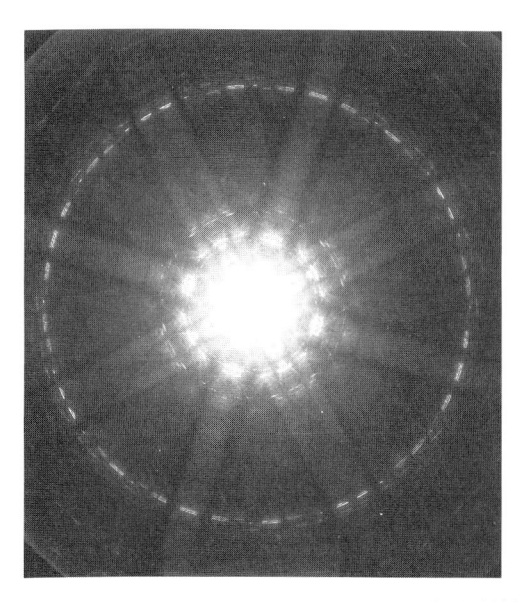

FIGURE 1. Convergent-beam electron diffraction pattern of GaAs/AlAs superlattice.

TEM studies on different subjects in materials science were carried out in other universities and in industrial research laboratories as well. At the Department of Applied Physics at the University of Groningen, Radelaar (later succeeded by De Hosson) and co-workers carried out studies on lattice defects (dislocations, antiphase boundaries) in the ordered alloy Cu_2NiZn using TEM. At the University of Twente, Verbraak continued his work on recrystallization textures in f.c.c. metals that he had started in Delft in the group of Burgers. His earlier work in Delft mainly concerned modeling, but in Twente TEM was employed in the determination of recrystallization textures. Viegers and collaborators at the Philips Research Laboratories in Eindhoven worked on dislocation structures in epitaxial semiconducting layers as well as on high-resolution electron microscopy (HREM, see below) of interfaces in heterostructures based on III–V alloys and other semiconductors.

HREM is also important for materials science. In classical TEM either bright-field or dark-field images are employed, with the image formed by a single diffracted beam. The contrast obtained is called diffraction contrast; it can be considered as an artefact, in the sense that it is not related to the perfect crystal structure. In order to obtain a true (projected) image of the structure, the image should be formed by recombining all diffracted beams with their correct mutual phase relationships. Unfortunately, this is not possible in a classical electron microscope. Structural images with a limited resolution can be obtained only in terms of a number of factors such as the number of beams. However, the reliability of structural images can be tested by comparing experimental images with computer-simulated images until a satisfactory match is obtained. In the past five years, the possibilities in this respect in HREM have much improved, partly due to the better resolution of microscopes and partly to more sophisticated image simulation techniques.

HREM work has been performed by Viegers and co-workers at the Philips Research Laboratories in Eindhoven. Until recently, electron microscopes capable of performing HREM work were not available at the universities in the Netherlands. Consequently, the work of Zandbergen on HREM of high-temperature oxide superconductors was carried out in collaboration with RUCA in Antwerp and partly in the United States at the Berkeley Center for HREM. In 1989, a Center for High-Resolution Electron Microscopy was founded in Delft, and this center is operated as a national facility for HREM. The center forms part of the former group of Burgers (now led by Mittemeijer); the simultaneous availability of high-quality electron microscopical, X-ray diffraction, and surface analytical techniques and expertise allows a multidisciplinary approach in investigations which is essential to materials research.

B. Electron Microscopic Research on Vapor-Deposited Metal Films and Solid Catalysts[4]

During the 1960s, vapor-deposited metal films were extensively investigated by Kiel, Nadorp, and Geus of the Central Laboratory of DSM. Tungsten, nickel, and iron films were deposited in ultrahigh vacuum. The BET surface area of the films was calculated from the extent of adsorption of xenon, while the electrical resistance measured at a range of temperatures provided information about the defect density and the geometry of the electrically conducting phase. The macroscopic data thus obtained complemented the electron microscopical results very nicely. A fairly complete description of the growth phenomena of vapor-deposited metal films could be developed, of which an extensive report has been published (Geus, 1971).

Earlier, Haanstra, Kooy, and Nieuwenhuizen (1966) had published an important contribution to the understanding of the elementary processes occurring during growth of vapor-deposited metal films. These authors, working at the Philips Research Laboratories in Eindhoven, observed the growth of acicular metal crystallites during vapor deposition of aluminium. A well-designed replica technique clearly showed the presence of the acicular metal crystallites. The axis of the metal crystallites was pointing to the source of the metal atoms. When the direction of the beam of metal atoms was changed without intermediate exposure of the film to atmospheric air, the original direction of the crystallites was maintained.

The evidence obtained by electron microscopical investigation of vapor-deposited metal films added greatly to the elucidation of the phenomena proceeding during thermal treatment of supported metal catalysts. Van Hardeveld and Hartog (van Hardeveld and van Montfoort, 1966) developed a description of the surface structure of small nickel and platinum particles (dimensions 2–7 nm) based on work on supported nickel and platinum catalysts by Kiel. The authors developed models indicating the presence of "B5" sites, sites that are able to adsorb molecular nitrogen in an infrared active state.

A strong increase of the BET surface area of silica catalyst supports after application of nickel compounds indicates reaction of the silica with nickel ions to nickel hydrosilicate. Kiel and Geus established the reaction of the silica from electron microscopical investigations on nickel-on-silica catalysts prepared by deposition-precipitation. At sufficiently high nickel loadings, the silica completely reacts to nickel hydrosilicate, as was clearly evident from micrographs of the silica before and after loading with nickel.

---

[4] Authored by John W. Geus.

The above investigations were initially performed on Siemens Elmiscopes and later on Philips EM 300, 400, and 420 electron microscopes. Modern electron microscopical investigations of solid catalysts call for the ability to assess accurately the chemical composition of very small active particles. An example is the determination of the Pt/Sn or Pt/Pd ratio of supported alloy particles of about 1.5 nm. For this work, electron microscopes equipped with a field-emission electron source, such as the Vacuum Generators instrument, are indispensable. Due to the lack of these instruments in the Netherlands, Dutch workers in catalysis have to use instruments in France or the United States for these investigations.

Another important item in the investigation of catalytic materials is the investigation of the atomic structure of zeolites and of the shape of small Pt crystallites in zeolites by lattice imaging. The lack of suitable electron microscopes in the Netherlands during the 1980s caused Dutch workers to call on the hospitality of the institute of Professor Van Landuyt of the University of Antwerp in Belgium. Using the facilities in Antwerp, Zandbergen was, for instance, able to elucidate the effect of lanthanum oxide on the thermal stability of τ-alumina. He found that the surface layer of τ-alumina is amorphous, which facilitates nucleation of the more stable α-alumina, which has a small surface area. The presence of a small amount of lanthanum oxide, significantly less than corresponding to a monolayer of lanthanum oxide, causes recrystallization of the amorphous surface layer to τ-alumina, thus impeding nucleation of α-alumina.

REFERENCES

Geus, J. W. (1971). In "Chemisorption and Reactions on metallic Films" (J. R. Anderson, Ed.), Vol I, p. 129 and 327. Academic Press, New York.
Kooy, C., and Nieuwenhuizen, J. M. (1966). In "Basic Problems in Thin Film Physics," R. Niedermayer and H. Mayer, Eds.), p. 181. Van den Hoeck and Rupprecht, Gottingen.
van Hardeveld, R., and van Montfoort, A. (1966). Surface Sci **4**, 396.

ELECTRON MICROSCOPY FOR BIOLOGICAL SPECIMENS[5]

A. Seeing Is Believing: 1950–1970

The breakthrough of electron microscopy for biological cells and tissues came with the introduction of ultramicrotomy. This technology was started

---
[5] Authored by Arie J. Verkleij.

by Elbers in Bretschneider's laboratory in Utrecht in 1952, but it took several years before the classical thin sectioning technique—i.e., fixation with OsO_4 and glutaraldehyde, dehydration, embedding in plastics, sectioning, and staining with lead and uranyl—became a reproducible and mature methodology. However, there was a great deal of scepticism among Dutch scientists, as expressed by the famous Dutch botanist, Professor V. Koningsberger, when he saw the first electron micrographs: "I cannot deny that these pictures are fascinating, but whoever admires them must bear in mind that he is like an archeologist, who is impressed by the beauty of an Egyptian princess . . . through observing her gilded mummy." Indeed, this remark quite accurately conveys the true nature of our electron microscopical observations in those days.

Besides the thin sectioning technique, very soon other technical approaches began to emerge in our country in both fundamental and medical research. In the 1950s and 1960s, chairs in submicroscopic analysis were founded in the science faculties, particularly in the field of molecular cell biology. In Groningen, Van Bruggen started his investigations on the structure of biological macromolecules: on hemocyanins with negative staining and on DNA with the Kleinschmidt method. In Utrecht, Elbers studied the Lymnaea egg cell and was successful in studying biological membranes in collaboration with the Department of Biochemistry of Van Deenen, who had started research on the biochemistry of lipids of biological membranes in our country. In Amsterdam, the structure of the bacterial cell, including the membrane, cell wall, and the genome, was the central theme in the Department of Mrs. van Iterson, later continued by Nanninga. In the 1970s, chairs at the Medical Faculties were founded in Leiden, where Daems began his studies of phagocytes and lysosomes, in Groningen, where Molenaar commenced his studies on the influence of vitamin E on cellular membranes, in Nijmegen, where Stadhouders started his studies on the pathobiology of neuromuscular disorders, especially mitochondrial myopathies, in Utrecht, where Geuze began with the unraveling of secretory and endocytotic transport routes in eukaryotic cells, in Amsterdam, where Mrs. Hoefsmit set up her studies of macrophages and dendritic cells in the immune response, and again in Amsterdam, where Leene began his studies of lymphocytes and the immune response. All these professors have been active in the Dutch Society for Electron Microscopy, many of them as members of the board of the society.

Although most of the institutes for electron microscopy at the different universities were centered around these chairs, there were many other centers for electron microscopy where the electron microscopes and the auxiliary apparatus of various types were brought together for financial and efficiency reasons to serve the whole scientific community: the universities of

Rotterdam, Wageningen, and Maastricht, and research institutes such as the Hubrecht Laboratory in Utrecht, the Netherlands Cancer Institute and the Netherlands Ophthalmic Research Institute in Amsterdam, and the industrial laboratories of Philips, Unilever, and DSM.

B. Do We Believe What We See? 1970–1980

At the end of the 1960s, it gradually became clear that we were looking at "mummies," which, although representing reproducible derivatives of the living state, were nevertheless dead and full of artifacts. At the same time, it was realized that the classical thin sectioning technique resulted in denaturation of proteins, oxidation of molecules, extraction of lipids and ions, beam damage in the microscope, and above all extraction of water, the basis of life.

Cryofixation, strongly advocated by Fernandez-Moran, contained the clue to more reliable images of biological specimen. The first method to demonstrate the reliability of cryofixation was freeze-fracture etching. This method, introduced by Steere, was applied at the end of the 1960s in the Philips Research Laboratories by Van den Berg, at the University of Amsterdam by Nanninga, and at the University of Utrecht by Ververgaert and Verkleij. This method, which was very successful with regard to the study of biological membranes, stimulated research enormously at the Institute of Molecular Biology in Utrecht, especially with regard to membrane structure, the discovery of lipidic particles as intermediates in membrane fusion (Verkleij et al., 1979), and lipid polymorphism. Cryofixation also became the basis of a new development in the thin sectioning technique as advocated by Kellenberger and Müller in Switzerland. Cryofixation was followed by cryosubstitution, after which embedding in lowicryl and polymerization with UV was carried out at low temperature as well. Finally, to rule out any compromise, cryofixation was used in combination with cryoelectron microscopy, a method developed by Dubochet in Heidelberg. This method was applied with a great deal of success by Van Bruggen of Groningen in his work on biological macromolecules and by Frederik of Maastricht in his work on lipid structures and small cells. Cryofixation not only yields a more reliable structural preservation, it also offers the opportunity for time resolution of biological processes, since fast dynamic biological processes can be trapped. Brakenhoff and Nanninga, concerned at having to study dead and dehydrated specimens in electron microscopy, made a search for an alternative approach to investigate hydrated biological samples. This led to the development of confocal scanning laser microscopy CSLM (Brakenhoff et al., 1979).

C. We See Everything We Want to Explore: 1980–1991

It has always been our wish to localize molecules in cells and tissues. First, many studies have been dedicated to the localization of molecules by cytochemical methods. One example is the visualization of phosphatases with the cerium method, in which the phosphate ion formed through enzymatic hydrolysis is precipitated by cerium, the precipitate being electron dense. This method was introduced in Groningen by Veenhuis and brought to general application by Hulstaert (Hulstaert et al., 1989). After introduction of the cryosectioning technique according to Tokuyashi and the introduction of antibodies, the dream of localizing molecules came true. Initially ferritin was used as an electron dense marker, but it is especially the combination of cryosectioning and labeling of molecules with immunogold labels which caused an immense breakthrough in cell biology (Geuze et al., 1989). The introduction of immunogold was a Dutch initiative by Geuze and Slot of Utrecht. They also introduced the double and triple labeling on cryosections (Fig. 2), which makes biochemical studies in a single cell possible. Immunogold labeling is now also used on sections of freeze-substituted resin-embedded tissues, where antigenicity appears to be preserved for many epitopes, and on freeze-fracture membranes, the so-called label fracture [for a complete overview, see *Labeling in Cell Biology*, edited by Verkleij and Leunissen (CRC Press, Boca Raton, FL, 1989)]. An improvement of immunogold labeling can still be expected by using smaller gold or 1-nm gold probes for a better penetration. These probes have been developed by Leunissen at Janssen Pharmaceutica.

In recent years the computer has also entered the field of electron microscopy. It tremendously increased the possibilities for processing and interpretation of image data on electron micrographs. In Groningen the software package IMAGIC was born for maximum information retrieval from both single and (artifically) ordered structures. CORAN (correspondence analysis, a multivariate statistical method) is of great importance for the classification and averaging of noncrystalline biomacromolecules such as the hemocyanins (Van Heel and Frank, 1981). Furthermore, the computer was used by the group of Mrs. Arnberg for the discovery and very accurate mapping of introns in DNA and for the structural analysis of RNA processing phenomena. With regard to the identification and localization of atoms, many centers have contributed to this technology, especially Rotterdam, Leiden, Nijmegen, and Utrecht. However, the application of this technology for biological specimens is still rather limited.

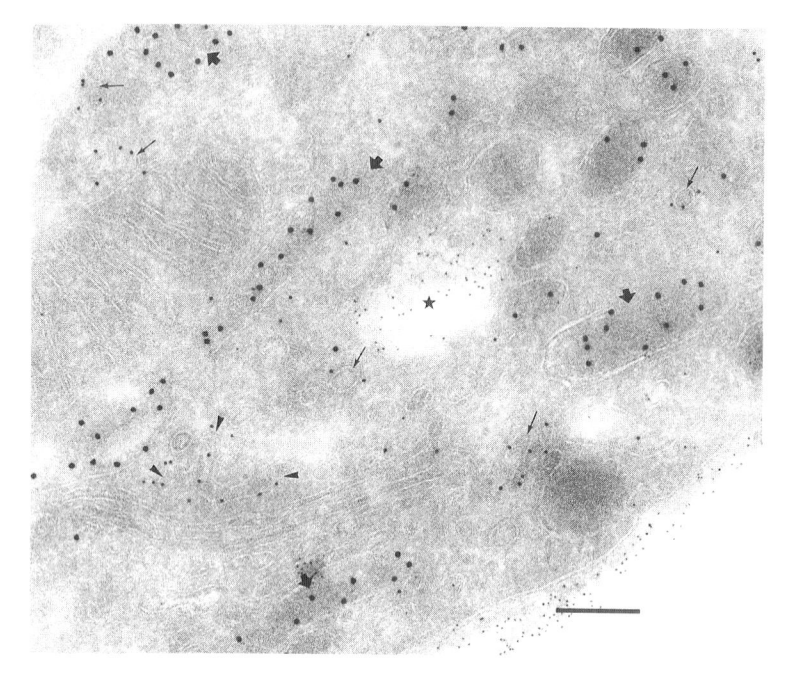

FIGURE 2. Ultrathin cryosection of brown adipose tissue from an underfed rat. The section was immunolabeled by the protein A-gold multiple staining protocol for three proteins: 10 nm gold = the insulin-regulated glucose transporter (arrowheads, in trans-Golgi reticulum; thin arrows, small cytoplasmic vesicles), 15 nm gold = cathepsin D (bold arrows, lysosomes), 5 nm gold = rat serum albumin (asterisk, endosomal vacuole; along the cell membrane). For further details, see Slot *et al.* (1991), *J. Cell Biol.* **113,** 123. Bar = 200 nm.

IV. The Dutch Society for Electron Microscopy[6]

In the early years of electron microscopy, the numerous difficulties encountered by electron microscopists in their practical work necessitated regular contact. Fourteen electron microscopists assembled on 1 July 1952, in the lecture hall of the Hortus Botanicus in Amsterdam, and decided to found a group of electron microscopy workers. The first elected board consisted of Le Poole (President), Van Dorsten (Vice President), Wiebenga (Secretary), Bretschneider (Treasurer), and Van Iterson (Secretary for Foreign Relations). In 1958 the society adopted a new name: Nederlandse Vereniging voor Electronenmicroscopie (Dutch Society for Electron Microscopy). Right from the beginning, the membership was open to technicians and technical staff, which led to a steady growth of the membership. At the

---

[6] Authored by Caesar E. Hulstaert.

end of 1995, the society had 510 members and 31 benefactors. The society has five sections: (1) Practical Electron Microscopy; (2) Scanning Techniques and Microscopy; (3) Cryo Electron Microscopy; (4) Clinical and Pathological Electron Microscopy; (5) Materials Science. The society has two meetings a year. In spring, a one-day meeting is organized, usually in cooperation with one of the benefactors. In autumn a two-day meeting is held with a trade exhibition, usually with more than 200 participants in attendance, and with sessions for medical biological sciences, materials sciences, and electron optics.

The society has organized the second and the Seventh European Congress on Electron Microscopy in, respectively, Delft in 1960 and The Hague in 1980. In 1978 the Stichting tot Bevordering van de Electronenmicroscopie in Nederland SEN (Foundation to Promote Electron Microscopy in the Netherlands) was founded. Every two years a prize of Dfl. 3,000 is awarded by the foundation to an electron microscopist in the Netherlands below the age of 35 for an outstanding contribution in the field of electron microscopy.

The Dutch Society for Electron Microscopy is a dynamic organization with members in many circles of society: universities, large and small industries, and public institutions. Throughout the years the society has remained of great importance for all its members belonging to very different scientific disciplines. The most recent proof of its dynamic nature is the opening of the society to practitioners of all forms of advanced microscopy, symbolized by changing its name, in 1995, to Nederlandse Vereniging voor Microscopie (Dutch Society for Microscopy).

REFERENCES

Brakenhoff, G. J., Blom, P., and Barends, P. (1979). *J. Microsc.* **117,** 219.
Geuze, H. J., Slot, J. W., Vanderley, P. A., and Scheffer, R. C. T. (1981). *J. Cell Biol.* **89,** 653.
Hulstaert, C. E., Kalicharan, D., and Hardonk, M. J. (1989). *In* "Techniques in Diagnostic Pathology" (G. Bullock, A. G. Leathem, and D. van Velzen, Eds.), Vol 1, p. 133. Academic Press, London.
Van Heel, M., and Frank, J. (1981). *Ultramicr* **6,** 187.
Verkleij, A. J., Mombers, C., Leunissen-Bijvelt, J., and Ververgaert, P. H. J. (1979). *Nature* **279,** 162.

2.11
The Development of Electron Microscopy in Scandinavia

ARVID B. MAUNSBACH

Department of Cell Biology, Institute of Anatomy
University of Aarhus,
DK-8000
Aarhus C, Denmark

BJÖRN A. AFZELIUS

Department of Ultrastructure Research
Stockholm University
S-10691 Stockholm, Sweden

I. Introduction

In this chapter we describe three aspects of the history of electron microscopy in the Nordic countries: the early development of a transmission electron microscope (TEM) by Manne Siegbahn, the dramatic increase of microscopy in the 1950s, and the contribution of the Scandinavian Society for Electron Microscopy (SCANDEM) for the development of structural research in biology and materials science. The subject is vast, and we have chosen to focus on the developments and events before the year 1960, but in the context of the society up to the present time. We have interviewed several colleagues about early events, but naturally there is some emphasis on aspects where we have first-hand information.

II. The Siegbahn Electron Microscope

Electron microscopy in Scandinavia began with Manne Siegbahn (Fig. 1). He moved in 1938 from a Chair in physics at Uppsala University to become Head of the Research Institute for Experimental Physics in Stockholm, now named the Manne Siegbahn Laboratory. The emergence of an electron microscope with a higher resolution than that of the light microscope, built by Ernst Ruska and described in his 1934 paper (Ruska, 1934), inspired Manne Siegbahn to explore the new possibilities for structural analysis by building one himself. As one of his research lines, he

FIGURE 1. Manne Siegbahn, Nobel Prize winner in Physics in 1924 for X-ray spectroscopy, built the first Scandinavian electron microscope in 1938.

therefore started electron optical experiments using an evacuated glass tube with a heated wire as cathode at one end, a glass window coated with fluorescent material at the other end, and with a metal grid mounted inside the tube as an object. The apparatus was mounted on an optical bench, and iron-shrouded solenoids constituted the lenses; these could be moved along the glass tube for focusing. Images of the metal grid were produced on the fluorescent screen in experiments that agreed with Ruska's observations. He then built in 1938 a transmission electron microscope that was based on the design principles of Ruska. The microscope was finished early in 1939, and the resolution was clearly better than that of the light microscope (Siegbahn, 1939a, 1939b). Manne Siegbahn was helped, almost from the start, by Fritiof Sjöstrand (Fig. 2), then a young assistant at the Karolinska Institute. During a later stage, Kai Siegbahn, son of Manne and Nobel Prize winner in Physics in 1981 for the development of the electron spectroscopic method for chemical analysis (ESCA), also became involved in the project.

FIGURE 2. Fritiof Sjöstrand, Swedish pioneer in biological electron microscopy.

The Siegbahn microscope had a horizontal column and air-cooled lenses (Fig. 3). A second prototype was begun in 1939 (Siegbahn, 1940) and was completed in 1941 (Fig. 4). Its high tension of 60 kV was provided from a commercial high-voltage unit for clinical X-ray equipment produced by the company Georg Schönander AB in Stockholm. The microscope had a tilting device that allowed stereo recording (Siegbahn, 1942). The microscope, with different modifications, was applied to several different investigations in the early and mid 1940s. Thus, Fritiof Sjöstrand (Fig. 5) analyzed the structure of skeletal muscle following preparation of the tissue with his newly developed sectioning method (Sjöstrand, 1943a, 1943b, 1943c), and Björn Ingelman and Kai Siegbahn studied the structure of dextran molecules (Ingelman and Siegbahn, 1944). Ivar Elvers applied the instrument to plant cytology in an attempt to depict plant pachytene chromosomes in 3-μm-thick paraffin sections of plant tissues (Elvers, 1941, 1943). In some cases he used a micro-grid as an object support instead of a supporting

FIGURE 3. First prototype of the Siegbahn microscope. (Reproduced from Siegbahn, 1939a.)

film. The grids were manufactured from transversely cut pinewood that had been impregnated with osmium tetroxide.

A commercial production of the Siegbahn microscope was begun around 1944 by the company Georg Schönander AB in Stockholm. The microscope was now referred to as the Siegbahn-Schönander Electron Microscope (Fig. 6). Its resolution was stated to be 20 Å, and the highest magnification was 30,000× (Bergqvist, 1946). The commercial version had facilities for stereo pictures and for recording diffraction patterns. A series of about 10 microscopes was produced and sold to various institutions, including the Karolinska Institute, the University of Lund, the Royal Institute of Technology in Stockholm, and at least two industrial customers, one of which was ASEA in Västerås. Additionally, one instrument was delivered

FIGURE 4. Second prototype of the Siegbahn microscope (From Siegbahn, 1941. Reproduced with permission from the Royal Swedish Academy of Science.)

FIGURE 5. Fritiof Sjöstrand operating the Siegbahn microscope. (Reproduced from the daily Swedish newspaper, *Svenska Dagbladet*, July 17, 1943.)

to Czechoslovakia and one to Poland. One of the instruments in Stockholm remained in operation until the end of the 1950s. In materials science, Nils Hast used this microscope to analyze the structural properties of clays and developed methods of manufacturing thin films of beryllium and aluminum

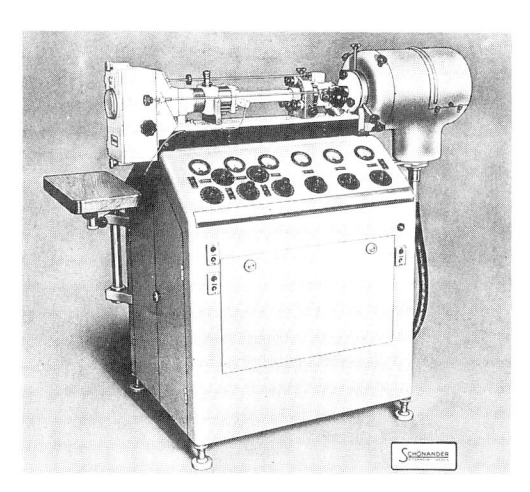

FIGURE 6. The commercial Siegbahn-Schönander microscope from the 1940s. (Reproduced from the manufacturer's brochure.)

as well as of other metals (Hast, 1948), and from the mid-1950s the Danish scientist Asger Lindegaard-Andersen, working at ASEA, Västerås, applied the microscope for the analysis of various materials, including zink oxide films (Lindegaard-Andersen, 1958), dust particles, and synthetic diamonds. In biomedicine the Siegbahn-Schönander microscope was used by Hans Engström and Jan Wersäll in their early studies on the structure of cilia and flagella and of ear epithelial cells (Engström and Wersäll, 1952; Wersäll 1954) and by Arvid Syrrist and Gösta Gustafson from the State Dental College in Malmö for analyses of dental enamel and dentine (Syrrist, 1949; Syrrist and Gustafson, 1951).

The first commercial electron microscope imported to Scandinavia was a Siemens Übermikroskop 100. It had been ordered by the laboratory of Arne Tiselius in Uppsala before the war, but was not installed until September 1943. It was used primarily for analyses of viruses and blood proteins (Leyon and Gard, 1950). However, prior to its installation, Arne Tiselius had visited Bodo von Borries and Ernst Ruska in Berlin, with the aim of studying polio viruses (Tiselius and Gard, 1942).

III. Developments in the 1950s

The early years of the 1950s represent a turning point for biological electron microscopy. Sanford Newman, Emil Borysko, and Max Swerdlow (1949) developed the methacrylate embedding method, James Hillier and Mark Gettner (1950) noted that sectioning and section collection were facilitated if the sections were allowed to float off the knife edge onto water in a trough on the knife, and Harrison Latta and Francis Hartmann (1950) introduced glass knives. The thin sectioning techniques improved significantly by these and other measures.

At the Department of Anatomy, Karolinska Institute, in Stockholm, Fritiof Sjöstrand first developed an improved sectioning technique using a modified Spencer 820 rotary microtome. He then constructed a new ultramicrotome (Fig. 7), which was capable of producing ultrathin sections with a thickness of less than 200 Å (Sjöstrand, 1953b). The knife consisted of carefully polished razor blades. With these instruments he performed, for example, the first ultrastructural studies at high resolution of mitochondrial membranes (1953c), and myelin sheath (1953a) (Figs. 8 and 9). The new microtome was produced commercially from the mid-1950s by the LKB Company, as the LKB-Sjöstrand Ultramicrotome; about 500 units were sold worldwide.

This breakthrough in biological electron microscopy resulted in a cascade of new information. Investigators inside or outside Scandinavia joined Frit-

FIGURE 7. Prototype of Sjöstrand's ultramicrotome from 1951, capable of cutting sections with a thickness below 20 nm (200 Å).

iof Sjöstrand and, together with him, published high-resolution studies of cells and tissues: Johannes Rhodin on renal proximal tubules (Sjöstrand and Rhodin, 1953) (Fig. 10), Aksel Birch-Andersen on bacteria (Birch-Andersen *et al.*, 1953), Erwin Steinmann on chloroplasts (Steinmann and Sjöstrand, 1955) (Fig. 11), Viggo Hanzon on the Golgi apparatus, and Ragnar Ekholm on the thyroid epithelium. Other major projects from his laboratory in the mid- and late 1950s were presented as doctoral dissertations: Björn Afzelius investigated the nuclear envelope, Johannes Rhodin renal proximal tubule cells, where he discovered microbodies (peroxisomes) (Rhodin, 1954), Hans Zetterquist the intestinal epithelium, Jan Wersäll the inner ear, Åke Holmberg the ciliary epithelium, Ove Nilsson the uterine epithelium, and Ebba Cedergren the heart muscle and its T system. The founding of a new journal in 1957 by Fritiof Sjöstrand can be regarded as a

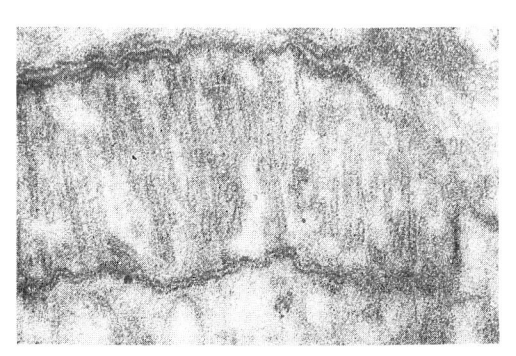

FIGURE 8. First high-resolution electron micrograph of a mitochondrion. X65,000. (From Sjöstrand, 1953c. Reproduced by permission from *Nature*.)

FIGURE 9. Electron micrograph of myelin sheath. X120,000. (From Sjöstrand, 1953a. Reproduced by permission from *Experientia*.)

spinoff product of this hectic activity: the *Journal of Ultrastructure Research* (recently renamed the *Journal of Structural Biology*), which ever since has been a leading journal in the field of biological electron microscopy.

The above studies were performed on an RCA EMU 2C electron microscope granted to Fritiof Sjöstrand in 1949 by the Wallenberg Foundation. For periods in the mid 1950s, this microscope was working virtually day and night and was exceptionally productive. The microscope was equipped with a mechanical stigmator that could be adjusted only by breaking the vacuum, removing the objective polepieces, and adjusting the iron screws of its stigmator—not an easy procedure compared to present-day compensation systems. A parallel RCA 3A in the same laboratory remained essentially nonfunctional until Holmes Halma, at that time an RCA engineer, spent a year extensively rewiring the microscope. As a result, this microscope was the only one of its model (3A) to work satisfactorily.

An additional group of biologists began their ultrastructural studies in Sjöstrand's laboratory in Stockholm in the late 1950s and continued in his laboratory in Los Angeles after his move there in 1959: Isser Brody working on the epidermis, Lars-Gösta Elfvin on peripheral nerves, Sven Erik Nilsson on retinal rods, Ulf Karlsson on neurons in the central nervous system, and Arvid Maunsbach on protein endocytosis and lysosomes in kidney tubules. These investigators then moved to other departments inside or outside Sweden and formed separate research groups.

Enzyme cytochemistry at the ultrastructural level had its origin in Sjöstrand's laboratory in Stockholm with a study by Huntington Sheldon, Hans Zetterqvist, and David Brandes in 1954 (Sheldon *et al.*, 1954). Another significant achievement was the development of the serial sectioning technique at the ultrastructural level by Fritiof Sjöstrand, who applied this

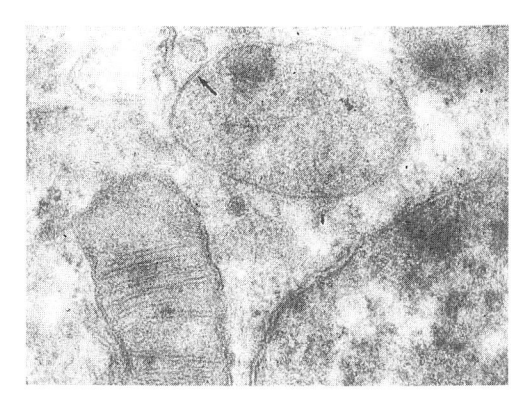

FIGURE 10. Cell organelles inside the cytoplasm of renal proximal tubule cell. Notice the triple-layered appearance (arrow) of the membrane that limits the granule. X50,000. (From Sjöstrand and Rhodin, 1953. Reproduced by permission from *Experimental Cell Research*.)

method to an analysis of the retinal rod synaptology. On the technical side, a new ultramicrotome, the Ultrotome (Hellström, 1960), was designed and produced by the LKB Company in Stockholm and became a standard instrument in biological laboratories all over the world (Fig. 12). More than 4000 of these instruments were sold over the next 20 years.

Other groups of biomedical electron microscopists were also started in Sweden in the 1950s, notably those of Ragnar Ekholm in Gothenburg and Hans Engström in Uppsala. In Stockholm, Anders Bergstrand was one of the first pathologists worldwide to analyze human kidney biopsies by electron microscopy (Bergstrand and Bucht, 1957). The Department of Cell Research at the Karolinska Institute had one of the first commercial electron

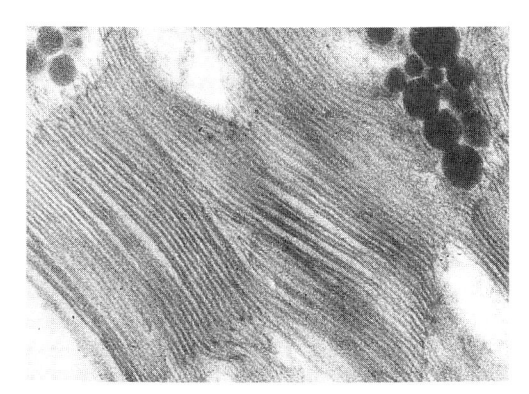

FIGURE 11. Electron micrograph showing the lamellated organization of a chloroplast. X60,000. (From Steinmann and Sjöstrand, 1955. Reproduced by permission from *Experimental Cell Research*.)

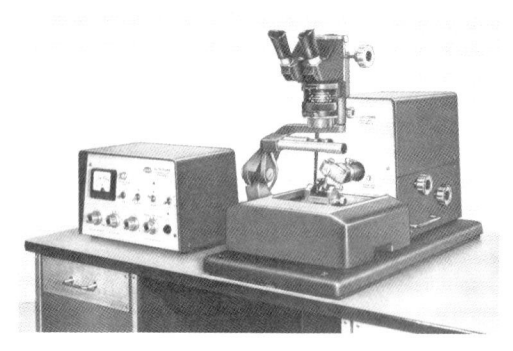

FIGURE 12. The well-known Ultrotome microtome manufactured by the LKB Company. More than 4000 units were manufactured.

microscopes in Sweden, an RCA EMU 2, installed in 1947, and attracted some foreign guest scientists. Thus, Humberto Fernández-Morán initiated work on cryoultramicrotomy and introduced the diamond knife for tissue sectioning (Fernández-Morán, 1953). The method of sharpening diamond for knives had previously been refined in the Siegbahn Laboratory, where similar knives were used in the preparation of spectroscopic gratings. Günther Bahr carried out basic *in vitro* studies of the interactions between osmium tetroxide and various biological substances (Bahr, 1954), and Elmar Zeitler, in collaboration with Günther Bahr, studied the quantitative interpretation and contrast in electron micrographs (Zeitler and Bahr, 1957, 1959).

In the field of physics and materials science, microscopes were installed in the early 1950s at the Royal Institute of Technology and at the Metal Research Institute in Stockholm, as well as at Chalmers University of Technology in Gothenburg. Some of this work focused on the analysis of steel alloys (Modin and Modin, 1955). The first million-electron-volt microscope was acquired by the Institute of Metal Research in Stockholm in 1969 and was used throughout the next decade. Already in the 1940s, three electron microscopes were installed in major industries. We have been unable to find out the type of initial work and the results obtained with most of these early microscopes, except for the one installed at ASEA. Subsequent microscopes applied to industrial work in Sweden were installed only in the early and mid-1960s.

A total of about 220 transmission electron microscopes (TEMs) had been purchased in Sweden by 1990. About two-thirds of these were involved primarily in biomedical research. The number of TEMs installed in Sweden is shown in the histogram in Fig. 13. Thus, 14 microscopes, five of which were Siegbahn-Schönander microscopes, were installed between 1943 and

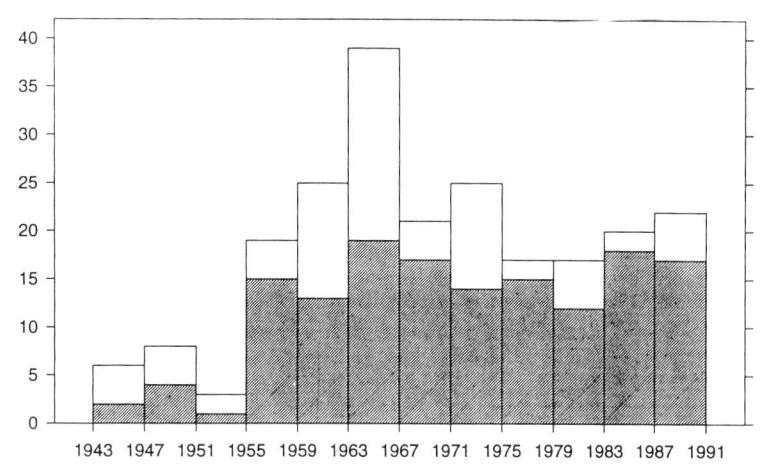

FIGURE 13. Histogram showing the total number of TEMs installed in Sweden from 1943 to 1990. Note the rapid increase in purchases in the mid-1950s. Shaded area, biology and medicine; white area, materials science and physics.

1950; the majority were directed toward studies of materials and industrial work. A dramatic increase in the purchase of electron microscopes took place in Sweden in the mid-1950s and reached a peak in the mid-1960s. Much of this increase was due to the establishment of new research groups in biology and medicine as a result of the breakthrough in techniques in the first half of the 1950s. New groups within materials science were also formed. The total purchase of transmission microscopes decreased somewhat in the 1970s, but sales for biological applications has remained at a fairly constant level since 1955.

It is noteworthy that scanning electron microscopy had a slow start in the Scandinavian countries. In the 1960s only a few microscopes were installed, some more during the 1970s, and only during the 1980s did a considerable increase in the sale of scanning microscopes take place. Most scanning electron microscopes were acquired by industrial laboratories, and only a few were used in basic research. This is reflected in the low numbers of contributions at SCANDEM meetings of scanning electron microscopy studies.

In the Nordic countries other than Sweden, developments in electron microscopy had a later start, due mainly to wartime events in these countries. In Norway, Joar Markali obtained funds from Marshall Plan aid and introduced electron microscopy into materials science in 1951 at the Center for Industrial Research. This organization had a similar character to that of a present-day Research Park, and electron microscopy was used in a number of different projects, particularly in studies on metal oxidation and the examination of planktonic organisms (Braarud and Nordli, 1952; Halldal et al., 1954). Slightly earlier, a transmission electron microscope was installed at the Department of Bacteriology of the University Hospital of Oslo. Another instrument was used at the Technical University of Norway for chemical research. In 1959, a Tesla instrument from Czechoslovakia was obtained by Theodor Blackstad for neuroanatomical research.

In Finland the first commercial microscopes were installed in 1956 in Helsinki and in 1957 in Turku, in the Anatomy Departments of both cities, and organized as separate laboratories within the Medical Faculties. Several years earlier, however, Alvar Wilska had begun to build low-voltage electron microscopes in the Physiology Department of the University of Helsinki. After a visit to Stockholm in 1947–1948, Wilska decided to build his own transmission electron microscope. The microscope was ready in 1952 and, contrary to the practice in commercial instruments, the acceleration voltage could be brought down to exceedingly low values, even to a few kilovolts, while retaining focus. Wilska was impressed by the improved specimen contrast at low voltages and, over the following decades, initiated the design of a series of low-voltage microscopes (Wilska, 1964a, 1964b). Some of these prototypes had accelerating voltages in the range of 6–10 kV (Fig. 14) and exhibited molecular resolution of protein molecules. From 1959, Wilska continued his work on low-voltage electron microscopy in the United States. It seems that, at that time, various technical difficulties prevented the commercial manufacture of these instruments, in spite of the fact that some well-resolved electron micrographs had been taken with the various prototypes. This situation is now changing as high-brightness field emission guns have become available, and some new instrumental designs incorporate low-voltage functions, thus illustrating the validity of Wilska's original concept.

In Denmark, transmission electron microscopes were installed in three laboratories in 1949. That of Aksel Birch-Andersen at the Serum Institute was the first serially produced Philips 100 electron microscope, and it was used largely for bacteriological work. In studies published in 1956, Birch-Andersen was the first to use epoxy resins for embedding (Fig. 15) (Maaløe and Birch-Andersen, 1956). At the Biophysical Laboratory, University of Copenhagen, H. M. Hansen and Frits E. Carlsen, also in 1949, installed an RCA electron microscope for biological and materials research, which in 1959 was supplemented with a Siemens Elmiskop I. In the Medical Faculty in Copenhagen, the first electron microscope, a Philips 100, was installed in 1957. A complete laboratory for ultrastructure research was then built by Harald Moe, who investigated in particular intestinal glands and capillaries (Moe, 1960). In 1960 Moe also set up a new laboratory at the Dental College in Copenhagen, and was joined by Olav Behnke and Jørgen Rostgaard, who

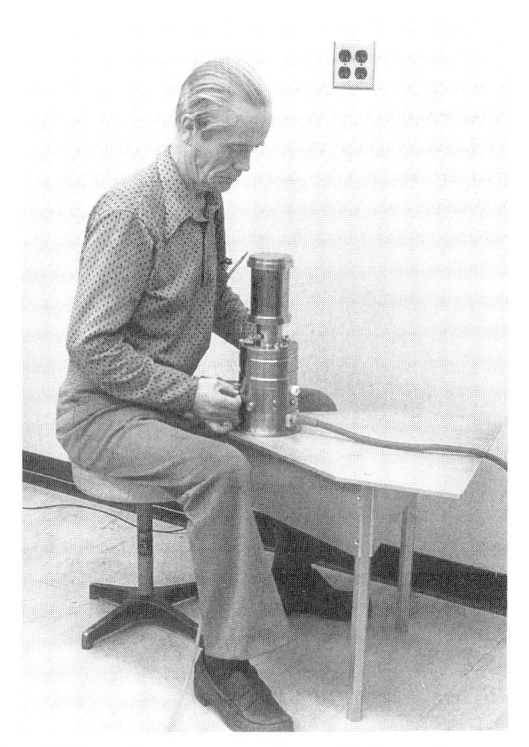

FIGURE 14. Alvar Wilska with a prototype of one of his low-voltage transmission electron microscopes. (Reproduced by permission of Mrs. M. Wilska.)

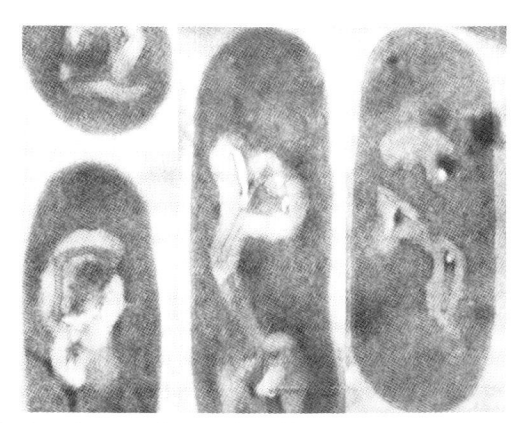

FIGURE 15. The first electron micrographs of biological material (coli bacteria) embedded in an epoxy resin. X25,000. (From Maaløe and Birch-Andersen, 1956. Reproduced by permission of A. Birch-Andersen.)

investigated microtubules and transporting epithelia, respectively. Shortly thereafter, electron microscope laboratories were also built in Aarhus for biomedical research, both at the Faculty of Medicine and at the Dental College.

Electron microscopy at the Danish Technical University was originally planned around a Philips EM 100 electron microscope, acquired in 1949 by Robert Asmussen at Chemical Laboratory B. In 1954, however, Ebbe W. Langer started a group in metallurgical research in the Department of Metallurgy and equipped his laboratory with the smaller and more versatile Philips EM 75 microscope. The work included studies of precipitates in metal alloys (Fig. 16) (Langer, 1955; Langer and Hansen, 1957) and this microscope was used for almost a decade, until it was replaced by a Hitachi HU-11A microscope. In the early 1960s, Ole Sørensen installed a microscope at the chemical engineering firm Haldor Topsøe for research in catalysts for chemical reactors, and Finn Grønlund at the Chemistry Laboratory at the University of Copenhagen initiated work on oxidation reactions on copper surfaces. Thus, with a start in the mid-1950s, electron microscopy became increasingly popular in Denmark within different fields of materials science, both in universities and industry.

Personal accounts from the above-mentioned early Danish laboratories should remind us of the hardships and frustrations encountered with some of these old commercial instrument models. Thus, in one case, the microscope was practically useless for a year due to specimen drift, which was finally diagnosed as being due to a manufacturing error in the mechanical support of the objective lens system. Another microscope had a porosity defect in its cast aluminum column which made it impossible to achieve high vacuum until it was sealed off externally with rubber pads and vacuum grease!

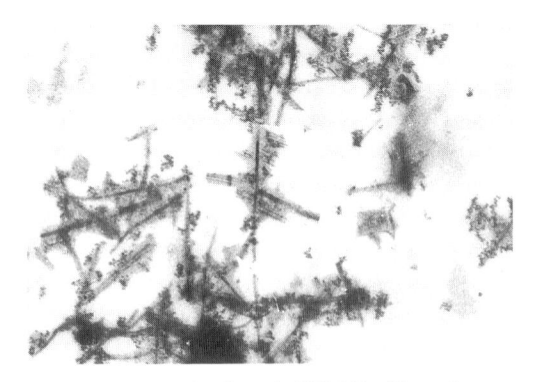

FIGURE 16. Electron micrograph of steel. X32,000. (From Langer and Hansen, 1957. Reproduced by permission of E. W. Langer.)

IV. THE FORMATION OF THE SCANDINAVIAN SOCIETY FOR
ELECTRON MICROSCOPY

The founding meeting of the Scandinavian Society for Electron Microscopy took place on Saturday, 16 October, 1948, at the Research Institute of Experimental Physics in Stockholm. The meeting was chaired by Kai Siegbahn and attracted 21 participants representing life sciences, physics, and materials science. Nineteen of the participants were from Sweden, one from Denmark and one from Norway. The sole lecture at the meeting was given by Allan Danielsson, who reported on his impressions from a study tour. The major point of the meeting was a discussion of whether or not to form a society for electron microscopy. It is evident from the minutes of the meeting that a certain scepticism was expressed about the relevance of founding such a society, considering the fact that the members would represent highly different fields of research. However, it was concluded that such a society might still be useful, since electron microscopy was a new technique, and a society could promote the spreading of information about specimen preparation techniques and the operation of the microscope itself. Some of the discussions of today's EM societies echo the same questions that were aired in 1948! The meeting elected a board of three members: Nils Hast of the Royal Institute of Technology in Stockholm as President, Fritiof Sjöstrand of the Karolinska Institute as Secretary/Treasurer, and Gösta Glimstedt, University of Lund, as an ordinary member of the board. The board was asked to prepare statutes for the society before the next meeting in 1949 and to contact the other Nordic countries and the EM Society in the UK.

Since Finland was not represented, either at the founding meeting or at the next few meetings, the name of the society became the Scandinavian Society for Electron Microscopy. As Finnish scientists joined the Society in the early 1950s, a better name would have been the Nordic Society for Electron Microscopy (Denmark, Norway, and Sweden are generally called the Scandinavian countries; Denmark, Finland, Iceland, Norway, and Sweden are the Nordic countries. The society has never had any members from Iceland.)

From the 21 participants at the founding meeting in 1948, the Scandinavian Society has grown to 717 members in 1993. Of these, 97 are from Denmark, 141 from Finland, 118 from Norway, 338 from Sweden, and 23 from non-Nordic countries, including the UK, Germany, France, and the United States. It is noteworthy that all branches of science using electron microscopy are represented in the society and, also, that the majority of all scientists using electron microscopy in the Nordic countries are members of the society. In the 1980s, the society introduced a special Company Membership for commercial companies, allowing a closer contact between the society and industry. While the official name is the Scandinavian Society for Electron Microscopy, the abbreviated form, SCANDEM, initially introduced by the Finnish organizers of the Annual Meeting in 1974, is also used both for the society and for its annual meetings. The total membership of 694 from the Nordic countries should properly be considered in relation to the total population of Denmark, Finland, Norway, and Sweden, which is 23 million.

The board of the society consists, in addition to the President (Fig. 17), Vice President, and General Secretary/Treasurer, of two representatives from each of the four participating Nordic countries, one being a biologist and one a physicist or a materials scientist. This composition, together with a gradual renewal of the board members, has functioned well over a long time. An important stabilizing factor of the society is the fact that the General Secretary has usually served for several years, sometimes up to

FIGURE 17. Seven consecutive SCANDEM Presidents, 1973–1995. Upper row, from left: Bjørn V. Johansen, Norway (1983–1986); Arvid B. Maunsbach, Denmark (1977–1980); Björn Afzelius, Sweden (1973–1976), Jorma Wartiovaara, Finland (1981–1982). Lower row, from left: Sven-Olof Bohman, Sweden (1987–1989); Kaarina Pihakaski-Maunsbach, Finland (1990–1993), Anders Thölén, Sweden (1994–present).

eight consecutive years, as is also the case in the International Federation of Societies for Electron Microscopy (IFSEM).

V. ACTIVITIES AND MEETINGS OF THE SCANDEM

In September 1956, the Scandinavian Society for Electron Microscopy, at the request of the International Federation of Electron Microscope Societies, organized the first European Regional Conference on Electron Microscopy in Stockholm. The conference was chaired by Fritiof Sjöstrand and held at the Karolinska Institute with 370 participants from 27 countries. With 176 papers read, this conference represented the largest meeting on electron microscopy up to that date and was one of the very first conferences with an instrument exhibition. It also had an extensive exhibition of electron micrographs and was thus, in a sense, an early forerunner of the present-day poster sessions.

The main ongoing activity of the society is the organization of its annual meetings. These meetings have been held every year since the founding meeting, at the beginning of June. At the founding meeting there was only one lecture; the next year there were seven. The number of lectures increased so rapidly that the meetings had to extend over two entire days, and by the beginning of the 1970s had expanded to three days. The venue rotates among the four member countries. During the first decades the rotation was irregular, but since the mid-1970s the fixed sequence of Denmark, Sweden, Finland, and Norway has been strictly adhered to. Furthermore, the meeting locations shift within the host countries and thus, over a number of years, SCANDEM participants come to visit most Nordic university cities. Almost all annual meetings have been organized on university campuses to reduce organizational expenses and registration fees. Likewise, student dormitories have often been available to reduce accommodation costs.

In the 1950s and 1960s, most presentations at the annual meetings were in Swedish, Danish, or Norwegian. Participants from Finland spoke either in Swedish or English. Since Finns and Swedes are notorious for having difficulties in understanding Danish—and sometimes vice versa—the discussions were sometimes confused, with many unexpected side effects leading to impaired scientific efficiency. In 1975, therefore, it was recommended that all lectures be in English, and this has been the official language of the society ever since. The few invited lectures had in any case been given in English or German from the very beginning of the society. For the first 20 years or so, the society was therefore a rather closed Nordic society, in which all presentations were oral and where senior scientists as well as younger ones presented their contributions in plenary sessions to all participants, biologists, physicists, and materials scientists alike. Many scientists who are now experienced congress veterans can remember their debut with their knees shaking in front of a large—but basically friendly— SCANDEM forum.

The "internationalization" of SCANDEM accelerated in the 1970s due to the increasing number of invited scientists from non-Nordic countries (Fig. 18). At about the same time, the size of the meeting expanded from 50–100 participants in the 1950s and 1960s to 200–400 in the 1970s and 1980s. This necessitated parallel sessions and, from 1975, also poster sessions. Instrument exhibitions were arranged from about 1970. Exhibitions of electron microscopes and auxiliary equipment are now highly valued aspects of the annual meetings.

This "internationalization" of the meetings has not come about without animated discussions. On the one hand, it was agreed that it is important to establish close contact with leading scientists. On the other hand, it has been emphasized that SCANDEM meetings should focus on a genuine Scandinavian exchange of ideas rather than on an imitation of large European or international congresses. In particular, it has been considered important to stimulate young scientists in the field and allow them to give their presentations in a more relaxed atmosphere than that of an international meeting. The same problem seems to face most national societies; they have to choose between arranging mini-international congresses or truly national meetings with richer possibilities for neophytes to appear in person and explain their results. The increasing tendency of some organizers to arrange mini-international congresses has also resulted in a dramatic in-

FIGURE 18. Two well-known participants at the SCANDEM meeting in Jyväskylä, Finland, in 1982: Elmar Zeither (left) and Alvar Wilska (right).

crease in the registration fee at the SCANDEM meetings, from about $15 in 1970 to $150 at present (representing a doubling of costs in real terms). This sum, together with high traveling costs within Scandinavia—remarkably high considering the short distances involved in many cases—has become a limiting factor for many participants.

The topics during the annual meetings of SCANDEM have undergone striking changes during its more than 40 years of existence. In the beginning, "preparation methods," together with "biomedical reports," dominated the meetings. The 1960s and much of the 1970s saw a dramatic expansion of ultrastructural cell biology after the invention of such methods as ultrastructural cytochemistry and autoradiography. In this period, physics and materials science were hardly represented, sometimes by only a handful of participants. However, in the 1980s, an increasing number of physicists and material scientists participated due to stronger programs in these areas. Presently, there is a corresponding decrease in the number of participating biologists, who apparently prefer to go to specialized meetings on cell biology, molecular biology, neurobiology, plant biology, microbiology, or other topic-centered congresses. Thus, the question asked in 1948, whether the society should deal primarily with preparatory methods and microscopy or with various specific fields of research, still remains a hot topic.

Another welcome tendency over the last decade has been increased participation by technical assistants. At some meetings, special sessions have been arranged for these participants on topics such as laboratory safety, and they have attracted much interest. Apart from the annual meetings, one-day winter meetings have been held in January and have usually focused on one specific technique. These have been appreciated by both technicians and researchers. Furthermore, the society has for several years sponsored two or three one-day courses on various aspects of electron microscope methods, such as immonocytochemistry. These "SCANDEM courses" have been held at different universities and have generally been well attended.

Some national electron microscopic societies have now dropped the specification "electron," to indicate a broader scope. However, the General Assembly of SCANDEM recently (1992) decided to retain its name, but to include in its programs also other aspects of microscopy, for instance, confocal microscopy and scanning probe microscopy.

VI. The Scientific Impact of the SCANDEM

Assessment of the importance and impacts of any scientific organizations is of necessity subjective. Nevertheless, in respect to SCANDEM we feel that the society has successfully filled several functions. We believe that it has been instrumental in improving the quality of electron microscopy research in the Nordic countries, mainly through its annual meetings, which have accelerated the spread of new techniques and findings and given examples of front-line research in various fields of electron microscopy. Through its regular annual meetings, the society has also stimulated collaboration among scientists in related fields throughout the Nordic countries.

REFERENCES

Bahr, G. F. (1954). Osmium tetroxide and ruthenium tetroxide and their reactions with biologically important substances. *Exp. Cell Res.* **7**, 457–479.

Bergqvist, A. (1946). Ett svenskt elektronmikroskop. *Industritidningen Norden* **33**, 3–7.

Bergstrand, A., and Bucht, H. (1957). Electron microscopic investigations on glomerular lesions in diabetes mellitus (diabetic glomerulosclerosis). *Lab. Invest.* **6**, 293–300.

Birch-Andersen, A., Maaløe, O., and Sjöstrand, F. (1953). High resolution electron micrographs of sections of *E. coli. Biochim. Biophys. Acta* **12**, 395–400.

Braarud, T., and Nordli, E. (1952). Coccoliths of Coccolithus huxleyi seen in an electron microscope. *Nature* **170**, 361–362.

Elvers, I. (1941). An electron-microscopic study of chromosomes and cytoplasm in *Lilium. Arkiv Botanik* **30B**, 1–8.

Elvers, I. (1943). On an application of the electron microscope to plant cytology. *Acta Horti Bergiani* **13**, 149–245.

Engström, H., and Wersäll, J. (1952). Some principles in the structure of vibratile cilia. *Annals Otol., Rhinol. Laryngol.* **61**, 1027–1038.

Fernández-Morán, H. (1953). A diamond knife for ultrathin sectioning. *Exp. Cell Res.* **5**, 255–256.

Halldal, P., Markali, J., and Naess, T. (1954). A method of transferring objects from a light microscope to marked areas on electron microscope grids. *Mikroskopie* **9**, 197–200.

Hast, N. (1948). Production of extremely thin metal films by evaporation on to liquid surface. *Nature* **162**, 892–893.

Hellström, B. (1960). The *Ultrotome* ultramicrotome—Basic principles and summarized description of construction. *Science Tools* **7**, 10–17.

Hillier, J., and Gettner, M. E. (1950). Sectioning of tissue for electron microscopy. *Science* **112**, 520–523.

Ingelman, B., and Siegbahn, K. (1944). An electron-microscopic study of dextran molecules. *Arkiv kemi, mineral. geol.* **18B**, 1–6.

Langer, E. W. (1956). Electron-microscopic observations on precipitation of the theta phase in an aluminium 4% copper alloy. *J. Inst. Metals* **84**, 471–473.

Langer, E. W., and Hansen, N. (1957). The precipitation of ε-carbide by ageing of soft steel. *J. Iron Steel Inst.* **186**, 422–424.

Latta, H., and Hartmann, J. F. (1950). Use of a glass edge in thin sectioning for electron microscopy. *Proc. Soc. Exp. Biol. Med.* **74**, 436–439.

Leyon, H., and Gard, S. (1950). Electron microscopy of Theiler's virus, strain FA. *Biochim. Biophys. Acta* **4**, 385–390.

Lindegaard-Andersen, A. (1958). Plastic deformation in zinc oxide films. *Acta Metallurgica* **6**, 306–308.

Maaløe, O., and Birch-Andersen, A. (1956). On the organization of the "nuclear material" in *Salmonella typhimurium. In* "6th Symp. Soc. Gen. Microbiol." Cambridge University Press, Cambridge, 1956, pp. 261–278.
Modin, H., and Modin, S. (1955). Pearlite and bainite structures in an eutectoid carbon steel. An electron microscopic investigation. *Jernkontorets Ann.* **139**, 481–515.
Moe, H. (1960). The ultrastructure of Brunner's glands of the cat. *J. Ultrastr. Res.* **4**, 58–72.
Newman, S. B., Borysko, E., and Swerdlow, M. (1949). New sectioning techniques for light and electron microscopy. *Science* **110**, 66–68.
Rhodin, J. (1954). "Correlation of Ultrastructural Organization and Function in Normal and Experimentally Changed Proximal Convoluted Tubule Cells of the Mouse Kidney." Aktiebolaget Godvil, Stockholm.
Ruska, E. (1934). Über Fortschritte im Bau und in der Leistung des magnetischen Elektronenmikroskops. *Z. Phys.* **102**, 417–422.
Sheldon, H., Zetterqvist, H., and Brandes, D. (1955). Histochemical reactions for electron microscopy: Acid phosphatase. *Exp. Cell Res.* **9**, 592–596.
Siegbahn, M. (1939a). Vetenskapsakademiens Forskningsinstitut för Experimentell Fysik. *Nordisk Familjeboks Månadskrönika* **2**, 571–576.
Siegbahn, M. (1939b). *Kungliga Svenska Vetenskapsakademiens Årsbok* **37**, 147–148.
Siegbahn, M. (1940). *Kungliga Svenska Vetenskapsakademiens Årsbok* **38**, 151–153.
Siegbahn, M. (1941). *Kungliga Svenska Vetenskapsakademiens Årsbok* **39**, 131–133.
Siegbahn, M. (1942). *Nordisk Tidskrift för Fotografi* **26**, 205–206.
Sjöstrand, F. S. (1943a). Electron-microscopic examination of tissue. *Nature* **151**, 725–726.
Sjöstrand, F. S. (1943b). Eine neue Methode zur Herstellung sehr dünner Objektschnitte für die elektronenmikroskopische Untersuchung von Geweben. *Arkiv Zool.* **35A**(5), 1–22.
Sjöstrand, F. S. (1943c). Fixering och preparering för elektronmikroskopisk undersökning av vävnad. *Nordisk Medicin* **19**, 1207–1223.
Sjöstrand, F. S. (1953a). The lamellated structure of the nerve myelin sheath as revealed by high resolution electron microscopy. *Experientia* **9**, 68–69.
Sjöstrand, F. S. (1953b). A new microtome for ultrathin sectioning for high resolution electron microscopy. *Experientia* **9**, 114–115.
Sjöstrand, F. (1953c). Electron-microscopy of mitochondria and cytoplasmic double membranes. *Nature* **171**, 30–32.
Sjöstrand, F. S., and Rhodin, J. (1953). The ultrastructure of the proximal convoluted tubules of the mouse kidney as revealed by high resolution electron microscopy. *Exp. Cell Res.* **4**, 426–456.
Steinmann, E., and Sjöstrand, F. S. (1955). The ultrastructure of chloroplasts. *Exp. Cell Res.* **8**, 15–23.
Syrrist, A. (1949). An introduction in electron microscopy with some results from histological investigations of enamel and dentine. *Odont. Tidskr.* **57**, 79–105.
Syrrist, A., and Gustafson, G. (1951). A contribution to the technique of the electron microscopy of dentine. *Odont. Tidskr.* **59**, 500–513.
Tiselius, A., and Gard, S. (1942). Übermikroskopische Beobachtungen an Poliomyelitis-Viruspräparaten. *Naturwissenschaften* **30**, 728–731.
Wersäll, J. (1954). The minute structure of the crista ampullaris in the guinea pig as revealed by the electron microscope. *Acta Oto-Laryngol.* **44**, 359–369.
Wilska, A. P. (1964a). Low-voltage electron microscopy. A 6-kV instrument. *J. Roy. Microsc. Soc.* **83**, 207–211.
Wilska, A. P. (1964b). Recollections and experiences from fifteen years of constructing electron microscopes. *Ann. Acad. Scient. Fennicae Series A. V. Medica* **106/23**, 1–14.
Zeitler, E., and Bahr, G. F. (1957). Contributions to the quantitative interpretation of electron microscope pictures. *Exp. Cell Res.* **12**, 44–65.
Zeitler, E., and Bahr, G. F. (1959). Contributions to quantitative electron microscopy. *J. Appl. Phys.* **30**, 940–944.

2.12
Electron Microscopy in Southern Africa

MICHAEL J. WITCOMB

Electron Microscope Unit
University of the Witwatersrand
WITS 2050, South Africa

SONIA A. WOLFE-COOTE

Experimental Biology Programme
Medical Research Council
Tygerberg 7505, South Africa

I. THE ELECTRON MICROSCOPY SOCIETY OF SOUTHERN AFRICA (EMSSA)

An electron microscopy society for the whole of Southern Africa had its beginnings in 1953 with informal meetings held by physicists interested in electron microscopy. These were organized by Dr. H. G. F. Wilsdorf second on the left in Fig. 1, then a member of the Council for Scientific and Industrial Research (CSIR), Pretoria, which had a Philips EM100 microscope; Dr. P. J. Jackson and J. H. Talbot, of the Chamber of Mines Research Laboratories, Johannesburg, which had an AEI EM3A microscope; and Dr. J. W. F. Hampton, of the Poliomyelitis Research Foundation laboratory, also in Johannesburg, which also had an AEI EM3. There were three meetings a year, held at each laboratory in turn. The group was joined in 1954 by Dr. J. T. Fourie, second from the right, in Fig. 1 (CSIR), and in 1956 by Prof. D. J. Fourie, of the University of Pretoria (UP), which had a Zeiss electrostatic microscope. These meetings addressed everyday problems such as how to get the microscope to work, how to prepare samples, and what advances were being made in electron microscopy overseas.

Although the first electron microscope in Southern Africa was an RCA Console transmission microscope delivered to the Veterinary Research Institute at Onderstepoort in Pretoria in 1946, it was the arrival of the first "high-resolution" transmission electron microscope (TEM), the double-condenser Siemens Elmiskop 1, at the University of the Witwatersrand, Johannesburg (Wits), that really galvanized the interest in electron micros-

FIGURE 1. From left to right, Dr. Neville Comins, Prof. Heinz Wilsdorf, Mrs. Judy Harris, Dr. Koos Fourie, and Mr. Jerry Thirlwall. The photograph was taken in 1981 with the group in front of the CSIR's JEOL JEM-200B.

copy. The microscope was placed in the Department of Physics, which subsequently delegated its installation and supervision to one of its laboratory assistants, Mr (later Dr.) J. W. Matthews (Fig. 2), a name that subse-

FIGURE 2. Dr. John W. Matthews.

quently became synonymous with epitaxy. With the addition of interested staff from Physics, Zoology, and other Wits departments, together with the financial support of commercial firms, this informal group soon blossomed into a formal society.

The first move to become associated with the International Federation of Societies for Electron Microscopy (IFSEM) was made when Dr. R. Yodiaken, then of the Wits Medical School, attended an international meeting in 1962 and suggested that the society be constituted with the aim of joining the IFSEM. As a result, in the same year, an inaugural conference/planning meeting of the EMSSA, attended by 30–40 interested participants, was held at Wits. The commercial firms feted the delegates, and it is rumored that this contributed to a most pleasant environment, conducive to both constructive planning and considerable partying! At the meeting, Dr. Fourie was elected the first Chairman of the society, and Dr. Talbot the Secretary/Treasurer. The most significant decisions taken were to hold regular conferences and to draft a nonracial constitution. Subsequently, the society applied for formal membership in the IFSEM in 1964; this was granted in 1966 at the Sixth International Conference in Kyoto, Japan.

South Africa is a multilingual country; the official languages are Afrikaans, a language derived from seventeenth-century Dutch, and English. This was reflected in the original EMSSA constitution, which states that conference abstracts and presentations could and still can be submitted in either language. In particular, the Chairman's Conference Welcoming Address and the Annual General Meeting alternate each year between the two languages. While many Afrikaaners speak and understand both languages, most English speakers do not. Hence, when Prof. B. I. Balinsky (Fig. 3), the society's second Chairman, presented his Welcoming Address at the 1965 Conference, most delegates thought, due to his strong accent, that he was speaking in his native Russian. They only realized that he was speaking in Afrikaans when he was well into his address! At the inaugural Boris Balinsky Lecture in 1985, the language problem reared its head yet again. On that occasion, one of the members of the Executive Committee, showing off his prowess in Russian, greeted Professor Balinsky in his native language. After a while, Professor Balinsky interrupted him with, "Are you speaking some language I am supposed to be able to understand?" Since then, the society has confined its linguistic adventures to the two official languages, although the constitution allows Chairmen and subsequently Presidents to conduct meetings in their language of choice, which has generally been English. However, questions can still be addressed to them in either official language.

The society's early annual conferences were held alternatively in Pretoria and Johannesburg, since the majority of members were located in these

FIGURE 3. Professor Boris I. Balinsky.

centers in the Transvaal. Invited speakers for the keynote addresses were drawn only from the national pool of electron microscopists until 1970, when both these traditions were broken. In that year, the Ninth Annual Conference was held at the University of Natal, Durban, and one of the invited speakers, Prof. M. J. Karnovsky, an ex-patriot of South Africa, was brought out from the Harvard Medical School for the occasion. Since that time, the EMSSA has continued to invite each year at least one, usually two, overseas guests of international repute. Some are remembered more than others for events, catastrophic or humorous, surrounding their visits. Some years ago, both invited speakers were single and set off on a trip around the local nightspots with two young ladies, one of whom had borrowed transport from her cousin, an undertaker. They recounted the next morning that the transport was, in fact, a hearse, which was lent on the proviso that the corpse it contained would be delivered to its destination on schedule! They have probably dined out on that story ever since.

The year 1971 was significant in that it marked the first independent publication of the society's own conference proceedings, each author being allowed two numbered pages, high-quality micrograph reproduction being guaranteed. Previous to that, abstracts of papers given at the conferences from 1966 to 1969 had been published in the *South African Journal of Science*, as roughly hundred-word summaries. This change in the method of publishing the conference presentations and particularly the improved quality of the product immediately resulted in a doubling of the contributions to 40.

Except on one occasion, all conferences have been held at universities, for financial reasons. Hotels are used mainly by the commercial representatives. A bond is quickly established between the rest of the delegates through the mutually interesting experience of staying in the halls of residence, some of which have been pretty spartan. It is surprising how one takes for granted shower nozzles and curtains, comfortable beds, reasonable food! One of us (MJW) was rather desperate at one conference for a hot shower. He heard a rumor that there was hot water at the far end of the women's floor above. So, midafternoon, when the other microscopists were at the scientific sessions, he located the place and proceeded to shower in truly wonderful hot water. To his surprise—or possibly delight—in came a very attractive young lady, who started to strip in the central area. The dilemma of this respectable Executive Committee member was resolved by humming loudly—singing would have driven her away!

Other factors influenced these educational experiences. There was a surprise one year when it snowed on the mountains overlooking the conference venue on the first night, and residents had only one sheet and one blanket. So much for typical summer weather! But the social events are designed to overshadow any deficiencies. One evening there is always a traditional South African *braaivleis* (barbeque), with cooked meat, *boerewors* (sausage), salads, and liquid delights. Since everything is totally informal, including dress, everyone is relaxed, and worries are forgotten. It is surprising, though, how often it rains soon after the *braai* has started!

After the death of Dr. Matthews (Fig. 2) in the United States in 1977 (see Sec. II. F), his family donated a sum of money to the society to be used to sponsor a named lecture in his honor. A distinguished speaker has presented the lecture each year since the 1978 conference, when Prof. F. R. N. Nabarro, FRS (Wits), delivered the first John Matthews Memorial Lecture in the Physical Sciences. Later, it was agreed to honor the internationally known embryologist and second Chairman of the EMSSA, Prof. B. I. Balinsky (Fig. 3) (see Sec. II. E). In 1985, Prof. Balinsky was present on the occasion of the first Boris Balinsky Lecture in the Life Sciences, which was delivered by Dr. O. Moestrup of the University of Copenhagen. Without the generous financial support over the years from both the CSIR and the Medical Research Council (MRC) toward the travel costs of the invited speakers, these lectures would not have been possible.

Since 1981, the EMSSA has successfully instituted, with the generous help of the commercial sector, a number of significant prizes pertaining to both the quality of the conference abstract and its presentation. This not only encourages students but acknowledges excellence in research on the different aspects of microscopy, techniques, and microanalysis work presented at the annual conference. In 1988, South African Philips offered a substantial annual award for the best published electron microscopy paper in the open literature. The first recipient of the prize was Dr. H. K. Schmid (CSIR), for his work on the microstructure of glass-ceramics in the system B-Na-Si-Ta oxides. In the following year, the Anglo-American Chairman's Fund generously provided funds for an initial two-year period to assist needy students to participate in the conferences. All these awards and prizes are presented at the annual dinner and, in recent years, a dance held on the last evening of the three-day conference. With no papers to present the following day, the event takes on a party atmosphere, especially as it is the end of the academic year and the beginning of the summer/Christmas holidays. The frustrations of the year fade, only to be replaced later by the frustration of being told to vacate the dance venue in the early hours of the following morning! Since at least 50% of the society membership of 350 attend the conference, the atmosphere is warm and friendly, an occasion for friends to renew old acquaintances. For members who are not able to come to the conference, and to keep everyone in touch with news and events throughout the year, the society has for the past six years published annually a very successful four-monthly newsletter under the able editorship of Mr. R. H. M. Cross of Rhodes University. This is particularly necessary since it must be remembered that there are considerable distances between most of the major centers—e.g., 1450 km between Johannesburg and Cape Town, and 1750 km between the University of the North and the University of Cape Town.

For the past 19 years, the society has successfully encouraged the organization of workshops in electron microscopy and related techniques. These are normally held in midwinter or run back-to-back with the conference so as to make use of the invited speakers. For example in 1991 there were three workshops held just before the conference, one each on environmental scanning microscopy, light microscopy techniques, and three-dimensional electron microscopy. Although the society had been unsuccessful in its attempts to establish a correspondence training program for technicians in electron microscopy, it actively supported the opening in 1984 of a Centre for Electron Microscopy at the Technikon Pretoria and assisted with the course structure. The first students entered this three-year diploma course in 1988. This intake has since been integrated into EM units throughout the country for the 18-month in-service training component of the diploma. The combined effect of the cutback in tertiary education funding, the relatively high cost of the course per student for the Technikon, the small student intake, and a recession reducing the number of jobs available has resulted in no students being allowed to register in 1991. It remains to be seen whether student exposure to electron microscopy will revert to being a component of other courses.

Contact between the EMSSA and electron microscopists in Angola and Mozambique has regrettably been curtailed for years, due to the civil wars in both those countries. It was only in 1990 that some educational contact was reestablished with Mozambique. Similarly, very limited contact was possible with microscopists in Rhodesia after UDI and the transition to a subsequent independent Zimbabwe. It is hoped that with improved international relations, the situation will be remedied in the near future so that the society will again become representative of Southern Africa rather than just South Africa.

Since the EMSSA became a member society of the IFSEM, increasing numbers of microscopists from Southern Africa have traveled to IFSEM conferences; see Table I. However, a deterrent to further participation has always been the ever-decreasing value of local currencies against those of the major industrialized world and the geographic isolation of the region,— for example, Johannesburg to London is 9000 km. There have never been cheap or charter flights available, the airfares always being high in order to support South African Airways, which for a long time could not overfly Africa, causing it to fly around the bulge of West Africa (direct flights were restored in mid-1991). The EMSSA has, however, been able to have a local representative attend the committee meetings at all the IFSEM conferences. It received support during all the sanction years, being permitted access to the international electron microscopy community. In particular, Prof. G. Thomas, both as Secretary and as President of the IFSEM, ensured that our members were able to present their papers at the Hamburg and Kyoto conferences. In 1975, the society was honored to be able to host the current IFSEM President, Dr. V. E. Cosslett, as one of its invited speakers to report on "Progress in High-Voltage Electron Microscopy" at a time when the

TABLE I

STATISTICS RELATING TO SOUTHERN AFRICAN DELEGATES ATTENDING IFSEM CONFERENCES

Year	Location	Number of delegates	Life sciences	Physical sciences	Total
1966	Kyoto	2	1	1	2
1970	Grenoble	8	3	5	8
1974	Canberra	6	—	4	4
1978	Toronto	11	4	3	7
1982	Hamburg	10	4	4	8
1986	Kyoto	7	10	6	16
1990	Seattle	13	6	4	10

Number of presentations given

possible establishment of a national HVEM facility was being considered. In the event, it was more practical to send people overseas to use such facilities.

Since 1962, when Dr. J. T. Fourie took the chair of the EMSSA, the position has been held successively by Prof. B. I. Balinsky, Prof. T. A. Villiers, Dr. N. R. Comins (the last Chairman of the society), Prof. R. N. Pienaar (the first President), Prof. B. B. Rawdon, Prof. H. C. Snyman, Dr. S. A. Wolfe-Coote (the first lady President, elected in 1988), and Dr. M. J. Witcomb. During most of the time, indeed from 1965, Dr. E. M. Veenstra has served the society, as Secretary for 26 years and as Treasurer for 23 years. With her wealth of experience, she has provided continuity to the society, and has quietly and efficiently assisted the Chairmen and Presidents through their periods of office. She has seen many changes, including the recent opening up of the society to include all forms of microscopy. Dr. Veenstra has watched the society grow from its small beginnings to a large, multidisciplinary group of which 72% of the members are now from the life sciences (including dental and medical), 22% from the physical sciences (including materials science and engineering), and 5% from the earth sciences. Many members have worked tirelessly for the society over the years; in recognition of their services, the following have been made Honorary Fellows: Prof. B. I. Balinsky, Dr. N. R. Comins, Dr. J. T. Fourie, Dr. R. J. Murphy, Dr. E. M. Veenstra, and Prof. T. A. Villiers.

II. ACTIVITIES AND ACHIEVEMENTS IN ELECTRON MICROSCOPY

Outlined below are some examples of the research carried out over the years involving electron microscopy in Southern Africa.

A. Instrumentation

In 1982, Prof. D. Crawford, Director of the Electron Microscope Unit, University of Cape Town (UCT), and Mr. D. A. Gerneke, developed SIMSEM, the first computer-based interactive electron microscope simulator for teaching SEM operation. The student controls the simulator by turning live knobs on a control panel, so that the instrument looks, feels, and acts just like a real SEM (Fig. 4). SIMSEM has provided novices with a self-tuition course that has allowed them to become familiar with SEM principles through text, diagrams, and references and, via interactive simulations, the operation of an SEM before they approach the real instrument. This has conserved productive research time on the SEM as well as preventing user-initiated accidents on the microscope during the initial teaching period. Since its completion in 1983, the simulator has been successfully

FIGURE 4. The prototype simulated SEM.

used at UCT by some 160 postgraduate students, 110 undergraduate students, and 75 technicians.

B. Specimen Preparation

Major contributions in optimizing processing methodology for plant TEM have been made at the University of Pretoria by Prof. J. Coetzee and Mr. C. F. van der Merwe. Their work on glutaraldehyde and buffer osmolarities has led to a better understanding of "effective osmolarity" during fixation of plant tissue.

Drs. S. A. Wolfe-Coote, J. Louw, and B. Day, of the MRC, have established a biological model from which the effects of tissue preparative procedures on animal tissue dimensions can be assessed in an objective manner. The model has been used successfully so far to investigate the effects on tissue dimensions of varying fixative/buffer osmolarities and of storage and drying resin-embedded blocks. Since 1976, this group has also been very active in the development, refinement, and teaching of immunolabeling procedures.

Dr. J. R. Lawton, at the University of Durban-Westville, has studied the use of dye-glutaraldehyde primary fixation, using dyes that have shown some chemical specificity by light microscopy but that do not contain heavy metals. Neutral red, for example, was found to make nucleic acids clearly resolvable, malachite green to retain lipids. The basis of the reaction between lipid and dye has been considered in relation to the structural formulas of the dyes and model lipid.

C. Botanical Sciences

Professor T. A. Villiers is considered to be the father of botanical electron microscopy in Southern Africa. His research interests until the mid-1970s at the University of Natal, Durban (UND), concerned seeds, a topic which still forms the basis of a very flourishing line of research at Durban. Professor Villiers was one of the major world figures in seed dormancy, and he made extensive use of the TEM in his correlative ultrastructural and physiological studies. Professor P. Berjak, the present Head of Biological Sciences at UND, was his first Ph.D. student to use the TEM as a major research tool in microscopical characterization of deteriorative changes in stored maize seeds. Subsequently, she and her team have contributed significantly to aspects of fungal-mediated deterioration of stored, air-dried (orthodox) seeds, with characterization of the fungal species involved, and to elucidating the basis of recalcitrant (dessication-sensitive) seed behavior. Another former student was Prof. R. N. Pienaar, Head of the Department of Botany, Wits. Professor Pienaar has gained worldwide recognition among phycologists for his TEM work on the characterization of unicellular organisms known collectively as nanoplankton, and particularly in the area of scale production. He has concentrated on describing and explaining the life cycles of new species, particularly those of marine origin.

D. Dental Sciences

Although dental research is active in South Africa, little, except that at the Dental Research Institute at Wits, is based on electron microscopy. Dr. L. Fleisch and Mrs. E. S. Grossman of that institute have investigated soft tissue problems associated with denture wearing, and their results have enabled a more accurate prediction to be made of the biological limits of tolerance of denture-bearing tissue. Another problem that has been addressed by the institute since 1980 is bacterial invasion of air–tooth surfaces which can take place at the restoration–tooth interface. They have shown that the fastest sealing occurs when low-copper amalgam and a calcium hydroxide base are used in the absence of varnish to produce a seal that is high in tin and chlorine. Use of such a procedure will ensure that a rapid-sealing, bacteria-resistant, restored tooth will be achieved with expected longevity.

E. Medical and Zoological Sciences

Professor B. I. Balinsky (Fig. 3) was born in Kiev and carried out pioneering work in the field of experimental embryology before he arrived in South Africa in 1949 to take up a post in the Department of Zoology at Wits. He subsequently became Professor and Head of the Department in 1954. In the mid-1950s, he visited Yale University, where he learned the techniques available at that time for transmission electron microscopy. Upon his return, he was the first to apply these methods to the study of embryology in South Africa. One of his numerous papers (B. I. Balinsky and R. J. Devis, "Origin and Differentiation of Cytoplasmic Structures of the Oocytes of *Xenopus laevis, Acta Embryol. Morph. Exp.* **6,** 55–108) was nominated as a Citation Classic by Current Contents in 1984. In 1960, Prof. Balinsky first published *An Introduction to Embryology,* which was based on the courses that he had been teaching at Wits. Four English editions of this textbook have been widely used throughout the world. The book has also been translated into Italian, Japanese, and Spanish. Not only did Prof. Balinsky pioneer the application of electron microscopy techniques to the study of embryonic structure, he also foresaw the relevance of molecular biology to the understanding of developmental processes. Through his teaching, his writing, and by his example of industry in research, Professor Balinsky has made an outstanding contribution to electron microscopy in South Africa.

A very active group within the society are the andrologists. Probably the largest unit in the country is that of Prof. van der Horst at the University of the Western Cape. He, with the assistance of 10 postgraduate students, tackles three areas of research: comparative spermatology, epididymal sperm physiology with special reference to male contraception, and clinical spermatology. Mr. J. T. Soley's results of studies on cheetah sperm, under the leadership of Prof. R. I. Coubrough at the Department of Anatomy, Faculty of Veterinary Science, Onderstepoort (UP), have been instrumental in improving breeding programme to such an extent that these animals are now bred successfully in captivity. This has enabled the cheetah to be reintroduced to the wild and to be supplied to zoos worldwide. Professor R. T. F. Bernard, Dr. A. N. Hodgson, and co-workers at Rhodes University have found that sperm ultrastructure may be used in the identification of species and the analyses of taxonomic affinities of species from the same genus or family.

Professor B. T. Sewell, appointed in 1981 to develop the electron microscopy of chromatin in the UCT Biochemistry Department, has significantly improved the technique for preparing chromatin for TEM. His work has permitted the characterization of reassembling histones and DNA into "synthetic" chromatin as well as the characterization of the reconstruction of the histone octamer from separated histones. A major outstanding problem in chromatin research is the structure of the so-called 30-nm fiber. This is a complicated, somewhat irregular structure in which there is no *a priori* reason to suppose that any two regions of the fiber are identical. Through collaboration with Dr. M. C. Lawrence, then of the MRC, it became possible to approach the problem by tomographic reconstruction of negatively stained fibers. The application of the maximum-entropy method of EM tomography has led to substantially improved images in which the nucleosomes can be well resolved.

Dr. A. Lochner of the University of Stellenbosch (US) and co-workers have made extensive use of TEM in their investigations into the critical factors affecting the recovery of the rat heart after periods of transient ischemia. Electron microscopy by Dr. D. A. Sanan, formerly of the MRC and now in the United States, and Mrs. E. L. van der Merwe (MRC), was particularly helpful in establishing that damage appeared first in the subendocardium (inner region) and later in the subepicardium. Such knowledge greatly facilitated the understanding of some of the biochemical assays, which generally do not take spatial heterogeneity into account.

Using lanthanum as an ionic probe for altered membrane permeability, Dr. I. S. Harper (MRC) and Dr. A. Lochner (US) have demonstrated reversible permeability changes during reperfusion after short periods of ischemia. Irreversible changes were associated with lethal injury. The results from this probe technique can be semiquantified and thus statistically evaluated. As a result, lanthanum probing has also been used to investigate cell membrane involvement in the development of ischemic contracture [Harper, van der Merwe (MRC), and Owen, Opie (UCT)] and lysophosphatidylcholine-induced myocardial damage [Lochner and Mouton (US), Harper (MRC)].

Electron microscopy has inevitably played an important role in virology. South Africa's first publication of negatively stained viruses was by Prof. W. B. Becker (US) in association with Nobel Prize winner Dr. A. Klug, FRS. Professor G. Lecatsas (Medunsa) has made significant contributions in the area of veterinary virology, in particular, with his classification of the viruses of African horse sickness and ephemeral fever, the Wesselsbron virus, and equine encephalitis virus. More recently, he has been involved in AIDS research, publishing locally the first South African micrographs of negatively stained AIDS virus. Mechanisms by which viruses attach to and infect cells have been elucidated with the electron microscope by Dr. L. Stannard (UCT) and Prof. B. D. Schoub, present Director of the National Institute of Virology, Johannesburg, and successor to Prof. O. W. Prozesky and Prof. J. Gear, the doyen of virology in South Africa. Professor Prozesky's early work in electron microscopy involved the characterization of the morphology of bacteriophages from sewage isolates. He and his co-workers were also the first to describe methods to monitor the elimination of hepatitis A virus from drinking water and so to ensure the safety in recycling sewage water for human use—a necessity in a country where water is a valuable and often rare commodity. Dr. L. Stannard has also worked on the hepatitis B virus with Prof. J. Alexander (then at Medunsa), who used electron microscopy to characterize her famous Australia antigen-excreting cell line PLC/PRF/S. More recently, utilizing immunocytochemical techniques, Dr. Stannard has made the unexpected finding of herpes virus antigens not only in the capsid protein coat, which was previously considered to hold the antigen determinants specific for each virus, but also in the virion.

The major contribution to microbiological research using electron microscopy has been from Electron Microscope Unit of the Red Cross Children's Hospital in Cape Town. Early work on Campylobacters associated with gastroenteritis by Dr. A. J. Lastovica and Mr. M. Emms has been extended to the reporting of some very unusual, previously unreported, forms of the bacterium by Dr. C. Sinclair-Smith, Mr. M. Emms, and Mrs. E. Le Roux, and to the first report internationally of an association between the now *Helicobacter pylori* and protein-losing enteropathy by Dr. I. Hill, Dr. C. Sinclair-Smith, Prof. M. D. Bowie, and Mr. M. Emms.

The low success rate of the human *in vitro* fertilization (IVF) and embryo transplantation program became the basis of a study by Prof. B. Kramer and Dr. B. Stein, both of Wits. They found that hormone hyperstimulation of the endometrium resulted in deleterious changes in the surface epithelial cells and the stroma, so preventing pregnancy. It appears that alterations in morphology interfere with attachment and implantation of the embryo and may explain the limited success rate in the human IVF program.

The recently retired Professor A. Andrew (Wits) and her co-workers are well known internationally for their work in developmental biology on the origin of the endocrine pancreas in the chick using TEM and immunocytochemistry at the light microscope level. In collaboration with Prof. B. B. Rawdon, then of Wits and now of UCT, and Prof. B. Kramer, the embryonic origin of gut and pancreatic endocrine cells has been studied, the results having a direct bearing on the APUD concept of A. G. E. Pearse: It was shown, using the quail/chick chimera technique, that neither pancreatic nor gut endocrine cells are of neural crest or neuroectodermal origin. Factors influencing the differentiation of the endocrine cells in the avian gastrointestinal tract have been examined immunocytochemically by light microscopy. These studies have shown that the mesenchyme plays an important role in determining the regional specificity of populations of the gut endocrine cells. In recognition of her research, Prof. Andrew was inducted in 1988 as a Fellow of the Royal Society of South Africa.

The gastro-enteropancreatic (GEP) system of the subhuman primate has been studied by an MRC group under the direction of Dr. S. A. Wolfe-Coote. In collaboration with Prof. D. F. du Toit, then of the Tygerberg Hospital (US), the effects of pancreas transplantation and immune suppression techniques on the pancreatic endocrine tissue were investigated. More recently, the group has been studying the ontogeny of the subhuman primate GEP system. Accumulating evidence from studies on the baboon and monkey suggests that both pancreatic primordial buds have the potential to produce all pancreatic endocrine cells and that an ontogenic relationship exists between A and PP cells.

F. Physical Sciences

Having just received a first-class honors degree in Physics in 1956, John W. Matthews (Fig. 2) was persuaded not only to take control of the first Wits TEM but to develop his own field of scientific research. The presence of Dr. J. T. Fourie at CSIR, with his experience of electron microscopy of thin metal foils, and that of Dr. now Prof., J. H. van der Merwe at the University of Pretoria, a pioneer in the theory of epitaxy, induced John Matthews to work on the electron microscopy of epitaxial deposition. He soon established a reputation for undertaking clean and elegant experiments. In 1962, he presented a Ph.D. thesis entitled "An Electron Microscopical Study of Defects in Evaporated Single-Crystal Films." His main interests then were in the formation of defects by the coalescence of growth nuclei and in the use of moiré fringes to study epitaxial misfits. The conditions for good epitaxial growth seemed to include rapid growth and the presence of contamination, which he reasoned into a need for many initial nuclei. He also studied the generation of defects when interdiffusion occurred between the substrate and deposit on annealing, and developed a special interest in the glide and climb of misfit dislocations. Years later, this led to a patent on the production of dislocation-free deposits by encouraging the dislocations present in the deposit to migrate into the interface.

Dr. Matthews worked closely with Prof. W. A. Jesser at the University of Virginia, each visiting the other's laboratory. Research involved the mechanical properties of thin films including spontaneous cracking when both substrate and deposit are brittle. For 15 months in 1966/1967, Prof. Jesser visited Wits, where he and Dr. Matthews used a Siemens Elmiskop 1A to investigate mechanisms of the epitaxial growth of gold, silver, and copper individually deposited in UHV onto 12 vacuum-cleaved alkali halides. They developed models for pseudomorphic epitaxial growth and misfit accommodation, by misfit dislocations and homogeneous elastic strain, from electron microscopy and diffraction of thin epitaxial bicrystals of fcc metals. The epitaxial layers were observed to copy the lattice constant and in some cases the crystal structure of the substrate. These observations were the first of their kind and provided pioneering work that is still quoted in the literature. In recognition of the caliber of his work, Wits appointed him to a Readership in Electron Microscopy in 1968. Dr. Matthews' connection with Jesser and the Wildorfs at Charlottesville led to an increasingly close collaboration with the IBM Watson Research Center at Yorktown Heights, New York. He finally joined IBM in 1969 and continued his innovative work there, laying the foundations for the production of multilayered electronic materials until his untimely death in 1977 at the age of 45. For an overview and general reference to Dr. Matthews' research, see Matthews (1975, 1979).

Epitaxial work has also been carried out at the Universities of Port Elizabeth and Pretoria. At the latter, Prof. H. L. Gaigher and Dr. N. G. van der Berg have been undertaking a systematic study of the epitaxial growth of vapor-deposited bcc metals on thin (111) fcc metal substrates. This is part of a program to determine the conditions favorable for any particular required mode of epitaxial growth, since present theories for such predictions have limited applicability, except for well-defined experimental conditions. Earlier work at Port Elizabeth studied not only the epitaxial growth of bcc metals on fcc metals (Fe/Ag and Fe/Au) but also fcc on fcc (Ag/Au). These represent cases where the atomic misfit is different in two perpendicular directions, and also where they are the same. Professor J. S. Vermaak and his students (UPE), in collaboration with Prof. D. Kuhlmann-Wilsdorf and Dr. C. A. O. Henning, published extensively during the period 1968–1974 on TEM studies of the growth and characterization of fcc metals on alkali halide crystals and amorphous carbon.

Little TEM work has been undertaken on compound semiconductors in Southern Africa. At Port Elizabeth, under the direction of Prof. Vermaak, Prof. H. C. Snyman, Dr. J. H. Neethling, and Prof. C. A. B. Ball have studied the surface modification of GaAs by proton bombardment. They reported for the first time for this material that the lattice defects consisted mainly of hydrogen platelets on (110) planes with effective Burgers vectors that vary with platelet size. Microcracks were found on the same planes, and these acted as sources of glissile dislocations that spread out on (111) planes.

In early 1950, Dr., now Prof., H. G. F. Wilsdorf (Fig. 1) joined the National Physical Laboratory, CSIR. He brought with him an authoritative knowledge of electron microscopy, having worked with the then-famous Prof. H. König of Berlin University. Dr. Wilsdorf's expertise in electron microscopy was unique in South Africa at that time, and at the CSIR it was to become the genesis of some excellent electron microscopy in future years. Dr. Wilsdorf acted as mentor to many budding electron microscopists, including Dr. J. T. Fourie (Fig. 1). Some months after his arrival at the CSIR, a Philips EM100 transmission microscope was delivered. This instrument had an almost horizontal column and the image was observed *through* the fluorescent screen. It was the second commercial electron microscope to leave the Einhoven factory. Work at 100 kV proved to be difficult, with many high-voltage cables shorting out and having to be replaced. Dr. Wilsdorf's initial work was carried out at 40–60 kV and was concerned with amorphous Al_2O_3 films. From electron diffraction patterns and a considerable amount of calculation, he determined a molecular configuration consisting of two Al_2O_3 molecules. In 1951, he reported that the basic building block was an octahedron of six closely packed oxygen ions with four Al ions in a tetrahedral arrangement. Most of his later research at the CSIR was concerned with the mechanisms of plastic deformation in metals and alloys. This he undertook with his wife, then Dr. D. Kuhlmann-Wilsdorf, a lecturer in the Department of Physics at Wits. Using a high-resolution (2.5–3.0 nm) replica technique that he had invented, he made the significant discovery that there was a fine structure of slip steps between the known glide planes. This observation subsequently influenced the theory of dislocation glide in fcc metals. With Dr. J. T. Fourie, in 1956 he reported, using the alpha brass system, that a large amount of slip could occur on a single plane rather than being spread over a number of planes. Since Prof. Wilsdorf's departure to the United States in 1956, Dr. Fourie has continued to make notable contributions to electron microscopy. His first was the excellent work on making repeated replicas of the same area to study slip lines, and this has since become a benchmark against which present-day micrographs used in reviews are judged. He then, with Prof. Wilsdorf, discovered dislocation dipoles in plastically deformed fcc metals. His other major contributions have been the first cinematographic recording of electron microscopic images of mechanical twinning in a metal, the discovery of the soft-surface effect in crystal plasticity, the proposal of an electric field effect in hydrocarbon contamination phenomena in electron microscopy, and recently he has been the first to recognize the possibility of using a crystal as an objective aperture in the optical system of an electron microscope. In recognition of his work, Dr. Fourie received the CSIR Merit Award in 1984 and was inducted as a Fellow of the Royal Society of South Africa in 1988.

Continuing Dr. Fourie's line of investigations, Dr. N. R. Comins (Fig. 1) (CSIR) elegantly investigated with the TEM the surface effects in the plastic deformation of copper and dispersion hardened copper–silica alloys. Detailed cross-sectional microscopy through the surface regions proved that the soft-surface effect was depth-influenced by the dislocation mean free paths, being limited to tens of micrometres in copper–silica as against about 1.5 mm in copper. Dr. Comins also carried out quantitative weak-beam studies of the three-dimensional configurations of multipole arrays in copper–aluminum single crystals. From these data, anisotropic elasticity dislocation equilibrium calculations were used to determine the mechanisms of the array formation. For this work, Dr. Comins was awarded the prize for the best paper published in the *South African Journal of Physics* in 1986.

Mr. J. T. Thirlwall (Fig. 1), of Dr. Fourie's Electron Microscopy Division, carried out novel work in the field of scintillators for electron detectors, where, based on a statistical model developed by Dr. Comins, a qualitative description of each quantum conversion stage in these detectors was obtained. This led directly to improvements in the performance of the Robinson-type backscattered electron detector. As a result of this work, CSIR for many years supplied scintillators to all the local electron microscopy suppliers, resulting in about 50% of the instruments in South Africa using them.

More recently, Dr. D. E. Jesson (then at the CSIR), in collaboration with the University of Bristol (1983–1989), studied and clarified new structure refinement, foil thickness determination, and defect characterization techniques (Jesson and Steeds, 1989, 1990) based on CBED, in particular, the diffraction to high-order Laué zones (HOLZs) when the projected atom columns (in 2Hb MoS_2) contain more than one atomic species; the characterization of transverse faults in layered structures using zero-layer contrast and computer simulations; and has demonstrated that a simple kinematic interpretation of HOLZ line splitting to obtain the 3D fault vector from a single pattern is inappropriate when the patterns are obtained from low-index crystal projections. For this work, the EMSSA awarded him the Best Physical Science Student prize in 1984, 1986, and the Best Innovative Technique prize in 1985, 1987.

Professor S. Kritzinger (US), using TEM during the 1970s and 1980s, contributed significantly to the understanding of the formation, characterization, and annealing behavior of secondary defects, especially in aluminum and aluminum alloys.

In the mid-1970s, Dr. M. J. Witcomb (Wits), using scanning microscopy, elucidated the development of ion-bombardment surface structures on different materials and in particular on stainless steel. Since 1980, in collaboration with Drs. U. Dahmen, M. A. O'Keefe, and K. H. Westmacott at the National Center for Electron Microscopy, Berkeley, he investigated the structural factors that govern phase changes and phase stability in interstitial alloys, especially in the Pt–C system. Vacancies were shown to be essential for the precipitate nucleation and growth sequence, due to the oversized nature of the carbon atoms. This has been characterized by conventional, high-voltage, and atomic-resolution microscopy; see Figs. 5 and 6. Detailed studies of the model fcc/bcc system, Cu–Cr, have finally explained the differences in morphology and orientation relationships reported over the years by different authors. The disagreements were recently found to be

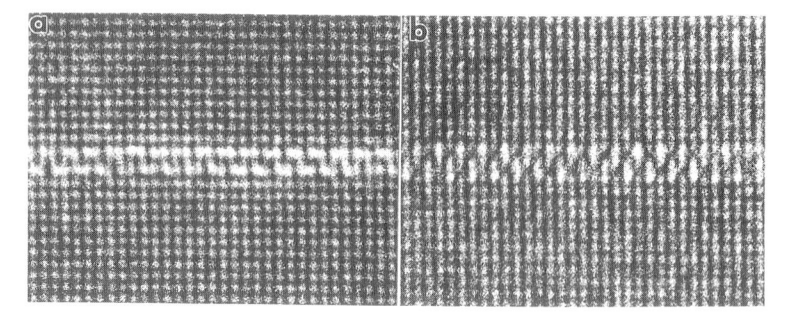

FIGURE 5. High-resolution images of {001} Pt$_2$C plates, imaged edge on in a ⟨100⟩ zone axis orientation. The images, photographed at 800 kV in a JEOL ARM 1000 (resolution 0.16 nm), were taken in (a) at −40 nm defocus, black atoms, and in (b) at −80 nm defocus, white atoms. The closest atom column separation is 0.196 nm. (Reprinted with permission from Witcomb, M. J., Dahmen, U., O'Keefe, M. A., and Westmacott, K. H. (1991). HREM imaging of single unit cell carbide precipitates in Pt-C alloys, *Proc. EMSA*, 572–573.)

a consequence of variations in the quenching conditions used and alloy phosphorus content.

In the 1970s, Prof. A. Ball, then of Wits, with Dr. G. S. Woods of De Beers Diamond Research Laboratories, Johannesburg, investigated the precipitation of nitrogen in chromium. With the aid of transmission microscopy, they were able to study, for the first time, the early stages of interstitial precipitation in terms of layer-by-layer thickening as the precipitate

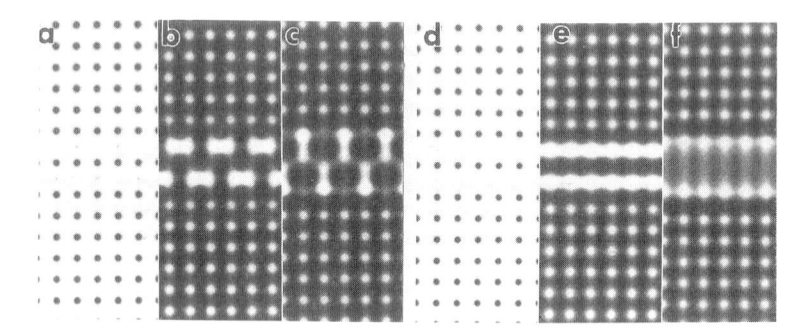

FIGURE 6. Image simulations for the conditions in Fig. 5 for a 4-nm-thick foil, 0.03 nm vibration. (a) is the projected potential, while (b) and (c) are the simulations at −40 nm and −80 nm defocus, respectively. (d)–(f) correspond to the same configuration but without any carbon atoms. Only the carbon-containing model matches the experimental images. Simulations show that a resolution of around 0.1 nm would be required to image the carbon atom columns directly. (Reprinted with permission from Witcomb, M. J., Dahmen, U., O'Keefe, M. A., and Westmacott, K. H. (1991). HREM imaging of single unit cell carbide precipitates in Pt-C alloys, *Proc. EMSA*, 572–573.)

changed its degree of coherency. After moving to UCT, Prof. Ball worked with Dr. S. White, then of Imperial College, on the mechanism of creep in quartzite, an area of significance in geological deformation. TEM analysis revealed that creep takes place by climb of dislocation loops in the basal plane, an observation in agreement with one of the predicted Nabarro mechanisms. More recently, in collaboration with Prof. C. Allen and Dr. J. Heathcock, both of UCT, SEM and TEM studies have been made of the surface deformation of materials that have been subjected to abrasive and erosive wear. The work has indicated the significant role played by phase transformations in metastable materials such as 304 austenitic steel and cobalt-rich stellite.

The microstructure and deformation of hard metals, important materials in, for example, the mining industry for rock drilling tools and for use for high-pressure components for diamond synthesis, have been studied extensively by Prof. S. Luyckx (Wits). She has determined the slip system, cleavage planes, and work-hardening mechanisms of WC crystals and, with Prof. F. R. N. Nabarro, has proposed a theory for the strength of WC-Co compacts on the basis of observations on TEM replicas of fracture faces. With scanning microscopy, she has established the role of porosity and inclusions in the fracture process of WC-Co and, more recently, has determined the effects of precompression and ion implantation on the mechanical properties of the same material.

Of note in the area of ceramics, Dr. H. K. Schmid (CSIR), using analytical TEM, has made important contributions to the understanding of segregation behavior and the redistribution of dopants and impurities in grain boundary regions during the processing for high-temperature applications of a variety of materials, e.g., silicon nitride, zirconia, and boron carbide for high-temperature applications.

G. *Electron Microprobes and the Earth Sciences*

With the development of the wavelength-dispersive (WDS) electron microprobe in the late 1950s, the mining and commercial research laboratories in South Africa quickly realized the potential of the technique. The first microprobe was purchased as a joint venture between the Atomic Energy Commission (AEC) and the then National Institute for Metallurgy (NIM), now the Council for Mineral Technology (Mintek). The instrument, an ARL EMX, was installed in South Africa in 1965 at AEC, Pelindaba. However, it soon proved not to be multi-user friendly and was relocated under a single operator at NIM in Johannesburg. Being the only microprobe available, a service was rendered to the whole country, entailing the investigation of an interesting range of projects including the identification of inclusions in lung tissue, analysis and study of kidney stones, the description of new minerals, investigations on meteorites, and studies of bomb shrapnel. The installation of a second microprobe, a JEOL JXA-3SM, soon followed at the Johannesburg Consolidated Investments (JCI) Laboratories. The first instrument delivered to a university was a Cambridge Microscan 5 in 1971 at the Department of Geochemistry, UCT. At around the same time, MRC installed an AEI EMMA-4, a combined TEM/WDS microprobe, at the National Research Institute for Occupational Diseases, now the National Centre for Occupational Health, for the study of fibers in the atmosphere, in particular, asbestos fibers in lung tissue—Pneumoconiosis.

By the mid-1980s, quantitative microprobe analysis had become a routine technique and as a result there are now over 20 instruments in laboratories in South Africa, with some South African organizations having installed their third instrument. The most recent acquisition is at the Anglo-American Research Laboratories in Johannesburg, where a Cameca SX-50 has been innovatively fitted with five WDS detectors and one EDS detector. There is only one microprobe in a neighboring country, a JEOL JXA-50A at the Institute of Mining Research in Harare, Zimbabwe, and this has been used primarily for contract work for mining companies. As to the distribution of the instruments, nine of the microprobes at universities are in geology/geochemistry departments and one is in a physics department; at parastatal organizations, eight are in geological and materials research departments, while the remaining five instruments are in the minerals research laboratories of mining houses. Thus it is evident that the most important application of the microprobes is in the mining industry, covering the range from basic research at universities to that of final added-value products in the extractive metallurgical field. In addition, they are used in actual mineral exploration. For example, quantitative analyses are performed on thousands of individual mineral particles on a routine basis, in order to locate markers for promising deposits of diamonds.

Worthy of mention is the Department of Geochemistry, UCT, which has gained a worldwide reputation for excellence over the years, their staff having carried out work in such areas as lunar rocks and kimberlites, while also discovering a number of new minerals using the microprobe, e.g., Lindsleyite and Mathiasite. It was at their suggestion, and with some input from them, that Cameca developed their geological computer package, which is now commercially available.

III. The Future of Electron Microscopy in Southern Africa

The number of electron microscopes in Southern Africa, that is, South Africa and its bordering countries, currently (late 1991) stands at 225; see Table II; the status of the instruments in Angola and Mozambique is currently unknown. From this resource, a record number of some 150 papers will be presented at the 1991 EMSSA conference, which, as a result, has been extended to three days. Interestingly, this growth over the years has come mainly from the life sciences, which have shown an average annual growth rate just over twice that of the physical sciences. Of worry in the latter category is that the contributions in the TEM field now basically originate from a decreasing number of authors. Part of the problem has been highlighted by a recent EMSSA survey of microscopes. This has revealed that the average age of TEMs is 14 years, almost double that for SEMs. Of more concern is the fact that there are only three fully analytical TEMs (STEM, standard windowed EDS, EELS) in South Africa, their ages being 14, 12, and 6 years. Not one of these instruments is at a university, the youngest being a Philips 420 sited at the CSIR. In addition to these, the physical sciences has only one instrument with STEM and EDS facilities that is less than 13 years old. Not surprisingly, the net consequence of this situation is that new physical science students and researchers are not being attracted to the TEM field. Additional negative influences, of course, also contribute to the situation. For example, the geographic isolation from mainstream microscopy centres and researchers, and greatly reduced library budgets, has severely restricted information transfer and to some extent suppressed both enthusiasm and progress, while the fact that electron microscopy is taught as a separate topic in only a couple of university physical science departments has meant limited student exposure to the subject. Further, prevailing economic conditions for the past few years have resulted

TABLE II

Distributions of Electron Microscopes in Southern Africa

Country	Microprobe	SEM	TEM
Angola	—	—	1
Lesotho	—	—	—
Mozambique	—	—	1
Namibia	—	—	—
South Africa	24	114	78
Swaziland	—	—	—
Zimbabwe	1	3	3

in educational institution subsidies being cut dramatically, by more than 30%, and the parastatal research laboratories being commercialized. The latter have almost stopped TEM-based work at some locations. These events, coupled with the real possibility that science and technology will become a low priority, have placed electron microscopy, and the physical sciences perhaps, in particular, in straitened circumstances. It can only be hoped that this trend will be reversed in the not-too-distant future as the political order is changed in South Africa, sanctions are lifted, foreign investment is restored, and the economy moves into meaningful growth. Since South Africa is considered the powerhouse of the region, this should have a ripple effect in stimulating the neighboring regions.

For all countries in Southern Africa, the biggest problem in the years to come will be how to fund new instruments, whether first-time purchases or the replacement of obsolete microscopes. One result of present economic policies must be a move toward centralization of facilities in order to optimize their use and contain costs. Only limited movement in this direction has been seen to date. In comparison to the physical sciences, the life scientists are not as severely affected, since they are somewhat less dependent, particularly in Southern Africa, on state-of-the-art microscopes. As a result, they have continued to prosper even if some of their microscopes are now quite elderly. In the area of microprobe analysis, the techniques are well established and the equipment is generally adequate. In 1991, South Africa saw the completion of proton microprobes having 1 μm diameter probe sizes at Wits and the National Accelerator Centre at Faure in the Cape Province. In both cases, the beams are obtained from Van der Graaff generators. Hopefully, at some point in the future, a high-resolution ion probe will be purchased so that microprobe analysis can be complemented with these techniques. Although an ARL Ion Microprobe Mass Analyzer (IMMA) was installed at the CSIR in the mid-1970s, the resolution was so poor that it has been used solely for depth profiling in materials.

Acknowledgments

The authors extend their grateful thanks to all those who have generously provided them with information and memories of the EMSSA, microscopes, people, and research covering the past 40 years. In particular, we wish to thank Prof. P. Berjak, Dr. J. T. Fourie, Mr. D. A. Gerneke, Dr. J. R. Lawton, Prof. F. R. N. Nabarro for allowing us to quote from his first John Matthews Memorial Lecture, Mr. R. S. Rickard, Dr. E. M. Veenstra, Mr. E. A. Viljoen, and Prof. H. G. F. Wilsdorf. Due to space limitations, we have had to be selective with respect to the information included in this article. We thus apologize to those whose work does not appear here.

REFERENCES

Balinsky, B. I., (1981). "An Introduction to Embryology," 5th ed., Saunders, Philadelphia, p. 768.

Matthews, J. W. (1975). Coherent interfaces and misfit dislocations. *In* "Epitaxial Growth" (J. W. Matthews, Ed.), Academic Press, New York, Part B, pp. 559–609.

Matthews, J. W. (1979). Misfit dislocations. *In* "Dislocations in Solids" (F. R. N. Nabarro, Ed.). North-Holland, Amsterdam, Vol. 2, pp. 461–545.

2.13
Highlights in the Development of Electron Microscopy in the United States: A Bibliography and Commentary of Published Accounts and EMSA Records

ROBERT M. FISHER

Department of Materials Science and Engineering
University of Washington
Seattle, Washington 98195, USA

I. INTRODUCTION

The history of the invention and development of the electron microscope and the growth of the broader field of electron microscopy now spans a period of more than 60 years. Consequently, the scope of the history of electron microscopy (EM) on the international level is enormous and even for the United States alone, and a late starter in the field, it is impossible to present more than the highlights in a brief account. It is not known just how and when news of the achievements of Knoll and Ruska reached North America and the United States. However, news of major scientific advances did get around the world quite quickly even before today's nearly nonstop schedule of conferences and, of course, the Internet. In Berlin itself, and other cities in Germany, news of the activities at the Technical University reached colleagues and competitors very rapidly and sparked intense rivalries and resentments that outlived the protagonists.

The pioneering work of Knoll and Ruska and the growing desire to see smaller things first attracted interest and then stimulated activity in other laboratories and in other countries including the United States. Although it is a fact of history that the development of electron microscopy in the United States was strongly influenced by the immigration of ideas and individuals, first from Europe and then from Canada, there are American threads woven into the historical fabric that can be traced back to the time of the seminal concepts and discoveries that culminated in the construction of the first electron microscope by Knoll and Ruska in 1931. The first event in the United States was experimental confirmation by Davisson and Germer (1927) at the Bell Telephone Laboratories (then in New York City) of the wave character of an electron beam (by demonstrating the existence of electron diffraction simultaneously with G. P. Thompson (in the UK in 1927). Several years later in the same Bell Telephone laboratory, Calbick and Davisson (1931) first estimated the properties of an electron lens. This work was contemporaneous with the early work on electron optics in Europe.

The primary history of electron microscopy in the United States resides, of course, in published records of the original scientific work in journals and in conference abstracts and proceedings. Bibliographies of publications are very valuable sources to the original work, since they list many papers together and generally include the titles of the articles. The history of EM as found summarized in the reference and textbooks and in review articles is more selective in emphasis, to fit the objectives of the author. However, these sources usually do give proper credit to the major accomplishments. Finally, personal accounts, although they may not be totally accurate or objective, do provide much insight into the chance occurrences, personalities, motivations, and rivalries that are always present in human endeavors. Such details are rarely included in scientific papers. It was not practical, for reasons of limited accessibility and time, to reexamine more than a few of the earliest articles that were published in the United States.

Understandably, reference books and texts properly consider the development of specific topics in electron microscopy as a whole and do not take a national perspective. However, the purpose of this "IFSEM history" is to focus strictly on the contributions of the microscopists in a particular country. In the case of the United States, with a large population and a sizable industrial economy, the history of electron microscopy is extensive and complex. Just 11 years after the construction of the Knoll-Ruska microscope, 75 people turned up at the first EM meeting in the United States to talk (informally) about their EM work using upwards of 25 RCA EMB production microscopes and several home-built instruments. The number of attendees and active laboratories essentially doubled for the second meeting in January 1944. Nearly 50% of the attendees at these first two meetings did not remain involved in the field or become microscopists; however, most of the others remained active for years, producing volumes of material for historians to consider. Since the earliest days of electron microscopy in the United States, upwards of 8000 individuals (including a substantial number of highly talented people who came from overseas and Canada) have been active to some extent, so the complete history, comprising at least 15,000 oral or written papers, is rich indeed. The published scientific record provides the "hard" evidence, but it is the "soft" personal accounts that tell later generations just how it all happened. Fortunately, some years ago the *EMSA Bulletin* initiated a "Reflections" section by John Reisner to focus on EM history, and he encouraged or, if necessary, cajoled some of the other pioneers to record their memories. These articles provide a rich resource for historical research.

Since much of the U.S. history has already been published, the primary purpose of this chapter is to draw attention to these original sources and indicate their contents (suggestions on how to obtain them are given along with the bibliography). In addition, some key "milestone events" in the United States are also listed. No attempt is made to specifically connect historical events in the United States with contemporary activities overseas or even across the border with Canada, although the interplay was substantial and enormously beneficial. For this account the history of EM in the United States is discussed in approximately 15-year interval: "The Opening Era: 1927–1942" (through the first Electron Microscopy Society of America meeting); "Period of Progress and Promise: 1943–1957" (period of rapid instrument and technique development), "World Watch and Welcome: 1958–1970" (major impact of ICEMs and overseas scientists); "Center Emphasis: 1970–1984" (formation of HVEM, HREM, and IVEM centers); and "New Microscopies and Media: 1985–1995" (new scanning probe and optical techniques evolve). The availability of comprehensive reviews, covering developments in electron microscopy in two- or four-year intervals, that have been published in *Analytical Chemistry* since 1950 proved to be invaluable.

II. Sources of Historical Information About EM Activities in the United States

Sources of information about the history of electron microscopy in the United States are quite abundant, and many are readily available through the libraries at major universities or through service agencies or technical societies for a small fee. The most important sources are described below in order of the date of publication (the dates refer to the year that the particular publication started or stopped carrying information of interest to electron microscopists).

Physical Review (1927–1941). Abstracts of early meetings of the American Physical Society and some relatively brief technical papers. The titles of the earliest articles are of historical interest because they provide the first reference to "electron diffraction, "electron lens," and electron microscope" in any U.S. publication. However, the abstracts do not include illustrations or references.

Journal of Applied Physics (1937–present). (1) Most of the early papers on instrumentation, techniques, and applications were published in this journal (8–15 per year from 1941 to 1950), and electron micrographs graced its covers on numerous occasions. (2) Abstracts and programs of annual EMSA meetings (1944–1966) were published promptly, (EMSA was, and still is, an Affiliate Society of the American Institute of Physics. In most cases, the abstracts contain useful details about the work as well as references to previous studies. (3) Papers presented at the inaugural meeting (1944) of the Division of Electron and Ion Physics of the American Institute of Physics by C. J. Calbick, R. C. Williams, F. Keller and A. H. Geisler, and C. S. Barrett. (4) Bibliographies of Electron Microscopy (1943–1950). Claire Marton, with Marton's Electron Optics group at Stanford, and Samuel Sass, University of Michigan librarian, prepared the 1943 compendium. Other issues appeared in 1944, 1945, and 1950. The 1950 version included more than 2000 references and was co-authored by Marton, Sass, Swerdlow, Van Bronkhorst, and Meryman. This was also issued as Circular 502 by the National Bureau of Standards. Later the EMSA, and then the New York Society of Electron Microscopists, tried valiantly to publish key-sortable bibliographies, but the large number of new papers coming out each month overwhelmed the mechanical technology of the day.

Electron Optics and the Electron Microscope, by Zworykin, Morton, Ramberg, Hillier, and Vance (1945). This was the "Whole Earth Catalog" of EM at the time, with hundreds of references and specific discussion of all important subjects in electron optics and microscopy in some 760 pages of text. Five years later it would have been impossible to cover every EM topic, from vacuum systems to lens design and aberrations to image contrast, in the same detail.

Microscopy, Techniques and Applications, by Ralph Wyckoff (1949). Contains a comprehensive and well-organized bibliography on all that was known or had been tried on specimen preparation and applications of EM to both biological and nonbiological materials.

Analytical Chemistry (1949–present). Literature reviews on developments on electron microscopy were initiated by C. J. Burton (1949) and continued by F. A. Hamm (1949–1951); they became biennial with Max Swerdlow in 1954. Each article lists and comments on from 100 to 600+ papers on EM that appeared during the previous two years. Books and major conferences are also described.

ASTM Special Publication (1950–1985). Proceedings of a series of symposia on "Techniques for Electron Microscopy" that provide excellent coverage of early work on replication and etching for studies of metals, especially STP No. 155 in 1953.

Introduction to Electron Microscopy, by Cecil Hall of MIT (1st ed. 1953). The first comprehensive textbook produced for courses on the theory and uses of the electron microscope. It contains virtually all of the important references prior to its publication.

Proceedings of the NBS Semicentennial Symposium on Electron Physics, Washington DC, 5–7 November 1951, edited by L. Marton (1954). This was the first international conference on electron optics in the United States, and it brought the work of Castaing, Cosslett, and others to the attention of American electron microscopists.

Proceedings of the Fifth International Congress for Electron Microscopy (1962). Records a fairly accurate snapshot of the state of the art in the United States as it existed at the time. Of the nearly 270 papers on the physical aspects of EM, less than 100 came from the United States, whereas nearly 150 were from Europe and just over 20 came from Japan. However, the situation for the biological sciences was essentially reversed, with more than 180 papers from the United States, 100 from Europe, and 30 from Japan. Less than 3% of all the papers originated in other parts of the world. (This situation had changed remarkably by the time of the 12th ICEM in Seattle in 1990.) The primary emphasis on biological studies with the EM persisted into the late 1980s. By then, materials science had become the dominant application of electron microscopy.

Fundamentals of Transmission Electron Microscopy, by Robert Heidenreich (1964). An excellent source of significant references relating to original work on electron optics, scattering, and diffraction contrast from 1939 to 1963.

EMSA Proceedings (1967–present). Published by Claitor, Baton Rouge, and San Francisco Press, San Francisco. The EMSA ceased publication of short abstracts of papers presented at the annual meetings in the *Journal of Applied Physics* in 1966. In 1967 the society began publication of illustrated, two-page abstracts as their Proceedings. The 1971 Proceedings contains an illustrated description of a magnetic transmission electron microscope constructed at Washington State University in 1935. The 1992 Presidential Symposium, organized by Patricia Calarco at the EMSA 50th Anniversary meeting, featured a historical perspective to EM in the United States. Contributed papers, "Reflections of the Masters," with a historical theme, are also included in the two volumes.

Early History of the Electron Microscope (San Francisco Press, San Francisco, 1968). A very personal account by L. L. Marton of his work in electron microscopy at RCA, Stanford University, and the National Bureau of Standards (now the National Institute of Standards and Technology, NIST). "New edition, 1994."

EMSA Bulletin (1971–1994). Provides a record of the activities of the society and the highlights of the annual meetings during this period. In 1981, the *Bulletin* established a section on "Reflections" that was organized by John Reisner, where EM pioneers recorded their reminiscences. Reisner was a member of (and later directed) the RCA electron microscope development department at RCA and contributed many "reflections" himself. Although not fully relevant to this article, personal accounts from microscopists from Europe and Canada, particularly James Hillier, John Watson, Ernst Ruska, Jan Bart Le Poole, Bernhard Reimann, and Martin Freundlich, as well as the descendants of Reinhold Rudenberg, also appear.

Advances in Electronics and Electron Physics, Supplement 16 and Volume 73 (1985 and 1989). These volumes, edited by Peter Hawkes, contain historical articles by Cecil Hall and John Reisner about early EM developments in the United States, along with personal accounts by pioneers from other countries.

EMSA and Its People—The First 50 Years, by Sterling Newberry (1992). Published by the EMSA to commemorate its 50th anniversary. The "family album," as Newberry calls it, contains numerous behind-the-scenes stories and photographs that he gleaned from many personal interviews, correspondence, and unpublished reports. Newberry himself has been active in electron microscopy since the late 1930s (beginning at Washington University in St. Louis) and attended most of the "memorable" meetings, so he knew many of the pioneers personally. As a supplement to the book itself, Newberry has prepared printed excerpts from "oral history" video interviews with 23 EMSA Charter Members, which he had recorded to show at the 50th Anniversary meeting.

III. THE OPENING ERA: 1927–1942

Any account of the vanguard events in electron microscopy in the United States is, somewhat surprisingly, largely a tale of three cities. At the beginning of this period the center of the action was New York City, the home of Bell Telephone Laboratories in the 1920s, and at the end of the era, it was Camden, New Jersey, the location of RCA research at the time. However, for several years in the late 1930s, St. Louis, Missouri, the site of Washington University, was really the most important center of EM work in the United States, and in fact, the first publications describing instrument design and construction and actual applications of EM in biology originated there.

The era opened in 1927 at Bell Telephone Labs, when Davisson and Germer (assisted by Chester Calbick) carried out a Nobel Prize-winning experiment in which they diffracted electrons from a single crystal of nickel and observed specific directions of reflection which almost, but not quite, matched the expectations of the de Broglie theory of the wave nature of electron beams. Davisson and Germer used low-voltage electrons (100–1000 eV), where the refraction effect of the inner crystal potential is quite significant, giving rise to a noticeable deviation from the predicted directions of the scattered electrons. The cause of the deviation was soon recognized, and the correct wavelength of the electron beam in the experiment determined. The following year (1928), Lester Germer (1928) described their diffraction observations as an "optical experiment in the sense of an interference phenomenon," coupling electron beams with optical wave effects for the first time. Work in electron optics at Bell Labs intensified toward improving sealed cathode-ray tubes. Early in 1931, Davisson and Calbick described the optical characteristics of an electrostatic pinhole lens. In 1934 they used the lens to produce enlarged shadow images of fine-mesh screen. This series of publications by Davisson and Calbick appear only as abstracts in the *Physical Review*. Later research in electron optics at Bell was directed more toward communications applications, and it was not until much later that Calbick (1944) published drawings of the 1934 microscope and examples of the low-magnification micrographs that he obtained with it.

By the middle of the 1930s, the idea of using electrons instead of light for microscopy was drawing considerable attention in the United States. Thus it was not entirely a coincidence that in 1935 efforts to construct electron microscopes were initiated independently in Pullman, Washington, at Washington State University (WSU), in St. Louis, Missouri, at Washington University (WU), and in Toronto, Ontario, at the University of Toronto (the events in this city, which determined the course of EM history in the United States, are outside the boundaries of this article). At WSU, Paul Anderson and Kenneth Fitzsimmons undertook to construct a transmission electron microscope patterned after the Knoll and Ruska instrument. Although it was put together quickly and was operable in 1936, the performance of the Pullman microscope was severely limited by technical problems with electrical and mechanical stability, which proved to be difficult to correct with the meager financial resources available. Unfortunately, the project was terminated in 1938 when Anderson and Fitzsimmons felt that others were outstripping their progress and that it was unlikely that biological materials could be studied effectively by EM anyway. The project was abandoned before significant successes were achieved or anything was published. The original microscope was discovered in storage at WSU in about 1967, restored to close to its original condition, and put on display. The column incorporated features that were very advanced for its day, including three magnetic lenses, externally adjustable beam shift controls, and an internal photographic chamber holding six plates.

In contrast, an electron emission microscope constructed about the same time by Scott and McMillan at Washington University in St. Louis did work quite well from the beginning. They published a description of the WU microscope in 1937, and Scott with D. Packer was able to show that emission electron microscopy could provide useful information about mineral localization in biological tissue in a series of papers that appeared in 1938 and 1939. Unfortunately, this project was also terminated, not for lack of success this time, but for lack of money (aspirations to convert their emission EM to a TEM would not be realized). The choice was made by the WU administration to put all available funds into cyclotron biomedical research rather than continue electron microscopy development, and Gordon Scott moved to Philadelphia.

An important episode of a different sort also began in the United States in the mid-1930s. In 1936 and 1937, U.S. patents were granted to Siemens Schukertwerke and Reinhold Rudenberg for an electron microscope concept and an electron lens, although neither the words "microscope" nor "lens" appeared in the titles. The German patent office had previously rejected both applications. The U.S. government seized the Siemens-Rudenberg patents as alien property in 1941. Surprisingly, they were not cited in the technical literature until six years after they were issued (Bachman and Ramo, 1943). Rudenberg successfully filed suit against the U.S. Attorney General in 1947 and was awarded sole ownership of the patents, and most major EM companies paid some royalties. Detailed articles about the case by people who were closely involved have appeared in the *MSA Bulletin:* Rudenberg and Rudenberg (1994) and Freundlich (1994). Insight about the case from another perspective can be found in a telling exchange of letters between M. Knoll and M. Steenbeck reproduced in "The Beginnings of Electron Microscopy" (*AEEP,* Suppl. 16).

Interest in the promise of EM for microstructural studies continued to grow in the United States, and before the end of the decade, Prof. Burton's students were being drawn away from Toronto to take up attractive positions in American companies and universities and construct TEMs. The émigrés included C. E. Hall, who went to Eastman Kodak, James Hillier to RCA, A. F. Prebus to Ohio State University, and W. Ladd to Columbian Carbon (the microscope was actually started in Toronto and completed in New York). Hall, Prebus, and Ladd constructed TEMs at great speed, and these proved to be productive for several crucial years before commercial instruments, based on Hillier's more thoroughly engineered design, became available.

Thus it was that the team that Zworykin assembled at RCA that had the most striking success. Zworykin had moved to RCA from Westinghouse in 1929, and reportedly exhibited interest in the electron microscope as early as 1933. However, he did not receive approval to initiate a program to develop a commercial electron microscope until 1938 (*The New York Times* carried an article about Siemens achieving a maximum magnification of 100,000× in May 1938). Zworykin immediately invited Ladislaus (Bill) Marton to move to Camden and undertake the project. By 1940, Marton's RCA EM "A" model was completed and described in various publications. However, Zworykin could see that this was not user-friendly nor sufficiently attractive in appearance to be a commercial success. He brought in James Hillier from Toronto at the beginning of 1940 to start over and design the EM "B" electron microscope. In December of that same year, the engineering prototype of the "B" model was delivered to American Cyanamid. This was the first sale of a commercial microscope in the United States. Seven regular commercial units were soon in production for delivery in late spring of 1941. At this point, Marton chose to leave RCA and go to to Stanford University to form his own EM group. Several production runs of EMBs would be delivered before the end of 1941. A total of 58 EMBs were produced and delivered in less than two-and-a-half years (10 of these were sent overseas).

With the EM "B" design work completed and the formation of a talented engineering group at RCA, Hillier turned his attention to consideration of image effects, particularly Fresnel interference fringes and basic optical and specimen limits to the attainable resolution. He also constructed a 300-kV microscope with Zworykin for exploratory studies of high-voltage effects.

Early on, Zworykin's entrepreneurial instincts told him that it would be necessary to demonstrate the advantages of the electron microscope over the light microscope to ensure its commercial success. He and Ramberg published a paper on the use of plastic replicas to demonstrate the utility of the EM for examining bulk materials. He encouraged other RCA people to work on techniques and applications, and invited the top managers of major companies and senior researchers to visit Camden to see the microscope in operation. To stimulate interest in EM by biomedical scientists, he arranged to set up a special National Research Council Fellowship for studies at RCA laboratories in Camden, aimed at developing biological applications of TEM. The Fellowship was awarded to T. F. Anderson, and in just two years, more than 50 landmark publications were completed, most of them by Anderson and colleagues. This turned out to be a good investment by Zworykin, as biological research (but not medical diagnosis) became a major market area for RCA microscopes for many years.

General Electric was also interested in producing an electron microscope for sale, and *The New York Times* article generated questioning internal memos (Newberry, 1992) but management hesitated and the project developed very slowly. Development work finally began (1939–1940) on a microscope with electrostatic lenses. C. H. Bachman and S. Ramo (1943) described the properties of electrostatic lenses in full detail in a series of papers in the *Journal of Applied Physics*. GE management recognized that RCA had taken the lead with "conventional" TEMs, but were persuaded by Sterling Newberry that a small desktop microscope would find a place in the market. A few were built and sold, but RCA continued to dominate the market and GE terminated the program along with several others aimed at producing laboratory instrumentation in 1949. (GE did somewhat better with a dedicated electron diffraction unit and an X-ray point projection microscope, but these products could not stand up against RCA's marketing muscle and were not commercial successes.)

EM development efforts at Farrand Optical Company followed about the same fate that they did at GE. Farrand also decided on electrostatic lenses, and several units were constructed following a well-considered design by Gertrude Rempfer (1947). However, the microscope was never put into commercial production. Curiously, the cover of the January 1949 issue of the *Journal of Applied Physics* carried a TEM image of a shadow cast Bacillus taken on a Farrand electrostatic microscope with no accompanying article or explanation. It is worth mentioning that the great Austrian electron optician, Walter Glaser, spent several months with the Farrand Co. in 1955; it was during this stay that he developed all the basic theory for quadrupole-octopole correction of spherical aberration.

As 1942 came to a close, any doubts about the utility or acceptance of the electron microscope had largely been put to rest. At least 30 RCA microscopes were in operation (or awaiting installation) in various industrial and university laboratories. The home-builds had served their purpose and were being shut down, and a backlog of orders was growing rapidly (in fact, defense-related priorities were required for a time). The specimen preparation methods that were available were adequate to examine powder and pigment dispersions on strong, electron-transparent substrates, as well as certain biological specimens such as butterfly wings. The concept of taking thin replicas of bulk or wet materials had evolved and stimulated a search for stronger, more stable materials that would be capable of replicating finer details. Quantitative analysis of the effects of lens geometry and excitation on aberrations was under study, and a device to permit use of a conventional TEM as a diffraction camera had been described. Along with the rapid expansion of EM activities, instrument developers, users, and promoters sensed a growing need for closer and continuing communications and were ready to welcome a scientific/technical society that could serve this need.

IV. Formation and Growth of the Electron Microscope Society of America (EMSA)

By the summer of 1942 there was intense interest throughout the scientific community in learning about the latest developments in all the phases of electron microscope, and RCA salesmen and the public relations department did their part to spread the news. In late September, efforts to organize a meeting on electron microscopy in conjunction with a national chemical exhibition in Chicago at the end of November were put in motion. With the help of the RCA sales department, Prof. George Clark (University of Illinois) sent out invitations to everyone thought to be active in electron microscopy. They were encouraged to bring micrographs illustrating their work and to be prepared to participate actively in an open meeting following an agenda of topics ranging from instrumentation to applications. The final item on the agenda was a proposal to form a permanent organization to arrange an annual conference. This idea was endorsed and a committee selected to establish a name for the society and to draw up a charter and a membership application form. Invitations to join the society as charter members were sent to everyone believed to be interested in electron microscopy. Accounts of these organizational activities have been published by Reisner (1990), and a more detailed record, including names of all of the individuals involved, is reprinted in the EMSA 50th Anniversary book by Sterling Newberry (1992).

Because of wartime research priorities and travel restrictions, the next national conference on electron microscopy did not take place until early January 1944 at Columbia University in New York City, some 14 months after the Chicago meeting. In addition to 30 technical papers on instrumentation and EM applications, the proposed charter for the new society was duly ratified, dues were collected and officers elected, and the Electron Microscope Society of America (EMSA) was formally brought into existence. (The society later changed its name to the Electron Microscopy Society of America so that it would appear less like a trade association.) The membership also voted to become affiliate societies of the American Institute of Physics and the American Association for the Advancement of Science. These connections have been maintained ever since.

Although the first "official" (scheduled, with a formal call for papers) EMSA meeting took place at Columbia University in January 1944, the date of founding of the society is taken as November 1942 and the 50th anniversary was celebrated in 1992. (In the same vein, Chairman George L. Clark is now listed as the first President, with footnotes to explain the circumstances.) A list of EMSA meeting sites and presidents is given at the end of this article. A second meeting was held in 1944 in mid-November, and for years the annual meetings generally took place in November or December. However, the summer school vacation time was favored by many members, and eventually the EMSA Executive Council agreed upon mid-summer as the scheduled date of the annual meeting. Membership in the EMSA has grown by 50–200 per year, reaching 2100 in 1971 and nearly 5000 by 1992. Similarly, the annual meeting program grew from 30 papers in 1944 to 290 in 1971 to over 900 at the 50th anniversary meeting in Boston in 1992. Another very important growth area has been the size and scope of a commercial exhibit of instruments, accessories, and books. Over 100 vendors have taken booths at recent meetings, and their participation has proven to be a strong attraction for attendees. In order to manage the growth in membership, meetings, and publications, the society opened a permanent business office in Woods Hole, Massachusetts, in about 1986.

EMSA international contacts began at the 1947 annual meeting with a report by Wyckoff of his experiences with European microscopes, descriptions of prototype microscopes at Delft and C.S.F. Paris, as well as two letter reports from V. E. Cosslett and C. F. Robinow about the activities of the British EM Group. In 1953, Wyckoff (then Scientific Attaché at the American Embassy in London) and T. F. Anderson attended the German EM annual meeting as EMSA representatives. They also attended the London ICEM in 1954. In 1955 (following initiatives first proposed at the 1954 London meeting), the international community of electron microscopists organized an "International Federation of Electron Microscope Societies" (IFSEM) to coordinate worldwide contacts through a regular series of international meetings. The EMSA was invited to join and send an official delegation to the 4th ICEM in Berlin in 1958 (as with the EMSA, ICEMs were back-counted to the "1st"—Delft, 1949). T. F. Anderson was elected IFSEM President in Berlin, and Philadelphia, Pennsylvania selected as the site of the 5th ICEM.

For many of the 50 or so U.S. microscopists who attended, the Berlin meeting was their first opportunity to see and meet the internationally renowned European microscopists as well as to initiate contacts with trans-Atlantic colleagues that have been maintained for years. European microscopists followed suit and came in large numbers to Philadelphia. Many of them managed to visit EM labs throughout the country as a part of their trip. The EMSA has sent delegates to each IFSEM General Assembly, and members of the society have played an active role in the IFSEM (Gareth Thomas served for almost 20 years as an Executive Committee member, General Secretary, and President). Over the years the EMSA expanded its international activities by joining CAPSEM (Pacific regional organization) and, more recently, CIASEM (Inter-American regional organization). A series of China Exchange Programs was conducted in the early 1980s that was very productive and of value to both sides. The society has since established in International Committee to coordinate its foreign affairs and set up an IFSEM-EMSA-MAS Travel Fund to facilitate attendance of students and microscopists from developing countries at the ICEMs. In recent years the focus has been on Mexico and South America.

The EMSA has always maintained close contact with the Microscopy Society of Canada (MSC), and has collaborated with the MSC on the organization of the 9th ICEM in Toronto in 1978 (many prominent EMSA members and officers came from Canada). A number of EMSA annual meetings have been joint meetings with the MSC. Cognizant of the rapid introduction of new "microscopies" such as confocal microscopy, near-field and STM/AFM techniques, and video microscopy methods, EMSA incorporated major symposia on these topics into the annual meeting program. To emphasize its commitment to the new microscopies, the society changed its name to the Microscopy Society of America (MSA) beginning in 1993. The EMSA was not so responsive to new instrumentation years back, so the microbeam analysts and the field ion microscopists elected to form their own societies when the EMSA failed to welcome them at the critical point in time (25+ years ago). Relations with the Microbeam Analysis Society (founded in 1966) improved with time, and joint meetings are now more the rule than the exception. It is quite conceivable that some future historian will note the merger of the MSA and the MAS as a "milestone" event. An annual spring "scanning" meeting at the Illinois Institute of Technology in Chicago was initiated shortly after by Ohm Johari in 1968. A separate SEM meeting is still organized, often in conjunction with local societies, although scanning electron microscopy is now an important theme at MSA meetings as well.

Since its inception, the EMSA has endeavored to ensure that members who cannot attend the annual meeting (70+% of the membership) can keep abreast of EM developments of interest to them. For many years the society published the abstracts of the meeting in the widely distributed *Journal of Applied Physics* (they appeared remarkably soon after the meeting). Subsequently, the EMSA began publication of illustrated abstracts in hard-bound *Proceedings* and then added a biennial *Bulletin* of society news, which in later years also carried review articles on important topics in microscopy. The scientific papers in the *Bulletin* were well received, with the result that, beginning this year (1995), the society has initiated the publication of the *Microscopy Society of America* (MSA) *Journal* for distribution to the membership, libraries, and other subscribers.

No doubt, change and, hopefully progress, will always be with us, and the Microscopy Society of America will not be immune to its impact. At the 1950 annual banquet at the Detroit meeting, outgoing EMSA President Ralph Wyckoff said that a scientific society based on a single instrument could not survive and that the members would drift over to the larger professional societies of chemists, biologists, and others. This remark surprised the audience and drew much comment in reports of the meeting. In a sense he was correct. By 1950, electron microscopists had gained much confidence in their results and were anxious to present them to their peers who were using other methods of observation and analysis as described by Hamm in his 1951 review of electron microscopy for analytical chemistry (Hamm, 1951). What Wyckoff did not foresee was that EMSA members would do both. They would attend the EMSA meeting to learn about new instrumental developments and techniques and to tour the ever-enlarging commercial exhibit. They would present application papers as their "ticket to travel." The other development that Wyckoff did not foresee was tremendous advances in all aspects of microscopy, reaching resolution levels and interpretative sophistication hardly imagined in 1950. The fact is that membership and participation in the annual meeting has grown more than a factor of four since Wyckoff expressed his doubtful view about the longevity of the society.

V. Period of Progress and Promise: 1943–1957

Substantial advances in EM, possibly more than in any of the other "eras" in this account, occurred during the 15-year interval between the first meeting of electron microscopists in Chicago in 1942 and the IVth ICEM in Berlin in 1958. Even by the time of the second EMSA meeting, just 14 months after the first, the initial tentative mentions of EM work or suggestions had been succeeded by confident reports of accomplishments. By then, RCA had produced and installed nearly 50 EMBs in the country, they were being heavily used, and "electron microscopy" was emerging as a recognized discipline for both academic and industrial research. The program of the January 1944 meeting was never published, but the titles of the presentations were included in the "year book" that was distributed to charter members and is reproduced in Newberry's book (1992). The authors and their topics had profound influence on EM developments for years. The most familiar names are Marton for electron lens, specimen holders, and microscope design; Hillier and Baker for electron microscope parameters; Stuart Mudd for bacterial structures; Tom Anderson for biological specimen preparation; Gordon Sharp for animal viruses; Francis Schmitt and Cecil Hall for collagen fibers; Robert Heidenreich for surface replicas;

Robert Picard for electron diffraction of thin films; Charles Burten and R. Bowling Barnes for polymer films; William Kissinger for resolution considerations; and Hillier for the microanalyzer. However, the most intriguing of the titles is "A Dynamic Method of Correcting Spherical Aberration in Electron Lenses," by Zworykin, Ramberg, and Hillier. Nothing further was ever heard of their attempt to grasp the avidly sought "Holy Grail" of every electron optician.

Progress was uneven, however, and not every instrument development proved to be successful. Just weeks after the inaugural EMSA meeting, the first of three papers on the electron optics and construction of the GE electrostatic EM appeared (Bachman and Ramo, 1943). Coincidentally, Ramberg (1942) had shown in the previous issue of the same journal that the aberrations of magnetic lenses were less than those of corresponding electrostatic lenses. Nevertheless, perhaps sensing a competitive threat, RCA quickly submitted a paper to the same journal describing the prototype of a new table model EM (the EMC). Although the general appearance of the GE and RCA instruments was somewhat similar, the EMC was smaller and gave superior performance (Zworykin *et al.*, 1943). Although nearly 100 EMCs were sold, mostly for quality control purposes, the choice of 30-kV accelerating voltage and a lack of a condenser lens limited its performance, and the EMC was dropped from production in 1948. (The EMC was the first EM to use a biased gun, which provided a sorely needed increase in brightness.) During the 1940s, RCA had experimented with both scanning electron microscopes and electron microanalysis. However, they did not have adequate technology for detecting scattered electrons or emitted X rays, so these approaches to materials characterization were not pursued further by RCA. GE never had any commercial rewards at all.

The EMBs were well received, but RCA was aware of their inherent limitations and moved rapidly to develop an upgraded model. By mid-1944 they were ready to deliver the first of exactly 300 EMU-series models that were produced through 1953. Although the resolving power of the EMU was better than that of the EMB, it fell short of the expected theoretical resolution of the design and seemed to vary considerably from instrument to instrument. Questions concerning resolution and how to measure it prompted the EMSA to appoint a committee to consider the matter. Several years later the committee issued a report that carefully side-stepped arguments about poor focus and bad specimen preparation and suggested that Fresnel fringe measurements might prove to be the best approach to standardizing resolution measurements. Efforts to improve resolution occupied Hillier and colleagues at RCA, and the EMU proved to be a good platform for their investigations. Their activities lead to the self-biased gun with more brightness and stability, objective lens compensation for astigmatism using adjustable iron set screws, a high-contrast aperture for thick specimens, an intermediate lens and electron diffraction adapter, quantitative understanding of Fresnel fringe phenonama, and recognition of the effect of contamination on practical attainment of good resolution. Although RCA had experimented with high-voltage (150-kV) microscopy, they did not see that the advantages would outweigh the cost and space requirements and so elected to use 50 kV for the EMUs. Their next new model, the EMU-3, did use 100 kV, but it was aimed at the large market for EMs in biology laboratories and lacked the features that scientists would soon demand for work on crystalline materials. The first EMU-3, destined for the Karolinska Institute in Stockholm, was unveiled with great fanfare at Rockefeller Center in New York City in 1953–1954. By that time, U.S. microscopists had become quite familiar with European instruments, which would soon start to arrive from Philips and then Siemens in substantial numbers. The electrifying news that dislocations could be seen in metal films, and actually moving at that, reached the United States in 1956 and created an enormous demand for these microscopes, and material scientists hastened to use them (in 1960, US Steel Research had two Siemens 1As that were used around the clock, seven days a week, by 10+ basic researchers).

Advances in specimen preparation techniques during this period kept pace with improvements in instrumentation. In fact, the same January 1943 issue of the *Journal of Applied Physics,* mentioned above, carried a classic paper on replication by Heidenreich and Peck (1943) and described at the Chicago EM meeting. The essential requirement for specimen substrates or replicas for transmission electron microscopy is an electron-transparent film. Several readily soluble polymers were tried (Schaeffer and Harker, 1942), but Formvar (polyvinal formal), first used by Lester Germer (1939) to support films for transmission electron diffraction, proved to be superior and continues to be widely used by microscopists. It was not ideal for high-resolution replication but Heidenreich and Peck had found that it was possible to evaporate silica and form thin, strong, and virtually structureless films that adhere very closely to the contours of the surface of the sample (many surfaces require the use of double-negative plastic-positive silica techniques). In addition to reproducing very fine detail, silica replicas are very stable, so they are suitable for stereo observations of surface relief (Heidenreich, 1943). An alternative method of observing surface structure came to hand after Williams and Wyckoff (1944) developed shadow casting as a method of determining the height of objects dispersed on substrate films. They were inspired by studies of evaporated metal film structures by Picard and Duffendack (1943) in the same laboratory and aided by the circumstance that Williams (trained as an astronomer) was familiar with interpreting shadows on the lunar surface. Many variations on the replica theme employing different materials and methods developed during the next 10 years, including an "extraction" replica process (Fisher, 1953), where prior etching is used to ensure that fine inclusions and precipitates come away with the replica and can be identified by electron diffraction, energy-dispersive X-ray spectroscopy, or electron energy-loss spectroscopy. The original extraction replica method generally used collodion, but after Bradley (1954), evaporated carbon was preferred.

The reservations and skepticism of metallurgists toward electron microscope studies of metals began to subside when highly regarded colleagues such as Charles Barrett (1944) and G. H. Geisller (1944) obtained good results on important problems and endorsed the technique. Nevertheless, shadow-cast replicas appeared too three-dimensional (often looking much like SEM images), and were confusing to those who saw metallurgical structures as stains under the optical microscope rather than relief. As a result, a joint committee (Subcommittee XI, ASTM E-4) was formed to develop a convincing correlation between optical and electron micrographs of important microstructures "Sub XI" issued its first report in 1950 (ASTM, 1950) and remained active for more than 30 years after.

Although etching and replication were reasonably satisfactory for observing grain structures and fine precipitates, this approach was totally inadequate for studying deformation structures and lattice defects. Heidenreich (1949) utilized electrolytic thinning to prepare thin sections of Al, but with only a 50-kV beam and no double condenser for good diffraction contrast, he did not see any features recognizable as dislocations. [Some years earlier, he and Lorenzo Sturkey (1945) had correctly interpreted contrast effects in images of crystals correctly in terms of dynamical diffraction concepts.] By the end of the decade, other areas of materials and chemical science were also receiving increasing attention, such as catalysts, pigments, and magnetic fringe fields. The pace quickened rapidly, and by the end of this "era," major developments in instrumentation, electron scattering and diffraction phenomena, specimen preparation techniques, and professional society activities had totally altered the world of electron microscopy. Methods of preparing thin foils had been developed in Europe, and the physical basis of diffraction contrast of lattice defects had been recognized. Electron microscopes with 100-kV accelerating voltage, goniometer stages, and precision diffraction capabilities were available and facilitating sophisticated studies that would have been hard to imagine 15 years earlier.

VI. WORLD WATCH AND WELCOME: 1958–1969

In the world of materials science at least, this era opened with tremendous excitement over the newly developed ability for direct observations of defects in crystals by thinning interesting samples for transmission electron microscopy and interpreting the observations by diffraction contrast theory. Information about the newest advances reached a receptive U.S. audience through publications and presentations at EMSA meetings and several Gordon Conferences. In addition, a large group of U.S. microscopists went to the Berlin ICEM and toured the leading EM centers as well. There is a striking parallel between the situation regarding EM in the United States at the end of the 1930s and the end of the 1950s and into the early 1960s. In the first case, it was instrument developments at Toronto that attracted attention, and many of the graduate students involved were soon induced to move to the United States and continue their work here. In the second instance, the largest single center of leading-edge EM work on materials and methods was at Cambridge University, and again many of the graduate students from several departments were attracted to industrial and academic positions in the United States. The very earliest group included Gareth Thomas (Berkeley), Peter Swann, Alan Baker, and Peter Salmon-Cox (US Steel), John Silcox (Cornell), and Oliver Wells (Westinghouse). In addition, some North Americans studied in Cambridge and elsewhere and returned to U.S. positions, including Tom Everhart, J. S. Lally, and R. M. Fisher. Biologists and other EM users were also attracted to the overseas work that was coming to their attention at about the same time.

ICEM V was held in Philadelphia in 1962, and the *Proceedings* (preprinted for the first time) are very informative about the state of EM work in the United States at that time. In keeping with the pattern that had been in vogue for nearly 15 years, more than 66% of the papers were on biological techniques and applications. The only paper in the opening general session was by M. Beer on staining, although a number of notable and well-regarded presentations were made in the other sessions. The Vth ICEM program included 22 sessions on physical topics and nonbiological applications. The United States was a minor participant in much of this half of the Congress, with no presentations at all in sessions on electron optics, scattering, and diffraction. The only paper from RCA was by Reisner, on charge neutralization of contamination on objective apertures.

Participation in the sessions on applications of EM to materials was also small, considering that the meeting took place on the populous east coast of the United States (the presence of new arrivals from overseas was apparent in the program). However, the meeting stimulus, availability of federal and industrial funding, and the newcomers spurred an active and long period of EM activity in the United States.

The next ICEM took place in Kyoto in 1966, and a relatively large contingent of American microscopists managed to attend. This meeting had about the same effect as Berlin had in 1958, and initiated personal contacts that lead to special topical meetings and exchange visits as well as greater acceptance of Japanese microscopes.

Choosing the years to begin and end the historical subdivisions selected for this article is obviously rather arbitrary. Progress in science moves steadily forward, and most changes in pace are difficult to detect. However, a well-defined era did end in 1969, when RCA closed its electron microscope division after 30-plus years of active and mostly successful efforts to design, produce, and market instruments for the United States and world markets. No doubt international economic factors such as labor costs and government financial policies were the primary reasons that forced RCA to drop EMs as a product line. These factors affected RCA no less than they had impacted the steel industry, automobile manufacturers, television production, and all the rest. However, another negative factor was the very small size of the electron microscope division within a huge corporation. At no time did the EM division equal as much as 1% of RCA's annual business, and most years it was much less. The managers of such tiny divisions have little power to promote their business objectives at the corporate level, and talented managers tend to be drawn away to greener departments. This same size disparity likely influenced Siemens management to eventually drop electron microscope production. Similarly, pessimistic market projections caused GE to drop an instrument division before they were even ready to make a serious effort to market electron microscopes.

RCA did misjudge the market opportunity with both the electron probe analyzers and the scanning electron microscopes. Even with the transmission electron microscope, RCA was slow to follow their initial technical successes, which included highly stable power supplies, convenient vacuum systems, high-brightness self-biased guns, astigmatism compensation, and the intermediate lens. Significant improvements such as 100-kV accelerating voltage and externally adjustable apertures were slow in coming. Many potential buyers felt that marketing considerations influenced RCA designers to make their microscopes excessively user friendly, even to the point of providing built-in ashtrays on the final series of the EMU-4s. Ironically, RCA sold the electron microscope division to ForgeFlow of Sunbury, Pennsylvania, in June 1969. This was the same month that the first HVEM Conference was held at the US Steel Research Center in Monroeville, where the RCA Million Volt Electron Microscope stood proud and tall (literally, four stories from floor to ceiling) and was working well for distinguished visitors, including Cosslett, Dupouy, Kobayashi, and Hashimoto, and others, to see. In addition, a smaller, 500-kV RCA instrument had been installed at the University of Virginia for use by Kenneth Lawless and colleagues. ForgeFlow tried valiantly to regain a foothold in the market with a U.S.-made TEM (including submission of bids to supply the RCA-design MVEM), but to no avail, and the RCA era quietly came to an end.

VII. CENTER EMPHASIS: 1970–1984

Despite regrets over the disappearance of RCA from the world of microscopy, this era began on an upbeat note with several easily defined landmark events and a promised flowering of recent EM innovations. One of the important chapters for the era was about to open with the successful retrieval of the first load of moon rocks by *Apollo 11* astronauts. High-visibility EM studies of lunar samples during the next several years drew the attention of mineralogists, geologists, and ceramists to the use of the electron microscope to observe internal microstructural features that pertain to the genesis of rocks and soil as well as the processing of man-made materials.

Interest in high-voltage electron microscopy (HVEM) reached perhaps its highest level in 1970. HVEMs were in operation at the US Steel Research Center (R. M. Fisher), University of Virginia (K. Lawless), and at the University of California at Berkeley (G. Thomas). With funding from the Division of Research Resources of the National Institutes of Health (NIH), 3 1-million-volt electron microscopes were on order for the University of Wisconsin (Hans Ris), the University of Colorado at Boulder (K. Porter), and the State University of New York (SUNY) at Albany (D. Parsons). While waiting installations of their own microscopes, these and other biologists were able to use the US Steel MVEM through an NIH contract. Interest in the HVEM would get a further boost at the VIIth ICEM in Grenoble, where exciting work from France, the UK, and Japan would be featured. An important area of application of HVEMs worldwide was *in situ* studies of radiation displacement damage, which leads to serious swelling of stainless steels used for cladding of reactor fuel. Although radiation damage experiments could be done in a few hours in an HVEM, eliminating months or more of reactor exposure, it was later found that electron beam simulation was not an exact parallel to neutron radiation, and the Westinghouse-Hanford HVEM, purchased for just this purpose, was shut down. Other *in-situ* experiments continued to be important, however, and the U.S. Department of Energy installed a 1.5-mV "conventional" HVEM for this work at the Lawrence Berkeley Laboratory at the end of the decade under the direction of Gareth Thomas and Ken Westmacott.

Despite all of the attention given to the HVEM, significant changes were occurring in high-resolution and conventional scanning electron microscopy. By 1972, there were 400 SEMs in use in the United States. Most of these were employed in the physical sciences, but biologists who were aware of the SEM's potential were busily devising fixation methods, cold stages, and other techniques to facilitate its use. To meet the need for people trained in advanced SEM methods, summer school courses were initiated at Lehigh University. In a review for analytical chemistry, McAlear (1972) predicted that pathologists would be reluctant to accept SEM diagnosis at first, but felt that work at the NIH using ETEC SEMs would dispel their doubts (it is not clear in 1995 that this mission was satisfactorily accomplished). McAlear also speculated that chemical analysis in the SEM using EDS would become important to biologists. In addition, the success of Crewe and Wall (1970) in resolving DNA molecules by field-emission STEM prompted McAlear to suggest that STEMs would supersede the TEM for many such applications. The advent of micro recording and image processing to improve TEM images and aid in their interpretation was also noted.

During the remaining 1970s, much attention was focused on electron–specimen interactions in terms of both damage and information content. This included energy-loss processes, X-ray generation and displacement, and ionization radiation damage. This emphasis is reflected in the 1974 and 1976 reviews on electron microscopy prepared for Analytical Chemistry by Michael Beer. As high-resolution TEMs became available, much attention was placed on structure images of crystals and atomic rearrangements at surfaces and interfaces. As a measure of the diversity that had evolved in electron microscope techniques and instrumentation, Cowley, in his review for Analytical Chemistry in 1978, elected to provide a very useful list of definitions of nearly 40 acronyms and other terms common in contemporary EM nomenclature. He also noted the need for more advanced training in HREM experimentation and interpretation. Later, Arizona State University was to provide such training through annual winter (January) schools and application workshops.

The intense interest in high-resolution electron microscopy stimulated efforts to establish national centers at major laboratories. As a result, the Department of Energy established the National Center for Electron Microscopy at the Lawrence Berkeley Laboratory at UC Berkeley (G. Thomas), and the National Science Foundation set up a similar facility at Arizona State University (Tempe, AZ, J. Cowley). The Berkeley instrument, installed in 1983, was capable of 1.6 Å resolution at 1 meV and was properly designated the ARM (atomic resolution microscope). Because of the high cost of the ARM and the special building requirements, intermediate 300- to 400-kV instruments were favored, and 15–20 were installed in university and industrial laboratories throughout the United States. As this era closed, the million-volt microscopes at US Steel and Westinghouse-Hanford had already been shut down.

VIII. New Microscopies and Media: 1985–1995

The cycle of IFSEM congresses brought ICEM XII back to the United States (Seattle) in 1990. It was the largest at that time, with nearly 1700 papers and 5000 participants. Times were generally good around the world, which favored a satisfactory economic outcome for the meeting. Through association with the Microbeam Analysis Society, analytical methods received more attention than ever before, and the new scanned probe microscopies were also prominent additions to the program. Although the U.S. contributions were again more numerous in the applications areas than the physical topics, the disparity was not so large as in 1962 and the country presented a creditable account of its work in the now broader field of microscopy.

From a technical perspective, the past 10 years is clearly too immediate to be examined as a historical era, but changes in emphasis have occurred that already are a part of the record. One obvious change that affects almost all facets of daily life is the invasion of the computer. Computation has been important to microscopy from the earliest days, but the giant blue-gray computer console used to be far away in an air-conditioned room with uniformed attendants on hand to protect the unknowing. Now the smaller, largely tan-colored computers have pushed into the laboratory, moved right up against the column, often crowding the microscopists' work space or even crawling inside the console itself. They respond without emotion to tactile instructions, and some even accept voice commands. The microscopes respond readily to digital direction and will produce a micrograph to suit your fancy, even in color if you wish. If you do not have a microscope, no matter, feed the computer some tasty software and it will create an image for you. If you insist on doing experimental work, send your sample to San Diego and the people there will let you so some "virtual" microscopy on line (Ellisman et al., 1994). As yet, not many microscopists have come to this, but computers are reducing the time spent compensating for astigmatism or focusing or adjusting the exposure and increasing the time available for creative work such as making good specimens or interpreting the observations.

The satirical commentary given above not withstanding, the availability of fast, high-capacity computers has been crucial to all of the important advances of recent years. In the case of what might now be called conventional TEM microscopy, computer technology has provided the ability to observe and to record "structure" images of atom columns in crystals electronically, and often to interpret the observations on-line, which has provided much important new information. This is especially true for studies of the atomic arrangements at interfaces with respect to bonding and adhesion as well as semiconductor device performance. Nevertheless, future historians will note that the large electronics companies decommissioned their HREMs in the early 1990s and donated them to university laboratories. But the university centers themselves are not immune to budget pressures and can no longer readily accept "free" offers of surplus microscopists. In fact, the HVEMs installed in 1971–1973 have well reached the final years of their useful life. None will be replaced in kind.

Computers have also affected the media of scientific exchange. In addition to storing images and data on disk instead of film, research results, including micrographs, are also going out to colleagues on disk and, in a few cases, by ethernet. CD-ROMs are ideal for storing large numbers of micrographs, and at some point large laboratories will archive their images in CD atlases. Conventional journal publishers are adapting computer technology as fast as possible to keep costs competitive and reduce publication delays. Electronic publication will eventually largely displace books and journals.

The late Chuck Fiore totally devoted his 1988 *Analytical Chemistry* review of EM to "The New Electron Microscopy: Imaging the Chemistry of Nature." His emphasis properly reflected the importance that quantitative analysis had assumed and, in fact to many, modern electron microscopy and analytical microscopy are synonymous (Fiore, 1988).

From a purely scientific point of view, computer technology has done much more for microscopy than simply reducing the microscopist's workload of tedious tasks. Computer technology has truly made totally or somewhat new microscopies feasible and affordable. The scanned probe microscopies, which include STM and AFM (scanning tunneling and atomic force microscopies), and near-field, confocal, and fluorescent optical microscopy have become a very important part of the ultrastructure researchers toolkit and taken their place on conference platforms. Developments in this area may well prove to be milestone events of the 1985–2000 era in microscopy. It should be noted that *Analytical Chemistry* has not carried a review on electron microscopy per se since 1988, whereas STM/AFM and other surface techniques have been featured. Electron microscopy is still by far the most productive tool for research, but it has far more sibling rivals to cope with than it did in its youth.

IX. Concluding Remarks

In this record of the history of electron microscopy in the United States, I have focused primarily on the events of the early years. They are rather

FIGURE 1. Keith R. Porter, in his capacity as President of the Electron Microscopy Society of America, addressing the Opening Ceremony at the Twelfth International Congress for Electron Microscopy in Seattle, Washington, 13 August 1990. Dr. Porter has had the unique distinction of serving as President of EMSA during the two ICEMs that have been held in the USA (Seattle '90 and Philadelphia '62). Keith Porter was a great source of inspiration and guidance to this author, stemming perhaps from our common Canadian ancestry, but certainly enhanced by many exciting events and experiences that followed a still-remembered question he put to me in Philadelphia in 1962 after Gaston Dupouy had presented an eloquent description of the pioneering work underway in Toulouse: "Shouldn't we have an HVEM or two in the US?"

few in number and their importance well established. As the field grew over the years, an increasing number of individuals made some contribution to virtually every development, and the importance of each contribution is sometimes difficult to assess. Innovators do not always come up with the optimum implementation of their own idea. People use and remember what works best. In history, the first is not always foremost. Even Hannibal, renowned in history for crossing the Alps, followed some foot sloggers who were carrying back-breaking loads.

The recent eras are discussed in much less detail in view of the huge number of events and contributions that occurred. Later historians will have to put this material into perspective on a topic-by-topic basis, e.g., advances in contrast theory, electron sources, vcacuum technology, etc. The author apologizes for any excess emphasis on TEM and reliance on historical events as recorded in EMSA programs and *Proceedings,* as well as for the errors and omissions that knowledgeable readers will detect.

APPENDIX I: ELECTRON MICROSCOPY SOCIETY OF AMERICA PRESIDENTS AND
MEETING SITES

1981	John Hren	Atlanta, GA
1942	G. L. Clark[a]	Chicago, IL[b]
1943	R. Bowling Barnes[c]	No meeting
1944	R. Bowling Barnes[d]	New York, Columbia University[e]
"		Chicago, IL, La Salle Hotel[f]
1945	James Hillier	Princeton, NJ, Princeton University
1946	David Harker	Pittsburgh, PA, Mellon Institute
1947	William G. Kinsinger	Philadelphia, PA, Franklin Institute
1948	Perry C. Smith	Toronto, Canada, University of Toronto
1949	F. O. Schmitt	Washington, DC
1950	Ralph W. G. Wyckoff	Detroit, MI
1951	Robley C. Williams	Philadelphia, PA, Franklin Institute
1952	R. D. Heidenreich	Cleveland, OH
1953	Cecil E. Hall	Pocono Manor, PA
1954	Robert G. Picard	Highland Park, IL
1955	Thomas F. Anderson	State College, PA, Penn State University
1956	William L. Grube	Madison, WI, University of Wisconsin
1957	John H. L. Watson	Boston, MA, Massachusettes Institute of Technology
1958	Max Swerdlow	Santa Monica, CA
1959	John H. Reisner	Columbus, Ohio State University
1960	D. Gordon Sharp	Milwaukee, WI Marquette University
1961	D. Maxwell Teague	Pittsburgh, PA
1962	Keith R. Porter	Philadelphia, PA (Vth ICEM)
1963	Charles Schwartz	Denver, CO
1964	Sidney S. Breese	Detroit, MI
1965	Virgil G. Peck	New York, NY
1966	Walter Frajola	San Francisco, CA
1967	Joseph J. Comer	Chicago, IL
1968	John H. Luft	New Orleans, LA
1969	W. C. Bigelow	St. Paul, MN
1970	Russell Steere	Houston, TX
1971	Robert M. Fisher	Boston, MA
1972	Daniel C. Pease	Los Angeles, CA
1973	Benjamin Siegel	New Orleans, LA
1974	Russell J. Barrnett	St. Louis, MO
1975	Gareth Thomas	Las Vegas, NV
1976	Etienne de Harven	Miami Beach, FL
1977	T. E. Everhart	Boston, MA
1978	Myron Ledbetter	Toronto (IXth ICEM)
1979	John Silcox	San Antonio, TX
1980	Michael Beer	Reno, NV
1982	Lee Peachey	Washington, DC

(continues)

APPENDIX I: (*Continued*)

1983	David Wittry	Phoenix, AZ
1984	J. David Robenson	Detroit, MI
1985	Dale Johnson	Louisville, KY
1986	Robert Glaeser	Albuquerque, NM
1987	Linn W. Hobbs	Baltimore, MD
1988	Jean Paul Revel	Milwaukee, WI
1989	Ray Carpenter	San Antonio, TX
1990	Keith R. Porter	Seattle, WA (XII ICEM)
1991	Charles Lyman	San Jose, CA
1992	Patricia Calarco	Boston, MA
1993	Michael Isaacson	Cincinnati, OH
1994	Robert R. Cardell	New Orleans, LA
1995	Terence E. Mitchell	Kansas City, MO

[a] Chairman of the First National EM Conference.
[b] In conjunction with a National Chemical Exposition.
[c] Elected interim, preconstitution president.
[d] First elected president.
[e] In conjunction with the 1944 annual winter meeting of the American Physical Society.
[f] Second meeting in 1944.

APPENDIX II: SIGNIFICANT EVENTS AND ACHIEVEMENTS IN ELECTRON MICROSCOPY IN THE UNITED STATES

Date	Event or achievement
1927	Demonstration of reflection diffraction of low-energy (100+ eV) electrons from a nickel crystal by Davisson and Germer at the Bell Telephone Laboratories, confirming the wave nature of an electron beam.
1928	Description, by Lester Germer, of electron diffraction observations as "optical interference" experiments in a journal on chemical education.
1929	Vladimir Zworykin leaves Westinghouse Electric to join RCA to work on television tubes.
1930	P. M. Morse publishes a wave mechanical treatment of reflection electron diffraction (the Bragg case) to explain intensity details in the Davisson-Germer experiments.
1931–1932	Optical properties of an electrostatic pinhole "electron lens" described by C. J. Calbick and C. J. Davisson at a summer meeting at Pasadena. Apparent first use of the term "electron lens" in English. The full paper was published in 1932.
1933	Zworykin writes about the idea of an electron microscope in the *Journal of the Franklin Institute*.
1934	Calbick and Davisson describe an electrostatic emission-type "electron microscope."
1935	Efforts to construct emission electron microscopes initiated independently at the University of Toronto and at Washington University in St. Louis,

(*continues*)

APPENDIX II: (*Continued*)

Date	Event or achievement
	Missouri. Construction of a TEM started at Washington State University (WSU) in Pullman, Washington. The WSU microscope was completed by Ken Fitzimmons and Paul Anderson the same year. A few low-resolution images were produced, but nothing was ever published and all work with the instrument was terminated in 1938.
1936	V. K. Zworykin and G. A. Morton publish a paper on "Applied Electron Optics" pertaining to image conversion devices.
1937	C. E. Hall completes thesis on electron emission microscopy in 1936 and leaves the University of Toronto at the beginning of the next year to join Eastman Kodak in Rochester, New York, with expectations to build an electron microscope. Approval to begin construction was not received until 1938. Nevertheless, Hall's EM was apparently the first to begin operation in the United States.
	G. H. Scott and J. H. McMillan at Washington University in St. Louis construct "A Magnetic Electron Microscope of Simple Design." This was an electron emission microscope using magnetic lenses to study chemical elements in biological specimens.
1938	*The New York Times* carries a report of a Siemens EM reaching $100,000\times$ magnification.
1938	Zworykin initiates program to develop a commercial electron microscope and invites L. Marton to join RCA Laboratories and begin construction of an electron microscope. Description of the RCA EMA microscope is published in 1940.
1939	First reports of the Toronto microscope by E. F. Burton, J. Hillier, and A. F. Prebus.
1939	Lester Germer initiates studies of thin films by electron diffraction and notes "anomalous secondary scattering effects." A series of papers on this subject was published through 1942.
1939	Scott and Packer (Washington University, St. Louis) describe their use of "The Electron Microscope as an Analytical Tool for the Localization of Minerals in Biological Tissues." Pronounced effect of Mg and Ca on electron emission provides strong image contrast.
1940	James Hillier leaves Toronto to join RCA and begin construction of RCA type "B" electron microscope. The prototype instrument was delivered to American Cyanamid in December. A. F. Prebus leaves Toronto for Ohio State University and begins construction of an electron microscope.
1940	Fresnel fringes due to phase shifts along edge contours discovered simultaneously by Hillier (USA) and Boersch (Germany).
1941	William Ladd moves from Toronto to Columbian Carbon Corp. in Brooklyn, New York. He takes along an electron microscope that he constructed with guidance from Hillier.
1941	Siemens-Rudenberg patents seized by U.S. government as alien property.
1941	C. E. Hall publishes TEM studies of photographic exposure and development phenomena. L. Marton leaves RCA to move to Stanford University in Palo Alto, California.

(*continues*)

APPENDIX II: (*Continues*)

Date	Event or achievement
1941	Sixteen production instruments delivered during the first nine months of 1941 Electron microscope image contrast interpreted on a mass-thickness scattering basis by Marton and Schiff.
1941	Zworykin, Hillier, and Vance publish "A Preliminary Report on the Development of a 300 kV Magnetic Microscope."
1941	James Hillier discusses the "fundamental limit of performance of an electron microscope."
1941	Zworykin and Ramberg report on the use plastic impressions to observe the structure of the surface of bulk materials in a transmission electron microscope.
1941	T. F. Anderson takes up two-year special National Research Council Fellowship funded by RCA to expedite development of biological applications of the EM. Work was done largely at Camden, but some was carried out at Woods Hole. Fellowship program resulted in more than 50 publications in two years.
1942	*The Electron Microscope,* by E. F. Burton and W. Kohl, published.
1942	The first meeting of the Electron Microscope Society of America convened in Chicago under the chairship of George L. Clark (University of Illinois), with 75 registrants in attendance. More than 25 of the attendees undertook long careers in electron microscopy. The organizational structure of the society and the format of the annual meetings that were established have been maintained ever since. Professor Clark is considered to be the first EMSA president.
	At the meeting, V. J. Schaefer of GE (Schenectady, NY) discussed the merits and limitations of various types of plastic replicas, with Formvar as the preferred material. Heidenreich, from Dow Chemical Co., described the development (with Virgil Peck) of a polystyrene-evaporated silica replica technique with sufficient stability and resolution to observe fine slip steps on the surfaces of deformed metals.
	Various replica techniques evolved over the next 10 years, including two-step positive-negative methods and the shadow-cast and thermal oxide replicas and extraction replicas. Virtually all of these methods were superseded when carbon evaporation was developed by Bradley in Britain.
1942	First studies of the structure of collagen by TEM by F. O. Schmidt, Marie Jakus, and C. E. Hall at MIT.
1942	Detailed description of a "scanning electron microscope" published by Zworykin, Hillier, and Snyder.
1942	Hillier and Baker publish "micrographs of hydrated alumina exhibiting Bragg crystalline reflections."
1942	Studies of electron diffraction contrast effects in MgO crystals in the TEM reported by Robert Heidenreich. (In 1945, Heidenreich and Lorenzo Sturkey discussed interference.
1943	Wartime travel restrictions prevented holding a second EMSA meeting. R. B. Barnes of American Cyanamid served as interim, pre-constitution, President.

(*continues*)

APPENDIX II: (*Continued*)

Date	Event or achievement
1943–1944	Papers on "Microanalysis by Means of Electron Beams and Microspectroscopy" appear (Hillier, 1943; Marton, 1944).
1944	Second EMSA meeting takes place at Columbia University in New York City. James Hillier is elected under the newly ratified constitution to serve in 1945.
1944	Review chapter by A. F. Prebus on the achievements and promises of "The Electron Microscope" published in *Colloid Chemistry*.
1944	Use of two-direction shadow casting by R. C. Williams and R. W. G. Wyckoff to determine "The Thickness of Electron Microscopic Objects."
1945	*Electron Optics and the Electron Microscope* by Zworykin, Marton, Ramberg, Hillier, and Vance appears. This comprehensive treatise covers electron optics including aberrations, instrument design, electron scattering phenomena, image characteristics and recording, and specimen techniques, and forestalls the appearance of another book of this type for 18 years.
1945	Use of a TEM to achieve very long camera length for electron diffraction reported by G. L. Simard, C. J. Burton, and R. B. Barnes.
1945	E. F. Burton presents a paper on electron microcopy of butterfly wings and relates ridge spacing to interference colors at EMSA meeting at Princeton. (This was the only EMSA meeting he attended).
1945	K. R. Porter, A. Claude, and E. F. Fullam report on first electron microscope studies of tissue culture cells.
1945	Use of shadow casting by R. C. Williams and R. W. G. Wyckoff to enhance contrast for the "Electron Micrography of Virus Particles" and other objects.
1945	Electron microscope studies of tissue culture cells initiated by Keith Porter with associates A. Claude and Ernest Fullam. The program continued for more than 10 years.
1945	Mary Schuster Jaffe, first woman to present a paper at the EMSA, describes preparation methods for powdered materials.
1946	Heidenreich, Sturkey, and Woods report on studies of age-hardening alloys using silica replicas and note a connection between the scale of the substructure and mechanical stength.
1946	L. Marton leaves Stanford University to go to the National Bureau of Standards.
1946	EMSA Committee on "Resolution" defines the problem but is unable to agree on an acceptable method to determine resolution.
1946	Review chapter on "Electron Microscopy" published by L. Marton in *Progress in Physics*.
1947	Gertrude Rempfer describes her design of the Farrand electrostatic TEM at the EMSA meeting in Philadelphia. This microscope was never put into commercial production.
1947	Claude introduces osmium fixation.
1947	The effects of electron bombardment causing damage to the specimen are noted by J. H. L. Watson, E. F. Burton, R. S. Sennett, and S. G. Ellis.

(*continues*)

APPENDIX II: (*Continued*)

Date	Event or achievement
1947	Rudenberg sues U. S. Attorney General in District Court in Massachusetts to regain the Siemens-Rudenberg patents. Judge upholds his case and grants sole ownership to Rudenberg.
1948	*Further Researches on the Electron Analyzer* published by S. G. Ellis of RCA.
1948	Cecil Hall demonstrates the use of dark-field electron microscopy to examine crystalline materials.
1948	*Analytical Chemistry* inaugurates biennial reviews of EM literature.
1948	"Phase Contrast in Electron Microscope Images" is analyzed by E. G. Ramberg.
1949	Robert Heidenreich publishes results of efforts to see slip bands using imaging and diffraction studies of thin metal sections. Image features characteristic of dislocations were noted later by M. J. Whelan in unpublished Heindenreich micrographs.
1949	Discovery of exceptionally uniform latex spheres by R. C. Backus and R. C. Williams.
1949	Cecil Hall describes a method for "measuring spherical aberration in an EM objective lens."
1949	Methacrylate introduced as an embedding medium by Sanford Newman, Emile Borysko, and Max Swerdlow at the National Bureau of Standards. Ultimately replaced by epoxy in 1956.
1950	ASTM Subcommittee XI-E4 on the "Electron Microstructure of Steel" issues its first progress report. This committee succeeded in gaining acceptance of the results of "electron metallography" by the initially skeptical metallurgical community and sponsored a long series of symposia on EM techniques and applications.
1950	K. R. Porter with J. Blum at the Rockefeller Institute for Medical Research build a microtome to cut ultrathin sections for TEM examination. Design details were published in 1953.
1950	Hillier introduces a removable intermediate lens for the RCA EMU2 to extend the magnification range. (This author modified the intermediate lens power supply to obtain selected area diffraction patterns from extraction replicas in 1952.)
1950	Robley Williams leaves Michigan to join Wendall Stanley at the Virus Laboratory at the University of California in Berkeley.
1950	The glass knife introduced for ultra microtomy by Harrison latta and J. Francis Hartmann.
1950	In addition to its many pioneering scientific reports, the 1950 EMSA meeting in Detroit is also noteworthy for a widely quoted pronouncement by outgoing President Ralph Wyckoff that a society based on a single instrument was not sustainable. Actually, EMSA grew by almost a factor of 10 in both membership and the number of papers presented during the next 40-plus years.
1951	Marton organizes a truly milestone "National Bureau of Standards Semi-Centenial Symposium on Electron Physics." For most attendees, this was

(*continues*)

APPENDIX II: (*Continued*)

Date	Event or achievement
	their first close encounter with EM activities in Europe, including the construction of an "electron microanalyzer" by Raymond Castaing." The *Proceedings* were not published until 1954, but information about the papers was widely circulated and prompted Robert Ogilvie at MIT, Verne Birks at the Naval Research Laboratory, Paul Duwez and David Wittry at Cal Tech, and Robert Fisher at US Steel to undertake the construction of microprobes based on the Castaing concept during 1953 and 1954.
1951	Critical point drying and other "Techniques for the Preservation of Three-Dimensional Structure in Preparing Specimens for the Electron Microscope" described by Tom Anderson.
1952	George Palade shows that a buffering system can be used to fix tissue in osmium tetroxide.
1953	*Introduction to Electron Microscopy*, by Cecil E. Hall (MIT), is published.
1953	Ernest F. Fullam, Inc. (a consulting company and supplier of EM accessories and materials), established. William Ladd, Inc., established in 1955.
1953	Porter and Blum introduce popular Sorval microtome. Fernandez-Moran demonstrates superiority of diamond for microtomy.
1955	Benjamin Siegel inagurates the first summer school on electron microscopy at Cornell. The series continued for about five years.
1956	EMSA becomes a member of the International Federation of Electron Microscope Societies (IFEMS).
1956	Palade and Siekevitz publish classic paper on "Pancreatic Microsomes: An Integrated Morphological and Biochemical Study."
1958	M. L. Watson describes "Staining of Tissue Sections for Electron Microscopy with Heavy Metals" (using lead and barium).
1960	Migration of British microscopists to America begins. Gareth Thomas to Berkeley, where he eventually established National Center for Electron Microscopy; Peter Swann to US Steel and eventually to found Gatan, Inc.; John Silcox (1961) to Cornell to establish electron microscopy and spectroscopy program; Oliver Wells (somewhat earlier) to Westinghouse, then to IBM. T. E. Everhart also joins Westinghouse after completing thesis on the development of an efficient detector for SEMs.
1961	J. H. Luft publishes on first embedding resins forming covalent bonds with the tissue itself.
1961	EMSA membership exceeds 1000.
1961	Berkeley conference on "Electron Microscopy and the Strength of Crystals" organized by G. Thomas and J. Washburn. The conference and the widely distributed *Proceedings* drew much attention to the power of the TEM to study crystal defects. Studies of ceramics, polymers, and magnetic materials were featured at the subsequent conference in 1971.
1962	Fifth International Congress on Electron Microscopy held in Philadelphia. T. F. Anderson and John Reisner, Co-Chairs, T. F. Anderson EMSA President. Abstracts preprinted for ICEM Proceedings for the first time. 650 papers were presented, about equally divided between physical and biological sciences. Some of the many highlights: keynote lectures by

(*continues*)

Appendix II: (*Continued*)

Date	Event or achievement
	Ernest Ruska on the theoretical limit to the resolution of the TEM, and Gaston Dupouy on megavolt electron microscopy. The presentations and informal interactions with attendees from abroad had a major impact on developments in the United States, such as stimulating strong interest in developing domestic HVEMs.
1963	The availability of reliable, high-performance electron microscopes spurred intensive work in biological, electronic, and engineering materials, with steadily increasing sophistication of image interpretation and quantitative microscopy. The links forged with foreign microscopists at the Philadelphia ICEM proved to be extremely productive.
1964	*Fundamentals of Transmission Electron Microscopy,* by Robert Heidenreich, is published.
1966	Workshop on "High Voltage Electron Microscopy" at Argonne National Laboratories—R. K. Hart, Chair. Announcement by A. V. Crewe of intentions to build an "atomic resolution" scanning electron microscope.
1967	The 25th EMSA anniversary meeting is held in Chicago, featuring the prepublication of abstracts as the EMSA *Proceedings* for the first time. Notable keynote projections of things to come are featured. Papers from RCA and US Steel on construction and performance of the just installed USS Million Volt electron microscope. Other high-voltage microscopes soon came into operation at the University of Virginia and UC Berkeley.
1967	First of three US–Japan Conferences on High Voltage Electron Microscopy takes place in Honolulu.
1969	First of seven International Conferences on High Voltage Electron Microscopy held at US Steel Center. R. D. Heidenreich predicted that resolution, penetration, and radiation damage considerations will establish 400 kV as the optimum accelerating voltage.
1970	Thorium atoms attached to DNA strands observed by Albert Crewe with a field-emission STEM. Sample was prepared by Michael Beer.
1970	Lehigh University's summer "Short Courses" inaugurated.
1970	John Cowley arrives at Arizona State University to initiate an EM program that evolved into a major center for high-resolution microscopy.
1971–1972	High-voltage EM centers established with NIH funding at Boulder (K. Porter), Madison (H. Ris), and Albany (Parsons).
1973	EMSA Radiation Safety Committee reports on X-ray radiation surveys and recommends maximum level of 0.5 mR/h measured at 5 cm from the microscope.
1974	George Palade and Albert Claude receive the 1974 Nobel Prize in Medicine for their pioneering applications of electron microscopy to cell biology.
1975	EMSA inaugurates Distinguished Physical and Biological Scientist Awards. Robert Heidenreich and Keith Porter were the first recipients.
1978	EMSA annual meeting takes place in Toronto as part of the Ninth International Congress on Electron Microscopy.
1986	ARM (atomic resolution microscope) installed in Berkeley in National Center for Electron Microscopy (G. Thomas).

(*continues*)

APPENDIX II: (*Continued*)

Date	Event or achievement
1990	Twelfth International Congress for Electron Microscopy held in Seattle; repeat performance by Keith Porter as President of the EMSA; R. M. Fisher and Dale Johnson as ICEM XII Co-Chairmen.
1992	50th anniversary of the EMSA takes place in Boston, with Patricia Calarco as the first woman President and Mort Maser as Chairman of the Organizing Committee. EMSA membership approaching 5000. Book by Sterling Newberry, *EMSA and Its People—The First 50 Years*, is published by EMSA and distributed to attendees.
1993	EMSA becomes the Microscopy Society of America (MSA) to reflect the full breadth of the various microscopies employed by its members.
1994	Final issues of the *MSA* (EMSA) *Bulletin* appear.
1995	Publication of the *MSA Journal* begins with Jean-Paul Revel of Cal Tech as Editor. *MSA Proceedings* will be published as special issue of the Journal.

Selected References

American Society for Testing Materials (1950). Subcommittee XI of Committee E-4 Electron Microstructure of Steel. *Proc. ASTM* **50,** 1–49.

Bachman, C. H., and Ramo, S. (1943). Electrostatic electron microscopy I, II, III. *J. Appl. Phys.* **14,** 8–18, 69–79, 155–160.

Backus, R., and Williams, R. (1949). Small spherical particles of exceptionally uniform size. *J. Appl. Phys.* **20,** 224–225.

Backus, R., and Williams, R. (1950). Use of spraying methods and of volatile suspending media in the preparation of specimens for electron microscopy. *J. Appl. Phys.* **21,** 11–15.

Baker, R., Ramberg, E., and Hillier, J. (1942). The photographic action of electrons in the range between 40 and 212 kilovolts. *J. Appl. Phys.* **13,** 450–456.

Barnes, R., Burton, C., and Scott, R. (1945). Electron microscopical replica techniques for the study of organic surfaces *J. Appl. Phys.* **16,** 730–739.

Beer, M. (1974). Electron microscopy. *Anal. Chem.* **46,** 93–117R.

Beer, M. (1976). Electron microscopy. *Anal. Chem.* **48,** 88–94R.

Brown, A., and Jones, W. (1947). A methyl methacrylate-silica replica technique for electron microscopy. *Nature* **169,** 635–636.

Burton, E., Sennett, R., and Ellis, S. (1947). Specimen changes due to electron bombardment in the electron microscope. *Nature* **160,** 565–567.

Ceren, B., and McCulloch, D. (1951). Development and use of the Minot rotary microtome for thin sectioning. *Exp. Cell Res.* **2,** 97–102.

Claude, A., and Fullam, E. (1946). The preparation of sections of guinea pig liver for electron microscopy. *J. Exp. Med.* **83,** 499–503.

Claude, A., Porter, K., and Pickles, E. (1947). Electron microscopy study of chicken tumor cells. *Cancer Res.* **7,** 421–430.

Columbian Carbon Company Research Laboratories (1940). The particle size and shape of colloidal carbon as revealed by the electron microscope. *Columbian Colloidal Carbons* **2,** 5–53.

Comer, J., and Hamm, F. (1949). Modified silica replica technique. *Anal. Chem.* **21,** 418–419.
Cowley, J. M. (1978). Electron microscopy. *Anal. Chem.* **50,** 88–94R.
Crewe, A. V., and Wall, J. (1970). Field emission STEM of DNA molecules. *J. Mol. Biol.* **48,** 375–388.
Draper, M., and Hodge, A. (1949). Submicroscopic localization of minerals in skeletal muscle by internal "microincineration" within the electron microscope. *Nature* **169,** 576–580.
Ellis, S. (1948). Further researches on the electron microanalyser. *J. Appl. Phys.* **19,** 1191.
Ellisman, M. H., Desoto, G. E., and Martone, M. E. The merger of microscopy and advanced computing: a new frontier for the 21st century. *Proc. Microscopy Society of America* (G. W. Bailey and A. J. Garratt-Reed, Eds.). San Francisco Press.
Fiore, C. (1988). Electron microscopy. *Anal. Chem.* **60,** Review section.
Fischbein, I. (1950). Electron microscopy of wet biological tissues by replica techniques. *J. Appl. Phys.* **21,** 1199–1204.
Fisher, R. M. (1953). Extraction replica technique for electron metallography. *J. Appl. Phys.* **23,** 113 (1952 EMSA Annual Meeting).
Fullam, E., and Gessler, A. (1946). A high speed microtome for the electron microscope. *Rev. Sci. Instrum.* **17,** 23–35.
Germer, L. (1939). Electron diffraction studies of thin films. I. Structure of very thin films. *Phys. Rev.* **66,** 58–71.
Gerould, C. (1947). Preparation and uses of silica replicas in electron microscopy. *J. Appl. Phys.* **18,** 333–343.
Gessler, A., and Fullam, E. (1946). Sectioning for the electron microscope accomplished by the high speed microtome. *Am. J. Anat.* **78,** 245–280.
Grey, C., and Kelsch, J. (1948). Use of the electron microscope and high speed microtome in medicine. *Exp. Med. Surg.* **6,** 368–389.
Gross, J., and Schmitt, F. (1948). The structure of human skin collagen as studied with the electron microscope. *J. Exp. Med.* **88,** 555–568.
Hall, C., and Sehoen, A. (1941). Application of the electron microscope to the study of photographie phenomena. *J. Opt. Soc. Am.* **31,** 281–285.
Hall, C. E. (1948). Dark-field electron microscopy. I. Studies of crystalline substances in dark-field. *J. Appl. Phys.* **19,** 198–212.
Hall, C. E. (1949). Electron microscopy of fibrinogen and fibrin. *J. Biol. Chem.* **179,** 857–864.
Hall, C. E. (1949). Method of measuring spherical aberration of an electron microscope objective. *J. Appl. Phys.* **20,** 631–632.
Hall, C. E. (1950). Low temperature replica method for electron microscopy. *J. Appl. Phys.* **21,** 61–62.
Hall, C. E. (1951). Scattering phenomena in electron microscope image formation.
Hall, C. E., Hauser, E., Le Beau, D., Schmitt, F., and Talalay, P. (1944). Natural and synthetic rubber fibers; Electron microscope studies. *Ind. Eng. Chem.* **36,** 634–640.
Hall, C. E., Jakus, M., and Schmitt, F. (1945). The structure of certain muscle fibrils as revealed by the use of electron stains. *J. Appl. Phys.* **16,** 459–465.
Hall, C. E., Jakus, M., and Schmitt, F. (1946). An investigation of cross striations and myosin filaments in muscle. *Biol. Bull.* **90,** 32–50.
Hamm, F. A. (1951). Electron microscopy. *Anal. Chem.* **23,** 17–20.
Hawn, C., and Porter, K. (1947). The fine structure of clots formed from purified bovine fibrinogen and thrombin: A study with the electron microscope. *J. Exp. Med.* **86,** 285–292.
Heidenreieh, R., and Sturkey, L. (1945). Crystal interference phenomena in electron microscope images. *J. Appl. Phys.* **16,** 97–105.
Heidenreieh, R. (1942). Electron reflections in magnesium oxide crystals with the electron microscope. *Phys. Rev.* **62,** 291–292.
Heidenreich, R., and Peck, V. (1943). Fine structure of metallic surfaces with the electron microscope. *J. Appl. Phys.* **14,** 23–29.

Heidenreich, R. (1943). Interpretation of electron micrographs of silica surface replicas. *J. Appl. Phys.* **14,** 312–320.

Heidenreich, R. (1949). Electron microscope and diffraction study of metal crystal textures by means of thin sections. *J. Appl. Phys.* **20,** 993–1010.

Highberger, J., Gross, J., and Schmitt, F. (1951). The interaction of mucoprotein with soluble collagen; An electron microscope study. *Proc. Natl. Acad. Sci.* **37,** 286–291.

Hillier, J., and Baker, R. (1944). Microanalysis by means of electrons. *J. Appl. Phys.* **16,** 663–675.

Hillier, J., and Gettner, M. (1950). Improved ultra-thin sectioning of tissue for electron microscopy. *J. Appl. Phys.* **21,** 889–895.

Hillier, J., and Vance, A. (1941). Recent developments in the electron microscope. *Proc. Inst. Radio Eng.* **29,** 167–176.

Hillier, J. (1943). On microanalysis by electrons. *Phys. Rev.,* 318–319.

Hillier, J. (1949). Some remarks on the image contrast in electron microscopy and the two-component objective. *J. Bacteriol.* **67,** 313–317.

Hillier, J. (1950). A removable lens for extending the magnification range of an electron microscope. *J. Appl. Phys.* **21,** 785–790.

Hodge, A. (May 1952). Studies on paramyosin: The in vitro reconstruction and transformation of periodic structure. Ph.D. thesis, Biology Department, Massachusetts Institute of Technology, Cambridge, MA.

Kaye, W. (1949). Aluminum-beryllium alloy for substrate and replica preparations in electron microscopy. *J. Appl. Phys.* **20,** 1209–1214.

Kellenberger, E. (1948). New replica method for electron microscopy. *Experientia* **4,** 449.

Latta, H., and Hartmann, J. (1950). Use of a glass edge in thin sectioning for electron microscopy. *Proc. Soc. Exp. Biol. Med.* **74,** 43–439.

Mahla, E., and Nielson, N. (1948). Oxide replica technique for the electron microscope examination of stainless steel and high nickel alloys. *J. Appl. Phys.* **19,** 378–382.

Marton, L. (1944). On electron microspectroscopy. *Phys. Rev.* **66,** 159.

McAlear, J. H. (1972). Electron microscopy. *Anal. Chem.* **44,** 97–127.

Meryman, H. (1950). Replication of frozen liquids by vacuum evaporation. *J. Appl. Phys.* **21,** 68.

Newberry, S. P. (1992). EMSA and its people—the first fifty years. *Electron Microscopy Society of America.*

Newman, S., Borysko, S., and Swerdlow, M. (1949). New sectioning techniques for light and electron microscopy. *Science* **110,** 66–68.

O'Brien, H., Jr., and McKinley, G. (1943). New microtome and sectioning method for electron microscopy. *Science* **98,** 455–456.

Packer, D., and Scott, G. (1942). A cryostat of new design for low temperature tissue dehydration. *J. Tech. Meth. Bull. Int. Assoc. Med. Museums* **22,** 85–96.

Pease, D., and Baker, R. (1948). Sectioning techniques for electron microscopy using a conventional microtome. *Proc. Soc. Exp. Biol. Med.* **67,** 470–474.

Picard, R. G., and Duffendack, O. S. (1943). Studies of the structure of thin metallic films by means of the electron microscope. *J. Appl. Phys.* **14,** 291–305.

Porter, K., and Hawn, C. (1947). The culture of tissue cells in clots formed from purified bovine fibrinogen and thrombin. *Proc. Soc. Exp. Biol. Med.* **66,** 309–314.

Porter, K., and Thompson, H. (1947). Some morphological features of cultured rat sarcoma cells as revealed by the electron microscope. *Cancer Res.* **7,** 431–438.

Porter, K., Claude, A., and Fullam, E. (1945). A study of tissue culture cells by electron microscopy. *J. Exp. Med.,* **81,** 233–246.

Prebus, A. (1942). The electron microscope. *Ohio State Univ. Eng. Exp. Sta. News* **14,** 6–32.

Prebus, A. (1944). The electron microscope. In "Colloid Chemistry" (J. Alexander, Ed.), Vol. 5, pp. 152–235. Reinhold, New York.
Ramberg, E. G. (1942). Variations of axial aberrations of electron lenses with lens strength. J. Appl. Phys. **13,** 582–589.
Ramberg, E. (1948). Phase contrast in electron microscope images. J. Appl. Phys. **19,** 1190.
Reisner, J. H. (1990). Reflections. EMSA Bull. **20,** 49–53.
Richards, A., Anderson, T., and Hance, R. (1942). A microtome sectioning technique for electron microscopy illustrated with sections of striated muscle. Proc. Soc. Exp. Biol. Med. **61,** 148–152.
Schmitt, F., and Gross, J. (1948). Further progress in the electron microscopy of collagen. J. Am. Leather Chemists' Assoc. **43,** 658–675.
Schwartz, C., Austin, A., and Weber, P. (1949). Positive-replica technique for electron microscopy. J. Appl. Phys. **20,** 202–205.
Scott, G., and Packer, D. (1939). The electron microscope as an analytical tool for the localization of minerals in biological tissues. Anat. Record **74,** 17–29.
Schaefer, V., and Harker, D. (1942). Surface replicas for use in the electron microscope. J. Appl. Phys. **13,** 427–433.
Schuster, M., and Fullam, E. (1946). Preparation of powdered materials for electron microscope. Ind. Eng. Chem. **18,** 653–657.
Turkevich, J., and Hillier, J. (1949). Electron microscopy of colloidal systems. Anal. Chem. **21,** 475–485.
Williams, R., and Backus, R. (1949). Macromolecular weights determined by direct particle counting. I. The weight of the bushy stunt virus particle. J. Am. Chem. Soc. **71,** 4052–4057.
Watson, J. (1947). An effect of electron bombardment on carbon black. J. Appl. Phys. **18,** 153–161.
Williams, R., and Wyckoff, R. (1946). Applications of metallic shadow-casting to microscopy. J. Appl. Phys. **17,** 23–33.
Williams, R., and Wyckoff, R. (1945). Electron shadow micrography of virus particles. Proc. Soc. Exp. Biol. Med. **68,** 265–270.
Williams R., and Wyckoff, R. (1944). The thickness of electron microscopic objects. J. Appl. Phys. **16,** 212–715.
Williams, R., and Wyckoff, R. (1945). Electron shadow micrography of the tobacco mosaic virus protein. Science **101,** 594–596.
Williams, R. (1953). A method of freeze-drying for electron microscopy. Exp. Cell Res. **4,** 188–201.
Wyckoff, R. (1946). Frozen-dried preparations for the electron microscope. Science **104,** 36–37.
Wyckoff, R. (1949). "Electron Microscopy." Interscience, New York.
Zworykin, V., Hillier, J., and Vance, A. (1941). A preliminary report on the development of 300 kV magnetic microscope. J. Appl. Phys. **12,** 738–742.
Zworykin, V. K., and Hillier, J. (1943). Compact high resolving power electron microscope. J. Appl. Phys. **14,** 658–673.
Zworykin, V., Morton, G. A., Ramberg, E G., Hillier, J., and Vance, A. W. (1945). "Electron Optics and the Electron Microscope." John Wiley, New York.

Part III

HIGHLIGHTS OF THE IFSEM CONGRESSES

3.1
Biology

BJÖRN A. AFZELIUS

Department of Ultrastructure Research
Stockholm University
S-10691 Stockholm, Sweden

I. Is It Worth Going to International Conferences?

A large international conference is a strenuous and sometimes chaotic event, a stress for the participants and probably a penance for the organizer. So why go to such a big meeting? After having attended over a dozen international meetings on electron microscopy, I can answer without hesitation: There are many good reasons. As a participant you see the latest microscopes and learn about their advantages. You may even try your hands on one. You will be able to see the latest auxilliary equipment, from huge and complex instruments that drink liquid nitrogen (or liquid helium), like calves drinking milk, to small things such as new types of copper grids or "rub-on" micron marks. You will contact colleagues, and maybe a contact will lead to a former competitor becoming a future collaborator. You will meet the pioneers in the field—or, to misuse a well known metaphor—you will have the privilege of sitting side by side with the giants on whose shoulders you stand. You may get in touch with the author of some complex technique and get a first-hand account of the "tricks of the trade." You also have the opportunity to make yourself and your work known to a large audience. You may get a response to your lecture or to your poster that is useful to the progress of your work. You may gain new insights into your own field and are likely to get a broader vista. You have the opportunity to learn new techniques and to update your own methods. You may well make new friends. You will probably have great fun.

II. Personal Recollections

This piece contains personal recollections from the international and European meetings on electron microscopy. Emphasis will be mainly on biological contributions and also in other respects will be colored by my own interests and prejudices.

A. European Regional Meetings (ERMs)

European Regional Meetings have been held regularly every fourth year since 1956, in the same years as the Olympic Games. (No comparison is implied!) The international meetings have been held on the even years between the Olympic years and, from 1950 on, with the same regularity. Some Asia-Pacific Regional Meetings have also been organized—Tokyo in 1956, Calcutta in 1965, Singapore in 1984, Bangkok in 1989—but these meetings were smaller than the European or international meetings. There have also been some Latin American meetings of this kind: Maracaibo, Venezuela, 1972; Sao Paulo, Brazil, 1974; Santiago, Chile, 1976; Mendoza, Argentia, 1978; Bogota, Columbia, 1981; Venezuela, 1984; Barcelona, Spain, 1987; and Havana, Cuba, 1989. The yearly Electron Microscopy Society of America meetings served the same purpose in the United States and Canada.

B. International Meetings[1]

The history of the international meetings really starts with the International Conference on Electron Microscopy in Delft in 1949. Nearly half of the 208 participants came from the Netherlands, the others from 10 European and 3 non-European countries (India, the United States, and Uruguay). No one came from Germany. Forty-nine papers were presented. The proceedings contain photographs of most of the participants, and anyone with a desire to see the young Cosslett, Le Poole, Bernhard, Sjöstrand, or Kellenberger can consult that volume. For those interested in biological ultrastructure, there is less to find. Useful thin sections could not be produced at that time, and metal shadowing of whole-mount preparations was one of the few techniques that gave useful specimens. In a survey paper on specimen preparation techniques, D. G. Drummond reported on the status of ultrathin sectioning and maintained that the "cyclone microtomes" were promising although "the speeds of cut attained so far are hardly high enough for the full theoretical requirement." These microtomes had cutting speeds of up to 1425 m/s (as compared with today's speeds of 0.1–2 mm/s—seven or eight orders of magnitude slower). It must have been a dramatic experience to section with a cyclone microtome!

---

[1] A list of the international conferences is to be found at the end of this volume.

The "First International Congress on Electron Microscopy," as it was then styled, held in Paris in September 1950, is now generally reckoned as the Second International Congress, with the Delft meeting being reckoned as the first. Thomas F. Anderson gave a paper on his ingenious method of critical point drying. He also showed that the bacteriophage attaches to the bacterium with its shaft rather than with its head, as was thought previously.

It was at the Third International Meeting, held in London in 1954, that Ernst Ruska introduced his Siemens Elmiskop I, a model that was to become an outstanding scientific success and a commercial bestseller. About a thousand of these instruments were subsequently sold. As for specimen preparation, large strides had been taken since the preceding congress. Ultramicrotomy, which in 1950 was so imperfect that no really useful sections were obtained, had been enormously improved. Thus Fritiof S. Sjöstrand could produce ultrathin sections on a regular basis and show high-resolution micrographs of the various organelles. Furthermore, he was able to make serial sections and thereby to reconstruct the complex architecture of retinal synapses. Palade lectured on "a small particular component of the cytoplasm," later to be named variously ribonucleoprotein particles, RNP particles, or Palade particles. They are now termed ribosomes.

C. The ERM at Stockholm

The European Regional Meetings (ERMs) got off to an excellent start with the Stockholm conference in 1956, organized by Fritiof S. Sjöstrand and Johannes Rhodin. It attracted 370 participants from 27 countries. Seven well-known participants are shown in Fig. 1. From left to right they are Jan Bart Le Poole, inventor of the wobbler, the first to introduce selected-area electron diffraction mode into a transmission electron microscope, and one of the team that manufactured the first 400-kV microscope; Johannes Rhodin, who was the first person to describe peroxisomes, then called microbodies; Fritiof S. Sjöstrand, the first to take high-resolution micrographs of biological sections and to resolve the mitochondrial membranes, as well as the first to use isosmotic fixatives; V. Ellis Cosslett, the guiding spirit behind the Cambridge University High Resolution Microscope, in which atomic resolution was first obtained routinely; James Hillier, who patented the electron-probe x-ray microanalyzer in 1943 and in 1946 recognized that astigmatism, and not spherical aberration, was the limiting factor in commerical electron microscopes of the day, who devised a "stigmator" to correct it, and who also equipped the ultramicrotome with a water trough; Thomas F. Anderson, inventor of the critical-point drying technique; and Ernst Ruska, the first to design and construct a TEM that surpassed the

FIGURE 1. Seven of the 370 participants at the Stockholm Conference of 1956. Left to right: Jan LePoole, Johannes Rhodin, Fritiof Sjöstrand, Ellis Cosslett, James Hillier, Tom Anderson, and Ernst Ruska.

light microscope in resolving power and who, with Bodo von Borries, invented the iron polepiece—Nobel Prize winner 1986.

The Stockholm ERM was the largest EM meeting held until then. At least three important new techniques were presented (although not described under their present names): section staining (I. R. Gibbons and J. R. G. Bradfield), negative staining (H. E. Huxley), and gelatin encapsulation (V. P. Gilëv). Phase-contrast effects due to defocusing were demonstrated with some of Sjöstrand's high-resolution micrographs in a paper read by F. Lenz. Dislocations in a platinum phthalocyanin crystal seen at a magnification of 1 million times was another major achievement (J. W. Menter). At this conference, 176 papers were read.

D. The Berlin Conference of 1958

The International Meeting in Berlin in 1958 was also memorable for some technical firsts. The LKB company introduced the Ultrotome ultramicrotome; 4000 units were sold over the next 20 years. I. R. Gibbons lectured about a water-soluble embedding medium, the first of its kind, apart from gelatin. He had produced it simply by water extraction of the now commonly used epoxy resin Epon 812; he named his resin Aquon. Also, the third group of embedding substances, the polyesters, was introduced with a paper on Vestopal W by A. Ryter and E. Kellenberger.

At the 1960 meeting in Delft, J. David-Ferreira made use of thorotrast as an electron-dense marker substance and showed that the canaliculi in blood platelets are open to the external environment. H. E. Huxley and G. Zubay showed that ribosomes consist of two parts, a large and a small subunit.

E. Philadelphia, 1962

The International Meeting in Philadelphia 1962 was the first meeting in this series in which the proceedings were distributed to the participants during the meeting (rather than some years later). The organizers had scrutinized over 600 contributions to the congress over a period of a few months before the meeting. A few highlights and "firsts": A high-voltage microscope with an accelerating voltage of 1–1.5 MeV (G. Dupouy and P. Ferrier); an anticontamination device, now usually termed a cold finger (H. G. Heide); negatively stained actin filaments with actin subunits visible (J. Hanson and J. Lowy); and mitochondrial DNA in plants (H. Ris). (Mitochondrial DNA in animals had been documented elsewhere a few months earlier by M. K. Nass and S. Nass.) The first demonstration of the clathrin coat came with a presentation by T. F. Roth and K. R. Porter.

F. The Prague Meeting of 1964

The European Meeting in Prague in 1964 attracted around 500 participants and was one of the first occasions when electron microscopists from the East could meet their colleagues from the West. This was probably the first meeting where R. Weibel lectured about morphometric methods. Immunoelectron microscopy was in its infancy, but J. Baxandall introduced an important improvement, the use of two antibodies for the demonstration of a certain antigenic site: first an unlabeled one prepared in a rabbit, and then a second anti-rabbit antibody labeled with ferritin.

G. Kyoto, 1966

The International Meeting in Kyoto 1966 was larger still, with more than 700 contributed papers. As in the other meetings, new instruments and techniques were introduced. The Philips 200 microscope was shown here for the first time. Strioscopy, a form of dark field microscopy, was demonstrated by G. Dupouy and F. Perrier. D. C. Pease lectured on his concept of inert dehydration. A. Monneron showed that it is possible to perform a simple chemical analysis of proteins in standard ultrathin Epon sections by removing osmium, so as to permit digestion of the proteins with proteolytic enzymes.

H. The ERM at Rome, 1968

The European Meeting in Rome in 1968 attracted 575 contributed papers. R. Valentine gave a tutorial lecture on the symmetries of biological molecules—a paper that even today can be recommeded as an introduction to this important aspect of molecular biology. A new commercial microscope, JEOL's JEM 100B, was introduced by M. Watanabe *et al.*, and an improved specimen fixative consisting of a mixture of osmium tetroxide and potassium ferrocyanide was introduced by W. C. Bruyn.

I. Grenoble, 1970

The European Meeting in Grenoble in 1970 was huge, with over 1000 contributed papers (1140, to be exact) in 12 parallel sessions. It was often difficult to get to the proper lecture at the proper time in the proper lecture hall. Those who managed it got to hear Crewe's lecture on "single atom visibility" in his STEM and will probably remember the atmosphere of amazement and the background of scepticism at Crewe's claim to be showing single atoms! Whether this kind of dedicated scanning transmission electron microscope is better for high resolution than the traditional TEM was something that was discussed long afterward; the debate continues even to the present day. The potential advantages and disadvantages of microscopes with superconducting lenses constructed by, among others, H. Fernández-Morán were also extensively discussed.

J. Manchester, 1972

The organizers of the European Meeting in Manchester in 1972 restricted the presentations to papers concerned with "the electron microscope as an instrument and electron microscopy as a physical technique." As a result, the number of papers presented was below 250, although the number of registered participants was still of the order of 1000. From this meeting, I remember best T. Mulvey's introductory lecture on the early history of electron microscopy, J. Russ' tutorial lecture on the possibilities of energy-dispersive X-ray microanalysis, the new Reichert Ultracut microtome, and, also at the commercial exhibition, a cross-sectioned sperm tail, with all tubulin protofilaments visible, fixed with the tannic acid-containing fixative of V. Mizuhira and Y. Futaesaku.

Many European or international meetings on electron microscopy have been held since 1972 and have, without doubt, been as important and interesting as the early ones. However, the relatively short time separation of these meetings has made it difficult to select presentations that contain new information with lasting influence on biological electron microscopy. Furthermore, the subject areas covered by the meetings has proliferated to such an extent that it is difficult to evaluate the importance of the new methods. Examples of relatively new areas include immunoelectron microscopy, image analysis of periodic biological molecules, investigations on frozen hydrated material, perhaps enclosed in vitreous ice, as well as competing new microscopical techniques at atomic resolution, which are performed with instruments other than conventional electron microscopes.

In addition, since 1976 there have been International Meetings on Cell Biology, which compete with the electron microscopical congresses and which pose a dilemma to those of us who can afford to go to one international meeting per year but not to two or more.

Let me conclude this rehearsal of recollections by highlighting one that is for me unique in more than one respect: a lecture which was never presented and which was made public only two years after the death of its author. Wilhelm Bernhard, who died in 1978, had prepared a lecture for a cancer congress in Buenos Aires and entitled it "Spirituality and Insubstantiality in Science." It appears in the Proceedings of the meeting at the Hague in 1980 and is an appeal to scientists to regard their research not only as an enjoyable adventure but also as an occasion to go forward with humility, respect, and love of creation.

It has been a privilege to enter the field of biological electron microscopy in the period of its infancy and to witness its development and growth. The congress volumes can be regarded as the historical documents of the evolution that electron microscopy has undergone, and as a proof that the development has been both amazingly rapid and has followed unpredicted routes. They also show that we still work in an expanding phase of development. These Proceedings can be regarded both as historical documents of important scientific breakthroughs and as still-useful sources of information.

3.2
Materials Science

JOHN L. HUTCHISON

Department of Materials
University of Oxford
Oxford OX1 3PH
United Kingdom

I. INTRODUCTION

My interest in international meetings in electron microscopy began in the mid-1960s, when, as a final-year chemistry undergraduate, I was introduced to electron microscopy by Dr. Ian Dawson. I can still recall the wonder of seeing particles of carbon black in the Cyclopean viewing screen of Glasgow University Chemistry Department's elderly Philips EM100 instrument, with a resolution around 20 Å, and pondering whether one day we might ever be able to see atoms. As a naïve research student I decided that atomic resolution (particularly of chemically interesting systems) would be a worthy goal, and it was one that I have been privileged to see fulfilled over the past 25 years. Not knowing anyone else in the field of "high resolution" in these early days, I soon found the value of those thick volumes of papers and abstracts from the various International and Regional conferences, and as I worked my way back through them from the 1966 Kyoto meeting to the early meetings in Berlin (1958) and Stockholm (1956), a remarkable global picture of electron microscopy and its development gradually emerged. Names began to accumulate; new instruments came (and some disappeared just as quickly); pet projects were described; kites flown. It was soon clear that here was a worldwide club of enthusiasts from many different disciplines, who met regularly to share a common interest: electron microscopes and their use. For a young student trying to gain an insight into how the field developed and what were currently the key questions, these volumes provided many of the answers, and if not, the people to contact.

In attempting to provide something of an overview of these conferences from a materials scientist's point of view, we can look at several aspects:

1. Commercial developments
2. "Home-built" projects/"large" projects

3. Interpretation of image contrast
4. Holography
5. Applications
6. New developments
7. Workshops

II. Commercial Developments

As makers and suppliers of the instruments on which these conferences were centered, the various manufacturers were evident in two ways: first, at the exhibitions, where on display were the latest instruments and attachments, usually hurriedly assembled and put into working order during the frantic 2 or 3 days before the conference opened! Here was a unique opportunity to meet not only the sales representatives, who were ever eager to strike a deal, but also to gain genuine hands-on experience of the latest instruments, into which you might even be allowed to insert your own specimens! There was also the valuable opportunity of talking to the design engineers, who were ultimately responsible for new developments. Thus the London conference of 1954 saw the introduction of the legendary Siemens Elmiskop I, an instrument which was to remain at the forefront of research for more than a decade. The exhibition at the Stockholm (1956) conference had instruments from no fewer than five manufacturers on display: a calculated risk, bearing in mind the potential loss of future sales if a machine failed to work on the day! Nowadays the exhibitions have become a major focus for the meetings—new instruments are announced, new attachments offered. Indeed, today's conferences depend to a large extent on significant financial support from manufacturers and other commercial exhibitors to ensure overall financial viability.

A second aspect of the manufacturers' input to these meetings was the presentation of papers describing the capabilities of their new instruments. This put flesh onto the dry bones of sales brochures, with the opportunity to hear the authors' replies to adverse criticism from other designers and researchers.

High-voltage electron microscopes (HVEMs) had been developed commercially, first in Japan and then also in the United Kingdom. The Proceedings of a series of HVEM conferences, usually, but not always, organized under the auspices of the IFSEM, document the significant contributions of these instruments, and their applications, particularly in metallurgical research. Sadly, many of these instruments have now succumbed to financial (and operational) pressures, and are no longer in operation. Almost all new high-voltage installations, i.e., in the range above 500 kV, are, with a few notable exceptions, in Japan. Many of the commercial HVEM developments stemmed from "home-built" projects, and again the accounts of these are to be found in the Proceedings of the international and European conferences—see below.

III. "Home-Built," In-House Projects

For me one of the most fascinating "instrumentation" aspects of these conferences and their Proceedings were the reports of "home-built" instruments, pet projects that ultimately met with varying degrees of success. Particularly during the 1960s and 1970s, many such projects were being pursued, some in the quest for ever higher resolution, others having more radical approaches to instrumentation. While space does not permit a comprehensive review of all of them, it is all there in the various Proceedings, which still make fascinating reading. Outstanding examples of the pioneering spirit of that time include the following: the amazing dual-gun electron microscope built in Cambridge by Nixon (popularly known as "two-gun Nixon" at the time) and described at Stockholm (1956), in which the electrons traveling *down* the column were used to form the image, while those traveling *up* the column from the second gun, located close to the image viewing screen, could exploit the demagnifying effects of the imaging lenses to form a small probe for analytical purposes. Successors to this novel instrument were described at later meetings (e.g., Delft, 1960). Other notable examples include the early 1–1.5 MeV instruments, such as that described in the Philadelphia meeting (1962) by Dupouy, a remarkable achievement which probably catalyzed other developments in Japan, the United States, and the United Kingdom, along with several other "one-off" megavolt projects such as the incredibly compact 1 MeV TEM designed and built by Le Poole and his colleagues at the University of Delft. This ingenious instrument employed a Van der Graaf generator, and featured a uniquely folded electron optical column to give an overall height of just over 2.9 m (Grenoble, 1970); the 1–1.5 MeV developments were (logically, but at enormous expense) followed by two much larger projects, 3 MeV instruments in Toulouse and also in Osaka. This latter instrument (built by Hitachi) was described in the 1970 Grenoble meeting. After more than 20 years of use, this instrument is now being replaced by one operating at 3.5 MeV, details of the "renewal project" being given at the 1994 Paris Congress.

Crewe's pioneering field-emission STEM instruments appeared during the late 1960s (Kyoto, 1966; Rome, 1968; Grenoble, 1970). Predictions of atomic resolution appeared later (Manchester, 1972), and these were confirmed in some remarkable—and controversial—images (Canberra, 1974). Opinions at that time were strongly divided on the relative merits of STEM and HVEM for achieving high resolution, and this can be discerned in, for example, some of Cosslett's contributions at these conferences. Crewe and Zeitler's high-voltage STEM in Chicago was again a unique concept, in which the entire lens column was to be suspended from a pressure vessel containing the high-voltage generator and power supplies (Canberra, 1974); Strojnik's horizontally mounted, modular 1-MeV STEM was also a radical departure from conventional design concepts. Sadly, it failed to survive beyond early trials (Manchester, 1972). Siegel's high-coherence, high-stability, 150-kV instrument incorporated a liquid helium-cooled objective lens and specimen area and was designed for UHV operation with minimum contamination and radiation damage of biological specimens (Canberra, 1974). The idea behind it (Kyoto, 1966) was to observe biological molecules directly.

Fernandez-Moran's elaborate superconducting lens instruments and sophisticated installations were, among others, ones that also aroused great interest at the Kyoto (1966), Grenoble (1970), and Canberra (1974) meetings.

The conferences also provided platforms for new ideas on lens design, such as the iron-free minilenses introduced by Le Poole at the Prague conference in 1964. This type of lens was later to find widespread use in conjunction with the condenser-objective lens of Riecke and Ruska (Kyoto, 1966). Mulvey built on Le Poole's early ideas by exploiting the miniexcitation coil approach to produce the single-pole lens of low aberration and a large solid angle for an array of analytical detectors (Canberra, 1974). In the continuing quest for higher resolution, there have been several different approaches, such as Ruska's ultrastable instrument (Canberra, 1974), suspended on ropes from the roof of a tower as a means of reducing vibrations from the nearby U-Bahn. The new, commercially produced high-voltage/high-resolution installations with a point resolution approaching 1 Å, such as has recently been installed in Stuttgart (Paris, 1994), may represent the physical point resolution limit for TEMs with uncorrected objective lenses. But note that this is not the end of the story. Accurate amplitude and phase transfer from the specimen is not possible in the presence of spherical aberration. Until this is achieved, it is not possible *a priori* to identify image structure at high resolution. Furthermore, if spherical aberration were completely corrected, phase contrast would disappear from the TEM image. In an aberration-corrected TEM, electron beam holography would then be essential to retrieve phase contrast. Attempts to design and build aberration correctors continue with unflagging optimism on the part of their designers.

Their incorporation into an operational TEM surely cannot be very far off! However, the advent of objective lenses of remarkably low spherical aberration has to some extent postponed the need for corrected lenses in all but some specialized instrument development projects, where it is necessary to attempt to determine an unknown atomic structure.

Specimen stage development can also be traced back to some of the first conferences: Heated specimen holders and tilting devices were both presented at the London (1954) conference, specimen cooling devices being described two years later at Stockholm. Stage design was initially the territory of a relatively small band of researchers with good design ideas (H. G. Heide, G. Lucas, J. Mills, P. Swann, U. Valdré, P. Ward, and others). Ever more elegant solutions to problems of specimen tilting, heating, cooling, straining, etc., were produced as the needs arose, and the sections devoted to stages and attachments in the Proceedings, particularly during the 1970s, show many examples of great ingenuity. Interestingly, many of these early designs were subsequently developed commercially and are incorporated into the plethora of stages and holders which are now available as standard, "off-the-shelf" items by the microscope manufacturers.

IV. Image Contrast and Its Interpretation

While image contrast from crystalline specimens—both perfect crystals and also defect structures—is now more or less understood, the early Congress Proceedings (London, 1954; Stockholm, 1956) give a flavor of the debate and the different approaches with which problems of contrast interpretation were eventually solved.

Thus we see in the Stockholm meeting of 1956 the first tentative interpretation of diffraction contrast from dislocations in metal foils, at the same time as Menter's famous "lattice image" micrographs of platinum phthalocyanine crystals, also showing dislocations. During the 1960s the Congresses charted the steady improvement in both theoretical underpinning of image interpretation, and also in TEM resolution, with Japanese workers clearly setting the pace for resolution: Hibi with his pointed tungsten filaments (London, 1954), which with much improved spatial coherence later proved capable of producing remarkable "half-spacing" fringes with very fine resolution. Pointed tungsten filaments were to remain the best practical electron sources for high-resolution imaging until the development of the LaB_6 cathodes in the early 1970s (Manchester, 1972); Despite their relatively high costs, LaB_6 emitters are now standard items on most modern TEMs, their long life and improved performances making them cost-effective. In the 1960s, Komoda [and also Dowell (Philadelphia, 1962)] devised the use of tilted illumination to produce one- or two-dimensional lattice images of metal foils (some containing defects). These developments were followed in North American in the work of Parsons *et al.* (Grenoble, 1970) and others, who were analyzing defect structures by direct interpretation of lattice fringes in terms of atomic planes. At the Grenoble Congress a paper by Cockayne *et al.* pointed out that, in general, there need *not* necessarily be a one-to-one correspondence between lattice fringes and atomic planes in the specimen. The development of the "weak-beam" techniques, again by Cockayne *et al.* (Grenoble, 1970; Mancheter, 1972), was regarded at that time as presenting a safer and more rigorous approach to studying defects in thin crystals. On the other hand, a paper by the present author (Manchester, 1972) demonstrated an unambiguous correlation between planar "chemical" faults and lattice fringes in HREM images of a complex niobium oxide. Two years later the underlying theory and computational techniques for image simulation were sufficiently developed to confirm this initially pragmatic approach (Canberra, 1974).

V. Holography

The late Denis Gabor announced to the young IFSEM at the Paris meeting of 1950 his invention of holography; at the same meeting (see the Proceedings, which appeared in 1952), the first electron holograms were presented by Haine and Mulvey, and their optical reconstructions by Wakefield and Dyson. This was only the start. It took many years of work by many people, Hanssen's group at Braunschweig making seminal contributions. Gabor did not live to see his ideas brought to full fruition: it was to take nearly another 40 years before Lichte succeeded in obtaining subatomic resolution by holography (Granada, 1992; Paris, 1994), this being made possible by the development of high-coherence, high-stability FEG electron microscopes. It is now possible to capture the wave leaving the specimen in amplitude and phase, from which the amplitude and phase information in the image can be digitally computed and displayed as a high-resolution image. This is an essential prerequisite for eventually determining unknown crystal structure. Alternative methods of "focal series reconstruction" by Van Dyck and his group at Antwerp, in collaboration with the Philips research team in Eindhoven, are now also highly developed, as shown in the Paris (1994) Proceedings. Results from the Japanese Electron Wavefront Project are also highlighted in this volume, in a series of papers by Tonomura and his team, especially in the field of fundamental research in magnetic materials and quantum physics.

VI. Applications

Almost all areas of solid state were covered in the "applications" sessions of the conferences, although some meetings (such as Manchester, 1972) deliberately concentrated more on the techniques themselves. It is impossible to mention all the applications, but one or two interesting trends may be discerned: first, the ongoing and growing interest in metallurgical problems (these were the first to benefit from the development of the underlying theory for diffraction contrast); the relative sparseness of metallurgical applications in the Stockholm (1956) conference (8 papers) and its predecessors (Paris, 1950, 17 papers; London, 1954, 19 papers) reflected the severe difficulties of preparing thin foils; this lay behind the early attemps to build high (i.e., 300–500 kV) instruments, such as were reported in the Berlin (1958) meeting, but chemical polishing and ion-bombardment techniques capable of producing suitably thin metal foils (and, for biologists, the development of good ultramicrotomes) soon removed that particular driving force and the first "HVEM" prototypes thus passed off the scene. The growth of the field was evident at the Philadelphia meeting (1962), with well over 60 metallurgical papers presented.

Complex oxides began to be studied by HREM in the early 1970s, when instruments with stable tilting stages became available. For more than a decade these materials (as opposed to ceramics such as alumina and zirconia) were of interest only to a relatively small band of microscopists, until the discovery of high-T_c superconducting oxides, which triggered an explosive growth in electron microscopic studies of these and related materials. The Proceedings of the 1990, 1992, and 1994 conferences all contain sections devoted to superconducting oxides, with Volume 4 of the Seattle 1990 congress containing no fewer than 56 papers in that area.

Electron microscopy has made a very significant contribution to mineralogy, and examples of this appeared as early as Berlin (1958) (airborne coal and rock dust) and Delft (1960) (study of the mineral Tobermorite). Early EM studies of diamond (using both TEM and SEM) also appeared at Berlin (1958) and Philadelphia (1962), severe difficulties of specimen thinning having been solved by high-temperature, controlled oxidation of diamond flakes. Perhaps the most dramatic of all the mineralogical EM investigations were those in which lunar material, brough back in 1969 by the *Apollo* astronauts, was carefully studied by teams of electron microscopists. The Grenoble (1970) Congress contained for the first time a session devoted entirely to minerals, and Volume II of the Proceedings includes reports of their first findings—less than a year after those unique samples of "moon-rock" arrived on earth.

It is almost three decades since electron microscopy of carbon black particles was first reported at the Kyoto (1966) conference; for many years subsequently, graphitized carbon blacks were employed mainly as standard resolution test specimens, the 3.4 Å interlayer spacing being a readily achievable target. Microscopists have now returned to this material with renewed enthusiasm since the discovery of carbon fullerenes and nanotubes; at the Paris (1994) meeting, a whole session (over 20 contributions) was devoted entirely to these novel carbon materials. Even more intriguing in this session were papers describing *inorganic* fullerene-type structures, of compounds such as boron nitride or, even more surprisingly, tungsten and molybdenum disulfides. An example of closed-shell structures of tungsten disulfide (presented by the author at ICEM XIII in Paris) is shown in Fig. 1.

FIGURE 1. High-resolution TEM image of a new form of tungsten disulfide (WS_2) taken by the author in a 400-kV TEM (point resolution 0.16 nm). This compound normally occurs in the form of sheets, but here they have curled up to form closed shells. The dark contrast in the concentric shells corresponds to layers of tungsten atoms, seen edge-on. Variations in the relative stacking of successive tungsten layers can also be discerned in such images. This new structure is an unexpected analog of the well-known "buckeyballs" and "buckey-onions."

VII. OTHER DEVELOPMENTS

In the immediate aftermath of World War II, the considerable effort in development of electron microscopy in Germany was not reported at these international conferences, until the London conference of 1954. This was clearly rectified by the time of the Stockholm meeting two years later, with no fewer than 112 German delegates among a total of 370!

Two other large regions were relatively underrepresented at these international conferences: the former Soviet Union and Eastern Europe, and China. Thus there is very little in the Proceedings up to Hamburg (1980) to provide clues of the considerable effort in development and applications of electron optical equipment in either of these regions. Two of the European conferences (Prague, 1964, and Budapest, 1984) gave what were at that time unique opportunities for electron microscopists from all parts to meet and exchange information and ideas. Now that international travel is in some ways easier, these countries are now better represented—although economic factors now appear to have replaced political ones.

VIII. WORKSHOPS AND SATELLITE MEETINGS

A relatively recent development at the Congresses has been the informal "workshop," a session organized at short notice, on a new or controversial topic. Satellite meetings, either before or after the "main event," are now also regular features, being run by local organizers. The Spring School on Lattice Imaging which followed the Canberra (1974) meeting was a first get-together for many HREM enthusiasts sharing new ideas and discoveries. "Buying a New Microscope" (Jerusalem, 1976), "The Impact of Intermediate Voltage Microscopy" (Seattle, 1990), "Relative Merits of EDX and EELS" (Seattle, 1990), and "Reaching 1 Å: Why and How?" (Paris, 1994) were all memorable workshops. Unfortunately, not all the spontaneous contributions (or heated exchanges) at these lively sessions are recorded in the Proceedings! Other "open-lab" workshops (notably at the Paris 1994 Congress) offered hands-on experience with specimen preparation procedures, computer image processing and simulation, etc., with experts on hand to guide the novice.

IX. CONCLUSIONS

Where are we now? At the 1994 Paris meeting we saw convincing evidence that 1 Å point resolution has genuinely been achieved, by more than one route: the "direct route" of high voltage and a low-aberration objective lens (recent installations in Germany and in Japan), and that involving sophisticated reconstruction methods, such as developed by Van Dyck and colleagues in Belgium and the Netherlands and by Lichte and colleagues in Germany. The former will inevitably have problems of specimen damage due to the high-energy electron irradiation, not to mention high instrument costs, while the latter requires elaborate computing and an expensive field-emission gun (FEG) instrument with excellent image recording facilities. Interestingly, both approaches appear to have succeeded at around the same time. With the growing need for simultaneous chemical information on the subnanometer scale and structural information on the atomic scale, it is evident that FEG–TEMs will become increasingly widely used, as researchers try to push the limits of chemical detection and spatial resolution even farther. It is in this area that the current commercial developments in FEG instruments are likely to be especially important, and the combination of high spatial point resolution with a small, coherent probe will allow us to exploit a much improved "information limit" to extract much more information than has hitherto been possible. It was encouraging to see (and try out) new FEG-HREM instruments displayed by some of the major manufacturers at the Paris (1994) Congress. They represent a dramatic achievement on the part of the designers and manufacturers, and will open up many new areas of research.

X. What Has Been Achieved?

Concerning the imaging of atoms, my early hopes have been fully realized. Black dots (or white dots) in a suitable image can in some instances be interpreted in terms of projected atomic columns in a crystal lattice; those particles of carbon black which fascinated a young student have now been shown by direct atomic imaging to contain some of the most intriguing structures (buckeyballs, "buckey-onions," and fullerenes) of carbon so far discovered. Even more surprising is the discovery of metal sulfide equivalents of these structures. Figure 1 shows an example of a fullerene-type structure of WS_2. Examples of these novel materials were presented at the 1994 Paris Congress (Volume 2) by the present author. It was therefore appropriate that André Guinier entitled his introduction to the 1994 Paris volumes, "Peeping at aToms." We can now do what the early pioneers of electron microscopy could only dream about and X-ray crystallographers do only indirectly: we can "see" the internal atomic structure of (optimally thinned and correctly oriented) real, defective crystals. Over the last decade, scanned probe microscopies such as scanning tuneling microscopy (STM) have revealed with stunning clarity "atomic-resolution" images of crystal surfaces. It is appropriate that the emerging "scanned probe (albeit lensless) microscopies" are now included in the conferences; this nicely complements the information about internal crystal structures, giving us fresh insight into what "real" crystals look like on the atomic scale, both inside and on their surfaces.

XI. A Postscript: "Deadlines"

The Proceedings of the conferences up to 1960 were correctly designated as such, since they were records of the papers that had been presented, and for which manuscripts were later collected, edited, and published, more than a year (sometimes two) after the meeting. The Delft (1960) conference started a tradition which has to some extent been maintained: distributing preprints of the majority of the papers before the meeting started. This put a novel pressure on prospective authors, as is evident in the Preface to the full Proceedings of that meeting, which appeared a year after the conference had taken place:

> In order to have the pre-prints ready and sorted out in time (156 out of 245 final articles) the prospective authors were begged to submit their papers not later than June 15th—less than $2\frac{1}{2}$ months before the conference. Quite many of the authors deserve our thanks for keeping to this deadline. In doing so they did enable those colleagues, for whom the date proved rather too early, to submit their manuscripts one to several weeks later.

Times have changed—late or overlong submissions are now liable to be rejected altogether! Since Delft, the *only* Proceedings are collated abstracts (2 or 4 pages), submitted, from 1962 onward, in camera-ready format, several months beforehand and distributed as bound volumes to delegates upon arrival at the conference, a considerable feat for today's large gatherings (the 1994 Paris Proceedings fill no fewer than five thick volumes, with more than 1800 papers). What appears in the Proceedings was thus prepared up to six months prior to the conference—giving an element of risk to the submission of an abstract, when the necessary micrograph or diffraction pattern had perhaps not yet been obtained.

As requested by the Editor, this review is highly subjective; it is based on my own impressions and feelings gained from attending many of the conferences and from reading the Proceedings volumes of the others. There are many omissions; other electron microscopists would probably tell a very different story. Space has not allowed me to discuss the developments in SEM, microanalysis, STEM, image processing, etc., which are also key elements in any IFSEM conference. The solution, of course, is to locate these books and find out for yourself!

Finally, it seems to me that the IFSEM has clearly succeeded in promoting, on a global scale, electron microscopy as an interdisciplinary science, charting its development over more than four decades, and fostering fruitful contacts between electron microscopists in many countries. Long may it continue!

Acknowledgments

I am indebted to Prof. Tom Mulvey for his very helpful comments and insights, without which this article could not have been written, and would also like to thank several colleagues for helping me to locate copies of the early volumes of Proceedings.

Note Added in Proofs. In keeping with the latest developments in information technology, the Proceedings of the next (XI) European Congress on Electron Microscopy (Dublin, 1996) will be distributed to all delegates in the form of a CD-ROM. Printed volumes will be available *only* if specially requested!

3.3
Electron Optics

PETER W. HAWKES

*CEMES/Laboratoire d'Optique Electronique
du Centre National de la Recherche Scientifique
31055 Toulouse, France*

"What one might call 'orthodox electron microscopy' has practically reached the limit of its development three years ago when Hillier and Ramberg compensated the astigmatism of their objective, and to all intents and purposes realized the theoretical resolving power," declared D. Gabor at the Delft conference in 1949. Although the IFSEM did not yet exist, this is an appropriate starting point, for the Proceedings of that meeting, rare though copies may be, mark the beginning of the series of international congresses now held quadrennially under the aegis of the IFSEM. A glance at subsequent proceedings volumes shows that, important though the development of the stigmator was (and the names of Rang and Bertein should not be forgotten here), theoretical and experimental electron optics were by no means at an end in 1949 and, indeed, many of the new developments were first presented at these Congresses and first appeared in print in their volumes of extended abstracts. In the following pages, we draw attention to some of these highlights.

Before leaving that early international meeting in Delft (which was truly international, with participants from as far afield as India, the United States, and Uruguay, as well as many European countries), we should just mention that several papers were devoted to parasitic aberrations, including an introduction to the theoretical foundations by P. A. Sturrock; a first attempt to use an electron phase plate was described; and Gabor considered the practical problems of (electron) holography, proposed by him a year earlier.

That pioneering meeting attracted 209 participants. A year later, more than 600 electron microscopists met in Paris and the Proceedings record 33 papers in the section on electron optics. All the giants of the early days were there, with numerous papers by W. Glaser, three contributions on holography, a study of scattering and contrast by K. Kanaya and K. Kobayashi, and another on energy losses by G. Möllenstedt. R. Castaing discussed probes, G. Liebmann and E. M. Grad published curves of magnetic lens properties, while E. Regenstreif considered the same quantities for electrostatic lenses. There were descriptions of many types of microscope. And we must not forget the contribution by O. Scherzer on aberration correction, Fig. 8 of which represents, if only schematically, "Un microscope sans aberrations sphérique et chromatique." It was not until the 1994 International Conference in Paris that such a microscope appeared to be close to realization. Moreover, this by no means exhausts the list.

Four years later, the first International Congress to be held "under the auspices of the Joint Commission for Electron Microscopy" met in London. The electron optics section contains 12 papers, but if we add those on new electron microscopes and on the attainment of high resolution, this figure increases to 19. It is invidious to single out particular papers for mention, but those by K. Kanaya *et al.* and R. Uyeda on the use of wave theory in image formation, by J. C. Burfoot and G. D. Archard on quadrupoles, and by M.-Y. Bernard and P. Ehinger, by A. Septier, and by W. Lippert on electrostatic lens properties should not be forgotten. The contributions on high resolution are memorable for the long account of the Elmiskop I by E. Ruska, for the attempt by G. Möllenstedt to correct spherical aberration by means of quadrupoles and octopoles, and for the discussion by M. E. Haine and T. Mulvey of "very high resolving power" (1 nm or better).

By 1956, the International Federation of Electron Microscope Societies had not only come into existence but had also defined the pattern of meetings to which we still adhere: international meetings every four years and regional meetings in between. In that year, therefore, Regional Conferences were held in Stockholm and Tokyo. At the former, F. Lenz set out clearly what is meant by asymptotic aberrations, already considered briefly by P. A. Sturrock in his book the year before; K. W. J. Picht examined what today we would call an inverse problem, namely, the calculation of lens fields that have specific imaging characteristics; and P. Durandeau discussed unified lens curves. Many new developments in microscope design were reported, and V. E. Cosslett and P. Duncumb described a scanning microscope capable of giving electron and X-ray signals. Among the papers on electron optics at the first regional conference in Asia and Oceania in Tokyo were discussions of permanent-magnet lenses, a 300-kV microscope equipped with a van de Graaff accelerator, an energy filter, and a new pointed filament, improving Hibi's recent design.

With the Berlin International Congress of 1958, the scale of these meetings increased sharply. The Proceedings now occupied two large and expensive volumes, the physical sciences alone occupying 853 pages. Individual sections were devoted to cathodes, lenses and deflection systems, transmission microscopes, interference microscopy and interferometry, reflection and emission microscopes, and scanning microscopes and other microprobe instruments. The average standard was high, but unlike some other meetings, no one group of papers or theme stands out; so far as electron optics is concerned, Berlin saw the consolidation of many earlier ideas but no major novelties.

In 1960, the European Regional Electron Microscopy Congress returned to Delft, and the Proceedings contain a long section on electron optics and shorter sections on microscope developments and attachments and on scattering. The account by A. P. Wilska of low-voltage microscopy provides a rare opportunity to study the ideas of this pioneer of the use of low voltages.

The international meeting of 1962, held in Philadelphia and thus the first to be convened in the American continent, was noteworthy for the first conference papers by G. Dupouy *et al.* on the 1.5-MV transmission microscope constructed in Toulouse and for the discussion by W. D. Riecke of the practical aspects of condenser-objective imaging. The survey by E. Ruska in which he asks, "What is the theoretical resolution of the electron microscope and when will it be reached?," could be republished today with only a few changes! Two years later, at the Prague meeting, we note the description of the first commercial scanning electron microscope, the Cambridge Instrument Co. Stereoscan, to be marketed a year later, and the first work by F. Thon on the granularity of high-resolution images of amorphous specimens; the optical diffraction pattern is not yet mentioned, but we perceive the germ that led to Thon's experimental demonstration of the contrast transfer function. Attempts to correct spherical aberration, which had occupied a few pages in the preceding Proceedings, continued to be described, and A. D. Dymnikov, T. Ya. Fishkova, and S. Ya. Yavor presented the configuration that we now call the "Russian quadruplet" for the first time at an IFSEM congress. Prague was also regrettably noteworthy for the number of postdeadline papers, the abstracts of which were distributed as a separate booklet; this contains the first account by J. B. Le Poole of his miniature lens, repeated a year later at the Second Regional Conference on Electron Microscopy in the Far East and Oceania, in Calcutta. Also at that meeting, E. Ruska described a project for a transmission microscope with a single-field condenser objective in his invited survey lecture.

In 1966, the International Congress was held in Japan, and several papers introduced ideas that attracted considerable interest and excited much research activity. W. D. Riecke and E. Ruska described "A 100-kV transmission electron microscope with single-field condenser objective," already announced a year earlier, as we have seen. F. Thon showed the optical diffraction patterns of amorphous carbon with the rings corresponding to the transfer function clearly visible. T. Ichinokawa and Y. Kamiya presented the magnetic analog of the Möllenstedt analyzer. R. Castaing, A. El Hili, and L. Henry showed filtered images obtained with the Castaing–Henry analyzer. The Cambridge high-voltage electron microscope was described by K. C. A. Smith, K. Considine, and V. E. Cosslett in a session on high-voltage microscopes that contained 15 contributions, mostly from Japan. R. Lauer and K.-J. Hanszen described an electron-gun model that explained most of the observed behavior of triode guns. For the first time, a section was included on superconducting lenses, with papers by H. Fernández-Morán, S. Ozasa *et al.*, B. M. Siegel *et al.*, and A. Laberrigue *et al.* Finally, among the many other papers that could be mentioned, we must draw attention to "Digital Computer Techniques in Electron Microscopy," in which J. W. Butler introduced the "Butler gun" and foreshadowed the development of the STEM by A. V. Crewe, adumbrated in the section of the Proceedings entitled "Supplements."

The next few years represent a period of intense activity in electron optics, and the space devoted to this subject in the various proceedings volumes reflects this dramatically. In Rome, in 1968, there were sessions on high-voltage electron microscopy, scanning microscopes, mirror microscopes, high resolution, image formation, phase and interference microscopy, lens aberrations, superconducting lenses, instrumentation in general, and energy analyzers. With so many papers to choose from, it is unfair to single out a small number, but the paper by A. V. Crewe, J. Wall, and L. M. Welter announcing the construction of a STEM must be mentioned. Another landmark was the paper by P. Schiske on image reconstruction from focal series, though the effect of noise was not yet fully understood. K.-J. Hanszen surveyed the uses of the optical bench for reconstruction, thereby announcing the themes of his contributions to many future conferences.

By 1970, when the International Congress was held in Grenoble, the conference Proceedings had grown to three volumes, bound in the colorful livery of the Dauphiné, and electron optics is to be found in both Volumes I and II. A number of themes that were to recur at many later conferences appeared here: Atomic resolution had of course been discussed earlier, but was now seriously envisaged by W. D. Riecke. Zone plates and phase plates too were not new, but they became a major topic at Grenoble with papers by F. Thon, with D. Willasch and B. M. Siegel, and by L. Reimer and H. G. Badde, among others. The papers by J. Frank *et al.*, R. Langer *et al.*, and W. Hoppe *et al.*, all from the Munich laboratory of W. Hoppe, launched many years of research on digital image processing. H. Boersch and B. Lischke measured the quantized magnetic flux trapped in superconducting tubes by means of the effect discovered by Ehrenberg and Siday that is known as the Aharonov–Bohm effect. Volume II has long sections on aberration theory, with a subsection on multipole systems, on guns and filaments, on superconducting lenses, and on energy analyzers. A paper by

E. Munro on the "Computer-Aided Design of Magnetic Electron Lenses Using the Finite-Element Method" marks the beginning of an activity still in lively development. Papers by H. Ahmed and A. N. Broers and by K. H. Loeffler introduced the LaB_6 cathode. I. Dietrich and R. Weyl described the shielding-lens design of superconducting lens for the first time at such a conference.

At Manchester in 1972, many of the themes already evoked were consolidated and gradual progress was announced. An ingenious attempt to correct spherical aberration by means of a ring-shaped beam was described by W. Kunath, and R. W. Moses established the best field shape for coma-free magnetic lenses by means of the calculus of variations. The Manchester Proceedings occupy only a single volume, but the full papers delivered in the specialist symposia on "Applications of Image Processing" and "Computer-Aided Design of Electron Optical Systems" were published separately (Hawkes, 1973). This volume contains a long account of "Contrast Transfer and Image Processing" by K.-J Hanszen, a description of the Gerchberg–Saxton algorithm by its authors, the first announcement by P. Schiske of a Wiener filter for focal series, and a survey of reconstruction procedures by W. Hoppe *et al.* In the CAD section are full papers by E. Munro, by R. W. Moses, and by H. Rose and E. Plies, who designed a system with a curved optic axis free of certain aberrations. The optics of high-frequency lenses, already considered by N. C. Vaidya at Grenoble, was pursued more fully by L. C. Oldfield.

The 1970s and their European and International Congresses were the decade during which digital image processing, for three-dimensional reconstruction and for overcoming the phase problem in particular, acquired major importance. Contrast-transfer theory was extended, to include the effects of partial spatial and temporal coherence and to include unconventional modes of illumination. Superconducting and various other lenses of unusual shape—pancake and snorkel lenses—were studied. Field-emission guns became better understood. All these themes are to be found in the Proceedings of the meetings in 1974 (Canberra), 1976 (Jerusalem), and 1978 (Toronto). At the latter, an important new development was reported, namely, off-axis holography using a microscope equipped with a field-emission source and a Möllenstedt-Düker biprism (A. Tonomura, T. Matsuda, and T. Komoda).

At the Hague in 1980, the preoccupations of the new decade were anticipated in the survey papers. Thus E. Kasper and F. Lenz reviewed "Numerical Methods in Geometrical Electron Optics," destined to attract much research with the widespread availability of large, fast computers and, in the following decade, of extremely powerful PCs. W. O. Saxton provided "A Survey of Motivations and Methods" in "Digital Processing of Electron

Images." Papers by A. Tonomura *et al.* showed that holography was useful for real specimens and demonstrated the possibility of phase amplification by creating interference between opposite side bands in off-axis holography. More unconventional magnetic lenses were described by T. Mulvey *et al.* At Hamburg in 1982, there were many more papers on these topics, and several sessions on image processing, in which J. Frank and M. van Heel described the use of correspondence analysis for classification. B. Lencová and M. Lenc proposed a way of applying the finite-element method to very open structures without making unreasonable assumptions about the boundary conditions, a problem to which they returned at the 1984 meeting in Budapest. At Budapest, too, W. O. Saxton and W. M. Stobbs revived the old idea of bright-field/dark-field subtraction for phase determination, and on-line image processing was clearly a growing area.

Also in 1984, the Asia-Pacific Regional Conferences were resumed, the third such meeting being held in Singapore. This was devoted largely to applications, though S. Maruse gave an invited survey of high-voltage microscope development and Y. Kondo *et al.* described applications of hollow-cone illumination. After this, Asia–Pacific regional meetings have been held at four-year intervals, and hence coincide with European Regional Congresses.

The International Congress of 1986 met in Kyoto and, in a plenary lecture, A. Tonomura surveyed the achievements of electron holography, including an irrefutable experimental proof of the reality of the Aharonov–Bohm effect, a subject of vigorous controversy over the years. N. Mori *et al.* introduced the Fuji imaging plate, and advances in the familiar themes of electron optics continued to be reported. A short note on the spherical aberration of a combined electrostatic and magnetic lens by K. Yada is noteworthy. Automatic microscope control took a step forward with the proposal by A. J. Koster *et al.* to exploit the beam-tilt-induced image shift for automatic focusing. This was extended at the European Regional Meeting of 1988, held in York. There, too, R. Degenhardt and H. Rose spelled out clearly for the first time the relations between system symmetry (and antisymmetry) and aberrations, relations that have led to the design of analyzers with very desirable properties. The construction of multichannel STEM detectors was reported by M. Haider *et al.*, a step toward the STEM image processing technique subsequently devised by J. Rodenburg and R. H. T. Bates.

Electron holography in a STEM equipped with a biprism was reported by T. Leuthner *et al.*, and digital rather than optical reconstruction in holography gained ground (G. Ade). The Asia–Pacific meeting the same year was held in Bangkok, and several papers on electron optics were presented, notably a survey by K. Yada of "Electron Optical Elements and

Image Processing for High Resolution Works" and a discussion of statistical electron noise in STEM and its reduction.

The organizers of the International Congress in Seattle in 1990 adopted a novel classification of the papers presented, with the result that "Electron Optics" no longer appeared as a separate heading and the articles that might have appeared there were scattered among "High-Resolution TEM," "Evolutionary Developments in EM Instrumentation," "Revolutionary Developments in EM Instrumentation," "Advances in SEM Instrumentation," and several sections on various aspects of image processing. The symbiosis of microscope and computer was now complete, with several contributions on microscope control and many on on-line image processing. In holography, T. Matsuda *et al.* showed individual fluxons in a superconductor, while in interference, F. Hasselbach and M. Niklaus reported observation of the electron equivalent of the Sagnac effect.

In 1992, two regional conferences were again held, in Granada and in Beijing. At the former, J. M. Rodenburg and B. C. McCallum explained how phase can be determined directly in the STEM if the current distribution in the detector plane is recorded for every object-element. Both geometric electron optics and image processing inspired by wave optics figure in the Beijing Proceedings.

We thus arrive at the most recent international meeting, held in Paris in 1994. This marked a real revival of electron optics, for some 400 pages of the volume on interdisciplinary developments and tools are devoted to one or other aspect of this topic. Miniaturized optical elements, the extension of the focal-depth extension method to hollow-cone illumination, microscope control for very high resolution, highly corrected imaging filters, comb probes, bigger imaging plates, new developments in holography, and standing-wave illumination, all generated papers of high interest. Aberration correction was at last successful, for J. Zach and M. Haider described a low-voltage SEM equipped with a working corrector. Moreover, the correction of a high-resolution TEM, although not yet proven, seemed to be imminent.

The Proceedings of these international and regional conferences, held under the auspices of the IFSEM for many years now, thus provide a fascinating record of electron optical progress over the past 45 years, from the slender and rather rare volume produced after the 1949 Delft meeting to the five heavy volumes generated by the Paris conference of 1994. Some themes—such as the study of the properties of guns and lenses—are perennial, though the arrival of the computer revolutionized the methods of studying them. Others—digital image processing and autotuning—were not even dreamed of at Delft. It is a happy chance that holography, which occupied an important place at Paris, was invented just in time to be mentioned at Delft.

What can we conclude from all this? That the international and regional congresses have been regarded as sufficiently important and representative of the electron microscope community for electron opticians to be sure to present their new ideas there, and often to announce them. In the foregoing account I have emphasized papers in which new material was presented for the first time, but occasionally, of course, the two-year interval was too long and the announcement was made in print or at a national meeting. Even then, a further announcement has often appeared in the ICEM or EUREM Proceedings too. Future historians of the subject will surely use these volumes as hilltop beacons, before turning to the richer but less well illuminated material to be found in journals and compendia.

REFERENCES

Hawkes, P. W., Ed. (1973). "Image processing and computer-aided design in electron optics." Academic Press, London, New York.

Part IV

INSTRUMENTAL DEVELOPMENTS

4.1
The Story of European Commerical Electron Microscopes

ALAN W. AGAR[1]

Agar Scientific, Ltd., Stanstead, United Kingdom

I. Introduction

The early history of electron microscopes, from the point of view of scientific discoveries and developments, has been extensively cataloged (Ruska, 1980; Hawkes, 1985). There is a good review of developments in Japan and of the Japanese electron microscope companies (Fujita, 1986), which also includes sections on the Philips and Zeiss commercial developments. Reisner (1989) has recorded the early history of microscopes in the United States, and Doane *et al.* (1993) the story of the remarkable early work in Canada. The history of electron microscopy in Switzerland (Günter, 1990) has also been brought together in a book. The progress of European commercial microscopes generally has not, however, been fully documented, and this is an attempt to remedy that omission. It is believed that the account is complete in broad detail, but not all the relevant work has been published in accessible journals, and some of the statements are based on various personal recollections which, like my own, are now blurred by the passage of time. I therefore apologize in advance to anyone who feels that some of his work is inadequately acknowledged.

This record sets out to be historically correct as far as information could be obtained and verified. However, it is not perhaps a true history in the strictest sense in that I was personally involved in the affairs of both Metropolitan-Vickers (later AEI) and also Siemens and Halske GmbH. The balance of comments about some design and policy decisions must inevitably be biased toward the affairs that I knew of personally, but to have omitted them, as well as some personal reminiscences, would, I feel, have detracted from the general interest of the account.

It should be made clear at the outset that this review is confined to commercial instruments offered for sale (not prototypes, except for one or two exceptions which are identified). Any dates mentioned refer to the date when a particular intrument type was first delivered, and any resolution quoted refers to the manufacturer's guaranteed figure for point-to-point

---

[1] Present address: Hinds Cottage, Sproxton, Helmsley (Yorks), YO6 5EF, United Kingdom.

resolution at the date of introduction, even though it may later have been improved. In order to avoid a mere catalog of instruments, the main text attempts to feature the important developments, relating them to any relevant research papers, and to the progression of techniques of specimen preparation and of instrument operation. It would have been impracticable to have included a picture of every instrument made. Apart from including those which were most important historically, the selection has been biased in favor of those instruments which are not widely known or extensively illustrated.

Since it is just 60 years[2] since Ruska announced the construction of the first electron microscope, it seemed convenient to divide the period into six decades and to treat the different types of instruments in separate sections within each decade. Particularly in the later years, no attempt has been made to mention all the instruments appearing, but a listing of all the European manufacturers, together with a list of the microscopes they produced, appears as Appendix I. In order to present a smooth chronology of events, a listing of all the instruments by date of first delivery, and with brief details of their main features, is set out in Appendix II. Appendix III summarizes the total production numbers of the main instrument types. A glossary of abbreviations used is included immediately before the references.

II. 1931–1940, THE FIRST DECADE

This period starts with the first electron microscope, with a magnification of 17, which was designed and constructed by Ernst Ruska and Max Knoll in 1931. By 1933, Ruska, then working alone, had constructed an instrument comfortably exceeding the resolution of a light microscope. Intensive activity over this period proceeded in three groups in Germany: Knoll, Ruska, and von Borries (see Ruska, 1980); von Ardenne (von Ardenne, 1941, 1985) and Brüche, Johannson, and Scherzer (Ramsauer, 1943). A broad survey of the early developments in Germany and in the other countries where excitement over the new possibilities arose is given by Mulvey (1962). This surge of interest resulted in the first commercial order for an electron microscope in 1935. It was ordered by Prof. Martin of Imperial College, London, from the Metropolitan-Vickers Electrical Company in Manchester. It was designed to enable a specimen to be compared successively by light and electron optics. However, it did not surpass the resolution of the light microscope.

[2] *Editor's note:* This statement was true at the time at which most of this text was written. Publication has been delayed while waiting for other contributions to the book. It was felt, however, that the history should not attempt to include the instruments launched in the subsequent years, and the listing is discontinued after 1991.

The involvement of the Siemens Company in electron microscopy started early in 1937. It is interesting that the commercial backing for this development was to some extent influenced by the positive results obtained on biological material by Marton (1934) on an instrument that he built himself in Belgium. Marton's contribution to electron microscopy has been summarized by Susskind (1985). These results were seized upon by Helmut Ruska, who felt strongly that the potential of electron microscopy for medical research should be pursued. He asked his professor (Prof. Siebeck) to write a strong recommendation that electron microscopes should be produced, because of their potential importance in medicine, and this was finally successful. For several years, Ruska and von Borries (Fig. 1) had been trying to persuade the Siemens Company and Carl Zeiss to start building electron microscopes, without success. Following Prof. Siebeck's intervention, both companies finally agreed. Ruska and von Borries decided that Siemens was more involved in electrotechnical expertise, which would be important for the building of the very stable supplies required by the

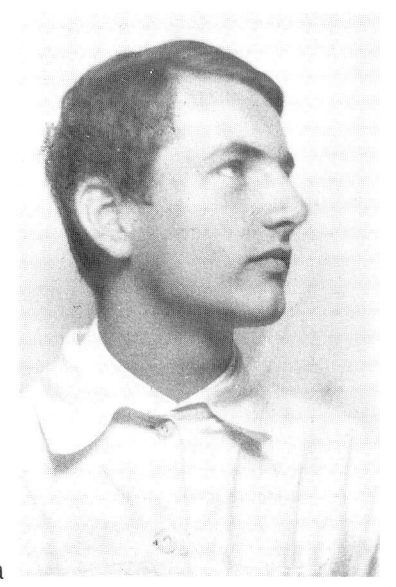

FIGURE 1. (a) Ernst Ruska, 1906–1988. Photographed 1928. Developed the first electron microscope. Jointly with von Borries, developed the world's first serially produced electron microscope. Awarded the Nobel Prize for his contribution, 1986. (b) Bodo von Borries, 1905–1956. Photographed 1931. Joint developer of the ÜM and active in forming first the German Electron Microscope Society, and later the International Federation of Electron Microscope Societies. (Reprinted with permission from Ruska, E. (1980).)

instrument. They therefore agreed to join Siemens, and they were given full control of the project. They in turn recruited H. O. Müller and Helmut Ruska, and a few young co-workers, to the team. They also secured the assistance of W. Glaser as advisor. In the spring of that year, 1937, they started the new laboratory in a disused building in Spandau. It should be noted that from the end of 1933 until this time, neither Ruska nor von Borries had been able to do any practical work on electron microscopy because of lack of any funding.

They made several very important decisions at this stage. Perhaps the most important for the future of electron microscopy was the decision to run applications work in parallel with the development program, and eventually to provide a separate instrument devoted to use in testing the range of possible applications in different fields of science for the new instrument. This work was put under the direction of Helmut Ruska, the brother of Ernst, who was a medical doctor and who had been instrumental in securing the very positive assessment of Prof. Siebeck which helped to start the project. He was assisted in the first instance by C. Wolpers and G. A. Kausche. Wolpers (1993) recalls that, through 1938, the applications team had access to the microscope only between the hours of 8 p.m. and 6 a.m.! The rest of the time, it was in use for instrument development.

The technical decisions made were hardly less significant. One was to construct the whole microscope column in steel, rather than incorporating brass connecting sections for the necessarily long distances between the lenses. This gave much better protection against stray magnetic fields. They decided from the first that both specimen and camera airlocks were required for reasonable convenience of working. Following the 1934 design, the column was vertical, as opposed to the horizontal arrangement favored by some other workers. The other critical decision was that the length of the imaging section of the microscope should be minimized, so as to keep the screen at a convenient height for the operator, without placing the upper alignment controls too high to reach comfortably. This seems to have been a continuation of an idea by Knoll and Ruska (1932) that the focal length of the magnetic lens must be kept short if an unacceptable length of the electron microscope was to be avoided. This criterion, together with a desirable maximum magnification set at about 25,000 meant that the focal lengths of both objective and projector lenses had to be small. The first prototype had an objective focal length of 5.4 mm and a projector lens with the remarkable focal length of only 1.1 mm. These figures resulted in a distance of 800 mm for the imaging length of the microscope, and a practical resolution of 13 nm (Ruska, 1980). It should be noted that in this paper, and the succeeding ones to be quoted, the short objective focal length was achieved specifically as a result of the requirement for a shorter column, while retaining a high top magnification.

The first microscope was completed by 2 December 1937 (H. von Borries, 1991). The initial design had a cold-cathode electron gun, shown in cross-section diagram, Fig. 2 [Fig. 36 in Ruska's (1980) book]. Since they had started with an empty building, had to equip the workshop before starting

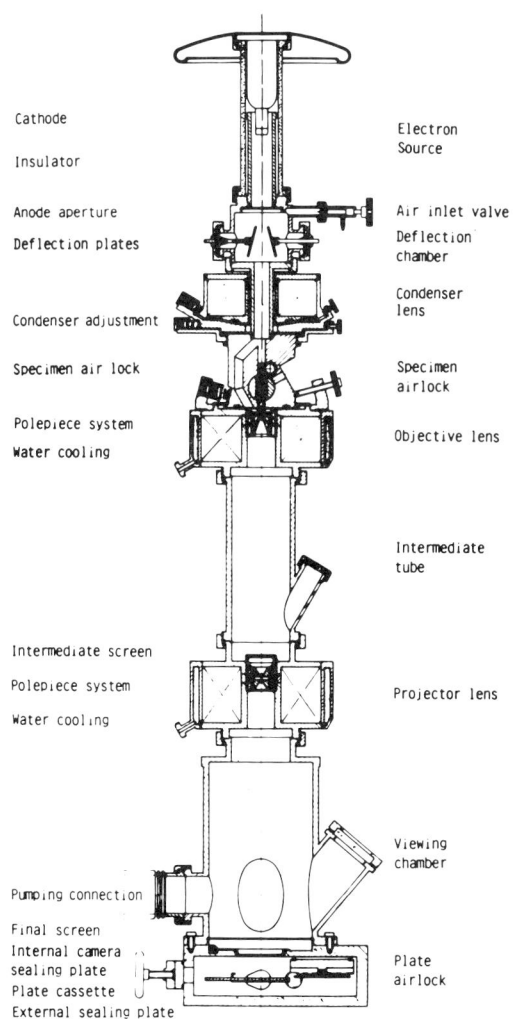

FIGURE 2. The first model of the Siemens experimental prototype, drawn in cross section, showing a cold-cathode gun. (Reprinted with permission from Ruska, E. (1980).)

any construction, and according to von Borries (H. von Borries, 1991), there were no machinists available for work on the electron microscope as late as June 1937, progress had indeed been rapid.

A paper by von Borries and Ruska (1938a), which was received for publication on 25 February 1938, described the instrument and showed a number of micrographs of bacteria, which were said to have been recorded using a hot-cathode gun. The original cold-cathode gun had therefore been replaced very soon after tests were started. The authors had included in the design a beam-blanking system above the condenser lens to control the amount of irradiation of the specimen. They pointed out the great saving in irradiation of the specimen by making full use of the superior resolution of the photographic plate compared to the eye (a factor of 10), and keeping the electron optical magnification low. They also remarked that in order to focus satisfactorily, a magnifier was needed for viewing the image screen. Measurements taken on two of the micrographs in this paper led to a claim of a 10 nm resolution. A footnote records that since submitting the paper, it had been found possible to reduce the focal length of the objective lens to 2.8 mm. Furthermore, a companion paper submitted on the same date, by von Borries, E. Ruska, and H. Ruska (1938), included micrographs of eight different bacteria, which in those days might well have taken some considerable time to prepare and photograph. This represented remarkable progress in two-and-a-half months.

By September 1938, a paper presented in the *Transactions of the Society for Natural Science and Arts* (von Borries and Ruska, 1938b) showed that the instrument appeared substantially unchanged from that shown earlier, except for the thermionic cathode (Fig. 3). The new paper included a number of micrographs of different specimens from a range of applications, and cited four papers which had been published, describing these results.

The applications work had in fact proceeded so well that the results obtained on some synthetic fibers, dyestuffs, and pigments had convinced I. G. Farben-Werke to order four microscopes for research work at four of their works. This order was received by Siemens at the end of 1938, before I. G. Farben had even seen a production prototype. This caused Siemens to uprate the first production batch to 12, even at this very early stage in the program.

Probably toward the end of 1938 (no exact date has been established), the experimental instrument had been further refined to include the "bathtub" around the electron gun, which also held the battery supplies for the bias and filament heating (Fig. 37 in Ruska's book). However, it was still mounted on an open metal stand, and it had the lens controls on a separate control box alongside the column. This was the production prototype (Fig. 4).

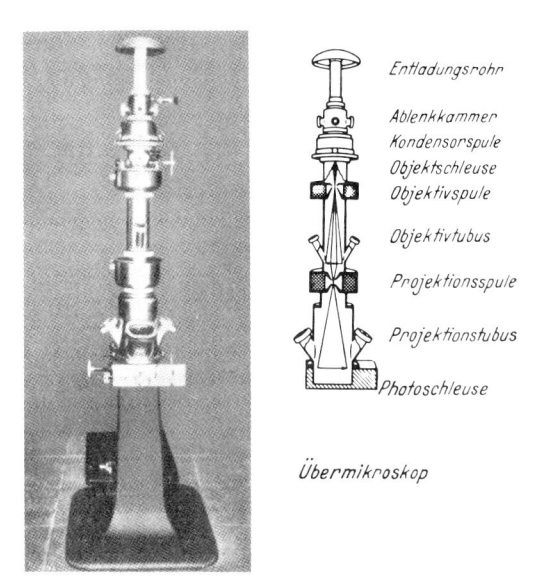

EUROPEAN COMMERCIAL ELECTRON MICROSCOPES   421

FIGURE 3. Siemens experimental prototype, February 1938, modified with a thermionic cathode.

Comparison with Fig. 3 will show that most of the column layout is unchanged at this stage. Early in 1939, the first instrument devoted exclusively to applications work was delivered, and it seems probable that it was this production prototype.

Further developments proceeded at an incredible speed, as shown by a paper submitted for publication in April 1939 (von Borries and Ruska, 1939a). The work done by the applications team was said to have shown the need for the microscope to be more compact, more reliable, and simpler to operate. These needs were addressed in the redesign, which resulted in the actual production instrument which was illustrated in this paper. As an example of improved simplicity of operation, they mention that the earlier design had a specimen movement driven by two push rods at 120° to one another, working against a return spring. The difficulties of steering the specimen with such an arrangement can be imagined! The new instrument had acquired a proper desk, which also incorporated the lens controls, and all the vacuum equipment as well. A viewing telescope was incorporated for enlarging the image on the final screen by a factor of 4. It had a multiplate airlocked camera. Furthermore, it had been found possible to introduce the specimen through the airlock even closer to the objective lens, and thereby to decrease its focal length to 3.8 mm for normal working [foreshadowed in their earlier paper (1938a)]. This gave a final magnification of 20,000 to be achieved with an imaging length reduced to

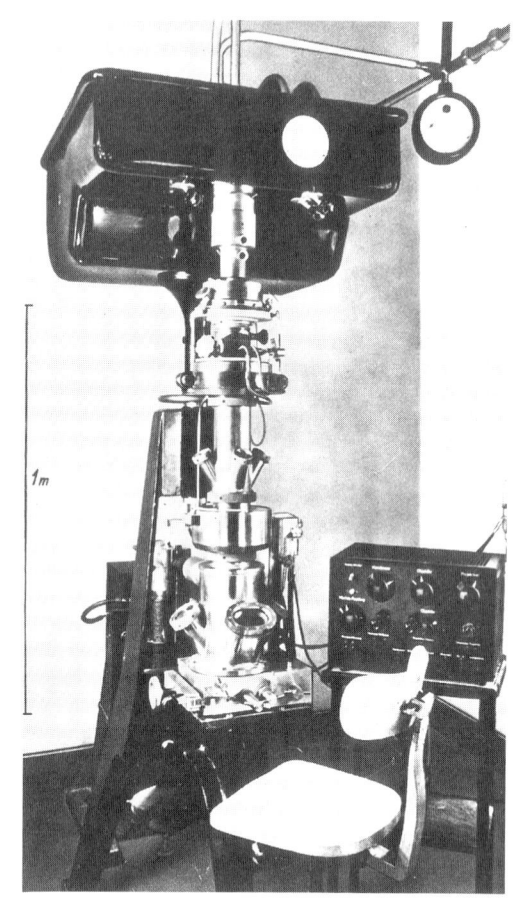

FIGURE 4. Siemens production prototype of the ÜM, showing "bathtub" around the electron gun. The imaging length is still 800 mm. Late 1938. (Reprinted with permission from Ruska, E. (1980).)

580 mm. By working at a specially reduced objective focal length of 2.5 mm, the top magnification could be increased to 30,000 without changing the length of the imaging system, and the resolution was improved to 10 nm. The authors make the interesting comment that the lens was designed so that the polepieces might be easily interchanged, should it be found possible in the future, through a further development of the polepiece system, to achieve a higher resolution through a reduction in lens aberrations. This statement suggests that they did not know at that time what geometric factors in the lens polepiece design affected the aberrations. More details of the instrument are given in von Borries and Ruska (1939b).

This microscope was called the Übermikroskop (ÜM) (Fig. 5), and it initially had an accelerating voltage of 70 kV, and a guaranteed resolution of 7 nm. In a paper submitted for publication on 6 June 1939, a resolution of 7 nm had been achieved with this lens (von Borries and Ruska, 1939c), and in a footnote added in proof, it is recorded that a particle separation of 5 nm had been achieved by H. O. Müller on 12 July 1939. The column components were joined by greased cone joints, providing vacuum sealing and mechanical alignment. It was a well-engineered instrument; the fact that only six years after the first experimental model had just exceeded the resolution of the light microscope, this first production instrument should have improved on that resolution by a factor of 100 was a stunning tribute to this young team of Siemens engineers. It is remarkable how many features of present-day microscopes were already to be found on this first production

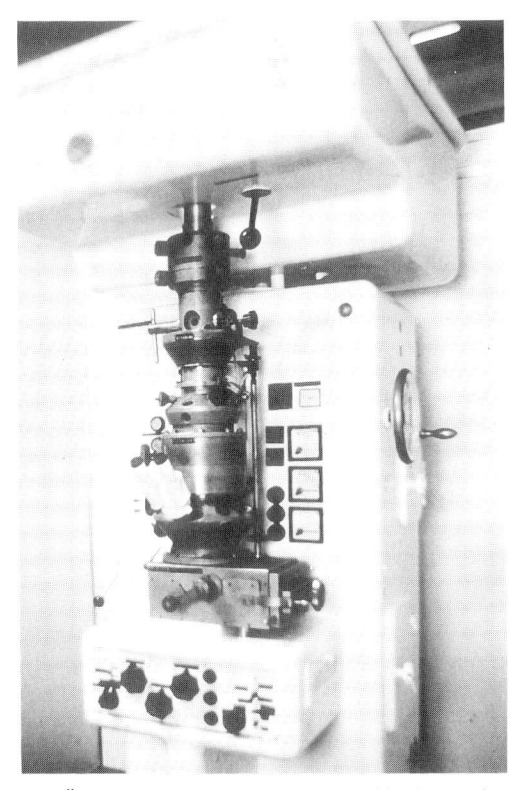

FIGURE 5. Siemens ÜM electron microscope, the world's first series production instrument. First production of polepiece lenses. Initially worked up to 70 kV, resolution 7 nm. Equipped with desk, short-focal-length objective, specimen and camera airlocks, and viewing telescope.

model, and the features of this microscope are due to the foresight in devoting intensive effort to applications work.

The paper (von Borries and Ruska, 1939a) shows that the production instrument existed as early as April 1939; it had in fact been exhibited at the Leipzig Trade Fair in March 1939 (H. von Borries, 1991). This information seems to be at variance with the previously generally accepted date of late 1939 for the delivery of the first production-line electron microscope to the Hoechst laboratories of I. G. Farben. It is now apparent (Kehler, 1993) that the first three instruments from the production line were all delivered to the customer laboratory of the Siemens Company in Berlin. The instrument shipped to Hoechst was serial no. 4. It remains, of course, the first serially produced microscope shipped to a customer.

A booklet published to celebrate the formal opening of the Electron Microscope Laboratory at Siemens & Halske in Berlin in April 1940 gathers a remarkable collection of electron micrographs already recorded from the fields of biology, botany, and chemistry and already employing stereomicroscopy and reflection microscopy for the examination of surfaces (Siemens and Halske, 1941). In the introductory chapter is reproduced the layout of the Siemens applications laboratory, of no less than 30 rooms. It was operational from April 1940, with three production ÜM (confirming the information from Kehler) and an experimental instrument, and equipped with a range of preparation laboratories for different disciplines. This very substantial effort devoted to applications demonstrates the enormous commitment made by Siemens to the rapid exploitation of the new microscopes. The book also includes a micrograph obtained by von Borries and Ruska (1940) in which a resolution of 2.5 nm is demonstrated—fantastic progress in under three years of work. Over 3000 micrographs had been recorded by the applications team by the end of 1939, and some 25 papers on different applications had been published.

It is fascinating to record that in all the publications concerning the development of the instrument, there is no mention of the enormous technical advantage of the short focal length of the objective in bringing corresponding improvements in the spherical and chromatic aberrations, although it seems to have been Ruska's idea from his first experiments that a very short field length was a good thing. In fact, in a paper published in June 1939, von Borries and Ruska (1939c) do detail a theory of the effects of aberrations on resolution, which they found yielded results in agreement with their experimental measurements of resolution. They showed that they were familiar with the quadratic sum of the spherical, chromatic, and diffraction aberrations as an indication of the resultant resolution limit. But they concluded in the paper that the most important limit to resolution lay in the spherical aberration, and that the diffraction and chromatic aberrations were relatively unimportant. In fact, at that time they could not have measured the effect of spherical aberration accurately, as the illumination aperture actually used could not be well defined. This analysis of the order of importance of the aberrations is correct, as far as it goes, but it perhaps illustrates a mentality found in other researchers on electron microscopes—they calculated the expected resolving power of the instrument entirely in terms of the aberration coefficients, but neglecting the effect of the specimen. So, while they were arithmetically correct in choosing spherical aberrations as the major defect, and the conclusions would be valid for very small particles on an extremely thin support film, von Ardenne (1938b) was more practical in realizing that with the relatively thick specimens they used, the chromatic defect was more important. (His analysis was criticized by von Borries and Ruska.)

There was no correlation in their paper between the size of the aberration coefficients and the focal length of the lens, although, on the other hand, they clearly recognized the value of increased accelerating voltage in increased gun brightness and increased screen efficiency. They noted that this also improved the energy loss in the specimen (always referred to in this paper as loss in speed of electrons) relative to the accelerating voltage. If one examines their micrographs, some of them seem to indicate a directional structure, which suggests that astigmatism might be present. On the other hand, it could equally have been due to vibration or supply instabilities. In addition, there was a very poor definition of the illuminating beam. In this respect von Ardenne seems to have had a much clearer idea of the sources of resolution limitation. His paper (von Ardenne, 1938a), gives a good account of the various aberrations, though he, too, ignores astigmatism. In his book (von Ardenne, 1940), he linked for the first time the focal length of the objective with the magnitude of the aberrations. Von Ardenne deduced that multiple scattering of the electrons started for film thicknesses of greater than 0.1 μm, and suggested a loss of 200 eV in beam energy for 50-kV electrons traversing a 1 μm section.

It is interesting to speculate why astigmatism was not detected at this time. It could well be due to the very poor quality of the illumination system, but it may also be a tribute to the high skills of the machinists of the polepieces. Their task was made more difficult by the very small dimensions of the short-focal-length lenses, and they would have had no idea of the exacting tolerances actually required for an astigmatism-free lens. The fact that neither Ruska nor von Ardenne apparently noticed the presence of astigmatic focal lines while focusing certainly suggests that the machining of the polepieces was of a very high standard.

It appears, therefore, that the short-focal-length lenses introduced for the excellent reason of making the instrument more user-friendly (a word not then dreamed of!) were not at that time connected by the authors with the prime reason (lower lens aberrations) for the superior performance of the Siemens microscopes over those of all other manufacturers for the next 15 years.

The Allgemeine Elektrizitäts Gesellschaft (AEG) started experimental work on electrostatic electron microscopes at much the same time as Siemens. After a preliminary model in 1936, it was further developed by Mahl (Fig. 6), and completed in 1939 (Mahl, 1939). This was engineered into a production instrument, the EM 5, by Golz and Schliebe, the first model of which appeared in 1940. It had an objective and a double projector lens, with the voltage switched to one lens or the other to provide magnifications of 1000, 3000, and 10,000. By switching off both the lenses, a diffraction pattern could be obtained from a specimen located below the projectors. It was equipped with airlocks for the specimen and camera, and, incred-

FIGURE 6. Hans Mahl, 1909–1988. Photographed 1936. Pioneer of the electrostatic electron microscope of AEG. He also developed the important plastic replica technique for surface examinations. (Reprinted with permission from Ruska, E. (1980).)

ibly for so early a date, a 24-plate camera. It used a mains frequency transformer to generate the high voltage, as was used in X-ray machines at that time. Focusing was by physical movement of the specimen in the z direction. It had an optional tilting specimen holder for stereo working. A book by Ramsauer (1943) gives details of the developments leading up to this instrument. The book also contains an impressive collection of electron micrographs covering a wide variety of disciplines. They were of very high quality, considering their early date; some show a resolution of about 5 nm.

A. Electron Diffraction Cameras

One of these instruments was built by G. I. Finch in 1932 and 1933, and was described in Finch and Quarrell (1934). Finch used a cold cathode gun at an accelerating voltage of 40–70 kV, a fine anode aperture, and a focusing coil to bring the electron beam to a focus approximately in the plane of the single plate camera. The vacuum in the system was obtained with a mercury diffusion pump. A vacuum of about 10^{-3} Torr was maintained in the gun by differential aperturing and a controlled leak.

A delightful feature, reminiscent of the "string and sealing wax" days, was the gun insulator. This insulator required a glass cylinder with a narrow neck. Finch found the gun insulator could conveniently be made from a straight-necked wine bottle (he recommended Sauternes) with the bottom removed and then ground flat with carborundum to form a vacuum sealing surface. This instrument was made to his design by Cambridge Instruments, who later made further models based on this design that were sold in the 1930s.

III. 1941–1950, THE SECOND DECADE: NEW COMMERCIAL INSTRUMENTS

The Siemens ÜM instrument continued in production through the war years, and although the applications laboratory was destroyed during an air attack in 1944, the factory itself remained intact, only to be dismantled and transported to Russia after the war. A recent review by Wolpers (1991) lists the destinations and in some cases the subsequent fate of the 38 microscopes listed by Ruska as being manufactured by Siemens during the war years. It appears from this information that while there was nominally a single design, and all the instruments were called "Übermikroskope" (Supermicroscope), and abbreviated as ÜM, the operating voltage increased from 60 kV for the first two experimental instruments, through

70 kV for the first six commercial models, and 85 kV for eleven. The run ended wtih sixteen 100-kV microscopes. If Siemens had adopted the numbering convention which was adopted by all companies after the war, there would have been three distinct models launched in this period, to distinguish the changed specification. Since they were first in the market, there was no commercial pressure to adopt this course at the time. In the course of the production, three microscopes operating above 100 kV were also built for experimental purposes. One, built in 1939, had an accelerating voltage of 120 kV, a 220-kV model was made in 1940, and a further 120-kV instrument in 1942. The grand total was 38 microscopes. Nearly all these instruments worked throughout the war in Germany, though one was shipped to Uppsala University in 1943. After the war, five of them were brought to England and gave useful service there for some further years. The remaining postwar distribution was 10 in West Germany, 3 each in Austria and France, 1 each in Sweden and the United States, and 13 in the USSR. One had been destroyed in an air raid, and one was destroyed by hammer in Vienna at the end of the war. This was brought about by a disagreement between the three scientists in the laboratory concerning the handing over of the microscope to the Russian troops. Two of them wished to preserve the instrument, and one to destroy it. The destruction was successful, but all three young men lost their lives, two were shot; the other committed suicide (Wolpers, 1991).

Like Siemens, the AEG Company had its factory in Berlin, but this was destroyed in an air raid in 1943, after they had manufactured four instruments. The latest of these, the EM 5, was installed in Frankfurt-am-Main by Dr. Rang (Rang, 1992). The electron microscope work then moved to Schönberg (near Dresden) in Oberlausitz, and after many difficulties, a further instrument was built there, the EM 6, which had an HF-driven cascade generator for the high-voltage supply (Panzer, 1950). With the approaching end of the war, the Eastern front came too close to Schönberg for comfort (Rang, 1991), and the activity moved again, first to Helmbrecht, and finally to Mosbach, near Heidelberg. The sheer determination of the team to go on developing the microscope in the face of all these obstacles is awe-inspiring. One other event of the war years should be mentioned here: AEG and Carl Zeiss had contracted to establish a joint development of electron microscopes, with Zeiss responsible for the marketing. This was to have important long-term effects on the electron microscope market.

Before dealing with the new postwar instruments, we have to add a postscript to the story of the Übermikroskop of Siemens. Some 13 of the instruments installed in the Eastern regions of Germany were seized by the Russian forces and shipped to the USSR as reparations. One of the first of these instruments was taken to Moscow, where it was disassembled and studied in detail. Complete working drawings and full technical documentation were prepared. A small series of these instruments (called the EM-100), was put into production in the experimental plant BEU (All-Union Electrotechnical Institute) in Moscow. Production continued for several years after the war, at a rate of about 12 microscopes per year. This microscope was anonymously described in the book of Lebedev (1954). It led to a fairly rapid advance in the technology of electron microscopes in the USSR, and gave the Moscow group a considerable lead over others in the country (Kiselev, 1992).

Meanwhile, the VEB Werk für Fernsehelektronik in Berlin (abbreviation WF) (the Television Works) started to manufacture copies of the Siemens design, and between 1945 and 1950 produced a further 12 instruments. There were minor changes in detail in these microscopes, particularly in the last three to be made, which were given the works identification OSW 2752 (Fig. 7). The first nine instruments were all sent to the USSR, but the last three were installed in Germany, in Insel Reims (Tierseucherinstitut), Jena (Physik Institut), and Berlin (Institut für Silicatforschung) (Schramm, 1994). In 1949, Prof. Jung, Berlin-Buch, received one of the original Siemens models.

One of the earliest new instruments to appear outside Germany in this period was the Swedish Siegbahn-Schönander electron microscope (Bergq-

FIGURE 7. The last three production copies of the ÜM, designated type OSW 2752, made by the WF Company.

vist, 1946). It was a two-stage microscope, with a horizontal column. Perhaps because of this, it resembled an electron optical bench rather than a production instrument. Users at the time commented that it was rather awkward to adjust because of the distance between the viewing screen and the alignment controls (Afzelius, 1988). Nevertheless, it is believed that at least 15 of these instruments were produced, and some were shipped to Poland and Czechoslovakia. After this initial effort, however, all work on the production of electron microscopes in Sweden seems to have been abandoned.

A considerable effort was expended in France during the war years by Grivet and his colleagues in the Compagnie Générale de Télégraphie sans Fil (CSF). After an initial prototype electrostatic electron microscope, three laboratory "production" instruments were constructed in 1943. The work was summarized by Grivet (1946). Further improvements were incorporated in a production instrument in 1945, improved in 1946 in the model M III (Fig. 8). This was a two-lens instrument, with a telefocus gun providing the illumination. The resolution was 10 nm and the top magnification 45,000. This was a very well engineered instrument for that time.

The next serial production of electron microscopes in postwar Europe was by the Metropolitan-Vickers Electrical Company ("Metro-Vick"), which

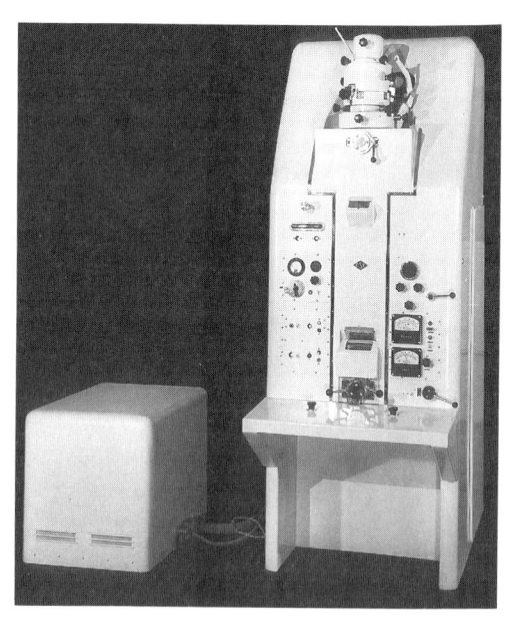

FIGURE 8. CSF Electrostatic electron microscope type M III. A very early postwar series production instrument. Accelerating voltage 50 kV, with a telefocus gun.

FIGURE 9. M. E. Haine. Seen with a group at the St. Andrews Conference, 1951. Left to right: J. B. Le Poole, V. E. Cosslett, D. G. Drummond, G. Liebmann, M. E. Haine.

made the EM 2 microscopes. These were designed largely by M. E. Haine (1947a) (Fig. 9) from ideas he acquired from the Metrovick EM 1 and in the United States during the war. All work on electron microscope design and development had ceased in the United Kingdom for the duration of the war. Haine himself had partly been employed on radar research and later, in the United States, on atomic research. During his stay there from June to December 1944, he had been able to meet Marton, then working at Stanford University in California, and they discussed their ideas on electron microscope design. Marton had designed the RCA model A microscope while working for that company between 1938 and 1940, but they had rejected his design in favor of the model B, designed by Hillier. It is clear that Marton did not bend easily to the production-oriented atmosphere at RCA, and Zworykin did not feel that the suspension of the lenses in a wide-bore vacuum tube (a basic feature of the model A) was an acceptable basis for production (Reisner, 1989). To what extent Haine was influenced by Marton we do not know, but the EM 2 design (Fig. 10) started by Haine immediately on his return to England bore some striking resemblances to the model A, notably in having the lenses suspended inside a wide-bore vacuum vessel. This might, however, trace its antecedents to the prewar high-voltage oscilloscopes made by the company. Nevertheless, production of the early models was fast enough for it to be demonstrated at the Physical Society Exhibition in January 1946, just 12 months after Haine's return from the United States.

The EM 2 was a reliable workhorse but not particularly innovative. The high-voltage supply was based on a 50-Hz mains transformer, but it did

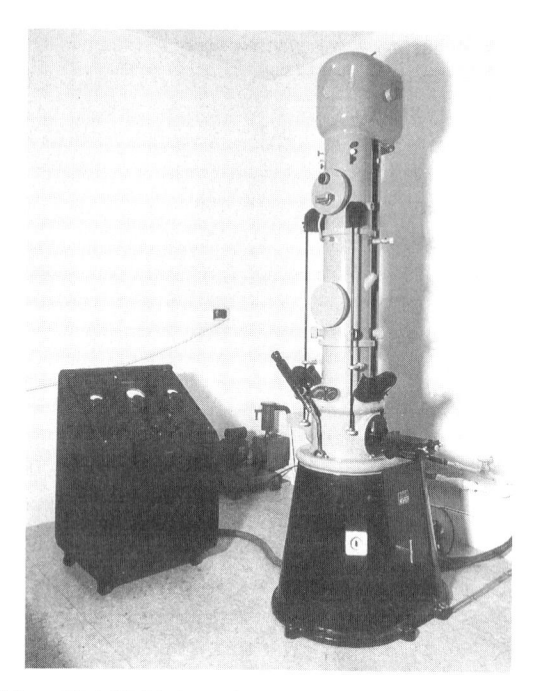

FIGURE 10. Metro-Vick EM 2. Note the very long column enclosing the lenses, and the separate console with lens supplies. The monocular viewer was a nonstandard addition.

have a quite advanced electronic stabilization and the lens supplies were also electronically stabilized. It had a thermionic filament, and the vacuum seals were rubber cord seals rather than greased joints. The vacuum pumping was controlled by electrically operated valves. The nominal resolution was 10 nm, although with much care and many photographs one could coax much better performance from it. One feature of importance was the alignment system for the illumination. Haine had designed this so that the three independent alignment controls could achieve a logical alignment without any of the iterative hit-and-miss procedures needed in other contemporary microscopes (Haine, 1947b).

My personal recollection of operating the instrument was of my good fortune to be tall—a minimum height of 6 ft. (1.85 m) was required if one was to be able to adjust the gun and condenser controls while watching the viewing screen. This was partly due to the 6-mm focal length of the objective lens—the value of a short focal length had not been recognized. The specimen stage was prone to drift, and one had to wait a long time before the movement became slow enough to allow for the photographic exposure. A sharp kick on the front panel of the base of the microscope was often useful in jerking the stage into a temporarily stable position. These comments, which would be echoed by operators of other instruments of the period, are included less as a criticism of a particular microscope, but more as an indication of the state of the art at that time. A stable specimen stage has to combine a subtle mix of high-quality mechanical movements, balanced thermal expansions, and just enough friction. The mix was never right in the early instruments, which was hardly surprising when one considers the technical specification: ability to move 3 mm in two orthogonal directions, positioning accuracy of, say, 0.1 μm, and no drift of more than 1 nm in 5 s.

The Trüb-Täuber Company (TTC) in Switzerland introduced the KM microscope in 1947, based on the earlier work of Induni (1945a, 1945b), who described a similar instrument but which was limited to 40-kV operation. The production microscope operated at 50 kV, and had a cold-cathode electron gun, an electrostatic objective lens, and electromagnetic condenser and projector lenses.

In their new home at the Süddeutsche Laboratorien (SDL) in Mosbach, which was under the direction of G. Möllenstedt, the microscope development team led by Rang and Seeliger, with valuable ideas contributed by Scherzer, produced a further modification of the Mahl design. This was launched as the AEG-Zeiss EM 7 in 1947. This was achieved in the face of enormous difficulties—supplies of needed materials had been almost completely disrupted by the war and its aftermath, and compromise and ingenuity were needed in order to produce an instrument at all (Rang, 1991). The EM 7 had to rely on a prewar type of high-voltage generator, a mains transformer system, which was all that could be obtained. There was no possibility of obtaining the steel needed for the desk supports, and it had to be constructed in wood, with metal panels.

In 1949, the Dutch electron microscopists hosted a conference in Delft which was to be the forerunner of a long series of international conferences on electron microscopy. That first meeting, which attracted some 200 delegates, was a great success. The meeting was used to launch the Philips EM 100 instrument (Fig. 11) (van Dorsten et al., 1950) (Fig. 12), which was the descendant of the Mark II experimental microscope of Le Poole (1947), who described his development efforts in an earlier review (Le Poole, 1985). This Philips instrument was sensational in many respects. It operated at voltages up to 100 kV, and had a resolution of 5 nm. It included four imaging lenses (one for selected-area electron diffraction) and was very compact. It was slightly angled to the horizontal to meet the line of sight of a seated operator, and it had a very large viewing screen. It was the first commercial microscope to have a focusing aid in the form of the focus wobbler developed by Le Poole (Le Poole, 1947). The specimen was intro

434 ALAN W. AGAR

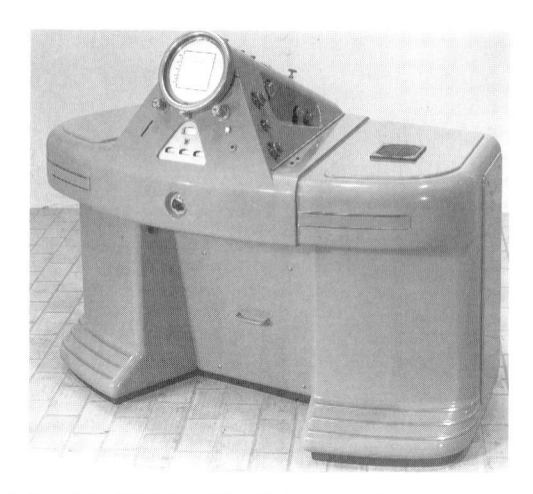

FIGURE 11. Philips EM 100. First 100-kV electron microscope with four imaging lenses, selected-area diffraction facilities, and a focus wobbler, 35-mm camera.

duced into the objective lens gap on a rod which could be rotated to achieve stereo tilt. Since Le Poole had worked through the war years without any contact with workers in other countries, he had had to develop all these design features independently. He had indeed heard a lecture in 1942 by

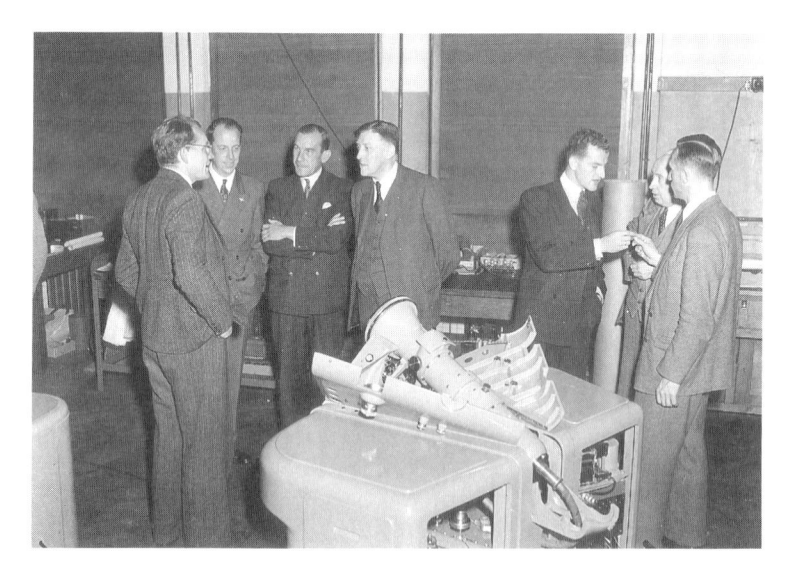

FIGURE 12. Ir. van Dorsten and his colleagues of the Philips Research Laboratory in Eindhoven showing the EM 100 to professors from the Laboratory of Technical Physics in Delft. Left to right: Prof. Casimir, Dr. Ir. Oosterkamp, Ir. Alting, Prof. Dorgelo, Ir. Le Poole, A. Verhoeff, Ir. van Dorsten.

von Borries on the early Siemens microscope, but this would hardly have affected his concepts of multilens imaging.

At the time that the Metro-Vick EM 2 was launched, work had already started on a new instrument, EM 3, a 100-kV, 5 nm-resolution microscope, which incorporated many new ideas known to be desirable but which had to be left to one side in order to get the EM 2 quickly on the market. These included multilens imaging to reduce column length [already employed by Marton (1945) in his experimental microscope at Stanford University]. The third imaging lens also permitted selected-area electron diffraction, which had been proposed and realized as early as 1936 by Boersch (1936) in an electron optical bench. The idea had been realized independently by Le Poole in the Delft electron microscope during the war. The EM 3 microscope was probably the first to have completely logical alignment facilities for the whole electron optical column. This arose because Mike Haine was the recipient of one of the Siemens instruments brought over to England after the war. This particular instrument had a bad misalignment between the objective and projector lenses. (The whole instrument was prealigned on its greased cone joints.) Haine solved the problem by sawing the column in half and inserting a traverse mechanism between the lenses. The double projector required something more sophisticated, and Haine adopted a shift and a tilt adjustment, the tilt being centered on the middle of the second projector. Another feature of this instrument was that the pumped volume was greatly reduced, both because of the shorter column and because the lens windings were now outside the vacuum. The mains frequency HT set of the EM 2 was replaced by an electronic generator (Haine *et al.*, 1950) similar to that already used by RCA. The first instrument was shipped in 1949, soon after the Delft conference. It is worth recording that the spectacular decrease in pumping time for the EM 3 column, compared with the EM 2, convinced Haine that there was no longer a need for a specimen or camera airlock, and this resulted in a considerable simplification of the vacuum design. This view was not shared by those of us who had to examine many specimens before finding a good one, or who frequently exhausted the stock of plates in the camera (just two!). The other effect was that the column was hardly ever properly degassed. It is perhaps worth noting here that RCA, in the design of the EMU 2, introduced in 1944, had also abandoned specimen and camera airlocks in favor of simplicity of the vacuum valving system (Reisner, 1989). But two wrongs do not make a right!

It is interesting to note that Haine, Le Poole, and van Dorsten had described the main features of the Metro-Vick and Philips instruments at a conference on electron microscopy held at Oxford in 1946 (see Cosslett, 1947). The two groups had therefore arrived at very similar instrument specifications independently. What is striking about the reading of these papers is that the two companies were publishing the main features of the specification of instruments which were to arrive on the market some two-and-a-half years later. It illustrates the very open atmosphere at the time, and may well have contributed to the very fast rate of development which ensued. It was shortly to become apparent that the Siemens group had also chosen a similar specification when the ÜM 100 was launched in 1951. All were 100-kV, multilens instruments with selected-area diffraction and stereo facilities. The Siemens instrument claimed better resolution (1.5 nm). The Siemens as well as the Metrovick instrument used a three-lens imaging system, as it was found possible to achieve electron diffraction without the special lens employed by Philips. However, the Siemens design involved interchangeable projector polepieces, which were each operated fairly close to the minimum-focal-length condition. Here again, this instrument had been developed and produced under the very difficult conditions prevailing in Germany after the war.

The selection of 100 kV as the upper operating voltage must be attributed at least in part to the pioneering work of the Siemens team during the war, when they steadily increased the maximum voltage from 70 to 100 kV through the production run of the ÜM. The distribution of numbers of these instruments around Europe in the immediate postwar years would have propagated the idea. The initial incentive for choosing this voltage would undoubtedly have been the need for penetration of the relatively thick specimens which could then be prepared. It happens that 100 kV is also near the maximum voltage which can be used reliably in a single stage of acceleration in the electron gun. But the continuing work to obtain reliable operation at this voltage was undoubtedly of great importance when metallurgists became involved, as they could not have obtained useful results at lower voltages. This proved valuable commercially to the European companies later, when entering the market in the United States, because the RCA philosophy, much more closely geared to biological requirements, had been to settle for 50-kV operation.

In each of these companies, the design team was led by one man with very clear, and often rigid, ideas of his own, and an inflexible determination to carry them through. Furthermore, the design teams were very small, which contributed to their efficiency and close integration. The designs were accomplished with short lead times compared with current practice. Each instrument therefore carried the unmistakeable stamp of the team leader, so that, although the actual specifications from the different companies were similar, the physical appearance and design philosophies were quite different. The influence of these giants—Ernst Ruska at Siemens, Jan Le Poole, and later, van Dorsten at Philips, and Mike Haine, followed by Dick Page, at Metropolitan-Vickers—was to continue, to a diminishing extent, through the decade. (In mentioning Ruska in this context, it is appropriate to recognize the very substantial contribution of von Borries to the early Siemens instruments.) Some of the interesting results of their design decisions will be mentioned in the context of later designs.

Yet another new microscope design was described by CSF at the Delft conference and in a later publication (Grivet and Regenstreif, 1952) (Fig. 13). This was the M IX instrument, which had a revolutionary design in that the three electrostatic imaging lenses were constructed in a single rigid block, which also incorporated the anode of the telefocus gun. The instrument was therefore capable of selected-area diffraction. It was fitted with an electrostatic astigmatism corrector after the design by Bertein (1947, 1948). It was unfortunate that the CSF team split up in 1950 due to its dispersal among a number of important postwar reconstruction projects in France. All electron microscopy work at CSF ceased at that time.

The first instruments to be produced in quantity in the USSR, apart from the copies of the Siemens instrument, were the EM 3 models, designed by

FIGURE 13. P. A. Grivet, the leader of the French team which developed the CSF electron microscopes.

Lebedev, Vertzner, and Zandin (Figs. 14 and 15), following the work of the Leningrad group. These first instruments were very crude, with an open 50-kV gun, and condenser, objective, and projector lenses. The resolution was 10 nm. There were no mechanical alignments, apart from a movement of the filament in the horizontal and vertical planes. There were no airlocks for the specimen or camera. They were produced from 1949 by the experimental plant of the Krasnogorsk (near Moscow) electromechanical school. A considerably improved model, the UEM 100, was designed by Sushkin, Popov, Kushnir, and Plakhov in 1950–1951, and put into production from 1952. This model had an open 100-kV gun, the same lenses as before, but full alignment of the illumination system and with airlocks for specimen interchange and plate removal from the 12-plate camera. The resolution was improved to 5 nm.

Development continued at the SDL in Mosbach. The AEG-Zeiss EM 8-I appeared in 1949 (Fig. 16). The electronics had been modernized, with the battery supply for the filament replaced by HF generation, and a RF generator and cascade replaced the old high-voltage transformer and rectifier. This was a return to the Panzer design first used in the EM 6. Steigerwald (1949) had developed the long-focus electron gun, and this gave a much-needed brightness increase. The EM 8-I also dispensed with the requirement to move the specimen physically to focus the picture. The adjustment was made by a small change to the high voltage applied to the center electrode of the objective lens. The projector lenses were redesigned to operate at minimum focal length to eliminate distortion in the image. The vacuum system was also significantly improved, and a substantial steel desk replaced the temporary wooden construction of the EM 7. The camera, holding 24 plates of 65 × 90 mm, gave a much more generous photographic provision than in any other microscope at that time.

FIGURE 14. An interesting photograph taken about 1946. On the left is S. I. Vavilov, a specialist in physical optics and the President of the Soviet Academy of Sciences. In the center is Academician A. A. Lebedev (1893–1969). On the right is Dr. V. N. Vertzner.

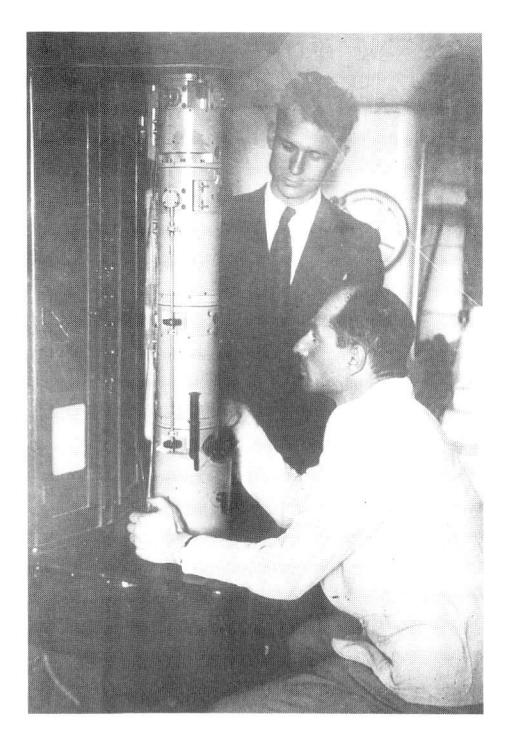

FIGURE 15. V. N. Vertzner, 1908–1980, seated at the EM 3, with N. G. Zandin, a fellow designer of the instrument, in Leningrad, 1946.

The Delft conference was memorable not only for the launch of the Philips EM 100 but also for the first paper by Castaing and Guinier (1949) on the electron probe microanalyzer. I can remember being most impressed by the idea, which unfolded as I helped Castaing to translate the paper into English, the official language of the conference. I am less impressed by my subsequent failure to press Metro-Vick to follow up the idea immediately. In extenuation, one could perhaps cite our very limited engineering strength at the time and our concentration on transmission electron microscopes. At that time the only way of examining a metal surface was by a plastic replica; even extraction replicas were five years in the future; and thin metal foils nearly 10 years away.

Specimen preparation techniques were at an early stage of development in this period. The normal support film was of Collodion or Formvar (Drummond, 1950). Biological specimens had to be viewed intact, sometimes after shadow casting to reveal surface detail and height (Williams and Wyckoff, 1946). Metal specimens were normally examined by replica techniques, either with a plastic replica (Mahl, 1942; Schaeffer and Harker, 1942) or

FIGURE 16. Ernst Brüche and Otto Rang with the first AEG-Zeiss EM 8-1, 1949.

by a two-stage polystyrene/silica replica technique (Heidenreich and Peck, 1943). Very flat surfaces could be examined by reflexion microscopy (von Borries, 1940). Until 1946, specimen supports were punched out of woven mesh. It was only after Smethurst-Highlight started making the electroformed Athene grids in 1946 that a flat and stable support for the thin films was available commercially.

A. Early High-Voltage Electron Microscopes

High-voltage electron microscopes flickered into life briefly in these early years. There had been a 200-kV microscope built by von Ardenne in 1941, and then one by Zworykin et al. (1941). Van Dorsten, working with Oosterkamp and Le Poole (1947), built a 400-kV microscope and had it operating in 1946. It was used to look at yeast cells, but in general the results were disappointing. In the early 1950s, Page at Metrovick started to build a

500-kV microscope to be designated the EM 5. The air-immersed high-voltage set was completed and operated on a prototype electron optical column. However, the successful development of ultramicrotomes capable of cutting thin sections of biological material, which occurred at about this time, removed most of the incentive for these projects, and they were discontinued.

B. Electron Diffraction Cameras

The design evolved by Prof. Finch, which had earlier been manufactured by Cambridge Instruments, was produced after the War by W. Edwards & Co. (later Edwards High Vacuum). Edwards improved the instrument by using O-ring seals with elastomer rings in place of ground joints, and the mercury diffusion pump was replaced with an oil pump. They provided ports for a 250 mm and a 500 mm camera length. The high voltage range was 40–70 kV and the high voltage, generated remotely from the camera column, was transmitted on an open line suspended from the ceiling by high voltage insulators.

Edwards found that a smooth transition from wide diameter to narrow diameter in the gun insulator was essential for stable operation of the gun. It is fascinating to learn that they could find nothing more suitable than the wine bottles originally specified by Prof. Finch—a tribute to the genius of the original design decision.

Improved instruments based on the EM 3 electron microscope, using a high-voltage cable and thermionic emitter, were produced by Metropolitan-Vickers from about 1950. After Edwards and Metrovick had ceased producing these designs, however, no further specialized electron diffraction cameras were produced in Western Europe. From then on the diffraction facilities in electron microscopes were used.

IV. 1951–1960, THE DECADE OF THE HIGH-RESOLUTION EXPLOSION

A. Microscopes for Routine Operation

This decade started quietly, partly because of the limited number of microscopes in service, but also because of the limitation imposed by the available specimen preparation techniques.

This did not deter the enthusiasts who worked for the electron microscope manufacturers in the West. We could already imagine the enormous potential usage in many different scientific fields, and were encouraged by the launch by RCA and GEC in the United States of simple microscopes for routine applications. The Metrovick Company developed the EM 4, launched in 1952, which was a console model with a horizontal column operating at 50 kV (Fig. 17). The projector lens had two sets of polepieces energized by the same coil, enabling the column length to be shortened. The specimens were introduced on a rod which held three interchangeable grids. This was the first commercial use of an idea developed first by Liebmann (1948). The EM 4 had a 70-mm roll-film camera allowing 40 exposures per loading. The vacuum system was controlled by a single interlocked knob, which prevented faulty operation. The resolution was 10 nm (Page, 1954) (Fig. 18). When this microscope was at an advanced stage of development, Dick Page asked Revell and myself in the Applications Laboratory to try it out for operational convenience, and we expressed ourselves satisfied. When we had the instrument in production, we were called upon to instruct the first customer in its operation, who turned out to be a diminutive lady. It was at this stage that we discovered that it was

FIGURE 17. Metro-Vick electron microscope type EM 4. Simplified 50-kV console instrument, with double-projector lens and triple specimen holder. At the controls, M. Venner, author of the paper, "Where to Kick the Electron Microscope." Power supply cabinets of the EM 3 instruments appear in the background.

FIGURE 18. R. S. Page, 1917–1985, Chief Design Engineer for Metropolitan-Vickers and AEI electron microscopes, in a New York hotel room, while at an EMSA meeting.

well-nigh impossible for someone with short arms to sit at the front of the microscope and adjust the gun alignment controls. A stark lesson in ergonomics! Page, Revell, and Agar were all about 6 ft (1.85 m) in height, with arm lengths to match.

In the same year, Siemens introduced the Elmiskop 2 (Müller and Ruska, 1955) (Fig. 19), which had no condenser lens, but an objective lens and a projector with a turret of four interchangeable polepieces to give a range of magnifications from 300 to 30,000. The lens currents were supplied from a bank of batteries to avoid the need for electronic stabilization. It operated at 50 kV, with a resolution of about 2.5 nm. It had a 12-plate camera.

In 1954 Philips introduced the EM 75 (van Dorsten and Le Poole, 1955), another radically different design (Fig. 20). It operated at 75 kV and had a resolution of 5 nm. It was designed with a very short focal length objective, which had a very small chromatic aberration coefficient, so that it could give good performance with a cheap stabilizer. The projector lens was excited at a fixed strength near its minimum focal length, but with a subsid-

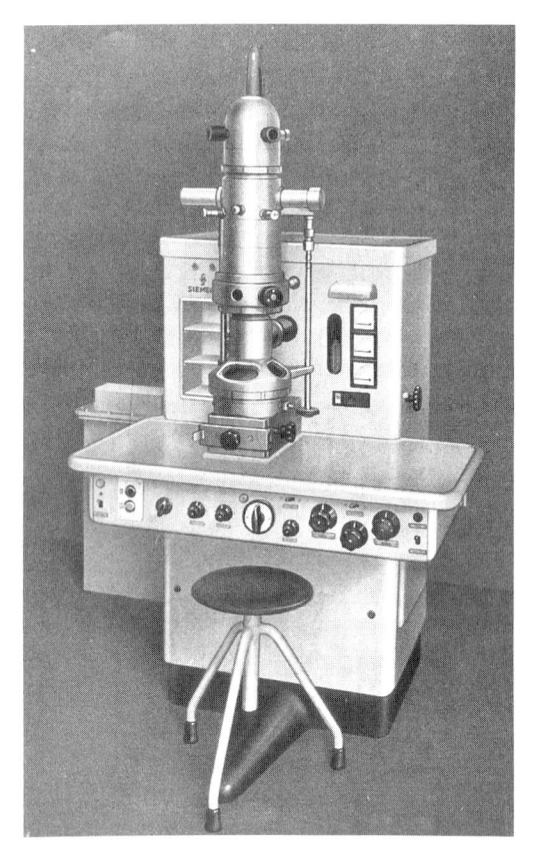

FIGURE 19. Siemens simplified electron microscope type Elmiskop 2. 50 kV, with battery-driven lenses.

iary polepiece which could be moved vertically over a distance of 15 mm by rotating a ring around the top of the lens. The effect of moving this polepiece was to sink it progressively into a soft iron tube, which therefore short-circuited the gap and rendered the extra lens ineffective. This scheme enabled the magnification to be changed by a factor of 9, from 1200 to 11,000, and the magnification change involved no rotation of the image (Fig. 21).

Although all the manufacturers had decided that there was a need for a simple, routine microscope, none of these instruments sold in quantities which made them commercially viable. The very different conceptual designs of the three available instruments illustrated a high degree of ingenuity, but, interestingly, contrasted sharply with the high degree of convergence in the designs of the top-line instruments. What now seems apparent is that the main objective for each team had been to make an instrument

FIGURE 20. Philips electron microscope type EM 75. Simplified 75-kV instrument with very-short-focal-length objective and a fixed excitation projector lens unit with a subsidiary variable polepiece gap. Objective lens supplied by a power pack with selenium rectifiers and smoothing condensers.

significantly cheaper than the main instrument in its range, and hopefully easier to use. With the benefit of hindsight, it appears obvious that reliable techniques of specimen preparation were insufficient to justify any widespread demand for these simple microscopes at this early stage, however clever the design, or however much cheaper they were.

AEG-Zeiss introduced the EM 8-II electrostatic microscope in 1951 (Rang and Schluge, 1951) (Fig. 22). The objective lens was now fitted with an electrostatic stigmator, after a study by Scherzer and a practical realisation by Rang and Seeliger (Rang, 1949). This work had been carried out quite independently of that of Bertein (1947), but was essentially the same idea. The earliest mention of a corrector for astigmatism was actually in a patent filed by Mahl in 1942 or 1943. It consisted of a diaphragm with a long narrow slit in it, eccentric to its center, and which could be physically

FIGURE 21. A. C. van Dorsten with the Philips EM 75, in 1953.

rotated. It was never actually built, though it is believed that the Philips Company approached AEG concerning the patent during the war. This was the first instrument fitted with an electrostatic astigmatism corrector, though RCA in the United States had in fact started to incorporate a corrector in their microscopes in about 1948, based on the use of eight soft-iron screws inserted into the polepiece gap (Hillier and Ramberg, 1947). Though effective for a fixed-kV instrument, this was a less elegant and convenient system. In the same year as the EM 8-II was launched, Siemens had introduced a magnetic corrector in their ÜM 100. If the CSF had continued their operations, the M IX would also have appeared with a stigmator at this same time. As a result of the astigmatism corrector, AEG was able to offer an instrument with a resolution close to the best available on any instrument at that time. It is worth reminding ourselves that up to this time electrostatic instruments seemed a viable alternative to the electromagnetic ones. The great attraction of the electrostatic design was that, since the center electrode of each lens was connected to the same high-voltage supply as the electron gun, the system was effectively achromatic, at

FIGURE 22. AEG-Zeiss 50-kV electron microscope type EM 8-II. An electrostatic electron microscope launched in 1951 with the highest resolution (2 nm) attained at that time by an electrostatic microscope. First commercial instrument with an electrostatic stigmator.

least at relatively low accelerating voltages, where the relativistic correction was small. Since in those days it was very difficult to achieve reliable high stability in both voltage and current supplies, this was a major advantage. What was less advantageous were the decreasing dimensions needed for a strong lens and the consequent problems of HT discharge. The trend toward higher accelerating voltages compounded the difficulties. The EM 8-II offered as an accessory an intermediate accelerator before the object, so that from a gun voltage of 50 kV, the specimen being raised to a potential of +50 kV, the effective penetration of the electron beam was 100 kV. The beam was decelerated again after passing through the specimen, before falling on the fluorescent screen or photoplate. This concept was due to Möllenstedt (1955) (Fig. 23). The instrument was also equipped with a new viewing telescope of high aperture, designed by Zeiss. It enabled the image to be viewed at an additional 20× without loss of brightness.

FIGURE 23. G. Möllenstedt, photographed in 1938. A contributor of many inventive features of the AEG microscopes. He built and exploited the advantages of the first convergent-beam electron diffraction camera. He also invented the electron biprism while at Mosbach, which did not find immediate application but was later exploited by Lichte to achieve electron-beam holography at atomic resolution in a modified Philips TEM.

In 1953, the electron microscope business was taken over completely by Zeiss, Oberkochen from the joint venture with AEG. Most of the microscope team at SDL moved to Oberkochen, with the exception of Prof. Brüche and Dr. Rang. In 1954 the EM 8-III (Schluge, 1954) was introduced. The EM 8-III, operating up to 60 kV, had prealigned lenses. It had an optional top magnification of 25,000, although Zeiss believed that 16,000 was entirely adequate, and this was retained as the normal top magnification. There were survey magnifications of 200 and 800. The best resolution remained at the 2 nm level achieved earlier.

In parallel with this development in electrostatic microscopes, an instrument was developed and produced in the former German Democratic Republic by VEB Carl Zeiss Jena (later known as Jenoptic Carl Zeiss Jena). The instrument in prototype form in 1951 was named the ELMI A and was a three-stage electrostatic microscope. This instrument progressed through the type B to the ELMI C, which in 1953 had a resolution of 5 nm (Recknagel, 1952). However, by 1954 they had made a number of significant improvements, resulting in the ELMI D (Fig. 24) (Guyenot, 1955; Schulze, 1960), which had a resolution of 1.5 nm., the best performance claimed for any electrostatic instrument. It had five electrostatic imaging lenses and an octupole electrostatic stigmator.

It was in 1951 that the first results of a program of research initiated by Oatley were presented at a meeting of the Institution of Electrical Engineers in London by Dennis McMullan. This paper, on the design of a scanning electron microscope, was published in 1953 (McMullan, 1953). The first commercial scanning microscope was, however, still many years in the future.

In this same period, an improved Trüb-Täuber microscope KM 4 appeared (Wegmann, 1952), again with an electrostatic objective and with

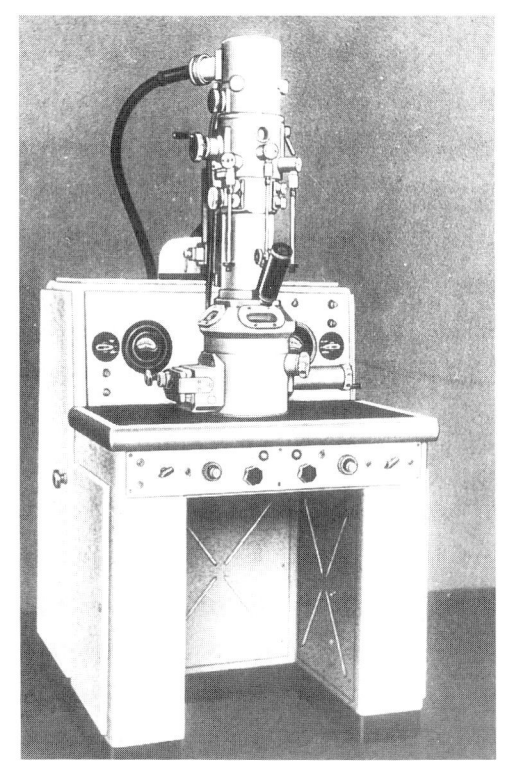

FIGURE 24. Zeiss Jena electrostatic electron microscope ELMI D, which achieved the best resolution (1.5 nm) for an electrostatic instrument. It was the last such instrument to be made commercially.

electromagnetic condenser and projector lenses. The new instrument had a much improved high-voltage set and gun to give stable operation at 50 kV. It was found, however, that thick specimens gave very poor-quality images, due to the high chromatic aberration of the objective lens.

In 1953 the first commercial Czech instrument, the Tesla BS 241, appeared (Fig. 25). This was a 50-kV microscope with two imaging lenses, based to the recognizable extent on the RCA EMU 2A, and achieving about 10-nm resolution (Delong and Drahoš, 1951). It had a magnetic astigmatism corrector.

The upgrading of the Metrovick EM 3 to the EM 3A also occurred in 1953. With a new high-voltage stabilizer and objective lens, the resolution was improved to 2.5 nm, and it was given a new tilt/translation stage beneath the condenser lens to permit the gun to be tilted over to an angle of 8°, for use in reflexion microscopy. It gave the microscope an extraordinary

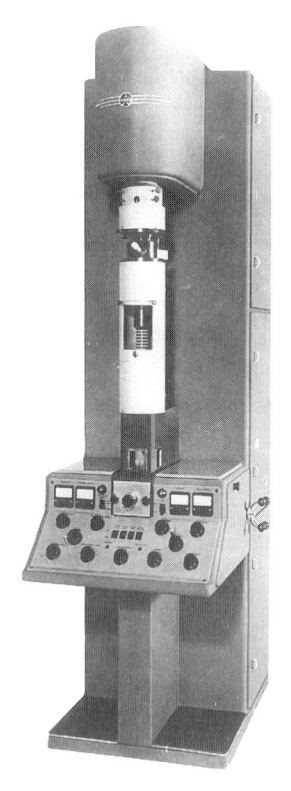

FIGURE 25. Tesla electron microscope type BS 241. The first Czech microscope, showing its evolution from the RCA EMU 2A.

appearance (Fig. 26). This was at the time still the only way of directly examining the surface of a solid specimen (at glancing incidence), having been originally described by von Borries (1940). It was finding application in a number of laboratories (see, e.g., Menter, 1952; Fert, 1956), and was also used in the Research Laboratories of the Metrovick Company in a study of the wear process in metals. The EM 3A was fitted with a four-pole electrostatic stigmator, which could be rotated inside the back bore of the objective lens. There was an interesting background to this development. In 1950, I had been given a project to assess the soft-iron screw system of Hillier and Ramberg (1947), except that the soft-iron screws had been inserted through the top face of the pole piece of a EM 3, so that one had to dismount the top of the instrument to gain access to the screw heads. I found that it was indeed possible to correct the astigmatism, but it was an impossibly tedious task, taking some three days to achieve a correction! In 1952, Mulvey, having completed a redesign of the magnetic circuit of the lens to eliminate the stray fields arising from inadequate iron in parts of the polepiece and yoke, equipped the lens with eight iron rods protruding into the air gap. He found to his surprise that the rods now had

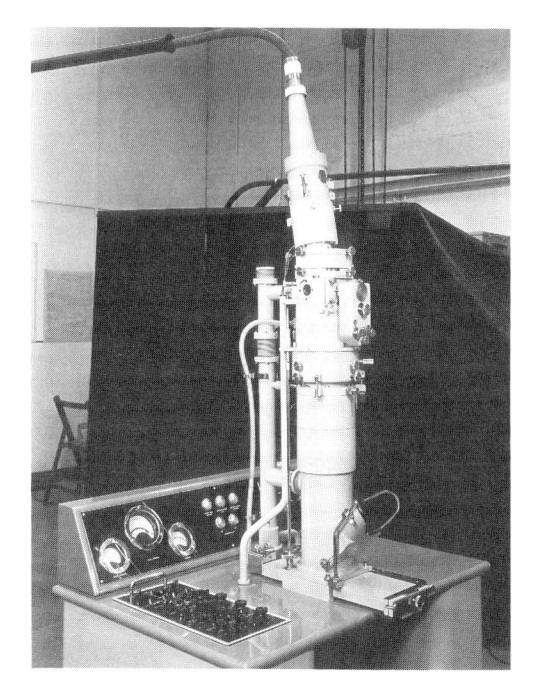

FIGURE 26. Metro-Vick electron microscope type EM 3A, showing the gun and condenser tilted at 8° for reflection microscopy, and the reflection stage in position above the objective.

no effect on the astigmatism, and turned instead to the early design of Rang for an electrostatic corrector (Mulvey, 1992). This was the one adopted for the EM 3A. It may seem strange that we started using the Hillier system rather than the superior systems which had been developed in Europe at the same time (Bertein, 1947; Rang, 1949). The only explanation I can offer was that the United Kingdom for all the war years and the immediate aftermath looked to the United States for technical cooperation, and the European links were slow to be reforged, in spite of the attendance of a few European colleagues at EM meetings in the United Kingdom. There had also been the very good meeting in Delft in 1949, but the German scientists had not been invited to attend.

In the WF Company, following the production of copies of the Siemens ÜM, Eckart and Langbein in 1949–1950 constructed a prototype magnetic electron microscope, OSW2748, called the SEM 2 (Standard Elektronenmikroskop). It was a simple design with single condenser, objective and projector lenses, but operating up to 100 kV. There were problems with the high-voltage supply, resolved by Schramm when he joined the team in 1951. This led to the production instrument SEM 2-1 (Fig. 27) by Langbein, Beissse, and Schramm, to be followed by a further improved high-voltage design by Rassmus, leading to the SEM 2-2. The instrument was very elegantly packaged, clad in a white aluminum cabinet, and was jokingly referred to as Die Weisse Frau (The White Lady) (Schulze, 1992). Of the 16 microscopes of this design produced in the period 1952–1955, two were shipped to China.

In 1955, this company produced a small microscope, the KEM 1 (Klein Elektronenmikroskop) (Walz and Schramm, 1962). This instrument had three electromagnetic imaging lenses and operated in the voltage range 40–60 kV. It had a single condenser illuminating system but with electromagnetic alignment coils for tilting the beam from the gun. With a resolution of 5 nm, the magnification range was 1200–30,000 in seven fixed steps. It had selected-area electron diffraction facilities, and stereo tilt was possible with a special specimen holder.

In 1956, Tesla introduced the table-top microscope, BS 242 (Fig. 28) (Drahoš and Delong, 1958). This small, 60-kV instrument had four imaging lenses and a rod airlock for specimen entry. It became a considerable commercial success, no fewer than 847 of the various models being made. Two years later, Tesla introduced an improved version, the BS 242a, in which current stabilizers replaced the original battery supply. It was to turn out that this was the only table-top electron microscope to be produced anywhere in Europe, although RCA had produced such an instrument earlier.

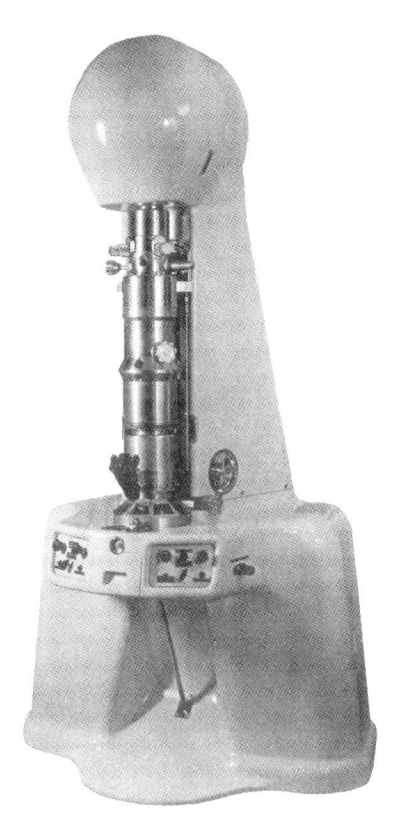

FIGURE 27. WF electron microscope SEM 2-1, the "White Lady." First instrument designed by B. Schramm and co-workers.

1958 also saw the introduction of the KM 5 by Trüb-Täuber (Gribi *et al.*, 1959). This was a 70-kV cold-cathode model, again with a mix of magnetic and electrostatic lenses, and with a resolution of 2 nm. The new design of electrostatic lenses was very stable at 70 kV, and the magnetic lenses were now provided with transistorized stabilizers for the lens current. Nevertheless, the retention of the cold-cathode source meant that the performance was limited to magnifications below 35,000. According to Villiger (1990), the company technical staff were enthusiastically in support of a recommendation from the University of Basel that the instrument be upgraded by employing electromagnetic lenses and a thermionic cathode. However, the management of TTC apparently "insisted for prestige reasons on the use of the cold cathode" and would not accept the technical imperative of a thermionic emission source for the future success of the line.

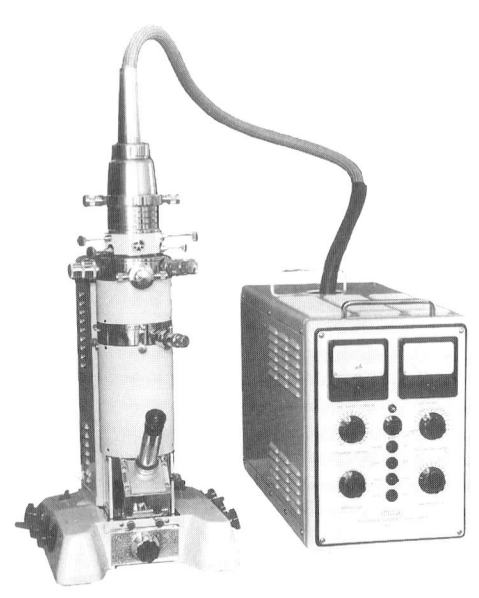

FIGURE 28. Tesla type BS242. First (and only) European tabletop microscope, 60 kV, four lenses. It was in widespread use, particularly in Eastern Europe.

They therefore withdrew from further development, and production of transmission electron microscopes by TTC ceased at the end of 1960.

Toward the end of this decade, a new instrument, the EM 5 (Vertzner et al., 1961) was produced in the USSR. This was an improvement on the EM 3 instrument which had appeared almost 10 years earlier. The EM 5 (Fig. 29) had a condenser system giving a small demagnification of the gun spot size, and with an intermediate lens added. An eight-pole electrostatic stigmator was also fitted, allowing 2.5-nm resolution to be attained. There was still no specimen airlock. This instrument was designed and built at the Krasnogorsk Mechanical Plant (KMZ) from 1959. The designs were passed to a new production facility at Sumy, in the Ukraine, which also started to build this model in 1959. This factory soon had a large development team, and later became the principal source of all the electron microscopes in the USSR.

B. The First High-Resolution Electron Microscopes

The year 1954 was a landmark year because it featured the launch of the Siemens Elmiskop 1 (Fig. 30), the first high-resolution (1-nm) electron microscope (Ruska, 1956; Ruska and Wolff, 1956). For the first time in a

FIGURE 29. An early microscope from the USSR. The EM 5 instrument was manufactured both in Krasnogorsk and in Sumy. It had three imaging lenses but a single condenser. The objective was fitted with an octupole stigmator.

commercial instrument, one had reliable 100-kV operation of the electron gun (with a choice of 40–60–80–100 kV). The double condenser lens appeared for the first time in a commercial instrument, complete with stigmator. Thus a controllable illumination was available, and much specimen damage avoided. The specimen stage was controllable and stable enough to allow the routine attainment of the guaranteed resolution of 1 nm (improved later to 0.8 nm). The three imaging lenses (including the turret of four projector polepieces) gave a very wide magnification range, from 200 to 160,000. The camera contained a loading of 12 plate cassettes, making the productivity of the instrument vastly superior to all its predecessors (except for the electrostatic microscopes, which were well equipped in this respect). It was indeed a giant step forward, which ensured that it became the standard against which all other microscopes were to be judged for years to come.

Other manufacturers had not, of course, been idle. The powerful research team at the AEI Research Laboratories at Aldermaston had established important design parameters for electron guns (Haine and Einstein, 1952),

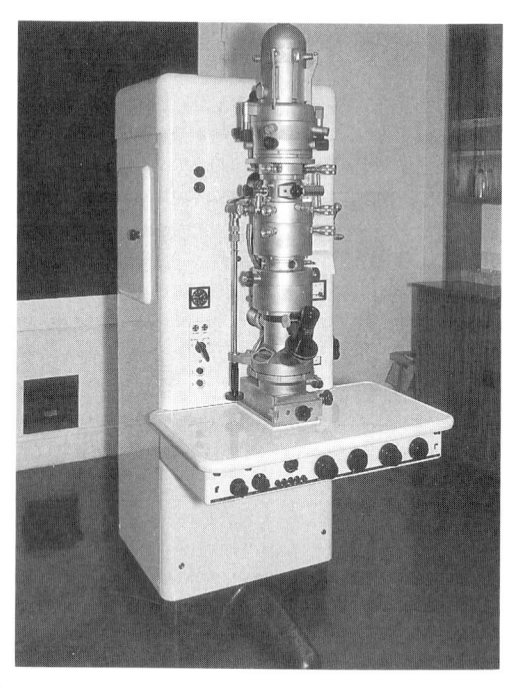

FIGURE 30. Siemens electron microscope type Elmiskop I. The first high-resolution electron microscope (1 nm). Perhaps the most important single development in the history of commercial instruments. The first instrument with a double condenser and two stigmators.

electron lens design (Haine, 1956), anticontamination measures, and alignment systems. [See extended reviews by Mulvey (1985) and Haine and Mulvey (1956).] This work was incorporated in the design of the AEI (formerly Metrovick) EM 6 microscope, which appeared in 1956 (Haine and Page, 1957). It had the same important design features as the Siemens Elmiskop, except that its astigmatism correctors were electrostatic and therefore much easier to use. The lenses were of a new design, since they had integral polepieces instead of separate inserts, and this contributed to the good magnetic design. This integral-polepiece design did, however, have a disadvantage—Haine designed the lens with a 6.25 mm bore and gap, the main idea being to simplify the machining and to make easier the obtaining of a truly round bore. It seems that the perceived manufacturing problems may have outweighed the desirability of reducing the objective focal length further. In fact, the focal length was 4 mm, which gave the instrument good contrast and very good general purpose performance, but which made the highest resolution more difficult to obtain. The AEI design retained the single intermediate lens and projector which were featured in

EUROPEAN COMMERCIAL ELECTRON MICROSCOPES 457

the EM 3. This involved a compromise in that, in order to achieve a reasonable balance of barrel and pincushion distortions at low magnifications, as well as using the intermediate lens for diffraction, the final projector had to be operated at low excitation in the lower part of the magnification range. It was therefore operating far from its optimum condition at this point. The importance of the minimum-focal-length condition had been the subject of a paper by Liebmann (1952) of AEI and had also been recognized by van Dorsten and Le Poole (1955). This design economy was a perfectly acceptable solution at the time because of other instrumental and specimen limitations. It was only at a later date that the disadvantages of this approach became apparent.

In 1958, Philips produced the EM 200 microscope (Fig. 31), which again incorporated all the important features of the Siemens design but included simpler astigmatism correction, and the focus wobbler system which had first appeared on the EM 100. All four imaging lenses had alignment of

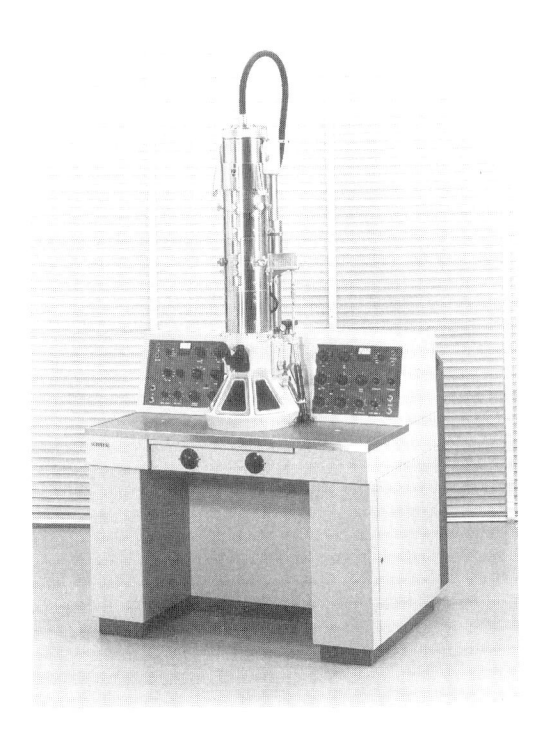

FIGURE 31. Philips EM 200. Solid column construction with lens alignment carried out by movements of polepieces.

the polepieces by screws operated by removable Allen keys, so that the column design was very stable mechanically as well as having a neat appearance.

1956 saw the introduction of the EM 8-IV by Zeiss Oberkochen. It was similar to the EM 8-III, but had the additional very interesting facility of a filter lens. Rang had observed that the minimum-focal-length condition for a projector gave distortion-free working. If one made the potential of the center electrode more negative still, then the focal length went to infinity. If one made it still more negative, a new focal point appeared, followed once again by an infinite focal length. Eventually, one arrived at a condition where the lens became a mirror. It was pointed out by Möllenstedt and Rang (1949) that, just before the mirror condition, the lens achieved a condition of very high sensitivity as an energy analyzer for the electrons. One could arrange that the lens pass only those electrons which had suffered no energy loss, or some small defined energy loss, thus excluding inelastically scattered electrons from the image. The filtering of these electrons increased the contrast of the image, and the monochromaticity of the beam reduced the chromatic aberration. The instrument therefore anticipated by some 30 years the later Zeiss interest in the filter lens. This instrument also turned out to be the last high-resolution electrostatic electron microscope. The earlier advantage over the magnetic instruments was being eroded by better stabilizers for voltage and current, and the tendency toward higher accelerating voltages accentuated the problems of voltage stability in the lenses; the problem of the greater spherical aberration in electrostatic lenses remained, so the line had to be abandoned. Zeiss had meanwhile designed a small console-type electrostatic microscope, the EM 70-E, described by Mahl, Volkmann, and Weitsch (1956), but it was not put into series production.

In 1957, the WF Company (Television Factory) first produced the SEM 3 (Fig. 32), as a high-resolution instrument, being a development from the SEM 2-2. It operated in the voltage range 40–100 kV, and had four electromagnetic imaging lenses but with a single condenser illuminating system with electromagnetic alignment of the beam from the gun. It was equipped with an electromagnetic stigmator with alignment coils in the objective lens. There were 12 fixed magnification steps, and a top magnification of 100,000 on the viewing screen (nearly 120,000 on the photographic plate). This was an advanced design from Schramm and Rassmus (Fig. 33), who were certainly not as well known as the scientists involved in the other instruments launched at this time. There appears to be no scientific publication covering the overall design of this instrument, only a commercial brochure. Papers by Schramm and Walz (1962) and Schramm (1964) described the stigmator of this instrument.

EUROPEAN COMMERCIAL ELECTRON MICROSCOPES 459

FIGURE 32. B. Schramm (left) with the SEM 3 at the 3rd European Conference on Electron Microscopy in Prague, 1964.

FIGURE 33. B. Schramm (right) and M. Rassmuss photographed at the Czech EM meeting in 1959. They designed the WF SEM 3 instrument.

In 1955, the French Company Optique et Précision Levallois (OPL) launched the instrument series MEU. It arose from a long program of development work carried out at the University of Toulouse. The channel for communication was Prof. Fert (Fig. 34), who devoted considerable effort to this project. He collaborated with M. Selme of OPL in the engineering of the first microscope (Fert and Selme, 1956). The MEU IA in 1955 was followed by the MEU IB in 1956. Both were single-condenser 100-kV instruments with three imaging lenses, and thus with microdiffraction facilities. The series was upgraded in 1960 to the MEU IIA, when it was equipped with a double condenser system, bringing the specification in line with the leading instruments from other suppliers. The objective lens was fitted with an electrostatic astigmatism corrector.

In 1958/1959, the Vyborg Instrument Making Plant in the USSR tried to organize the serial production of the UEMB-100 (Stoyanov *et al.*, 1958), which was designed to be a high-resolution microscope, the predecessor of the UEMV 100 to be produced in 1961 in Sumy. The design parameters were not achieved in this batch, and no further transmission electron microscopes were produced by Vyborg. They did, however, make a number of scanning microscopes over the next decade.

C. Electron Microprobe

1958 was important because the first electron microprobe, MS 85, was introduced by Cameca following on from the pioneer work of Castaing and Guinier (1949). It was a fixed-probe instrument, following with modifi-

FIGURE 34. C. Fert (front row, right) at a conference in the 1950s. Also in the front row, left, is R. Castaing, and behind Fert is E. Ruska. Prof. Fert, who was based at Toulouse University, collaborated closely in the development of the OPL electron microscopes.

FIGURE 35. R. Castaing. Photographed in 1951. Pioneer of the technique of electron probe X-ray microanalysis.

cations the ideas presented at the London Conference by Castaing (1956) (Fig. 35).

D. Specimen Techniques

The decade saw a great change in the available specimen preparation techniques. After the first publication of a method for preparing carbon films (König and Helwig, 1951), a much simpler method was devised by Bradley (1954a), and this provided a very thin but strong support which could be used for high-resolution preparations. Major advances were made in biological specimen preparation. The most important advance was the development of ultramicrotomes, which permitted the routine cutting of very thin sections (Porter and Blum, 1953; Huxley, 1956). The studies of Sjöstrand on the sharpening of knives for ultramicrotomes assisted the development of the LKB ultramicrotomes in Sweden. The cutting of thin sections also depended on suitable embedding media. The first was a methacrylate formulation (Newman *et al.*, 1949), and then in 1956 a medium based on the epoxy resin Araldite (Glauert *et al.*, 1956; Glauert and Glauert,

1958) was introduced. These developments, coupled with suitable stains, opened up the huge field of biological research.

At the same time the application of the carbon evaporation technique to the formation of carbon replicas (Bradley, 1954b) enabled a much wider range of surfaces to be examined than was practicable with plastic replicas. The extension of the carbon replica process by Smith and Nutting (1956) to yield extraction replicas from metal surfaces enabled precipitate particles to be removed and examined by electron diffraction. This gave a first possibility of using an analysis technique in conjunction with morphological information.

Later in the decade, the first successful thinning of metal foils for direct examination in the electron microscope was reported (Hirsch *et al.*, 1956a, 1956b; Bollman, 1956; Brandon and Nutting, 1959), to be followed by a flood of other papers giving methods for different materials.

Menter (1956a, 1956b), using a Siemens Elmiskop I, published the first high-resolution micrographs of crystal lattice planes. He demonstrated that one could see individual defects in the crystal lattice, and so opened up a whole new field of investigation.

It can truly be said that this decade was a golden age for new techniques in electron microscopy. These developments on one hand exploited the benefits of the new instruments. Menter's work would not have been possible without the high resolution of the Elmiskop 1. Likewise, the extraction replicas would have been ineffective without the selected-area electron diffraction facility. On the other hand, the possibilities of preparing thin sections for biological research opened the flood gates of demand for new instruments. The total number of microscopes in service in Western Europe at the beginning of this decade cannot have been more than 250, with under 200 in the Eastern Bloc. By 1960 there were perhaps 1400 in the West and a further 1400 in Eastern Europe.

The electron microscopes of the 1950s and early 1960s were very versatile and were capable of all the resolution that could be exploited by the available specimen techniques. It was, however, necessary to have quite a fundamental understanding of these instruments if good results were to be obtained. It needed an intensive course of at least one week, with a good deal of individual tuition, to teach anyone the basics of how to align the lenses and correct astigmatism and to operate a Siemens Elmiskop. This was before the complications of high-resolution dark-field microscopy had been developed, before weak-beam techniques were used, and before the microscopy of radiation-sensitive materials had been seriously attempted.

E. Designers' Luck?

It is worth recording at the end of the consideration of this important decade some of the idiosyncrasies of the famous designers. To take the most famous, Ernst Ruska: He designed the Elmiskop 1 to be operated in a way which ensured that the magnification be accurately known. Thus, the specimen was to be mounted on a platinum disk containing seven holes of 70 μm diameter. This would ensure that the specimen was truly flat. The image magnifications were then to be achieved by combining the objective with one of the four projector polepieces, contained in a turret, with the intermediate lens either on or off—a total of eight magnifications. In this mode, the intermediate lens is excited from the objective lens supply, to ensure a known excitation. These magnifications were to be standardized by fitting the image of a field-limiting aperture in the projector polepieces into a circle inscribed on to the final viewing screen. This was the only attempt I know of to define the electron optical magnification other than by lens currents, which as is well known are but an approximate guide to magnification. It is only recently that concern has arisen for the provision of accurately certified magnifications. A tribute to Ruska's foresight! He did mention in the original paper that the intermediate lens could be independently controlled, to yield a wider magnification range, particularly at lower kilovoltages. This was, however, considered to be a secondary mode of operation by Ruska.

For most operators of the Elmiskop, the "standard" method of operating was unduly restrictive, and it was indeed fortunate that the current control of the intermediate lens was by a stepped switch with accurate wire-wound resistors at each step. The result was that for a number of years, operators of the Elmiskop 1 would calibrate the magnifications with intermediate lens in the free mode, and thereafter take pictures at "Click 7" or "Click 11." It was only some six years after the initial launch of the series that Siemens provided a magnification meter based on the intermediate lens current.

Another feature of the Elmiskop 1 was that there were four ports giving access into the objective lens gap. Two of these accommodated the drives for the objective lens aperture. The other two bores had blanking plugs in them. I can only assume that they were initially put in for the sake of symmetry in the iron circuit. However, after some seven years of production, one of these bores came in very useful for an anticontaminator, and many years later, the other accommodated the first attempts at analysis of X rays emitted from the specimen. Perhaps that is what a really good designer is all about—he deserves his luck!

A further example of designer "luck" comes from the AEI Company, and concerns Mike Haine. He employed quite a lot of effort, both theoretical and practical, in developing the electrostatic/electromagnetic beam-deflector system for the EM6, which enabled the incident electron beam to be tilted through an angle of up to 12°, for reflection microscopy. This device had the advantage of being extremely compact. Reflection microscopy gradually fell into disuse, and with it the need for such beam tilting. However, within another year or two, the materials scientists required a high-angle dark-field tilt facility, and the beam-deflector unit needed only to be mounted on a turntable to fulfill this requirement. This enabled AEI to be the first to provide this particular facility.

V. 1961–1970, THE DECADE OF THE SCANNING PROBES: HIGH NOON FOR THE TEM

A. Electron Microprobes

Electron probe microanalysis started with the classic paper by Castaing and Guinier (1949). The earliest production model by Cameca, noted earlier, had followed the concept of a fixed electron probe located on the specimen by observation under a light microscope. Furthermore, a panel established by the BISRA (British Iron and Steel Research Association) in the United Kingdom to look into the technique had recommended a similar design, and the development work was carried out by Mulvey (1960) at the AEI Research Laboratories at Aldermaston. This design was duly produced in 1960 by AEI in Manchester as the SEM 1. It had a fully focusing spectrometer and two crystals. In the meantime, Cosslett and Duncumb (1956) had started on a very different tack by developing an instrument based on a scanning electron beam. This was to have a decisive advantage, in providing an electron image or an image formed by the characteristic X rays from the area being studied. This was in spite of the disadvantage of only a semifocusing spectrometer. The concept was developed into a working instrument, described by Duncumb (1959). This instrument had a single semifocusing spectrometer and had gas proportional counters. The prototype instrument had aroused the interest of Melford of the Tube Investments Research Laboratories as a potential metallurgical research tool, and as a result of the collaboration, an improved design was evolved (Duncumb and Melford, 1960). The improved instrument had X rays entering the counter with an enhanced solid angle, and the area to be analyzed could be moved from observation under a light microscope to the electron probe by a 180° rotation of a turntable.

When the design was adopted by Cambridge Instruments, they incorporated two semifocusing spectrometers in an instrument which was launched as the Microscan in 1960 (Fig. 36). This was the first commercial scanning electron probe microanalyser (EPMA). An improved version appeared in

EUROPEAN COMMERCIAL ELECTRON MICROSCOPES 465

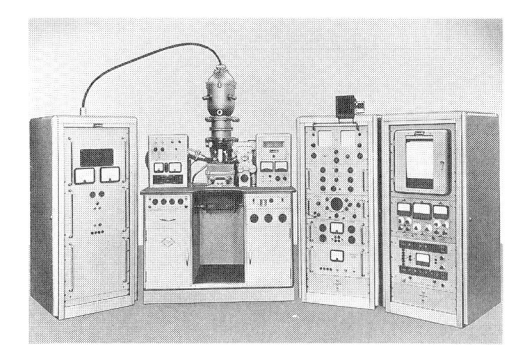

FIGURE 36. Cambridge Instruments Microscan 1. The first scanning electron probe microanalyzer. Two semifocusing X-ray spectrometers. Note the bulky electronics required at that date.

1962 as the Microscan II. It had an element detection range from $Z = 11$ to 92.

Following the success of the scanning concept, the AEI instrument was improved by the inclusion of a scanning system for the electron probe and was launched as the SEM 2 in 1961. It operated at 50 kV and had a resolution of 250 nm (Page and Openshaw, 1960).

Publications by Kushnir (Fig. 37) and his co-workers in the USSR (Kushnir *et al.,* 1961) show that they had also developed a prototype scanning microanalyzer/microscope at about the same time. Their paper quotes the various papers already published in the United Kingdom, and a photograph of the prototype appears to show an inverted column and a crystal spectrometer in an arrangement similar to the Mulvey design. The first instrument had electrostatic lenses, but when designing a production instrument, they adopted electromagnetic lenses in the REMP-1. This instrument, manufactured in the Vyborg plant, first appeared in 1964. An analytical version of the REMP-1, called REMMA, was produced in small numbers by the Sumy factory in 1968–1969. These instruments are perhaps difficult to classify as microprobes, since they might more properly be called microscopes with X-ray analytical facilities, and in this respect they were conceptually very advanced.

By 1964, Cameca had introduced a much improved microanalyzer, the MS 46, which had an operating range from 3 to 40 kV, a resolution of 200 nm, and no less than four spectrometers. It also introduced scanning, though still retaining the optical viewing of the specimen. (In case anyone is confused by the model numbers used by Cameca, they are simply the inverse of the year of issue.)

Cambridge Instruments, following the work of Agrell and Long (1960), who built a prototype analyzer to suit the particular needs of a geology

FIGURE 37. J. M. Kushnir, 1906–1971. SEM designer and leader, Moscow group.

laboratory, produced the Geoscan in 1964. This had a horizontal column, two fully focusing spectrometers, and, significantly, a high take-off angle of 75° through the lens to reduce X-ray absorption effects in the specimen. It appeared in improved form as the Microscan 5 in 1968. Thus, in the course of only 10 years, electron probe microanalyzers moved from a few prototypes to a state where there were some 400 production instruments in operation.

B. Scanning Electron Microscopes

Following the early pioneering work by Knoll (1935), by von Ardenne (1938a) and by Zworykin, Hillier, and Snyder (1942), little attention was paid to the scanning electron microscope until Oatley in the Engineering Laboratory at Cambridge University inspired a series of talented research students in a round of classic investigations into the subject (McMullan, 1953; K. C. A. Smith, 1956; Everhart and Thornley, 1960; and others). The potential fields of application were reviewed early by Smith (1955). In 1961, Smith (Fig. 38) completed the building of an instrument (SEM 3) which had been commissioned by the Pulp and Paper Research Institute of Canada (Smith, 1961). This instrument was based on a column from an AEI EM 4 microscope. It was fitted with a Liebmann pinhole lens (Liebmann, 1951), designed in collaboration with Page of AEI, which was used as the final probe-forming lens, with the specimen outside the magnetic field. It was the first instrument to include the new Everhart-Thornley detector. It was designed so that the camera and viewing chamber of the EM 4 could be fitted in place of the final lens and stage, to enable it to be used as a transmission electron microscope. Alternatively, by fitting a special target

FIGURE 38. K. C. A. Smith. Built the first scanning electron microscope to be delivered to a customer. In collaboration with V. E. Cosslett, built the 750 kV high-voltage microscope in Cambridge. He was also the pioneer of computer-controlled scanners.

holder in the upper polepiece of the final lens, it could give X-ray projection images. In this case, a special specimen stage with provision for adjustment of the target to specimen distance replaced the scanning reflexion stage. This extensive specification reflected the uncertainty of success with the fiber specimens to be examined. As it turned out, the scanning pictures were a great success, and the workload was so heavy that the TEM and X-ray facilities were never tested in practice. This may have been unfortunate for posterity, because the additional components needed for the X-ray projection facilities were minimal, and the complementary information might have been valuable in some investigations if the facility had been incorporated in future production instruments.

In the meantime, AEI had designed a scanning microscope in house, and the prototype went to the Physical Chemistry Department at Cambridge University. Perhaps because this work made very little use of the expertise of the Engineering Laboratory in Cambridge, the design was never very satisfactory, and the management of AEI decided that the considerable effort needed to redesign this instrument could not be sustained, in view of the limited engineering effort available in the department. If this now seems incredible, it has to be set against the fact that the development team for the TEM consisted of just three professional engineers at that time. Additional staff were employed on the electron probe microanalyzer, and it seems that there was not the will or vision to invest the needed effort in the scanning instrument. The major defect in the AEI effort was that they did not adopt many of the design features already evolved by Oatley's team (the NIH, or "not invented here" syndrome). A review of the detailed

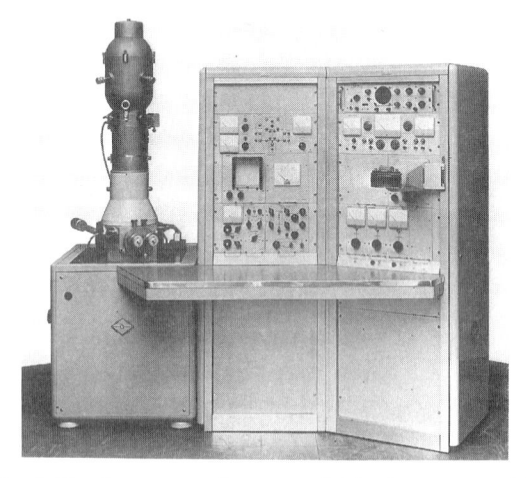

FIGURE 39. Cambridge Instruments Stereoscan 1. The first commercial scanning microscope in series production: 20 kV, resolution 50 nm.

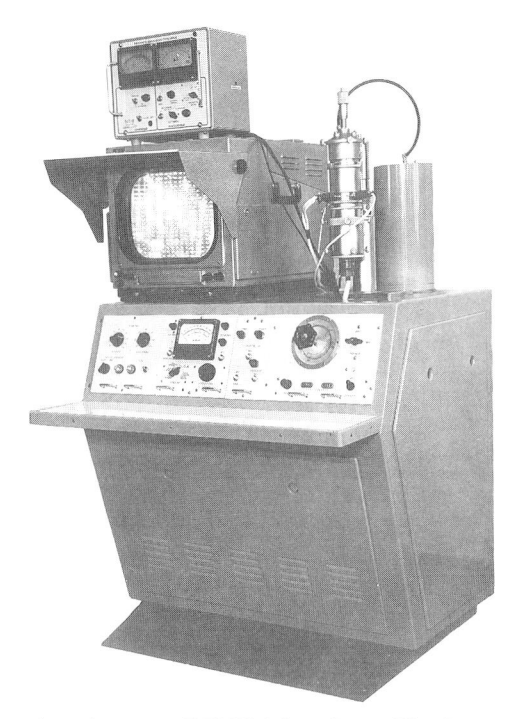

FIGURE 40. Scanning microscope REMN-1 from Sumy. The first scanning microscope manufactured in the USSR. Operated at the very low voltages of 0.5–2.3 kV, it had only absorbed current detection.

history of the events leading up to the production of scanning microscopes by the Cambridge Instrument Company is recorded by Jervis (1971) and Oatley (1982). At all events, this decision by AEI was probably the most significant factor in their later withdrawal from electron microscopy altogether.

When the Cambridge Instrument Company decided to take up the manufacture of scanning microscopes, they made the critical decision to place a postgraduate, Gary Stewart, from the Engineering Department, to head the team. This ensured a close collaboration with the university, and a minimum delay in launching the product. The first series production scanning electron microscope was produced by the Cambridge Instrument Company in 1965, launched as the Stereoscan (Fig. 39) (Stewart and Snelling, 1965). This had an accelerating voltage of 20 kV and a resolution of 50 nm. The world now began to appreciate that the scanning microscope offered quite distinct advantages over the transmission instrument, even

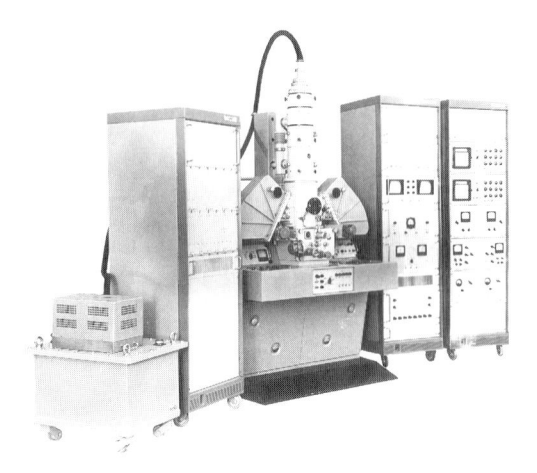

FIGURE 41. REMMA from Sumy. Based on the REMP-1, which was built in Vyborg. Its main imaging mode was by secondary electrons, and the voltage range was 10–50 kV. Elements in the range 5 to 92 were analyzable.

though it could not compete in resolution. The main advantages were: little or no specimen preparation; direct examination of surface structure; large depth of field, tending to give an apparent three-dimensional picture; and the possibility of observing large specimens. The market was in fact much more closely related to light microscopy than to the area covered by transmission electron microscopes.

An improved instrument, the Stereoscan 2, was produced in 1966 with an operating range up to 25 kV and a resolution of 50 nm. The Stereoscan S.2A of 1968 had a voltage range up to 30 kV and a resolution of 25 nm. Two years later, the new Stereoscan S 4 had a further improved resolution of 15 nm, partly attributable to a more efficient gun, and with facilities for fitting an energy-dispersive X-ray detector (EDX).

An early scanning microscope was produced in 1967 in Sumy for the Russian market. This was the REMN-1 (Fig. 40), a low-voltage (0.5–2.3kV) scanner using specimen current imaging. It appears that this was intended for use in examining semiconducting devices. In 1968, the Vyborg plant introduced the REMP-2, a scanning microscope with a three-crystal spectrometer, operating at up to 50 kV. In the same year, another ambitious design appeared from the Sumy works—REMMA (Kisel *et al.,* 1970), also with a conventional voltage range of 10–50 kV and a 50-nm resolution (Fig. 41). It was based on the REMP-1 scanning microscope/analyzer, which had earlier been made at Vyborg. It was equipped with a wavelength spectrometer for analysis of elements between boron (5) and uranium (92).

1969 saw the founding of the Cambridge Scanning Company (later known by its abbreviation, Camscan), which initially produced a simple SEM for routine visualization, and which therefore concentrated on features to help the operator. The most important idea was the provision of TV rate viewing as standard, and with an automatic airlock sample exchange. The instruments were sold by Bausch and Lomb as Balscan instruments.

In 1970, Cameca introduced the MEB 07, which operated over the voltage range 1–50 kV, having a 20-nm resolution and a goniometer tilt stage.

By the end of this decade, there were probably 900 of these new microscopes in service, and the new instrument had established a clear position in the market, ready for a spectacular climb in the 1970s.

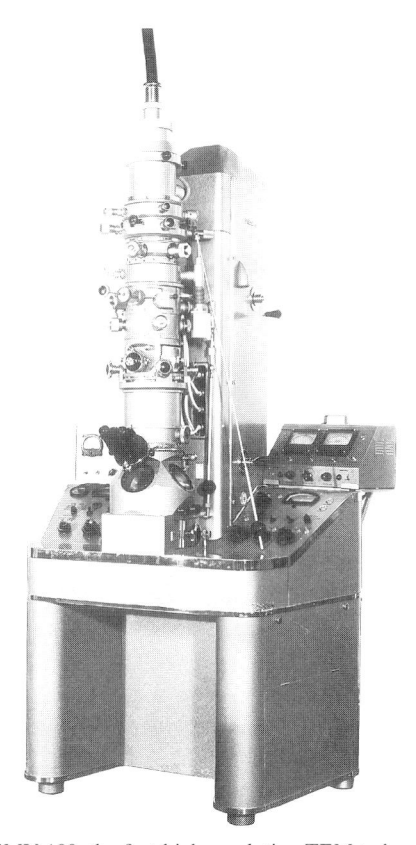

FIGURE 42. The UEMV-100, the first high-resolution TEM to be produced in the USSR, operating at up to 100 kV and within 1-nm resolution. A number of different models were produced to meet specialized requirements; the most striking perhaps was the UEMV-100B, which accommodated a wet stage for examination of biological specimens.

FIGURE 43. P. A. Stoyanov, designer of some 16 different versions of TEM.

C. Transmission Electron Microscopes

The rapid spread of numbers of electron microscopes in biological laboratories encouraged the development of instruments suitable for this market. At the same time, increased specialization for materials science applications was provided on other models.

In 1960, the Sumy factory "Electron" in the USSR started to launch the UEMV-100 (Fig. 42), the first high-resolution microscope to be made in that country. In view of this, they ceased production of the EM 5 the following year, though production continued at Krasnogorsk (KMZ) until 1962, when production was phased out there in favor of the EM 7 (Vertzner *et al.*, 1961). The UEMV 100 was a 100-kV, 1-nm instrument incorporating double-condenser illumination, with a stigmator for the second condenser as well as for the objective. The illumination could be tilted to yield high-resolution dark-field pictures. It had a 24-plate camera. The UEMV 100A in 1964 was provided with specimen heating and straining stages and an anticontaminator. In 1965, a version of this instrument, the UEMV-100B (Stoyanov *et al.*, 1965) (Fig. 43), was introduced for investigating living biological specimens in a gas cell, which could operate up to atmospheric pressure (Stoyanova, 1958). (Audrey Glauert later, perhaps more accurately, described it as a study of dying biological specimens, due to the radiation received). Nevertheless, Sumy produced no fewer than 96 of these instruments, something not attempted elsewhere in the world.

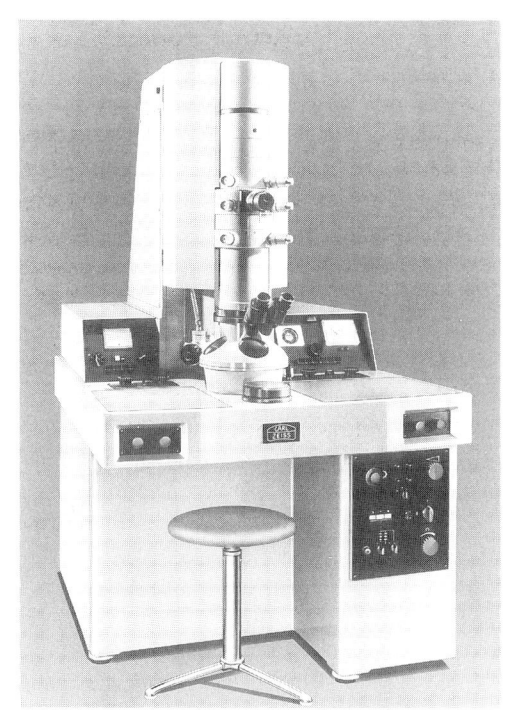

FIGURE 44. Zeiss EM 9. A very successful simplified electron microscope: 60 kV, 1.2 nm. It had automatic focusing and an automatic camera.

The model was further developed in 1965 as the UEMV-100V with an improved resolution of 0.8 nm; a metallurgical instrument (UEMV-100K) followed in 1967, fitted with a goniometer stage, though tilting only ±10°.

In 1961, Zeiss Oberkochen, having abandoned electrostatic lens microscopes, brought out their first electromagnetic microscope, the EM 9 (Fig. 44) (Gütter, 1961; Gütter and Mahl, 1962). This was a simplified microscope operating at 60 kV with a 1.2-nm resolution. By using a telefocus gun, it dispensed with any condenser lenses. It had a rod specimen holder, an auto-focus facility, and an auto camera. The top magnification was 35,000. This was the first simple instrument to appear with the necessary user friendliness, coupled with reasonable performance, which could wean some customers away from the apparently insatiable demand for the maximum possible resolution and maximum facilities. The reward was a steady demand for many years. Although it made no special claim as such, it could have been labeled as a "biologists" instrument. It had upgrades in 1966

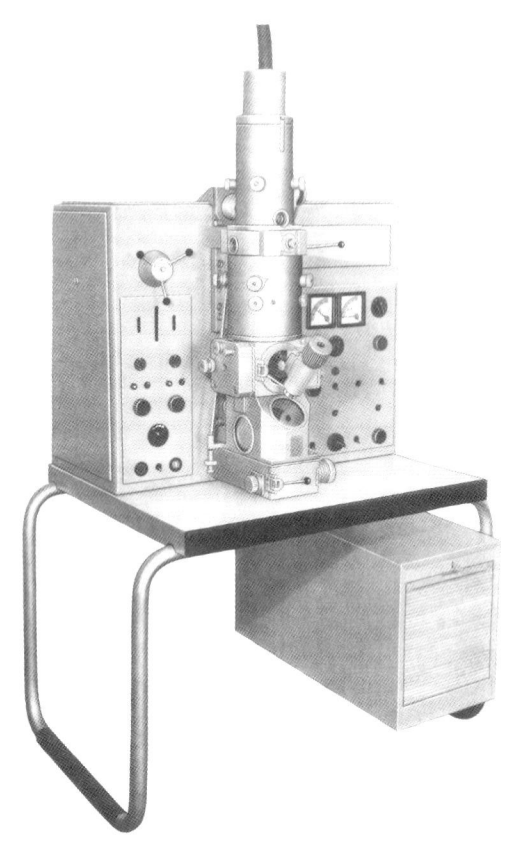

FIGURE 45. Carl Zeiss Jena model EF 4 modular microscope for transmission and reflexion microscopy and diffraction.

(EM 9A), 1968 (EM 9S) (Gütter and Mahl, 1967), and 1970 (EM 9S-2).

In 1963, VEB Carl Zeiss Jena introduced the model EF4, which had accelerating voltages of 35, 50, and 65 kV, and a resolution of 2 nm (Fig. 45). It had three magnetic imaging lenses. It included an electrostatic/electromagnetic beam centering arrangement, and the beam could be tilted up to 30° for reflection working. Apart from selected-area electron diffraction, there was a facility for mounting the specimen beneath the projector to achieve high-resolution diffraction in both transmission and reflection. The instrument was designed in modular form, and in 1964 two new modules appeared: The EF5 offered heating of the specimen to 1000°C, evaporation of materials onto the specimen, and etching of the specimen surface by ion

FIGURE 46. A. Delong and V. Drahoš (seated), with an early version of the Tesla BS413.

beam etching. An additional feature was a heating stage for large-reflection diffraction specimens, for temperatures up to 900°C. The other module, the EF6, was an emission microscope system which will be described later, in the section on emission microscopes. The instruments were described as aiming for the physical and materials science markets, although able to be used for life science research. The fact that of 180 EF4 instruments sold, 125 were also sold with the EF6 module, suggests that there was a very heavy bias toward the physical sciences at that time in the former East Zone of Germany and in the USSR, where a number of these instruments were sold. The explosion of interest in biological studies in the West had clearly not been reflected in the East. It also appears that, with a top accelerating voltage of 65 kV, the pressures from materials scientists for voltages high enough to penetrate thin metal foils had not then developed in that country. It should be recorded that this was the last electron microscope produced by Carl Zeiss Jena. In 1970 they stopped this production to concentrate on electron-beam control and measurement instruments, particularly for microelectronic circuits.

Tesla launched a high-resolution instrument, the BS 413, in 1964, based on the prototype described by Delong, Drahoš and Zobač (1961) (Fig. 46).

In 1964 AEI finally launched an instrument aimed specifically to attract biologists—the EM 6B (Page, 1964) (Fig. 47). It used a very asymmetric arrangement of the bores of the objective lens to achieve very low aberration coefficients, combined with a limited prefield. This was effectively a Lieb-

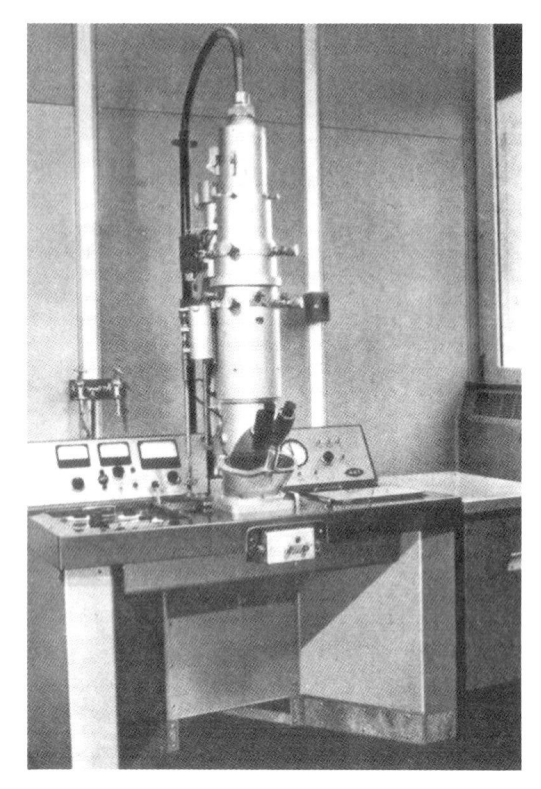

FIGURE 47. AEI EM 6B, the first microscope designed specifically for biology. It had very simple controls—the first zoom magnification, with the full magnification range covered by one control without realignment. It also had compensation of objective and condenser excitation for change of magnification. A two-stage astigmatism corrector gave very accurate compensation.

mann pinhole lens (Liebmann, 1951) slightly modified to allow space for a robust specimen rod and for the upper and lower plates of the anticontaminator. It also avoided the alignment problems associated with the condenser-objective lens of Riecke and Ruska. This lens, together with extensive magnetic screening of the column, combined to result in a spectacular improvement in resolution, coupled with a useful increase in image intensity caused by the low-aberration prefield. It became very easy to observe phase effects at focus, and to achieve a regular 0.5-nm resolution. It had a two-stage objective stigmator and a built-in anticontaminator. Thus it was the first microscope to be fitted with an anticontaminator at its introduction, although the principles had been known for a long time (En-

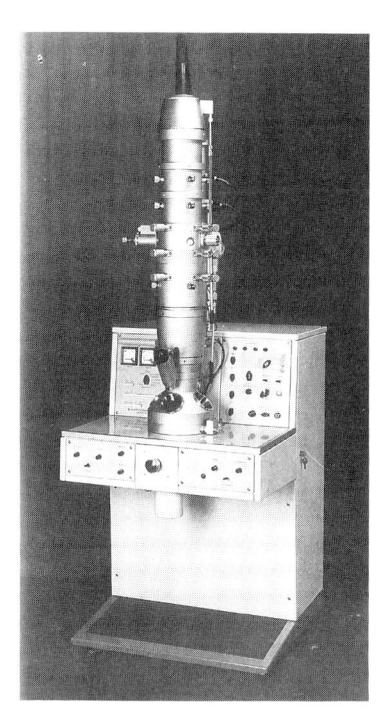

FIGURE 48. WF (TV factory) SEM 3-2. An advanced design of TEM with four imaging lenses and a built-in anticontaminator. Options for a heating stage and a high-resolution electron diffraction stage.

nos, 1954; Schott and Leisegang, 1956; Heide, 1958). It had a single knob controlling magnification over the whole range from 1500 to 250,000. This control also simultaneously adjusted the focus and controlled the condenser lens excitation to maintain illumination with change of magnification. The triple-specimen rod holder saved time in specimen changing and enabled comparative studies to be made easily. As a further contribution to ease of use, the microscope was designed so that most lens alignments could be preadjusted in the factory and did not need attention from an operator. This was a remarkably innovative instrument and remains the best tribute to Dick Page, a very talented but underrecognized engineer. The whole instrument was evolved by Page and two other engineers, Ray Coles and Reg Beadle.

1965 saw the introduction of an improved version of the SEM 3, the SEM 3-1, by the WF Factory in Berlin (Schramm, 1966; Butzke *et al.*, 1966). The instrument was now equipped with a double-condenser lens system, individual mechanical alignments for each lens, and a stigmator in the

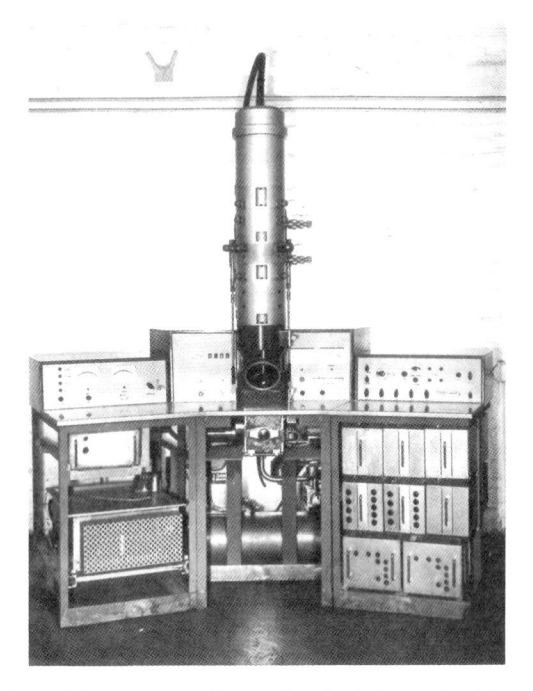

FIGURE 49. WF SEM 4 prototype, with transistorized electronics. This model was withdrawn in 1970.

second condenser. There had been significant improvements in the electronic stabilisation of high voltage and lens currents, bringing the resolution to 1 nm or better. There was also an anticontaminator in the objective lens, facilities for a heating stage to 1000°C, and for a high-resolution diffraction stage. Apart from being used for microdiffraction, the first intermediate lens was used only to demagnify the image in the lower part of the magnification range. The second intermediate lens was used to magnify the image in the higher range of magnification.

The instrument was further improved in 1967 to the model SEM 3-2 (Fig. 48), with a magnification range up to 200,000 and a reduced illumination spot size of 2 μm. The instrument was now a very good high-resolution instrument, even though not achieving the ultimate resolutions available from some of the Western European microscopes. It was a tribute to the excellent design work of Schramm and his colleagues. Some 69 of the SEM 3 series instruments were manufactured. At the conclusion of the production run of the SEM 3-2, a new instrument, the SEM 4, had been developed. This had a fully transistorized power supply, an accelerating voltage range from 40 to 100 kV, and a resolution of 0.5 nm (Fig. 49). It

FIGURE 50. Philips EM 300. High-resolution general-purpose instrument. Excellent specimen tilt made it suitable for metallurgy as well as biological applications. The best-selling electron microscope; 1850 units were produced.

was not, however, put into commercial production because the management of the factory decided in 1970 to concentrate production on microelectronics. It was interesting that in the same year, Zeiss Jena also discontinued production of electron microscopes in favor of electron-beam control and measurement instruments. These may have been ordinary commercial decisions, as there is no documentary evidence to say otherwise. However, it may be signficant that, also in 1970, the production of electron microscopes in the USSR was sharply reduced at all sites except at Sumy, where production was greatly expanded at this same time. This makes it at least plausible that a policy decision had been made to concentrate all electron optical work in Comecon at this unit.

In 1966, AEI introduced the metallurgical electron microscope EM 6G, with a built-in electrical beam tilt capable of high angle (up to 3°) beam tilt for high-resolution dark-field experiments and up to 12° for reflection microscopy. It also had a high-tilt specimen holder operating at high resolution. This was thus another instrument which had facilities to meet the needs of a developing special market—in this case for materials science. Following the pioneering work of Haine *et al.* (1958; Haine and Einstein, 1960), who built an early image intensifier inside the microscope vacuum, a commercial form of image intensifier was developed to fit to the EM 6, EM 6B, and EM 6G instruments (Anderson, 1968). In this design, a transmission screen beneath the microscope camera was linked by a wide-aperture lens to the input of a four-stage intensifier, whose output was in turn picked up by a vidicon camera to yield a picture on a monitor screen. It was possible to photograph Fresnel fringes from an image at the input of the intensifier with a beam intensity of only 50 fAcm^{-2}.

In the same year, 1966, Philips introduced the extremely successful EM 300 (Fig. 50), the first fully transistorized TEM, with excellent viewing facilities provided by the large fluorescent screen (Rakels et al., 1968). It had the best styling yet seen for a European electron microscope and a very good ergonomic design. It has to be said, however, that in this design aspect, Europe lagged far behind the United States. The RCA Company had as early as 1952 employed an industrial design expert—an exceptionally good one—in the early planning of the EMU 3A microscope (which appeared in 1954). The desk and controls evolved in this design looked not so very different from the 1966 EM 300 (allowing for the introduction of transistorized circuitry in the meantime).

The gun airlock of the EM 300 was another innovation. Crucially, Philips managed to solve the very difficult mechanical problem of incorporating a tilt mechanism within the specimen rod, and thereby made the instrument suitable for metallurigcal as well as biological use. It was therefore truly a universal instrument, and became the best-selling electron microscope of all time.

A very important new concept was published by Riecke and Ruska (1966). This was a new high-resolution microscope fitted with a symmetrical condenser-objective lens. This lens has become an essential feature of very-high-performance TEM instruments today. Its other significance, perhaps unremarked at the time, was that the condenser system had been carefully designed to integrate with the image-forming system. Previous to this publication, the condenser system had been very much the "Cinderella" part of the electron optical system, and seemed not to have received adequate study.

In 1968, Siemens brought out the Elmiskop 101 (Asmus *et al.,* 1968), which had improved electronic stabilization compared with the Elmiskop 1A, and guaranteed a resolution of 0.5 nm. It was optimized for efficient X-ray output from the specimen.

In 1968, AEI replaced the EM 6B and EM 6G with the EM 801 and EM 802. The EM 801 corrected a defect in the EM 6B which had not been at first apparent. Low-magnification pictures of thick specimens showed unacceptable loss of definition around the extremities of the image. The effect was due to chromatic change of magnification, a defect not previously recognized as being specially important. It had been discussed by Liebmann (1952), who showed that the coefficient of chromatic change of magnification was zero at the minimum focal length of the projector. He assumed that it would normally be negligible for other excitations by calculating the size of the defect for the expected variation in the high-voltage supply. This calculation overlooks the very high energy losses suffered in a thick specimen, which are most obvious at low magnifications, when the field of view is large. Consequently, one really needs to run the final projector at minimum focal length for all magnifications. There then remains a much smaller component of this error due to the weaker intermediate lens. Accepting that the final lens strength cannot be changed significantly, it now becomes imperative to provide an additional intermediate lens to achieve the required magnfication range. AEI did not at the time accept this logic, but by juggling the lens parameters for the EM 801 succeeded in reducing this chromatic defect to a level where it was scarcely apparent, without appreciably compromising other lens properties. (This was obviously not an ideal solution, and Philips later introduced a more complete solution by taking advantage of their more flexible projector system with its extra lenses.) The whole episode is another illustration of lens designers using the raw data for lens performance, without checking for the effects which might be introduced by thicker specimens. The EM 801 also had a six-position specimen holder and high-angle specimen tilt, now becoming important for interpretation of structures in biological sections. There was an alternative specimen holder with a long slit which could be continuously scanned to examine a ribbon of serial sections.

Philips introduced the first European-based scanning transmission (STEM) attachment for their EM 300 microscope in 1968. This was based on the work of Ong (1968), who showed that by operating an EM 300 microscope with increased objective excitation, so that the specimen lay in the midplane of the lens, one had achieved the symmetrical condenser-objective operation which resulted in a very small illumination probe, as well as high resolution from the objective lens. This enabled the lens to be used in the STEM mode, with a very good resolution. This had particular relevance for thicker specimens, which would have high chromatic losses in the normal TEM mode. In the STEM mode, by contrast, the aberrations of the objective lens have no effect on the resolution of the image. The probe could also be used to yield normal scanned images from a solid specimen of small enough dimensions to fit inside the polepiece gap. In an independent piece of research, K. C. A. Smith and Considine

(1968) reported at the Rome conference that they had used a symmetrical condenser-objective in their high-voltage electron microscope. In addition, they had constructed a crude energy-analyzing system. The work had been presented orally at an earlier meeting of the EM Group of the Institute of Physics, but was mentioned only as an abstract.

In 1969, Sumy produced the EMV-100L (Stoyanov et al., 1970). (The suffix "L" was in celebration of the 100th anniversary of the birth of Lenin.) It was a 100-kV multipurpose instrument with a resolution of 0.3 nm and with a magnification range from 400 to 600,000. All the lenses in this instrument were constructed in μ-metal (an alloy of 78% Ni-Fe and 50% Ni-Fe) in order to minimize hysteresis and stray magnetic fields. The specimen stage had thermocompensators, and was designed to be vibration-proof. This was an instrument of considerable personal interest. I attended a Trade Fair in Moscow that year, following which I was privileged with an invitation to visit the microscope factory in Sumy and to lecture on the then new AEI EM7 high-voltage microscope. I saw a very impressive range of electron optical instruments in Sumy, including a special scanning microscope for examining semiconducting devices, as well as diffraction cameras, neither being at that time manufactured in the West. I was then led to the "clean room" of the test area to operate this EMV-100 instrument. There was astonishment at the size of my feet, as they could find no overshoes big enough to fit! Some resourceful improvization soon had me suitably clad for access to the clean area, where I found the instrument performing impressively. Figure 51 shows a version of this microscope, which was the first instrument in the USSR to guarantee a resolution of 0.3 nm.

D. Simplified Instruments

The interest in simplified instruments seems to have been rekindled at this time. It is not difficult to imagine why the various manufacturers should market such instruments. It was quite apparent that for almost all biological work, the resolution of the micrograph would be limited by the specimen itself to a figure significantly worse than the resolution limit of the top-line instruments then available. For some particle examinations also, an instrument of more modest performance would be adequate.

Siemens started with the Elmiskop 51, which had a turntable specimen stage carrying 15 specimens available for interchange (Müller and Ruska, 1958). It was a 50-kV microscope with a cold-cathode source, and with permanent magnet lenses, which provided a fixed magnification of 9200 at the photographic plate. It had a magazine of 24 plates. It was clearly designed for routine repetitive work with a large throughput of specimens.

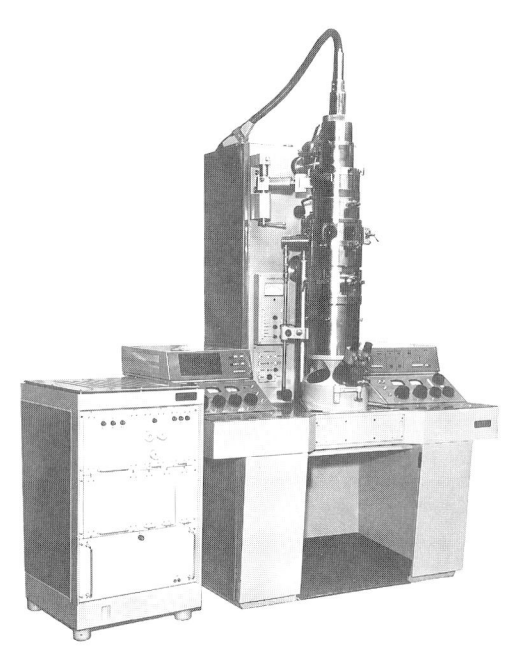

FIGURE 51. Model EMV-100B from Sumy. The first microscope in the USSR to guarantee a resolution of 0.3 nm.

Because of the large number of specimens which could be loaded at one time, it dispensed with a specimen airlock.

The WF Company improved the small microscope KEM 1 in 1963 by the addition of an octupole electomagnetic stigmator, electrically aligned, and the instrument became the KEM 1-1 (Rassmus, 1966). A further change in 1967 brought a plate camera in addition to the film camera, the type number being changed to KEM 1-2 (Fig. 52). It now had a resolution of 1.5 nm and an increased top magnification of 50,000.

Russian interest in simple instruments continued. The EM 7 was an improvement on the EM 5, being now provided with a specimen airlock. Production started in Krasnogorsk in 1961. A paper at about this time (Kushnir, 1962) classified instruments then available into three classes. Class 1 instruments had a resolution of 0.8–1.5 nm; class 2 instruments had a resolution between 2 and 3 nm; and class 3, between 5 and 15 nm. The paper reviewed the instruments Kushnir knew of at that time; the above-mentioned EM 7 was a class 2 instrument. An instrument of the third class produced throughout the decade after its launch in 1959 was the EM 6. It was a highly simplified model with objective and single projector lens

FIGURE 52. Model KEM 1-2 from WF (TV factory), East Berlin. This was a routine microscope, but it had an electromagnetic stigmator, and electromagnetic alignment of the beam from the gun. It had a choice of plate camera or a small-format, 50-exposure camera with picture size 24 mm × 24 mm.

producing an image on a single-crystal screen. This was then examined with a ×20 light microscope, the image being recorded on film. It had a rod-type specimen airlock. A later variant (EM 9) provided an option for in-vacuum photography. These instruments were also produced in Krasnogorsk (KMZ).

Three companies produced their first—and last—instruments at this time. In 1967, the successor to the OPL company (now called Société d'Optique, Électronique et Méchanique—SOPELEM SA) introduced the model Micro 75, working at 30–75 kV. It was intended as a routine instrument, with a magnification range of 2000–128,000.

Polaron in England in 1969 launched the MR 60, operating at a fixed 60 kV and with a resolution of 2 nm. The illumination was from a telefocus gun, with no condenser lens. A second version of the instrument, the MR60C, had a normal gun and a double-condenser lens system. In this form, the instrument had a resolution of 1.2 nm. These instruments were later taken over and marketed by Miles Engineering, but did not long survive.

In 1969, the Jugoslav Iskra LEM 4C (Fig. 53), a 50-kV, 1.5 nm instrument appeared. In an introduction to the concept of the instrument (Millen, 1969), adapting a paper prepared by Strojnik, analyzed the 1500 micrographs presented at the Third European Regional Conference, and showed that about 70% were at a magnification of 40,000 or less, even after photographic enlargement. The LEM 4C therefore aimed to satisfy this part of the market. In order to avoid the complication of a double-condenser system, use was made of a specimen partially immersed in the objective lens field, so that the prefield acted as an additional condenser. It was equipped with a single-

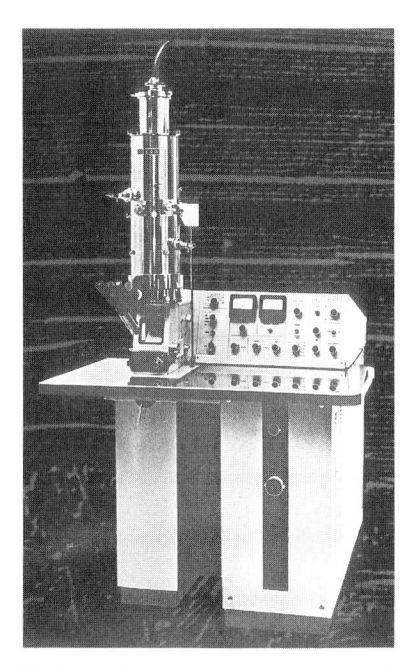

FIGURE 53. Iskra LEM 4C, a medium-power instrument from Jugoslavia, using the objective prefield to reduce the size of the illumination spot at the specimen.

knob magnification control and a wobbler alignment system, and depended on a 35-mm camera system. This instrument was intelligently conceived, but probably suffered from an insufficient market base, and was eventually discontinued. It was based on designs by Strojnik (1964) and Strojnik and Kralj (1964). It turned out that none of the above companies could market enough instruments to cover the substantial costs of the development.

In 1970, Philips produced the EM 201, an instrument designed to be easy to use, have good resolution, and yet be economically priced. It had a guaranteed resolution of 0.5 nm, with an accelerating voltage range of 40–100 kV. It had a single minicondenser lens in lieu of the double-condenser system of the EM 300, and an electromagnetic alignment system for simplicity. Some of the lenses were factory prealigned. It was reasonably successful, but in no way matched the popularity of the EM 300, perhaps because the price differential for the reduced specification could not be made sufficiently attractive. This microscope was one of many examples of the difficulty of designing a simple instrument acceptable to the market. The problem is that many of the expensive components of a universal microscope are still required for a simpler one for it to have an acceptable specification. The components one can omit often do not make for major savings in the cost of the instrument. The perceived need for a simple instrument is illustrated by the number of models produced. The most successful turned out to be the Zeiss EM 9 and its successors. The design was conservative and more limited in performance than some of its competitors. What its designers got right was the need for the utmost simplicity in operation to help those who did not want to become experts in electron microscopy. Substantial numbers of these instruments were produced.

E. Emission Microscopes

In 1961, Trüb-Täuber introduced an emission electron microscope, designated the Metioscope KE, and evolved through the collaboration of Trüb-Täuber with Möllenstedt and Düker (1953) at Tübingen University. In this instrument the surface of the specimen was bombarded with an ion beam to give secondary electron emission. This was accelerated in an electrostatic immersion objective and the image magnified by a magnetic projector lens. The specimen temperature could be varied between room temperature and 1200°C. The specimen surface could be modified during examination by ion-beam etching procedures. This instrument was improved, and in 1964 the KE 2 was launched. In the period from 1961 to 1967, 20 of these instruments were shipped. The first 13 of these were manufactured by Trüb-Täuber, after which the business was taken over by Balzers in 1965, and they shipped all future production. A survey of the instrument and its applications was published by Wegmann (1972).

The Metioscope KE 3, which was a photoemission electron microscope, evolved from this in 1965 (Wegmann, 1968, 1972), after the work of Engel (1966). In order to release the photoelectrons, the specimen surface was irradiated with ultraviolet light from a mercury pressure lamp. The light collected from the specimen surface by the quartz lens system was focused from an anode acting as a mirror. The photoelectrons were accelerated from a very low velocity in an electric field of 8–10 kV/mm between the specimen surface and the anode. After passing through the anode aperture,

---

FIGURE 54. (a) Trüb-Täuber Metioscope KE3 photoemission electron microscope, in basic operation. (b) KE3 in operation in Leeds University School of Materials using the high-temperature straining stage, which sticks out at the top of the instrument and which contains the displacement dial G (out of sight in the roof recess). Operator A has the job of reading this displacement gauge. Operator B, largely hidden in front of A, is recording the temperature on a galvanometer just behind his head. C is focusing the instrument, and also controls the specimen traverse. D is operating/controlling the load applied to the specimen (and who has to interact with A so that the displacement/strain rate is slow and uniform). E is out of sight and controls the room light switch.

EUROPEAN COMMERCIAL ELECTRON MICROSCOPES 487

a

b

the image was magnified by a magnetic projector lens system. The instrument was provided with a treatment chamber, where the specimen surface could be cleaned by ion bombardment before examination. The specimen could also be treated by heating in the presence of a vapor or gas during examination. In addition, it could be strained while being examined. The instrument operating in a basic mode is shown in Fig. 54a. When used for more complicated experiments, the operation became complex. Figure 54b shows staff from Leeds University School of Materials using the high-temperature straining stage—clearly a team effort was needed!

In 1964, as mentioned earlier in the section on transmission electron microscopes, VEB Carl Zeiss Jena introduced the EF 6 (Fig. 55), which was one component of the EF electron optical system, providing emission microscope facilities. The specimen could be operated at temperatures between 700 and 2000°C, heating being by means of electron beam dissipating 100 W. Temperature measurement was by built-in pyrometer or a

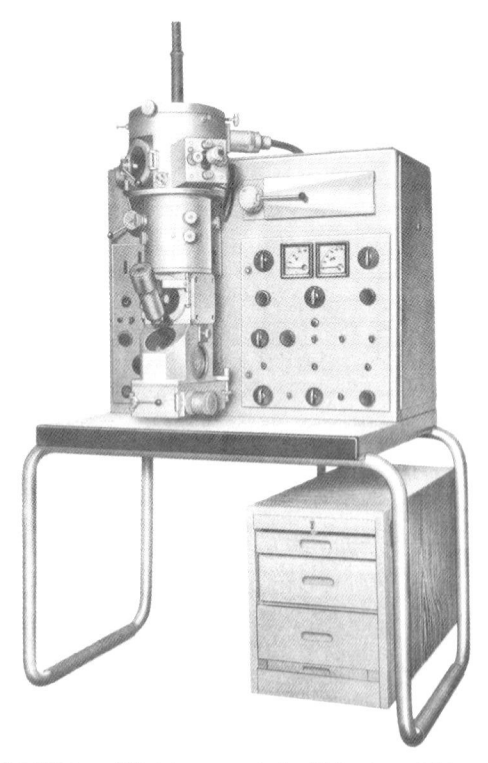

FIGURE 55. Model EF 6 on EF 4 base, made by Zeiss Jena. This emission microscope could operate at temperatures up to 1000°C, and had facilities for secondary electron imaging from electron or ion beams.

thermocouple. Secondary electron imaging could be generated either from an ion beam or a primary electron beam. The striking feature of this instrument is the large number that were sold—125—far more than the total of TTC and Balzers units in Western Europe. It transpires that a number of these instruments were bought for use in the USSR, since their own emission microscope design, the EEM 50, was less successful, only two models being built.

F. Electron Diffraction Cameras

In contrast to the lack of interest in electron diffraction cameras in Western Europe, production started in Sumy in 1960 and within the decade of the 1960s at least 250 instruments were produced. The first universal model EG-100A was launched in 1960, and had voltages available from 40 to 100 kV and a resolution of 0.001 nm for atomic spacings $d = 0.05$ nm and a camera length of 755 mm. Ellipticity of the diffraction rings was less than 0.4%. Shortly after came the EVR-1, a diffraction camera with a chart recorder (Fig. 56), designed for structural investigations of solid crystals and amorphous materials. It had accelerating voltages of 40, 50, and 60 kV. Ellipticity of the rings was better than 0.3%.

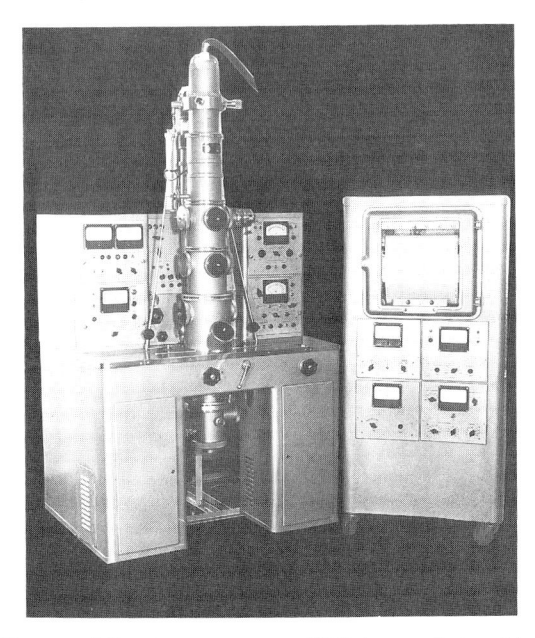

FIGURE 56. Electron diffraction camera type EVR-1, manufactured in Sumy. Operated at 40–60 kV. Ports for three camera lengths.

FIGURE 57. Diffraction camera ER 100, introduced in 1967. Voltage range 25–100 kV.

A later (1967) version of the EG 100A, the EG-100M, allowed examination of materials heated to 1000°C or cooled to −140°C, and heated vapors and gases in the range 300–400°C. Also in 1967, a further new instrument, the ER-100 (Fig. 57), was introduced, with accelerating voltages of 25–100 kV and a stability of the emission current better than 2%. Specimens up to 0.1 μm in thickness could be examined in transmission, and reflexion specimens up to 10 × 10 × 3 mm in size.

Electron diffraction was not entirely neglected in the West; AEI, for instance, produced a high-resolution diffraction attachment for its transmission microscopes. This fitted into a port on the side of the viewing chamber, and employed the very fine electron beam formed by the projector system to yield diffraction patterns of very high resolution. Of course, the camera length was limited by the dimensions of the viewing chamber, but it was adequate for a wide range of specimens.

G. Analytical Microscopes

In 1969 AEI had adopted the design of the electron microscope/microanalyzer (EMMA) which had been developed by Cooke and Duncumb (1968) at the Tube Investments Research Laboratories. This was launched as the first analytical electron microscope EMMA-4 (Fig. 58) (Cooke and Openshaw, 1969). It used a minilens to reduce the illuminating spot size on the specimen to 0.13 μm. It was fitted with two fully focusing spectrometers with interchangeable crystals. There was an excellent high take-off angle of 45° for the X rays, and the electron image remained available during the analysis. While the instrument was undoubtedly a technical success, it was not continued for very long in production. This was because the rapid introduction shortly after of improved solid-state X-ray detectors

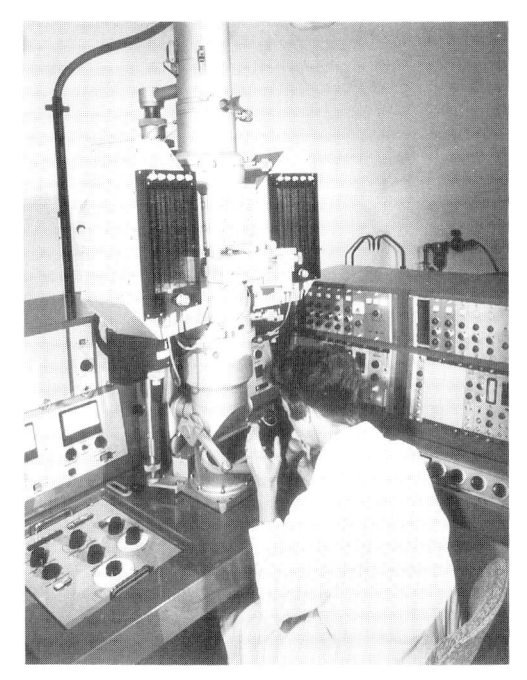

FIGURE 58. AEI EMMA-4 electron microscope-microanalyzer. The first analytical TEM designed as an integrated instrument. Two fully focusing spectrometers. Spot size reduced to 0.13 μm by use of a minilens as condenser 3.

made it feasible to fit them to conventional transmission instruments, and to offer some kind of X-ray analysis as an accessory. Only those who had problems requiring the extra sensitivity of X-ray detection of the EMMA 4 and the smaller volume analyzed found it worthwhile to buy a special instrument rather than having an add-on to an existing microscope.

One of the earliest attempts at X-ray analysis from the specimen was described by Neff and Herrmann (1964). With considerable ingenuity, they arranged that the specimen be tilted at about 45°, so that the characteristic X rays generated could be extracted along the counterbore opposite the objective aperture rod, and picked up on a solid-state detector at the edge of the column. There was therefore a very small acceptance angle for the X-ray beam, and the detector was farther from the specimen than desirable. The minimum area of the specimen illuminated was also relatively large. At the state of development of the detectors at that time, the resolution was fairly poor compared with a fully focusing spectrometer. With all these drawbacks, it is remarkable that solid-state detectors were used so soon. It happened, however, that there was a rapid improvement in resolution, with new developments in the detectors, and the potential for analysis was so great that it was not long before the design of the TEM itself came to be modified to produce a very small illuminated area at the specimen by the adoption of a symmetrical condenser-objective, or a close relation of it. The magnetic circuit was changed to allow a moderately high take-off angle and a much closer approach of the detector to improve the acceptance angle, while retaining the specimen horizontal to allow normal imaging during analysis. Once these technical problems had been solved, and the energy resolution of the detectors had been significantly improved, the energy dispersive X-ray detector became a necessary accessory for any serious analytical work.

In 1970, an EMMA instrument was produced at Sumy. It was based on a UEMV-100 microscope, and had a resolution of 0.7 nm in normal operation. It was fitted with a wavelength spectrometer, analyzing elements between magnesium (12) and uranium (92). The magnification range using the microanalysis equipment was 300–20,000 and the resolution 5 nm.

H. High-Voltage Electron Microscopes

The pioneer work of Dupouy *et al.* (1961) in Toulouse in building a 1-MV electron microscope was followed by a 3-MV model (Dupouy *et al.*, 1969). This helped to encourage Smith, Considine, and Cosslett (1966) in the development of an experimental 750-kV instrument at the Cavendish Laboratory in Cambridge and in turn signaled the strong rise in interest in being able to image thicker specimens than had hitherto been possible. The Japanese had already started to build commercial high-voltage microscopes. The Hitachi Company had a 500-kV model working (Tadano *et al.*, 1966), as had Shimadzu (Shimadzu *et al.*, 1966), and JEOL had a 1000-kV instrument (Watanabe *et al.*, 1966). Popov (1959) in the USSR had also built a 400-kV microscope/diffraction camera. It was decided that in view of the strong demand in the United Kingdom, a development of the Cavendish instrument should be built. A committee of potential users monitored development and fed in design requirements. This finally resulted in the AEI EM 7 microscope (Fig. 59), operating at 100–1000 kV (Agar *et al.*, 1970). Most of the 12 instruments manufactured were kept in intensive use for nearly 20 years, though many have now succumbed to the harsh economic climate in science today. The EM 7 had remote electronic control of specimen and aperture movements. The convenience of stereo photography was greatly enhanced by the use of the four imaging lenses to provide nonrotational imaging through the magnification range for the first time in any commercial microscope.

The results obtained with the aid of these instruments initially proved to be somewhat disconcerting. One of the earliest successes was the discov-

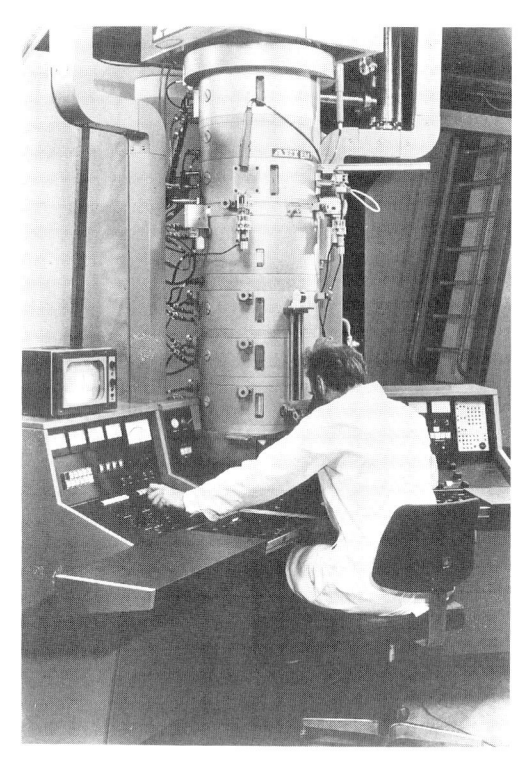

FIGURE 59. AEI EM 7. 1000-kV high-voltage electron microscope. First with nonrotational image with change of magnification. Many special *in situ* treatment stages.

ery by the scientists at the Atomic Energy Research Establishment (AERE) in Harwell that the rate of radiation damage sustained by the specimen in the EM 7 was far greater than in any of their normal radiation facilities. They were therefore able to carry out radiation studies in a few hours which would have taken months in an atomic pile, so this particular instrument paid for itself in an incredibly short time! The next discovery was that after comparative experiments on penetration of thick specimens at different accelerating voltages, the penetration at 300–400 kV was better than had been thought, and it was concluded that many useful experiments with thick specimens would be possible at 300 kV. The third item was the shift of emphasis away from the penetration of thick specimens to the improved resolution that could be possible because of the shorter wavelength of the high-energy electrons. This eventually led to the building of the HREM (High-Resolution Electron Microscope) at Cambridge specifically for this purpose (Nixon *et al.*, 1978; Cosslett *et al.*, 1979; Cosslett, 1980). This instrument was also based on the EM 7 column. Apart from these observations, these very-high-voltage instruments have, of course, also shown the value of being able to conduct dynamic experiments in the microscope (Butler and Hale, 1981). There is a continuing role for very high voltages in this context. Nevertheless, it is now probable that most needs can be met by microscopes operating at 300–400 kV, which have the great merit of being able to operate in a room of normal size, and hence not to need a very expensive special building.

The first move in Europe to produce a routine microscope operating at a voltage higher than 100 kV was made by Sumy. The EMV 150 was first produced in 1964, as a small batch of three instruments, operating at 150 kV. As a result of experience with these instruments, an improved model EM 150 was produced in small quantities from 1968. It was regarded as a transitional design leading to the EM 200, which appeared in the next decade.

I. Specimen and Operating Techniques

The 1960s saw a huge expansion in the use of the TEM in both biology and metallurgy. In this period, the biologists' concern about the effects of preparation techniques had led to the development of cryo techniques so as to avoid the need for chemical fixation and embedding (Steere, 1957; Bernhard and Leduc, 1967; Bullivant, 1965). The metallurgists, for their part, had always been aware of the potentially destructive nature of radiation, but had not had the means for observing the effects (Pashley and Presland, 1961). The widespread use of thin metal foils emphasized the need for precise tilting of the specimen and the ability to control beam tilt as well. The extreme sensitivity of plastics to electron-beam irradiation was recognized, and minimum exposure techniques were needed to enable some organic materials to be imaged at all (Agar *et al.*, 1959). The use of scanning microscopes for many metallurgical problems began to be favored, especially because large specimens could be examined directly, and the addition of X-ray detectors enabled identification of inclusions in fracture surfaces.

VI. 1971–1980, THE DECADE OF ANALYTICAL MICROSCOPY

A. Transmission Electron Microscopes

Zeiss Oberkochen moved into the research class of transmission microscopes with the launching of the EM 10 (Fig. 60) in 1971. This was a 100-kV, 0.5 nm instrument having a pushbutton-operated, fully automatic

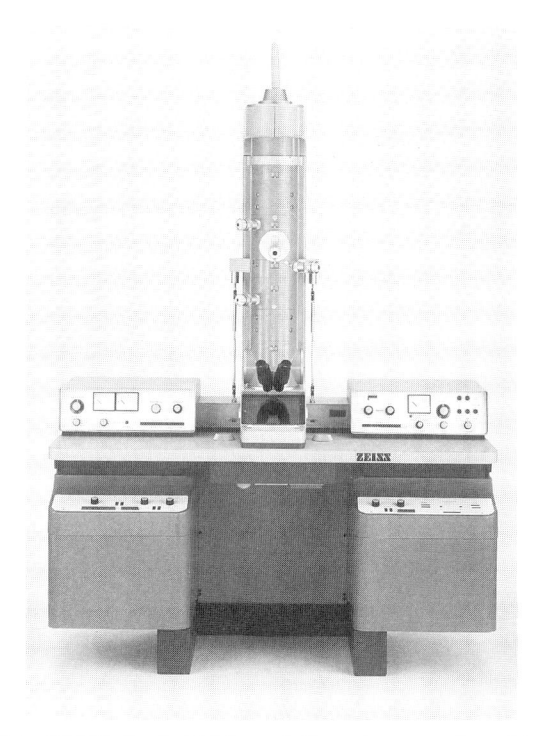

FIGURE 60. Zeiss EM 10, the first high-resolution microscope from this company.

camera system, which represented a move to simplify operation of even sophisticated instruments.

There was a move to increase the maximum high voltage on research instruments. The earlier EM 150 of the Sumy factory was followed by the EM 200, a 200-kV instrument, in 1974, and over 100 were produced. The first move to higher voltage in Western Europe was taken by Siemens, which launched the Elmiskop 102 (Müller and Schliepe, 1973), operating at 125 kV, in 1972. This instrument, however, had a resolution matching the best available at 100 kV, so it had a slightly different philosophy from the Sumy instrument, which was exploring further into the intermediate-high-voltage zone. The Siemens example was followed by Tesla, which in 1974 launched the BS 540 (Fig. 61), a 120-kV microscope designed for materials science applications.

The other significant instrument of this class was a completely new design of Siemens, the CT 150, with operating voltages up to 150 kV, introduced in 1976. This represented a complete break from the family of instruments running from the Elmiskop 1 to the Elmiskop 102. The Elmiskop CT 150

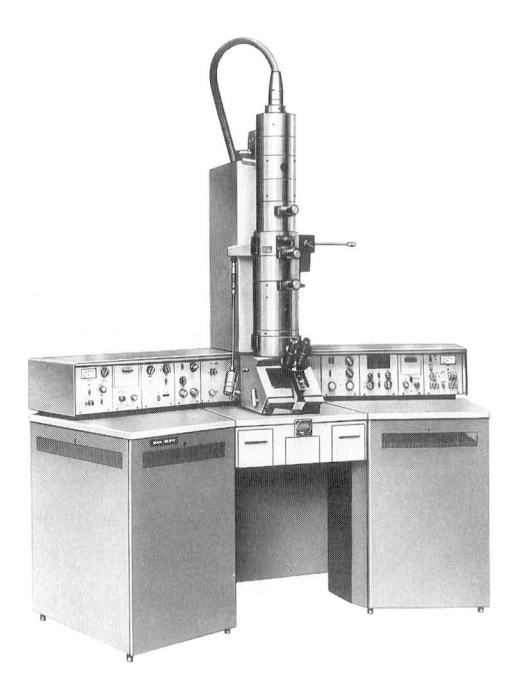

FIGURE 61. Tesla BS540, 120-kV materials science microscope.

had a very high vacuum in the gun and specimen area. The full 0.3 nm resolution (point to point) could be obtained while using the eucentric goniometer stage. There was a TV camera fitted beneath the column, with a small display monitor visible at the back of the viewing chamber, giving an image of the center of the field at a 10× increased magnification. The relatively large space around the specimen gave room for efficient extraction of X rays for analysis, and for secondary electron detection. The lens alignment system was preprogrammed by microprocessors. This very advanced instrument, unfortunately, fell prey to the withdrawal of Siemens from electron microscopy in 1979.

The first step toward true medium-high-voltage instruments had been taken in the USSR by Stoyanov and Renskii (1977). They designed a 300-kV microscope, but it did not enter production, being perhaps somewhat ahead of its time.

Another significant technical advance in this period was the improvement of the vacuum system. This has already been mentioned in the context of the CT 150. It was also realized in the Philips EM 400, launched in 1975 (Fig. 62). This instrument entailed an extensive redesign to achieve a clean

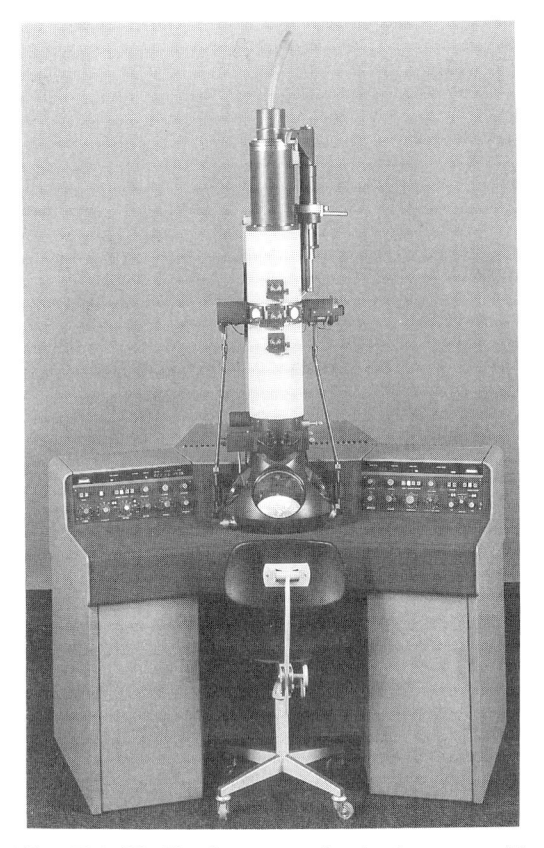

FIGURE 62. Philips EM 400. The first conventional microscope with a clean vacuum system. Differential pumping aperture in the final projector. First with five imaging lenses.

vacuum system, so that specimens could be examined over a period of time without suffering contamination, even without the use of the anticontaminator. The gun and microscope column were pumped by an ion getter pump, and the specimen region also had a cryopump near to the specimen. Stainless steel liners were fitted to the condensers, and through the imaging lenses, all motions being transmitted through bellows drives. This ultrahigh-vacuum region was separated from the viewing chamber and camera by a small 200-μm differential pumping aperture in the final projector lens. The camera region was pumped with a conventional diffusion pump and backing pump, but with a large reservoir to avoid continuous running of the backing pump, and therefore minimizing backstreaming. To ensure that the image diameter at the pumping aperture was always less than 200 μm in any operating mode, an additional imaging lens had to be fitted, making five

FIGURE 63. AEI Corinth 275. A console-type microscope with inverted column for routine applications. It had a four-specimen holder in a simple airlock, and large screen viewing.

in all, in addition to the two condenser lenses. Philips took advantage of this redesign to correct the chromatic change of magnification error in the projector system to a level that was imperceptible at any magnification. The biological version of this microscope was the EM 410 (Hax, 1982).

A further instrument aimed at the biological market was the Corinth 275, produced by AEI in 1971 (Fig. 63). It was unusual in having an inverted column design with the gun near the floor and a large viewing screen at desk height. Thus it did not depend on the rigidity of the desk to give it mechanical stability. It had a holder for four interchangeable specimens, and simple controls. The resolution was 1 nm, improved to 0.5 nm in the Corinth 500 the following year. In 1976, an analytical version of this microscope (Cora) allowed a beam probe size down to 25 nm to be obtained, and X rays to be extracted from a level specimen (Anderson *et al.*, 1976). The elemental detection limit was of the order of 10^{-18} g. The design was technically ingenious in producing an economic solution to the technical specification. Unfortunately, it did not take into account the likely customer reaction to the unconventional design and it did not make sufficient impact on the market to support future developments. AEI withdrew from electron microscope production in 1978.

The other main technical advance was the development by Philips of the so-called TWIN lens in order to optimize the instrument for both high resolution and for STEM operation (van der Mast *et al.*, 1980). This continued to be based on the Riecke and Ruska (1966) symmetrical condenser-objective, as earlier proposed by Ong (1968) and by Smith and Considine (1968). The strong field below the specimen ensured a large detection angle for diffracted electrons, combined with a stationary diffraction pattern in the STEM mode. However, in order to retain a large field of view and freedom to adjust the second condenser lens (C2) for TEM work, an auxiliary condenser lens was required; by an ingenious piece of design, this was built into the objective lens as a minilens. This enabled the electron source to be imaged in the plane of the objective aperture, or at choice, with a larger crossover at the specimen itself. This was the new feature in the TWIN lens. The design also enabled the achievement of a high tilt of the specimen (±60°), leaving space for efficient detectors for electrons and X rays. It also allowed a spot size of only 2.5 nm to be achieved, which greatly improved the precision of the area selected for analysis.

An optional field emission gun was announced for the EM 400 which permitted an increase of gun brightness by a factor of 1000. This improved the detection limits for analysis and also provided a very coherent beam for high-resolution microscopy.

Another new idea for photography was introduced by Zeiss in the EM 109 in 1978. This was the first transmission fiber optic system to keep the photographic material out of the microscope vacuum. The principle of external photography had been demonstrated some years earlier by Anderson and Kenway (1967), but using a 35-mm camera focused on a transmission screen. The Zeiss system was obviously more user-friendly, but it had benefitted by the development in the meantime of fiber-optic plates of adequate resolution, which brought the performance up to the standard attained by direct exposure of the photographic material to electrons.

The Sumy factory further developed the high-resolution instruments of the EMV 100 series first launched in 1969. Various models to suit particular requirements were made. By 1976, the model EMV 100AK, with quantitative computer analysis of electronic images, was produced. An instrument with a STEM attachment appeared in 1978 (EMV 100-BR).

The development of X-ray spectrometer attachments during the 1960s ushered in a period when an analytical function was added to what had been purely a morphological source of information. Interest developed in other available quantitative techniques. As X-ray techniques at that time seemed unlikely to yield useful results for very light elements, the possibilities of electron energy loss spectroscopy (EELS), which is most sensitive for light elements, were investigated. It was also in this decade that the technique of convergent-beam diffraction became accepted as a powerful tool in structural analysis. The old technique of stereoscopy achieved a new level of interest, and stereology began to be actively promoted in electron microscopy applications.

Biologists increasingly turned to cryo techniques in an effort to avoid the artefacts generated by the chemical treatments involved in fixation, staining, and embedding for sectioning. There were some indications that low temperatures might also reduce the susceptibility of biological materials to radiation damage. Microscope manufacturers began to include low-dose irradiation modes into the operating procedures and to design reliable cryo stages to accommodate frozen specimens.

B. Electron Microscope-Microanalyzers

Although production of electron microscope-microanalyzers as a special design ceased in the West by about 1975, because of the modification of conventional TEM adapted for EDX, production continued at Sumy. The EMMA 2 (Fig. 64) appeared in 1974, based on the UEMV 100 instruments, and after a transitional design (EMMA 3), the EMMA 4 (Fig. 65) (Zelev *et al.*, 1977) was produced in 1980, based on the EMV 100L design. Both of these instruments had a cylindrically shaped spectrometer of the semifocusing type and with a take-off angle of 15°. Improvements were thus made both of resolution, and of X-ray sensitivity and range of elements detected (down to boron). Instruments of this type were made until 1985. Analytical microscopy, of course, continued to advance in the West as well, but as a part of the broadening of the functions of both TEM and SEM.

C. Electron Diffraction Cameras

New models continued to appear from Sumy. They showed steadily improving performance, and in particular a much wider range of camera lengths, and hence of interplanar spacings which could be investigated. This was achieved by introducing a lens between the specimen and the viewing screen to increase the effective camera length as in the EMR 100 (Fig. 66) (Alekseev *et al.*, 1978). The normal operation without lenses was of course

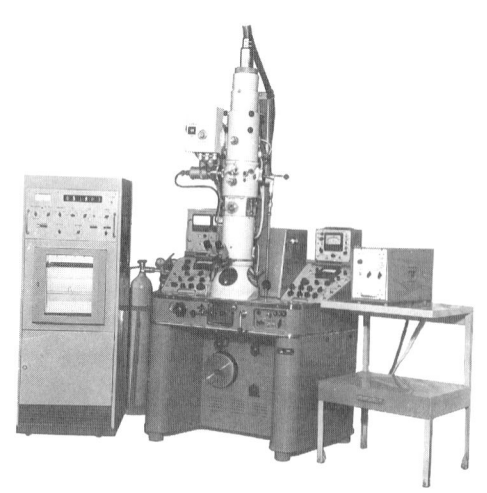

FIGURE 64. Electron microscope/microanalyzer EMMA 2, based on the Sumy TEM UEMV-100.

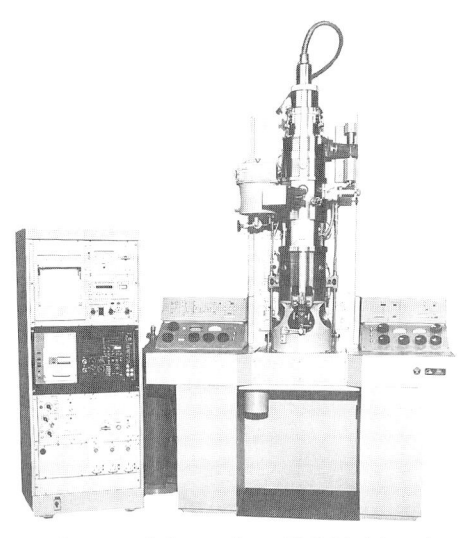

FIGURE 65. Electron microscope/microanalyzer EMMA 4, based on the EMV-100B. This was the last dedicated EMMA to be produced in Europe.

retained for the shorter camera lengths. Toward the end of this period, auxiliary lenses were added to increase sensitivity, and facilities for a computer link were added (EMR 100M).

D. Scanning Electron Microscopes

There was a great proliferation of scanning microscopes in this decade. Several companies made their first entry into scanning microscopes. Philips

FIGURE 66. Electron diffraction camera EMR 100. Besides the physically determined camera lengths, a lens between the specimen and screen allowed for much longer effective camera lengths for the examination of large diffraction spacings.

FIGURE 67. Philips scanning electron microscope PSEM 500. The first SEM from this company, and incorporating a motor-driven eucentric stage for this specimen.

introduced the PSEM 500 (Kuypers and Tiemeijer, 1975) (Fig. 67) in 1972, with good resolution of 10 nm and with the spectacular opening gambit of a eucentric stage driven by stepping motors inside the vacuum of the large specimen chamber.

After an earlier analytical scanning microscope (REMMA), produced in small numbers, Sumy began to produce the REM 200 (Fig. 68), a SEM

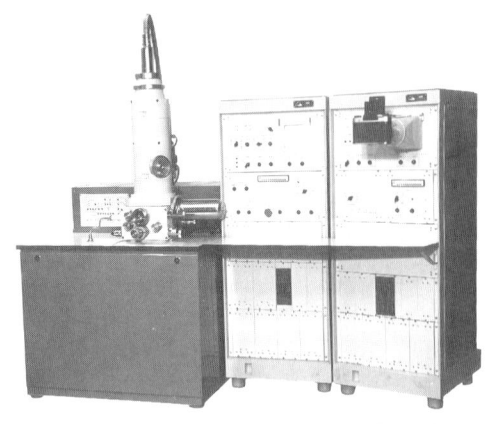

FIGURE 68. The Sumy REM 200 scanning microscope with voltage range 5–40 kV, and magnification range 40–80,000. The first SEM with large-scale production in the USSR.

operating in the 5- to 40-kV range, in 1974, on a mass production basis. It should be noted that the Sumy scanning microscopes, when adapted with two or more spectrometers, were designated REMMA. There is always a close resemblance to the current model of scanning microscope, e.g., the REMMA 200 and REM 200. Improved instruments appeared at intervals during the decade.

Tesla brought its first scanner, the BS 300 (Fig. 69), to the market in 1977.

The first simpler instruments appeared—the first being the Stereoscan 600 with all solid-state supplies (Gibbons *et al.*, 1971). A few years later, Philips built the SEM 501, which had very good resolution but simplified electronics and a limited stage.

In 1975, Camscan started to market its instruments under its own name, and it continued with the horizontal-column design previously marketed by Bausch and Lomb and now called the Series 3. It was now, however, mounted on a dynamic self-leveling system which minimized vibration. The voltage range of the Series 3 was 2–30 kV and the resolution 10 nm. The display incorporated two large, 210 × 170 mm CR tubes—considerably bigger than those of other contemporary instruments. The basic design of the Series 3 was modular, so new features could be added without the need to alter other parts of the microscope. In particular, Camscan developed a number of special stages, processing facilities, and specimen chambers to meet particular customer requirements. One such chamber was large enough to require the help of a crane to load the specimens (Fig. 70).

As a result of these early experiments with large chambers, the instruments quickly found acceptance in forensic laboratories, where large samples frequently had to be examined. This distinctive design philosophy

FIGURE 69. The first Czechoslovak scanning microscope: the Tesla BS 300, operating between 1 and 50 kV, and with secondary, back-scattered, and absorbed current imaging modes, slow scan and TV-rate imaging, and twin-screen display.

FIGURE 70. Camscan Series 3 with designer Dick Paden, with a special large specimen chamber for bulky specimens. Pioneer of modular construction offering special facilities.

served the Camscan well. Each such special-featured instrument was given a subnumber to distinguish it. This concept of organic growth has persisted to the present day (1992), and, as a result, the nomenclature has changed infrequently—in 16 years, there have apparently been only two research instruments, the Series 3 and the Series 4. In fact, the changes introduced would correspond to a considerable number of models if they adopted the nomenclature policies of some other companies.

Cameca built on its initial lead in EPMA by producing an advanced scanner with both energy and wavelength spectrometers in 1972. By 1974, it moved to the Camebax (Fig. 71) instrument, which was the first combined microprobe/scanning microscope with full performance in both modes of operation. It seemed a logical combination, in the light of the much-improved analysis facilities becoming available on scanning microscopes. Cambridge Instruments indeed ceased making specialized microprobes after the Microscan 9, but as we shall see, Cameca eventually split the two instruments again at a later date.

The research-level microscopes developed quite quickly during the period, and there was a steady improvement in the attainable resolution, so that new models appeared at relatively short intervals. The other development that is most noticeable was the availability of both energy and wavelength spectrometers to enhance the analytical value of the microscopes.

Cambridge Instruments were the first to incorporate alphanumeric micrograph notation as a standard feature, on the S180 (Paden *et al.*, 1973). This fairly soon appeared on all the instruments. Toward the end of the period, dual display tubes were becoming common, and the first backscatter detec-

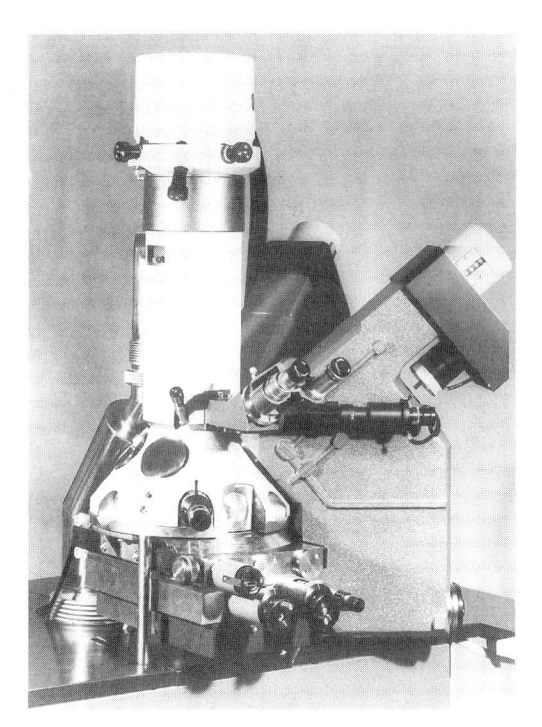

FIGURE 71. Cameca Camebax. First microanalyzer/SEM with full performance in each mode. The photograph shows the electron optical column with two vertical spectrometers.

tors made an appearance. The backscatter detector employed in the Philips SEM 505 was interesting, in being made up of four doped yttrium silicate scintillator detectors beneath the final lens, the signals being transmitted in two pairs of flexible fiber optics to photomultipliers outside the chamber.

In 1980, Cambridge Instruments introduced the S 250, with a resolution of 6 nm. This instrument was the first to be equipped as standard with turbomolecular pumping, which greatly improved the vacuum cleanliness. Another new feature on the S.250 was the Optibeam program. This was a system of controlling the three probe-forming lenses to give the maximum beam current density for a given setting of probe current. The beam-limiting aperture, which is normally located in the final lens, was in this instrument moved up the column above the objective lens. A projection of the physical aperture could now be imaged into the normal aperture plane and the effective aperture angle changed electron-optically. The microprocessor system noted the operating voltage and the working distance and, for any given setting of the probe current, adjusted the excitation of the three lenses so as to maximize the probe current density, and hence minimize the spot size for the current working conditions. It turned out that the optimum aperture for given working conditions changed only slowly with spot size, so the calculation of the required excitations was not very stringent. The program could also be set to achieve maximum depth of field in given conditions.

A somewhat similar data link was provided in the new Philips SEM 505 at about the same time.

As early as 1973, Vacuum Generators (VG) had entered the market for scanning microscopes, but specializing in ultrahigh-vacuum columns. The HB 50 microscope (Griffiths *et al.*, 1973) exploited the clean vacuum environment for the specimen by offering Auger electron analysis facilities.

In 1978, the HB 50A was developed as a UHV field-emission-source SEM, operating at 60 kV, and with an Auger spectrometer for surface analysis at high spatial resolution.

Tesla made very rapid technical advances to introduce the BS 350 in 1980 (Fig. 72), which also had a UHV column equipped with ion getter pumps and a field-emission gun to achieve much greater brightness and therefore improved resolution.

E. Scanning Transmission Electron Microscopes

Scanning transmission electron microscopes (STEMs) arose from the pioneering work of Crewe *et al.* (1968), who as outsiders from nuclear physics approached microscopy from an entirely new viewpoint. (Crewe actually

FIGURE 72. Tesla BS 350 scanning microscope based on a UHV column and using a field emission gun to give greater brightness and resolution (5 nm). Ion getter pumps; ion-beam cleaning and etching of the specimen surface.

described the instrument first at the 1965 Conference of the Electron Microscope Group of the Institute of Physics.) They employed a field-emission electron gun to create a very small source of extremely high brightness. This overcame at a stroke the limit on performance of a scanning microscope due to the limited number of electrons in a spot small enough for high resolution. Crew *et al.* also demonstrated the value of Z-contrast images, and the potentialities of an energy analyzer as part of the system. After some years, Crewe solved the problems of obtaining stable emission from his guns and began to show the advantages of different imaging modes. His pictures of individual atoms aroused intense interest, and it was natural that commercial development should follow.

The earlier instruments were all produced by small specialist firms in the United States, but both Siemens and AEI started the development of a field-emission STEM as being likely to be the microscope of the future. Siemens completed a production instrument, the ST 100F (Fig. 73, Krisch *et al.*, 1976) in 1975. This was a very well engineered instrument, which had a resolution of 0.2 nm. It had the expected bright- and dark-field detectors, and a magnetic-sector energy analyzer beneath the column. A minilens in the upper bore of the objective lens gave the possibility of large area imaging. Some 10 of these instruments were sold, which was probably a considerable success in the light of the state of development of the market at that time. However, this can in no way have repaid the cost of development of the instrument. It is also possible that it was in some ways rather too integrated a design for the early stages of the technique, since the desk adopted might have limited the scope for modification of the column. The AEI team also completed a first microscope of this type at about the same time (Ray *et al.*, 1976). These developments proved fatal

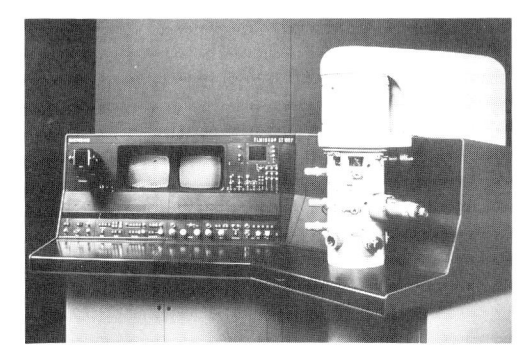

FIGURE 73. Siemens ST 100F. The first European field-emission STEM; with bright- and dark-field secondary electron detectors, X-ray analyzer, and electron energy-loss analyzer with magnetic-sector spectrometer.

for both companies, because the STEM development was done at the expense of the conventional instruments which still formed the bulk of the electron microscope business. Both companies had heavy overhead in the form of factory capacity, sales forces, and demonstration facilities, and would have needed a substantial injection of funding for this new development. Such funding was not forthcoming, and the result was that both companies withdrew from electron microscopy altogether in 1978 and 1979.

It seems that Philips made quite a different assessment, and decided to concentrate instead on the STEM facility within a normal TEM, based on the use of the symmetrical condenser-objective. This naturally had reduced potential performance, but on the other hand offered many imaging enhancements, without any of the major risks of a completely new development. In this, it has to be said that the company made an excellent commercial decision.

The advent of the new instrument, which had to be dependent on UHV (ultrahigh-vacuum) techniques, was very timely for the Vacuum Generators Company, which was looking to diversify its high-vacuum business into separate companies dealing with instruments employing this technique. It were happy to be able to acquire some of the team from AEI to help in the project.

This was in fact a more promising way of starting, because at this stage, the way in which the requirements would develop was not entirely clear. It made sense to build the instruments in modular form so that modifications and additions would be relatively simple, and it was easier to incorporate special features at the request of each customer—in other words, a market-led approach. The company was also free of the heavy overhead of the larger established companies, and was able to develop the business more as a one-off rather than on a mass production basis.

The first instrument, a STEM, launched in 1974, was designated the HB 5 (Wardell *et al.*, 1973) and operated at 80 kV. It achieved a line resolution of 0.34 nm. Although this instrument was fairly basic, it proved to be easy to make additions and modifications, as it was equipped with a number of standard UHV flanges on which additional items could be mounted.

The HB 5 was replaced in 1980 by the HB 501, which gave a much improved analytical performance as a result of attention to detail in the design of the X-ray interfacing. It achieved a current of 1 nA into a probe of 1-nm diameter. This instrument incorporated a number of features which had been developed or suggested by earlier customers, which is often the case in a quickly moving field of research, when the company develops the respect and confidence of its users. The most notable addition was a new energy-loss spectrometer which was later further modified to include a parallel detection system.

VII. 1981–1990, The Decade of Computer Control

A. Scanning Electron Microscopes

The title of this decade reflects the important changes wrought by computers in this period. It is appropriate therefore to pay tribute to the earlier work by K. C. A. Smith and his collaborators at the Engineering Laboratory, Cambridge. They established direct computer control of some of the microscope functions (e.g., automatic focusing and astigmatism correction), and they demonstrated the acquisition of images and some of the image processing treatments which could be applied (Unitt and Smith, 1976). They also established a computer measurement of height by stereo methods (Holburn and Smith, 1979). All of these things have appeared in production instruments, made much simpler by the very rapid increase in capacity of the computer and storage memories in the few years since the first work was done. Smith's computer had just 48 kB of RAM!

In the first few years of this decade, the main trend in the output of the various manufacturers was to launch a range of instruments rather than just the top-of-the-line microscopes which had so far been the norm (e.g., Fig. 74). Both Cambridge Instruments and Philips settled for three instruments to cover the range of price and facilities which seemed attractive. The simplification mainly took the form of restricting the size of specimen which could be accommodated on the stage, and streamlining the electronics. There was an increasing concern to optimize the X-ray analysis facilities on the more expensive models. A fairly common feature appearing at this time was a conical polepiece for the final lens, so that larger specimens could be accommodated without losing too much resolution when the specimen was tilted.

As time went on, there was a clear development of options offering a lanthanum hexaboride gun to increase the current density in the beam,

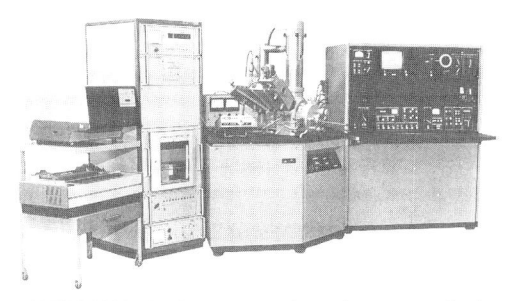

FIGURE 74. Sumy REM 100U. Universal scanning microscope with heating and straining stage, STEM operation, and facilities for characterizing dielectric specimens. Eucentric stage.

and so increase the available resolution. By the end of the period, field-emission guns were also available, with even greater brightness increases. These were particularly useful on specimens where the nature of the specimen was not a resolution-limiting factor.

A whole new area of interest arose as a result of advances in microchip technology, since the structures became too small to be seen by optical techniques, and dedicated scanning electron microscopes had to be used. Several different models were launched in this decade. A very valuable feature was the voltage contrast effect, discovered many years previously by Oatley and Everhart (1957), but now invaluable in visualizing the microcircuits in operation, particularly when used in conjunction with a sophisticated beam-blanking system.

Another development was the provision of special features for forensic laboratories, such as a gunshot residues package, and a comparator stage. All manufacturers steadily increased the size of the specimen which could be accommodated, and the size accepted on a eucentric stage.

As a part of the search for ways of examining biological specimens in a state more approaching natural conditions, the use of very-low-voltage scanning microscopy was tried. This results in a great diminution in surface charging, so that the coating of the specimen by sputtering heavy metal on to it could often be avoided. There is now a considerable literature on low-voltage results. However, it appears that to obtain the best results without coating, a voltage of 1.1 kV or less is required, and this has been feasible only since the development of field-emission guns. With a normal filament, the gun is very inefficient at these voltages, and both brightness and resolution suffer. More recently, instruments have been built which have differential pumping of the object chamber, which permits relatively high pressure (up to about 3 kPa) in the region of the specimen, and thus the examination of tissue in the wet state (Danilatos, 1988). Such instruments are termed ESEM (environmental SEM).

In this decade, the concept of the analytical microscope and the quantitative microscopy spread rapidly, and the REM 101 from Sumy (Fig. 75) is typical. The analytical version, REMMA 202M (Kisel *et al.,* 1988) appeared later in the period. Owing to the rapid developments in small computers, it became easy to attach image analyzers which could count and size particles. To match this facility, accessory manufacturers, and later the instrument makers themselves, produced electrically driven stages which could be programmed to scan a specimen automatically, and to log the position of many fields of interest for later examination.

By 1985, computer-controlled instruments started to appear, such as the Cambridge S 360 (Fig. 76). At first, some control functions were provided by microprocessors, so that the operating conditions could be optimized

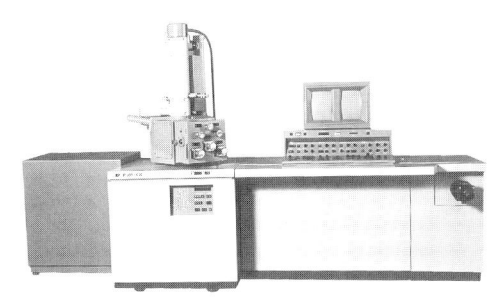

EUROPEAN COMMERCIAL ELECTRON MICROSCOPES   511

FIGURE 75. Sumy REM 101 SEM, 0.2–40 kV, 5 nm. A range of instruments for different applications, and when fitted with two fully focusing spectrometers, known as the REMMA 202M.

against some variable, such as beam current. The scope of such control rapidly widened, so that by the end of the decade full computer control of vacuum, electron optics, and image processing and control were provided. There was some divergence of opinion between the manufacturers as to the form these controls should have. The most radical solution was adopted by Philips, which went for full software control of all functions by a keyboard and mouse, using Windows-based software in the XL 20 (Fig. 77). Zeiss preferred to retain some control knobs for specified functions. It will be interesting to see if one of these systems gains preference in the eyes of the public. The result might turn on the actual cost of providing such systems. One might expect that the Philips system would be cheaper to produce than the hybrid system of Zeiss. (In the time that has passed since

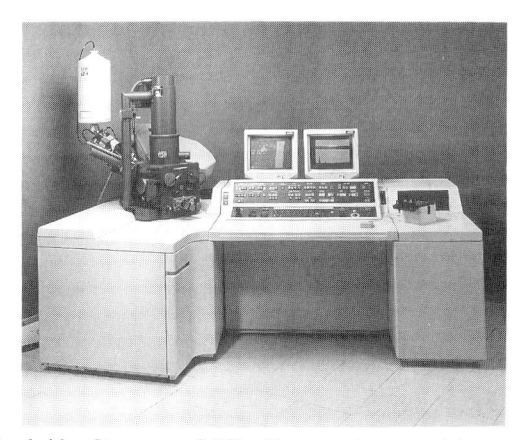

FIGURE 76. Cambridge Stereoscan S.360 with computer control, image store, and central image processing. The column is fitted with an X-ray solid-state detector and a focusing spectrometer.

FIGURE 77. Philips XL 20 desktop scanning microscope. The logical conclusion of computer control—all normal control knobs are replaced by a mouse-operated menu system.

this part of the history was first written, several new scanning instruments have appeared, from all parts of the world. Opinion still seems to be rather equally divided between the Windows/mouse format and knobs controlling a computer-driven microscope. It seems that the younger generation which has grown up with computers tends to favor the mouse, while those old enough to be writing historical surveys favor the retention of a few knobs, which seem to them easier to control. An overwhelming victory for Windows does not yet seem to be assured.

The inclusion of computers to control the microscope itself left the possibility of using some of the storage capacity to do image processing. Computer techniques also made possible stereoscopic height measurements by measurement of parallax between features in fields recorded before and after specimen tilting. Thus, the technique no longer depended on the ability of the operator to fuse stereoscopic images—one of the main obstacles to progress in the past. Another useful device was the dynamic stereo viewing offered by using the color display tubes to produce red and green images for observation through standard stereo viewers.

Image stores started to be a built-in feature rather than an add-on, and the choices for image recording and archiving widened. Images could now be printed out directly to a video recorder, or archived on to an optical disk.

A quite different development came from Tesla, which introduced a simple instrument, the BS 343. This was a miniscanner which was designed to be portable, being independent of water supplies and with a very low power requirement. It had a unique feature—a suction attachment, which enabled the electron-optical column to be sealed to any large object which could not be carried to the laboratory or inserted into a conventional scanning microscope. It was equipped with a range of useful features. (See Appendix 2, serial no. 217.)

There were several technical developments of a more specialized nature. Digital scanning became a standard feature. Improvements were made to the final lens to reduce the spherical aberration, and so improve the resolution of electron channeling patterns. Later, a dynamic correction of this aberration gave further improvement in ECP performance (Fig. 78). A feature of the early electron probes was reintroduced—namely, the use of an optical microscope to target the area to be analyzed. Significant improvements in the performance of backscatter detectors has resulted in very extensive application of this imaging mode. The introduction of windowless detectors has extended X-ray analysis into the light element range ($Z = 4$).

B. Conventional Transmission Electron Microscopes

The developments in transmission instruments in this decade mirrored to a large extent those in the scanning field. In the middle of the period, large computers were installed to handle all the operating routines. This is particularly useful with a TEM because of the need to move between normal imaging, selected-area diffraction, or dark-field imaging in selected diffraction orientations, convergent-beam diffraction, and STEM-mode imaging. These involve considerable time in adjustment and alignment, which could involve significant electron beam damage. The ability to transfer from one mode to another by preprogrammed switching is therefore a major advantage. Frame stores made possible both the examination of a

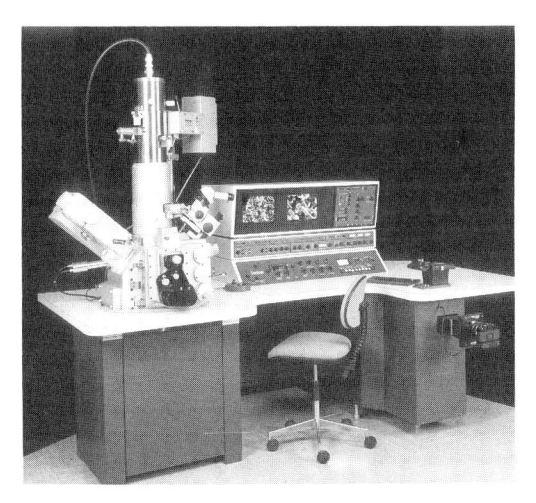

FIGURE 78. Camscan Series 4 SEM with targetted optical viewing of the specimen, X-ray spectrometer, and an ion-pumped gun.

field of view at leisure after the image had been captured under conditions of low irradiation, and noise reduction on otherwise uninformative images. We should not forget that there were also pressing economic arguments for using computers and microprocessors in electron microscopes. The individual components required for the highly stabilized electronic supplies were extremely expensive, and had to be duplicated for each lens stabilizer. Likewise, the mechanical adjustment controls were very expensive, and computer-controlled systems saved many costs.

Philips converted the 400 series microscopes into computer-controlled versions. The CM12/STEM (Gross et al., 1987) was the 120-kV materials instrument derived from the EM 420. The pursuit of ultimate resolution continued in parallel with the continued interest in the intermediate-high-voltage instruments of 200 kV (Otten, 1990; Otten and Bakker, 1991) and 300 kV. The latter instrument, the Philips CM 30, is shown in Fig. 79. The potential of the condenser-objective was further pursued, by Philips with their TWIN, Super-TWIN, and Ultra-TWIN lenses, and by Zeiss with the launch of the Köhler illumination system (Benner et al., 1990), also based on the Riecke and Ruska symmetrical lens. Both companies used a third condenser in the form of a minilens to control the area of illumination on the specimen, which was located in the center of the symmetrical lens. This type of lens has been found excellent in optimizing conditions for STEM operation, and in allowing space for eucentric goniometer stages of high tilt angle (60°) at high resolution. Where the ultimate resolution is required, the Ultra-TWIN lens (Bakker and Asselbergs, 1990) for the CM 20 achieves the remarkable figures of 0.5 mm for C_S and 1 mm for C_C, while allowing space for tilting the specimen to 20° to the horizontal in two axes. There is therefore some loss in the available tilt and in accessibility for X-ray extraction when the ultimate resolution is required.

The decade saw much more widespread use of the electron microscope as an analytical tool, employing several different analysis modes to supplement the morphological information. Apart from normal imaging, X-ray analysis and EELS systems are regularly employed. The use of digital image stores and of interfaces to image analysis systems has further strengthened the array of techniques available. The use of high-brightness sources, discussed under the heading of scanning electron microscopes, is also valuable in improving the sensitivity of the analytical techniques. It is also essential for those seeking to use holographic techniques.

The development of YAG single-crystal screen meant that there was now a high-resolution input available for a television system. The vidicon cameras used earlier began to be replaced by CCD cameras, which, by the end of the decade, when cooled and operated in slow scan mode, could match the resolution of the photographic plate. The digital form of the

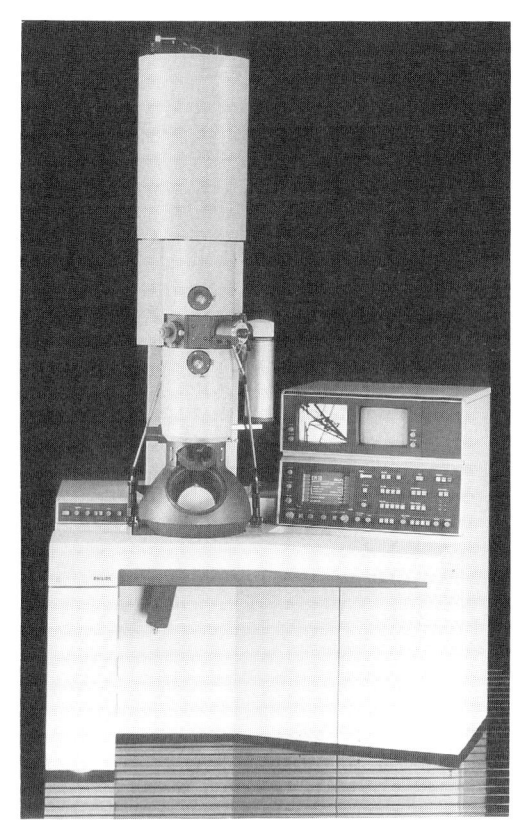

FIGURE 79. Philips CM 30 300-kV medium-high-voltage microscope with very high resolution (0.2 nm) when using the lanthanum hexaboride source. Four condenser and five imaging lenses. EELS detector available. Full computer control.

signal also allowed a more accurate quantitation of the image intensities, so that a move to video printers instead of photo plates, and scanned optical disks for archival storage of images, is now in prospect as the preferred method for the future.

One major new technique in this period was introduced by Zeiss, which included an integrated electron energy spectrometer in the column of the EM 902 (Egle *et al.*, 1984). It also had an image intensifier, a cryosystem, and a TV adapter. The long gap between research and production is interesting. Castaing and Henry (1962) had developed an energy-loss spectrometer more than 20 years earlier. AEI engineers had a laboratory model in 1968, but it did not reach production. The Zeiss engineers originally employed an electron mirror system similar to the Castaing and Henry design. This imposes limitations on the usable accelerating voltage, but was the basis of the EM 902, which introduced an entirely different operating mode from all other available microscopes. Figure 80 shows the CEM 902, which is the computer-operated version of the EM 902. The ability to exclude all inelastically scattered electrons from the image was very important for the examination of biological sections, particularly where thicker sections were to be examined. Energy-loss electrons were spatially separated in the spectrometer and selected at the energy-selecting slit below it. The elastically scattered electrons are normally imaged, but by increase of the gun voltage, electrons with any desired energy loss could be selected to pass through the slit. Since the energy loss was element-specific, these electrons could be used to give chemical information about the specimen (electron spectroscopic imaging). This also enabled biological specimens to be examined in high contrast without the use of heavy-metal staining. The instrument therefore opened up all kinds of new possibilities for electron microscopy. This instrument has now been significantly improved by the adoption of the omega-spectrometer design in the EM 912, having four magnetic 110° deflectors in a symmetrical array. This enables the instrument to work up to a voltage of 120 kV.

In a discussion of the possibilities of miniaturizing a magnetic lens, Mulvey (1982) showed that a very small lens could have low aberration if the excitation could be confined in a small space. This calls for very efficient cooling of the windings, and he proposed that one possibility would be to use the latent heat of vaporization to effect the cooling. Mulvey's suggestion was to use water converted to steam. His ideas have been converted into commercial use by the engineers at Sumy, who used miniature lenses cooled by the vaporization of an organic liquid in the PEM 100 (Fig. 81) (Lyalko et al., 1986a, 1986b). It can be seen that the column (140 mm diameter) is significantly slimmer than conventional instruments designed for 100-kV

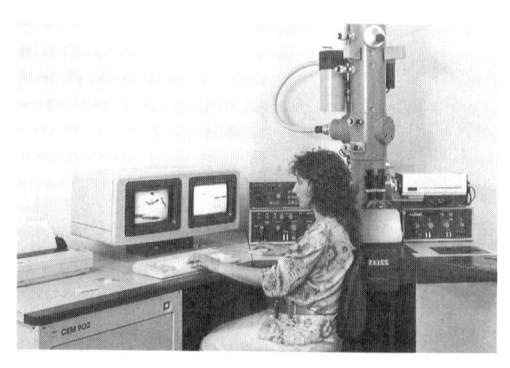

FIGURE 80. Zeiss CEM 902. Computer-controlled version of EM 902, the analytical microscope with energy-loss imaging and energy spectroscopic imaging.

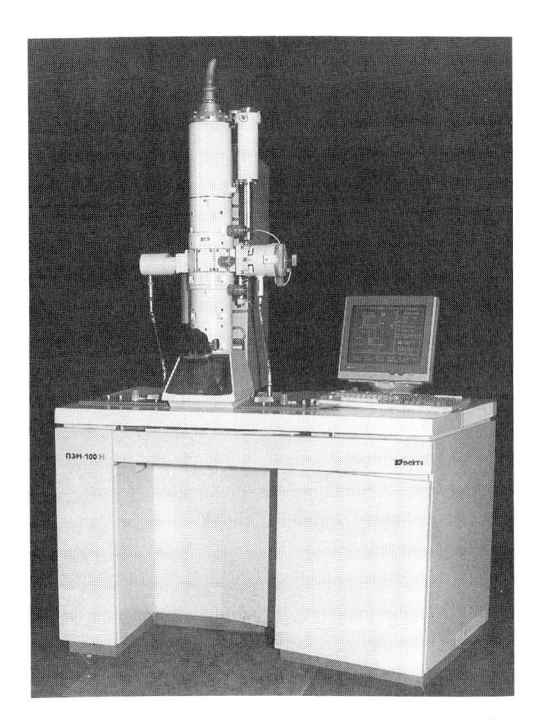

FIGURE 81. Sumy PEM 100, a computer-controlled TEM with miniaturized lenses and a special vapor cooling system for the lenses.

working. This gives it a lowered sensitivity to stray fields, and good mechanical stability. It has external photography on large-format roll film against a high-resolution fiber optic plate coated with a fine grain phosphor. The large latent heat of vaporization of the organic liquid avoids the substantial water flows of conventional cooling systems, and permits a current density in the windings of 10 A/cm^2. The temperature stabilizes at 60°C in less than 3 min.

A very-high-resolution materials science instrument, the EM 125K (Derbenev et al., 1990) (Fig. 82), features a 125-kV gun on a column designed to accept a 200-kV gun without further modification. The four-projector system allows for rotation-free imaging, and for a small differential pumping aperture in the final lens. Computer control enables a small number of multifunction knobs to control all alignments and operations.

One other interesting feature arises from the literature from Sumy. The company has taken a license to use the Zeiss side-entry stage as fitted on its TEM. It also obtained X-ray detectors from Tesla. All the microscope companies bought in their energy-dispersive detectors from companies specializing in that field. Perhaps the high costs of development will acceler-

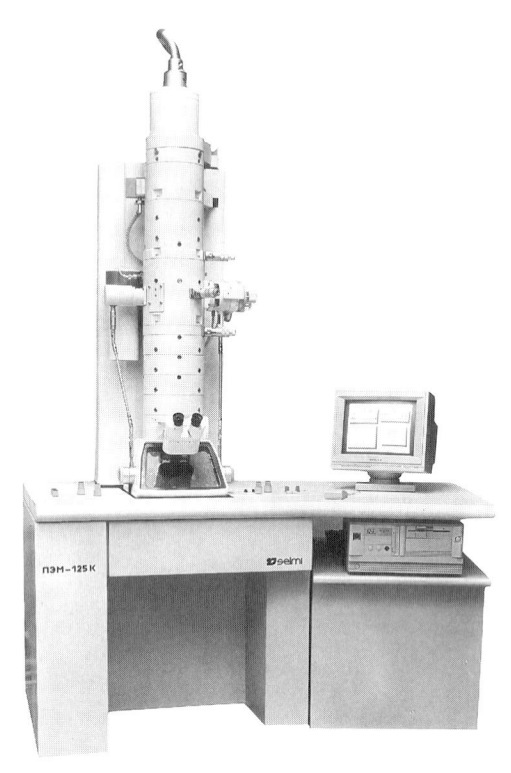

FIGURE 82. Sumy PEM 125K, a computer-controlled high-resolution microscope for materials science. Resolution 0.35 nm at 60° tilt.

the trend to sublicensing of particular specialized parts of the microscope in future. There is some evidence that this is the path chosen by Tesla for the future—providing a subcontracting service for components or complete instruments.

Old techniques—such as reflexion microscopy—reappeared, but this time on specimens immersed in the objective lens field, and therefore with very low aberration compared with the technique of the 1950s. Materials scientists found uses for ultramicrotomy of their specimens, thus reusing a technique first developed in the early 1960s.

As with the scanning electron microscopes, turbopumping was introduced to help clean up the vacuum system.

The reliability of the high-resolution instruments made possible a more systematic study of crystalline chemical compounds and the study of interface layers where the structure varied. These studies involved using off-line mainframe computers to generate simulated images for a range of values of out-of-focus and of specimen thickness, since owing to the strong dependence of the phase contrast on these factors, it was otherwise impossible to guess which of the structures to believe. Most of this work can now be done on desktop computers, due to the very rapid increase in speed and capacity of these machines.

With the widespread adoption of computer control of electron microscope operation, many of the traditional skills involved in electron microscopy are thereby rendered redundant, and one can concentrate on the manipulation and interpretation of the image in the light of whatever science is being studied.

As a postscript to the microscope story, the use of a computer and microprocessor is perhaps even more important in X-ray microanalysis. Here it is necessary to have equipment working for long periods under very stable conditions, and it may involve quite detailed instructions to analyze for many elements over a number of specimens (for example, 500 analysis points over 24 hours). This is achieved in the Cameca SX microprobe. Another feature is the ability to read the digital circuitry remotely via a modem, the so-called telediagnostic system. This permits quick diagnosis from an engineer at base with a great saving in traveling time—and hopefully, in downtime in a busy laboratory.

C. Scanning Transmission Electron Microscopes

The use of ultrahigh vacuum for STEM instruments, which was at first inevitable to ensure stable operation of the field-emission gun, gave these instruments a particular advantage in studies of surface effects, where operation at a very high vacuum was an essential if contaminating surface layers were to be avoided. These working conditions allowed the detection of Auger electrons, and in this period, an Auger spectroscopy facility was built in to the VG HB 501A.

The most significant development in this field was the use of high-angle, incoherently scattered electrons to form an image in atomic number contrast, which does not alter with change of focus, and is therefore directly interpretable, in contrast to the normal phase-contrast images, which are not (Jesson et al., 1991). They used the newly available HB 501UX, a very-high-resolution instrument, which had been launched in 1988. A further advantage of this operating mode is that the actual resolution limit is slightly improved because the contrast k in the equation $d = kC_s$ is 0.43 instead of 0.66 for coherent imaging. This new imaging mode is being further exploited in the 300 kV instrument, HB 603 (Fig. 83).

D. Electron Diffraction Cameras

The main features of electron diffraction cameras as far as facilities for the specimen, and resolution and circularity of the diffraction rings, were

FIGURE 83. Vacuum Generators HB 603. The first field-emission STEM instrument operating at 300 kV. Ultrahigh resolution; with wide-angle detector for incoherent imaging giving atomic number contrast.

concerned, were relatively unchanged in this period. However, useful improvements were made in data handling, and in information processing. Digital scanning and electron filters contributed to improved performance, and the EMR 102 typifies these developments (Fig. 84).

E. Then and Now

As a measure of the progress in instrumentation, a micrograph recorded on a Metro-Vick EM 2 in 1949 is shown in Fig. 85. This is a Formvar replica of the etched surface of a pearlitic steel. The electron optical magnification was about 3000×. For comparison, Fig. 86 shows a micrograph recorded by Jesson *et al.* (1991) of a superlattice of $(Si_4Ge_8)_{24}$ showing interfacial ordering. This was recorded in Z contrast by detection of wide-angle elastically scattered electrons in a VG STEM with a coherent electron beam of diameter 9.22 nm. The contrast between the micrographs is striking—the replica yielded only an indirect detection of different phases in a metal structure, with no indication of chemical constitution; the STEM permits a study on an atomic scale of a complex interfacial compound, and includes chemical information.

EUROPEAN COMMERCIAL ELECTRON MICROSCOPES 521

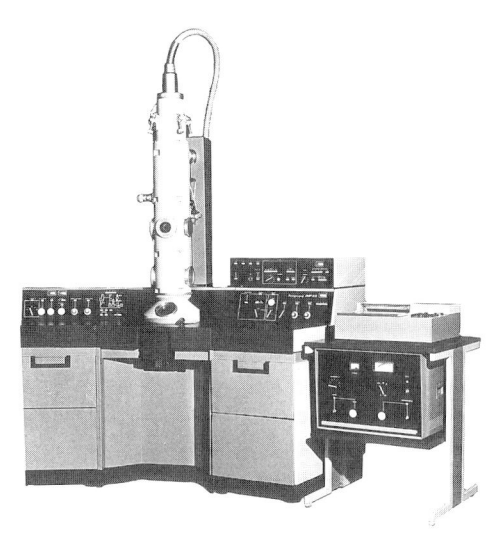

FIGURE 84. Electron diffraction camera EMR 102 with digital recording of information. A later modification included automatic measurement of the information and computer processing of diffraction patterns.

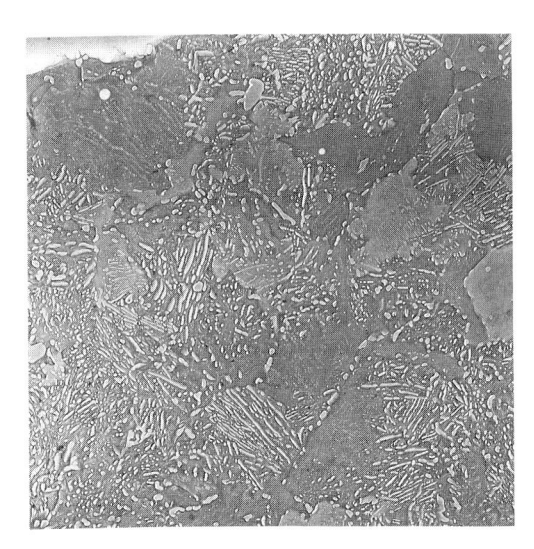

FIGURE 85. Micrograph recorded on a Metro-Vick EM 2 electron microscope, 1949. Plastic Formvar replica of etched surface of pearlitic steel. Electron optical magnification about 3000. Reprinted with permission from Jesson, D. E., Pennycook, S. J., and Baribeau, J.-M. (1991). Direct imaging of interfacial ordering in ultrathin $(Si_mGe_n)_p$. *Phys. Rev. Lett.* **66,** 750–753.

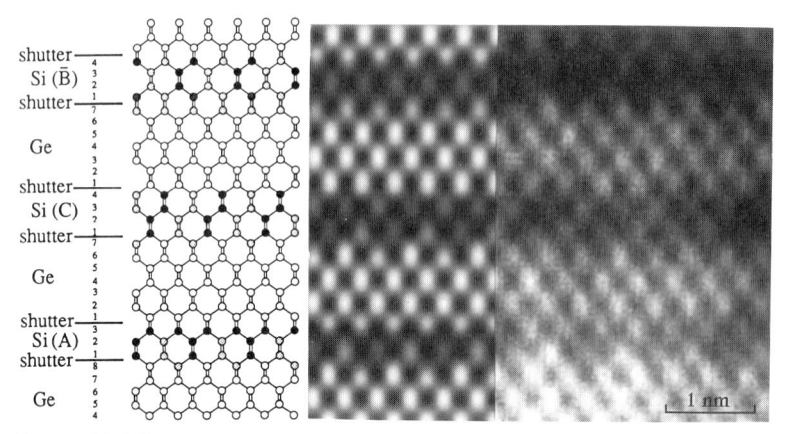

FIGURE 86. Micrograph recorded on a VG HB 501 STEM by Jesson *et al.* (1991). The image was recorded with wide-angle scattered electrons. The contrast and positions of intensity maxima are independent of objective-lens focus or of specimen thickness; the contrast is Z-dependent. The image shows a nominal $(Si_4Ge_8)_{24}$ superlattice with interfacial ordering. The authors' interpretation of the superlattice structure based on image simulation indicates the sequential deposition of Si and Ge layers together with the ordered structures B, C, and A resulting from the atom pump mechanism. Open circles indicate Ge columns; solid circles, Si columns; and shaded circles, alloy columns. Simulation parameters are $C_S = 1.3$ mm, convergence angle = 10.3 mrad, and defocus = -69.4 nm.

VIII. Summary and Conclusions

A listing of all the European manufacturers and of the instruments they made is given in Appendix I. In order to give an overview of all the information in summary form the different instruments have been arranged in chronological order in Appendix II. The date in each case is that at which first shipments of a particular instrument type commenced. In any given year, the instruments are arranged in alphabetical order by manufacturer's name, since the launch dates are not known to sufficient accuracy to make a valid date listing within any one year. Each instrument type is serially numbered, to facilitate cross-reference to other instruments in the appendix. Again, in the interests of standardization, the resolution figure is for point resolution only and is the manufacturer's guaranteed figure, except in a very few cases where this figure is not available. It may well have been possible to achieve better performance under special conditions with many of the instruments, but such results would not have formed a consistent basis for comparison. Most microscopes were upgraded in performance during the production run, but this is recorded here only if other significant changes were also incorporated. The most important technical features of each instrument are shown. Where known, the number of instruments of a given type which were manufactured has been indicated in a separate column. In cases where an instrument developed through several subtype numbers, the production numbers of each subtype are not generally given; in such a case, the whole production quantity is given against the first model of the series. The production information for the more recent years was not provided by Philips, so the detailed information received from the other manufacturers for these years has not been entered in Appendix II, to maintain equality of treatment among manufacturers. However, the totals produced are all included in the figures appearing in Appendix III, which summarizes the production quantities of the various instrument types. A glossary of abbreviations used in the text and the appendixes follows the appendixes. Appendix IV lists the principal Russian contributors to their microscope development program (Kiselev, 1992), since it was felt that they might not be well known in the West.

In Figs. 87 and 88, the guaranteed resolution of an instrument is plotted against the first year of its production, for transmission and scanning microscopes, respectively. This is interesting in showing the rapid progress in the earlier years and the asymptotic approach to the best performance for each type of instrument. In the case of transmission microscopes (Fig. 87), the results for the early Siemens instruments and for the AEG electrostatic microscopes are plotted separately. This figure shows that in the early years the Siemens instruments consistently outperformed other electromagnetic instruments in terms of resolution. It also shows that the electrostatic microscopes initially matched the performance of the best electromagnetic ones, but failed to improve their performance after 1951. The instruments have been divided rather arbitrarily into three classes: (1) high resolution, which includes any instrument approaching the state of the art at the time of launch; (2) simple or multipurpose, which covers instruments designed for simplicity rather than ultimate performance, or instruments such as EMMA, which may compromise resolution for improved X-ray performance; (3) basic, where the utmost simplicity has been aimed for, and resolution can be counted a secondary consideration. In some cases, it is not obvious which class to assign, and this leads to some points of discrepancy from the general trends. The plots of simplified instrument performance show that initially there was an appreciably poorer resolution, but that this difference narrowed significantly by about 1970.

In order to examine what has happened in more recent years, the plots have been placed on a resolution scale expanded by a factor of 10 from 1970 for TEM (Fig. 89). From this date we lose the strong correspondence between date and resolution, though there is still a slow improvement in ultimate resolution with time both for the most advanced instruments and the simpler ones. The trend of best and poorest resolution is indicated by

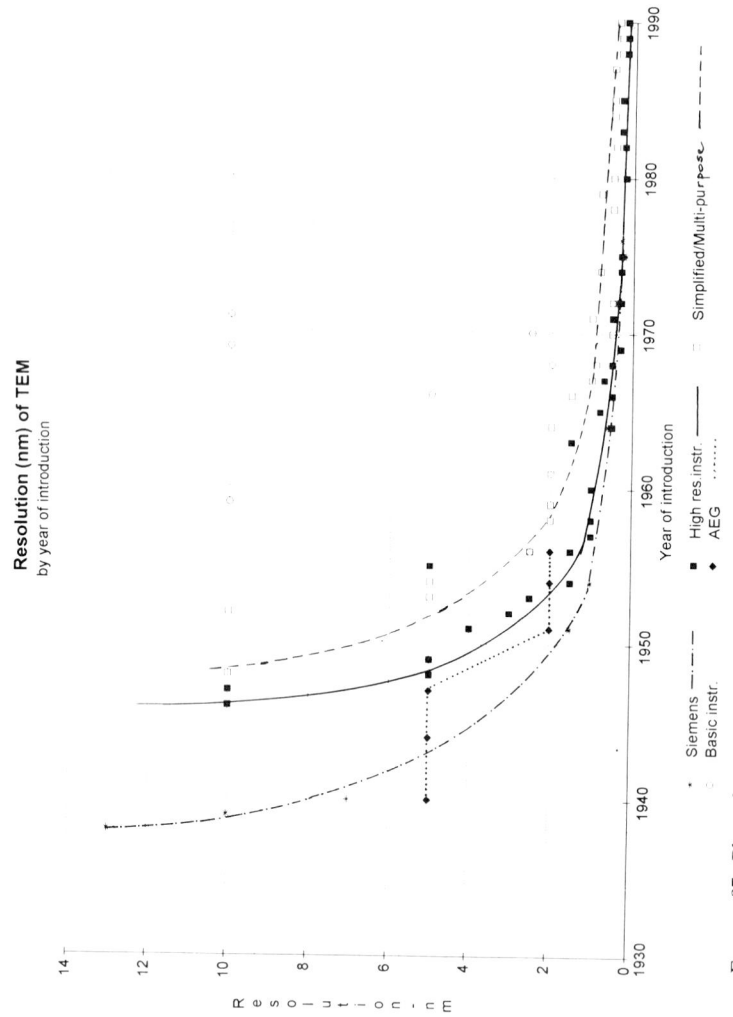

FIGURE 87. Plot of guaranteed point resolution at the date of introduction of transmission electron microscopes, against first year of shipment. Note the plot for AEG electrostatic microscopes (dotted line) and for Siemens microscopes (dot and dashed line). The state-of-the-art instruments (high resolution) are plotted as solid squares, and simplified and multiuse instruments as open squares. The basic instruments are shown as open circles.

FIGURE 88. Plot of guaranteed resolution at date of introduction of scanning electron microscopes against first year of shipment. Note that the resolution scale is different from that in Fig. 87, being contracted by a factor of 5. High-resolution instruments are plotted as open circles, mid-level as open squares, and simple as asterisks. The distinction has become more blurred in recent years, as the mid-range and simple instruments tend now to have a good resolution, but fewer additional facilities.

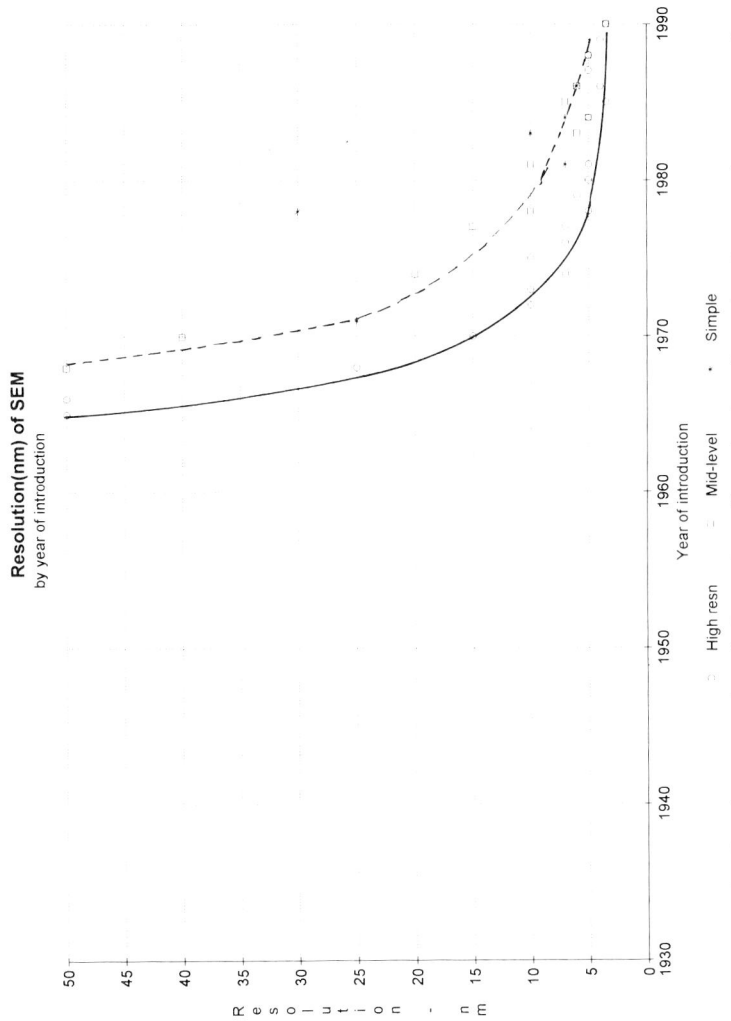

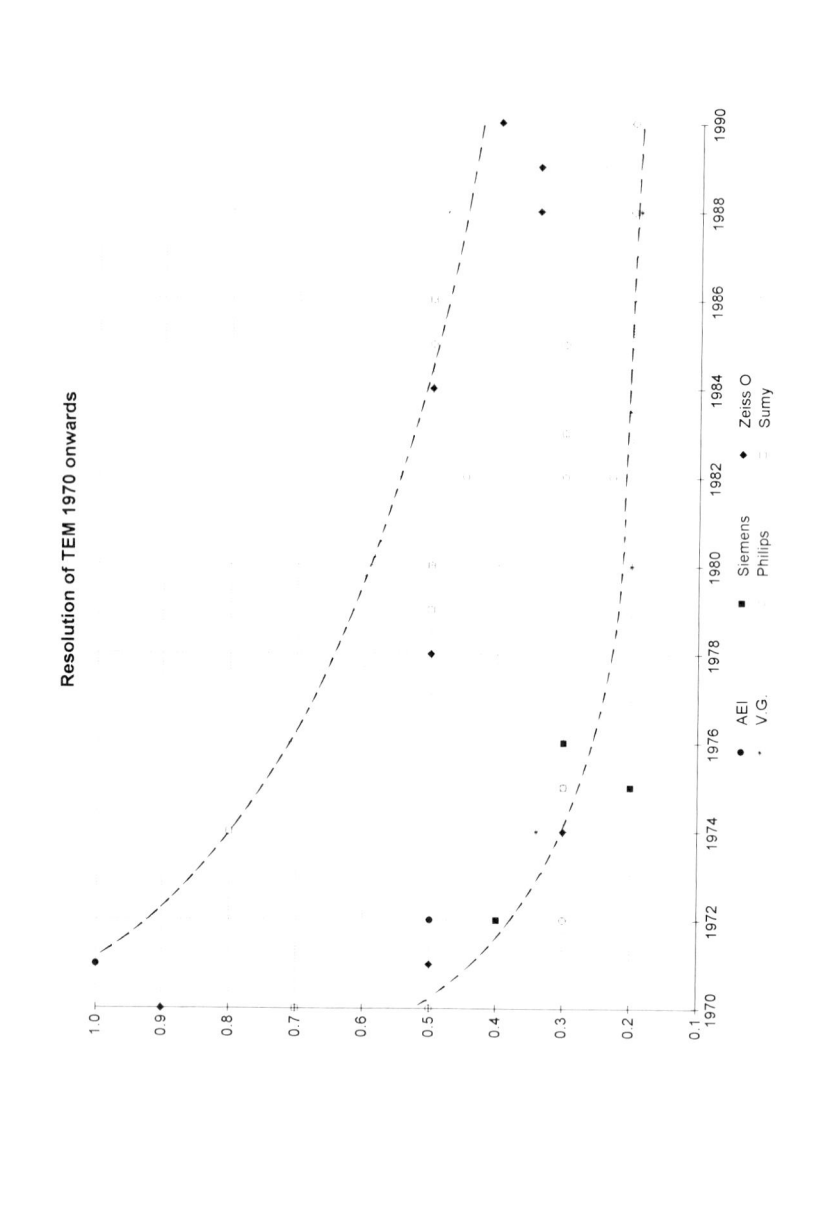

Resolution of TEM 1970 onwards

● AEI	◆ Zeiss O
· V.G	Sumy
■ Siemens	
Philips	

dotted lines. However, the differences now lie much more in the desired function of the microscope. This part of the plot includes instruments of up to 300-kV operation, and also includes the UHV STEM, which begins to cover a similar part of the analytical function. For all these reasons, the points plotted have not been used to define a continuous line plot, but the instrument makes have been distinguished by different symbols. A key below shows which instruments appear on the plot. It is interesting to note that even the simple instruments now have a resolution which equals the best available 25 years ago.

The plot of resolution against date for scanning microscopes in Fig. 88 starts with the date scale in the same place as in Fig. 87, so that comparison may be made between the two plots. It should be noted, though, that the vertical scale of resolution is different in the two figures, being condensed by a factor of 5 for the SEM. It is striking, though, that the plots follow very similar shapes, if one uses the curve for high-resolution TEM for the comparison. The slope of the curve for scanning microscopes is steeper than that for transmission instruments. This can readily be understood, since the slower developmental period for TEM included all the basic work on electron optics and lens design, and much of this work was relevant for the SEM also. Furthermore, vacuum techniques had greatly improved in the 25-year gap, as had electronic stabilizers, which were such a problem

---

FIGURE 89. Plot of resolution against year of introduction for TEMs from 1970 on. The resolution scale has been expanded by a factor of 10 compared with Fig. 87. The instruments are plotted by manufacturer, identified in the key. The individual instruments are listed below, with the instruments listed from the top of the plot downwards (i.e., improving resolution), for each year.

1970	1971	1972	1974	1975	1976
EM 9S-2	Corinth 275	Corinth 500	EM 200	EMV-100B	CT 150
EMMA	EM 10	Elmiskop 102	EM 10A	EM 400	
EM 201		EM 301	HB 5	ST 100F	

1978	1979	1980	1982	1983	1984
EM 109	PREM 200	EMMA 4	EM 410	EM 125	EM 902
		HB 501	EM 420		
			EM 430		

1985	1986	1988	1989	1990
CM 10	PEM 100	EM 902	EM 900	EM 910
CM 12		CM 30	CM 20	CM 20
		HB 501UX		UltraTwin

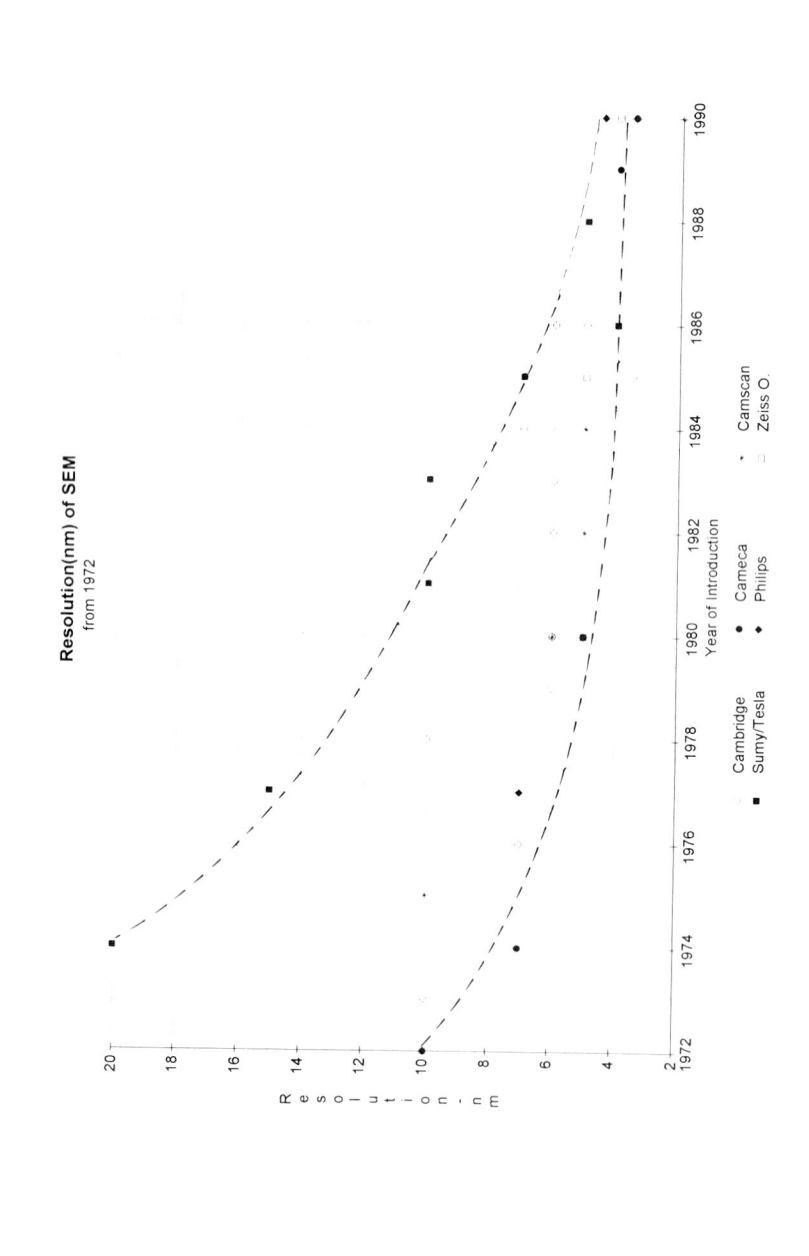

Resolution(nm) of SEM
from 1972

EUROPEAN COMMERCIAL ELECTRON MICROSCOPES   529

in the early years of the TEM. The resolution scale for the scanning microscopes is also expanded after 1982 (Fig. 90) so as to show more clearly the spread of results. These are in part due to the choice of high-brightness sources becoming available in this period.

These graphs illustrate well what can be deduced in other ways, namely, that the first 25 years or so were a time of a race for resolution as the main distinguishing feature of instrument performance and value. The end of this period corresponds roughly to the time when the theory of electron lens design achieved a high degree of accuracy and reliability, and when the main problems of magnetic design and stage design had been solved. In this period, the instrument design for much of the time was well ahead of the specimen preparation techniques which were required to exploit the instrument possibilities. The 1960s began to throw up demands for new facilities from the users, such as high-angle tilt stages; heating, cooling, and straining stages; and X-ray analysis facilities. The early impetus had come from university research into electron optics, in the different European countries. Outstanding among these were the Fritz Haber Institute in Berlin, the University of Tübingen, the Technological University of Delft, the Cavendish Laboratory in Cambridge, and Moscow University in the USSR.

←──

FIGURE 90. Plot of resolution against year of introduction of scanning electron microscopes from 1972 on. The resolution scale is expanded by a factor of 10 compared with Fig. 88. The instruments are plotted by manufacturer identified in the key. Individual instruments are listed under the year of introduction, reading downward on the plot.

1972	1973	1974	1975	1976	1977
Camsem II	S 180	REM 200	Series 3	S 150	BS 300
S 4-10		Camebax			SEM 501
PSEM 500					

1978	1979	1980	1981	1982	1983
S 604	S 150Mk.II	S 250	REM 100U	S 250Mk.II	REMMA-202
		Series 3	S100	Series 4	S 200
		SEM 505			
		BS 350			

1984	1985	1986	1988	1989	1990
S 90	Camebax SX 50	S 120	REM 102E	SEMPROBE	DSM 960
Series 2	BS 340	S 240			S 260
Phil. 515	DSM 85	S 360			Series 4FE
	S 250Mk.III	REMMA 202M			

The other source was from forward-looking research laboratories in industry, typified by the Tube Investment Laboratories of Hinxton Hall near Cambridge, which developed instruments to help to solve problems in materials science. Later, the pressures began to come from users with less direct interest in the instrument itself, and more in its problem-solving possibilities. As the bulk of the microscopes began to be in the hands of biologists, so the demand arose for simpler operation and greater productivity. The user no longer expected to have to service the instrument, and serviceability came imperative.

All these pressures increased the scale of the operation required to operate successfully in the market. More careful and detailed engineering was needed to ensure reliability, bigger teams of engineers to develop the range of special attachments, and larger service organizations to give the faster service needed. The substantial sales needed to pay for all this of course involved bigger sales staffs and expensive demonstration laboratories. This was a factor behind the withdrawal of Siemens and AEI from the electron microscope market. The smaller companies, such as Polaron and Miles Engineering in the United Kingdom, CSF and SOPELEM in France, and Balzers in Liechtenstein had already given up because of lack of funds. Carl Zeiss Jena left microscopy in 1970, to concentrate on electron beam lithography.

There were other factors at work, too. Initially, all the companies were making transmission electron microscopes, which were dominant until the 1970s. From then on, the market in scanning microscopes grew very rapidly, because, although they were not competitive in resolution terms, they had the great advantage of minimum specimen preparation, and could accommodate much larger samples. They were also spectacularly successful in the magnification range of the light microscope because of their great depth of field. So the balance of sales slowly moved over to the point where scanning microscopes predominated. At the same time, the TEM became much more of an analytical tool, and had X-ray and EELS attachments (and sometimes a STEM attachment as well). So although there was a continuing demand in biological work for a basic microscope, the materials scientists and physicists tended to need a wide range of facilities, the TEM became a very large investment, and fewer were purchased. With the shrinking of the mass market it became important for any major company to include all kinds of microscope in their range, and this is what both Philips and Zeiss did. Tesla also maintained a wide range of instruments. Cambridge Instruments still deal only with scanning instruments, but through a series of mergers, have diversified into related instrumentation. They have finally diversified out of their company name, and now trade as Leica.

It is worth considering in a little more detail why both Siemens and AEI should have withdrawn from the market. Siemens were the market leaders for a number of years after the introduction of the Elmiskop I, which had the best resolution and was an extremely reliable instrument. It was not an easy microscope to use, however, as both lens alignment and astigmatism correction required operator skill. This was not a major consideration while performance was the main criterion. However, over the years, the instruments made by Siemens' competitors achieved the same high standards of resolution and reliability, but became significantly easier to use. The upgrades made by Siemens, through the Elmiskop IA, 101, and 102, retained the main structural features of the first instrument over a period of 18 years, and it became progressively less attractive relative to competing instruments. Although Siemens had a large customer base in most countries of the world, it gradually lost market share, and was unable to exploit the very real merits of the CT 150 when it appeared. Siemens then produced the STEM ST 100F, which was probably too early into this new market to hope to achieve any considerable number of sales. Thus, the expenses of the development of two new instruments, set against sharply reduced sales, must have contributed to the decision to withdraw from the market.

The case of AEI was somewhat different. The microscope designs were in general advanced in concept, but in the earlier years, were insufficiently engineered for production, so that minor faults caused difficulties in production and irritation to the users. This was due mainly to the very small engineering strength allocated to the business, and limited production facilities reduced the exploitation of the designs. This was all part of a strange company policy which allowed a substantial research effort to be expended in Aldermaston, but which grossly underfunded the engineering, production, and sales effort. This continued until the early 1960s, when a new production facility was opened in Harlow, and when the engineering team was greatly expanded also.

The company continued to be very innovative in design ideas, though a good deal of engineering effort at first had to be diverted to sorting out production difficulties in a factory unused to making instrumentation of this degree of sophistication. Later in the 1960s, the engineering was at last adequate to put the required amount of detailed attention into design to ensure reliability and good production results. There were major investments in engineering the EMMA and the high-voltage microscope EM 7 as well as the conventional line. Unfortunately, sales began to fall behind production capability, after years of long delivery times. This was probably due both to insufficient investment in the marketing activity and to the small customer base inherited from the earlier years. The problem had been masked for too long by the production backlog, which for years determined the number of instruments sold.

Although a new biological microscope was almost completed in design and a new 300-kV, high-resolution instrument had been mapped out as an engineering proposal, the company declined to invest any further and withdrew from the market. It has to be said that the end was accelerated by a decision to make a STEM instrument, instead of continuing with the conventional high-resolution TEM, and in this, the company made the same error as Siemens. Even this might not have been fatal if the company had gone ahead in the 1960s with a much-discussed merger with Cambridge Instruments, which would have resulted in major savings in both engineering and sales for both companies. It should be mentioned for the record that AEI had been taken over by GEC in the mid-1960s, which in the early 1970s sold the scientific instrumentation business to the Kratos Company. This was fairly irrelevant to the present story, as it was clear that Kratos had no long-term interest in the electron microscope business, since it immediately sacked a substantial part of the engineering department.

This has left three smaller companies—Camscan, Cameca, and VG. They have wisely adopted a different philosophy, and have gone for niche marketing. Camscan, by making a very modular form of instrument, have been able to offer special facilities which have not been economic for the larger companies to attempt. By retaining a small scale it has a much lower expense level and can afford to be very responsive to customer needs and thus maintain its independence. Cameca has retained a reputation as a leader in electron probe microanalysis, which is the historical legacy of the work by Castaing. Even Cameca's scanning microscopes try to offer something special in the way of X-ray analysis. Vacuum Generators has been more adventurous in EM terms, though strictly inside the company specialty of UHV instrumentation. VG also deals in such relatively small numbers that it can offer almost a one-off service, while maintaining a high degree of standardization of parts. It has stayed firmly in the van of technical developments, by dint of using ideas and developments from users.

The case of the USSR is somewhat different. Here, the overall numbers of instruments were centrally determined, and distribution was by allocation rather than by market forces. Despite a rather late start, the Sumy factory managed to produce a wide range of instruments of rapidly improving performance, and with some very ingenious technical innovations of its own. What is very interesting is the much greater emphasis on instrumentation for the physicist apparent in the production emanating from East Germany and the USSR. The large numbers of electron diffraction cameras produced is especially striking, as are the emission microscopes.

As suggested earlier, it looks as if Tesla decided that market penetration on an adequate scale would prove too expensive for it to remain competitive, and it seems to have decided that in future it should concentrate on production of microscopes and components for other companies.

Appendix III summarizes details of the production quantities in both Western and Eastern Europe, and it will be seen that the combined production of the USSR and Czechoslovakia achieved a level which represented a significant proportion of the numbers of the total production of Western Europe, particularly in TEM. The production of scanning microscopes was considerably less in the USSR than in the West. The production of all kinds of electron optical instruments in the USSR reached a level of 250–300 instruments a month in the mid-1970s, but later fell away sharply. It seems that something similar has occurred in the West, though on a much reduced scale, and SEM sales seem to have continued to rise. The figures do illustrate the very rapid expansion of the technique, and the more recent falling off in demand, as new instruments with alternative imaging processes compete for available funds, and as the direction of research has changed.

The production numbers of the various microscope manufacturing companies of course include all the instruments that they exported, so that the numbers of their instruments actually used in European laboratories were always significantly smaller. At the same time, however, there were considerable numbers of imported electron microscopes, first from the United States and later from Japan, and it could be that the imports and exports were roughly in balance for much of the time. A "snapshot" of the position in the United Kingdom in 1970 shows that of 864 instruments in use, 141 were imports from outside Europe. A typical manufacturer in later years seems to have exported about half its output outside Europe. Making these rather broad assumptions, and those outlined in Appendix III concerning the effective operational life—a figure confirmed by the quoted experience of RCA (Reisner, 1989) that over two-thirds of its instruments were still in use more than 20 years after delivery—one can deduce that there are at least 10,000, and a best-guess figure of about 12,500, electron microscopes operational in Europe at the present time. However, the continued rise in relative strength of the Japanese companies might result in the 1990s in an excess of Japanese imports over exports from European manufacturers. This would mean that the estimates of operational instruments in Europe given here are low.

As far as can be seen at present, we should be served in the 1990s by three major Western European companies: Philips, Leica (formerly Cambridge Instruments), and Zeiss. The niche markets, needing fast reaction times and close customer relations, seem likely to be filled by Camscan, Cameca, and VG. At the final stage of revision of this chapter, I learn that the Sumy factory has suffered a severe drop in production levels as a result of the enormous changes in Russia and the Ukraine. It has, however, continued to make important technical advances to maintain a competitive position in the world market. Tesla has also suffered major changes, though apparently a new company, TESCAN Ltd, has been formed to carry on some of the work, and Dr. Delong has also entered the electron optics field with a new company. Of course, since electron microscopy is a global activity, future events will be influenced in a major way by the big Japanese companies, but that discussion lies outside the scope of this chapter.

ACKNOWLEDGMENTS

This compilation would not have been possible without the very helpful cooperation of the various microscope manufacturing companies. I would like to acknowledge the individual assistance of Mrs. Cilly Weichan, formerly of Siemens and Halske, and of Prof. Hermann of Tübingen University; the late Prof. Jan Le Poole, formerly of the Technological University of Delft, and David Roebuck of Philips Analytical, Cambridge; Dr. Rudolf Partsch of Zeiss Oberkochen, and Mr. Rodney Setterington, Mr. David Woodward, and Mr. Tom Kermeen of Zeiss U.K., Dr. Dick Paden of Camscan, Dr. Julian Davey of Cambridge Instruments, Mr. Colin Helliwell of Cameca, Dr. A. Delong, formerly of Tesla, Brno, and Monsieur Brian of SOPELEM S.A. Prof. Kellenberger was kind enough to send me a copy of the book on the history of electron microscopy in Switzerland, and M. Vastel a photograph of the CSF microscope.

I had much helpful information on the East German microscopes from Herr Feistel and Herr Sonnefeld of Carl Zeiss Jena though the good offices of Dr. Werner Noli of Plano W. Plannet. Prof. D. Schulze kindly provided photographs and information on the WF instruments from East Berlin, and Prof. Heydenreich sent brochures of the microscopes made by that company. Ing. B. Schramm has been most helpful in correcting my contribution concerning the WF Company. Dr. B. Tesche kindly provided a copy of the Wolpers paper. Prof. G. Möllenstedt, Dr. H. Schluge, and Dr. O. Rang have provided valuable details about the early AEG microscopes through a long correspondence. Dr. Wolpers has helped me with his personal memories of the early days in Siemens, and with additional reprints. Dr. Peter Hawkes provided me with some very helpful pointers to difficult sources of information. Dr. Ken Smith kindly read the text and made very helpful comments for improving it.

The important information about the production of electron optical instruments in Sumy is thanks to the comprehensive efforts of Prof. G. D. Kisel, of the "Electron" factory in Sumy, Ukraine. Prof. N. Kiselev and Dr. P. A. Stoyanov contributed valuable information on the instruments developed and manufactured near Moscow. Dr. Audrey Glauert employed her usual meticulous reading to clarify parts of the text. I am deeply indebted to Prof. Tom Mulvey, not only for his helpful and stimulating comments on the draft text, but for his inestimable help in translating the extensive information on the USSR production sent by Prof. Kisel. Finally, I should acknowledge the help and patience of Margaret, my wife, who over a period of many months has been a computer widow, as successive drafts of the work were extensively amended.

The illustrations have been kindly supplied by the various microscope companies or their successors. Dr. Chris Hammond of Leeds University provided Figs. 54a and 54b, and Fig. 85 is reproduced by kind permission of the authors and the American Physical Society. Figures 1(a) and 1(b), 2, 4, and 6 are reproduced by kind permission of Frau I. Ruska, Hirzel Verlag, Berlin.

APPENDIX I

Listing of Manufacturers and Model Numbers of Electron Microscopes and Microprobes

The instruments manufactured by a particular maker are listed in the fourth column of the table. Where large numbers of models were made by one manufacturer, the first to be made is listed in the appropriate position in the fourth column followed by the remaining model numbers in two or three columns. The order of manufacture is to be read down the first column, then down the second, and so on.

Manufacturer	Abbreviation	Country	Instruments	
Allgemein Elektrizitäts Gesellschaft	AEG	Germany (West)	EM 5	EM 6
Later: Süddeutsche Laboratorien-AEG	SDL-AEG	Germany (West)	EM 7	
AEG-Zeiss Opton	AEG-Zeiss	Germany (West)	EM 8-I	
			EM 8-II	
Carl Zeiss Oberkochen	Zeiss Oberkochen	Germany (West)	EM 8-III	
			EM 8-IV	DSM 950
			EM 9	CEM 902
			EM 9A	CSM 950
			EM 9S	DSM 940
			EM 10	EM 900
			EM 10A	DSM 960
			EM 10B	DSM 962
			EM 10C	EM 910
			EM 109	DSM 940A
			EM 109 Turbo	DSM 960A
			EM 902	EM 902A

(continues)

APPENDIX I (*Continued*)

Manufacturer	Abbreviation	Country	Instruments	
Balzers	Balzers	Liechtenstein	KE 2	
			KE 3	
Cambridge Instruments	Cambridge Inst.	United Kingdom	Microscan	S 150
			Microscan 2	S 604
			Geoscan	S 150 Mk II
			Stereoscan	S 250
			Stereoscan 2	S 100
			S. 2A	S 250 Mk II
			Microscan 5	S 200
			Stereoscan 4	S 90
			Stereoscan 600	S 250 MkIII
			S 4-10	S 120
			S 180	S 240
			Microscan 9	S 360
Later:				
Leica	Leica		S 260	
			S 360 FE	
Cambridge Scanning Company	Camscan	United Kingdom		
(Initially marketed by Bausch and Lomb)	Balscan		Camsem	
			Camsem II	
As Camscan:	Camscan		Series 3 plus variants	Series 4FE
			Series 4 plus variants	
			Series 2	
Cameca	Cameca	France	MS 85	
			MS 46 and ME 76	
			MEB 07	
			Camebax	
			Camebax Micro-beam	
			Camebax SX 50	
			SEMprobe	
Carl Zeiss Oberkochen	Zeiss Oberkochen	Germany (West)	See after AEG	
Compagnie Générale de Télegraphie sans Fil	CSF	France	M II	
			M III	
			M IX	
Iskra	Iskra	Jugoslavia	LEM 4C	
Metropolitan-Vickers Electrical Company	Metro-Vick	United Kingdom	EM 1	
			EM 2	
			ED 2	
			EM 3	
			EM 3A	
			EM 4	

(*continues*)

APPENDIX I (*Continued*)

Manufacturer	Abbreviation	Country	Instruments		
Later: Associated Electrical Industries	AEI	United Kingdom	EM 6 EM 6B EM 6G EM 801 EM 802	EMMA 4 EM 7 Corinth 275 Corinth 500 Cora	
Optique et Précision Levallois	OPL	France	MEU 1A MEU 2A MEU 2C		
Philips	Philips	Netherlands EM 75 EM 200 EM 300 EM300/ STEM EM 201 EM 301 PSEM 500 EM 400 PSEM 500X SEM 501	EM 100 SEM 505 EM 410 EM 420 EM 430 SEM 515 SEM 525 SEM 535 CM 10 CM 12	CM 12/STEM CM 30 CM 20 SEM 555 XL 20 XL 30 XL 40 CM 20 UltraTWIN	
Polaron	Polaron	United Kingdom	MR 60	MR 60C	
Siegbahn-Schönander	Siegbahn-Sch.	Sweden	EM		
Siemens and Halske	Siemens	Germany (West)	ÜM ÜM 100 Elmiskop 1 Elmiskop 2 Elmiskop 1A Elmiskop 101	Elmiskop 51 Elmiskop 102 Elmiskop ST 100F Elmiskop CT 150	
Société d'Optique, Électronique et Méchanique	SOPELEM	France	Micro 75		
Sumy, "Electron" factory	Sumy	USSR EG 100A UEMV 100 EVR 1 EEM 50 UEMV 100A EMV 150 UEMV 100B UEMV 100V EG 100M	EM 5 REMN 2 EMMA 2 EMV 100LM EM 200 REM 200 EMR 100 EMV 100B EMV 100A EMV 100AK REM 100 EMV 100BR REMMA 200 PREM 200	EM 125 REMMA 202 DE 120-1 EMR 102 PEM 100 REMMA 202M REM 102E REM 101M REM 101E EMR 102M PEM 100C PEM 100M EM 125K REM 101	

(*continues*)

Appendix I (*Continued*)

Manufacturer	Abbreviation	Country	Instruments	
		UEMV 100K	EMMA 4	
		ER 100	EMR 100M	
		REMN 1	REM 100U	
		REMMA		
		EM 150		
		EMV 100L		
		EMMA		
Tesla Brno	Tesla	Czechoslovakia	BS 241	
			BS 242	BS 540
			BS 242 A-E	BS 300
			BS 413	BS 350
			BS 513A	BS 301
			BS 613	BS 340
			BS 500	BS 343
Trüb-Täuber Company (TTC)	Trüb-Täuber	Switzerland	KM	KE
			KM 4	KE 2
			KM 5	
USSR (other than Sumy) Krasnogorsk Electro-Optical School Vyborg Instrument Plant Krasnogorsk Mechanical Plant	USSR	USSR		
		EM 100	EM 5	EM 9
		EM 3	EEM 75	REMP 2
		UEM 100	EM 6	EM 11
		UEMB 100	EM 7	EM 14
Vacuum Generators	VG	United Kingdom	HB 5	HB 501A
			HB 50A	HB 501UX
			HB 501	HB 603
VEB Carl Zeiss Jena (Jenoptic Carl Zeiss Jena)	Zeiss Jena	Germany (East)	ELMI B	EF 4
			ELMI C	EF 5
			ELMI D	EF 6
VEB Werk für Fernsehelektronik	WF (TV Factory)	Germany (East)	OSW 2752	KEM 1-2
		SEM 2-1	KEM 1	SEM 3-1
		SEM 2-2	KEM 1-1	SEM 3-2
			SEM 3	

APPENDIX II

Chronology of European Electron Microscopes

Date shows when commercial instrument first shipped. Within each year, manufacturers are listed alphabetically, as available information is not precise enough to attribute a date order to the nearest month.

Each instrument has a reference number (col. 2) to aid cross-reference.

Column 5 (Qty) shows number of instruments manufactured, if known. In some cases the number shows total production of the type, including later modifications listed separately here with no production quantity.

Date	No.	Maker	Model	Qty	Technical details
1936	1	Metro-Vick	EM 1	1	Combined optical/electron microscope. Cold-cathode electron source. Resolution not better than optical.
1939	2	Siemens	ÜM	38	70 kV (later 100 kV), 7 nm. First production-line instrument. First polepiece lens in a commercial instrument. Condenser and two imaging lenses. Steel column construction for improved screening. Column units joined by greased cone joints. Short-focal-length lenses for convenient column length. HT from mains transformer. Lenses battery-powered.
1940	3	AEG	EM 5	4	10–50 kV, 5 nm. Electrostatic microscope with objective and two alternative projectors. Magn. 1000, 3000, 10,000. HT from mains transformer. Stereo tilt holder. Diffraction from below projector.
1944	4	AEG	EM 6	1	As EM 5(3), but with HF cascade generator for HT.
1945	5	CSF	M II	2	60 kV. Electrostatic. Two imaging lenses.
	6	Siegbahn-Sch	EM	15	50 kV, two-stage, with horizontal column. Magnetic lenses.
	7	WF (TV Factory)	ÜM (OSW 2752)	12	45–100 kV, 5 nm. Copies of Siemens ÜM(2) completely manufactured in WF factory. Minor modifications to the last three models built.

(continues)

Appendix II (*Continued*)

Date	No.	Maker	Model	Qty	Technical details
1946	8	CSF	M III	60	60 kV, 10 nm. Telefocus gun, two electrostatic imaging lenses. Magn. to 45,000.
	9	Philips	HVEM	1	400 kV, 20 nm. Experimental.
	10	Metro-Vick	EM 2	26	50 kV, 10 nm. Thermionic gun. Condenser lens and two magnetic imaging lenses. Auto vacuum system.
1947	11	SDL-AEG	EM 7	3	10–50 kV, 5 nm. Two electrostatic lenses. Stereo tilt option. Magn. 1000, 3000, 10,000. Mains transformer for HT supply; wooden desk, monocular telescope.
	12	Trüb-Täuber	KM	3	50 kV, 10 nm. Cold-cathode gun. Electrostatic objective; magnetic condenser and projector with battery supplies. Molecular vacuum pump. Max. magn. 10,000.
1947	13	USSR	EM 100	60	Russian copy of Siemens ÜM(2).
1948	14	USSR	EM 3	112	30–50 kV, 10 nm. Open gun. Single condenser; objective and single projector. No airlock for specimen or camera.
1949	15	AEG-Zeiss	EM 8-I	56	10–50 kV, 5 nm. Steigerwald gun. Electrostatic; improved lens insulation, new minimum-focal-length projectors. HF cascade generator for HT. 24-plate camera. Steel desk. Magn: 1600, 3000, 16,000.
	16	Metro-Vick	EM 3	65	100 kV, 5 nm. Single condenser, three imaging lenses; magn. 1000–30,000 at 100 kV; selected area diffraction; stereo tilt; reflection microscopy accessory.
	17	Philips	EM 100	350	100 kV, 5 nm. First 5-lens microscope (4 imaging lenses); selected-area diffraction; stereo tilt; wobbler focus.

(*continues*)

Appendix II (*Continued*)

Date	No.	Maker	Model	Qty	Technical details
1950	18	CSF	M IX	2	60 kV. Telefocus gun; three e/s imaging lenses in solid block. Electrostatic astigmatism corrector.
	19	Metro-Vick	ED 2	6	100 kV. Electron diffraction camera.
1951	20	AEG-Zeiss	EM 8-II		20–60 kV, 2 nm. e/s with objective stigmator. Intermediate accelerator to give 100-kV beam through the specimen. Diffraction lens. High-res. electron diffraction option. 24-plate camera. Magn. 1600, 5000, 16,000.
	21	Siemens	ÜM 100	28	100 kV, 1.5 nm. Three imaging lenses; projector with turret of four polepieces. Specimen and camera airlock; selected-area diffraction; stereo tilt.
1952	22	Metro-Vick	EM 4	30	50 kV, 10 nm. Simple console model; two projector pole systems in single yoke; 3-specimen rod; 40-exp., 70-mm. roll-film camera; interlocked vacuum controls.
	23	Siemens	Elmiskop 2	30	50 kV, 2.5 nm. Simple instrument. No condenser; objective and single projector with a turret of four polepieces. Magn. range 300–30,000. Battery-driven lenses; 12-plate camera.
	24	Trüb-Täuber	KM 4	15	50 kV, 3 nm. Improved HT set and gun for stable operation; cold-cathode; e/s objective; e/m condenser and projector with battery supplies.
	25	USSR	UEM 100	320	40–100 kV, 5 nm. Open gun; same lenses as EM 3(14), but with full alignment of illumination. Airlocks for specimen and 12-plate camera.

(*continues*)

APPENDIX II (*Continued*)

Date	No.	Maker	Model	Qty	Technical details
1952	26	WF	SEM 2-1 SEM 2-2	16	40–100 kV, 4 nm. Single condenser, objective, and projector magnetic lenses. Max magn. 50,000. Polepiece exchange under vacuum for diffraction. 2-2 had more compact HT set.
	27	Zeiss-Jena	ELMI B	3	Preproduction model, 3-stage electrostatic lens system.
1953	28	Metro-Vick	EM 5	1	Experimental 500-kV TEM.
	29	Metro-Vick	EM 3A	15	100 kV, 2.5 nm. Improved HT and objective lens with e/s 4-pole stigmator. Mechanical gun tilt of 8° for reflection microscopy.
	30	Tesla	BS 241	25	50 kV, 10 nm. Single condenser, two imaging lenses. Magnetic astigmatism corrector.
	31	Zeiss Jena	ELMI C	10	50 kV, 5 nm. Electrostatic microscope with 3 imaging lenses.
1954	32	Philips	EM 75	60	75 kV, 5 nm. Simple microscope; very strong objective with low C_c; solid-state electronics with very simple stabilizers; fixed excitation projector, adjustable by movement of polepiece.
	33	Siemens	Elmiskop 1	1100	100 kV, 1 nm. First high-resolution microscope. Double condenser with stigmator; objective stigmator (magnetic); intermediate and projector lenses—the latter with four interchangeable polepieces. Magn. 200–160,000×; 12-plate camera.
	34	Zeiss Jena	ELMI D	100	50 kV, 1.5 nm. Telefocus gun; 5 electrostatic lenses—objective and four projectors.
	35	Zeiss Oberkochen	EM 8-III		20–60 kV, 2 nm. Electrostatic. Prealigned lenses; new auto vacuum system. Dark field

(*continues*)

APPENDIX II (Continued)

Date	No.	Maker	Model	Qty	Technical details
			EM 8-III (continues)		from defined crystal reflections. Survey magn. 200 and optional 25,000 high magn.
1955	36	OPL	MEU 1A	10	100 kV, 5 nm. Single electromagnetic condenser and 3 imaging lenses.
	37	WF (TV factory)	KEM 1	52	40–60 kV, 5 nm. Magn. 1200–30,000 in 7 steps. Three imaging lenses. e/m coils for tilting beam from the gun. High-frequency HT generator.
1956	38	AEI	EM 6	120	100 kV, 1.5 nm, later 1 nm. Double condenser with stigmator; electrostatic objective stigmator; magnetic/electrostatic beam shift and tilt; full magn. range 600–120,000 without realignment. 6-plate camera; reflection and stereo facilities.
	39	Tesla	BS 242	827	60 kV, 2.5 nm. Desktop EM. Rod specimen holder; 4 imaging lenses; selected-area diffraction; Steigerwald gun; battery supply for lenses. Objective stigmator.
	40	Zeiss Oberkochen	EM 8-IV		20–100 kV, 2 nm. Magn. 200–1600; 1600–30,000. Electrostatic filter lens for chromatic aberration correction. Wobbler for focus and stigmator.
1957	41	WF (TV factory)	SEM 3	28	40–100 kV, 1.5 nm. TEM with single condenser, 4 electromagnetic imaging lenses, alignable e/m stigmator for objective. Top magn. 100,000. 12-plate camera.
	42	USSR	UEMB 100	10	40–100 kV, 1.5 nm. Double condenser and 4 imaging lenses giving microdiffraction.

(continues)

APPENDIX II (*Continued*)

Date	No.	Maker	Model	Qty	Technical details
1958	43	Cameca	MS 85	22	10–35 kV. First electron microprobe analyzer. Fixed probe of approx. 1 μm diam. Area for analysis selected by optical mirror microscope. X ray take-off angle 15–18°. Two wavelength spectrometers, Z range 11 to 92.
	44	Philips	EM 200	250	100 kV, 1 nm. Double-condenser lens with stigmator and focus wobbler. Objective with stigmator and 3 further imaging lenses. Rod specimen holder.
	45	Tesla	BS 242-A-E		Like BS 242(39) but with current stabilizers instead of batteries.
	46	Trüb-Täuber	KM 5	7	70 kV, 2 nm. Cold cathode; new e/s objective design very stable at 70 kV, but still with high chromatic aberration; e/m condenser and projector with transistorized electronic stabilizers. Water-cooled diffusion pump instead of molecular pump.
1959	47	USSR	EEM 75	4	75 kV, 50 nm. Emission microscope.
	48	USSR	EM 5	400	40–60 kV, 2 nm. Single condenser; objective with octupole corrector, intermediate lens and projector.
	49	Sumy		79	
	50	USSR	EM-6	150	35 kV, 10 nm. Simplified instrument with line-focus source, two imaging lenses on to a single-crystal screen. The image on this screen was magnified a further 20 times by an optical microscope and recorded on film.
1960	51	AEI	SEM 1	6	30 kV. Fixed-probe microanalyzer, fully focusing Johannson spectrometer, with range $Z = 22$ to 92.

(*continues*)

Appendix II (*Continued*)

Date	No.	Maker	Model	Qty	Technical details
1960	52	Cambridge Inst.	Microscan	80	5–50 kV. First scanning electron probe microanalyzer; two semifocusing spectrometers with X-ray take-off angle of 20°. Light microscope for selection of the area for analysis.
	53	OPL	MEU IIA	60	50–100 kV, 1 nm. Double-condenser, e/m beam alignment, 3 imaging lenses. Magn. 3000–100,000 at 100 kV. e/s astigmatism corrector. Heating, cooling, and straining stages. Exposure measurement by beam current on screen.
	54	Sumy	EG 100A	158	Electron diffraction camera. 40–100 kV. Resolving power: 0.001 for $d =$ 0.05 nm and $L = 755$ mm. Ellipticity of rings: not more than 0.4%.
	55	Sumy	UEMV-100	215	50–100 kV, 1 nm. Universal instrument; double-condenser; objective and condenser stigmators; dark-field tilting of illumination; magn. 300–200,000. First high-resolution TEM in the USSR.
1961	56	AEI	SEM 2	18	5–50 kV, 250 nm. Scanning electron probe microanalyzer. Vacuum fully focusing spectrometer. Z range 12 to 92.
	57	Trüb-Täuber	KE	6	Emission electron microscope. Specimen temperature between 15 and 1200°C. Ion-beam etching facilities.
	58	Zeiss Oberkochen	EM 9		60 kV, 2 nm. Simple EM. Telefocus gun, no condenser. Electromagnetic lenses, autofocus, auto camera with selected-area electron densitometer, rod-type specimen holder with simple airlock.

(*continues*)

Appendix II (*Continued*)

Date	No.	Maker	Model	Qty	Technical details
1962	59	Cambridge Inst.	Microscan2		4–50 kV, 1 μm. EPMA with two lenses; probe diam. 0.2–10 μm. Two semifocusing spectrometers, each with 4 crystals. Flow-type proportional counters. Element range $Z = 11$ to 92. Light microscope viewing. Stage traverse $\pm$ 3 mm; anticontaminator.
	60	Sumy	EEM 50	2	50 kV. Emission electron microscope.
	61	Sumy	EVR-1	29	40–60 kV. Electron diffraction camera. Resolving power for $d = 0.1$ nm and $L = 600$ mm: 0.003. Chart recorder output.
	62	USSR	EM-7	349	40–60 kV, 1.5 nm. Like EM-5(50) but with specimen airlock.
1963	63	OPL	MEU IIC		Like MEU IIA (53) with automatic plate numbering, and high-resolution diffraction adapter. 18-plate camera.
	64	WF (TV Factory)	KEM 1-1	23	40–60 kV, 5 nm. Single condenser, electromagnetic gun alignment. 3 imaging lenses. Octupole e/m stigmator. Stereo tilt. 35-mm camera between projector and screen. 7 fixed-magn: steps to 30,000$\times$. Compare KEM 1 (37).
	65	Zeiss-Jena	EF 4	180	35–65 kV, 3 nm. Universal TEM in modular form. 3 electromagnetic lenses. 6-plate camera.
1964	65	AEI	EM 6B	250	30–80 kV, 0.5 nm. First biological EM. Immersion objective; double objective stigmator; 3-specimen rod; anticontaminator; full magn. range 1500–250,000 on one knob with auto brightness and autofocus adjustment. Electrical shutter.

(*continues*)

APPENDIX II (*Continued*)

Date	No.	Maker	Model	Qty	Technical details
1964	67	Cambridge Inst.	Geoscan		50-kV microprobe for geology. Horizontal column. 2 fully focusing spectrometers; high take-off angle of 75° for X rays.
	68	Cameca	MS 46	218	3–40 kV, 500 nm. Second-generation electron microprobe with 4 wavelength spectrometers. Beam diam. 0.2–300 μm. Optical mirror microscope with objective N.A. 0.48. Scanning electron beam to give a specimen current scanning image. Take-off angle 18°, Z range 5 to 92.
	69	Siemens	Elmiskop 1A		100 kV, 0.6 nm. Improved HT stabilizer compared to Elmiskop 1 (33).
	70	Sumy	UEMV-100A	132	Like UEMV-100 (55) but with specimen heating and straining stage and anticontaminator.
	71	Sumy	EMV-150	3	150 kV, 1 nm. TEM. Anticontaminator-fitted.
	72	Trüb-Täuber	KE 2	14	Improved emission microscope compared with KE (57).
	73	USSR	REMP-1	10	1–20 kV. Microanalyzer with magnetic lenses and X-ray analysis by WDX.
	74	Zeiss Jena	EF 5		Reflection electron diffraction attachment for EF 4 (65).
	75	Zeiss Jena	EF 6	125	20–40 kV. Emission microscope module for EF 4 (65). Resolution in thermal or secondary electron emission, 15 nm. Secondary electron imaging generated by ion beam or primary electron beam. Ion etching and cleaning of specimen surface. Specimen temperature 700–2000°C.
1965	76	Balzers	KE 3	10	40–50 kV, 15 nm. Photoemission microscope. Specimen irradiated by UV.

(*continues*)

Appendix II (Continued)

Date	No.	Maker	Model	Qty	Technical details
			KE 3 (continues)		Many specimen treatment facilities.
1965	77	Cambridge Inst.	Stereoscan	440	1–20 kV, 50 nm. First commercial scanning microscope in serial production.
	78	Sumy	UEMV 100B	96	Like UEMV 100 (55) but with gas cell working to 760 mm pressure for wet biological specimens. 24-plate camera.
	79	Sumy	UEMV-100V	234	50, 75, 100 kV, 0.8 nm. Universal EM. Magn. range 300–200,000. 18-plate camera for 6.5 × 9 cm plates.
	80	Tesla	BS 413	358	100 kV. Double condenser; 3 imaging lenses; 2 stigmators.
	81	WF (TV factory)	SEM 3-1	12	Like SEM 3 (41) but resolution improved to 1 nm by higher stability in electronic supplies. Double condenser with stigmator. Anticontaminator-fitted. Heating stage to 1000°C. High-resolution electron diffraction attachment.
1966	82	AEI	EM 6G	100	30–100 kV, 1 nm. Metallurgical EM. Electrical dark-field tilting and high-resolution tilt stage. Reflection microscopy with electrical beam tilt. Auto shutter.
	83	Cambridge Inst.	Stereoscan2		1–25 kV, 50 nm. SEM.
	84	Philips	EM 200 addnl		Goniometer stage.
	85	Philips	EM 300	1850	100 kV, 0.5 nm. First fully transistorised EM. Gun airlock; motorized camera. Rod-type specimen holder, later adapted for high tilt angles. Most successful commercial microscope with universal application.
	86	USSR	EM-9	500	Like EM-6 (50), 35 kV, 5 nm. Magn. 0.5–8000× on to single-crystal screen. This image further magnified 20×

(continues)

APPENDIX II (*Continued*)

Date	No.	Maker	Model	Qty	Technical details
			EM-9 (*continues*)		by optical microscope. In-vacuum photography optional.
1966	87	Zeiss Oberkochen	EM 9A		Like EM 9 (58), but resolution 1.5 nm with normal gun and single condenser permitting higher magnifications.
1967	88	Cameca	MS 46/ME 76		Accessory electron microscope attachment to MS 46 (68). Objective and projector insertable under the specimen stage to give transmission images up to 10,000× at resolution of 5 nm. A further attachment gave secondary and back-scattered electron detection for the scanned image.
	89	SOPELEM	Micro 75	20	30–75 kV, 1 nm. Single condenser; 3 magnetic imaging lenses; magn. 2000–128,000; anticontaminator. 18-plate camera with auto numbering. Goniometer stage option.
	90	Sumy	EG-100M	65	Like EG-100A (54), diffraction camera with facilities for cooling specimens to −140°C or heating to 1000°C. Also for gas or vapor molecules up to 350°C.
	91	Sumy	ER 100	84	25–100 kV. Recording electron diffraction camera for specimens up to 0.1 μm in transmission or size 10 × 10 × 3 mm in reflection. Ellipticity of rings < 0.3%. Emission current stability better than 2%.
	92	Sumy	UEMV-100K	570	Materials science instrument developed from UEMV 100 (55), with goniometer tilting ± 10°.

(*continues*)

APPENDIX II (*Continued*)

Date	No.	Maker	Model	Qty	Technical details
1967	93	Tesla	BS 513A		50, 80, 100 kV, 0.7 nm. Double condenser; 3-stage imaging. Stigmators for condenser 2, objective and intermediate lenses. Magn. 2500–250,000. Anticontaminator.
	94	WF (TV factory)	KEM 1-2	22	40–60 kV, 1.5 nm. Improved from KEM 1-1 (64). Better stabilities; 9 fixed steps of magnification to max. 50,000×. Plate camera in addition to film camera.
	95	WF (TV factory)	SEM 3-2	29	Like SEM 3-1 (81) but magn. range now 2000–200,000 on plate. Spot from double condenser 2 μm. Faraday cage for autoexposure system.
1968	96	AEI	EM 801	80	80 kV, 0.5 nm. Biological microscope; 6-specimen holder; motorized specimen tilt. Serial section holder. Autoexposure system.
	97	AEI	EM 802	50	40–100 kV, 0.5 nm. Metallurgical; high-resolution gonio stage, 360° rotation. New electromagnetic-beam tilt control up to 3.5°. Autoexposure system.
	98	Cambridge Inst.	Stereo.S2A		1–30 kV, 25 nm. Improved resolution compared to (83).
	99	Cambridge Inst.	Microscan5	60	Replacing the Geoscan (67)—horizontal column. Manual spectrometer.
	100	Philips	EM 300/STEM		First STEM attachment for a European TEM.
	101	Siemens	Elmiskop 101	150	100 kV, 0.5 nm. Improved high-voltage stability and vacuum control compared to Elmiskop 1A (69). Special lift cartridge giving very short focal length for spot size 0.5 μm for X-ray analysis. Optional liquid helium stage.

(*continues*)

APPENDIX II (*Continued*)

Date	No.	Maker	Model	Qty	Technical details
1968	102	Sumy	REMN-1	33	0.5–2.3 kV. First SEM in USSR. This was low-voltage model with magn. 100–1,000. Resolution in absorbed electron mode, 1 μm.
	103	Sumy	REMMA	4	10–50 kV, 50 nm with secondary electrons. Goniometer stage. Microanalysis with WDS, $Z = 5$ to 92.
	104	Sumy	EM 150	8	150 kV, 1 nm. Two-stage accelerator for gun. Iron-nickel alloy column. 24-plate camera for 90 × 120 mm plates.
	105	Zeiss Oberkochen	EM 9S		60 kV, 0.9 nm. Cf. EM 9A (87). Projector lens strength now doubled, increasing magn. range and improving electron diffraction patterns. Thin-film apertures fitted to reduce contamination, 70-mm camera, increased sheet-film capacity of 60 sheets, and TV adapter.
1969	106	AEI	EMMA-4	6	100 kV, 1 nm. First analytical microscope. Two fully focusing spectrometers. Spot size 0.13 μm by using minilens as third condenser in EM 802 column (97). X-ray take-off angle 45°.
	107	AEI	EM 7	12	100–1000 kV, 1 nm. High-voltage, high-resolution microscope. Magn. range 63–1,600,000, first with nonrotating image for magnification change. Camera lengths 400–6300 mm and 54–914 m. Joystick control of servo-operated specimen stage. Many treatment stages.

(*continues*)

Appendix II (*Continued*)

Date	No.	Maker	Model	Qty	Technical details
1969	108	Iskra	LEM 4C	10	60–75 kV, 1.5 nm. Routine microscope. Top magn. 70,000. Objective prefield used in lieu of a separate condenser lens. Wobbler alignment. Electronic exposure control.
	109	Polaron	MR 60	10	60 kV, 2 nm with telefocus gun. Magn. 400–50,000; fitted stigmator and anticontaminator.
		Polaron	MR 60C		60 kV, 1.2 nm. Like MR 60, but with double condenser and stigmator. Routine microscope for biological work.
	110	Siemens	Elmiskop 51	50	50 kV, 10 nm. Routine microscope. Cold-cathode source. Permanent magnet exciting objective and projector lenses. Total fixed magn. 9200. Focusing by variation of HT. 15 specimens on rotary stage.
	111	Sumy	EMV-100L	227	50–100 kV, 0.3 nm. Multipurpose high-resolution TEM. 400–600,000×, with mu-metal lenses and vibration-proof specimen stage with thermocompensation. Fitted eucentric stage with ±30° tilt and full rotation.
	112	USSR	REMP-2	50	20–50 kV, 50 nm. SEM with X-ray analysis in a 3-crystal spectrometer. Z range 12 to 92. Probe current 10^{-7} to 10^{-6} A. SE detection.
1970	113	Camscan	Balscan Camsem		2–25 kV, 40 nm. SEM with horizontal column. TV rate display. Pushbutton auto specimen change. Simple controls.
	114	Cambridge Inst.	Stereoscan S4	300	1–30 kV, 15 nm. Improved gun; energy-dispersive X-ray available.

(*continues*)

APPENDIX II (*Continued*)

Date	No.	Maker	Model	Qty	Technical details
1970	115	Cameca	MEB 07	35	1–50 kV, 20 nm. SEM with goniometer tilt stage. Magn. 20–100,000.
	116	Philips	EM 201		40–100 kV, 0.5 nm. Simplified TEM. Minicondenser lens for brightness control. Electromagnetic alignment system.
	117	Sumy	EMMA	48	100 kV, 0.7 nm as TEM. Based on the UEMV 100 series (55). Resolution using X-ray analysis equipment: 5 nm. Detection range: $Z = 12$ to 92.
	118	Tesla	BS 613		Like BS 513 (93) but resolution 0.9 nm with improved screening and mechanical stability. Focus wobbler. Improved field field of view and distortion. Electrical beam tilt $\pm 1.5°$.
	119	Zeiss Oberkochen	EM 9S-2		Cf. EM 9S (105). Resolution improved to 0.45 nm. Improved stigmator; top magn. now 60,000. Autofocus wobbler. Low-angle electron diffraction possible. Four beam-alignment coils instead of mechanical alignment.
1971	120	AEI	Corinth 275	40	80 kV, 1 nm. Simple console-model TEM with inverted column. Three imaging lenses. Four specimens in rod holder.
	121	Cambridge Inst.	Stereoscan 600		30 kV, 25 nm. First solid-state simplified SEM. Optional EDX. Signal processing accessories available. TV display available.
	122	Sumy	REMN-2	151	0.5–5 kV. Low-voltage SEM. Resolution 0.5 μm in absorbed electron mode.
	123	USSR	EM-11	37	40–60 kV, 1.2 nm. Like EM-7 (62) but also with airlock for photo plates.

(*continues*)

APPENDIX II (*Continued*)

Date	No.	Maker	Model	Qty	Technical details
1971	124	Zeiss Oberkochen	EM 10		20–100 kV, 0.5 nm. Research TEM. Single-button, fully automated exposure with choice of three different cameras. Simple specimen airlock.
1972	125	AEI	Corinth 500	65	Resolution improved to 0.5 nm from Corinth 275 (120).
	126	Cambridge Inst.	Stereoscan S4-10		Stereoscan 4 (114) upgraded to 10-nm resolution. EDX, WDX available.
	127	Camscan	Balscan Camsem II		2–25 kV, 10 nm. Resolution upgrade compared to (113).
	128	Philips	EM 301		100 kV, 0.3 nm. Magn. 100–1,000,000 and large diffraction camera length range.
	129	Philips	PSEM 500		1–50 kV, 10 nm. First Philips SEM. Fully eucentric goniometer stage driven by stepper motors. Large specimen chamber.
	130	Siemens	Elmiskop 102	200	125 kV, 0.4 nm. First routine high-resolution instrument over 100 kV in mass production from European manufacturer.
	131	Tesla	BS 500	399	60, 90 kV, 0.7 nm. Biological TEM. Double condenser, cold trap; 3 imaging lenses. Octupole stigmators in condenser and objective. Autoexposure and numbering of plates.
1973	132	Cambridge Inst.	Stereoscan S 180		20–60 kV at 10 nm resolution; 1–10 kV also available. Dual display. First SEM with alphanumeric notation on to micrograph. BS, SC modes. Tilt correction and scan rotation facilities. Selected area, large-area channeling pattern. Fully compensated magn. for any working distance.

(*continues*)

APPENDIX II (*Continued*)

Date	No.	Maker	Model	Qty	Technical details
1974	133	Cameca	Camebax		1–50 kV, 7 nm. First simultaneous microprobe analysis and scanning microscope imaging without compromising performance. Electron signals: SE, BSE, SC, CL, and Kikuchi. Z range, 5 to 92. Four WD and one ED spectrometer. X-ray take-off angle 35°.
	134	Sumy	EMMA 2	170	Like EMMA (117), but with improved sensitivity of X-ray analysis using a semifocusing spectrometer at take-off angle of 15°.
	135	Sumy	EMV-100LM	171	Similar to EMV 100L (111).
	136	Sumy	EM 200	106	25–200 kV, 0.8 nm. Two-stage accelerator for gun. First European intermediate-voltage TEM. Magn. range 300–200,000.
	137	Sumy	REM-200	468	5–40 kV, 20 nm in secondary electron mode. Magn. 40–80,000.
	138	Sumy	EMR-100	73	25–100 kV. Electron diffraction camera. Camera length between 360 mm and 40 m in large-angle mode for spacings up to 50 nm by adjustment of lens between specimen and screen. Normal camera length 600 mm without lens in the path, giving resolution of interplanar spacing of 0.002 nm at 0.1 nm.
	139	Tesla	BS 540	incl. in (131)	120 kV. Metallurgical TEM in BS 500 (131) series, with gonio stage.
	140	Vacuum Generators	HB 5		80 kV. FE STEM. Line resolution 0.34 nm. Microanalysis at 1-nm resolution.
	141	Zeiss Oberkochen	EM 10A EM 10B		20–100 kV, 0.3-nm line resolution with a semi-eucentric goniometer to 60°

(*continues*)

APPENDIX II (*Continued*)

Date	No.	Maker	Model	Qty	Technical details
			EM 10A EM 10B (*continues*)		tilt. High-resolution stage for 0.2 nm. Derived from EM 10 (124).
1975	142	Cambridge Inst.	Microscan 9	12	1–60 kV. Automated microprobe 0.3 μm beam. Two PDP 11 computer-controlled, fully focusing spectrometers. Electronics like Stereoscan 180 (132).
	143	Camscan	Series 3		2–30 kV, 10 nm. Horizontal column on dynamic self-leveling suspension. Split screen, dual-magn. displays on large CRTs.
	144	Philips	EM 400		100 kV, 0.3 nm. Ultraclean vacuum system. Extra projector to allow for differential pumping giving 5 imaging lenses. Full electromagnetic alignment of the column. Optional lanthanum hexaboride gun.
	145	Siemens	Elmiskop ST 100F	10	10–100 kV. FE gun, 0.2 nm. High-resolution STEM with eucentric stage at full resolution. Light- and dark-field detectors. Magnetic sector energy analyzer.
	146	Sumy	EMV-100B	76	10–100 kV, 0.3 nm. Magn. range 400–600,000×.
1976	147	AEI	CORA	10	Analytical version of Corinth 500 (125) using EDX with beam probe size down to 25 nm, and X-ray extraction from a level specimen. Detection limit: 10^{-18} g.
	148	Cambridge Inst.	S 150		1–30 kV, 7 nm. New 3-lens column. Alphanumeric print out on photo. Twin display tubes, dual magnification. Optional indirectly heated LaB$_6$ gun and autofocus.
	149	Philips	PSEM 500X		WD spectrometer in addition to ED on PSEM 500 (129).
	150	Siemens	CT 150	8	20–150 kV, 0.3 nm in high-resolution TEM with eucentric goniometer at full

(*continues*)

APPENDIX II (Continued)

Date	No.	Maker	Model	Qty	Technical details
			CT 150 (continues)		resolution. UHV in specimen chamber. Auto control of lens parameters. Minimum exposure system. Special TV viewing of enlarged image in chamber. Preprogrammed alignment system. Efficient extraction of X-rays from specimen.
1976	151	Sumy	EMV-100A/AK	181	Like EMV-100L (111), but with quantitative computer analysis of electronic images.
	152	Sumy	REM-100	23	SEM. Resolution 10 nm.
1977	153	Philips	SEM 501		1.8–30 kV, 7 nm. Simple SEM. Manually operated eucentric goniometer, tilt from $-15°$ to $+75°$.
	154	Tesla	BS 300	142	1–50 kV, 15 nm. Three probe-forming lenses. Slow scan + TV. Two screens plus record tube. SE, BS, AE imaging. Eucentric stage for small specimens. Automatic vacuum control.
1978	155	Cambridge Inst.	S 604		Replacing S 600 (121). 10-nm resolution.
	156	Philips	EM 400T		TWIN lens version of EM 400 (144) for optimizing TEM/STEM operation. Optional FE source to optimize resolution in TEM and SEM modes. Secondary electron detector added to existing BS detector.
	157	Sumy	EMV-100BR	110	Like EMV-100B (146) but with STEM attachment fitted, the first from the USSR.
	158	Sumy	REMMA-200	38	5–50 kV, 30 nm. Fitted with two wavelength spectrometers, $Z = 5$ to 92. 40–100,000×.
	159	V G	HB 50A		UHV FE SEM. 60 kV, 5 nm with Auger spectrometer for surface analysis at high

(continues)

Appendix II (*Continued*)

Date	No.	Maker	Model	Qty	Technical details
			HB 50A (*continues*)		resolution. Specimen preparation chamber on side of instrument.
1978	160	Zeiss Oberkochen	EM 10C		EM 10 (124) with ion getter pump EDX. Optional facilities: STEM attachment, TV adapter EDX, image analysis, image intensifier, cryo system.
	161	Zeiss Oberkochen	EM 109		50–80 kV, 0.5 nm. First TEM with transmitted fiber optics for external photography. Ion getter pumping; minimum-dose focusing. Factory-aligned imaging lenses. Lanthanum hexaboride gun available. TV adapter.
1979	162	Cambridge Inst.	S150 Mk II (cf 148)		1–40 kV, 6 nm.
	163	Sumy	PREM-200	43	25–200 kV, 0.5 nm (normal TEM), 0.45 nm (STEM). Scanning transmission electron microscope. Development of (136).
	164	USSR	EM-14	50	2–80 kV, 0.8 nm. Single condenser plus objective prefield to give spot demagnification of 5:1. Magnetic objective with electron retardation in front of the specimen and electron acceleration after it.
1980	165	Cambridge Inst.	S 250		1–40 kV, 6 nm. First SEM with turbo pumping standard. Optibeam-fitted for automated optimized resolution for given beam current. Full alphanumerics. Large visual displays. Split-screen dual magnification. EDX option.
	166	Camscan	Series 3 upgrade (143)		2–30 kV. Resolution improved to 6 nm. Very-large-movement goniometer stage ± 76 mm and objects up to 200 mm length. Viewing

(*continues*)

Appendix II (Continued)

Date	No.	Maker	Model	Qty	Technical details
			Series 3 upgrade (continues)		windows and chamber illumination. WDX, EDX available.
1980	167	Philips	SEM 505		1–30 kV, 5 nm. YAG high-sensitivity BS detectors fitted. TV rate on all modes. Data link for automatic compensation of operating parameters.
	168	Sumy	EMMA 4	78	10–100 kV, 0.5 nm. Based on EMV 100B (146). X-ray analysis, semifocusing spectrometer. $Z = 5$ (boron) to 92.
	169	Sumy	EMR-100M	36	Like EMR-100 (138) but with small-angle lens now allowing maximum interplanar spacings up to 60 nm. Auxiliary lenses act as electron filters to increase sensitivity for direct recording. Information output to punched cards or computer.
	170	Tesla	BS 350	18	1–25 kV, 5 nm. FE gun. Slow scan and TV rate. UHV chamber. BS, AE, SC modes. Ion getter pumps. Ion-beam cleaning/etching of specimen surface.
	171	V G	HB 501		80 kV, 0.2 nm. UHV STEM. Replacing the HB 5 (140), with much improved analytical performance. 1 nA of current in 1-nm probe. New energy-loss spectrometer.
	172	Zeiss Oberkochen	EM 109R		Simplified by omission of diffusion pump. Cold trap in manifold. Lower top magnification than EM 109 (161).
1981	173	Cambridge Inst.	S 100		2–25 kV, 7 nm. Simple SEM, cheapest of the Cambridge series. Three-lens prealigned optics, turbo pumped. Line measurement cursor. Large specimen chamber, 90° tilt

(continues)

APPENDIX II (*Continued*)

Date	No.	Maker	Model	Qty	Technical details
			S 100 (*continues*)		on 100-mm specimen. Interface for image analyzer.
1981	174	Sumy	REM-100U	291	5–40 kV, 10 nm with secondary electrons. Fitted wavelength spectrometer. First eucentric stage on USSR SEM; max. specimen size 60 mm diam. on this stage. Also heating and straining stage, STEM operation, ECP, and facilities for characterizing dielectric specimens.
1982	175	Cambridge Inst.	S 250 MkII		1–40 kV, 6 nm. Directly heated lanthanum hexaboride gun option. Optional image store. Upgrade of (165).
	176	Cameca	Camebax Microbeam		Microprocessor-controlled Camebax (133).
	177	Camscan	Series 4		0.5–40 kV, 5 nm. Vertical three-lens column. Standard 100-mm eucentric goniometer on a stage with $\pm$ 100 mm X and Y movements. Viewing port and internal illumination in specimen chamber. Microanalytical facility. Two 280-mm viewing CRTs.
		Philips			Specialized derivations of EM 400 (144).
	178		EM 410		100 kV, 0.45 nm. Dedicated biological microscope.
	179	Philips	EM 420		100 kV, 0.3 nm. Applied research instrument.
	180	Philips	EM 430		300 kV, 0.23 nm. First medium-high-voltage European TEM. Compact instrument with gonio stage. Resolution 0.2 nm when fitted with Super-TWIN lens.
1983	181	Cambridge Inst.	S 200		0.3–30 kV, 6 nm. Mid-range instrument with new electronics. Final lens now a

(*continues*)

APPENDIX II (*Continued*)

Date	No.	Maker	Model	Qty	Technical details
			S 200 (*continues*)		45° minilens to improve specimen tilt angle with large specimens. Large chamber to accept specimens up to 250 mm diam. Text keyboard standard.
1983	182	Camscan	Series 4 addnl (177)		Built-in image store. High-resolution electron channeling pattern introduced. Specialized variants for forensic work—a comparator, and gunshot residue package.
	183	Sumy	EM-125	98	25–125 kV, 0.3 nm. 100–850,000×. Automated vacuum, image control and photographic control system. Alphanumeric coding on photo.
	184	Sumy	REMMA-202	47	5–40 kV, 10 nm. A version of REM 100U (174), but with two wavelength spectrometers, $Z = 5$ to 92 for better analytical facilities.
	185	Sumy	DE-120-1	44	10–30 kV. Fast electron diffractometer for continuous monitoring of structures of thin films for molecular epitaxy. Diameter of beam at the screen: <0.3 mm.
	186	Tesla	BS 301		1–30 kV. Adapters for microelectronic measurements. Option for lanthanum hexaboride gun.
	187	Zeiss Oberkochen	EM 109 Turbo		Turbo-pumped EM 109 (161). First high-precision electronic stage positioning; microprocessor memory for 100 x, y stage coordinates; on-line image analysis.
1984	188	Cambridge Inst.	S 90		2–25 kV, 7 nm. Successor to S 100 (173) as base-range model.

(*continues*)

APPENDIX II (*Continued*)

Date	No.	Maker	Model	Qty	Technical details
1984	189	Camscan	Series 2		1–30 kV, 5 nm. Column like Series 4 (177) but with simplified electronics to provide a mid-range instrument with simplified control.
	190	Philips	SEM 515		1–30 kV, 5 nm. Microprocessor-controlled. Autofocus and astigmatism correction.
	191		SEM 525		Like 515, with large specimen traverse of 100 mm in X and Y. Second secondary electron detector available. EDX can be fitted.
	192		SEM 535		Like 525, with facility for WD spectrometer as well.
	193	Sumy	EMR-102	49	10–100 kV. Electron diffraction camera with digital recording system with two-coordinate digital scans. Electron filter operates up to 75 kV.
	194	Zeiss Oberkochen	EM 902		50, 80 kV, 0.5 nm. First commercial TEM with integrated energy-loss spectrometer for electron spectroscopic imaging, EELS, ESD. Image intensifier; TV adapter; cryo system; on-line image analysis. Energy resolution in image < 25 eV; in the spectrum: 1.5 eV.
1985	195	Cambridge Inst.	S 250 MkIII		0.5–40 kV, 3.5 nm. With optional lanthanum hexaboride gun. Cf. Mark II (175).
	196	Cameca	Camebax SX 50		1–50 kV, 7 nm. Camebax (133) with fully digital electronics, and software control. Electron signals: SE, BSE, SC, CL, Kikuchi. Light microscope fitted. Z range 4 to 92. Accommodates up to 5 WD spectrometers and 1 ED spectrometer.

(*continues*)

Appendix II (*Continued*)

Date	No.	Maker	Model	Qty	Technical details
1985	197	Philips	CM 10		40–100 kV, 0.5 nm. First TEM with full digital control by microprocessor. All lens controls by software, enabling operational controls to be set up in advance and changed instantaneously. Eucentric goniometer with 60° tilt. Five imaging lenses.
	198	Philips	CM 12		20–120 kV, 0.34 nm or 0.3 nm according to objective lens fitted. Minimum TEM probe size 1.2 nm. Software controlled as for CM 10 (197).
	199	Philips	CM 12/STEM		Like CM 12 (198) but with wide choice of STEM modes (Gross et al., 1987).
	200	Philips	SEM 525-IC & EB		Similar to SEM 525 (191) but designed specifically for nondestructive examination of semiconductor wafers. Optimized for low-voltage operation (200 V–5 kV). Conical end lens for high resolution under high-tilt condition. Motorised eucentrically tilting stage with 150-mm travel in X and Y. Linewidth measuring system.
	201	Tesla	BS 340		1–40 kV, 7 nm. Eucentric gonio with specimen size up to 140 mm. Stepper-driven stage with 99-position memory. Split-screen viewing. Scan rotation, tilt correction, gamma correction. Alphanumeric notation on micrograph. EDX available. BS, CL optional.
	202	Zeiss Oberkochen	DSM 950		0.5–30 kV, 5 nm. Fully digital SEM with integrated frame

(*continues*)

APPENDIX II (*Continued*)

Date	No.	Maker	Model	Qty	Technical details
			DSM 950 (*continues*)		store. All electron optical functions digitally controlled. New BSdetector; new zoom condenser system. Automatic choice of optimum beam diameter for given current setting. Take-off angle for EDX and WDX is 35°.
1986	203	Cambridge Inst.	S 120		0.3–30 kV, 6 nm. New, simple SEM to replace S.90 (188). Large displays, large chamber. Small step sizes at low-kV settings.
	204	Cambridge Inst.	S 240		0.3–30 kV, 6 nm. Like S 120, with more ports. New mid-range instrument replacing S200 (181).
	205	Cambridge Inst.	S 360		0.2–40 kV, 5 nm. Research SEM. Centrally controlled by computer with many automated setup and operating procedures. Image store, standard image processing function. Integral EDX color display.
	206	Sumy	PEM-100	212	25–100 kV, 0.5 nm. Magn. 30–400,000. TEM with miniaturized lenses for minimizing column dimensions to 140 mm diam. Lens cooling by boiling of organic liquid. Special temperature-stabilized stage. Anticontaminator. Auto exposure system; external photography by fiber optic plate on to 60-mm film. Microprocessor control of functions.
	207	Sumy	REMMA-202M	25	0.2–40 kV, 4 nm. Based on REM 101/102 series (214/215). Combined SEM-microanalyzer with SE, BS, AE, ECP. 100-mm specimen size. Two wavelength

(*continues*)

APPENDIX II (Continued)

Date	No.	Maker	Model	Qty	Technical details
			REMMA-202M (continues)		spectrometers and one EDX with automated microanalysis capability. Take-off angle 40°. Magn. 10–400,000.
1986	208	V G	HB 501A		HB 501 (171) modified for Auger spectroscopy simultaneous with STEM in UHV condition.
1987	209	Camscan	Series 4 addnl (177)		Digital scanning.
	210	Zeiss Oberkochen	CEM 902		80 kV, 0.34 nm. Like EM 902 (194), with applications-oriented computer control for image processing and analysis and with alphanumeric documentation. Software control of program operation and high voltage. High-resolution goniometer, 360° rotation, 60° tilt.
	211	Zeiss Oberkochen	CSM 950		Like DSM 950 (202) but with integrated computer for program-controlled operation. Linked with an IBAS image processing computer for automatic and interactive on-line image analysis. Precision 6-axis motor-driven stage. Specimen holder with Faraday cage.
1988	212	Camscan	Series 4 addnl (177)		First with optical disk storage and archiving. New chamber with 150-mm movement of specimen on fully eucentric stage. Take-off angle 40° for WDX.
	213	Philips	CM 30		50–300 kV, 0.23 nm with TWIN, 0.2 nm with Super-TWIN. Medium-high voltage like EM 430 (180) but with full digital control by central processor. Five imaging lenses plus three condensers

(continues)

APPENDIX II (Continued)

Date	No.	Maker	Model	Qty	Technical details
			CM 30 (continues)		in addition to the objective prefield, yielding electron probes down to 2 nm in diameter. X-ray take-off angle in objective is 20°. An EELS detector can be fitted under the column.
1988	214	Sumy	REM-102E	4	0.2–40 kV, 5 nm. SEM. 150-mm-diam. specimen. SE, SC, BS, and detection of induced charge currents in microelectronic circuits. Heating and cooling facilities on stage. Magn. 10–400,000. Image processing. EDX fitted.
	215	Sumy	REM-101M	19	Like REM-102E (214). Universal model SEM, with EDX, but display and image processing equipment not supplied.
	216	Sumy	REM-101E		Like REM-102E (214), but no EDX system.
	217	Sumy	EMR-102M	1	Like EMR-102 (193), diffraction camera with improved recording system giving automatic measurement and data handling by software control. Subsequent processing of diffraction patterns by computer. Range of recorded impulses: 11 kHz–1 MHz.
	218	Tesla	BS 343	49	15 kV, 16 nm. Mini-SEM, mobile model. Magn. range 10–50,000. Two-stage beam adjustment. e/m beam blanking. Double condenser lens with stigmator. Objective lens with dynamic focusing. Optional vacuum attachment to large objects, or minichamber to perform as conventional SEM. No water cooling required. Power input 750 VA.

(continues)

Appendix II (Continued)

Date	No.	Maker	Model	Qty	Technical details
1988	219	V G	HB 501UX		100 kV. FE STEM using high-angle scattered electrons giving incoherent beam for easily interpretable atomic number contrast independent of focus. Point resolution in this mode 0.192 nm.
	220	Zeiss Oberkochen	DSM 940		490 V to 30 kV. Basic SEM, but computer-controlled, turbo-pumped instrument. Eucentric stage standard. Built-in frame store.
1989	221	Cameca	Semprobe		0.2–30 kV. Analytical SEM with optimized X-ray facilities. Resolution 4 nm in SE image, 15–400,000×. TV-rate BS image. ECP. Digital image frame store. Eucentric gonio stage for large specimens. X-ray take-off angle 62°; long-term beam current stability of <0.5% in 12 h; fully integrated WDX and EDX; same working distance for SEM and EPMA. Full computer control with mouse-driven menu operation for all functions except tracker-ball operation of magnification and focus.
	222	Camscan	Series 4 addnl (177)		Fast-Track stage microprocessor control. Digital magnetic encoders on stage movement, with 250-position store.
	223	Philips	CM 20		20–200 kV, 0.27 nm (TWIN lens), or 0.24 nm (Super-TWIN) for optimum STEM. Back-scatter detector; high-resolution eucentric goniometer. Coma-free alignment program for ultimate resolution work.

(continues)

Appendix II (Continued)

Date	No.	Maker	Model	Qty	Technical details
			CM 20 (continues)		Convergent beam diffraction available. Many hybrid TEM/scanning modes.
1989	224	Philips	SEM 555		New automated wafer inspection and measurement SEM.
	225	Sumy	PEM 100C		Like PEM 100 (206) but with computer control of alignment and operation. Biological and general-purpose model, with two intermediate and one projector lenses.
	226	Sumy	PEM 100M		Like PEM 100C (225) but with side-entry goniometer for 60° tilt. Three intermediate and one projector lenses. Choice of direct recording or external photography via a fiber optic plate.
	227	Zeiss Oberkochen	EM 900		50, 80 kV, 0.34 nm. Routine microscope with double condenser and four imaging lenses. Specimen movement by tracker ball with 100 memory positions from digital readout. Translation speed is magnification-compensated. Microprocessor control of vacuum and photography. Camera holds 40 sheets of film.
1990	228	Camscan	Series 4 addnl (177)		Enhanced SAD by dynamic chromatic correction giving selected areas of less than 2 μm diameter. Stereo microscope with laser beam targeting of specimen for selected-area analysis. High-sensitivity CL with monochromator facility for analysis available simultaneously with EDX and WDX.

(continues)

Appendix II (Continued)

Date	No.	Maker	Model	Qty	Technical details
1990	229	Camscan	Series 4FE		300 V–25 kV, 2 nm. With thermionic-assisted FE gun and long-term stability of beam.
	230	Leica	S 260		0.3–30 kV, 4.5 nm (W), 3.5 nm (LaB$_6$). Medium SEM, digitally controlled. Autofocus, brightness, contrast, astigmatism correction. X, Y movements 100 mm, Z motion 70 mm. Alphanumeric keyboard. Integral digital image store. Dynamic real-time stereo option.
	231	Philips	XL 20		0.2–30 kV. Desktop microscope for ultimate resolution. 4 nm (W), 3 nm (LaB$_6$). Complete computer control in Windows system with mouse. Eucentric stage with 20-mm. movement. Continuous spot size control; Rotation-free focusing; auto astigmatism correction; electronic image shift ± 20 μm; TV-rate scan rotation. Magnification accuracy better than 3% over range 20 to 400,000. SCSI-bus digital interface for communicating with PC or downloading of digitized images. Option for video hard copy unit or digital optical recorder, or a range of cameras. Image store.
	232	Philips	XL 30		Like XL 20 (231), but resolution 4.5 nm (W). Eucentric stage for 50-mm movement in larger specimen chamber. Both EDX and WDX optional at same working distance. Full

(continues)

Appendix II (*Continued*)

Date	No.	Maker	Model	Qty	Technical details
			XL 30 (*continues*)		system automation with high-speed motor drives for the stage. Integrated X-ray analysis and macrofunction software.
1990	233	Philips	XL 40		Like XL 30 (232), but eucentric stage has 150-mm movement and chamber is designed for really large specimens. Multiple electrical lead-throughs for semiconductor work. WDX and EDX detectors for full analytical facilities. Highly automated.
	234	Sumy	EM-125K	41	EM-125 (183) with $\pm 60°$ goniometer for materials science with 0.37-nm resolution at full tilt. Four-projector lens system giving nonrotational magnification change. Small number of multifunction control knobs under computer control. Wobbler for current and voltage centers and for focusing. Differential pumping of specimen and gun.
	235	Sumy	REM-101	13	Like REM-102E (214) for the examination of specimens with hard phases by SE, BS, SC electrons.
	236	Zeiss Oberkochen	DSM 960		490 V to 30 kV, 4 nm. Research model. Completely digital microscope with computer control. Image-processing interfaces.
			DSM 962		Like DSM 960, but resolution 3.5 nm with optional LaB_6 gun. Includes dedicated image processing.
	237	Zeiss Oberkochen	EM 910		40–120 kV, 0.4 nm. Analytical microscope. First with Kohler illumination into symmetrical condenser-objective by addition of

(*continues*)

Appendix II (Continued)

Date	No.	Maker	Model	Qty	Technical details
			EM 910 (continues)		minilens. High-precision eucentric goniometer stage operating at full resolution up to 60° of tilt. Four-lens projector system; instrument controlled by AT computer with 30-Mb hard disk.
1991	238	Leica	S.360 FE		500 V–25 kV, 2 nm with W/ZrO thermal field emitter. Eucentric stage with 50-mm movement and 360° rotation Optical disk storage option.
	239	Philips	CM 20 Ultra-TWIN		20–200 kV, 0.2 nm with Ultra-TWIN lens (Bakker and Asselbergs, 1990). Specimen tilt up to 20° on gonio stage. Special heat insulation of specimen region, and a coma-free alignment regime for optimal performance.
	240	V G	HB 603		300 kV. FE STEM. Improved resolution and increased peak-to-background ratios for microanalysis.
	241	Zeiss	DSM 940A		Instruments with suffix A have a redesigned desk and control-knob layout. No change to the electron optics. Images can be output as a photo or as video print or output direct to an external computer in digital format.
	242	Oberkochen	DSM 960A		
	243		EM 902A		

Appendix III

Electron Microscope Production

Period	1939–50	1951–60	1961–70	1971–80	1981–90
Transmission electron microscopes					
W. Europe	227	1,212	2,665	2,731	2,065
E. Europe	12	620	1,190	399	—
USSR	172	688	1,833	1,712	642
Total	411	2,520	5,688	4,842	2,707
Scanning electron microscopes and microprobes					
W. Europe	—	45	889	1,418	2,780
E. Europe	—	—	—	160	49
USSR	—	—	85	564	527
Total	—	45	974	2,142	3,356
Electron diffraction cameras					
W. Europe	8	14	—	—	—
USSR	—	22	266	130	120
Total	8	36	266	130	120
Grand total	419	2,601	6,928	7,114	6,183
Cumulative total	419	3,020	9,948	17,062	23,245

Production in the USSR

Period	TEM—other	TEM—sumy	TEM—total	EDC	SEM/EPMA	Grand total
1941–1950	172	—	172	—	—	172
1951–1960	609	79	688	22	—	710
1961–1970	816	1017	1833	266	85	2184
1971–1980	551	1161	1712	130	564	2406
1981–1990	32	610	642	120	527	1289

Electron Microscopes in Use in the United Kingdom, 1970

AEI	EM 6	103	Siemens		Elmiskop 1	91
	EM 6G	63			Elmiskop 1A	21
	EM 6B	107			Elmiskop 2	4
	EM 802	8			Elmiskop 101	6
	EM 801	20			Elmiskop 102	1
Philips	EM 7	3	Zeiss		Elmiskop 51	3

(continues)

ELECTRON MICROSCOPES IN USE IN THE UNITED KINGDOM, 1970 (*Continued*)

Philips	EM 100	43	Zeiss	EM 9	1
	EM 75	25		EM 9A	12
	EM 200	27		EM 9S	2
	EM 300	47			
			Cambridge I.	Stereoscans	138
RCA	7				
Hitachi	32				
JEOL	102				
Grand total 864, of which, imports from outside Europe: 141					

The figures shown in the table at the top of page 572 are believed to be an accurate record of the production quantities of electron microscopes in Europe. However, the figures included for the production of Philips microscopes are estimates, because the company declined to furnish the information. The details of the production figures for the various Western production models have been deleted from Appendix II for the more recent years to maintain uniformity of treatment for the different manufacturers.

There were no AEI microscopes of types EM 2, EM 3, EM 3A, EM 4 still registered as operating in 1970. The operational life of these instruments is therefore taken conservatively as 10 years. On the other hand, a number of Philips instruments from the late 1940s and early 1950s were still in service. The operational life of these instruments and of the Siemens microscopes is therefore taken as 15 years. It seems likely that later instruments have an operational life of at least 20 years, and this is taken as the effective life of all instruments produced after 1960.

Using the above criteria, we can approximate to the operational microscopes of European manufacture as follows:

Period	Instr. produced	Total to date	Total obsolescent	Residual
1939–1950	419	419	6	413
1951–1960	2,601	3,020	367	2,653
1961–1970	6,928	9,948	1,837	8,111
1971–1980	7,114	17,062	5,790	11,272
1981–1990	6,183	23,245	11,013	12,232

A sampling of the proportion of microscopes of European manufacture which have been exported outside Europe suggests that an increasing proportion have been exported, and that the proportion may have grown from an initial 15% to nearer 60%. At the same time, however, there has been a considerable rise in imported Japanese instruments, which may roughly balance the exports. It is possible that the service life of the more recent instruments has extended to more than 20 years. It seems certain, therefore, that the number of instruments in service in Europe at present exceeds 10,000, and is more probably near 12,000.

Appendix IV

Russian scientists contributing to electron microscope design and development in the former USSR:

Anashkin, I. F.	Modell, N. M.
Barzilovitch, P. P.	Moseev, V. V.
Fetisov, D. V.	Polivanov, V. V.
Ivanov, M. G.	Popov, N. M.
Kabanov, A. N.	Schulyak, E. A.
Kaplichni, V. N.	Stoyanov, P. A.
Kisel, G. D.	Stoyanova, I. G.
Kiselev, N.	Sushkin, H. G.
Kushnir, Y. M.	Tregubov, M. I.
Lyalko, I. S.	Vertzner, V. N.
Michaelovski, G. A.	Voronin, Y. M.
	Zelev, S. F.

Glossary of Abbreviations

AE	Auger electrons
BS(E)	Back-scattered (electrons)
BSD	Back-scattered detector
CL	Cathodoluminescence
CRT	Cathode-ray tube
ECP	Electron channeling pattern
EDC	Electron diffraction camera
EDX	Energy dispersive X-ray analysis
EELS	Electron energy-loss spectroscopy
e/m	Electromagnetic
EM	Electron microscope
EPMA	Electron probe microanalyser
EMMA	Electron microscope/microanalyzer
e/s	Electrostatic
ESD	Electron spectroscopic diffraction
ESI	Electron spectroscopic imaging
FE	Field emission
HT	High tension (or high voltage)
HVEM	High-voltage electron microscope (generally >500 kV)

LaB$_6$	Lanthanum hexaboride
μm	Micrometer
nA	Nanoampere
nm.	Nanometer
Pa.	Pascal (newton/m^2) [1 torr = 133.3 Pa = 1.333 millibar]
REM	Rasterelektronenmikroskop (scanning EM)
REMMA	Scanning electron microscope/microanalyzer
SAD	Selected-area diffraction
SC	Specimen current
SE	Secondary electron
SEM	Scanning electron microscope
STEM	Scanning transmission electron microscope
TEM	Transmission electron microscope
TV	Television, and television scanning rate
UHV	Ultrahigh vacuum
ÜM	Übermikroskop
W	Tungsten
WDX	Wavelength dispersive X-ray analysis
YAG	Yttrium aluminum garnet

References

(*See note at end of References.)

Afzelius, B. (1988). Private communication.

Agar, A. W., Browning, G., Williams, J. L., Davey, J., and Heathcote, K. (1970). A new 1000kV electron microscope. *Proc. Int. Cong. Electron Microsc.*, Grenoble, *1970*, pp. 115–116.

Agar, A. W., Frank, F. C., and Keller, A. (1959). Crystallinity effects in the electron microscopy of polyethylene. *Phil. Mag.* **4**, 32–55.

Agrell, S. O., and Long, J. P. V. (1960). The application of the scanning x-ray microanalyser to mineralogy. *Proc. Int. Symp. on X-ray Microanalysis*, Stockholm, *1959*, pp. 391–400.

Alekseev, A. G., Boyandina, L. G., Vertzner, V. N., Verkhovskaya, T. A., Kisel, G. D., and Ovchinnikov, O. P. (1978). Commerical electronograph EMR-100 (in Russian). *Optiko-Mechanical Production* (OMP), No. 6, pp. 21–24.

Anderson, K. (1968). An image intensifier for the electron microscope. *J. Phys. E*, Series 2, Vol. 1, pp. 601–603.

Anderson, K., Brookes, K. A., and Finbow, D. C. (1976). The Corinth analytical electron microscope. *Proc. EMAG 1975, Developments in Electron Microscopy and Analysis* (J. A. Venables, Ed.). Academic Press, London, pp. 69–70.

Anderson, K., and Kenway, P. B. (1967). External photography of the microscope image. *Proc. 25th EMSA Meeting*, Claitor's Publishing Division, Baton Rouge, LA, pp. 244–245.

Asmus, A., Herrmann, K.-H., and Wolff, O. (1968). Elmiskop 101, a new high-power electron microscope. *Siemens-Z.* **42**, 609–619.

Bakker, J. G., and Asselbergs, P. E. S. (1990). An ultra high resolution objective lens for a 200kV TEM. *Proc. Int. Cong. Electron Microsc.*, Seattle, *1990*, pp. 132–133.

Benner, G., Bihr, J., and Prinz, M. (1990). A new illumination system for an analytical electron microscope using a condensor objective lens. *Proc. Twelfth Int. Cong. for Electron Microsc.*, Seattle, *1990*. San Francisco Press, San Francisco, pp. 138–139.

Bergqvist, A. (1946). Det svenska elektronenmikroskopet (in Swedish). *Tek. Tidskr.* **76**, 649–655.

Bernhard, W., and Leduc, E. H. (1967). Ultrathin frozen sections I. Methods and ultrastructural preservation. *J. Cell Biol.* **34**, 757–771.

Bertein, F. (1947). On some defects in electron optical instruments and on their correction (in French). *Ann. Radioelectr.* **2**, No. 10.

Bertein, F. (1948). Influence of electrode deformation in electron optics (in French). *J. de Phys.* **9**, 104–112.

Boersch, H. (1936). On the primary and secondary image in the electron microscope. Concepts of the diffraction image and its effect on imaging (in German). *Ann. der Phys.* (Leipzig) **26**, 631–644.

Bollman, W. (1956). Interference effects in the electron microscopy of thin crystal foils. *Phys. Rev.* **103**, 1588–1589.

Bradley, D. E. (1954a). Evaporated carbon films for use in electron microscopy. *Br. J. Appl. Phys.* **5**, 65–66.

Bradley, D. E. (1954b). An evaporated carbon replica technique for use with the electron microscope and its application to the study of photographic grains. *Br. J. Appl. Phys.* **5**, 96–97.

Brandon, D., and Nutting, J. (1959). Technique for preparing thin films of α-Fe. *Br. J. Appl. Phys.* **10**, 255–256.

Bullivant, S. (1965). Freeze substitution and supporting techniques. *Lab. Invest.* **14**, 440–457.

Butler, E. P., and Hale, K. F. (1981). Dynamic experiments in the electron microscope. *In* "Practical Methods in Electron Microscopy, Vol. 9" (Audrey M. Glauert, Ed.). Elsevier, Amsterdam.

Butzke, G., Schmidt, B., and Okraffka, J. (1966). "Eine Belichtungsautomatik für das Elektronenmikroskop SEM 3-1 vom VEB WF Berlin."

Castaing, R. (1956). État actuel du microanalyseur à sonde électronique. *Proc. Int. Conf. Electron Microsc.*, London, *1954*, pp. 300–304.

Castaing, R., and Guinier, A. (1949). Application des sondes électronique à l'analyse métallographique. *Proc. Int. Cong. Electron Microsc.*, Delft, *1949*, pp. 60–63.

Castaing, R., and Henry, L. (1962). Filtrage magnétique des vitesses en microscopie électronique. *C. R. des Seances Acad. Sci. Paris* **255**, 76–78.

Cooke, C. J., and Duncumb, P. (1968). Performance analysis of a combined electron microscope and electron probe microanalyser EMMA. *Proc. 5th Int. Conf. on X-ray Optics and Microanalysis,* Tubingen, pp. 245–247.

Cooke, C. J., and Openshaw, I. K. (1969). A high resolution electron microscope with efficient x-ray microanalysis facilities. Paper no. 64, Proc. Electron Probe Society of America 4th Natl. Conf., Pasadena (a 40-line summary in a ring-backed collection of 2-page papers).

Cosslett, V. E. (1947). Summarised Proceedings of Conference on the Electron Microscope, Oxford, 1946. *J. Sci. Instrum.* **24**, 113–119.

Cosslett, V. E. (1980). Principles and performance of a 600kV high resolution electron microscope. *Proc. Roy. Soc. London, Series A* **370**, 1–16.

Cosslett, V. E., Camps, R. A., Saxton, W. O., Smith, David J., Nixon, W. C., Ahmed, H., Catto, C. J. D., Cleaver, J. R. A., Smith, K. C. A., Timbs, A. E., Turner, P. W., and Ross, P. M. (1979). Atomic resolution with a 600kV electron microscope. *Nature* **281**, 49–51.

Cosslett, V. E., and Duncumb, P. (1956). Microanalysis by a flying-spot X-ray method. *Nature* **177**, 1172–1173.

Crewe, A. V., Eggenberger, D. N., Wall, J., and Welter, L. M. (1968). Electron gun using a field emission source. *Rev. Sci. Instrum.* **39**, 576–583.

Danilatos, G. D. (1988). Foundations of environmental electron microscopy. *Adv. Electronics and Electron Phys.* **71**, 109–250.

Delong, A., and Drahoš, V. (1951). Ceskoslovenský electronový mikroskop. (Czechoslovak electron microscope). *Sb. VST. Brno* **20**, 334–348.

Delong, A., Drahoš, V., and Zobač, L. (1961). An experimental high performance electron microscope. *Proc. Eur. Cong. Electron Microsc.* Delft, *1960*, Vol. 1, pp. 89–91.

*Derbenev, A. F., et al. (1990). Electron microscope EM-125R. Presented at *Proc. 15th All-Union Conf. Electron Microsc.*, Moscow, 1990. Private communication from N. Kiselev.

Doane, F. W., Simon, G. T., and Watson, J. H. L. (1993). Canadian contributions to microscopy. Microscopical Society of Canada, Toronto.

Drahoš, V., and Delong, A. (1958). A small universal electron microscope. *Br. J. Appl. Phys.* **9**, 306–312.

Drummond, D. G. (1950). The practice of electron microscopy. *J. Microsc. Soc.* **70**, 1–141.

Duncumb, P. (1959). The x-ray scanning microanalyser. *Br. J. Appl. Phys.* **10**, 420–427.

Duncumb, P., and Melford, D. A. (1960). Design considerations of an X-ray scanning microanalyser used mainly for metallurgical applications. *Proc. Int. Symp. for X-ray Microanalysis*, Stockholm, *1960*, pp. 358–364.

Dupouy, G., Perrier, F., and Fabre, R. (1961). An electron microscope operating at very high voltage. *C. R. Hebd. Séances Acad. Sci.* **252**, 627–632.

Dupouy, G., Perrier, F., and Fabre, R., Durrieu, L., and Cathelinaud, R. (1969). Microscope électronique 3 Millions de volts. *C. R. Acad. Sci. Paris* **269**, 867–874.

Egle W., Kurz, D., and Rilk, A. (1984). The EM 902, a new analytical TEM for ESI and EELS. *Zeiss MEM* **3** (3), 4–9.

Engel, W. (1966). Emission microscopy with different kinds of electron emission. *Proc. Int. Cong. Electron Microsc.*, Kyoto, *1966*, pp. 217–218.

Ennos, A. E. (1954). The sources of vacuum-induced contamination in kinetic vacuum systems. *Br. J. Appl. Phys.* **5**, 27–31.

Everhart, T. E., and Thornley, R. F. M. (1960). Wide band detector for micromicroampere low energy electron currents. *J. Sci. Instrum.* **37**, 246–248.

Fert, C. (1956). Observation directe des surfaces métalliques par réflexion. *Proc. Eur. Cong. Electron Microsc.*, Stockholm, *1956*, pp. 8–12.

Fert, C., and Selme, P. (1956). Le microscope électronique O. P. L. *Bull. Microsc. Appl.* **6**, 157–164.

Fitch, G. I., and Quarrell, A. G. (1934). Crystal structure and orientation in zinc-oxide films. *Proc. Phys. Soc.* **46**, 146–162.

Fujita H. (Ed.) (1986). History of electron microscopes. Publication of 11th Int. Cong. Electron Microsc., Kyoto, 1986.

Gibbons, R., Paden, R. S., and Kynaston, D. (1971). Engineering and design considerations for a simple scanning microscope. Proc. 25th EMAG meeting. *Inst. of Physics Conf. Ser. No. 10*, pp. 154–157.

Glauert, A. M., and Glauert, R. H. (1958). Araldite as an embedding material for electron microscopy. *J. Biophys. Biochem. Cytol.* **4**, 191–194.

Glauert, A. M., Rogers, G. E., and Glauert, R. H. (1956). A new embedding material for electron microscopy. *Nature*, **178**, 803.

Gribi, M., Thürkauf, M., Villiger, W., and Wegmann, L. (1959). Ein 70kV-Elektronenmikroskop mit kalte Kathode und elektrostatische Linse. *Optik* **16**, 65–86.

Griffiths, B. W., Jones, A. V., and Wardell, I. R. M. (1973). An ultra-high vacuum scanning electron microscope with Auger analysis facilities. Scanning electron microscopy: systems and applications. 1973. *Inst. of Phys.*, 42–45.

Grivet P. (1946). Industrial realisation of an electrostatic electron microscope. *Rev. Opt. Theor. Instrum.* **25**, 129–160.

Grivet, P., and Regenstreif, E. (1952). On a new electrostatic microscope with three magnifying stages. Paper 32, Paris, 1950. *Rev. Opt. Theor. Instrum.* **31**, 230–236.

Gross, U., Mescher, F. J. M., and Tiemeijer, J. C. (1987). The microprocessor-controlled CM 12/STEM scanning-transmission electron microscope. *Philips Tech. Rev.* **43**, 273–291.

Günter, J. R. (Ed.) (1990). "History of Electron Microscopy in Switzerland." Birkhäuser Verlag, Basel, Boston, Berlin.

Gütter, E. (1961). The new Zeiss Electron Microscope EM 9. *Zeiss Werkzeitschrift* (English ed.) **42**.

Gütter, E., and Mahl, H. (1962). Ein neues Elektronenmikroskop. *Z. Instrumentenkunde* **70,** Heft 5.

Gutter, E., and Mahl, H. (1967). The Zeiss EM 9-S Electron Microscope. *Zeiss Werkzeitschrift* **70,** 85–89.

Guyenot, E. (1955). Über die Entwicklung eines fünfstufigen elektrostatischen Elektronenmikroskops. *Jenaer Jahrbuch* **1955/1,** 28–34.

Haine, M. E. (1947a). The design and construction of a new electron microscope. *J. Inst. Elec. Eng.* **14,** 447–462.

Haine, M. E. (1947b). The electron optical system of the electron microscope. *J. Sci. Instrum.* **24,** 61–66.

Haine, M. E. (1956). Some simplified magnetic lens design features. *Proc. Int. Conf. Electron Microsc.,* London, *1954,* pp. 92–97.

Haine, M. E., and Einstein, P. A. (1952). Characteristics of the hot cathode electron microscope gun. *Br. J. Appl. Phys.* **3,** 40–46.

Haine, M. E., and Einstein, P. A. (1960). Image intensifier. *Proc. Eur. Reg. Conf. Electron Microsc.,* Delft, *1960,* Vol. 1, pp. 97–100.

Haine, M. E., Ennos, A. E., and Einstein, P. A. (1958). An image intensifier for the electron microscope. *J. Sci. Instrum.* **35,** 466–467.

Haine, M. E., and Mulvey, T. (1956). The regular attainment of very high resolving power in the electron microscope. *Proc. Int. Conf. Electron Microsc.,* London, *1954,* pp. 698–705.

Haine, M. E., and Page, R. S. (1957). A new universal microscope of high resolving power—Metrovick type EM6. *Proc. Eur. Cong. Electron Microsc.,* Stockholm, *1956,* pp. 32–37.

Haine, M. E., Page, R. S., and Garfitt, R. G. (1950). A three stage electron microscope with stereographic, dark field and electron diffraction facilities. *J. Appl. Phys.* **21,** 173–182.

Hawkes, P. W. (Ed.) (1985). "The Beginnings of Electron Microscopy." Academic Press, New York.

Hawkes, P. W., and Kasper, E. (1989). "Principles of Electron Optics—Vol. 2: Applied Geometrical Optics." Academic Press, London and San Diego.

Hax, W. (1982). EM 410: An instrument dedicated to biological applications. *Proc. Int. Cong. Electron Microsc.,* Hamburg, *1982,* pp. 381–382.

Heide, H. G. (1958). Die Objektverschmützung und ihre Verhütung. *Proc. Int. Cong. Electron Microsc.,* Berlin, *1958,* pp. 87–90.

Heidenreich, R. D., and Peck, V. G. (1943). Fine structure of metallic surfaces with the electron microscope. *J. Appl. Phys.* **14,** 23–29.

Hillier, J., and Ramberg, E. G. (1947). The magnetic electron microscope objective: Contour phenomena and the attainment of high resolving power. *J. Appl. Phys.* **18,** 48–71.

Hirsch, P. B., Horne, R. W., and Whelan, M. J. (1956a). A kinematical theory of diffraction contrast of electron microscope images of dislocations and other defects. *Phil. Trans.* **252,** 499–529.

Hirsch, P. B., Horne, R. W., and Whelan, M. J. (1956b). Direct observations of the arrangement and motion of dislocations in aluminium. *Phil. Mag. (8th Series)* **1,** 677–684.

Holburn, D. M., and Smith, K. C. A. (1979). On-line topographic analysis in the SEM. *Scanning Electron Microscopy* **1979,** 47–52.

Huxley, H. E. (1956). An improved microtome for ultra-thin sectioning. *Proc. Int. Conf. Electron Microsc.,* London, *1954,* pp. 112–114.

Induni, G. (1945a). Das schweizerische Übermikroskop. *Neue Zürcher Zeitung* **No. 357,** 28 Feb.

Induni, G. (1945b). Das schweizerische Übermikroskop. *Vierteljahresschrift der Naturf. Gesellschaft in Zürich Jahrg.* **90,** 181–195.

Jervis, P. (1971). Innovation in electron-optical instruments—Two British case studies. *Research Policy. North-Holland.* **1,** 174–207.

Jesson, D. E., Pennycook, S. K., and Baribeau, J.-M. (1991). Direct imaging of interfacial ordering in ultrathin $(Si_mGe_n)_p$ superlattices. *Phys. Rev. Lett.* **66,** 750–753.

Kehler, H. (1993). Private communication.
Kisel, G. D., Lyalko, I. S., Pavlenko, P. A., and Spinov, I. I. (1988). A modern scanning electron microscope microanalyser. Proc. Eur. Cong. Electron Microsc., York, 1988. *Inst. Phys. Conf. Ser.* **93,** 99–100.
*Kisel, G. D., Postnikov, E. B., Kalichnii, V. N., Peleshuk, L. P., Tregubov, M. I., and Fetisov, D. V. (1970). Commercial scanning electron microscope-microanalyser REMMA. *Izv. Akad. Nauk. SSSR. Ser. Phys.* **34** (7), 1452–1454.
Kiselev, N. (1992). Private communication.
Knoll, M. (1935). Charging potential and secondary emission of bodies under electron irradiation (in German). *Z. Tech. Phys.* **16,** 467–475.
Knoll, M., and Ruska, E. (1932). Das elektronenmikroskop. *Z. Phys.* **78,** 318–339.
König, H., and Helwig, G. (1951). Über dünne aus Kohlenstoffen durch Elektronen-oder Ionen-beschüss gebildete schichten. *Z. Phys.* **129,** 491–503.
Krisch, B., Müller, K. H., Schliepe, R., Thon, F., and Willasch, D. (1976). Elmiskop ST100F—Ein Durchstrahlungs-Rasterelektronenmikroskop höchster Leistung. *Siemens Z.* **50,** 47–50.
Kushnir, Y. M. (1962). Radiotechnics in electronics (in Russian). *Izv. Akad. Nauk SSSR* **5,** 747–781.
*Kushnir, Y. M., Fetisov, D. V., Raspletin, K. K., Pochtarev, B. I., Spector, F. U., Kabanov, A. N., and Anisimov, V. F. (1961). Scanning electron microscope-X-ray microanalyser (in Russian). *Izv. Akad. Nauk. SSSR Ser. Phys.* **25** (6), 695–704.
Kuypers, W., and Tiemeijer, J. C. (1975). The Philips PSEM 500 scanning electron microscope. *Philips Tech. Rev.* **35,** 153–165.
Lebedev, A. A. (1954). "Electron Microscopy" (in Russian). Ed. Government Theoretical Literature, Moscow, pp. 123–125.
Le Poole, J. B. (1947). A new electron microscope with continuously variable magnification. *Philips Tech. Rev.* **9,** 33–46.
Le Poole, J. B. (1985). Early electron microscopy in the Netherlands. In "The Beginnings of Electron Microscopy" (P. W. Hawkes, Ed.). Academic Press, Orlando, pp. 387–416.
Liebmann, G. (1948). A new experimental electron microscope. *J. Sci. Instrum.* **25,** 37–43.
Liebmann, G. (1951). The symmetrical magnetic microscope objective lens with lowest spherical aberration. *Proc. Phys. Soc. B* **64,** 972–977.
Liebmann, G. (1952). Magnetic electron microscope projector lenses. *Proc. Phys. Soc. B* **65,** 94–108.
Lyalko, I. S., Kisel, G. D., Vorinin, Y. M., Shchetnyov, Y. F., and Udaltzev, V. I. (1986a). A small-sized TEM. *Proc. 11th Int. Cong. Electron Microsc.*, Kyoto, *1986*, pp. 313–314.
Lyalko, I. S., Kisel, G. D., Voronin, Y. M., and Shchetnyov, Y. F. (1986b). Optical system of a small-sized TEM. *Proc. 11th Int. Cong. Electron Microsc.*, Kyoto, *1986*, pp. 315–316.
Mahl, H. (1939). Über das elektrostatische Elektronenmikroskop hoher Auflösung. *Z. Tech. Phys.* **20,** 316–317.
Mahl, H. (1942). Die übermikroskopische Oberflächendarstellung mit dem Abdruckverfahren. *Naturwissenschaften* **30,** 207–217.
Mahl, H., Volkmann, H., and Weitsch, W. (1956). Über ein neues elektrostatisches Gebrauchs-Elektronenmikroskop. *Proc. Eur. Cong. Electron Microsc.*, Stockholm, *1956*, pp. 34–37.
Marton, L. (1934). Electron microscopy of biological objects. *Nature* **133,** 911.
Marton, L. (1945). A 100kV electron microscope. *J. Appl. Phys.* **16,** 131–138.
McMullan, D. (1953). An improved scanning electron microscope for opaque specimens. *Proc. Inst. Elec. Eng.* **100,** 245–259.
Menter, J. W. (1952). Direct examination of solid surfaces using a commercial electron microscope in reflexion. *J. Inst. Met.* **81,** 163–167.
Menter, J. W. (1956a). The resolution of crystal lattices. *Proc. Eur. Cong. Electron Microsc.*, Stockholm, *1956*, pp. 88–93.
Menter, J. W. (1956b). The direct study by electron microscopy of crystal lattices and their imperfections. *Proc. Roy. Soc. A* **236,** 119–135.

Millen, D. (1969). A semi-automatic transmission electron microscope for medium power. *Lab. Electronic Equipment,* Feb. 1969, pp. 7–10.

Möllenstedt, G. (1955). 100keVolt-Elektronen im elektrostatischen Elektronenmikroskop (Zwischenbeschleuniger). *Optik* **12,** 441–466.

Möllenstedt, G., and Düker, H. (1953). Emissionsmikroskopische Oberflächenabbildung mit Elektronen, die durch schrägen Ionenbeschuss ausgelöst werden. *Optik* **10,** 192–205.

Möllenstedt, G., and Rang O. (1949). Die elektrostatische Linse als hochauflösendes Geschwindigkeitsfilter. *Z. Angew. Phys.* **3,** 187–189.

Muller, K., and Ruska, E. (1955). Ein vereinfachtes elektromagnetisches Durchstrahlungsmikroskop für Elektronen von 40 bis 60kV. *Z. Wiss. Mik.* **62,** 205–219.

Müller, K., and Ruska, E. (1958). Ein Hilfselektronenmikroskop für Kurs- und Routinebetrieb. *Int. Conf. Electron Microsc.,* Berlin, *1958,* pp. 184–187.

Müller, K.-H., and Schliepe, R. (1973). Elmiskop 102, ein neues Hochleistungs-elektronenmikroskop. *Siemens Z.* **47,** 471–475.

Mulvey, T. (1960). A new X-ray microanalyser. *Proc. Symp. X-ray Microanalysis,* Stockholm, *1959,* Elsevier, Amsterdam, pp. 372–377.

Mulvey, T. (1962). Origins and historical development of the electron microscope. *Br. J. Appl. Phys.* **13,** 197–207.

Mulvey, T. (1982). Unconventional lens design. *In* "Magnetic Electron Lenses" (P. W. Hawkes, Ed.). Springer-Verlag, Berlin and Heidelberg, pp. 359–449.

Mulvey, T. (1985). The industrial development of the electron microscope by the Metropolitan-Vickers Electrical Company and AEI Limited. *Adv. Electronics and Electron Phys* suppl. 16, 417–442.

Mulvey, T. (1992). Private communication.

Neff, H., and Herrmann, K.-H. (1964). Microanalysis in the Siemens electron microscope. Read at ACHEMA-Tagung, 25 June 1964.

Newman, S. B., Borysko, E., and Swerdlow, M. (1949). Ultra-microtomy by a new method. *J. Res. Natl. Bur. Std.* **43,** 183–199.

Nixon, W. C., Ahmed, H., Catto, C. J. D., Cleaver, J. R. A., Smith, K. C. A., Timbs, A. E., and Turner, P. W. (1978). Electronic, mechanical and electron optical engineering design features of the Cambridge University 600kV high resolution electron microscope. *Proc. 9th Int. Cong. Electron Microsc.,* Toronto, *1978,* pp. 10–11.

Oatley, C. W. (1982). The early history of the scanning electron microscope. *J. Appl. Phys.* **53,** R1–R13.

Oatley, C. W., and Everhart, T. E. (1957). The examination of *p-n* junctions in the scanning electron microscope. *J. Electron.* **2,** 568–570.

Ong, P. S. (1968). A combined conventional and scanning electron microscope. *Proc. 5th Int. Conf. of X-ray Optics and Microanalysis,* Tübingen, *1968,* pp. 84–85.

Otten, M. T. (1990). The CM 20/STEM: Design and applications. *Philips Electron Opt. Bull.* **127.**

Otten, M. T., and Bakker, H. G. (1991). The CM 20-Ultratwin: A 200kV analytical ultrahigh resolution microscope. Paper read at German EM Society Meeting, Darmstadt, 1991.

Paden, R. S., Tillet, P. I., and Upton, J. M. (1973). New display techniques for the scanning electron microscope. *Scanning Electron Microscopy/1973 (Part 1). Proc. Sixth Annual SEM Symp.,* IIT Research Institute, Chicago, pp. 211–216.

Page, R. S. (1954). A compact console type electron microscope. *J. Sci. Instrum.* **31,** 37–43.

Page, R. S. (1964). A new electron microscope for biological applications. *Proc. Eur. Conf. Electron Microsc.,* Prague, *1964,* Vol. A, pp. 31–32.

Page, R. S., and Openshaw, I. K. (1960). The Metropolitan-Vickers X-ray microanalyser. *Proc. Int. Symp. X-ray Microanalysis,* Stockholm, *1959,* Elsevier, Amsterdam, pp. 385–390.

Panzer, K. (1950). Hochspannungsanlagen nach dem Hochfreqenz-Kaskaden-Prinzip. *Optik* **7,** 290–293.

Pashley, D. W., and Presland, A. E. B. (1961). Ion damage to metal films inside an electron microscope. *Phil. Mag.* **6**, 1003–1012.
*Popov, N. M. (1959). A 400kV combined electron microscope and electron diffraction camera (in Russian). *Izvest. Akad. Nauk SSSR Ser. Fiz.* **23** (4), 436–441.
Porter, K. R., and Blum, J. (1953). A study in microtomy for electron microscopy. *Anat. Rec.* **117**, 685–710.
Rakels, C. J., Tiemeijer, J. C., and Witteveen, K. W. (1968). The Philips electron microscope EM 300. *Philips Tech. Rev.* **29** (12).
Ramsauer, C. (1943). "Elektronenmikroskopie. Bereich Über Arbeiten des AEG Forschungs-Institut 1930 bis 1942." Springer Verlag, Berlin.
Rang, O. (1949). Der electrostatische Stigmator, ein Korrectiv für astigmatische Elektronenlinsen. *Optik* **5** (8/9), 518–530.
Rang, O. (1991). "Zur Elektronenmikroskopie kurz nach 1945—Erinnerung eines Zeitzeugen." Der Festvortrag auf der Eröffnungssitzen der 25 Tagung der Deutschen Gesellschaft für Elektronenmikroskopie in Darmstadt am 2 Sept. 1991.
Rang, O. (1992). Private communication.
Rang, O., and Schluge, H. (1951). Aufbau des AEG-ZEISS-Elektronenmikroskops EM8. *AEG-Mitteilungen* **7/8**, 3–9.
Rassmus, W. (1966). Der derzeitige technische Stand des Elektronenmikroskops KEM 1-1 des VEB WF. *Arbeitstagung Elektronenmikroskopie*, Erfurt, *1966*, Programmheft.
Ray, I. L. F., Drummond, I. W., and Banbury, J. R. (1976). A high resolution scanning transmission electron microscope with energy analysis. In "Developments in Electron Microscopy and Analysis" (J. A. Venables, Ed.). Academic Press, London, New York, pp. 11–14.
Reisner, J. H. (1989). The early history of the electron microscope in the United States. In "Advances in Electronics and Electron Physics, Vol. 73, Aspects of Charged Particle Optics." Academic Press, New York, pp. 133–231.
Recknagel, A. (1952). Über ein elektrostatisches Elektronenmikroskop. *Wiss. ZS der TH Dresden* **2**, 515–522.
Riecke, W. D., and Ruska, E. (1966). A 100kV transmission electron microscope with single-field condenser objective. *Proc. Int. Cong. Electron Microsc.*, Kyoto, *1966*, pp. 19–20.
Ruska, E. (1956). Ein Hochauflösendes 100kV Elektronenmikroskop mit Kleinfelddurchstrahlung. *Proc. Int. Conf. Electron Microsc.*, London, *1954*, pp. 673–693.
Ruska, E. (1980). "The Early Development of Electron Lenses and Electron Microscopy." S. Hirzel Verlag, Stuttgart (English transl. T. Mulvey).
Ruska, E., and Wolff, O. (1956). Ein hochauflösendes 100kV Elektronenmikroskop mit Kleinfelddurchstrahlung. *Z. Wiss. Mik.* **62**, 465–509.
Schaeffer, V. J., and Harker, D. (1942). Surface replicas for use in the electron microscope. *J. Appl. Phys.* **13**, 427–433.
Schluge, H. (1954). Das AEG-ZEISS Elektronenmikroskop. *Zeiss Werkzeitschrift*, 15 Oct., pp. 105–112.
Schott, O., and Leisegang, S. (1956). Objektkühlung im Elektronenmikroskop. *Proc. Eur. Conf. Electron Microsc.*, Stockholm, *1956*, pp. 27–30.
Schramm, B. (1964). Über einen elektrisch centrierbaren elektromagnetisch Stigmator. Arbeitstagung Elektronenmikroskopie, Feb. 1964, Jena, Programmheft.
Schramm, B. (1966). Der derzeitige technische Stand des Elektronenmikroskops SEM 3-1 vom WF. Arbeitstagung Elektronenmikroskopie, Nov. 1966, Erfurt, Programmheft.
Schramm B. (1994). Private communication.
Schramm, B., and Walz, H. (1962). Ein Elektromagnetischer Stigmator für das Elektronenmikroskop SEM 3. Arbeitstagung Elektronenmikroskopie, Feb. 1962, Dresden, Programmheft.
Schulze, D. (1960). Elektronenmikroskop Elmi D2 aus Jena in der Festkörperphysik. *Jenaer Rundschau* **5**, 131–134.
Schulze, D. (1992). Elektronenmikroskope aus dem Osten. *Elektronenmikroskopie* **6**, 32–40.

Shimadzu, S., Iwanaga, M., Kobayashi, K., Suito, E., Taoka, T., and Fujita, H. (1966). Instrumental features of 500kV electron microscopes. *Proc. Int. Cong. Electron Microsc.,* Kyoto, *1966,* pp. 101–102.

Siemens and Halske (1941). "Das Übermikroskop als Forschungsmittel." Walter de Gruyter & Co., Berlin.

Smith, B. A., and Nutting, J. (1956). Direct carbon replica from metal surfaces. *Br. J. Appl. Phys.* **7,** 214–217.

Smith, K. C. A. (1955). The scanning electron microscope and its fields of application. *Br. J. Appl. Phys.* **6,** 391–399.

Smith, K. C. A. (1956). The scanning electron microscope and its fields of application. Ph.D. dissertation, Cambridge University.

Smith, K. C. A. (1961). A versatile scanning electron microscope. *Proc. Eur. Cong. Electron Microsc.,* Delft, *1960,* pp. 177–180.

Smith, K. C. A., and Considine, K. (1968). Scanning transmission microscopy at high voltages. *Proc. 4th Eur. Reg. Conf. Electron Microsc.,* Rome, *1968,* p. 73.

Smith, K. C. A., and Considine, K., and Cosslett, V. E. (1966). A new 750kV electron microscope. *Proc. Int. Cong. Electron Microsc.,* Kyoto, *1966,* pp. 99–100.

Steere, R. C. (1957). Electron microscopy of structural detail in frozen biological specimens. *J. Biophys. Biochem. Cytol.* **3,** 45–60.

Steigerwald, K. H. (1949). Ein neuartiges Strahlerzeugungs-System für Elektronenmikroskope. *Optik* **5,** 469–478.

Stewart, A. D. G., and Snelling, M. A. (1965). A new scanning electron microscope. *Proc. Eur. Cong. Electron Microsc.,* Prague, *1964,* pp. 55–56.

*Stoyanova, I. G. (1958). Environmental cell in EM (in Russian). *Proc. 2nd All-Union Conf. Electron Microsc.,* Moscow, 1958.

*Stoyanov, P. A., Mikhailovitch, G. A., and Moseev, V. V. (1958). UEMB-100 Electron microscope with two condenser lenses (in Russian). *Proc. 2nd All-Union Conf. Electron Microsc.,* Moscow, 1958.

*Stoyanov, P. A., Moseev, V. V., Rozorenova, K. M., and Renskii, I. S. (1970). High resolution electron microscope EMV-100L (in Russian). *Izv. Akad. Nauk. SSSR, Ser. Phys.* **34** (7), 1388–1395.

*Stoyanov, P. A., and Renskii, I. S. (1977). EM 300 electron microscope. *Proc. 10th All-Union Conf. Electron Microsc.,* Tashkent, 1976. Izvest. Akad. Nauk SSSR, Moscow.

*Stoyanov, P. A., Schulyak, E. A., Gurin, V. S., *et al.* (1965). UEMV-100B Electron microscope. Presented at *Proc. 5th All-Union Conf. Electron Microsc.,* Moscow, 1965. Private communication from N. Kiselev.

Strojnik, A. (1964). Ein Bestrahlungssystem mit kleinem Fleckdurchmesser für Elektronenmikroskope. *Proc. Eur. Cong. Electron Microsc.,* Prague, 1964, pp. 49–50.

Strojnik, A., and Kralj, A. (1964). A low frequency 60kV power supply for electron microscopes. *Proc. Eur. Cong. Electron Microsc.,* Prague, *1964,* pp. 33–34.

Susskind, C. (1985). L. L. Marton 1901–1978—A review. *In* "The Beginnings of Electron Microscopy" (P. W. Hawkes, Ed.). Academic Press, Orlando, pp. 501–523.

Tadano, B., Kimura, H., Katagiri, S., and Nishigaki, M. (1966). 500kV electron microscope and its accessories. *Proc. Int. Cong. Electron Microsc.,* Kyoto, *1966,* pp. 103–104.

Unitt, B. M., and Smith, K. C. A. (1976). The application of the minicomputer in SEM. *Proc. Eur. Cong. Electron Microsc.,* Jerusalem, *1976,* pp. 162–167.

Van der Mast, K. D., Rakels, C. J., and Le Poole, J. B. (1980). A high quality multipurpose objective lens. *Proc. Eur. Cong. Electron Microsc.,* The Hague, *1980,* Vol. 1, pp. 72–73.

van Dorsten, A. C., and Le Poole, J. B. (1955). The EM 75kV, an electron microscope of simplified construction. *Philips Tech. Rev.* **17,** 47–59.

van Dorsten, A. C., Nieuwdorp, H., and Verhoeff, A. (1950). The Philips electron microscope EM100. *Philips Tech. Rev.* **12,** 33–64.

van Dorsten, A. C., Oosterkamp, W. J., and Le Poole, J. B. (1947). 400kV high voltage electron microscope. *Philips Tech. Rev.* **9**, 193–201.
von Ardenne, M. (1940). "Elektronen-Übermikroskopie." Verlag von Julius Springer, Berlin.
von Ardenne, M. (1938a). Das Elektronen-Rastermikroskop. Praktische Ausführung. *Z. Tech. Phys.* **19**, 404–416.
von Ardenne, M. (1938b). Die Grenzen Für Auflösungsvermögen des Elektronenmikroskops. *Z. Phys.* **108**, 338–352.
von Ardenne, M. (1941). Über ein 200kV-Universal-Elektronenmikroskop mit Objektabschattungsvorrichtung. *Z. Phys.* **117**, 657–688.
von Ardenne, M. (1985). On the history of scanning electron microscopy, of the electron microprobe and of early contributions to transmission electron microscopy. *In* "The Beginnings of Electron Microscopy" (P. W. Hawkes, Ed.). Academic Press, Orlando, pp. 1–21.
von Borries, B. (1940). Sublichtmikroskopische Auflösung bei der Abbildung von Oberflächen im Übermikroskop. *Z. Phys.* **16**, 370–378.
von Borries, B., and Ruska, E. (1938a). Vorläufige Mitteilung über Fortschritte im Bau und in der Leistung des Übermikroskopes. *Wiss. Veröff. aus den Siemens-Werken* **17**, Heft 1, 99–106.
von Borries, B., and Ruska, E. (1938b). Das Übermikroskop als Fortsetzung des Lichtmikroskops. Die Verhandlungen der Gesellschaft deutscher Naturforschung und Ärzte 95. *Versammlung zu Stuttgart vom 18 bis 21 September 1938*, pp. 72–77.
von Borries, B., and Ruska, E. (1939a). Aufbau und Leistung des Siemens Übermikroskopes. *Z. Wiss. Mikrosk.* **56**, 317–333.
von Borries, B., and Ruska, E. (1939b). Ein Übermikroskop für Forschungsinstitute. *Naturwissenschaften* **27**, 577–582.
von Borries, B., and Ruska, E. (1939c). Versuche, Rechnungen und Ergebnisse zur Frage des Auflösungsvermögens beim Übermikroskop. *Z. Tech. Phys.* **20**, 225–235.
von Borries, B., and Ruska, E. (1940). Mikroskopie hoher Auflösung mit schnellen Elektronen. *Erg. exakten Naturwiss.* **19**, 237–322.
von Borries, B., Ruska, E., and Ruska, H. (1938). Übermikroskopische Bakterienaufnahmen. *Wiss. Veröff. aus den Siemens-Werken* **17**, Heft 1, 107–111.
von Borries, H. (1991). Bodo von Borries: Pioneer of electron microscopy. *Adv. Electronics and Electron Phys.* **81**, 127–176.
*Vertzner, V. N., Voronin, Y. M., Vorobyev, Y. V., Bogdanovskii, G. A., and Chentsov, Y. V. (1961). Electron optics of the EM-5 and EM-7 electron microscopes (in Russian). *Izv. Akad. Nauk. SSSR, Ser. Phys.* **25** (6), 680–682.
Villiger, W. (1990). The instrumental contribution of Switzerland to the development of electron microscopy. A historical review. *In* "History of Electron Microscopy in Switzerland" (John R. Günter, Ed.). Birkhäuser Verlag, Basel.
Walz, H., and Schramm, B. (1962). Über eine Plattenaufnahmevorrichtung beim KEM 1 Arbeitstagung Elektronenmikroskopie, Dresden, 1962. Programmheft.
Wardell, I. R. M., Morphew, J., and Bovey, P. E. (1973). Results and performance of a high resolution STEM. *In* "SEM Systems and Applications," Inst. of Phys. Conf. Ser. No. 18, pp. 182–185.
Watanabe, M., Hinaga, Y., Someya, T., Goto, To, Nakamura, O., Konno, K., Yanaka, T., and Takahashi, N. (1966). 1MV electron microscope. Structural features and some practical considerations. *Proc. Int. Cong. Electron Microsc., Kyoto, 1966*, pp. 105–106.
Wegmann, L. (1952). Eines neues schweizerisches Elektronenmikroskop. *Neues Zürcher Zeitung*, No. 1562, 16 July.
Wegmann, L. (1968). Photoemission electron microscopy. *Proc. 5th Int. Cong. X-Ray Optics and Microanalysis*, Tübingen, pp. 356–360.
Wegmann, L. (1972). The photo emission microscope: Its technique and application. *J. Microsc.* **96**, 1–23.

Williams, R. C., and Wyckoff, R. W. G. (1946). Applications of metallic shadow casting to microscopy. *J. Appl. Phys.* **17**, 23–33.

Wolpers, C. (1991). Electron microscopy in Berlin 1928–1945. *Adv. Electronics and Electron Phys.* **81**, 211–229.

Wolpers, C. (1993). Private communication.

*Zelev, C. F., Kisel, G. D., Vasilev, B. N., Volnukin, B. K., Udaltsov, V. I., and Batalin, V. P. (1977). Electron microscope with the possibility of X-ray microanalysis based on the EMV-100L (EMMA 4) (in Russian). *Izvestia Akad. Nauk. SSSR Ser. Phys.* **41** (7), 1437–1439.

Zworykin, V. K., Hillier, J., and Snyder, R. L. (1942). A scanning electron microscope. *ASTM Bull.* **117**, 15–23.

Zworykin, V. K., Hillier, J., and Vance, A. W. (1941). A preliminary report on the development of a 300kV magnetic electron microscope. *J. Appl. Phys.* **12**, 738–742.

* All papers marked with an asterisk are available in English translation. By courtesy of Hawkes and Kasper (1989), the list of Russian Conference Proceedings available in English translation from volumes of *Izv. Akad. Nauk. SSSR (Ser. Phys.)* is given below:

Conference Location	Date	Reference
Moscow	15–19 Dec. 1950	Vol. 15 (1951), Nos. 3 and 4 (Not translated)
Moscow	9–13 May 1958	Vol. 23 (1959), Nos. 4 and 6
Leningrad	24–29 Oct. 1960	Vol. 25 (1961), No. 6
Sumy	12–14 Mar. 1963	Vol. 27 (1963), No. 9
Sumy	6–8 July 1965	Vol. 30 (1966), No. 5
Novosibirsk	11–16 July 1967	Vol. 32 (1968), Nos. 6 and 7
Kiev	14–21 July 1969	Vol. 34 (1970), No. 7
Moscow	15–20 Nov. 1971	Vol. 36 (1972), Nos. 6 and 9
Tbilisi	28 Oct.–2 Nov. 1973	Vol. 38 (1974), No. 7
Tashkent	5–8 Oct. 1976	Vol. 41 (1977), Nos. 5 and 11
Tallin	Oct. 1979	Vol. 44 (1980), Nos. 6 and 10
Sumy	1982	Vol. 48 (1984), No. 2
Sumy	Oct. 1987	Vol. 52 (1988), No. 7
Suzdal	Oct.–Nov. 1990	Vol. 85 (1991), No. 8
Chernogolovka	May 1994	Vol. 59 (1995), No. 2

4.2
My Early Work on Convergent-Beam Electron Diffraction

GOTTFRIED MÖLLENSTEDT

Institute of Applied Physics
University of Tübingen
D-72076 Tübingen, Germany

The idea of observing electron diffraction in a convergent beam was born over 50 years ago at the Technical University of Danzig, where Walther Kossel (Fig. 1), of the Institute of Physics, had, together with his co-workers, discovered the interference of X rays emitted from sources within the lattice, formed within the crystal as fluorescent radiation of the lattice atoms excited by electrons or primary X rays (Kossel, 1937). An intensive exchange of ideas with the discoverer of X-ray interference, Nobel Prize winner Max von Laue (von Laue, 1931) (Fig. 2), who called this the "Kossel effect," resulted in active attempts to interpret the details of this new kind of X-ray interference. Each vector in reciprocal space, i.e., each system of parallel lattice planes, is associated with its Kossel cone formed by all spatial directions for which Laue's conditions are satisfied. The diffraction diagram shows a system of curved lines, which were interpreted as the intersections of the Kossel cones with the photographic plate. One striking feature was that the background intensity in the Kossel diagrams sometimes had different values on each side of a Kossel curve.

A similar and even more striking effect was observed in S. Kikuchi's (Fig. 3) electron interference patterns (Kikuchi, 1928, 1930). They are formed in relatively thick crystals, if an initially monoenergetic and parallel electron beam gradually assumes a much wider angular distribution by multiple elastic and inelastic scattering (Fig. 4). The directions of some of the scattered electrons that happen to satisfy Laue's conditions form Kossel cones, whose intersections with the photographic plate become visible as Kikuchi lines. There is no dependence whatever of the position of the Kikuchi lines on the initial direction of the primary beam.

In order to obtain additional information on the elementary processes of electron scattering in monocrystals, W. Kossel asked me to study "electron interferences in a convergent beam" as a subject for a diploma thesis. Electrons with a common energy of 45 keV but different initial directions should hit a thin monocrystal. I was very happy about the opportunity to study such an interesting problem and started my experiment with juvenile

FIGURE 1. Walther Kossel (1888–1956).   FIGURE 2. Max von Laue (1879–1960).

naivety, unimpeded by too much experimental experience. I had given up studying aeronautical engineering and had changed to the faculty of mathematics and sciences. Now I had to considerably reduce my activities in competitive athletics in the academic sports club Allemannia within the

FIGURE 3. Seichi Kikuchi.

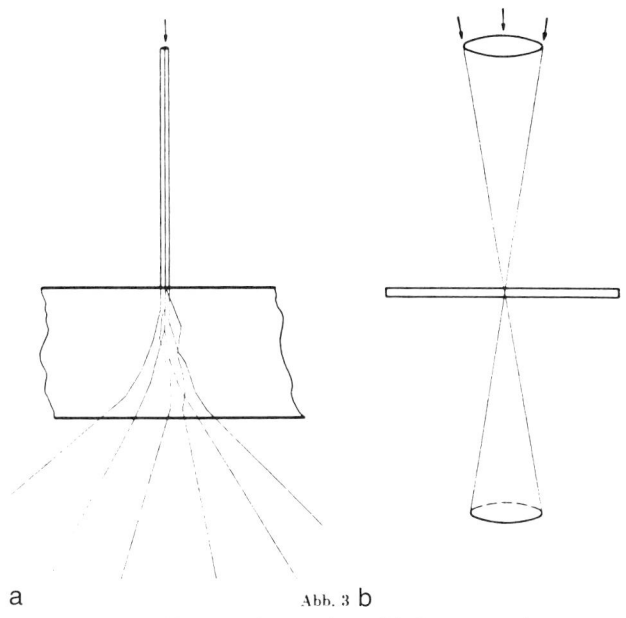

FIGURE 4. (a) Kikuchi procedure. (b) Convergent beam.

Akademischer Turnerbund (ATB). As a diploma candidate, I was at first supervised by an assistant, Dr. G. Borrmann, who later earned fame by his discovery of the anomalous absorption of X rays in ideal monocrystals. Very soon, however, Kossel instructed me to report directly to him as the director of the institute. He gave me a completely free hand in the construction of the instrument shown in Fig. 5, designed for convergent-beam electron interference. Only when I had to get his approval for the complete design did he cancel some constructional elements which were unnecessary from a physical point of view.

Some details of the diffraction apparatus are shown in Fig. 5. At that time, at the Danzig institute, preference was given to gas-discharge electron sources. According to G. I. Finch and H. Wilmann (1937), of Imperial College, London, an aluminum cone as a cathode was ground into the neck of a wine bottle and sealed with vacuum grease. The discharge volume could be shifted on a water-cooled brass block using a plane-ground surface and apiezon grease, in order to be able to adjust the electron beam precisely with respect to the first aperture. The beam intensity was controlled by varying the gas pressure in the discharge. The gas inlet to the discharge volume was connected to a glass balloon 25 cm in diameter, containing air at about 10 torr, by a flattened nickel tube. By bending this tube its flow resistance could be varied for sensitive control of the air supply to the

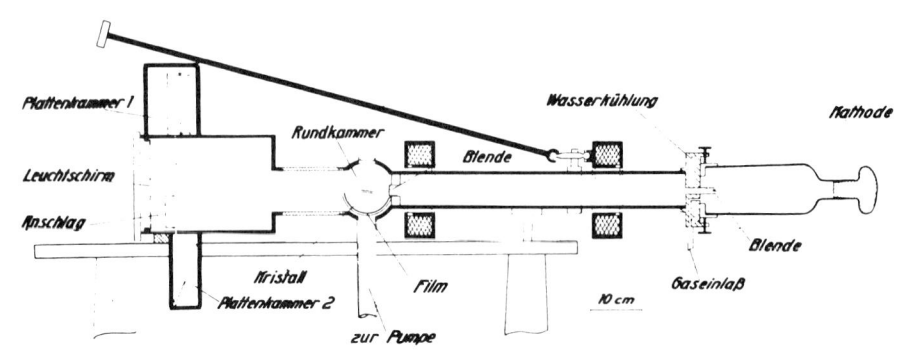

FIGURE 5. Instrument designed for convergent-beam electron interference, 1937.

discharge volume. The voltage was supplied by a high-voltage X-ray source from Seifert, Hamburg, with an additional high-voltage condenser for stabilization.

The convergent electron beam was focused by two iron-clad magnetic lenses positioned on special supports around a brass tube 60 cm in length and 7 cm in diameter. The first lens was held by adjustable ball joints and could be precisely adjusted with respect to the optical axis, producing a demagnified image of the anode aperture. The second magnetic lens operated at a focal length of about 2 cm to obtain a focus at the center of the round chamber, where the specimen was held by a specimen holder (Fig. 6) with two angular degrees of freedom (Schoon, 1937). A spring suspension with adjusting screws supplied two additional lateral degrees of freedom.

Joint 1, a ground cone, permitted azimuthal rotation over a Cardan drive; the angle between the surface normal and the optical axis could be varied using ground joint 2. The cylindric round chamber, 17 cm in length and 10 cm in diameter, was closed by brass plates on both ends. One of them served as a support for the specimen holder, the other held the mechanism for the observation of large-angle reflection patterns and was only rarely used. The round chamber was connected at its left to a photo camera for 9×12 cm plates by a tube that could be exchanged for a longer one in order to be able to extend the camera length. The camera permitted successive exposure of 12 plates without having to interrupt the vacuum. The upper part of the camera contained a carriage with 12 photographic glass plates. The photographic material, Agfa Kontrast and Agfa Normal, satisfied all demands with respect to vacuum usability and gradation of density. When an electron diffraction pattern on the fluorescent screen was to be recorded, it was first deflected by switching on a magnetic deflecting field. Then one of the plates, dropped through a slit from the upper chamber, fell into position, where it was held by an adjustable stop. After exposure, the stop was released, and the plate was dropped and collected in the lower chamber.

FIGURE 6. Specimen holder.

Later, after opening the whole system to air, the exposed plates could be removed by opening the lower lid.

A mercury diffusion pump was used to obtain a vacuum of about 10^{-3} torr against about 1 torr in a fore vacuum container consisting of a glass sphere 30 cm in diameter. A vacuum of 10^{-3} torr could be maintained for hours in the diffraction apparatus even with the rotary vacuum pump switched off.

In this instrument, the beam could be made to converge at any desired aperture angle over a wide range. The construction was sufficiently versatile to allow one to take diffraction patterns even at parallel irradiation in transmission or reflection. The fine adjustment of the convergent beam relative to the orientation of the crystal was achieved by turning the adjusting screws in the support of the first magnetic lens over two flexible drive shafts, one of which is shown in Fig. 5. I often wish that the fine adjustment in some modern high-performance electron microscopes would work with

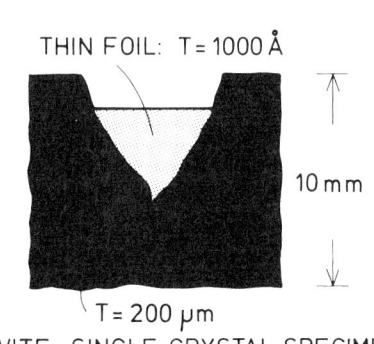

FIGURE 7. Plane mica lamella less than 100 nm in thickness.

FIGURE 8. Diffraction patterns of an Ag monocrystal about 10 nm in thickness, characteristic of a two-dimensional cross-grating taken at convergent and parallel irradiation.

the same precision and reliability! I never had any trouble with local charging even on highly insulating mica surfaces, probably thanks to the relatively poor vacuum of 10^{-3} torr. In addition, we did not suffer from specimen contamination 50 years ago, though we must have had a sufficient number of organic molecules in the residual gas. This can be easily understood with our present knowledge of the processes leading to contamination: I had a probe diameter of 40 μm, whereas nowadays probes 40 nm in diameter are used, which corresponds to a factor of 10 in the load per-unit area. Thus I had sufficient time to search the specimen and to adjust the different crystal orientations.

I enjoyed very much the construction and testing of the new instrument, which did not take too much time. I had much more trouble with the preparation of sufficiently large flakes of thin and even specimens. Since the probe diameter was of the order of several tens of micrometers, an even area of about 1 mm^2 was needed. Splitting the mica crystals by hand was a lengthy and tedious procedure, which, after many failures, resulted, in a few rare cases, in a specimen consisting of a thin, even lamella of angular shape held by thicker mica slabs on both sides, which had to be fixed to the specimen holder (Fig. 7).

Figure 8 (left) shows the diffraction of a convergent conical beam by a thin silver crystal about 10 nm in thickness. The central disk is formed by the primary beam. It is surrounded by a regular pattern of diffracted disks, each of which corresponds to one of the reciprocal lattice vectors in a plane perpendicular to the beam axis. The crystal was so thin that the diffraction pattern (Fig. 8, right) observed with parallel irradiation was formed only by the well-known diffraction spots characteristic of a two-dimensional cross-grating.

An excellent view of the elementary processes is shown in Fig. 9. The beam axis is adjusted to satisfy Laue's conditions for both the $(3\bar{3}1)$ and

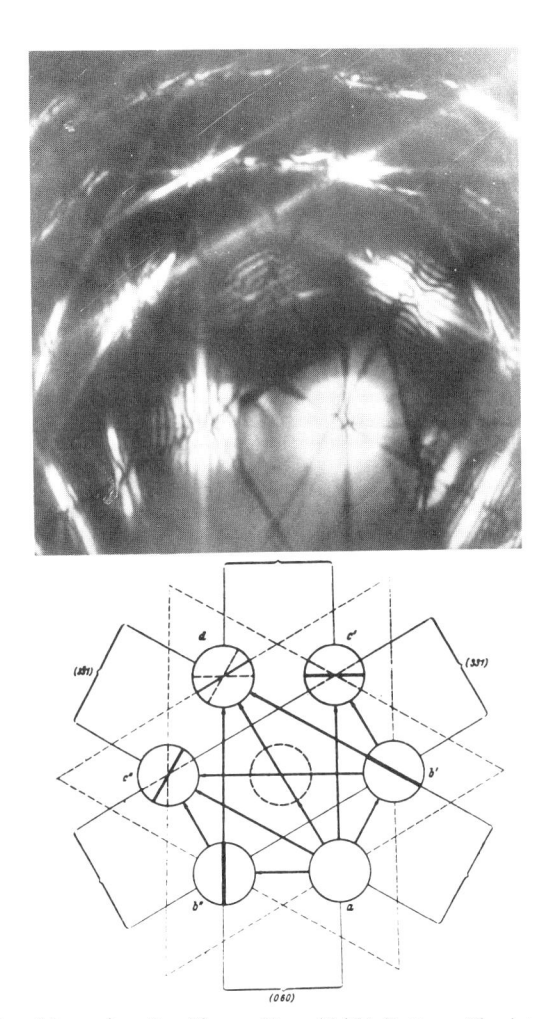

FIGURE 9. Top: Muscovite, $D = 80$ nm, $U_B = 45$ kV. Bottom: The intensity transfer is shown by arrows.

(060) lattice planes of the muscovite specimen. The schematic representation in Fig. 9 shows how parts of the beam intensity are transferred from a to b', from a to b'', etc.

A very interesting phenomenon in the reflected disks consists in a system of interference fringes parallel to the Kikuchi lines, whose spacing and intensity distribution depend on the thickness D of the crystal. According to the kinematic theory of diffraction, the fringes should have a central intensity maximum on the Kikuchi line, and equidistant minima at angular

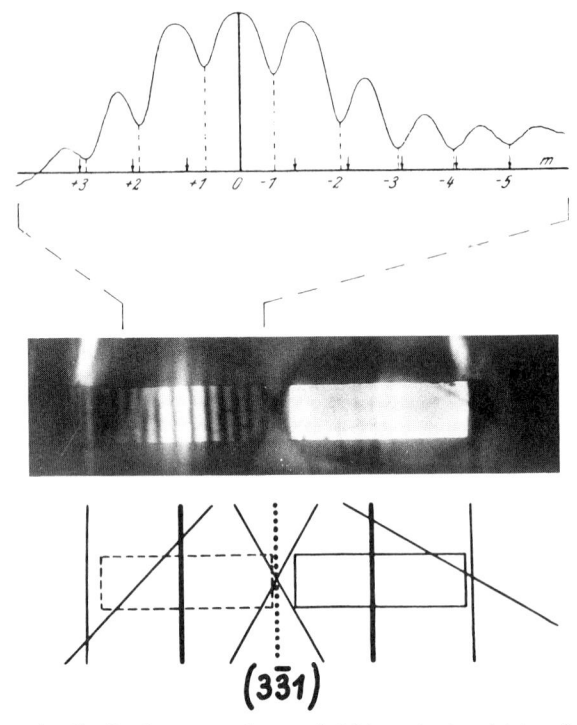

FIGURE 10. Intensity distribution across the parallel fringes in the vicinity of a Kikuchi line.

FIGURE 11. C. H. Mac Gillavry.

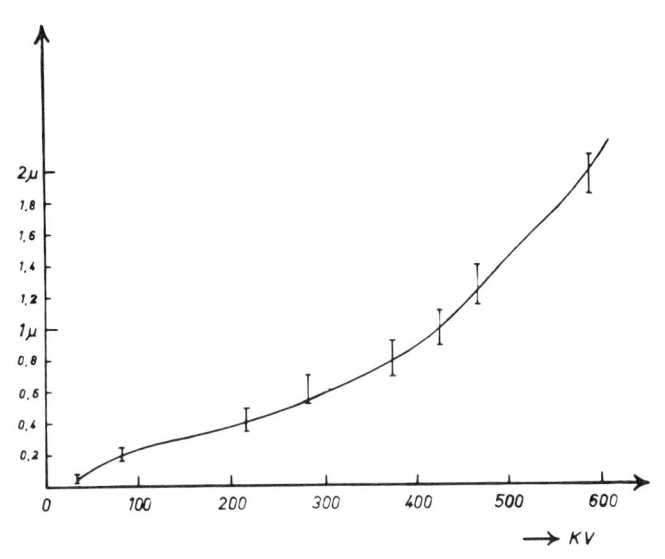

FIGURE 12. Dependence of penetrable thickness on beam voltages, measured with polycrystalline Al foils.

distances $\varepsilon_m = md/D$ ($m = 1, 2, 3, \ldots$) from the Kikuchi line, where d is the spacing of the lattice planes associated with the given reflection.

In our experiments we noticed deviations from the predictions of the kinematic theory. We found that the fringes were not equidistant, and that instead of a central maximum, even a minimum might occur (Fig. 10), a phenomenon giving the impression of a split Kikuchi line. Mac Gillavry (Fig. 11) showed later (1940) that this effect can be understood in terms of the dynamic theory of electron diffraction, and that the intensity minima should occur at

$$\varepsilon_m = d\sqrt{\left(\frac{m}{D}\right)^2 - \left(\frac{2em\lambda V_{hkl}}{h^2}\right)^2}$$

where V_{hkl} is the structure potential associated with the reflex defined by the Laue indices h, k, l. Careful measurements (Kossel and Möllenstedt, 1938, 1939, 1942; Möllenstedt, 1941) confirmed the excellent agreement between our experiments and Mac Gillavry's formula.

When later, in 1940, an improved design of the magnetic lenses allowed us to reduce the diameter of the electron probe from 40 µm to 1 µm at 65 kV (Möllenstedt and Ackermann, 1941), specimen preparation became easier, and the method could be applied to materials other than muscovite. After the construction of an electron diffraction instrument for 750 keV at the Institute at Danzig, it could be confirmed that the penetrable specimen

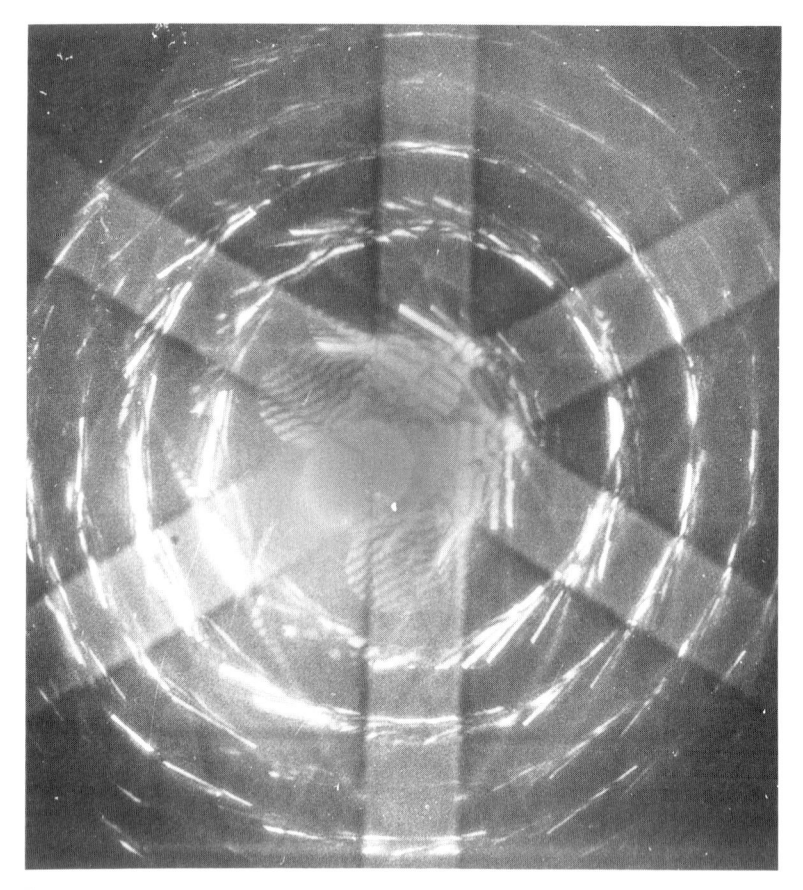

FIGURE 13. Convergent-beam electron diffraction pattern of a mica slab 110 mm in thickness, taken at 330 kV with the Danzig, 750-kV Van de Graaf generator in 1944.

thickness (Fig. 12) increases with beam voltage (Möllenstedt, 1946). Convergent-beam diffraction patterns of muscovite, biotite, and phlogopite were taken with excellent contrast at high voltages (Fig. 13). A detailed examination of the dynamic theory at high voltages, however, had to be postponed for later years, or even decades (Stumpp, 1983a, 1983b, 1983c, 1984; Stumpp et al., 1984), because the events of World War II brought an end to the projects at Danzig. A number of additional contributions from the team working at Danzig on convergent-beam electron diffraction are listed in the references (Ackermann, 1948; Kossel, 1949; Menzel-Kopp, 1951; Pfister, 1953).

References

Ackermann, I. (1948). *Ann. d. Physik* **2,** 19.
Finch, G. I., and Wilman, H. (1937). *Erg. exakt. Naturwiss.* **16,** 353.
Kikuchi, S. (1928). *Jpn. J. Phys.* **5,** 83.
Kikuchi, S. (1930). *Phys. Z.* **31,** 777.
Kossel, W. (1937). *Erg. exakt. Naturwiss.* **16,** 206.
Kossel, W. (1949). *Ann. d. Physik* **6,** 97.
Kossel, W., and Möllenstedt, G. (1938). *Naturwiss.* **26,** 660.
Kossel, W., and Möllenstedt, G. (1939). *Ann d. Physik* **36,** 113.
Kossel, W., and Möllenstedt, G. (1942). *Ann. d. Physik* **42,** 287.
Mac Gillavry, C. H. (1940). *Physica Vol. VII* **4,** 329.
Menzel-Kopp, C. (1951). *Ann. d. Physik* **9,** 259.
Möllenstedt, G. (1941). *Ann. d. Physik* **40,** 1.
Möllenstedt, G. (1946). *Nachr. d. Akadem. d. Wissensch. Göttingen,* 83.
Möllenstedt, G., and Ackermann, I. (1941). *Naturwiss.* **29,** 647.
Pfister, H. (1953). *Ann. d. Physik* **11,** 239.
Schoon, Th. (1937). *Z. Phys. Chem. Abs. B* **36,** 195.
Stumpp, H. (1983a). Thesis, University of Tübingen.
Stumpp, H. (1983b). *7th Int. Conf. on High Voltage Electron Microscopy,* Berkeley, CA, 89.
Stumpp, H. (1983c). Joint Meeting on Electron Microscopy, Antwerp, 106, 117a, 121.
Stumpp, H. (1984). *Optik '68,* No. 3, 193–207.
Stumpp, H., Lichte, H., and Möllenstedt, G. (1984). *Optik, '68,* No. 2, 147–152.
von Laue, M. (1931). *Erg. exakt. Naturwiss.* **10,** 133.

4.3
Atom Images and IFSEM Affairs in Kyoto, Osaka, and Okayama

HATSUJIRO HASHIMOTO

Okayama University of Science
Okayama 700, Japan

I. INTRODUCTION

Among the memorial lectures at the 30th anniversary of the IFSEM in Budapest, I presented a lecture with the title "Development of Atom Resolution Electron Microscopy" and showed the history, starting from the early theoretical prediction of the images of atoms and crystals by Boersch in 1946–1947 and by Scherzer in 1949 to the observations of crystal lattice fringes in 1956, single atoms in molecules in 1971, and the *in situ* observation of the movement of atoms in Si and Au crystals in 1979. In many cases in the past, the development of science was affected very strongly by the state of research, the social situation, and the possibilities for the exchange of knowledge. These are very important for the development of science, but in general they are *not* referred to in published papers.

In the case of electron microscopy, remarkable developments took place during and after World War II, when social conditions changed dramatically, by the exchange of knowledge among physicists, engineers, materials scientists, and biologists in many countries. After some discussion with the editor, I thought that instead of presenting yet another overall review of the field, it seemed more important to describe from an individual point of view, looking back over my own experience, the progress made during these 50 years.

I was involved in the affairs of the International Federation of Societies for Electron Microscopy (IFSEM) for 12 years as a committee member, President and Vice President, and also with the Committee of Asian Pacific Societies for Electron Microscopy (CAPSEM) for 12 years, as President and Vice President, and I experienced the enormous social political effects in these countries. An account of these affairs might be important for the future policy of the IFSEM.

Although the damage of the war disturbed research in Japan, I benefited in many ways from the Japanese environment, including our old tradition of electron diffraction dating from Kikuchi, to our more recent tradition in the manufacture of high-quality electron optical equipment.

It should also be clear from what follows that I have been enormously stimulated and helped by many contacts with scientists in other countries. In particular, I should mention the long and fruitful contact and collaboration that I have had with scientists in the United Kingdom, Germany, the United States, and France, to name only a few countries.

II. My First Experiments in Combined Microscopy and Diffraction

A. Electron Diffraction in Hiroshima

My first research was carried out in 1944 in the laboratory of Prof. H. Tazaki at Hiroshima University, where we used an electron diffraction camera, designed by the famous Kikuchi and built by the Institute of Physical and Chemical Research, to study the materials formed inside the water-cooling brass pipes of the engines of warships by the corrosion of sea water. I was involved in the design and construction of an electron lens to focus the electron beam in this camera and improve the quality of the diffraction patterns. It was the first time I met von Ardenne's book on the electron microscope, which was translated into Japanese and gave me the impetus to design an electron lens. The original book was brought to Japan by a German submarine; it stimulated Japanese scientists to construct electron microscopes, which enable us to see, for example, the virus, invisible in the light microscope. This book, in its Japanese version, was used very effectively by many researchers on electron optics, and thus when von Ardenne visited Japan in November 1979, a Manifesto of Gratitude was issued to him by the Japanese Society of Electron Microscopy.

Our university lay about 1 mile from the epicenter of the atomic bombing of Hiroshima and I suffered serious bodily injury, but somehow I survived. Almost all the rooms and equipment were destroyed and burned by the fire after the attack. Iron-reinforced concrete floors and roofs of the rooms close to mine were broken apart by the heat of the fire, which suggests that the temperature of some rooms was raised to the melting point of iron (1535°C). Thanks, however, to a small water pool which I had placed in my room using bricks and cement and to some timely firefighting that I carried out, our equipment and facilities in the room survived and were afterward used successfully by Prof. S. Kuwabara. Some equipment and my personal belongings in our unburned room are displayed in a memorial room in the new building in Hiroshima University, together with a plaque describing the above incident.

B. The Electron Diffraction Camera and the Universal Electron Diffraction Microscope in Kyoto

After graduating from Hiroshima University in September 1945, I stayed at my native place near Kyoto and engaged in teaching in my Alma Mater high school for one year. In 1946, I moved to Kyoto Technical College (KTC) and also engaged in teaching. Fortunately, one year later, I was allowed to do research in the laboratory of Prof. K. Tanaka at Kyoto University (KU), besides teaching at KTC. The conditions in Kyoto, though primitive (with gas available only during the day and electricity only during the night, and a serious shortage of food), were incomparably better than those at Hiroshima. In Kyoto, however, accommodation was very limited (most of the other cities had been burned down), and I had to commute every day from my native place, taking 3.5 hours one way by a combination of bicycle, streetcar, and train, whose windows were covered by wooden planks; the smoke from the locomotive engine made passengers sooty in tunnels.

At Kyoto University, though I was put in a room with only one desk, no research equipment was available to me. Thus, following a suggestion of Prof. Tanaka, I started to make an electron diffraction camera. Since a university machine shop was available, I machined and built it by myself using antiaircraft shell cases and brass blocks, which were wartime surplus. It was hard work collecting metallic pieces suitable as parts of an electron diffraction camera. I had to teach myself how to machine accurately by operating the lathe, end mill, drill, grinder, etc. In Kyoto city at that time, the streetcars were not available regularly, so, using a bicycle for transport, I moved to a storehouse of my uncle's factory at the south end of Kyoto to get some pieces of metallic blocks, cycled back to KTC at the north end to give lectures, and went to KU in the east end like a racing cyclist to machine the collected metals at night in the machine shop, sometimes alone. After spending one year like this, I succeeded in making a diffraction camera (Hashimoto, 1950a). This camera is now preserved in the Department of Applied Physics, Osaka University. As soon as I saw the greenish-yellow light from the fluorescent screen in the diffraction camera after switching on the electron beam, my heart started to beat strongly with joy. I felt that these electrons were my very own electrons and not ones made by an instrument manufacturer! It was the afternoon of Sunday, 30 November 1947.

Selective oxidation and sulfurization processed at elevated temperature were studied in this diffraction camera (Hashimoto, 1950b; Tanaka and Hashimoto 1951), and I noted that some fine structure in the electron diffraction patterns was closely related to the morphology of the crystals which grew. I therefore felt that electron diffraction studies should be carried out together with electron microscope observations.

We were stimulated by the idea of Boersch (Boersch, 1936) for such an instrument but did not really believe that we could go from imaging to diffraction conditions by changing the lens excitation until I had carried out some preliminary trials in a home-made optical setup using two lenses and an electric lightbulb as an object. With one setting of the lenses I could project on the "Fusuma" screen the manufacturer's name on the glass envelope (specimen) and, with a change of position to a second lens, project the image of the filament (diffraction pattern). Since it was supposed that the diffraction pattern would be very small, a three-lens electron microscope which could also function as a high-resolution electron diffraction camera

FIGURE 1. Universal electron diffraction microscope constructed in 1953.

(Hillier and Baker, 1946) was constructed and called the Universal Electron Diffraction Microscope (Fig. 1) (Tanaka and Hashimoto, 1953). The three-lens electron microscope constructed by Le Poole (1947) was not yet known in Japan when we started to construct ours. About the year 1952, I became a member of the Japan Electron Microscope Society, with a registered number (562), and could meet Drs. Tani, Sugata, Sakaki, Hibi, Kanaya, Kobayashi, Maruse, Uyeda, Honjo, etc., who gave me much stimulating advice at conferences (Fig. 2).

C. Diffraction Contrast Experiments

Using this three-lens Universal Electron Diffraction Microscope, I was able to observe (Hashimoto, 1954) in crystals of a MoO_3 a variety of diffraction contrast effects such as extinction and other features which I called "feather-like" and "wavelike" patterns. It is now evident that some of the pictures contained dislocation images, although I did not recognize this at that time and identified them as "boundary lines of crystal sections." I can therefore claim to be a member of that select group including Heidenreich, Hibi, and Bollmann who obtained dislocation images before the celebrated observations in 1956 (Hirsch *et al.,* 1956; Bollmann, 1956), but did not realize it. The subsidiary maxima were also observed at that time in a high-resolution electron diffraction pattern and explained by the dynamical theory.

At the time of the discovery of the dislocation images just mentioned, and the classic work of Menter (1956) in resolving lattice planes, I made observations of dislocations in moiré patterns arising from two overlapping crystals of slightly different orientations. It was demonstrated (Hashimoto and Uyeda, 1957) that when either crystal contained a dislocation, the moiré fringe would also contain a dislocation, which could be studied in a

FIGURE 2. Japan Electron Microscope Society conference in Hakone, 9 May 1955. From left, G. Honjo, H. Watanabe, R. Uyeda, H. Hashimoto, K. Kobayashi, K. Kanaya, and T. Hibi.

relatively low-resolution microscope as shown in Fig. 3. Eventually, dynamical theories of crystal lattice and dislocation images, moiré fringes, and Fresnel fringes were developed with the collaboration of Mannami and Naiki (1958–1961) (Hashimoto and Mannami, 1960, 1962; Hashimoto et al., 1961) using Bethe's two-beam theory.

In August 1957, J. M. Cowley, who was the first foreign visitor in my research field, visited our laboratory at Kyoto University and gave a lecture about his recently developed theory of electron diffraction (Cowley and Moodie, 1957), which stimulated us Japanese electron diffraction workers, who were using Bethe's two-beam theory. Since he stayed some days in Kyoto, he (one year younger than I) and I had a very nice time for discussion by visiting Japanese gardens, temples, shrines, the Imperial Palace garden and Japanese-style restaurants. I told him that I had applied Bethe's dynamical theory to explain the image contrast of crystal lattice fringes; these were calculated in the commuter train, in the morning and afternoon (a total of 5 hours). I was very lean and thin around that time, due to shortage of sleep and food, as can be seen in Fig. 4.

About this time I also became interested in the different contributions of elastic and inelastic scattering to electron microscope images and diffraction patterns. An energy analyzer of the net-filter type developed by Boersch was incorporated in the Universal Electron Diffraction Microscope (Hashimoto et al., 1956), and we were able to show that the Kikuchi lines from zincblende crystals contained both an elastic and an inelastic component.

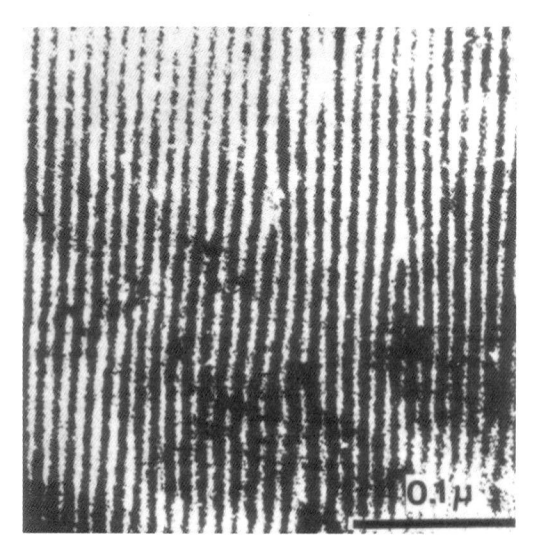

FIGURE 3. Moiré pattern in CuS crystal showing dislocation (1957).

FIGURE 4. J. M. Cowley and H. Hashimoto on the stage of Kiyomizu temple (1957).

III. Observation of Dynamic Processes of Crystal Growth

Various kinds of morphology of crystals formed by the selective oxidation and sulfurization stimulated me to carry out the direct observation of crystal growth by chemical reaction. After fitting the microscope just described with a specimen heating and gas reaction cartridge, we studied the processes of growth and evaporation of tungsten oxide crystals (Hashimoto, *et al.*, 1958).

I then designed a new microscope with an improved specimen gas reaction chamber and heating cartridge, constructed with the aid of the Shimadzu Co., and made a number of *in situ* observations of dynamic processes by recording with ciné film directly from the fluorescent screen, with the collaboration of colleagues (Hashimoto, *et al.*, 1959).

This microscope was actually for me the first microscope which was made by the engineers of a manufacturer; it gave quite a high resolution. Since, however, the highest sensitivity of ciné film was ASA 200, we could not record the images of growing crystals. Thus we used special "ultrahigh-power developer," which was not used by the cinema companies, and developed and processed the recorded film by ourselves in our own laboratory at KU, using home-made equipment.

The investigations which we made included the reaction between copper and sulfur vapor at 400°C, when a strange figure like a small hunter was seen and the growth on tungsten wire of tungsten oxide crystals in a form similar to banana leaves. The growth of tungsten oxide in the form of needles was also followed by heating ammonium tungstate at 700°C. A molten oxide drop forms at the top of the needle, and growth material comes from the vapor phase passing through the drop (Fig. 5). This process was called "drop growth" in 1960; a few years later, Wagner and Ellis of Bell Telephone Laboratories (1964) observed a similar phenomenon in the growth of Si and called it the VLS mechanism.

I brought the above ciné film to the first Thin Film Conference in 1959 in Lake George, New York, where I met many eminent scientists. My experience there will be described in Sec. V.

Later, I used a JEOL 7A microscope for these experiments (Hashimoto et al., 1968) and employed a TV orthicon camera (Hashimoto et al., 1970). Growth processes of needle crystals showing resolved lattice images (12 Å) were also recorded in movie film.

Some other observations were made of the reduction of cupric iodide to copper in hydrogen at 700°C. Reduction takes place at the surface of the molten CuI, and the metallic copper whiskers appear growing from the root to a length of a few micrometers before they are pulled down into the molten substrate and gradually change into pure copper.

All of this provided useful experience for the later experiments at the atomic level.

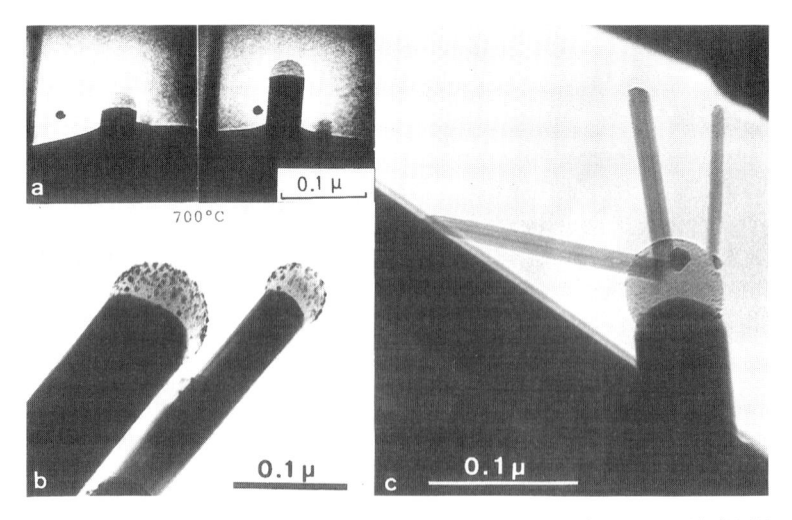

FIGURE 5. (a–c) Growth of γ-tungsten oxide needle crystals (drop growth) (1960).

IV. The First EM Regional Conference in Asia and Oceania in Tokyo (1956)

At the Third International Congress on Electron Microscopy, in London in 1954, the IFSEM was established and it was decided to hold Regional Conferences as well as International Conferences under the auspices of the IFSEM; it was suggested that Regional Conferences be held in Europe and Asia and Oceania regions every four years; in between, the International Congress also be held every four years. On 12 October 1956, the first Asia and Oceania Conference was held in Tokyo (Fig. 6), as well as the European Conference in Stockholm. This was actually the first International Conference for us in the Asia and Oceania regions. Delegates from Australia, China, France, Germany, India, Indonesia, Japan, Korea, the Soviet Union, and Taiwan attended, and 50 papers were read. The opening address by Prof Tani, Chairman of the organizing committee, and plenary lectures by Ruska and Bennett, were carried out with an interpreter, but other papers were presented using English (foreign participants) and Japanese. Though it was a very good chance for us Japanese to use oral English in a conference, most of the papers from Japan were presented in Japanese, except for a very few papers by authors who had stayed in Europe or the United States. I had the courage to present in English and presented our

FIGURE 6. Organizing Committee Chairman Y. Tani reading the opening address at the First Asia and Oceania Conference in Tokyo on 12 October 1956.

paper by using English memorized beforehand. I think this boldness came from my experience of meeting with Profs. A. H. Compton and H. Yukawa, even for several minutes [ref. Preface by M. J. Whelan in my *Festshrift in Ultramicroscopy* **54** (1994), Nos. 2–4].

In this conference, I met and made conversation with the Minister of Education, Mr. Kiyose, and the eminent Profs. Ruska, Bennet, Tani, Kaya, Sakaki, Sugata, etc. (Fig. 7). At this conference, I felt very strongly how poor my English was in listening and speaking; this did not disappear completely even after staying in England for a year and attending many International Conferences.

The delegates from the Peoples Republic of China (mainland China) and Republic of China (Taiwan) had a hard time with the arrangement of national flags on the stage at the opening ceremony due to the difference of their policies (see Fig. 6). The government of mainland China could not accept the existence of the Republic of China. Delegates from both countries and some of the Japanese Organizing Committee had to discuss all through the night; they decided to remove the flags as a tentative arrangement. Afterward, in any International Conference, national flag arrangements did not take place. But the problem was not solved completely by this arrangement; that had to wait until 1986, when I became the President of the IFSEM.

FIGURE 7. Front row, from the left: I. Kiyose (Minister), Y. Tani (Chairman), E. Ruska, Niricov, S. Kaya, K. Fukai, R. Uyeda, S. Ogawa, H. Hashimoto (12 October 1956).

It was a pity that a second Asia and Oceania Conference was not held in 1960 but in 1965 in Calcutta. Then came a pause of 20 years until the Third conference in 1984 in Singapore, with a change of name as the Asia-Pacific EM conference. Afterward APEM conferences were held regularly, in 1988 in Bangkok and in 1992 in Beijing.

After the conference in Tokyo, Prof. Ruska came to Kyoto and gave a lecture at Kyoto University and visited our laboratory (Fig. 8).

V. A Stay at Cambridge University After Attending a Conference in the United States

In 1959, I attended an International Conference on the Structure and Properties of Thin Films, which was held 9–11 September at Bolton Landing on Lake George, New York, and met many famous, active scientists. This was my first trip to abroad, where I could meet both old and young eminent scientists such as Drs. Marton, Müller, Schockley, Heidenreich, Siegel,

Figure 8. E. Ruska making comments on photographs taken at Kyoto University (1956).

FIGURE 9. (a) From the left, H. Fowler, H. Hashimoto, H. Reather, M. J. Whelan, R. Castaing; (b) H. Hashimoto, L. Marton; (c) R. D. Heidenreich.

Menter, Pashley, and Whelan. I presented movie film and slides on the growth of tungsten oxide (Hashimoto et al., 1959).

After the conference, I had enjoyable visits to several research laboratories at institutes and universities near the West Coast, and I stayed for nearly three weeks. It was an exciting time for me to join research groups in well-equipped laboratories. Dr. Marton gave me a chance to work in the National Bureau of Standards for 20 days together with Dr. Fowler. It was one of the highlights of my first stay in a foreign country (Fig. 9).

At the end of October, I visited Cambridge University, where I stayed until the end of the August of the next year as a Visiting Researcher of the Japanese government. I got a desk, by the arrangement of Dr. Menter and Dr. J. Nutting, in a room of the Department of Metallurgy, in which I was surrounded by young, active electron microscopists: P. Swann, R. Nicholson, R. M Fisher, M. Ashby, P. Kelly, D. Brandon, A. Baker, K. Williamson, etc. (Fig. 10). I could easily contact the researchers in the groups of Dr. P. B. Hirsch at the Cavendish Laboratory and Dr. J. W. Menter at the Tube Investment Research Laboratory, due to the close collaboration between these great leaders.

As soon as I got into the laboratory in the Department of Metallurgy, I saw many electron micrographs of electropolished metals and I got quite a shock, because the answer to a question which I had kept for a long time in mind in Japan was clearly answered in the micrographs. I said, "Oh!" quite involuntarily. When I was in Japan, I saw many extinction contours in the electron micrographs of naturally grown thin crystals of uniform thickness. The two parallel bend contours are due to the Bragg reflection from the same planes with opposite signs and sometimes showed dark contrast in the region between them but sometimes not. In the photographs of electropolished metals, whose thickness increases from the edges inward,

FIGURE 10. Young, active electron microscopists in the Department of Metallurgy, Cambridge University (1959).

the two parallel bend contours showed that the dark region between them becomes darker and darker with increasing thickness but at the thin edge no dark region was observed. I was convinced that this phenomenon was due to the appearance in the microscope images of absorption of Bragg reflectedlectron waves, which was observed in electron diffraction patterns by G. Honjo and K. Mihara in 1954 and by K. Kohra and H. Watanabe in 1959, and whose absorption coefficients were dervied by H. Yoshioka in 1957.

As I had some experience of the theoretical contrast calculation of crystal lattice fringes using Bethe's dynamical theory, including absorption (Hashimoto et al., 1961), it was easy to derive the equation representing the contrast of bend contours due to the thickness change of the specimen. The calculation was finished on 17 November and showed that the intensity asymmetry on either side of a bend contour increases with increasing crystal thickness in the bright field image, which agreed with the observation but, surprisingly, a symmetry in the intensity appears in the corresponding dark field image. I showed this calculation to Dr. Whelan, who was busy calculating the contrast of dislocation images (Howie and Whelan, 1961), and he showed me intensity anomalies of electron micrographs of dislocations and diffraction patterns in a thick crystal region by suggesting the absorption of electrons.[1] We both agreed that it would be worthwhile to investigate together the electron absorption effect in electron microscope images. I also showed these results to Dr. Nutting, who suggested that I investigate

[1] Prof. M. J. Whelan points out that the Cavendish Laboratory group were already aware the interpretation of these anomalies in terms of the electron equivalent of the Borrmann effect in X-ray diffraction. A paper by A. Howie and M. J. Whelan was read on this subject at the EM Group Conference at Exter in July 1959. [cf. *Br. J. Appl. Phys.* **11**, 31 (1960)]

the contrast anomalies of stacking faults in thick crystals to find out whether they are due to precipitates or to electron absorption. I then started to calculate the image contrast of stacking faults, considering absorption. But I thought I must make observations to confirm the above phenomenon directly. I took electron micrographs of electropolished thin films using the Siemens microscope on 25 November. This was the first experience for me in using the famous Siemens microscope. The next day, after developing and printing, I noted in the first two pictures, very clearly, that the bend contours showed asymmetry contrast in the bright field, and symmetry contrast in the dark field image which agreed with the calculated contrast. The next day, 27 November, I showed the photographs and calculated results to Dr. P. B. Hirsch, who told me that he was also thinking about the absorption effects on the contrast anomaly of the image in the thick crystal region and asked me to study the contrast of slip bands.

The intensity profiles of the images of stacking faults were calculated by using a mechanical Tiger calculator and took me a rather long time. Around that time, a high-speed electronic computer, EDSAC, was available at the Cavendish, and Dr. Whelan and his colleague Dr. A Howie deduced the profiles quite quickly. Though my calculation was slow, it was very effective for understanding the physics of electron wave fields in the crystal. It also showed up some errors in the electronic computation.

As a result of all these fortunate experiences, by 1960 I became quite familiar with the dynamical theory of diffraction contrast with the effects of anomalous transmission and absorption of electrons and its application to image computation in both perfect and imperfect crystals (Hashimoto *et al.*, Howie and Whelan 1960, 1962) (Fig. 11).

During my stay at Cambridge, I had the chance to meet and visit world-famous electron microscopists and made many good friends not only in

FIGURE 11. From the left: M. J. Whelan, A. Howie, H. Hashimoto.

England but also in European countries, too many names to be listed here (Fig. 12 and Fig. 13). Among them, Prof. Dupouy invited me to his laboratory in Toulouse, where he had constructed a 1.5-MeV electron microscope. When he accompanied me to the microscope on 15 June, he said, "You are the first foreigner to see this microscope." He had seen 1-MeV images the day before for the first time. As every electron microscopist knows, he produced beautiful photographs (see *Proc. 5th ICEM*) and was called father of high-voltage electron microscopy (*Proc. 3rd HVEM Conf.*) [Fig. 13(a)].

Toward the end of my stay, a very important task was imposed on me by Dr. Menter and the EM group studying defect structure of materials.

It was noted that the density of dislocations observed in electron micrographs is smaller than that estimated from the actual strength of the metals concerned. It was suspected to be due to the loss of dislocations from thin films. Since it was supposed that this phenomenon could be solved by observing thick metal films using a high-voltage electron microscope, for example, the 300-kV electron microscope at Kyoto University, which was made by Shimadzu Co. and installed by Prof. Kobayashi and Suito several years previously. I was asked to check the possibility of using this microscope by sending the specimens to Japan. After exchanging letters, specimens, and observed results with Kobayashi, I noted that I must make observations by myself in Japan because the Kyoto people did not know how to observe the inside structure of thick specimens using the condition of anomalous transmission of electrons, which we found at Cambridge.

FIGURE 12. (a) From the left: P. B. Hirsch, W. Bollmann, H. Hashimoto; (b) J. W. Menter, H. Hashimoto, D. Pashley.

FIGURE 13. I visited G. Dupouy in Toulouse (a) M. von Ardenne in Dresden (b) and H. Bethge in Halle (c).

VI. Relativistic Effects and High-Voltage Electron Microscopy in Kyoto in 1961–1962

Though it was necessary to spend three months to redesign the high-voltage electron gun system to elevate from 150 kV to 300 kV after I returned to Japan on 27 September, I was able to work for three months, from 10 January 1961, in the electron microscope room of Kyoto University at a temperature about 0°C, without any heating device, by wearing an overcoat and sometimes putting on gloves. In such a cold and dry room, the microscope could be operated rather nicely without high-voltage breakdown and I could get many bright and dark field images of wedge-shaped and bent Al films by changing the voltage from 100 kV to 200 kV and 300 kV. I could measure the energy dependence of extinction distance and absorption coefficient by measuring the spacings and intensity variation of bend contours (thickness fringes) in the exact Bragg reflecting condition. The measured energy dependence of extinction distance did not agree with the theoretical prediction of the dynamical theory at that time. This disagreement was made clear on 31 March 1961 at the Physical Society meeting in Japan. At that time, I presented my observed results and Dr. K Fujiwara showed that relativistic correction to the mass is necessary, not only in the wavelength, but also in the scattering potential, after checking the Dirac relativistic wave equation. I proposed to Fujiwara, who was on the stage, to check our observed results by his equation using our data. His calculation there and then on the blackboard agreed with my observed results exactly up to the third place! This agreement suggested that the extinction distance is proportional to v/c, where v and c are the speeds of electrons and light, and tends to constant value even at very high voltage.

The reciprocal of absorption coefficient of electron waves, which gives the transmissive power of the electron wave, was also known to be proportional to $(v/c)^2$ by Yoshioka's theory. My measured two absorption coefficients, the mean (ε_0) and the anomolous ($\Delta\varepsilon$) one, using effectively the same size of aperture, agreed with the theoretical prediction (Hashimoto, 1964). Thus it was known that even at very high voltages the transmissive power does not increase to a very high value but tends to the value of 3.35 times greater than the one at 100 kV, if the theory holds up to such high voltages.

After this meeting, Kyoto University decided to make a TEM change from 300 kV (Kobayashi *et al.*, 1963) to 500 kV (1964); this instrument was installed at the Chemical Institute of Kyoto University.

From August to September 1962, the U.K. Department of Scientific and Industrial Research sent two electron microscopists, Drs. T. Mulvey and M. J. Whelan, who were experts in the fields of electron optics and image contrast theory, respectively, to Kyoto to see the EM construction and the dislocation images in thick metallic specimens. It was a very hot and humid summer, and the microscope often suffered from discharge in the electron gun system, but they still got impressive and definitive results. I collaborated with them to get these good results (Fig. 14). Their experience became the stimulus for U.K. construction of six 1-MeV microscopes in 1969. It was also the beginning of a lifelong friendship with both of them.

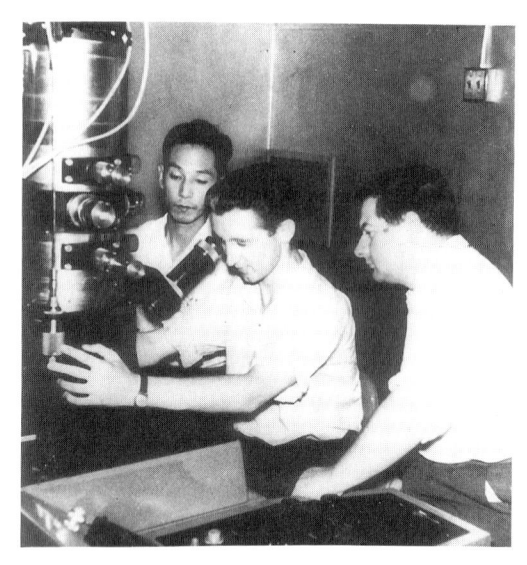

FIGURE 14. T. Mulvey (right) and M. J. Whelan (center) working with the Shimadzu Kyoto 300-kV electron microscope (1961).

On 25–30 September 1961, the same year, the International Conference on Magnetism and Crystallography (Electron Diffraction) took place. At this conference, relativistic effects appearing in high-voltage electron diffraction patterns (Miyake, Fujiwara, *et al.*) and our electron microscope images (Hashimoto *et al.*, 1962) were shown, together with the theory of Fujiwara and that of Howie, who also solved Dirac's equation and arrived at the same conclusion. At this conference, we showed several electron micrographs, showing inner fine structure of metal films of about 1 μm thickness and crystalline structure of high polymer crystals for the first time. It was also shown at this conference that the new developments of electron diffraction are also very important in the electron microscopy. They were many beam dynamical theories, the nature and behavior of inelastically scattered electrons, electron interference, structure analysis, low-energy electron diffraction, etc.

After these exciting days in 1961–1962, high-voltage electron microscopy began to play a very important role in the fields of physics and materials science and 500 kV–1 MV electron microscopes were installed in many active research laboratories in the world and many conferences related to HVEM were held. For example, in 1966, the Argonne National Laboratory organized the HVEM workshop to construct an 3–5 MeV electron microscope on 13 June–15 July 1966, to which Dr. B. Tadano of Hitachi and I were invited from Japan. The next year, Prof. DuPouy in France and Prof. E. Sugata *et al.* in Osaka started to construct 3-MeV electron microscopes.

The first U.S.-Japan Seminar on HVEM was held from 27 October to 5 November 1967 in Hawaii, of which the coordinators were Drs. C. E. Hall and R. Ogilvie (U.S.) and H. Hashimoto (Japan). The 2nd U.S.-Japan Seminar on HVEM was held 18–26 September 1971 in Hawaii, with coordinators Dr. G. Thomas (U.S.) and H. Hashimoto (Japan), and the 3rd one was 5–12 December 1976 in Hawaii, with coordinators Drs. R. M. Fisher (U.S.) and T. Imura (Japan).

The International HVEM Conferences were held first in Monroeville (USA, 1969), 2nd in Stockholm (Sweden, 1971), 3rd in Oxford (U.K., 1973), 4th in Toulouse (France, 1975) 5th in Kyoto (Japan, 1977) (Fig. 15), 6th in Antwerp (Belgium, 1980), 7th in Berkeley (USA, 1983), and 8th in Kyoto (Japan, 1986) (held with ICEM), whose proceedings clearly show the progress of HVEM which originated in 1961–1963.

VII. Early Atomic Imaging in Kyoto

A. Problems of Magnification and Contrast in Atomic Imaging

Though the size of the atom is well known, it may be useful to recall that when an atom is magnified 100 million times, it becomes as large as a ping

FIGURE 15. 5th HVEM Conference in Kyoto, 29 August–1 September 1977: opening ceremony.

pong ball; and when the ping pong ball is magnified by the same amount it becomes as large as the moon. Therefore our attempts to image the atoms in a ping pong ball may be compared with the problem of trying to observe in a terrestrial telescope a ping pong ball held by an astronaut on the moon. The problem is increased when we realize that atoms are mostly empty and do not absorb incoming electrons. They simply scatter them though small angles and indeed act like phase objects. A somewhat closer optical analog to observing an atom at a distance of 55 cm by illuminating 100-kV electrons can perhaps therefore be obtained if we imagine a 50-μm-diameter quartz ball on the top of a tower in Oxford to be viewed in the red light of the evening sun behind it from a distance of about 110 km, say, in Cambridge. In these bright-field imaging conditions, the ball would be essentially invisible if the telescope were focused exactly on it, but under suitable defocus conditions, it could appear as a very-low-contrast dark spot against a bright red background. Image contrast of single atoms and very thin crystals with atomic resolution was discussed by Boersch (1946, 1947), who approximated them as weak phase objects and predicted that some contrast would arise by the cutoff of the scattered waves by the aperture (amplitude contrast), and the images of thin crystals would become periodic fringes. Niehrs (1954, 1956) and Uyeda (1955) also discussed the imaging of crystals. Scherzer also discussed image contrast together with the phase shift which is produced by the spherical aberration of the imaging lens and compensated by a certain amount of underfocussing (Scherzer focus) (Scherzer, 1949) (Fig. 16). Theoretical image contrast of atoms considering the atomic scattering factor and spherical aberration of the lens was formulated by Heidenreich and Hamming (1965), by Kamiya (1965), and by Eisenhandler and Siegel (1966) in the bright-field image.

Returning to the optical analog, conditions equivalent to a tilted dark field could be obtained by waiting until after sunset and illuminating the quartz ball from behind the tower building, so that the ball scatters the light. If there were no fog or mist, the quartz ball would then appear bright, like a star in the night, with much better contrast. However, though the contrast is higher in dark field than in bright field, the intensity of scattered light is very small and thus a long exposure time would be needed for taking an image of the quartz ball.

B. Dark-Field Images of Single Atoms

In electron microscopy we cannot have isolated single atoms, so we consider the additional image contrast effects arising from the support film of the surrounding medium. Nevertheless, simple dark-field images of single heavy

FIGURE 16. (a) H. Boersch and (b) O. Scherzer.

atoms can be obtained when they are supported on a thin crystalline graphite substrate and an objective aperture is used to exclude the Bragg reflected beams from the substrate and to accept the electrons scattered at smaller angles by the heavy atoms. Figure 17(a) shows tilted dark-field images of Th atoms belonging to long-chain Th-pyromelitate molecules and separated by distances of 4 to 10 Å. Figure 17(b) is the image of a small ThO_2 crystal formed by the decomposition of Th-pyromelitate molecules shown in Fig. 17(a) (Hashimoto *et al.*, 1971, 1973) (Fig. 18). Images of heavy atoms supported by other crystalline or amorphous films have been reported, not only in dark-field, but also in bright-field images (Formanek *et al.*, 1971; Ottenmeyer *et al.*, 1972).

Since the scattering amplitude of electrons at large scattering angles is larger from heavy atoms than from light atoms, it is preferable to use a supporting film consisting of light atoms and an objective aperture setting to admit only large scattering; this inevitably needs long exposure time or strong illumination. Calculation suggests that the electron beam intensity passing through the aperture then becomes less than 1/100 of the illuminating beam intensity.

Strong illumination produces inherent radiation effects in the specimen, such as decomposition, movement of atoms and drift of the image as well as reduced image contrast, sometimes leading to incorrect interpretation of the images. Since it seemed interesting to see the movement of atoms

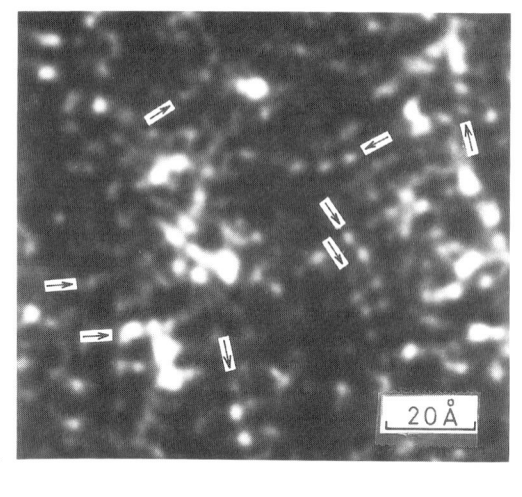

FIGURE 17. (a) Dark-field image of Th-pyromelitate molecules (arrows). (b) Dark-field image of ThO$_2$ small crystal. Th atoms appear bright (1971).

under strong-beam irradiation, the transition process from Th-pyromelitate molecules to small ThO$_2$ crystals was studied by illuminating the specimen with a beam current density of 100 A/cm^2 and recording on movie film (one frame per 0.08 s) (Hashimoto *et al.*, 1978).

By comparison, the dark-field atom imaging methods employed by Crewe (1973) in STEM use an annular detector which collects only the electrons scattered through rather high angles and give much higher contrast than in TEM. However, more than 15 s are needed for one frame of the image, which suggests that STEM is applicable to observe only rather slow motions of atoms (Crewe, 1979).

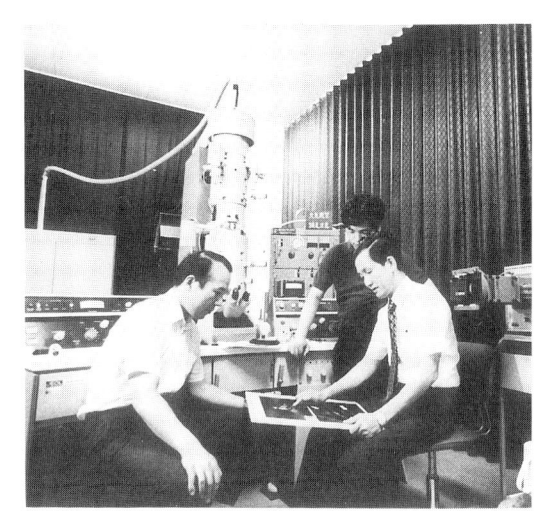

FIGURE 18. Research group at Kyoto Technical University. From the left: A. Kumao, H. Endoh, and H. Hashimoto (1971).

VIII. Lattice Defects and Their Motion at the Atomic Level, in Osaka

After moving to the Department of Applied Physics, Osaka University, in 1975, I realized that in order to apply high-resolution electron microscopy to the multitudinous problems in the field of materials science, I must employ many-beam imaging in both bright and dark field and consider in much more detail the problems of diffraction contrast in crystals. Fortunately, as indicated in the preceding sections, my previous experience in electron microscopy had involved me in the early development of diffraction contrast theory and had also given me some familiarity with the problems of dynamic observations in electron microscopy.

A. Stacking Faults, Twins and Their Motion

Around 1978 (Hashimoto *et al.*, 1978), we were able to obtain our first reliable atomic-resolution images of defects in metal crystals such as are shown in Fig. 19, and to observe their movement using a JEOL 120C electron microscope operating at 100 kV. We employed tilted-beam illumination with an intensity of 100 A/cm^2 and a divergence between 5×10^{-4} and 1×10^{-3} rad to form images under aberration-free focus conditions (Hashimoto *et al.*, 1978–1979). The images were transferred to a TV screen at a final magnification of 24 million (Hashimoto *et al.*, 1979, 1980). This

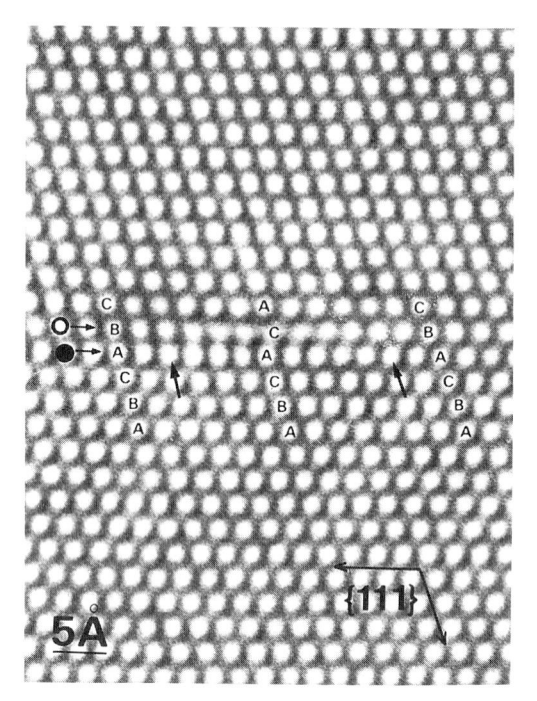

FIGURE 19. Gold single crystal in 100 orientation showing atomic structure of a stacking fault (1979).

video recording technique was later used by Sinclair *et al.* (1982) and by Marks *et al.* (1983) and then commercialized (1985).

In these situations, the images show details finer than the theoretical point resolution and can be interpreted only when detailed many-beam calculations based on the dynamical theory of electron diffraction, as developed by Bethe, are made, including aberration and defocus effects. Some of the more striking observations involved the motion of stacking faults and twins, probably generated as partial dislocations run in on neighboring slip planes from the edge of the specimen as a result of thermal stress there. Figure 20 is a series of TV images showing the dynamic rearrangment of Au atoms at the tip of a twin plate with an interval of 0.07 s between the images. Many further examples have been obtained (Hashimoto *et al.*, 1979).

B. Vacancies and Clusters

Vacancies can be formed by the displacement of atoms in material under irradiation from a nuclear fusion reactor and also by electrons accelerated

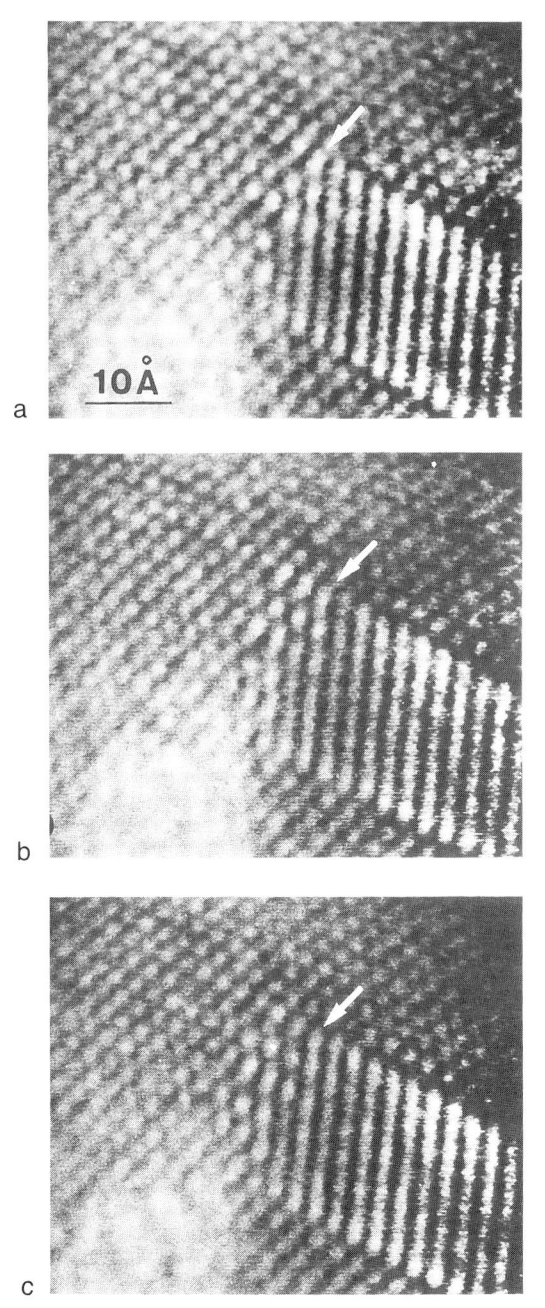

FIGURE 20. (a–c) Movement of atoms at the tip of the twin band in Au crystal (1979); interval 0.07 s.

to high voltages. In order to study the atomic process of displacement damage, gold films of 10 ± 1 nm thickness in 110 orientation were prepared by deposition and exposed to a 2-MeV electron beam in the electron microscope with a dose of 6×10^{21} electrons/cm^2 at room temperature and then transferred to the high-resolution electron microscope; atomic structure was studied at a magnification of 1,300,000 times on photographic film and at 26,000,000 times on the TV screen.

In the images of irradiated regions, some images of single atoms with less bright contrast and triangular regions with contrast anomalies were observed. By comparing with theoretical calculations, these seem to be the images of single vacancies and of stacking fault tetrahedra formed by the coagulation of vacancies, respectively. Adjoining pairs of black and white stacking fault tetrahedra (SFT) of various sizes were also observed.

In the vicinity of the SFT the contrast of the atom images changes during observation, which seems to be due to the migration of vacancies and of the remaining atoms in the stacking fault plane (Hashimoto *et al.*, 1980a).

IX. Atomic Image Processing by Fast Fourier Transform

The usefulness of the optical transform in high-resolution electron microscopy as a tool for measuring and correcting astigmatism, defocus, drift, vibration, etc., was well established (Thon, 1966; Erickson, 1973). For studies at atomic resolution and particularly for following dynamic processes of the type described in previous sections, it is particularly useful to have an on-line system. Since a computer-aided system (Saxton, 1977) can be operated on-line in the operating microscope, the Fourier transform can be obtained from a very small selected area of the specimen even if it changes very rapidly. Knowing that the Fourier transform of the atomic structure image can give crystallographic information such as electron diffraction patterns, I proposed to the government to fund me to construct a system in 1977. Around that time, no computer program or system of fast Fourier transform of two-dimensional structure images were available. With the help of the people in our department and the Matsushita Co., our ardent desires were realized and we published a paper at the Seventh European Congress on EM at the Hague on 25 August 1980. I projected a movie picture, showing the process of FFT of the images of moving atoms in a gold crystal, two or three times in the afternoon (Hashimoto *et al.*, 1980b).

It was also noted that processed images without random noise and defects in crystal structure can be obtained by the inverse Fourier transform after masking out the area corresponding to the random noise and the spread of diffraction

ATOM IMAGES AND IFSEM AFFAIRS 623

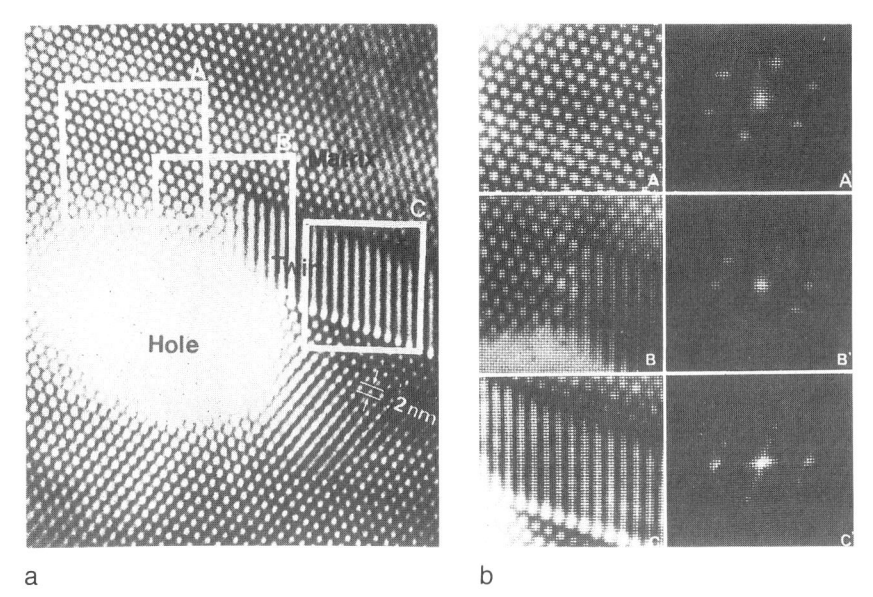

FIGURE 21. (a) EM image of Au film in 110 orientation containing a twin band. (b) A, B, C are digitized images in (a), and A', B', C' are Fourier-transformed patterns.

spots in Fourier space. The effect of masking aperture size on the elimination of defects was shown (Tanji *et al.*, 1982). Fast Fourier transforms of the images of amorphous carbon films were analyzed to obtain the defocus values of the objective lens. It was demonstrated that small selected-area diffraction patterns could be obtained from the images of a thin gold crystal containing an incoherent twin boundary and that Fourier transforms of the images of carbon film could be used for correcting the astigmatism of the objective lens.

X. Characterized Atom Images, in Kyoto, Osaka, and Okayama

When we took images of Th atoms in Kyoto in 1971 by the dark-field image method (Sec. VII, B), the characterization of the atoms recorded in the images had to be demonstrated. Making histograms of the intensity distribution of the bright spots in the image was used to indicate the most probable images of single atoms.

Theoretical contrast calculations using neutral and ionized single atoms and atomic chains, supported on the observed thin graphite, were also compared with the observed ones. It was noted from the contrast calculation that the intensity of the images of relevant atoms will change by the interference of neighboring atoms and the position of the aperture.

During the beam irradiation, long-chain molecules decomposed and became crystals, which was thought to be a good method of characterization. We constructed a continuous image recording system by modifying the orthicon TV camera so that the images accumulated for 1/20, 1/10, 1/5, 1, 2 s rather than the normal 1/30 s per frame. The recorded images in the tape were projected on the monitor intermittently, and were then recorded successively on movie film by the Aliflex movie camera operated by a synchronized pulse motor. These modification of the recording system made it possible to record the crystallization process of moving atoms (Hashimoto et al., 1979).

I was thinking that one of the most reliable methods for characterizing the atoms is to form the image of atoms by using the characteristic loss electrons by inner-shell ionization. This idea gradually came into my mind in the following way.

When I was staying in Cambridge in 1959–1960, I saw many times at Cavendish laboratory the instruments made by Nobel Prize winners such as Wilson, Aston, Bragg, Rutherford, Thomson, etc., and noted that the construction of these instruments was within my ability to machine. I noted from these instruments that the important contribution to science is not in constructing complicated fine machines, but in building original ones, even if they are very simple. I thought, at that time, that a unique contribution to the electron microscopy field from Japan could be made by constructing stable high-voltage electron microscopes and an energy-selecting electron microscope, which has a function similar to that of the heliostat for taking images of the Sun, using only the helium gas spectrum. At the beginning of 1960, I wrote about this idea from Cambridge to Dr. H. Watanabe at Hitachi, who had constructed a Möllenstedt-type energy analyzer. The answer was that he had started to construct such instruments with the collaboration of R. Uyeda, on which a paper appeared in 1962, the same year as the appearance of the paper by Castaing and Henry.

After I returned to Japan, Prof. Tanaka and I succeeded in getting funds to design and install such an instrument at Kyoto University with the aid of Hitachi. This instrument was not used by me due to the increase of my teaching load at Kyoto Technical University, but by M. Mannami and K. Ishida for the study of plasma loss electrons.

The idea of the characterization of materials by using the characteristic loss electrons was all the time in my mind and became very strong when I come to the idea of characterization of atomic images.

When I was discussing with Dr. R. F. Egerton at the EMSA conference in 1980, I noted that two-dimensional crystal structure images, corresponding to the selected energy spectrum, can be obtained using a simple sector type of magnetic energy analyzer attached below the column of a high-resolution electron microscope even without adding any electron lens.

Since the relation between the energy spectrum and energy-selected images corresponds to that between the electron diffraction patterns and images, a three-stage electron lens system was attached to the exit slit of a sector-type analyzer. After some preliminary observations of the images by plasma loss electrons and spectra in 1981–1983 (Ajika *et al.*, 1983, 1985), I asked the government to provide funds to construct a 500-kV analytical atom resolution electron microscope at Osaka University. After collaboration with JEOL engineers, I installed a microscope with accelerating voltages of 400 kV maximum in 1984 and a sector-type analyzer in 1985 (Hashimoto *et al.*, 1986).

Though this first 400-kV electron microscope produced sufficient resolution to see atomic images easily and thus produced many useful results after being installed in many laboratories in the world, the analyzer had heavy distortion and it was very difficult to produce energy-selected atomic images. I retired in March 1985 from Osaka University without taking characterized images of atoms.

In 1986, Drs. O. L. Krivanek and C. C. Ahn of Gatan Co. obtained atomic-resolution images of Si crystal using plasma loss electrons, which was the same kind as our images of Si taken in 1981, but the distortion of the images was corrected by using three quadrupole lenses, which was much better than our method of correcting the image distortion by computer after recording the images.

In 1988, three years later, I moved to Okayama University of Science. Fortunately, I had assistance from the university and the government to install a JEOL 400-kV electron microscope together with the improved new type of Gatan imaging filter (energy analyzer). Since mechanical, electrical, and electronic connections were necessary for interfacing the analyzer to the microscope, the JEOL company kindly sent a JEOL 200-kV electron microscope to the Gatan Co., and I could work with the engineers and my two colleagues in January 1992. After the successful interfacing of the imaging filter to the electron microscope, we could take images of Th atoms using zero loss electrons and energy loss electrons of the inner shell of $ThO_{4.5}$ (98 eV) with the help of Drs. Krivanek and Gubbens (Hashimoto *et al.*, 1992). After interfacing the analyzer to the 400-kV electron microscope in Okayama atomic-resolution images of Si crystal were recorded using 119-eV loss electrons which are the inner-shell loss of $Si-L_{2,3}$ (1994). The bright spot images were concluded to be the images of individual atomic columns containing the Si atoms which produced $L_{2,3}$ energy loss by referring the theoretical calculation on the Si crystal thinner than

15 nm (Endoh *et al.*, 1994) and not the structure images formed by the interference of Bragg reflected waves.

IV. Some International Affairs During My Presidency of the JEMS, IFSEM, and CAPSEM from 1978 Onward

Since I survived a rather long time, I could fortunately attend all the International Congress of EM (ICEM), the Asia-Pacific EM conferences (APEM), and the European EM congresses (EUREM) except the Prague conference for the 34 years since 1960. Though I had many important experiences at those conferences, especially at the 6th (Kyoto) ICEM in 1966, when I served as the Treasurer, I shall not write them here but confine myself to the affairs during my Presidency.

I was nominated to serve as Vice President of the Japanese Society of Electron Microscopy (JSEM) in 1977–1978 and as President in 1978–1979. As described in Sec. VI, the JSEM organized the 5th HVEM Conference in 1977, when I served as the Chairman of the Organizing Committee. One of the most difficult problems was to obtain a license from the government for collecting money without having to pay tax, which was essential for subsidy from industry and could be carried out only if the conference were to be organized under the auspices of the IFSEM. Unfortunately, there was no IFSEM Committee member from Japan, and thus to get such a title, I had to propose such to the IFSEM Committee meeting at the 6th EUREM in Jerusalem in 1976. Due to hijacking problems, it was difficult to reach Jerusalem, but it all worked out and we received permission from the IFSEM to go ahead. It was not easy to collect money from industry even though IFSEM support was given to the conference, but I was assisted by many people, especially by Drs. H. Fujita, T. Imura, S. Maruse, K. Ogawa, and others in organizing a successful conference.

At the general assembly in Toronto in 1978, I became a member of the Executive Committee of the IFSEM and became familiar with IFSEM business.

On 12 October 1978, I visited China for the first time, together with T. Taoka (Vice President, JSEM) and H. Yotsumoto, at the invitation of Mr. D. Lu, Vice Minister of Metallurgical Industry of China, and gave a series of lectures at Beijing, Shanghai, and Guanghou, where I met leading scientists in China, Profs. T. Ko, L. Y. Huang, F. H. Li, Lin Li, and others, who had the experience of staying in European countries. I showed images of atoms in molecules and crystals by slides and movie films showing the growth of tungsten oxides, movements of Th atoms and Au atoms in crystals, etc. I also presented several volumes of the proceedings of the 5th HVEM Conference at Kyoto, which were, I was told, distributed to the important laboratories in China, and read by many people.

On 22 May 1979, the JSEM, which was established on 13 May 1949, celebrated its 30th anniversary at Takarazuka and organized a lecture meeting, inviting Drs. A. V. Crewe, D. Fawcett, P. B. Hirsch, and E. Kellenberger, together with Japanese lecturers, Fukami-Murakami, Hashimoto, Ichinokawa, Imura-Fujita, Matsumoto, Saito, and Yamada. The lectures were published as a book entitled *Development of Electron Microscopy and Its Future* (Fig. 22). Already participants said that they had the impression that the electron microscope was now actually stepping into the atomic world.

Five delegates from China participated in this ceremony, later visiting several important laboratories in Japan. The next year, in 1980, on 4 November, China established the Chinese Society of Electron Microscopy in Chengdu, which was one of the capital cities in the Three Kingdom ages (1773 years ago), where I gave a lecture and contributed a Chinese-style poem to pay my respect. During my stay in Chengdu, Profs. T. Ko, K. H. Kuo, and I agreed to hold a Chinese Japanese Electron Microscopy Seminar the next year.

I worked as the Coordinator of the Japanese side and with the help of Prof. S. Kaya. We met in Dalian together with 19 Japanese participants

FIGURE 22. Proceedings of the 30th anniversary of the JSEM (1979).

(representing active laboratories in Japan) and 47 Chinese participants, on 27–31 July. It was quite a memorial meeting for both Chinese and Japanese participants, who had not met each other beforehand, even among the Chinese themselves. Presentations and discussions were carried out chiefly in English but sometimes by drawing Chinese characters on the blackboard, which were understandable to Japanese. After this seminar, exchange of researchers between the countries became more frequent than before.

Afterwards, the Japanese participants moved to Beijing and were invited to a dinner in the People's Grand Hall in Tian An Men Square as International VIPs (Fig. 23). The Chinese-Japanese seminar on EM was held afterwards, and then organized regularly every two years in Beijing, Hangzhou, Kunming, and Ulumuqi; the 6th one was organized in Okayama, Japan, in 1991 and the 7th in Zhang Jia Jie, China, in 1993. The proceedings of all seminars, except the first one published in Beijing, were published by the JSEM.

At the next General Assembly in Hamburg on 20 August 1982, I was elected to serve as the President of the IFSEM until the year of the Kyoto ICEM in 1986; and Prof. G. Thomas continued as General Secretary.

At this General Assembly, we agreed to propose the organization of the Asia and Pacific EM Conference (APEM Conf.), and then it was decided at the that the 3rd APEM Conference would be held together with a workshop under the auspices of the IFSEM in Singapore. The 3rd APEM Conference and Workshop was held successfully at the National University of Singapore from 29 August to 2 September 1984.

FIGURE 23. After dinner at the People's Grand Hall in Tian An Men Square in Beijing (1981).

In order to hold APEM conferences, regularly, I though we must form a Committee of the Asia-Pacific Society for Electron Microscopy (CAPSAM), similar to the already existing Committee of the European Society for Electron Microscopy (CESEM), which promote electron microscopy in Europe by organizing the EUREM congress every four years under the auspices of the IFSEM. After a discussion with General Secretary of IFSEM, Prof. G. Thomas, we sent invitation letters to the active electron microscopists in the countries in the Asia-Pacific region on 11 March 1983. However, the answer from mainland China indicated that they could not join CAPSEM as long as it belongs to the IFSEM. In other words, in order to invite mainland China to be a member of CAPSEM, CAPSEM had to be independent of the IFSEM until the political problem could be solved. The International Council of Science Unions (ICSU), which is the upper main body of the IFSEM, made a resolution and decisions on the membership of China, stating that the China Association for Science and Technology (CAST) is a National member acknowledging that there is only one China and Taiwan is a part of China.

The IFSEM had already accepted the proposal of membership from Taiwan as the Republic of China Electron Microscope Society at the General Assembly in Hamburg in 1982. Thus the answer from the Chinese delegates was that, as far as the IFSEM does not follow the principle of the IFSEM, CAST would not agree that the CSEM become a member of the IFSEM.

Thus, after some discussion with General Secretary Prof. G. Thomas, I wrote to all (CAPSEM) members as follows.

CAPSEM will remain totally independent of the IFSEM, at least until the satisfactory resolution of the political problem. I myself joined CAPSEM as an electron microscopist from Osaka and not at all in my capacity as President of the IFSEM. However, I will do my best as the President of the IFSEM to solve the current political problem.

Though I thought that finding a solution was beyond our power, we made every effort to collect information and opinions from both societies and governments by the exchange of letters, interviews with the Vice Chairman of the Science Council in Taipei, the Vice President of the Chinese Academy of Science in Beijing, and attendance at the Committee Meetings of both societies. Our activity was kindly helped by the CAPSEM Committees from Taiwan (Prof. C. K. Wu) and mainland China (Prof. L. Y. Huang, later by Prof. K. H. Kuo). I also sent the inquiry to the ICSU and its member International Unions, such as IUPAP (International Union of Pure and Applied Physics), IUPAC (Chemistry), IUIS (Immunology), IUB (Biochemistry), IUMB (Microbiology), IGU (Geography), all of which had a similar style, for example,

IUPAP
  The Chinese Physical Society
  The Physical Society located in Taipei, China
IUB
  Chinese Biochemical Society
  The Biochemical Society located in Taipei, China

Thus after many discussion with the General Secretary and all Committee Members, we proposed to both societies that the following would be acceptable:

  Chinese Electron Microscopy Society
  Electron Microscopy Society, Taipei, China

In order to accept two member societies from one country, the qualification of membership and related parts written in the constitution of IFSEM had to be changed. I thought that the idea of a nation would not become so important in the scientific field and International Congresses in Science are not a competition between countries or nations (like the Olympic games, for example,) but rather a forum to exchange and develop knowledge.

IFSEM Committee Members worked hard and changed the description and simplified the constitution for presenting to the General Assembly of the ICEM in 1988.

The President (myself) and General Secretary (Prof. G. Thomas) wrote a letter on 25 June 1986 to both societies on the above decision, emphasizing that this procedure was done simply to facilitate membership in the IFSEM and it was not intended that the name should be changed for any other purpose, nor to affect operations, communication status, etc. It is to be adopted, only within the IFSEM, in order to resolve these membership difficulties.

This proposal was confirmed at the IFSEM Committee Meeting on 31 August and accepted at the General Assembly of the IFSEM on 2 September 1986. Thus China became a member of the IFSEM and then CAPSEM became a member of the IFSEM as well as CESEM.

China invited the 5th APEM Conference to Beijing, which was held under the auspices of the IFSEM on 2–6 August 1992. I retained the Presidency of CAPSEM for two periods (8 years) and stepped down to the Vice Presidency at the end of 1992, when Prof. K. H. Kuo was elected the 3rd President of CAPSEM and Prof. D. J. H. Cockayne continued to serve as the General Secretary of CAPSEM for three periods.

Professor G. Thomas served the IFSEM for 20 years as General Secretary for three periods, President and Vice President, as well as former Presidents Dr. V. E. Cosslett and Prof. J. B. Le Poole, and moreover as a Committee

FIGURE 24. General Secretary G. Thomas (left).

Member of CAPSEM who helped my presidency from every point of view, which I appreciate very much (Fig. 24).

Acknowledgments

I miraculously survived the atomic bomb explosion in Hiroshima and fortunately I could work for 50 years in the fields of research and education. This is surely due to the fact that I was helped and supported continuously by many warm-hearted people, to whom I would like to express my sincere thanks.

References

Ajika, N., Hashimoto, H., Endoh, H., Yamaguchi, K., Tomita, M., and Egerton, R. F. (1983). *J. Electronmicrosc.* **32,** 250.
Ajika, N., Hashimoto, H., Yamaguchi, K., and Endoh, H. (1985). *Jpn. J. Appl. Phys.* **24,** L-41.
Boersch, H. (1936). *Ann. d. Phys.* **27,** 75.
Boersch, H. (1946). *Monatsheft Chem.* **78,** 163.
Boersch, H. (1947). *Z. F. Naturforschung* **2a,** 645.

Bollmann, W. (1956). *Phys. Rev.* **103,** 1588.
Cowley, M. J., and Moodie, A. F. (1957). *Acta Cryst.* **10,** 609.
Crewe, A. (1979). *Chem. Scripta* **14,** 17.
Eisenhandler, C. B., and Siegel, B. M. (1966). *J. Appl. Phys.* **37,** 1613.
Endoh, H., and Hashimoto, H. (1994). *Ultramicroscopy* **54,** 351.
Endoh, H., Hashimoto, H., and Makita, Y. (1994). *Ultramicroscopy,* **56,** 108.
Erickson, H. P. (1973). *Adv. Op. Electron Microsc.* **5,** 163.
Formanek, H., Miller, M., Hahn, M. H., and Koller, T. (1971). *Naturwiss.* **58,** 339.
Hashimoto, H. (1950a). *Mem. Kyoto Technical College* **7,** 4.
Hashimoto, H. (1950b). *Mem. Kyoto Technical College* **7,** 13.
Hashimoto, H. (1954). *J. Phys. Soc. Jpn.* **9,** 150.
Hashimoto, H., Yoda, E., and Maeda, H. (1956). *J. Phys. Soc. Jpn.* **11,** 464.
Hashimoto, H., and Uyeda, R. (1957). *Acta Cryst.* **10,** 143.
Hashimoto, H., Tanaka, K., and Yoda, E. (1958). *J. Electronmicrosc.* **6,** 8.
Hashimoto, H., Naiki, T., Mannami, M., and Fujita, K. (1959). "Structure and Properties of Thin Films," Wiley (Neugelauer, Newkirk, and Vermilyea, eds.), John Wiley, New York.
Hashimoto, H., and Mannami, M. (1960). *Acta Cryst.* **13,** 363.
Hashimoto, H., Howie, A., and Whelan, M. J. (1960). *Phil. Mag.* **5,** 946; (1962). *Proc. R. Soc.* **A269,** 80.
Hashimoto, H., Mannami, M., and Naiki, T. (1961). *Phil. Trans. R. Soc.* **253,** 459, 490.
Hashimoto, H., and Mannami, M. (1962). *J. Phys. Soc. Jpn.* **17,** 520.
Hashimoto, H., Tanaka, K. Kobayashi, K., Suito, E., Shimadzu, S., and Iwanaga, M. (1962). *J. Phys. Soc. Jpn.* **17,** suppl. B-11, Proc. Int. Conf. Magnet. and Crystallogr. II. Electron and Neutron Diffraction.
Hashimoto, H. (1964). *J. Appl. Phys.* **35,** 277.
Hashimoto, H., Naiki, T., Etoh, T., and Fujiwara, K. (1968). *Jpn. J. Appl. Phys.* **7,** 946.
Hashimoto, H., Kumao, A., Etoh, T., Fujiwara, K., and Maeda, M. (1970). *J. Cryst. Growth* **7,** 113; *7th ICEM Grenoble* **2,** 461.
Hashimoto, H., Kumao, A., Hino, K., Yotsumoto, H., and Ono, A. (1971). *Jpn. J. Appl. Phys.* **10,** 1115; (1973). *J. Electronmicrosc.* **22,** 123.
Hashimoto, H., Kumao, A., and Endoh, H. (1978). *Proc 9th ICEM,* Toronto, **3,** 244.
Hashimoto, H., Sugimoto, Y., Takai, Y., and Endoh, H. (1978). *Proc. 9th ICEM,* Toronto, **1,** 284.
Hashimoto, H., Endoh, H., Takai, Y., Tomita, H., and Yokota, Y. (1978–1979). *Chem. Scripta* **14,** 23.
Hashimoto, H., Yokota, Y., Takai, Y., Endoh, H., and Kumao, A. (1979). *Chem. Scripta* **14,** 125.
Hashimoto, H., Takai, Y., Yokota, Y., Endoh, H., and Fukada, E. (1980). *Jpn. J. Appl. Phys.* **19,** L-1.
Hashimoto, H., Takai, Y., Ajika, N., and Endoh, H. (1980a). *Proc. 6th Int. Conf. HVEM,* Antwerp, **4,** 240.
Hashimoto, H., Yokota, Y., Takai, Y., Tomita, M., Kori, T., Fujino, M., and Endoh, H. (1980b). *Proc. 7th EUREM,* The Hague, **1,** 118.
Hashimoto, H., Endoh, H., Tomita, M., Ajika, N., Kuwabara, M., Hata, Y., Tubokawa, Y., Honda, T., Harada, Y., Sakurai, S., Etoh, T., and Yokota, Y. (1986). *J. Electron Microsc. Technol.* **3,** 5.
Hashimoto, H., Makita, Y., and Nagaoka, N. (1992). *Proc. EMSA Boston,* **2,** 1194; (1992). *Optik* **93,** 119.
Heidenreich, R. D., and Hamming, R. W. (1965). *Bell System Tech. J.* **44,** 207.
Hillier, J., and Baker, R. F. (1946). *J. Appl. Phys.* **17,** 12.

Hirsch. P. B., Horne, R. W., and Whelan, M. J. (1956). *Phil. Mag* **1,** 677.
Howie, A., and Whelan, M. J. (1961). *Proc. R. Soc.* **A263,** 217; **A267,** 206.
Kamiya, Y. (1965). *J. Electron Microsc.* **14,** 83.
Kobayashi, K., Hashimoto, H., Suito, E., Shimadzu, S., and Iwanaga, M. (1963). *Jpn. J. Appl. Phys.* **2,** 47.
Menter, J. W. (1956). *Proc. R. Soc.* **A236,** 119.
Niehrs, H. (1954). *Z. Phys.* **138,** 570.
Niehrs, H. (1956). *Optik* **13,** 399.
Ottenmayer, F. P., Schmidt, E. E., Jack, T., and Pwel, J. (1972). *J. Ultrastructure Res.* **46,** 546.
Saxton, W. O. (1977). *Optik* **49,** 51.
Scherzer, O. (1949). *J. Appl. Phys.* **20,** 20.
Tanaka, K., and Hashimoto, H. (1951). *J. Phys. Soc. Jpn.* **6,** 406.
Tanaka, K., and Hashimoto, H. (1953). *Rev. Sci. Instrum.* **24,** 669.
Tanji, T., Hashimoto, H., Endoh, H., and Tomioka, H. (1982). *J. Electron Microsc.* **31,** 1.
Thon, R. (1966). *Proc 6th ICEM,* Kyoto, **1,** 23.
Uyeda, R. (1955). *J. Phys. Soc. Jpn.* **10,** 256.

4.4
Reminiscences on the Origins of the Scanning Electron Microscope and the Electron Microprobe[1]

MANFRED VON ARDENNE

*von Ardenne Institute for Applied Medical Research,
01324 Dresden, Germany*

Details of the first speculations and ideas are seldom recorded during the early stages of invention, because in the whirl of creative activity there is neither the time nor the ability to do so. The invention of the scanning electron microscope (SEM) and of the electron microprobe are exceptions to the rule. A first draft of both was made by the author on 16 February 1937; remarkably, this has survived to the present day, despite all the bewildering vicissitudes of the past. The handwritten sketches in Fig. 1 include the essential features of both the SEM and the STEM and of the electron microprobe, namely:

1. Inversion of the ray path normally used in transmission electron microscopes (TEM)
2. Production of a small-diameter scanning spot (electron microprobe)
3. Generation of a rasterlike pattern over the specimen surface by means of scanning coils incorporated in the final lens assembly
4. Imaging by means of reflected electrons (SEM) and transmitted electrons (STEM)
5. Amplification of the electron current, modulated by the specimen, by means of a photomultiplier
6. Synchronization of the beam scan of the cathode-ray tube with that of the probe by simple parallel connection of the relevant deflector coils

A ray diagram and the three principal modes for detecting electrons, modulated by the specimen, using the terminology of the time, are shown in Fig. 2.

When, in 1937, I tried to implement this invention at Siemens und Halske Berlin, Bodo von Borries and Ernst Ruska, both influential engineers with

---

[1] This is an updated version of a paper presented orally, but not published, at the European Conference on Electron Microscopy on the occasion of the 30th anniversary of the International Federation of Societies for Electron microscopy (IFSEM), Budapest, 13 August 1984.

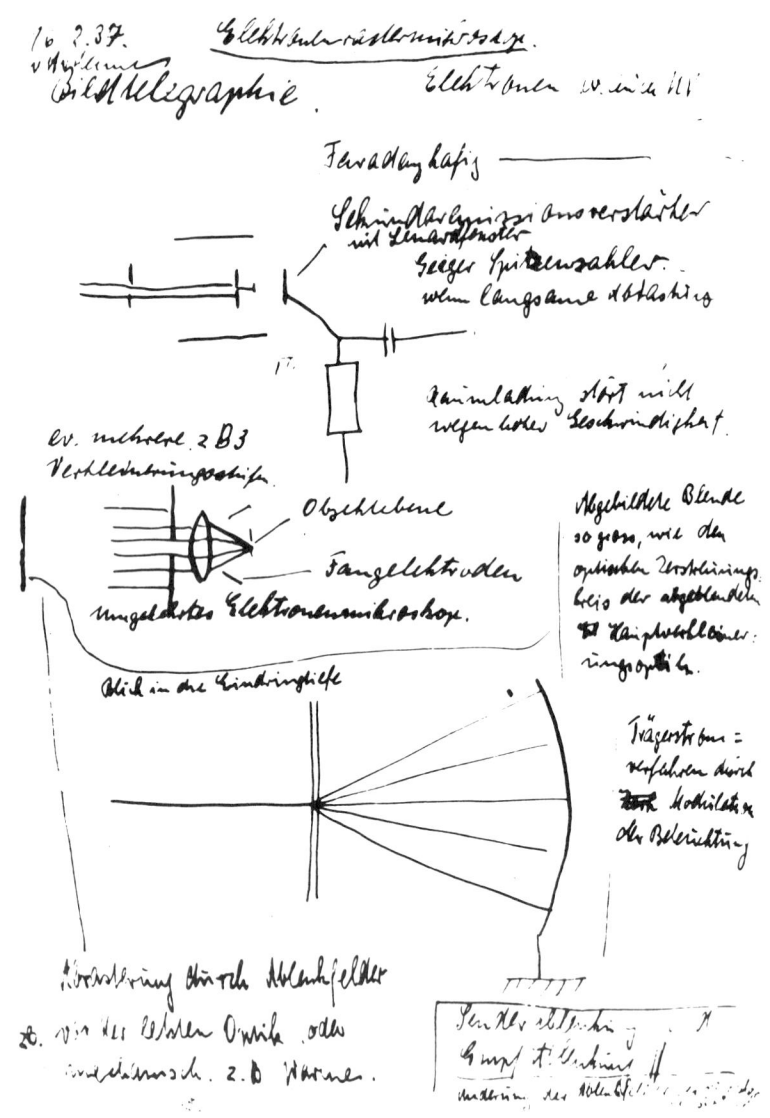

FIGURE 1. Very early ideas and sketches on scanning electron microscopy, taken directly from the laboratory notebook.

the company, regrettably gave a negative opinion about the basic principle of the scanning electron microscope. The first page of this confidential report is reproduced in Fig. 3. This document came into my hands quite accidentally, as late as 1946, when I was Head of Institute A near Sukumi on the Black Sea in the Soviet Union. There I had been asked by a government

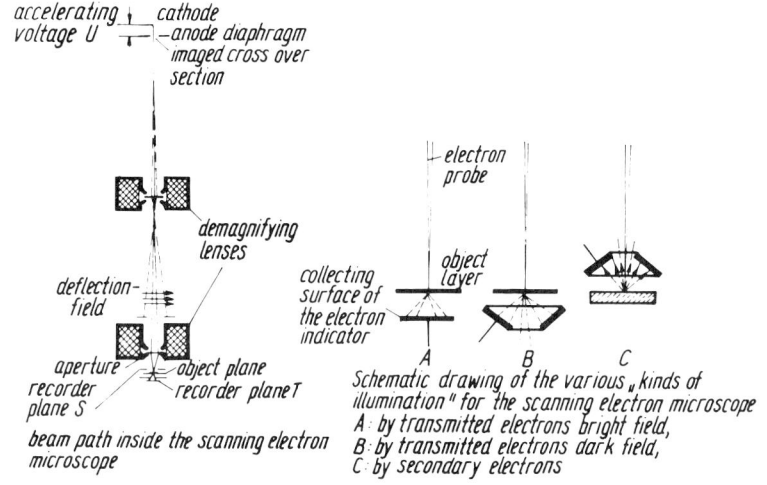

FIGURE 2. Ray diagram of an SEM and three principal modes for detecting electrons.

department to sort out the documents of the Siemens Research Laboratories. In translation, this curious memorandum on SEM by von Borries and Ruska reads as follows:

ZENTRAL LABORATORIUM
Spandau Nord, 30 June 1937
MEMORANDUM by Dr. v. Borries/Dr. Ruska (in German)
Confidential!
Comparison of the electron scanning microscope with the electron microscope
Summary.
The resolving power of the electron scanning microscope (SEM), on the purely energetic grounds of beam production, with present technology is limited to 0.05 micrometer for a one-off photograph or 1 micrometer for continuous observation. The latter rules it out completely. For one-off exposures one might, with successful development work and reduced requirements, achieve perhaps 0.01 micrometer. This limitation of the resolution by these difficulties of setting up the beam does not exist for the electron microscope (TEM). The reason for this difference lies in the fact that similar ray paths, geometrically inverted, are not energetically reversible. For the same resolution, the specimen loading, integrated over the exposure time, is 100 times as great in the SEM as in the TEM. The specimen loading in the SEM at the moment the beam crosses a given point on the specimen, which is likely to be decisive in damaging the specimen and disturbing the optics by local charging up effects, is, on the other hand, 1.5 million times as great as that at a specimen point in the TEM. The reason for these differences is that for all amplifier arrangements, a minimum input energy density is set by the noise level, while the photographic plate integrates over an arbitrary low energy density and moreover has a very low energy requirement. The resolu-

tion of the TEM, compared with that of the SEM, is also less affected by chromatic aberration because of these energy difficulties.

Despite the negative evaluation of SEM by von Borries and Ruska, but thanks to the far-sighted views of Prof. H. Küpfmuller, Chief Physicist at Siemens, a legal agreement, based on our SEM patents and giving further

---

ALLABORATORIUM

Ltw.

Verteiler:

Büro Dr.H.v.Siemens
Herr Dir.Dr. Lüschen
Herr Prof. Küpfmüller
Herren Dr.Mayer/Dr.Tamm
ZL. 44 2x
Akten
Herrn Prof.Dr.Hertz

Spandau-Nord, den 30. Juni 1937.
ZL. 188/44/v.Bs./Rk.

Bearbeiter: Dr.v.Borries/Dr.Ruska
Anruf : 01400
Anlagen: 1 Weißpause

V e r t r a u l i c h !

Aktenvermerk über

Vergleich des Elektronen - Rastermikroskops mit dem Elektronenmikroskop

Zusammenfassung:

Das Auflösungsvermögen des Elektronen - Raster - Mikroskops (E R M) ist schon aus rein energetischen Grunden der Strahlenerzeugung nach dem augenblicklichen Stande der Technik für eine einmalige Aufnahme mit 0,05 μ für dauernde Beobachtung mit 1 μ begrenzt. Letztere scheidet also vollständig aus. Für einmalige Aufnahme wird man bei erfolgreichen Entwicklungsarbeiten und geschmälerten Anforderungen vielleicht 0,01 μ erreichen. Diese Begrenzung des Auflösungsvermögens durch die Schwierigkeiten der Strahlherstellung besteht für das Elektronenmikroskop (E M) nicht. Der Grund für diese Verschiedenheit liegt darin, dass geometrisch umgekehrt ähnliche Strahlengänge energetisch nicht umkehrbar sind.

Für gleiche Auflösung ist die über die gesamte Aufnahmezeit integrierte Objektbelastung beim E R M 100 mal so hoch wie beim E M. Die im Augenblick des Überstreichens eines Gegenstandspunktes beim E R M auftretende Belastung, welche sehr wahrscheinlich für die Zerstörung des Objekts und die optischen Störungen durch örtliche Aufladung massgebend ist, ist dagegen $1{,}5 \cdot 10^6$ mal so hoch wie die Belastung eines Gegenstandspunktes im E M. Der Grund für diese Verschiedenheiten liegt darin, dass bei allen Verstärkeranordnungen eine Mindestenergiedichte durch den Rauschpegel bedingt ist, während die photographische Platte über beliebig geringe Energiedichten integriert, und überhaupt nur sehr geringen Energiebedarf hat. Das Auflösungsvermögen des E M ist durch den chromatischen Fehler erst später begrenzt als der des E R M durch die energetischen Schwierigkeiten.

FIGURE 3. "Confidential" report of 30 June 1937 from Siemens about the future prospects of SEM.

support for our research, was nevertheless negotiated with the Siemens Company. In fact, from 1936 to as late as 1945 there existed between Siemens and myself an agreement about patents and the development of the scanning electron microscope, which provided payments, if I remember correctly, to the extent of around 3000 Marks per month. I also received a monthly support at a similar level from the Deutsche Forschungsgemeinschaft, through the good offices of Prof. Thiessen of the Kaiser-Wilhelm-Institut für Physikalische Chemie at Dahlem. This support enabled me to pursue the development of both the scanning and the normal electron microscope, until the end of the war in 1945. It was indeed fortunate that Küpfmuller sensed the future importance of the scanning electron microscope and the agreement was kept going. One of the terms of the agreement was an understanding of the use of my patents by Siemens. Part of the document is shown in Fig. 4. In translation, the relevant clauses read as follows:

> Ardenne grants Siemens & Halske in the Agreement on protection rights an exclusive License for all application areas of the Agreement protection rights, with the exception of television and oscillographic measurement technology, with the right of granting subsidiary licenses.
> However, Ardenne reserves to himself:
>
> 1. . . .
> 2. . . .
> 3. To grant a simple licence for the purpose of production of electron scanning microscopes in the United Kingdom of Great Britain and Northern Ireland and of the marketing of these goods in any lands of the British Empire. . . .

Item 3 was concerned mainly with protecting my patent rights in the British Empire. In this connection, it is interesting to mention that the Stereoscan, developed some 20 years later by C. W. Oatley and his team, in conjunction with the Cambridge Instrument Company, was, in fact, the real starting point for the commercial development of the SEM.

It should perhaps be mentioned that in this agreement with Siemens, Ernst Ruska had inserted a clause that any development work on my part in connection with the transmission microscope was forbidden. Ruska wished, in this way, I suppose, to ensure his scientific monopoly. At the time, I found it an unethical and unfair part of my Siemens electron microscope agreement and, in fact, I disregarded this ban after 1939. By my ongoing inventions, I certainly enhanced the performance of the Siemens TEM. The Siemens agreement was so drafted, however, that in spite of the ban on my doing TEM research, ideas of mine about the transmission instrument could be, and were in fact, applied to the Ruska microscope. For example,

§ 2

A erteilt S & H an den Vertragsschutzrechten eine ausschliessliche Lizenz für alle Anwendungsgebiete der Vertragsschutzrechte - mit Ausnahme des Fernsehens und der oszillographischen Messtechnik - mit dem Recht zur Erteilung von Unterlizenzen.

Jedoch behält A sich vor,

1.) seinen etwaigen Lizenznehmern an den englichen Patenten, die gemäss § 1, Absatz 4, keine Vertragsschutzrechte sind, zum Zwecke der Ausfuhr der geschützten Elektronen-Raster-Mikroskope in andere zum Britischen Imperium gehörende Länder die Mitbenutzung der etwa in diesen Ländern bestehenden entsprechenden Vertragsschutzrechte zu gestatten;

2.) für die Zwecke der Herstellung und des Vertriebes von Elektronen-Rastermikroskopen in USA eine einfache Lizenz an den USA-Schutzrechten anderweitig zu vergeben, die den in Anlage 1 unter Ziffern 4 bis 6 aufgeführten Schutzrechtsanmeldungen oder der DRP-Anmeldung A 84 185 entsprechen;

3.) für die Zwecke der Herstellung von Elektronen-Raster-Mikroskopen im Vereinigten Königreich von Grossbritannien und Nordirland und des Vertriebes dieser Gegenstände in beliebige Länder des Britischen Imperiums eine einfache Lizenz an den etwa im Britischen Imperium bestehenden Schutzrechten anderweitig zu vergeben, die den in Anlage 1 unter Ziffern 4 bis 6 aufgeführten Schutzrechtsanmeldungen oder der DRP-Anmeldung A 84 185 entsprechen.

FIGURE 4. Extract from the "Terms of Agreement . . ." between von Ardenne and the Siemens Company.

by the reduction of the vibration sensitivity of the Ruska design, the resolution was improved by a factor of 4.

Eventually, in exasperation over the above situation, I decided in 1938, notwithstanding the ban in the Siemens agreement, to build at my Institute in Berlin Lichterfelde a TEM for bright-field, dark-field, and stereo operation. This was completed at the end of 1939 and produced the highest resolution at that time from many characteristic microobjects. These images were published in *Naturwissenschaften,* without previous communication with Siemens, at the beginning of 1940, and then soon after in my book, *Elektronenübermikroskopie* (1940a).

Through my constructional principle of "sideways" exchange of the objective-specimen system, I was able, in short order, to implement important improvements and thus succeed in competing with Siemens/Ruska. Stereo-electron microscopy came into being with images of extremely large depth of focus. Likewise, electron microscopy with specimen heating up to 2000°C, the biological cell objective for observing living processes, e.g., the imaging of the nucleation of spores, was achieved. There was also an objective for imaging specimen reactions and later the image cinematography of chemical reactions. All these important practical improvements brought benefits, within the terms of the agreement, to the Ruska design. I regret to say that E. Ruska neither mentioned these facts nor my name in his Nobel Prize lecture.

To realize and optimize both variants of the scanning electron microscope, SEM and STEM, only a few components and new constructional principles had to be created from scratch. Details, including a wide-ranging list of historical references, are given in my 1985 survey article, to which the interested reader is referred.

Of particular importance for SEM and STEM was the availability of the magnetic polepiece lens of short focal length of von Borries and Ruska (Ruska and von Borries, 1932; Ruska, 1934). A further improvement of this system, resulting in the single-field condenser-objective, gave increased resolution (von Ardenne, 1944). The system is shown in Fig. 5. Our construction allowed the entire polepiece unit to be changed through the side of the column. In addition to the system described, there were already at that time other polepieces having larger gaps or bores, whose dimensions were similar to those in use today. By means of such lenses, resolutions of 1.2 nm could be realized in TEM in 1943. Other construction units that were needed and the relevant technological principles could be taken over from related technology, such as high- and low-voltage electron beam oscillography, TV image convertors, and vacuum technology. The following devices should be listed here:

1. Three-electrode array with beam crossover
2. Electron beam scanning and image formation by horizontal line rasters and parallell deflection systems (von Ardenne, 1931)
3. Postacceleration of low-energy electrons (around 1 eV) from the object to about 10 keV and conversion of their energy by highly efficient phosphors
4. Photomultipliers for wide-band amplification of weak light (fluorescence) signals
5. Photographic chambers with airlocks

In several publications reviewing the development of SEM, for example, by Reimer and Pfefferkorn (1973) or Schneider *et al.* (1977), the simple

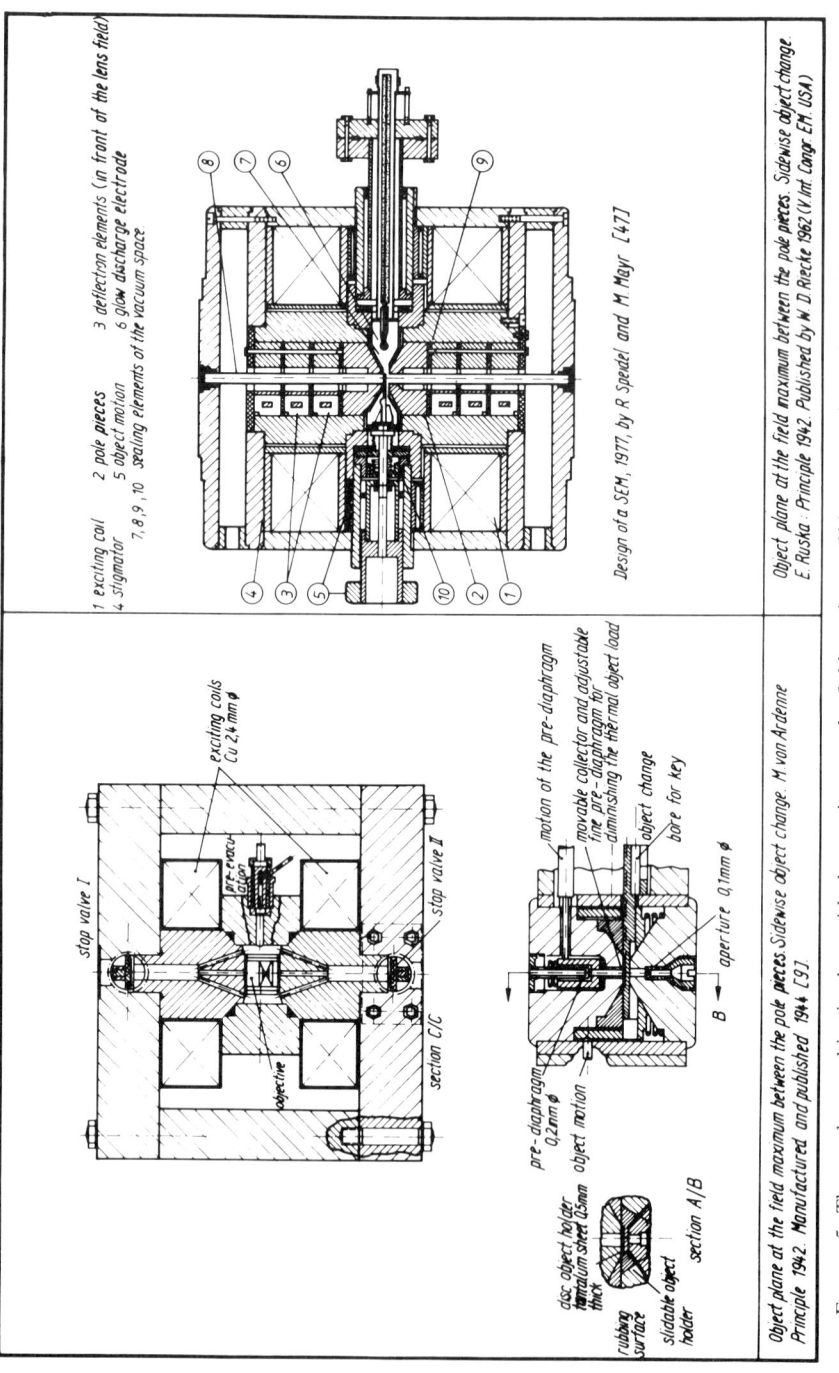

FIGURE 5. The condenser-objective lens with the specimen at the field maximum. Sideways specimen change. Manufactured and published by the author in 1944.

imaging of specimens by means of secondary electrons emitted from the surface during exposure to an electron beam is deemed to be the forerunner or even the main principle of SEM. One of our original micrographs of 1933 is shown in Fig. 6. At that time, surface structures such as scratches and lines inscribed on a light-sensitive semiconductor plate illuminated at glancing incidence were made visible (predecessor of the Vidicon camera).

In Fig. 7 the title page of the first contribution on SEM, published in *Zeitschrift für Physik* (von Ardenne, 1938) is shown. A photograph of the first SEM for imaging surfaces is shown in Fig. 8. A scanning electron beam of some 50–100 nm was produced by demagnifying the crossover with two magnetic polepiece lenses. The X–Y deflection took place immediately in front of the second magnetic lens. The video signal was obtained from a high-performance, low-capacitance collector system for the secondary electrons emitted from the specimen. The signal was further amplified by means of a wide-band TV amplifier. For direct viewing, a TV tube with a long-persistence screen was used, as seen on the left-hand side of Fig. 8. TV scanning rates were used for low resolution; for high resolution, strongly reduced numbers of lines, and hence frequencies, together with photographic recording, were used. By appropriately aperturing down the second demagnifying lens, its spherical aberration disk could be matched to the

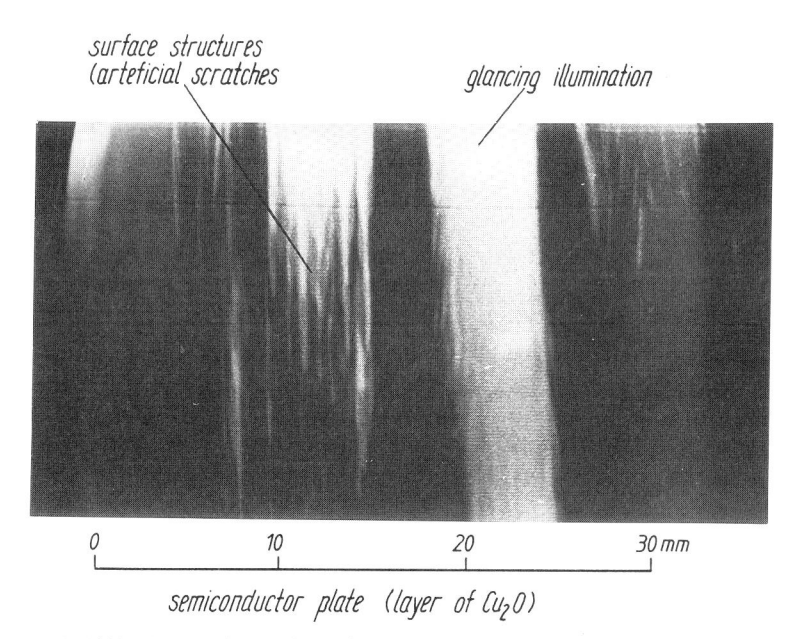

FIGURE 6. 1933 micrograph of a simple image formed by electrons emitted from a surface.

ZEITSCHRIFT FÜR
PHYSIK

HERAUSGEGEBEN UNTER MITWIRKUNG
DER
DEUTSCHEN PHYSIKALISCHEN GESELLSCHAFT

VON

H. GEIGER

Sonderabdruck 109. Band. 9. und 10. Heft

Manfred von Ardenne
Das Elektronen-Rastermikroskop.
Theoretische Grundlagen

VERLAG VON JULIUS SPRINGER, BERLIN

1938

FIGURE 7. Title page of the first contribution on SEM (1938).

small spot diameter required. This needs a small aperture, a feature distinguishing electron from light microscopy. Already in our earlier observations, when we had purposely selected diatoms as specimens because of their large axial extension, we were strongly impressed by the unusually large depth of focus of this method of imaging. Stimulated by the possibility of imaging spatial objects with extremely large depths of focus, I soon turned to stereo-electron microscopy (von Ardenne, 1940b, 1940c), which likewise concerns the sharply focused imaging of spatially extended objects. In Fig. 9 a tableau of the first stereo images taken in an electron microscope is set out. These were taken in the winter of 1940/1941 and were first published in my comprehensive book (von Ardenne, 1940a) (see Fig. 10). Unfortunately, studies on this method of imaging surfaces could not be continued, because, in fulfilment of other contractual commitments, we had to realize scanning transmission microscopy (STEM) using the same laboratory setup. Thus the first STEM was bult in 1938 (Fig. 11). At that time this type had priority for us because there was a chance that, even in

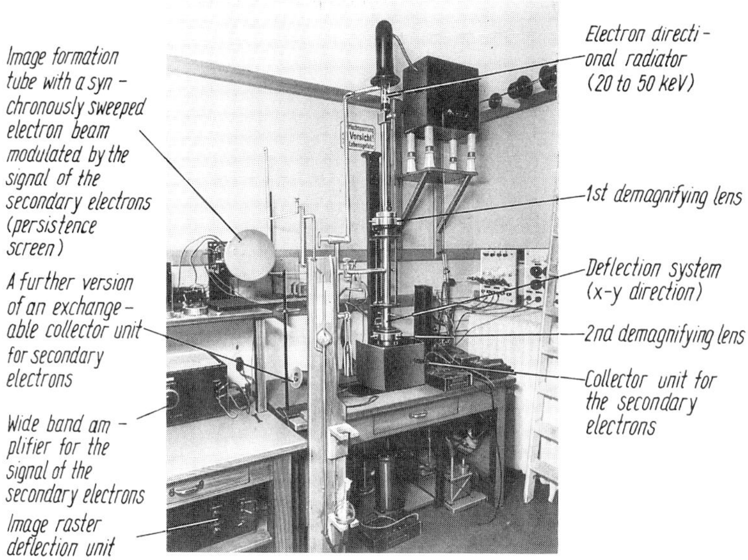
Image formation tube with a synchronously sweeped electron beam modulated by the signal of the secondary electrons (persistence screen)
A further version of an exchangeable collector unit for secondary electrons
Wide band amplifier for the signal of the secondary electrons
Image raster deflection unit
Electron directional radiator (20 to 50 keV)
1st demagnifying lens
Deflection system (x-y direction)
2nd demagnifying lens
Collector unit for the secondary electrons
Vorsicht!

FIGURE 8. The first SEM for imaging surfaces.

investigations of relatively thick specimens, such as standard microtome sections, chromatic aberration could be kept low. In the initial studies with STEM, the electrons leaving the specimen were recorded photographically. In the interests of high resolution, the integration time per pixel could be made suitably long by reducing the speed of the photographic drum. The mechanical stability needed was obtained by pressing the specimen holder firmly onto the polepiece of the second demagnifying lens. The photographic recording system was described in detail (von Ardenne, 1938b, 1940a). In later investigations, electron collectors with photomultipliers of different types were introduced or designed for bright- and dark-field operation (von Ardenne, 1938b, 1940a).

About that time, preliminary experiments were carried out aiming at the use of an electron probe for microanalytical purposes with simultaneous imaging of the specimen. Figure 12 shows a microbeam diffraction pattern of gold foil (von Ardenne *et al.,* 1942).

In 1939, matters concerning the offending clause in the agreement came to a head. The existence and the power of my TEM, known as the Universal Electron Microscope, for bright-field, dark-field, and stereo electron microscopy at the beginning of 1939 was a great surprise for the higher management of Siemens. Dr. Hermann von Siemens visited my laboratory in Lichterfelde, as I described in my autobiography, *Reminiscences.* During this visit, I put it to him that he must make a decision: Either accept my

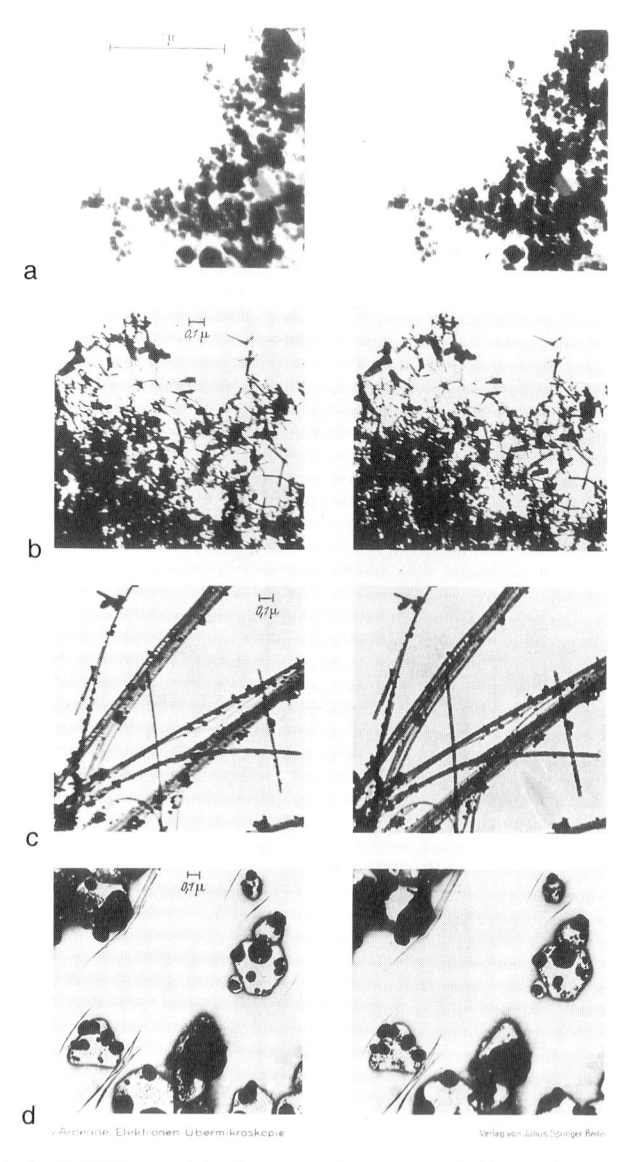

FIGURE 9. (a–d) Tableaux of the first stereo images recorded in an electron microscope (taken in the winter of 1940/1941).

infringement of the agreement by my working on TEM and continue the monthly payments, or, in view of the improvements made by me in my own instrument, take out a compulsory License (Zwangs Lizenz) for the

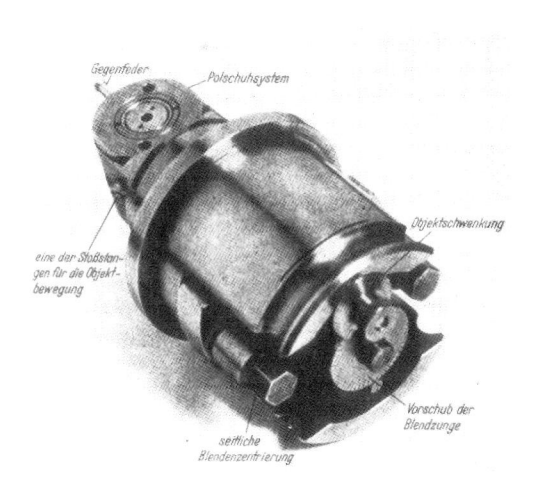

FIGURE 10. Stereo-attachment of the Universal Electron Microscope, published in 1940.

Borries/Ruska objective lens system. The Siemens management soon decided on a continuation of the agreement.

In the war years of 1940–1944, our electron microscope studies went at full speed. Supported by Siemens in Berlin and by Krupp at Essen, we worked all out in this promising field, as documented in Fig. 13. The outcome was our contribution to high-temperature EM (von Ardenne, 1941a), reaction-chamber EM (von Ardenne, 1942), electron microscopy of living matter (von Ardenne, 1941b), and electron vacuum microcinematography

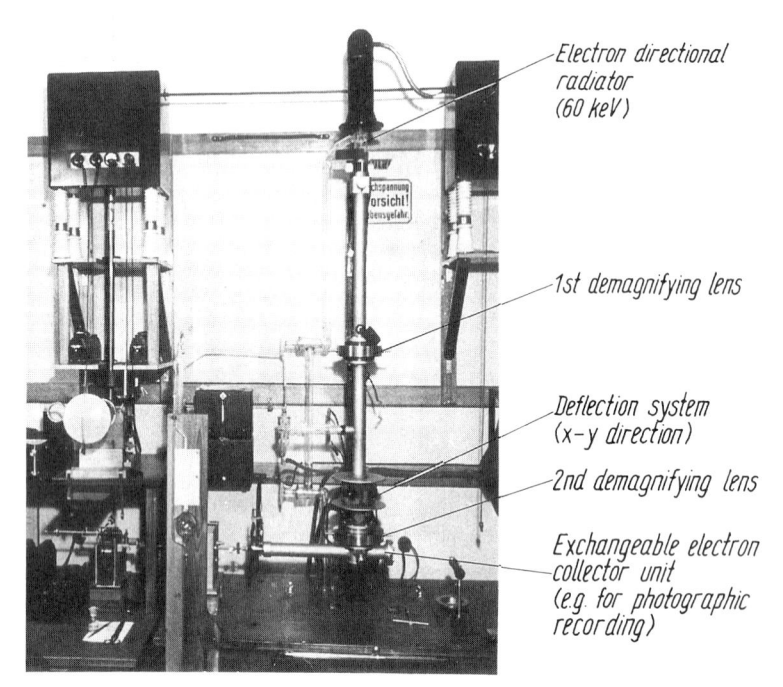

FIGURE 11. The first STEM, built in 1938. Electrons leaving the specimen were recorded directly on a film wrapped round a rotating drum, mounted below the specimen.

(von Ardenne, 1943a). Our developmental work and preparations for the construction of electron microscopes working at accelerating voltages of 200 to 1000 kV (von Ardenne, 1941c, 1943b) should be mentioned here. The wedge-cut microtome as a predecessor of the later ultramicrotomes (von Ardenne, 1939) was then a necessary preparative supplement for our work aiming at superhigh-voltage electron microscopy.

In those very busy years, several new electron microscopic investigations were carried out jointly with well-remembered friends such as H. Friedrich Freska, G. Schramm, and H. H. Weber. The precipitin reaction was studied by osmium tetroxide staining (von Ardenne et al., 1941). The filamentary structure of the muscle protein myosin (von Ardenne and Weber, 1941) and the fibrous structure of exposed and developed silver bromide grains (von Ardenne, 1940d) were detected. Especially important at that time was the discovery of the fibrous structure of developed silver bromide grains by stereo electron microscopy in my instrument. It was the discovery of a fundamental process in photography. Up to the end of the war there was a successful, close collaboration with the Dahlem Kaiser-Wilhelm-

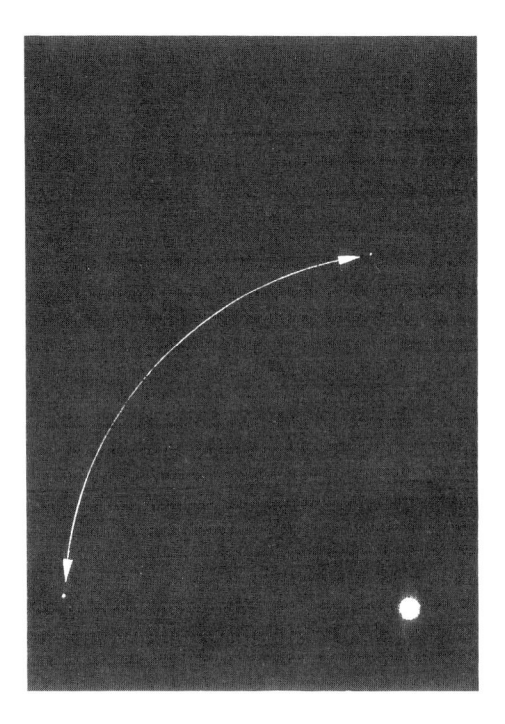

FIGURE 12. Microbeam diffraction pattern of a gold leaf with a 1-μm-diameter electron probe. Single crystal pattern; two of the reflections are marked. Accelerating voltage 140 kV. Exposure time 10 s. [Micrograph: M. von Ardenne, *Z. Phys.* **119**, 352 (1942).]

Institutes for Physical Chemistry, for Cell Physiology, and for Chemistry, which resulted in many publications. My book on electron microscopy (von Ardenne, 1940a) mentioned above, was published during the war, not only in the United States but also in Japan and the Soviet Union. Its appearance in Japan led to the sudden and rapid development of the Japanese electron microscope industry.

Sadly, these research activities in both variants of scanning electron microscopy, in the introduction of electron microprobes into microanalysis, and in superhigh-voltage microscopy, came to a violent end on the night of 25 March 1944. A British air raid destroyed all the devices mentioned above, including the 1000-kV generator, on that fateful night. When, in May 1945, on the initiative of Academician A. Joffe, the government of the Soviet Union invited me to establish and direct a research institute at Sukumi, it was decided that I should continue my work on the scanning electron microscope, the "universal" electron microscope, and the superhigh-voltage microscope, which had been interrupted by the bombardment of our Institute in Berlin-Lichterfeld. However, this situation was

FIGURE 13. Wartime (1940–1944) development of a high-voltage TEM. Prototype for TEMs working above 200 kV.

changed completely—within an hour, in fact—by the dropping of the American atomic bomb on Hiroshima in August 1945. By order of the highest-ranking authorities, the operational direction of the Sukumi Institute was newly defined and confined strictly to the development of industrial isotope separation, especially for the production of the fissionable material U-235. Many of my old friends from those early years of electron microscopy will have understood that, in 1945, to my great sorrow, I had to leave this field just when it began to bloom. Some members of the German Society of Electron Microscopy, however, misinterpreted my actions as an abandonment of my old speciality, because they could not be aware of the real, fateful situation. Their attitudes rendered the necessary changes even more difficult for me to make.

All these reminiscences lose their significance, however, in view of the fact that our attempts at the development of large-scale isotope separation were finally successful and hastened the attainment of nuclear parity between East and West, and hence made a contribution to nuclear world peace.

References

von Ardenne, M. (1940a). "Elektronen-Übermikroskopie." Springer, Berlin. (During World War II, Russian, English, and Japanese translations were published in the USSR, the United States, and Japan.)
von Ardenne, M. (1985). On the history of scanning electron microscopy, of the electron microprobe, and of early contributions to transmission electron microscopy. In "The Beginnings of Electron Microscopy" (P. W. Hawkes, ed.), p. 1. Academic Press,
Ruska, E., and von Borries, B. (1932). German patent application B. 154916.
Ruska, E, (1934). Über ein magnetisches Objektiv für das Elektronenmikroskop. *Z. Phys.* **89,** 90.
von Ardenne, M. (1944). Über ein neues Universal-Elektronenmikroskop mit Hochleistungsmagnet-Objektiv und herabgesetzter thermischer Objektbelastung. *Kolloid-Z.* **108,** 195.
von Ardenne, M. (1931). Über neue Fernsehsender und Fernsehempfänger mit Kathodenstrahlröhren. *Fernsehen* **2,** 65.
Reimer, L., and Pfefferkorn, G. (1973). "Raster-Elektronenmikroskopie." Springer, Berlin, Heidelberg, New York.
Schneider, V., Schwarz, W., Dunger, B., and Bahr, J. (1977). Rasterelektronenmikroskopie. *Pressedienst Wissenschaft FU Berlin* **2,** 306.
von Ardenne, M. (1938a). Das Elektronen-Rastermikroskop. Theoretische Grundlagen. *Z. Phys.* **109,** 55.
von Ardenne, M. (1940b). Stereo-Übermikroskopie mit dem Universal-Elektronenmikroskop. *Naturwissenschaften* **28,** 248.
von Ardenne, M. (1940c). Über ein Universal-Elektronenmikroskop für Hellfeld-, Dunkelfeld- und Stereobildbetrieb. *Z. Phys.* **115,** 339.
von Ardenne, M. (1938b). Das Elektronen-Rastermikroskop. Praktische Ausführung. *Z. Tech. Phys.* **19,** 407.
von Ardenne, M., Schiebold, E., and Cünther, F. (1942). Feinstrahl-Elektronenbeugung im Universal-Elektronenmikroskop. *Z. Phys.* **119,** 352.
von Ardenne, M. (1941a). Erhitzungs-Übermikroskopie mit dem Universal-Elektronenmikroskope. *Kolloid. Z.* **97,** 257.
von Ardenne, M. (1942). Reaktionskammer-Übermikroskopie mit dem Universal-Elektronenmikroskop. *Z. Phys. Chem. (B)* **52,** 61.
von Ardenne, M. (1941b). Elektronen-Übermikroskopie lebender Substanz. *Naturwissenschaften* **29,** 521 and 523.
von Ardenne, M. (1943a). Elektronenmikrokinematographie mit dem Universal-Elektronenmikroskop. *Z. Phys.* **120,** 397.
von Ardenne, M. (1941c). Über ein 200 kV-Universal-Elektronenmikroskop mit Objektabschattungsvorrichtung. *Z. Phys.* **117,** 657.
von Ardenne, M. (1943b). Über eine Atomumwandlungsanlage für Spannungen bis zu 1 Million Volt. *Z. Phys.* **121,** 236.
von Ardenne, M. (1939). Die Keilschnittmethode, ein Weg zur Herstellung von Mikrotomschnitten mit weniger als 10^{-3} mm Stärke für elektronenmikroskopische Zwecke. *Z. Wiss. Mikroskopie* **56,** 8.

von Ardenne, M., Friedrich-Freska, H., and Schramm, G. (1941). Elektronenmikroskopische Untersuchung der Präcipitinreaktion von Tabakmosaikvirus mit Kaninchenantiserum. *Arch. Ges. Virusforsch.* **II,** 80.

von Ardenne, M, Weber, H. H. (1941). Elektronenmikroskopische Untersuchung des Muskeleiweiβkörpers "Myosin." *Kolloid-Z.* **97,** 322.

von Ardenne, M. (1940d). Analyse des Feinbaues stark und sehr stark belichteter Bromsilberkörner mit dem Universal-Elektronenmikroskop. *Z. Angew. Photogr.* **2,** 14.

4.5
Electron Microscopes and Microscopy in Japan
4.5A Electron Microscope Development at Hitachi in the 1940s

TSUTOMU KOMODA

Central Research Laboratory
Hitachi, Ltd., Kokubunji
Tokyo 185, Japan

Electron microscope history began at Hitachi, Ltd., in 1939, when Hitachi took part in a development project of the Japan Society for Promotion of Science, to build electron microscopes in Japan. The first instrument was designed by K. Kasai, and was handed over to B. Tadano for experiment. They were experienced cathode-ray oscillograph engineers. This was coincident with Knoll and Ruska, who developed the first electron microscope in the 1930s. They also were cathode-ray oscillograph engineers.

The first electron microscope at Hitachi was completed in 1941 and was named H^{T}I-1. It had electromagnetic lenses and a horizontal column as shown in Fig. 1. It was built on an optical bench. There were no data or know-how to design an electron microscope. The engineers had to design and build the first instrument based on their knowledge of cathode-ray oscillographs. They ran into many technical as well as engineering difficulties before completing the microscope.

For example, the column was made of a bronze casting, so vacuum leak problems frequently occurred. The horizontal column posed a difficulty in aligning the beam and lenses, and it was quite sensitive to vibrations. One day, Tadano excitedly jumped out of a dark room, having succeeded in recording a sharp-quality image. It was already midnight and there was no one left in the entire building. It was through this experience that Tadano understood the problem of vibration, which was the primary cause of image disturbance!

Figure 2 is an enlarged part of a diatom shell recorded with the first microscope. It had a resolving power permitting a high magnification of about 20,000×.

Later, in 1941, K. Kasai and B. Tadano started designing the next electron microscope, based on the precious data and know-how obtained from the HU-1. A vertical column was employed to permit better alignment and

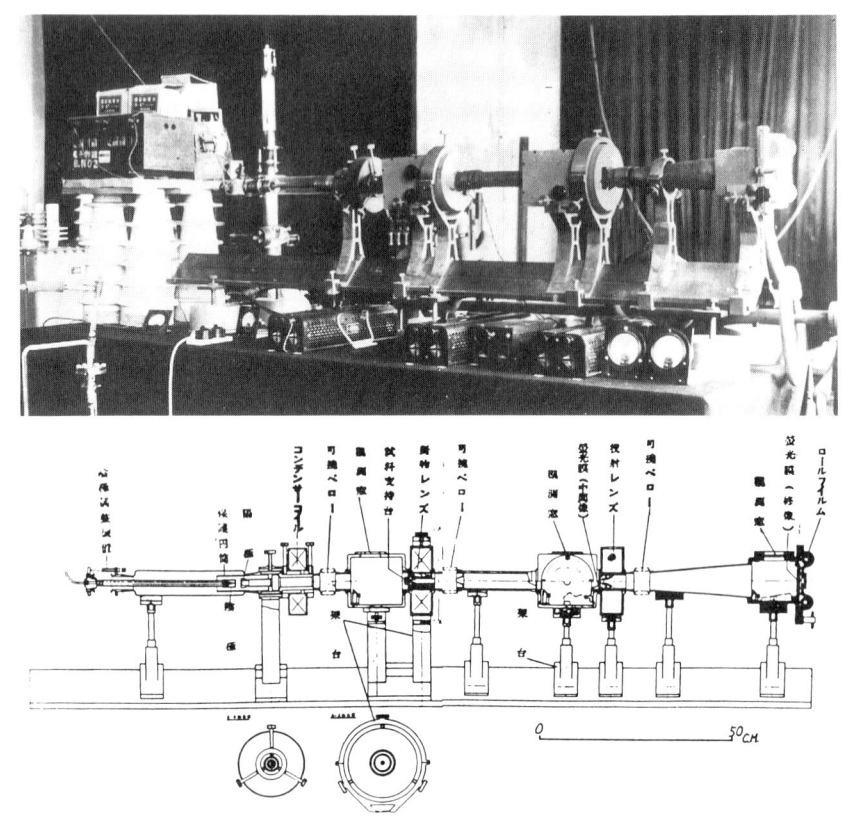

FIGURE 1. General view and cross section of the first electron microscope, HU-1, built at Hitachi (1941).

better stability against vibrations. A permalloy magnetic shield was placed along the electron beam path so that the column might be less susceptible to magnetic field disturbances. In order to achieve a better vacuum, sheet metal was used instead of cast materials.

This instrument was completed in 1942, and it was called HU-2 (Fig. 3). Two units of this model were manufactured. One was installed at the Central Research Laboratory of Hitachi, and the other at Prof. Sakaki's laboratory in the Engineering Department of Nagoya University, which was the first electron microscope installed at a customer's site in Japan.

The HU-2 had many superior points over the HU-1. High-voltage stability, mechanical stability against vibration, and electron optical lens performance were some of them. It was easy to record images at several ten thousand times. The objective lens was built with a very high mechanical accuracy and it had very low astigmatism even without any correcting

MICROSCOPES AND MICROSCOPY IN JAPAN 655

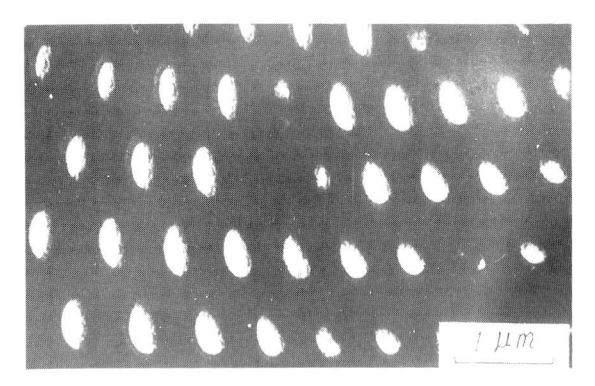

FIGURE 2. A part of a diatom shell taken with the HU-1 (1941).

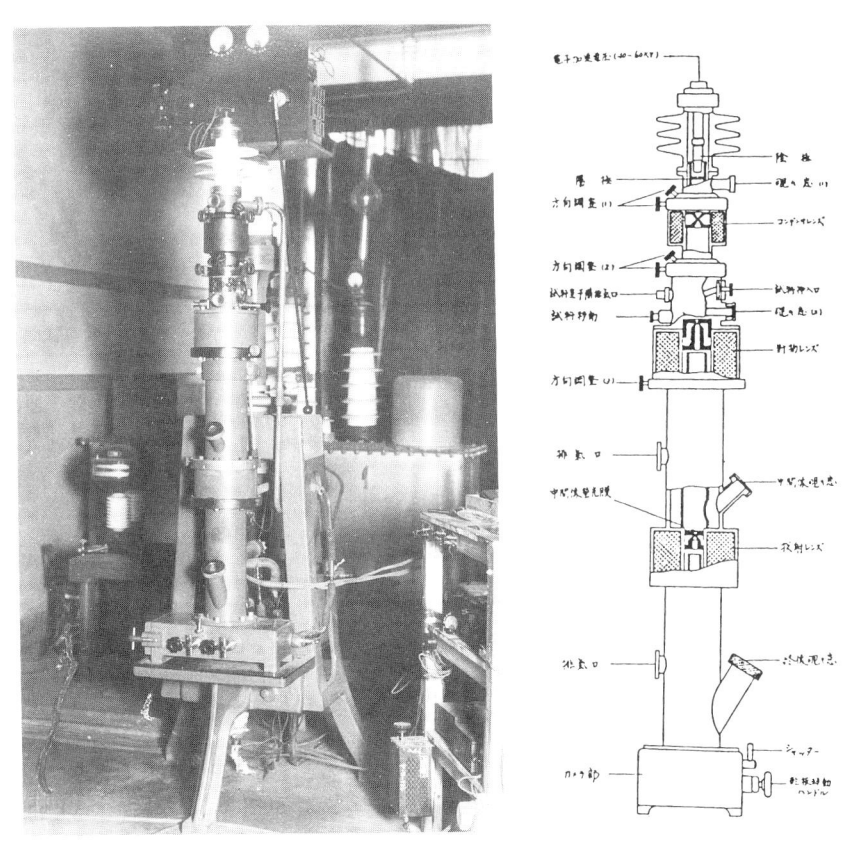

FIGURE 3. General view and cross section of HU-2 (1942).

devices. Many researchers at that time believed that spherical aberration was the most crucial factor for the microscope resolution. Tadano *et al.* already at this time believed that the astigmatism due to mechanical or machining inaccuracy was even more important. The objective lens polepiece of the HU-2 employed an inlaid metalwork in which all components were precisely machined and finely put together as a final structure. This precise machining accuracy for the lens laid the foundation for high-resolution electron microscopy at Hitachi for many years to come.

Figures 4, 5, and 6 are typical micrographs recorded with the HU-2. Specimens are cubic crystals of magnesium oxide, carbon black, which is used as a filler for rubber, and proteus vulgaris, a kind of bacteria. All of these micrographs are good-quality images even by today's standards. Figure 6 was in particular the most memorable one, in that it showed flagelli without staining, which gave many biological and medical scientists a very good impression about electron microscopy.

From 1944 to 1945, Tokyo was heavily attacked by the U.S. Air Force almost every night, and the Central Research Laboratory of Hitachi in Kokubunji, where the HU-2 was installed, was becoming a dangerous area. The engineers were advised to move to a mountainous area near the Japan Alps. In view of the anticipated delay in microscope research, Tadano vigorously rejected the move. Many sandbags were prepared and placed around the HU-2 to protect it from the bombing, and the researchers continued their work wearing steel helmets! Their correct decision and enthusiastic effort for developing technologies for electron microscopes made a great contribution to the uninterrupted development of Hitachi's electron microscopes even during the destructive period in and after World War II.

The present electron microscope technology at Hitachi is a great flower in bloom on the foundations laid by many forerunners during the 1940s.

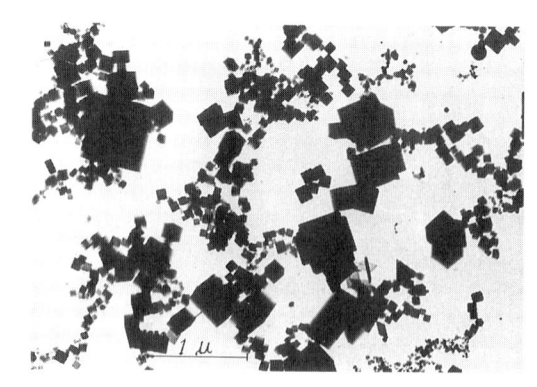

FIGURE 4. Magnesium oxide crystals taken with HU-2 (1943).

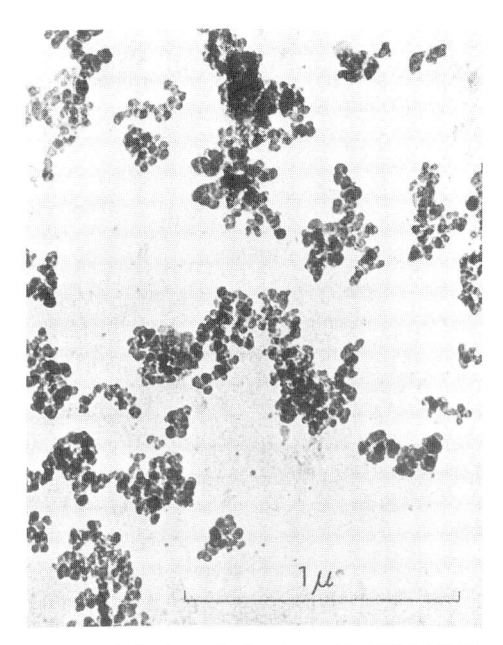

FIGURE 5. Carbon black taken with HU-2 (1943).

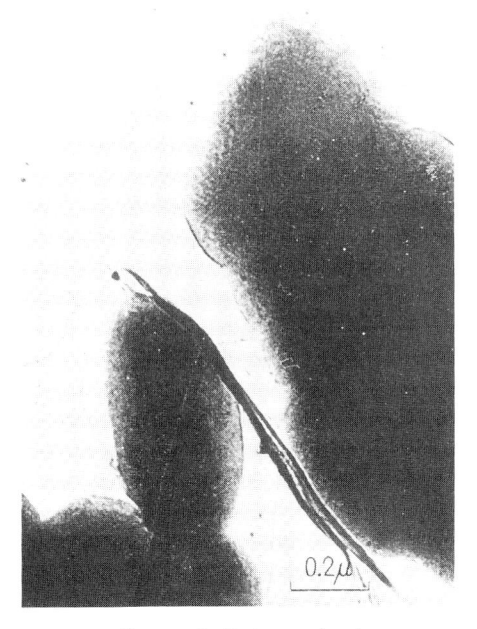

FIGURE 6. Proteus vulgaris.

4.5
Electron Microscopes and Microscopy in Japan
4.5B Development of the Electron Microscope at JEOL

KAZUO ITO

*JEOL, Ltd.
Nakagami, Akishima
Tokyo 196, Japan*

The founder of JEOL (Nihon Denshi Kabushiki-kaisha), Kenji Kazato, was one of the former Naval officers who were freed after the defeat in World War II. With painful memories of the war, he returned to his native town, Mobara, situated 60 km west of Tokyo. Although the entire society was extremely demoralized and quite at a loss as to what to do, his young and vigorous brain was not in this state. He searched continuously for a way to make use of his experiences in order to reestablish Japan from its devastated situation to an acceptable level, especially by means of a technological approach. Among numerous kaleidoscopic recollections during the war, the technological gap between the United States and Japan was very evident—for example, the United States' reliable radar, powerful airplanes, silent submarines, not to mention nuclear weapons—that it was inconceivable to build a new society without improving the technological base.

While sorting out his belongings, he came upon a small book titled *Denshikenbikyo* (*Electron Microscope*), written by Daisuke Kuroiwa, in 1942. This was like a gift from heaven, and he immediately realized that the electron microscope was the fundamental instrument for studying the physical characteristics of materials, which perfectly suited his purpose. This was the start of JEOL, now realizing nearly a half a billion dollars worth of business annually. At JEOL, this book by Dr. Kuroiwa is kept as a sacred souvenir of the founder.

Kenji Kazato started his business from scratch, with no money, no factory, and no people. The first things he did was to recruit key personnel. They came mainly from among his aquaintances in the Navy. Fortunately, his last assignment had been at the Naval Technical Research Center, where he was engaged in the early stage development of guided missiles.

Kazuo Ito, a physicist and later third President of JEOL, Kanichi Ashinuma, a mechanical engineer and former Executive Director, and

Kanjiro Takahashi, an electronics engineer and Former Executive director, were numbered among them. Though their backgrounds were all different, they had much in common. First of all, they knew nothing about the electron microscope before joining this group; however, they were all young, ambitious, extremely hard workers, and eager to learn; and they had nothing to lose.

When I recall the history of JEOL, I realize that the key to success is people, and JEOL was very fortunate to have these people together under the strong leadership of Kenji Kazato. At the same time, he found a building which had been used by the Naval Air Force. It was good both as a workshop and as lodging for his people. This was just after the war, and he got it under a cheap rental contract from the government. Actually, no one wanted this building at that time. He also got some machine tools from the Naval warehouse.

In order to get cash for day-to-day survival, we did many things, such as making saccharine, a substitute for scarce sugar, making imitation soya sauce, etc. This was April 1946, which was the start of JEOL.

As a physicist, my first task was to design the general specifications for the electron microscope. The only serious book available was *Ubermikroskopie,* written by von Ardenne; there was also a library operated by the Occupational Forces at Hibiya, Tokyo, where recent scientific journals and books were available to the public. I went there regularly with pen and notebook and copied them out by hand, word for word, in the notebook. Even with all this information, lens design was still very tedious and time-consuming work, because we had to do it all on a manual Tiger calculator! Today this can be done within a matter of minutes with much higher precision with the aid of a computer.

By trial and error, we arrived at the first commercial instrument, DA-1, shown in Fig. 1, only one and a half years after the start of development at Mobara. A photograph of the DA-1 is shown in Fig. 2.

This achievement—the success of electron microscope production by a few scientists in a remote town—became an object of public attention. It was picked up by newspapers as an example of a bright spot in a hard life after the defeat in war, and it gained the attention of the Royal Family. The present Emperor and then Crown Prince visited Mobara and saw our microscope when he was a middle-school student. Further, in July 1948, the DA-1 was exhibited at Tokyo, and the Emperor Showa was invited to study his sample, yeast, under the electron microscope.

The first commercial DA-1 was delivered to the Mitsubishi Chemical Industry Corporation at the end of 1946 and played a very important role in the study of ion-exchange resins. However, it was no better, in reality,

FIGURE 1. First transmission electron microscope, the DA-1, which was produced by Denshi Kagaku in 1947.

than the prototype, and the attainable resolution was 5 nm under the best conditions.

Our operation at Mobara lasted only three years, and Kenji Kazato with his key members had to leave Mobara and establish a new company, whose name was Nihon Denshi Kogaku Kenkyujo (Japan Electron Optics Laboratory Co., Ltd., or JEOL), in Tokyo. This was legally the start of today's JEOL Co., Ltd.

I believe that this event was a big blow to Kazato, because he had to leave his three years' work at Mobara, except for the brains, that is to say, the people. However, he became free from the local financial people from whom he had asked for help in the beginning. His decision was proved to be right, judging from his success.

Later the Japanese name of the company was changed to Nihon Denshi Co., Ltd. meaning Japan Electronics Corporation. However, the original English name JEOL was retained.

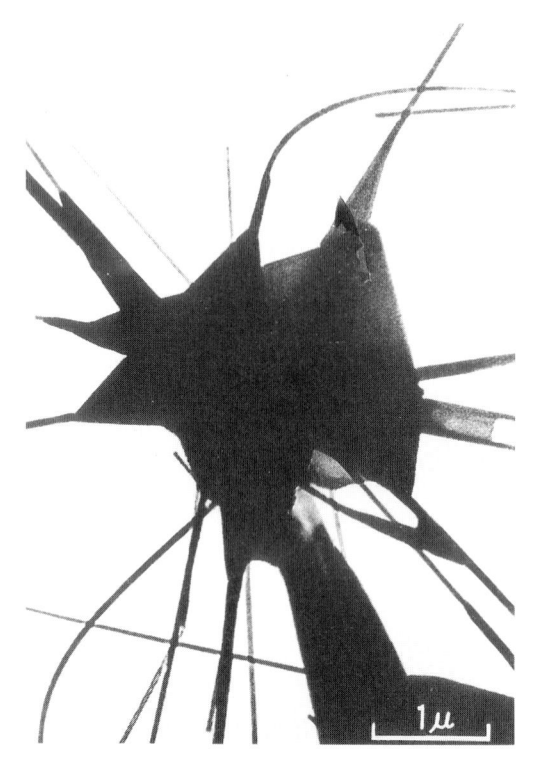

FIGURE 2. First electron micrograph taken with the DA-1.

Internal troubles did not restrain the eagerness of the young engineers. They started construction of a new electron microscope, the JEM-T1, which is shown in Fig. 3. This inverted type was thought to be quite natural because one could keep high-voltage parts, such as the high-voltage transformer and the electron gun, together under the desk safely without using a shielded high-voltage cable, which was not available at that time. We were not afraid of trying a new concept if we thought it right. This entrepreneurial spirit or engineer-oriented approach was the key to JEOL's spirit.

This continues to be true even today. Now, the Japanese electron microscope is considered to be No. 1 in the world. However, when we started the electron microscope company, the Japanese electron microscope was far behind other countries, such as German Siemens, U.S. RCA, British AEI, or Dutch Philips, etc.

Now, how could Japan become No. 1 in less than half a century is a very interesting question. I believe that this will also explain the history of JEOL. The first thing I have to mention is that even during the war, lots of activities concerning electron microscope were undertaken in Japan. As explained

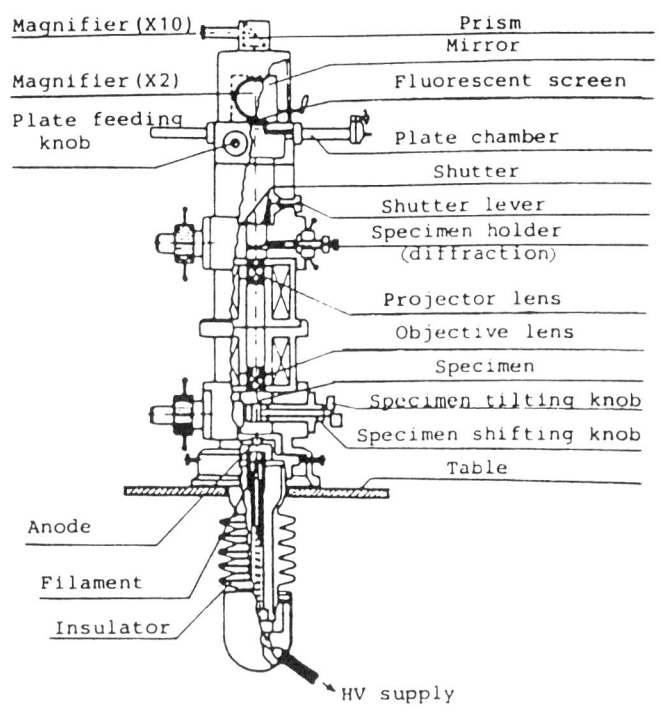

FIGURE 3. Cross section of JEM-T1 microscope column.

in this book, in 1938, after seeing the photographs of bacteria published by E. Ruska and B. von Borries, the famous 37 Subcommittee was created in the Japan Society for the Promotion of Science. The budget of this committee was 80,000 yen for three years. This is roughly equivalent to $1 million US today.

It is quite surprising that such an ambiguous scene created such enthusiasm and gave rise to such monetary appropriations during the tight and rather impoverished economy of the war. The effect of this encouragement was really astonishing. When the war ended in 1945, nearly 20 electron microscopes were working throughout Japan. All these microscopes were prototypes and not comparable to either Germany's or the United States'. However, the basic knowledge which is essential to appreciate the abundance of available information was there.

The next important fact is that in Japan, we considered the electron microscope and electron diffraction camera as a single unit, and not separate, right from the beginning. Optics theory says just this!

Many people in the world noticed this and published many papers concerning construction and application. However, it was only in Japan that the commercial instrument was oriented in this direction. All technologies available in electron diffraction were to be introduced into the electron microscope, such as specimen heating, specimen cooling, reflection method, etc. This approach has been pursued continuously until today and turned out to be a key to the analytical electron microscope.

Of course, this approach has not only advantages but also serious handicaps. It makes the electron microscope much more complicated and difficult to attain high resolution. We needed another 10 years to catch up with Siemen's world record of 10A.U. of the 1950s.

I have often wondered how Japanese electron microscopes would be today if we had simply followed the other microscopes then existing. After surviving those early days, the top scientists in the world had strongly requested that we achieve the highest resolution attainable. The continuous demands from all over the world and efforts to meet them pushed us step by step to the highest level in the world.

Instruments themselves or industry to produce them will progress only when there is a strong demand from customers. In this respect, I would like to express our thanks to all the people who have assisted JEOL in any way during nearly half a century of its existence.

4.5
Electron Microscopes and Microscopy in Japan
4.5C Development and Application of Electron Microscopes, Model SM-1 Series, at Shimadzu Corporation

SHIN-ICHI SHIMADZU

Shimadzu Corporation
Kawaramachi-nijo, Nakagyo-ku
Kyoto 604, Japan

I. A PROTOTYPE OF ELECTRON MICROSCOPE, MODEL SM-1

In 1938, B. von Borries and E. Ruska of Siemens Halske A.G. reported on the development of an electron super microscope which featured 100-kV accelerating voltage, a resolving power of 12 nm, and a magnification of 20,000 times.

Being stimulated by the news, the next year, in 1939, the Japan Society for the Promotion of Science decided to organize the 37th committee, which was to steer an integrating study on electron microscopes and was chaired by Prof. S. Setoh, University of Tokyo (Setoh, 1948). Several years later, in 1946, the author joined the committee as an active member.

The above news was of not a little interest at Shimadzu, which had been engaged in the manufacture of scientific instruments and had supplied them to universities and research institutes since its foundation.

In around 1939, Shimadzu began to develop an electron diffraction instrument under the guidance of Prof. K. Tanaka, Kyoto University, and the first product was installed at Kyoto University in March 1941. The instrument had a favorable reputation because it was equipped with a deflecting magnet for energy separation (Tanaka, 1942), as shown in Fig. 1.

In 1939, engineers of Shimadzu began to investigate the development of electron microscopes. However, it was not until July 1942 that construction of a prototype of an electron microscope was actually begun.

In parallel with this, Prof. N. Higashi, Faculty of Medicine, Kyoto University, was independently proceeding with the construction of an electron microscope at his laboratory and asked for our cooperation for manufacturing several main components such as electromagnetic lenses needed for his microscope, which substantially promoted the development of the proto-

FIGURE 1. Electron diffraction camera (1939).

type electron microscope at Shimadzu. Early the next year, 1943, the electron microscope was assembled. It was the realization of a scientist's dream, and was put to preliminary tests. Various tests were repeated on it and in April of the same year, it was completed and named model SM-1. A report on model SM-1 was promptly submitted to the 37th Committee (Shimadzu and Ohara, 1943). The features of the SM-1 electron microscope included resolving power of 5 to 10 nm, magnification of 20,000 times, and operation at 40 kV (Ohara, 1943). Figure 2 shows a general view of the SM-1, and a picture of zinc oxide crystals which was taken with the SM-1 is shown in Fig. 3.

The base supporting the microscope was made of wood, and parts that were available on hand were used for constructing electrical circuits. This was necessary due to the shortage of materials in Japan due to World War II, and the circumstances deteriorated day by day. Although many difficulties were encountered, the performance of the SM-1 was improved steadily through successive trials, and finally a satisfactory resolving power of 3 to 5 nm was attained in 1946.

II. Completion of Improved Models of SM-1A and SM-1B

Improvements of the prototype electron microscope SM-1 were made steadily on such components electromagnetic lenses, cathode filaments, and voltage stabilizers. In 1947, a new model SM-1A was put on sale and several sets were installed in universities and research centers including

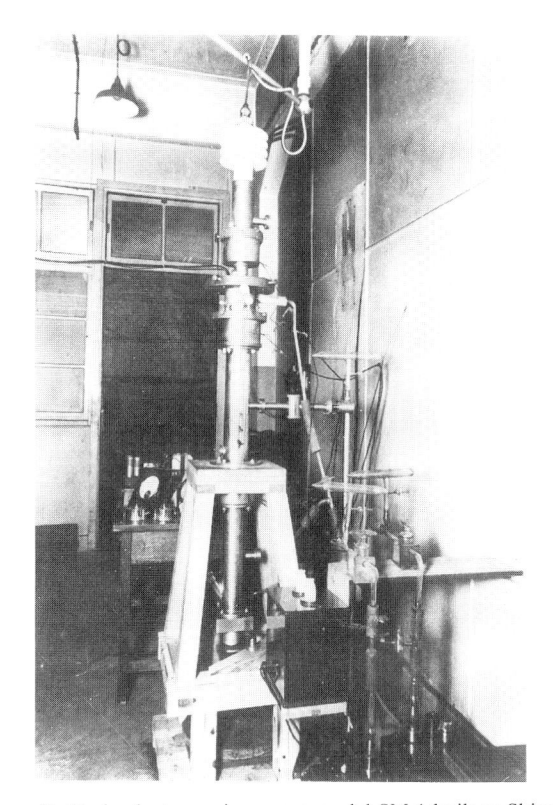

FIGURE 2. Early electron microscope model SM-1 built at Shimadzu.

the Electrotechnical Laboratory, Osaka University, and the Institute of Physical and Chemical Research, initiating electron microscopic studies in Japan. For the convenience of operation and maintenance, the SM-1 was designed so that the magnetic lens and the cathode filament were operated on an ordinary AC power supply. Furthermore, the self-biased electron gun was integrated into it, an innovation at that time (Hori, year unknown).

In 1948, an improved version, model SM-1B, was completed. It used a high-frequency, high-voltage power supply to accelerate the electron beam, i.e., the Cockroft quadrupling circuit, and was compact in design. Besides, to ensure safety, the high-voltage power supply and the electron gun were housed in a steel cabinet to protect the operator from electrical shock. It featured an accelerating voltage of 60 kV, magnification of 3000 to 12,000 times, and resolving power of 3 nm. It was also capable of taking stereo pictures. A general view of the SM-1B is shown in Fig. 4, a cross-sectional view in Fig. 5, and a picture of mycetozoan taken with the SM-1B in Fig. 6.

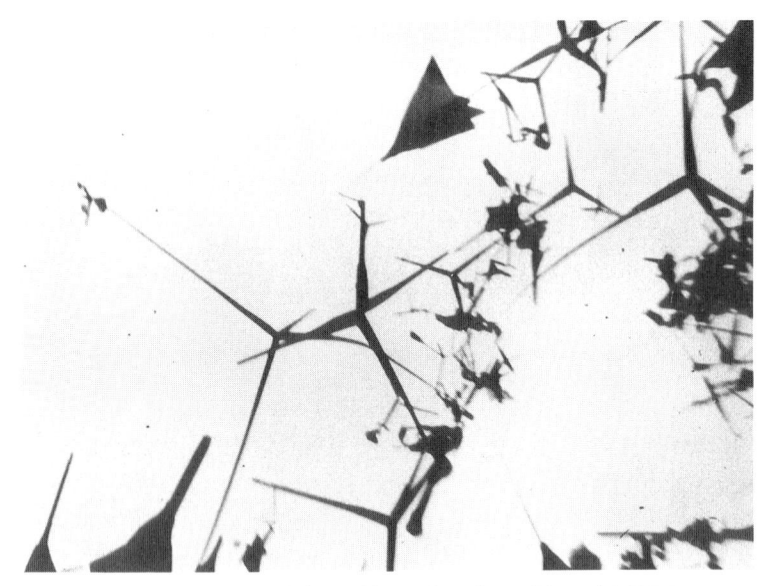

FIGURE 3. Crystal of zinc oxide smoke taken with model SM-1.

III. ELECTRON MICROSCOPIC STUDIES DURING THE EARLY PERIOD

Around 1943, the prototype electron microscope, just after its development, was used mainly for observing fine crystals of zinc oxide, magnesium oxide, carbon particles produced from benzene, and the fiber structures of asbestos.

By 1947, Japan had gradually recovered from the disruption after the war to such an extent that it became possible to hold scientific meetings. On 4, October 1947, a meeting on electron microscopic studies was held under the sponsorship of Shimadzu. The meeting attracted about 50 participants together with Prof. Y. Kimura, who was the Dean of the Faculty of Medicine, Kyoto University; Prof. S. Kato, a member of the 37th Commettee; and Prof. G. Yasuzumi, Osaka University. The participants discussed the progress of of electron microscope applications, performance, and history of the development of electron microscopes. At the end of the meeting, they visited the laboratory of Shimadzu to see the newly developed SM-1A being applied for practical study.

In January 1948, the *Shimadzu Review,* Vol. 5, No. 1, was issued as a special issue devoted to the reports concerning electron microscopes. In the special issue, Prof. Setoh, the chairman of the 37th Committee, presented an article titled "Electron Microscopic Study in Japan," together with reports on the advanced applications of the SM-1A in various research fields. The following articles are representative ones which appeared in the issue.

FIGURE 4. Electron microscope model SM-1B.

1. *Study on photoemulsion.* Mr. Y. Oyama of Mitsubishi Paper Mills Co. measured sizes of silver chloride and silver bromide particles in photoemulsion as well as sizes of silver colloid developed in the photos. Such quantitative comparison of particle sizes was correlated to the ones attained in pictures (Oyama, 1948).

2. *Study on genes.* Prof. G. Yasuzumi, Department of Anatomy, Faculty of Medicine, Osaka University, first observed fine structures of sperm of drosophilae and suggested that electron microscope images of genes be used in mutation study in biology (Yasuzumi and Takeda, 1948).

3. *Bacteriology.* Being guided by Profs. Y. Kimura and K. Sasagawa, Department of Bacteriology, Faculty of Medicine, Kyoto University, Prof. N. Higashi started electron microscope study of bacteria, rickettsias, and viruses (Kimura and Higashi, 1948).

4. *Metallurgy.* Profs. H. Nishimura and J. Takamura, Faculty of Engineering, Kyoto University, observed differences between the structures of aluminum alloy before and after rolling treatment. By such use of electron microscopes, they investigated the hardening of metals during physical

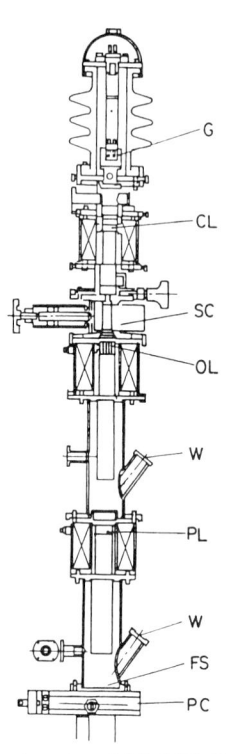

FIGURE 5. Simplified cross section of model SM-1B: G, electron gun; SC, specimen chamber; W, viewing window; FS, fluorescent screen; CL, condenser lens; OL, objective lens; PL, projection lens; PC, plate chamber.

processes such as rolling and bending and explained it in terms of location and orientation of microcrystallines in the bulk metals (Nishimura and Takamura, 1948).

5. *Taxonomy of the diatom.* Mr. H. Okuno, of the First Middle School of Kyoto Prefecture, inspected cell walls of diatom and observed fine structures of the cell, which had not been clarified by the conventional optical microscope but were characteristic of the individual species of diatom. When he had his first look at the diatom, he was surprised at the fine picture. His surprise deeply impressed the author. By thus utilizing electron microscopes to study diatoms, he contributed much to their classification (Okuno, 1948).

IV. Epilogue

Frankly speaking, the electron microscope, SM-1 series, which we developed in Japan nearly 50 years ago was not an elaborate product, but a

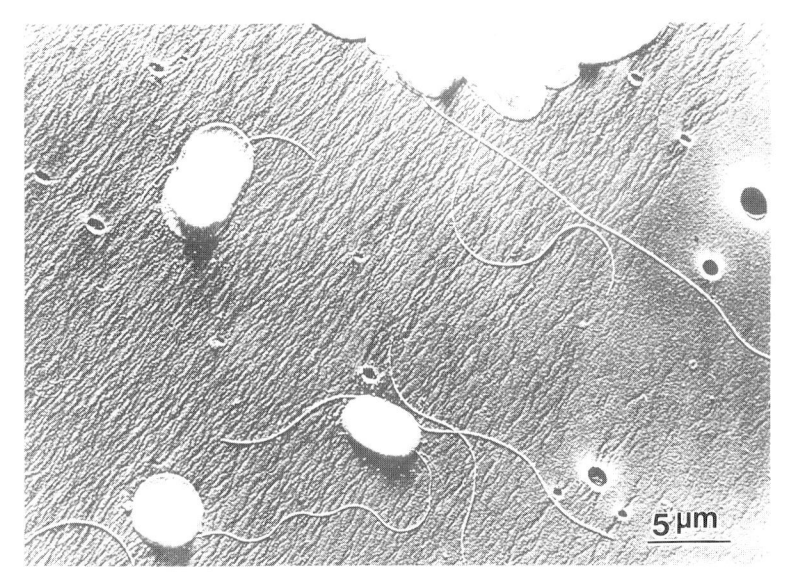

FIGURE 6. Mycetozoan taken with model SM-1B.

rather primitive one when compared with the highly sophisticated electron microscopes used currently. When we consider the technology and the shortage of materials at that time, however, this activity in R&D should be regarded as an excellent achievement.

It was natural that those who had seen the splendid views through the electron microscope could not diminish their strong urge toward the development of this fantastic instrument and the technology of electron microscopy in spite of the adverse conditions.

In November 1952, the Emperor of Japan, also a reputed biologist, visited the Sanjo Works of Shimadzu, and His Majesty himself observed biological specimens in the electron microscope, which was an honorable and unforgettable event for those who were concerned with the electron microscope.

REFERENCES

Hori, T. (year unknown). Japanese Patent 348451.
Kimura, Y., and Higashi, N. (1948). *Shimadzu Rev.* **5,** 43
Nishimura, H., and Takamura, J. (1948). *Shimadzu Rev.* **5,** 43.
Ohara, M. (1943). *Shimadzu Rev.* **4,** 105.
Okuno, H. (1948). *Shimadzu Rev.* **5,** 45, 100.
Oyama, Y. (1948). *Shimadzu Rev.* **5,** 17.

Setoh, S. (1948). *Shimadzu Rev.* **5,** 2.
Shimadzu, S. (1948). *Shimadzu Rev.* **5,** 5.
Shimadzu, S., and Ohara, M. (1943). The 37th Committee Report, No. 25-5.
Tanaka, K. (1942). *Shimadzu Rev.* **3,** 37.
Yasuzumi, G., and Takeda, Y. (1948). *Shimadzu Rev.* **5,** 28.

4.5
Electron Microscopes and Microscopy in Japan
4.5D Electron Microscope Research at the Toshiba Corporation

HIROSHI KAMOGAWA

Toshiba Research and Development Center
Komukai Toshiba-cho, Saiwai-ku
Kawasaki 210, Japan

The research and development of electron microscopes at the Toshiba Corporation (previously Tokyo Shibaura Electric Co., Ltd.) was begun by Soichiro Asao in 1939. This study was continued over 20 years by Hideo Inuzuka, Hiroshi Kamogawa, Hiroshi Yako, Kazuo Saito, and others. Since 1959 the study has been limited to use and application of electron microscopes and their technologies. The machines developed in those 20 years can be classified into three groups: electromagnetic (Toshiba No. 1), electro static (Toshiba No. 2–No. 6), and multifunctional (EUL-1, EUL-1B, EUL-2, and EUL-3). These 10 machines are described in *History of Electron Microscopes 1986* (HEMJ'86), pp. 64–79, (Kyoto, Japan, 1986). Among them, Toshiba No. 1, Toshiba No. 2, and EUL-1B are explained in detail in the following pages as representative of their respective groups.

I. Toshiba No. 1 (1940)

This electromagnetic electron microscope (EM) is one of the earliest successful EMs, and the TEM picture reported in the *American Mineralogist* in 1940 may be the first EM pictures from Japan.

Specifications:
  Trial fabrications: 1940
  Accelerating voltage: 30–45 kV
  Magnification: 10,000×
  Lens system: Electromagnetic type
    Condenser lens (solenoid)
    Objective lens (Ruska type)
    Projection lens (Ruska type)

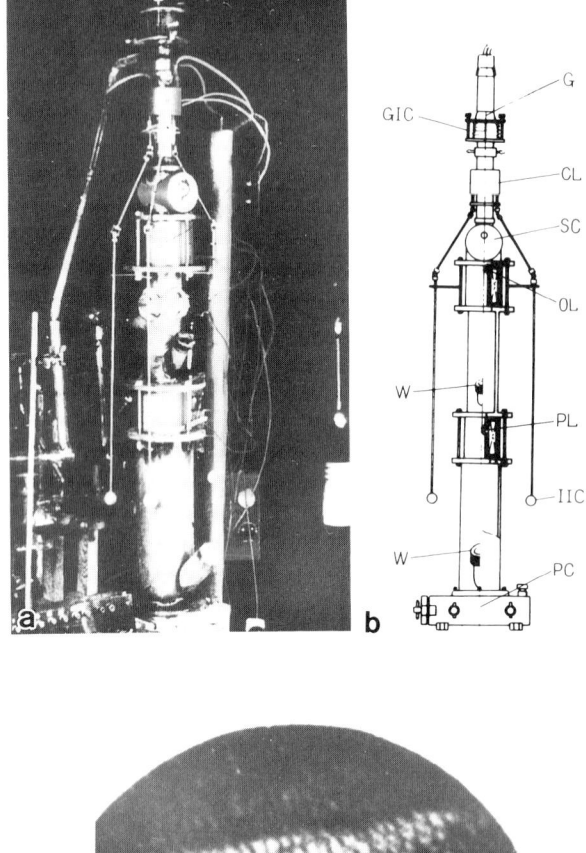

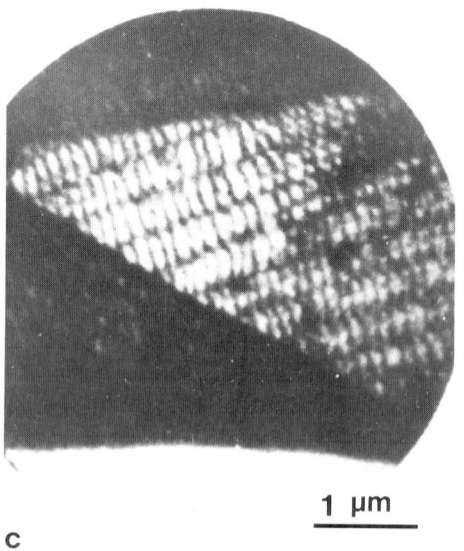

FIGURE 1. (a) Toshiba No. 1 (1940). (b) Sectional diagram. (c) Butterfly ramentum.

Pictures:
 Fig. 1(a): External view
 Fig. 1(b): Sectional diagram
 Fig. 1(c): Butterfly ramentum
Literature:
 H. Inuzuka (1940). Observation of clay minerals using electron micro scopes. *Am. Mineral.*, p. 448–1944.
 H. Inuzuka (1940). Super microscopic observation of clay minerals (in Japanese). *Matsuda Kenkyu Jiho* **16,** 171–176.
 HEMJ'86 (1986), pp. 64–65.

II. TOSHIBA NO. 2 (1941) TO NO. 6 (1943)

Toshiba No. 2 may be the earliest electrostatic EM completed in Japan. After the prototype fabrication, the improved Toshiba No. 2 was completed in 1941 as shown in Fig. 2 (a).

Specifications:
 Trial fabrications: 1941
 (*Specifications continued on page 676*)

FIGURE 2. (a) Toshiba No. 2 (1941). (b) Magnesium oxide smoke by Toshiba No. 2.

Specifications (*continued from page 675*)
   Accelerating voltage: 50 kV
   Resolution: 80 Å
   Magnification: 1000–6000×
   Lens system: Electrostatic type
      Objective lens (Einzel)
      Projection lens (Einzel)
Pictures:
   Fig. 2(a): External view of Toshiba No. 2
   Fig. 2(b): Magnesium oxide smoke by Toshiba No. 2
Literature:
   S. Asao, T. Watanabe, and S. Tomita (1942). Trial fabrication of electrostatic electron microscope (in Japanese). *Matsuda Kenkyu Jiho* **17,** 22–23.
   Toshiba No. 2, *HEMJ'86* (1986), pp. 66–67; No. 3, pp. 70–71; No. 4, No. 5, No. 6, pp. 72–73.

III. EUL-1 (1942) TO EUL-3 (1958)

The EUL series of electrostatic EMs has some original features, called multifunctional. This idea was registered as Japanese patent 161259, application date 5 October 1942, registration date 24 January 1944. Let us explain the principle by means of the EUL-1B. It has two specimen ports as shown in Fig. 3(*b*). The upper port, SC_1, is located just over the objective lens OL and is used for TEM; the lower port, SC_2, is just under the projection lens PL, and is used for shadow EM and electron diffraction of limited field.

Specifications:

	EUL-1	EUL-1B	EUL-2	EUL-3	
Fabricated:	1942	1947	1947	1958	
Accelerating voltage (kV)	50	56	56	80	
Lens system:	OL,PL(2Einzel)	OL,PL(2E)	OL,PL(2E)	OL,*IL,PL(3E)	
Magnification (×)	5,000	5,000	4,600	3,000 5,000 10,000
Resolution (Å)	80	60	60	50	
Literature	1	2	3	4	

* IL: Intermediate lens between OL and PL.

Pictures:
   Fig. 3(a): External view of EUL-1B
   Fig. 3(b): Sectional diagram of EUL-1B
   Fig. 3(c): TEM picture of zinc oxide smoke by EUL-1B
   Fig. 3(d): Shadow EM picture zinc oxide smoke by EUL-1B
   Fig. 3(e): Electron diffraction picture of zinc oxide smoke by EUL-1B

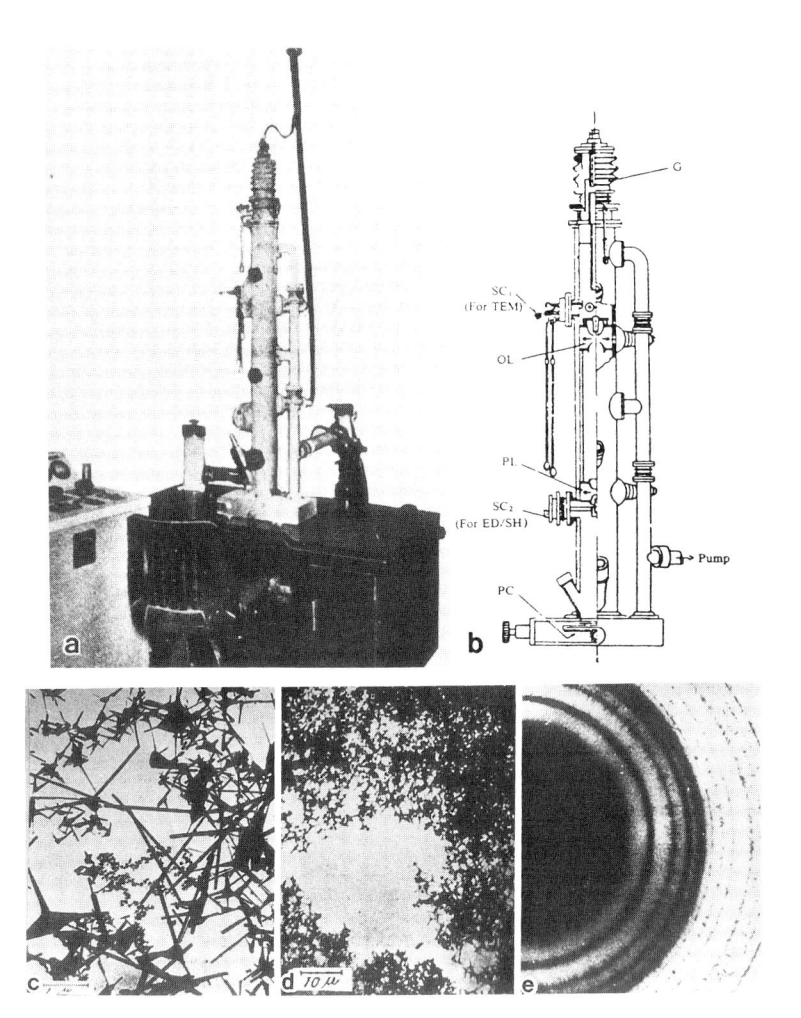

FIGURE 3. (a) External view of EUL-1B. (b) Sectional view of EUL-1B. (c) Zinc oxide smoke by EUL-1B. (d) Shadow picture of zinc oxide smoke by EUL-1B. (e) Electron diffraction picture of zinc oxide smoke by EUL-1B.

Literature:
1. H. Kamogawa and H. Yako (1943). 23rd Joint General Meeting of the Three Electrical Institutes of Japan, no. 15. *HEMJ'86* (1986), pp. 68–69.
2. H. Kamogawa and Y. Yokota (1942). The Matsuda electrostatic electron microscope (in Japanese). *Toshiba Rev.* **3**, 1–22. *HEMJ'86* (1986), pp. 74–75.
3. *HEMJ'86* (1986), pp. 76–77.
4. *HEMJ'86* (1986), pp. 78–79.

4.5
Electron Microscopes and Microscopy in Japan
4.5E Development of the Electron Microscope at Kyoto Imperial University Faculty of Medicine

YUTAKA TASHIRO

Department of Physiology
Kansai Medical University
Moriguchi-shi
Osaka 570, Japan

AKIO OYAMA

Department of Microbiology
Kansai Medical University
Moriguchi-shi
Osaka 570, Japan

I. Construction of Electron Microscopes and Their Application in the Early Period (1939–1950)

Electron microscope history at Kyoto Imperial University, Faculty of Medicine began in 1939 when Kyugo Sasagawa, Professor of Physiology, Osaka Medical Center, was selected as a member of the 37th Subcommittee of the Japan Society for the Promotion of Science, which was organized by Shoji Seto to develop electron microscopy in Japan. It was also in 1939 that Noboru Higashi, who graduated from Kyoto University in 1938 and was a graduate student in the Department of Microbiology (Prof. Ren Kimura), visited Sasagawa's laboratory to discuss how to construct an electron microscope for medical use.

Based on their earlier publications and memoranda (Sasagawa, 1951; Higashi, 1965) and on the records of the 37th Subcommittee by Ura (1983) and by Tadano (1986), the present article was written by Yutaka Tashiro and Akio Ohyama, who had been graduate students of the late Profs. Sasagawa and Higashi at Kyoto University Faculty of Medicine, respectively.

Sasagawa graduated from Kyoto University Faculty of Medicine in 1923 and studied General Physiology in the Department of Physiology (Prof.

Hidetsurumaru Ishikawa). In 1928 he moved to Osaka Medical College as a Professor of Physiology, where he worked enthusiastically on medical electronics in collaboration with a number of his students. Especially, he succeeded in recording the action potential of nerve fibers using a home-made cathode-ray oscilloscope in collaboration with Syunpei Watanabe of the Institute of Physical and Chemical Research (Riken) (Watanabe and Sasagawa, 1933). This was the first recording of the action potential by a cathode-ray oscilloscope in Japan.

In 1931, M. Knoll and E. Ruska succeeded in constructing an electron microscope by modifying a cathode-ray oscilloscope. Stimulated by this report, Sasagawa was very much interested in the construction and biological application of electron microscopy.

In 1939 Sasagawa decided to construct an electron microscope in collaboration with Higashi and Watanabe in order to apply it to biology and medicine. After Sasagawa was promoted to Professor of Physiology at Kyoto University on 30 January 1940, this construction was continued at Kyoto University Faculty of Medicine.

At the end of 1941, the original model of the electron microscope (KUM No. 1) was completed and installed in the Department of Physiology. At the 11th meeting of the 37th Subcommittee (13 January 1941), Sasagawa and Higashi presented four electron micrographs of polished alumina and chromium oxide taken by this microscope. At the 12th subcommittee meeting (29 March 1941), they presented electron micrographs of *Corynebacterium dipthteriae* and *Staphylococcus*. These micrographs were reported at the 15th meeting on bacteriology (4 April 1941, Kumamoto) and published by Kimura and Higashi (1941).

The original model of the electron microscope was modified extensively in collaboration with Shimadzu, Riken, Hitachi, and Yokogawa (YEW). Figure 1 shows the modified KUM, which was called KUM No. 2. Using this microscope, Sasagawa and Higashi reported the preparation of supporting film for electron microscopic observation and bacteriological study at the 20th subcommittee meeting (11 July 1942). The latter study was published by Kimura and Higashi (1942). This model was further modified to the final model, KUM No. 3. Using the microscope, Higashi observed anaerobic bacteria and *Corynebacterium diphtheriae* and succeeded in taking the first electron micrograph of a virus in Japan. These results were reported at the 24th meeting (2 April 1943) by Sasagawa and Higashi and published by Kimura and Higashi (1943). Electron microscopic observation of various kinds of bacteria was also reported by Kimura and Higashi (1944).

Fortunately, KUM No. 3 was neither damaged nor evacuated to the countryside during World War II (Kyoto was the only exceptionally large city in Japan which did not suffered from air raids). This machine was used

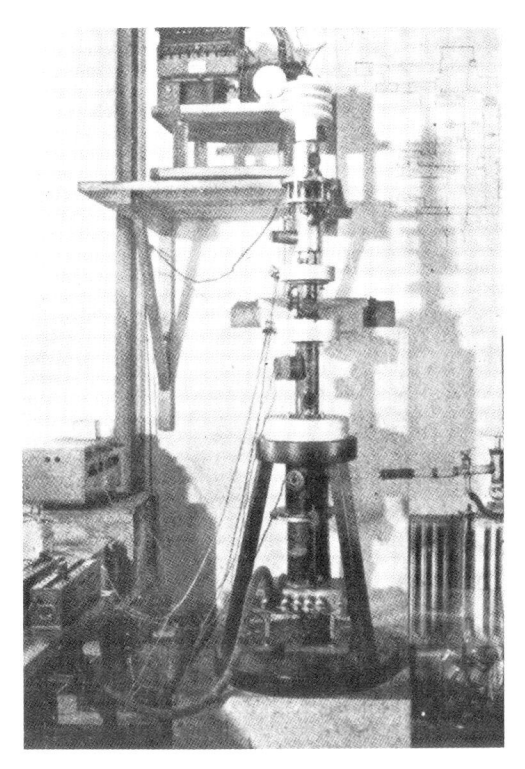

FIGURE 1. Kyoto University Medical School type 2 (KUM No. 2), trial fabrication 1942. Accelerating voltage 60 kV (max. 80 kV). Lens system: magnetic type, condenser, objective, and projection lenses.

even after the end of the war (August 1945), until good commercial electron microscopes became readily available in Japan in 1948–1950.

When I (Y. T.) started working in Sasagawa's laboratory in the Department of Physiology in 1950, four electron microscopes were in use: KUM No. 3, Toshiba EUL-1 (electrostatic type) (1942), Hitachi HU-4 (1948), and RCA EMC (1948). A JEM-5C was purchased (1954) for the Department of Microbiology. The HU-4 and JEM-5C were utilized extensively in the two laboratories, and various results were obtained as described in the following.

II. Application of Electron Microscopy to Microbiology by Higashi and His Collaborators (1951–1956)

Using the KUM No. 3, Higashi *et al.* observed various bacteria, virus, spirocheta, and rickettsia (Kimura and Higashi, 1943, 1944; Higashi, 1947).

However, the application of electron microscopy to microbiology in the early period was very limited, because the electron beam had low penetration power and bacteria are too thick to be observed in detail by electron microscopy. They should be cut into ultrathin sections before they are viewed under an electron microscope. Such a technique was not available in Japan at that time. Therefore Higashi visited the laboratory of Peace, University of California School of Medicine, Los Angeles, in 1950–1952 to learn ultrathin sectioning, and introduced this technique to Japan (1952). Using this ultrathin sectioning method, Higashi and his collaborators investigated proliferation and maturation of phage in bacteria (Higashi, 1953, 1955). Mumps virus (Higashi, 1954), vaccinia virus (Shimizu, 1951), ectoromelia virus (Ozaki, 1956a, 1956b), spirocheta (Asakura, 1951, 1952a, 1952b), etc., were also observed.

In 1956, Higashi was promoted to Professor of Biophysics, Institute for Virus Research, Kyoto University, where he continued electron microscopical research.

III. Application of Electron Microscopy to Cytology by Sasagawa and His Collaborators (1951–1957)

The application of electron microscopy to cytology was also very limited because of the lack of ultrathin sectioning technique. Erythrocyte membrane is an exceptional biological specimen which can be observed without ultrathin sectioning, and it was studied by Higashi *et al.* (1950) and by Mani (1951).

Sasagawa and Hosomi (1949) observed the ultrastructure of nerve, muscle, and collagen fibers of rabbit and frog after fixation with either Bouin fixative or 10% formalin. These fixed tissues were homogenized extensively in an agate mortar to disperse the fibers. They observed these fibers in the HU-4 electron microscope and found that the smallest fibrils, which they called protofibrils, are composed of small particles, 15–25 nm in diameter. Similar particles were also observed in the cytoplasm of various glandular cells. From these findings they suggested that all living materials are composed of small particles 15–25 nm in diameter, and they proposed to call them "elementary bodies of life." Later, ultrathin sectioning technique of plastic embedded cells and tissues became available and it was revealed that this proposal was an oversimplification. Tashiro and Ogura (1957) investigated rat liver microsomes and showed that, when the microsomes were incubated with RNase, electron-dense particles (Palade granules) embedded in the microsomal membranes are selectively digested. They suggested that these particles are ribonucleoprotein particles, now conventionally called ribosomes. Similar results were also reported independently by Palade and Siekevitz (1956).

IV. Further Application of Electron Microscopy in the Other Laboratories of Kyoto University Medical School and at the Institute for Virus Research (1955–1960)

In the mid-1950s, good electron microscopes became commercially available and a number of departments at the Kyoto University School of Medicine were interested in applying electron microscopy to solve their problems. The electron microscope at that time was too expensive for each department to buy one. As suggested by Sasagawa, therefore, an electron microscope center was established at the Kyoto University Hospital in 1955. Later, in 1960, another electron microscope center was established at the basic medicine campus. These centers were efficiently shared by all the researchers in the medical school, and a number of interesting papers were published.

At the Institute for Virus Research, not only Higashi's group but also Sigeyasu Amano's group engaged in electron microscopical investigation of normal and pathological cells.

References

Asakura, Z. (1951). *Saishinigaku* (in Japanese) **6,** 958.
Asakura Z. (1952a). *Jpn. J. Bacteriol.* (in Japanese) **7,** 335.
Asakura, Z. (1952b). *Jpn. J. Bacteriol.* (in Japanese) **7,** 567.
Higashi, N. (1947). *Saishinigaku* (in Japanese) **2,** 301.
Higashi, N. (1952). *Kagaku* (in Japanese) **22,** 480.
Higashi, N. (1953). *Virus* (in Japanese) **3,** 12.
Higashi, N. (1954). *Proc. Int. Conf. on Electron Microscopy,* p. 34.
Higashi, N. (1955). *Proc. 14th Medical Association of Japan,* pp. 253–265.
Higashi, N. (1965). "World of Electron Microscopy" (in Japanese), Iwanami, Tokyo.
Higashi, N., Wakisaka, K., and Mizukawa, I. (1950). *Electron-Microscopy* (in Japanese) **1,** 32.
Kimura, R., and Higashi, N. (1941). *Nihon Igaku* (in Japanese), no. 3234, p. 15.
Kimura, R., and Higashi, N. (1942). *Nihon Igaku* (in Japanese), no. 3285, p. 8.
Kimura, R., and Higashi, N. (1943). *Nihon Igaku* (in Japanese), no. 3338, p. 5.
Kimura, R., and Higashi, N. (1944). *Nihon Igaku* (in Japanese), no. 3366, p. 8.
Man-i, M. (1951). *Electron-Microscopy* (in Japanese) **2,** 53.
Ozaki, Y. (1956). *Acta Med. Univ. Kyoto* **33,** 159.
Ozaki, Y. (1956). *Acta Med. Univ. Kyoto* **33,** 168.
Palade, G. E., and Siekevitz, P. (1956). *J. Biophys. Biochem. Cytol.* **2,** 171.

Sasagawa, K. (1951). "Electron Microscopy" (in Japanese), Honda, Kyoto.
Sasagawa, K., and Hosomi, T. (1949). *Acta Schol. Med. Univ. Kyoto* **27,** 96.
Shimizu, T. (1951). *Acta Med. Univ. Kyoto* **29,** 145.
Tadano, B. (1986). In "History of Electron Microscopes 1986" (ed. H. Fujita), pp. 1–26. Komiyama, Tokyo.
Tashiro, Y., and Ogura, M. (1957). *Acta Schol. Med. Univ. Kyoto* **34,** 267.
Ura, K. (ed.) (1983). The records of the 37th Subcommittee of Japan Society for the Promotion of Science 1935–1947.
Watanabe, S., and Sasagawa, K. (1933). *Sci. Pap. Inst. Phys. Chem. Res.* **35,** 139.

4.5
Electron Microscopes and Microscopy in Japan
4.5F Instrumentation

TSUTOMU KOMODA

Central Research Laboratory
Hitachi Ltd.
Kokubunji
Tokyo 185, Japan

Development of electron microscopy in Japan, begun in 1939, reached the age of commercialization around 1950 with several manufacturers. The number of manufacturers was as many as six at one time. Intense competition took place among them, and consequently, the progress of electron microscopy techniques was greatly accelerated.

The history of the development of electron microscopy techniques after 1950 can be divided into two parts, before and after 1970. The former was the so-called growing period of electron microscopy, when many methods were investigated to improve the resolution and functional performance of the electron microscope. On the other hand, the latter was a ripening period, when problems such as the resolution and functionality were rigorously investigated, since the possibilities and limitations were more or less elucidated by 1970.

The history of technical developments in electron microscopy investigation in Japan therefore will be described into two parts, 1950–1970 and 1971–1990.

I. 1950–1970

A. The Study of Electron Lens Aberrations

Electron lenses had been designed and machined by trial and error at the beginning of the 1940s. Gradually it became clear that the machining accuracy of the lens polepiece has a great influence on the resolution. With this viewpoint, Tadano and Akeyama, of Hitachi, Ltd., introduced a new method for precision lens machining called a block lens (Tadano, 1957). In

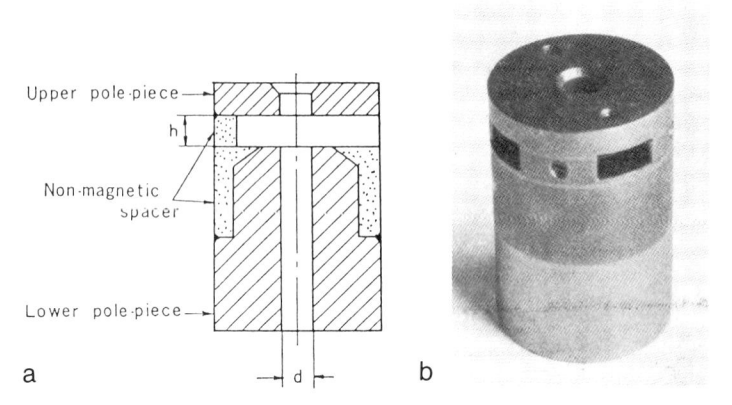

FIGURE 1. Sectional diagram (a) and photograph (b) of block lens.

this method, as shown in Fig. 1, two magnetic polepieces with a nonmagnetic spacer in between were soldered and then bored together, followed by lapping. This method assured nearly perfect alignment of the bore axis of the upper and lower polepieces, and the astigmatic difference was always kept less than 1 μm without a stigmator.

Around 1950, theoretical approaches to lens design, based on Glaser's theory, were made by Ito, Kanaya, Morito, and others. From their results, basic lens characteristics were adequately understood, although slight discrepancies still occurred between theory and experiment in the focal length or magnification. In those days, it was believed that the most important aberration of the lens was the spherical aberration, but it gradually turned out that the actual factors limiting the resolution are astigmatism due to imperfect machining of the polepiece and chromatic aberration due to instability of the power supply (high voltage and lens current). Thus, arguments on chromatic aberration were vividly discussed in the meetings of the Electron Microscopy Society of Japan, initiated by the study of chromatic aberration by Katagiri, of Hitachi, Ltd. (1953).

Chromatic aberration consists of an axial aberration and an off-axial one. Katagiri measured the amount of off-axial aberration upon varying the lens excitation for various shapes of polepieces. As a result, he found a lens condition under which the off-axial chromatic aberration became nearly zero (Fig. 2), by suitable combination of lens shapes (bore diameter and gap distance) of the objective and the projector, and excitation; this was called an achromatic lens system. Under this achromatic condition, where the excitation of each lens was fairly strong, with $NI/V^{1/2} = 16$–20, the spherical aberration and the axial chromatic aberration of the objective both became small, and the focal length also became small. Thus, it became

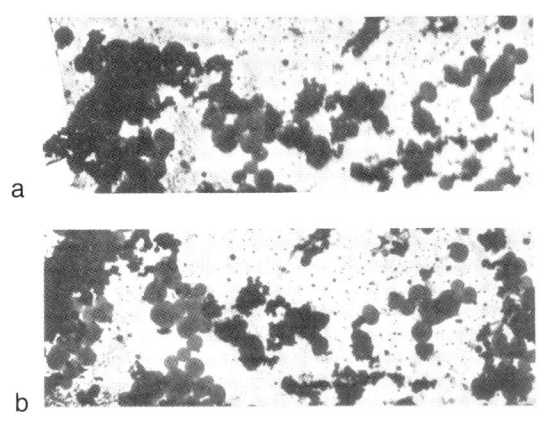

FIGURE 2. Compensation of off-axial chromatic aberration: (a) noncompensated image; (b) compensated image.

possible to use a larger-angle objective aperture, which resulted in a clear image of even a fairly thick specimen. Hitachi commercialized this idea of a chromatic aberration-free lens system with a strong excitation objective lens as the HU-9 in 1953, based on these experimental data.

Later, Katagiri et al. measured the astigmatic difference for various shapes of polepiece and showed that the astigmatism became remarkably small with increasing excitation strength (Morito et al., 1959), as shown in Fig. 3, which suggested that astigmatism depends not only on the machining accuracy of the lens bore but also on the magnetic property of the lens material. Thus, high excitation ($NI/V^{1/2} \geq 16$) for the objective was generally accepted in the middle of the 1950s in Japan.

B. Development of an Electron Microscope with a Permanent Magnet Lens

In the 1950s, Kimura (Hitachi, Ltd.) initiated the study of an electron lens excited by a permanent magnet, hoping for a miniaturized and low-cost electron microscope anticipated to have high stability and easy maintenance of the permanent magnet lens. However, there were some difficult problems in developing the permanent magnet lens, such as focusing and wide change of magnification because of the constant magnetic field. This needed a relevant invention to change the magnetic flux over the lens gap widely.

For this purpose, Kimura (Kimura and Kikuchi, 1956; Kimura,) developed a double-gap lens consisting of three polepieces and two lens gaps as shown in Fig. 4, where a part of the magnetic circuit between the permanent magnet and the intermediate magnetic pole was movable externally and

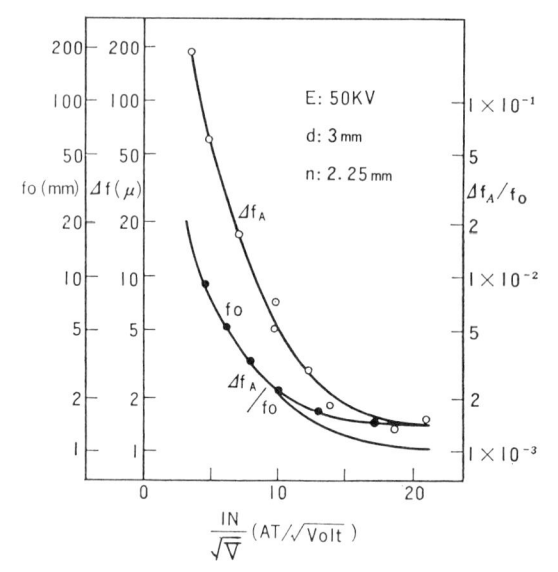

FIGURE 3. Relation for focal length (f_o) and astigmatic difference ($\triangle f_A$) versus lens excitation ($IN/\sqrt{V}$).

hence the magnetic resistance could be varied greatly to give a wide change of focal length. When this double-gap lens was used as a intermediate lens, a lens system which enabled a wide magnification range and electron diffraction capability was realized.

In addition to this, Kimura developed a special mechanism for objective focusing in which a magnetic bypass in parallel with the lens gap was changed mechanically. Very fine and stable adjustment of the focal length of the objective was thus realized within about 250 μm.

A permanent-magnet electron microscope was commercialized as the HS series in 1955 (Fig. 5). This type of microscope was welcome, at home and also abroad, as an easily handled, low cost microscope. The number of these microscopes produced totalled 850 by 1971.

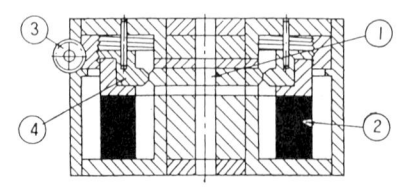

FIGURE 4. Cross section of double-gap lens: (1) double-gap lens, (2) permanent magnet, (3) gear, (4) movable piece.

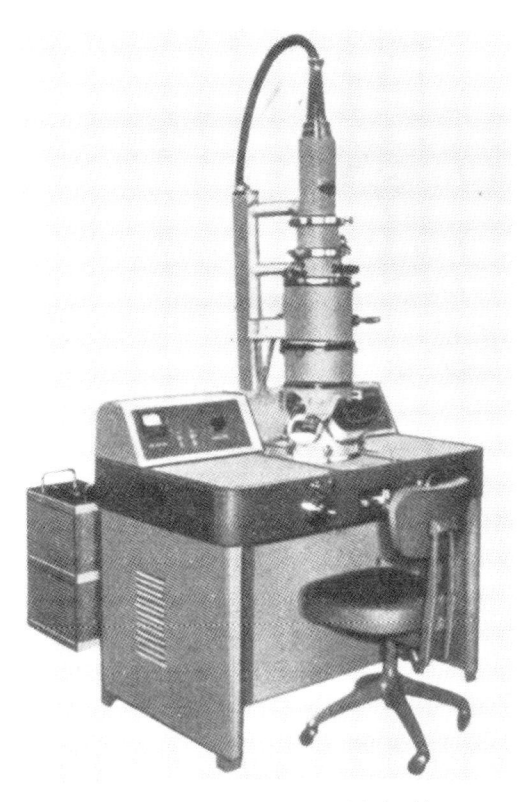

FIGURE 5. Model HS-6 electron microscope, provided with a permanent magnet lens system (1958).

C. Study of Pointed Cathodes

Studies of the electron gun had been made in the 1940s by Sugata (Osaka University) and others prior to the well-known work by Haine and Einstein, so the technical standard had reached a reasonable level around 1950. A little later, a creative and unique study of a pointed cathode was introduced. Hibi (1956) of Tohoku University, a pioneer, studied the pointed cathode to obtain a highly coherent electron beam for the realization of electron holography. He sharpened a tip of tungsten wire by hand polishing so as to have a radius of curvature less than 1 μm and made up a Müller-type pointed filament as shown in Fig. 6 (*a*). Thermionic emission from the tip of the cathode was used for illumination when it was heated by electric current. Figure 7 shows shapes of cathode tips and the electron image

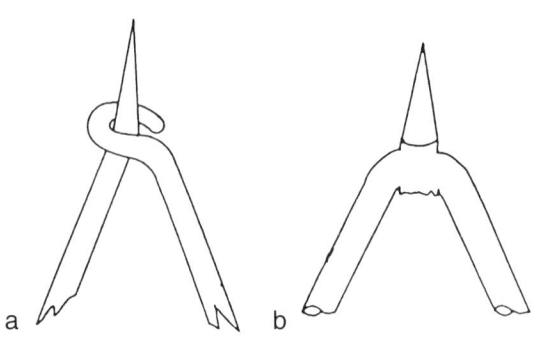

FIGURE 6. Pointed filament: (a) hand polished (by Hibi); (b) electropolished (by Sakaki).

obtained, where the coherence of the electron beam depends on the shape of the tip.

Hibi and Takahashi obtained about 20 Fresnel fringes on photographic film using an electron microscope equipped with this kind of pointed cathode in 1956. The coherence of this electron gun was not high enough to achieve electron holography, but it was found that the image contrast was enhanced when a normal electron microscope image was observed. They succeeded in observing highly magnified and bright images showing a resolution of 15 Å, at as much as 100,000 times with an electron microscope, the HU-6 (Hibi and Takahashi, 1959). This microscope, with a single condenser lens and two-stage imaging lenses, was a very old one, made in 1949, but its performance was comparable to that of a newer microscope with a double condenser and three-stage imaging lenses. Hibi and Yada (1964) eventually obtained lattice fringes of 3.8 Å spacing with further improvement in the two-stage microscope. This experimental data suggested that the pointed cathode provides an electron source whose brightness and coherence are higher.

Sakaki and Maruse also began to study pointed cathodes. They made a pointed cathode by electric etching as shown in Fig. 6(b) (Maruse, 1958). In 1959, they made clear the principal features of the pointed cathode from quantitative measurements of the emission characteristics, namely, that the pointed cathode can provide several times higher brightness even with the beam current as low as several microamperes, as compared with a normal hairpin cathode (Maruse, 1959).

Later, Komoda (Hitachi, Ltd.) improved the lattice image resolution greatly by using a pointed cathode in 1965 as described later. This result was due to the fact that sufficiently high brightness is provided at low beam current, so that the Boersch effect is minimized. Thus it was made clear that the pointed cathode has a smaller energy spread in addition to the

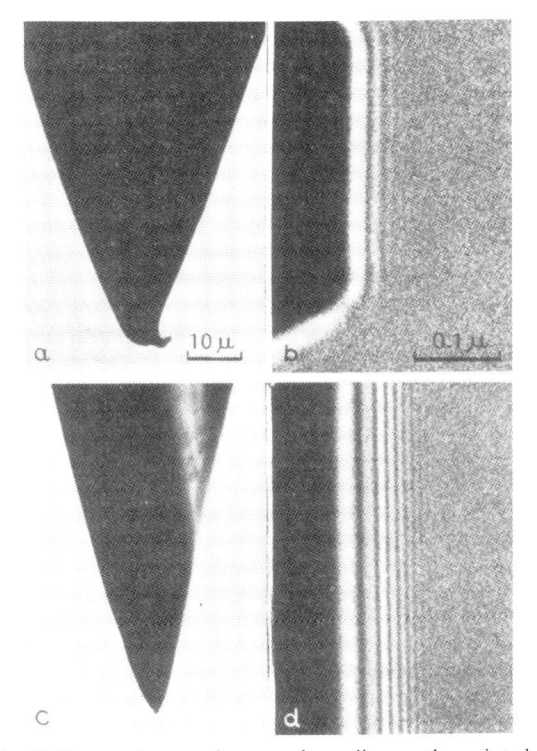

FIGURE 7. (a–d) Electron beam coherency depending on the pointed cathode tip.

known characteristics of pointed electron source, namely, high brightness and high coherency.

D. The Study of Electron Energy Loss

After 1948, the study of energy loss analysis of electrons transmitted by the specimen was begun by Ruthemann, Moellenstedt, and others. In Japan, Watanabe (1954), of Hitachi, Ltd., started to study electron energy loss around 1953. He modified a normal three-stage electron microscope to provide a fine slit and an energy analysis lens between the intermediate lens and the projector.

Figure 8 shows a schematic diagram, in which a Moellenstedt-type electrostatic lens having large chromatic aberration was used as the analyzing lens. A resolution of about 0.5 eV was obtained. Watanabe measured the characteristic energy-loss spectra of various kinds of thin films of metals and oxides, such as Al, Be, and Mg. Energy-loss spectra from part of the

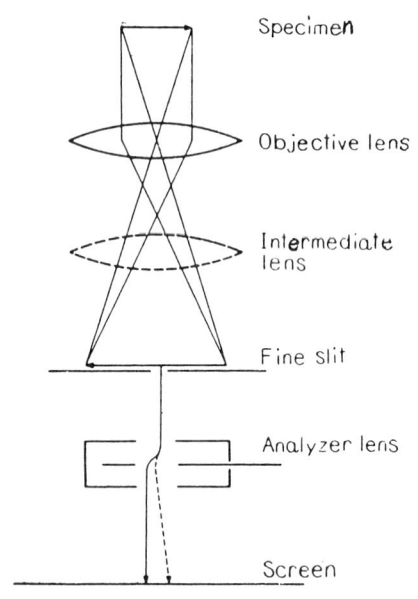

FIGURE 8. Ray diagram of electron velocity analysis in the electron microscope.

diffraction pattern was obtained by forming the electron diffraction pattern on the fine slit as shown in Fig. 9, where the relation between the scattering angle (ordinate) and the loss value (abscissa) was revealed. In this spectra of Al thin film, it is seen that the energy loss has an angular dependence in an archwise shape. Watanabe made it clear that the angular dependence arises from the plasma oscillation of electrons in metal, on the basis of Bohm and Pines plasmon theory (Watanabe, 1956). This was the first experimental proof of the plasma oscilation of free electrons in solids.

In 1961, Watanabe further developed this technique of electron energy-loss analysis in energy-selective electron microscopy (Watanabe and Uyeda, 1962). Figure 10 shows an energy-selecting electron microscope image thus obtained. Resolution was not much better than about 100 Å, but it was certainly in the lead in analytical electron microscopy for two decades.

E. Development of the Multifunctional Electron Microscope and Accessories

In the mid-1950s, electron microscopes of every Japanese manufacturer reached a standard level of 10-Å resolution at 100 kV. It was then desired to add other functions and accessories, such as elemental analysis capability

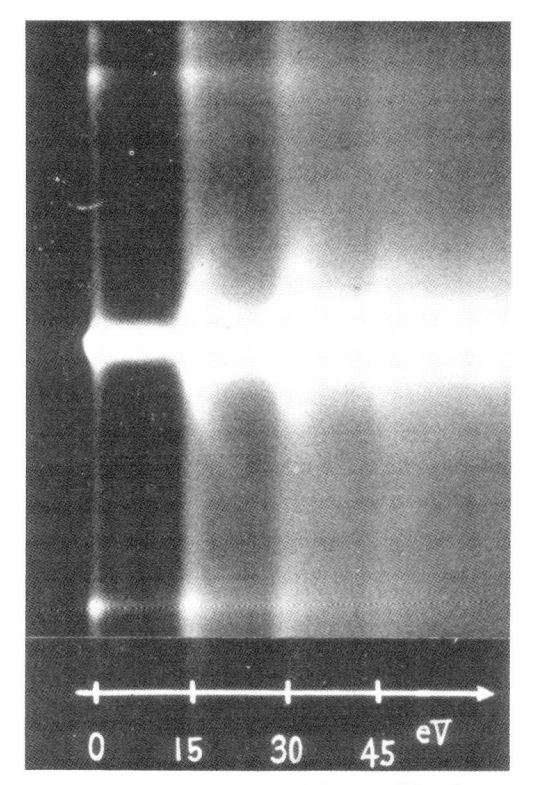

FIGURE 9. Energy-loss spectrum from a part of electron diffraction pattern of Al foil.

in addition to the normal image. The multifunction and accessory situation will now be described.

1. *Electron Diffraction*

Electron diffraction equipment had been made by several physicists around 1940. There were, however, few technical contacts between these physicists and electron microscope investigators in those days, in spite of common technical features in the electron beam instruments.

Around 1950, electron diffraction equipment was improved by incorporating multiple lenses. Especially, Tanaka and Hashimoto (1953) tried to combine electron diffraction and electron imaging, based on Boersch's proposal. They succeeded, but it was somewhat later than the first electron diffraction capability in an EM column, realized independently by Le Poole, based on the same principle. After this, communication among electron

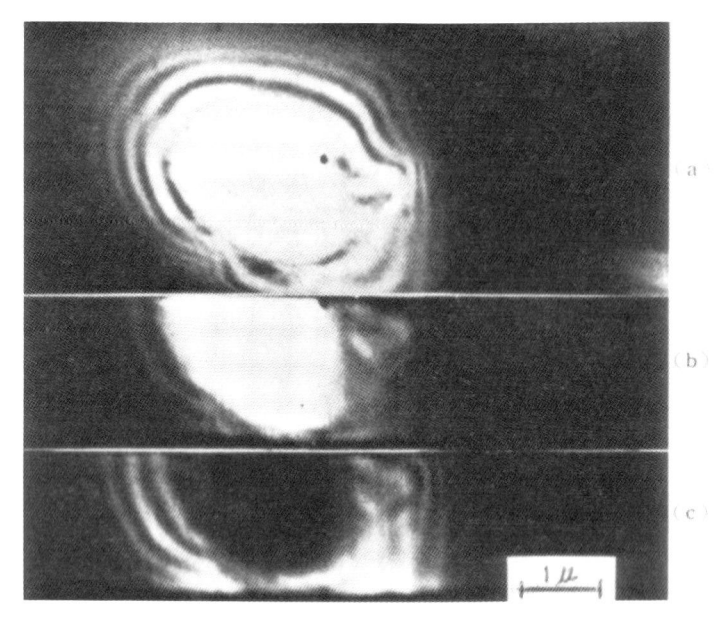

FIGURE 10. Energy-selected electron microscopic image of Al foil (equal thickness fringes): (a) bright-field image; (b) no-loss electron image; (c) energy-lossed electron image.

diffraction users and electron microscopists took place and every electron microscope manufacturer employed a three-stage lens system providing electron diffraction capability in 1950–1953.

2. Reflection Electron Microscope

The reflection electron microscope (REM) was developed for direct observation of the surface of bulk materials without using the established replica technique. Ito et al. (1954), of JEOL, Ltd., developed a REM with two pairs of deflection coils attached to the upper part of specimen chamber of JEM-5, as shown in Fig. 11, which provided reflection images of a bulk specimen by a beam deflection of more than 10°. The reflection image shown in Fig. 12 was obtained with the REM, which could take not only a reflection image but also the corresponding reflection electron diffraction pattern, because a three-stage lens system was employed. Later, a specimen heating device, up to 1000°C, was added to this REM, and the surface changes of Al-Mg-Si alloy were successively observed by Takahashi et al. (1956), of Yamanashi University.

3. Specimen Manipulation Devices

Strong demands to develop specimen handling devices for tilting, inducing thermal, mechanical, or chemical effects in the specimen, occurred around

FIGURE 11. Reflection-type electron microscope (JEM-5, 1954).

1955, and every electron microscope manufacturer was faced with solving these problems. For instance, a specimen holder which could change the temperature from liquid nitrogen temperature to 1000°C was developed.

As high-voltage electron microscopy developed, *in situ* observation of thicker specimens with various kinds of specimen treatments proved to be very effective for materials science; at the same time, the performance of these specimen treatment devices was greatly improved. In the early days,

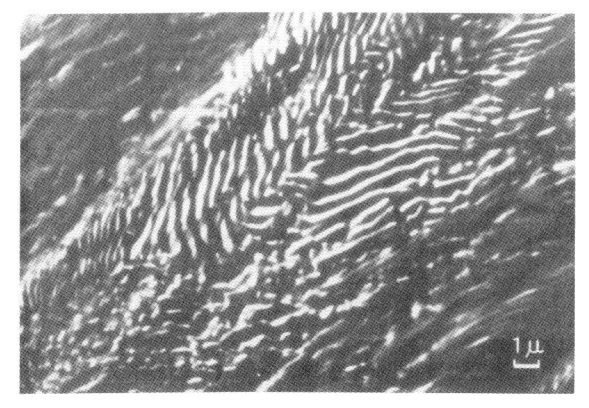

FIGURE 12. Reflection electron microscope image of pearlite with the beam deflection by 30°.

most specimens were inserted by top entry, but nowadays, side-entry devices have become preeminent because of easy specimen exchange.

For the successive observation and recording of a dynamically changing specimen, new recording media, besides photographic material, were demanded. For instance, Hashimoto *et al.* (1958), at Kyoto Technical University, studied the growth process of W-oxide crystals by using a gas reaction and heating specimen chamber and recorded the images with a 16-mm movie camera. After that, Takahashi *et al.* (1960), at Yamanashi University, added a movie camera directly to a JEM-6A. Later, as television technology developed, TV systems became standard for dynamic observations and recording.

4. *X-Ray Microanalysis*

The combination of X-ray analysis with electron microscopy, suggested by Duncumb in 1958, was realized by Katagiri and Ozasa (1967). They brought a wavelength-dispersive X-ray analyzer consisting of a crystal monochromater and a proportional counter into the specimen chamber of an HU-11B microscope as shwon in Fig. 13. With this instrument, they could obtain elemental analysis data in addition to the image and the corresponding electron diffraction pattern (Fig. 14).

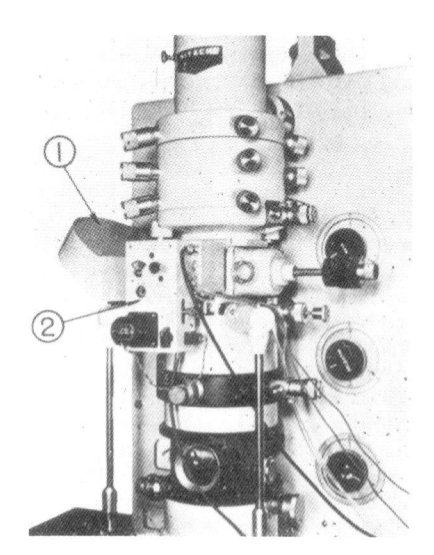

FIGURE 13. Model HU-11 electron microscope incorporated with X-ray microanalyzer (1965): (1) spectrometer, (2) driving unit.

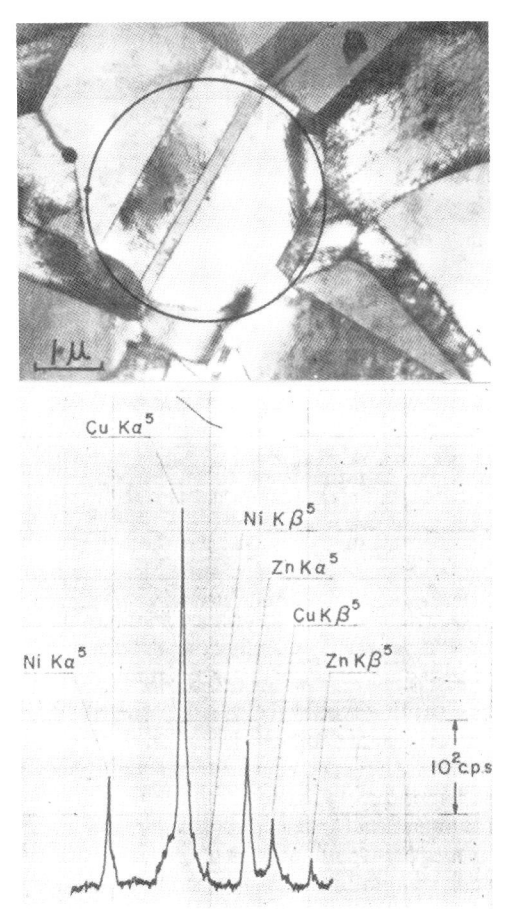

FIGURE 14. Electron micrograph of German silver film and X-ray spectrum obtained from the circular area on the image.

5. *Anticontamination Devices*

As the illumination spot was reduced by the double condenser lens, specimen contamination became a serious problem. An anticontamination system was introduced for the first time by Siemens. In Japan, Komoda and Morito (1960), at Hitachi, Ltd., studied the origins of the contamination and found practical ways to prevent it. Figure 15 shows a schematic drawing of their device, in which the specimen chamber was isolated from the vacuum and a cold trap chilled with liquid nitrogen was introduced into the chamber. Komoda and Morito obtained the results shown in Fig. 16

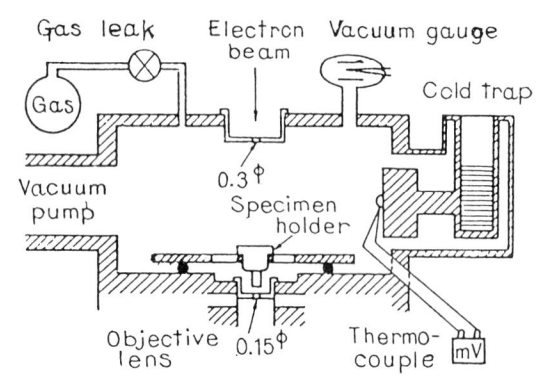

FIGURE 15. Schematic diagram of specimen chamber used for specimen contamination experiment.

from the measured contamination rate with temperature of the cold trap. It turned out that the contamination rate becomes almost zero when the cold-trap temperature goes down lower than −70°C.

Later, various kinds of anticontamination devices were supplied by every manufacturer on the basis of these experimental data.

F. Observation of Crystal Lattice Images

The report of a direct observation of the crystal lattice by Menter (1956) gave great stimulus to Japanese research workers to start the study of lattice image observation. At first, the aim of these studies was microscopic observation of the atomic arrangement, or lattice disorder and defects such as dislocation in crystals, but it changed to using the lattice image to estimate the performance of electron microscopes in Japan, because the lattice spacings were known constants.

It was desired to observe lattice images of increasingly narrower fringe spacing. Table I and Fig. 17 show the chronological records of the lattice spacings thus resolved, where the curve of the resolution shows a remarkable bend occurring about 1970, which suggests the following. (1) Before 1970, the resolution of electron microscope was limited by various disturbing factors due to electric and mechanical instabilities of the instrument, so that better resolution was obtained when some of them were removed. (2) Around 1970, these disturbing factors were almost completely absent, so that the theoretically expected resolution was achieved. (3) Rather slow progress thereafter was obtained by correction of essential problems concerning resolution, such as reduction of lens aberration, higher accelerating voltage, and use of field-emission electron guns.

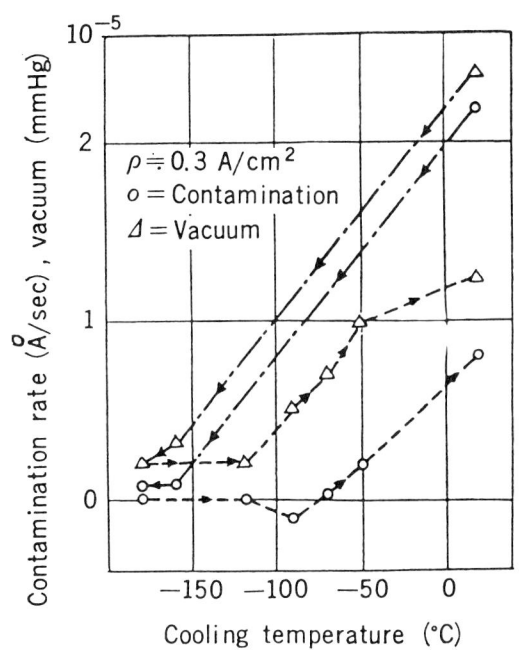

FIGURE 16. Contamination rate versus cold block temperature.

TABLE I

History of Crystal Lattice Image Observation

Year	Spacing (Å)	Specimen	Authors
1956	11.9	Pt-phthalocyanine	J. W. Menter
1957	9.8	Cu-phthalocyanine	E. Suito et al.
	6.9	MoO_3 (020)	G. A. Bassett, J. W. Menter
1958	5.8	Ni/Au (422) moiré	G. A. Bassett et al
	5.6	K_2PtCL_6 (111)	T. Komoda, S. Sakata
1959	4.9	1/2 Cu-phthalocyanine	K. Ito
1961	3.2	tremolite	W. C. T. Dowell
1963	2.35	Au (111)	T. Komoda
1964	2.04	Au (200)	T. Komoda
1965	1.81	Cu (200)	T. Komoda
1966	1.27	Cu (220)	M. Watanabe et al.
	1.18	Au 1/2 (111)	T. Komoda
1968	0.88	Ni 1/2 (200)	K. Yada, T. Hibi
1975	0.72	Au 1/2 (220)	P. Sieber, K. Tonar
1978	0.62	Ni 1/2 (220)	T. Matsuda et al.
1990	0.55	Au $(02\bar{2})(2\bar{4}2)$	T. Kawasaki et al.

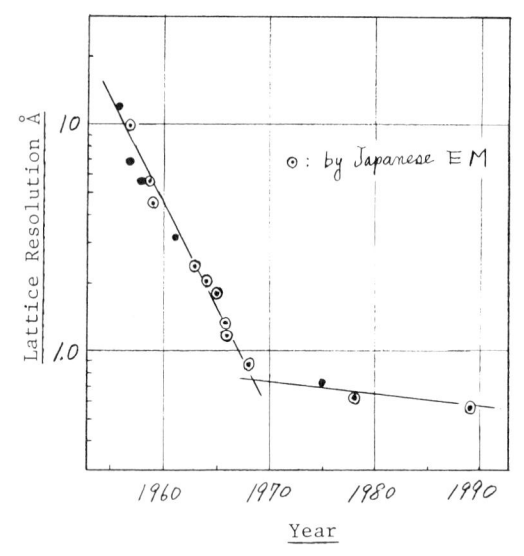

FIGURE 17. Historical trend of lattice resolution.

Lattice image observation began in Japan with the observation of Cu-phthalocyanine (12.5 Å), within a year of Menter's report, and then the 5.6-Å spacing of $K_2 PtCl_6$ (Komoda, 1958) and the 4.9-Å half-spacing of Cu-phthalocyanine (Ito, 1959) were observed.

In 1961, Dowell, at the Fritz Haber Institute, Berlin, resolved 3.2-Å fringes by means of tilted illumination. Taking advantage of the tilted illumination method to reduce the chromatic aberration nearly to zero, Komoda (1964) succeeded in observing the (111) planes of gold with a lattice spacing of 2.35 Å. He further developed the method to observe mutually crossed multiple fringes, such as the (200) and (020) planes of gold, as shown in Fig. 18 (Komoda, 1966).

Komoda improved the resolution record to 1.18 Å in 1966 by using a pointed cathode. In 1968, Yada (1969) succeeded in obtaining lattice fringes of 0.88 Å, which was the first record of a lattice image narrower than 1 Å. Improvement of the resolution of electron microscope was accelerated in Japan by this severe competition. Consequently, the resolving power of electron microscopes, even commercial ones, reached the theoretical value limited by spherical aberration and diffraction.

G. Development of High-Voltage Electron Microscopes

Some demands for making the accelerating voltage of electron microscope higher came from research workers in the medical and biological fields at

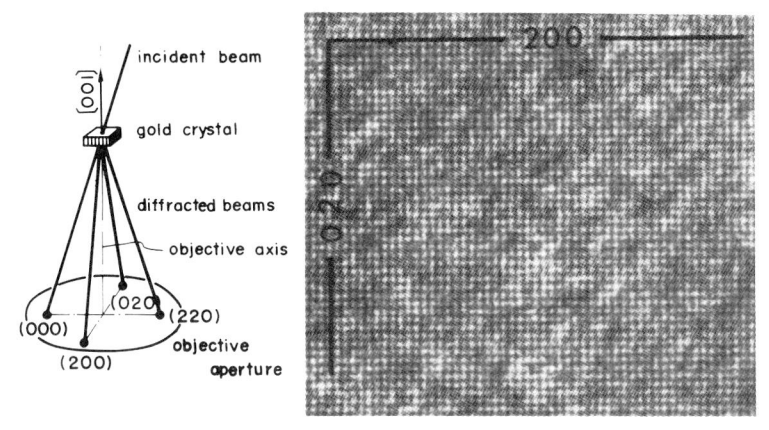

FIGURE 18. Tilted illumination method, and crossed lattice image of (200) and (020) planes in gold observed with tilted illumination.

the end of the 1940s, and high-voltage electron microscopes at the 300-kV level were planned to be built independently by two groups, Nagoya University and Hitachi, Ltd., and Kyoto University and Shimadzu, Ltd. The former employed a Van de Graaf-type belt generator, a high-energy accelerator, for the high-voltage power supply of the microscope (Tadano *et al.*, 1956). Figure 19 shows the principle, whereby electrons are supplied to a running belt by corona discharge and conveyed to the top electrode where high

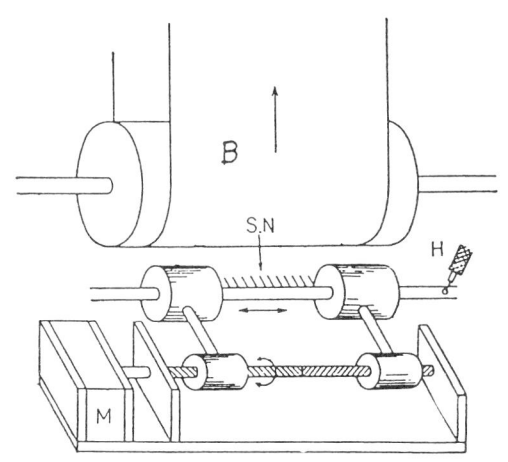

FIGURE 19. Schematic diagram of electron spray device in Van de Graaff generator: B, belt; SN, spray needles; H, spray voltage supplier.

voltage is produced. This apparatus was completed in 1954 (Fig. 20), and resolution of 25 Å was obtained. At that time, however, thin sectioning of biological samples was realized by the development of ultramicrotomy technique, so the necessity for a high-voltage electron microscope almost disappeared from the fields of biology and medicine.

On the other hand, the latter group made a 350-kV electron microscope by using three normal high-voltage transformers in a cascade connection. One of the researchers, K. Kobayashi, used this instrument, completed in 1957, and made it clear that electron radiation damage to organic materials is greatly reduced as the accelerating voltage becomes higher. In 1961, Hashimoto also measured the energy dependence of extinction distance and transmissive power for electron waves in crystals, and his findings became the theoretical basis for the construction of higher-voltage electron microscopes (Kobayashi *et al.*, 1963).

In the later half of the 1950s, as metal thinning techniques developed, it became possible to observe metals directly with a normal electron microscope. It became clear, however, that the metallurgical properties of samples sufficiently thin to be observable by 100-kV microscopes differ greatly from the those of the bulk sample, so that it is necessary to observe samples of a few micrometers thickness to know the true bulk state. Thus, the high-voltage electron microscope became indispensable for direct transmission observation of metal samples.

A user-friendly high-voltage electron microscope was thus urgently needed, and it began to be commercialized by the cooperation of Uyeda and Sakaki, Nagoya University, and Tadano *et al.*, Hitachi, Ltd., from 1962. They completed a 500-kV high-voltage microscope in 1964. This was such a compact type that the high-voltage generator and the electron accelerator

FIGURE 20. 300 kV high-voltage electron microscope incorporating Van de Graaff generator (1954).

are installed together in a high-pressure tank and the column is connected to the tank (Fig. 21) (Tadano *et al.,* 1965). Similar compact types of high-voltage electron microscopes were commercialized by every manufacturer, Hitachi, Shimadzu, and JEOL. In 1966, the accelerating voltage had risen from 500 kV to 1000 kV. Thereafter, the majority of high-voltage microscopes in the world were supplied by Japan.

In those days there was an argument, from a theoretical expectation based on relativistic laws, that the effectiveness of high-voltage electron microscopy, because of the penetrating power of electrons would approach a limit at around 500 kV. R. Uyeda (1967) tried to observe experimentally the image quality of samples of various thicknesses over a voltage range from 50 kV to 1200 kV; he showed that at the higher voltage, in fact, sharper images are observable. This result promoted the realisation of 3000-kV electron microscopes instead of negative discussion of 1000-kV microscopes.

In this context, Fujita *et al.,* (Osaka University) developed a 3000-kV high-voltage electron microscope in cooperation with Hitachi, Ltd. It was begun in 1968 and completed in 1970 (Fig. 22). The results obtained with the microscope were reported at the 7th International Conference on Electron Microscopy at Grenoble, France. This huge and highest-voltage instrument,

FIGURE 21. Compactly designed high-voltage electron microscope (1964).

FIGURE 22. 3000 kV ultrahigh-voltage electron microscope (1970).

12 m high and 70 tons in weight (Ozasa et al., 1972), was actively used thereafter at Osaka University, producing many remarkable results.

H. Development of the Scanning Electron Microscope

The X-ray microanalyzer had been studied in Japan in 1957 and commercialized around 1961. The functioning of this instrument—scanning of an electron probe over the sample, detection of characteristic X rays, mapping with a selected constituent element—are quite similar to those of scanning electron microscopes (SEMs), so that scanning electron microscopy was born as one of the electron probe microanalyzer capabilities in Japan.

Later, in 1964, Kimura et al. (1966), at Hitachi, Ltd., developed a SEM for observations of surface structure and electric potential distribution of semiconductors. More general samples, such as biological and mineralogical ones, were observed with commercial instruments produced by JEOL beginning in 1966. The SEM market gradually expanded, and the MINI-SEM developed by Akashi in 1972, was very compact and easy to handle so that it was accepted worldwide, almost like a light microscope, very rapidly.

II. 1971–1990

A. Reduction of Spherical Aberration

As described in Sec. I, the performance of electron microscopes in Japan had nearly reached the theoretical limit around 1970. Further efforts to improve the performance and function of electron microscope continued thereafter. The primary objective was the reduction of spherical aberration.

Around 1960, Maruse *et al.* (Nagoya University) started a study of a foil lens, which had been suggested by Scherzer in 1948 but not tried, for the purpose of correcting spherical aberration. The foil lens made by Maruse's group, shown in Fig. 23 (Hibino, 1986), consisted of a conductive foil and an adjacent electrode with a small hole. When a positive potential is applied to the electrode, the electric field consequently produced there shows a concave lens action, so correction of the spherical aberration becomes possible.

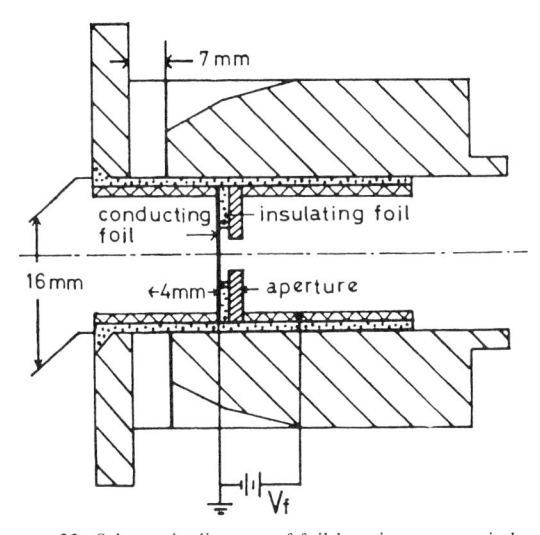

FIGURE 23. Schematic diagram of foil lens in a magnetic lens.

Figure 24 is an example of the results, showing that there is an actual possibility of correcting spherical aberration over a wide range (Hibino, 1986).

On the other hand, some efforts involving the shape and excitation condition of the lens polepiece continued to minimize the spherical aberration to its limit. For instance, Yanaka et al. (1967), at JEOL, Ltd., obtained $C_s \approx 0.5$ mm by strong excitation of a lens whose lower polepiece bore was 2 mm in diameter. In 1968, Suzuki et al. (1968), of Akashi, Ltd., also realized a lens of $C_s \approx 0.5$ mm, the so-called second-zone lens, by higher excitation than for the condenser objective.

On the other hand, Liebmann, as early as 1951, had suggested a lens called a pinhole lens, whose upper bore was much smaller, having a feature that a rather smaller value of the spherical aberration could be obtained with a fairly weak excitation. Yada and Kawakatsu (1976), at Tohoku University, made a systematic study of lens characteristics for various shapes of lens by employing a polepiece-in-polepiece system, whereby various dimensions of polepiece were made by inserting small polepieces into the bore of the

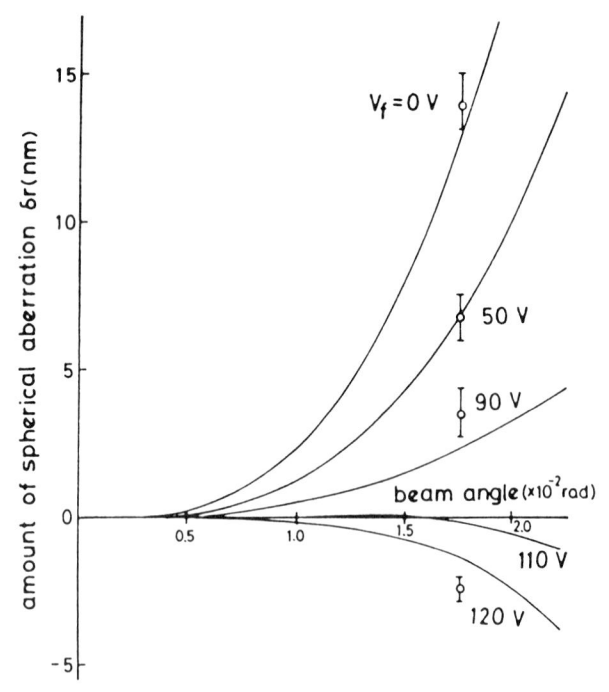

FIGURE 24. Spherical aberration of an objective lens compensated with the foil lens (parameter is electrode potential).

usual host objective lens (7ϕ-4h-7ϕ), as shown in Fig. 25. The use of a small bore in the upper polepiece was very effective for realizing small spherical aberrrations at the fairly weak excitation of $NI/V^{1/2} \leq 14$. By scaling down all the dimensions of the bore diameter and spacing, it was possible to realize a small C_s of 0.3 mm under the condition 1ϕ-1.5h-2ϕ, 100 kV, $NI/V^{1/2} = 23.4$. From these studies, the direction and limit of the reduction in spherical aberration of the objective lens were made clear.

In accord with these results, objective lenses of smaller spherical aberration were designed for commerical microscopes from the end of the 1970s. Above all, the EM-002A/002B microscope (Akashi, Ltd.) eventually realized $C_s = 0.3$ mm by scaling down the upper and lower bores and gap distance together (Yanaka et al., 1983). As the accelerating voltage of the microscope was 200 kV, its resolution was 1.8 A.U., which was realized by an extreme reduction of the thickness of the side-entry specimen holder to a practical limit.

B. Elevation of Accelerating Voltage

Improvement of resolving power depends on two factors, smaller spherical aberration and shorter wavelength. If we use higher accelerating voltage to shorten the wavelength, the other factor, spherical aberration, is apt to increase. On the whole, however, elevation of the accelerating voltage is more effective in improving the resolving power.

In Japan, a 500-kV electron microscope was developed for high resolution by N. Uyeda et al. (Kyoto University) at the beginning of the 1970s. The voltage of 500 kV was selected for the following reasons: (1) stabilization

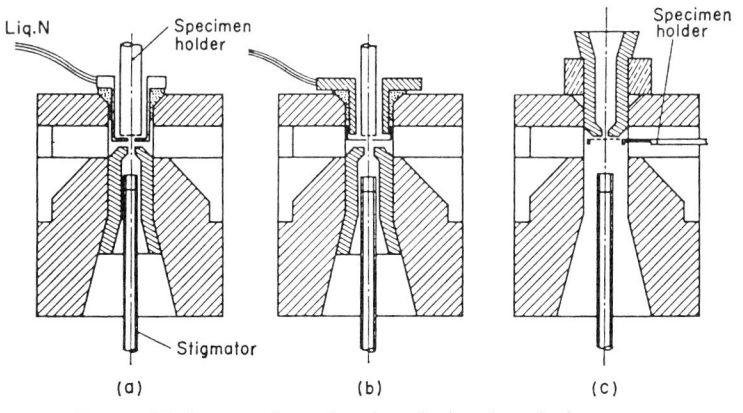

FIGURE 25. Lens configurations in polepiece-in-polepiece system.

of the accelerating voltage becomes more difficult as the voltage goes up; (2) radiation damage will be critically reduced as the voltage goes up but later increased with voltage higher than a certain value by another effect, knock-on collision causing the direct displacement of atoms. They observed a molecular image of chlorinated Cu-phthalocyanine crystal as shown in Fig. 26 (Uyeda *et al.*, 1979), where atoms such as Cl and Cu were clearly imaged.

On the other hand, Hashimoto (Osaka University) continued to observe the motion of individual atoms in a sample, directly, at the end of the 1970s. In parallel with this, techniques such as X-ray analysis and EELS analysis in the nanometer range had become possible at that time, so he extended his ideas to a new high-resolution electron microscope which was capable of identifying individual atoms and pursuing motion of atoms by combining those techniques. He actually tried to make a new microscope called an analytical atom-resolution EM (AARM) in cooperation with JEOL, Ltd. (Hashimoto *et al.*, 1983). The accelerating voltage of this microscope was selected to be 400 kV from consideration of the knock-on collision effect at higher than 500 kV. JEOL, Ltd., produced the microscope from 1984

FIGURE 26. Molecular image of chlorinated Cu-phthalocyanine crystal taken with 500 kV electron microscope.

and put a number of these microscopes on the market, adding the JEM-400 FX series with an analytical probe size of 2 nm. Hatachi, Ltd., produced a 300-kV high-resolution microscope in 1985.

In such an atmosphere of increased accelerating voltage, Bando *et al.* (1986), of the National Institute for Research in Inorganic Materials made it clear that spatial resolution and signal-to-noise ratio of the analysis improve as the accelerating voltage goes up. Therefore, the accelerating voltage of 300–400 kV was commonly accepted to be a standard for material researches with high-resolution and high-performance analytical function.

On the other hand, Horiuchi *et al.* (1991), of the National Institute for Research in Inorganic Mateials, developed a high-resolution, high-voltage microscope, normally 1300 kV and 1500 kV at maximum, in cooperation with Hitachi, ltd., according to their idea that a resolving power of 1 Å, necessary for resolving the light elements such as oxygen in the crystal, will be realized with accelerating voltages higher than 1250 kV. With this instrument completed in 1990, they succeeded in observing the array of oxygen atoms in ZrO_2 crystal as shown in Fig. 27. Thus, the resolving power of electron microscopes now reached 1 Å by elevating the accelerating voltage.

C. Development of Field-Emission Electron Guns and the Study of Electron Holography

As the resolving power of the electron microscope had nearly reached its theoretical limit by the end of the 1960s, Tonomura *et al.* (1968), at Hitachi,

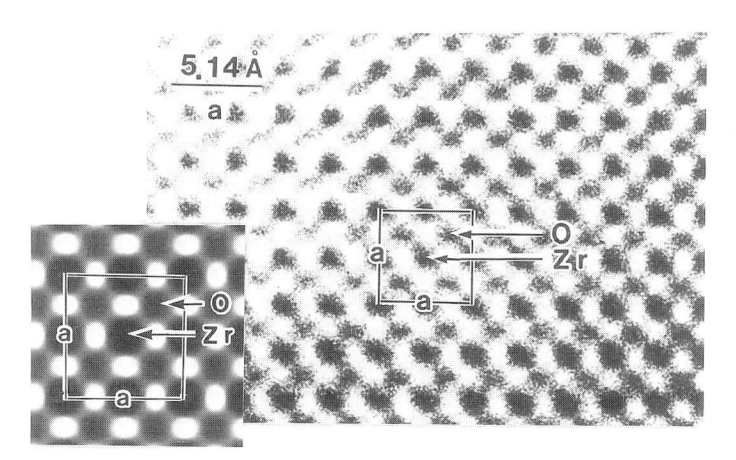

FIGURE 27. Structure image of ZrO_2 crystal observed with H-1500 high-resolution, high-voltage electron microscope.

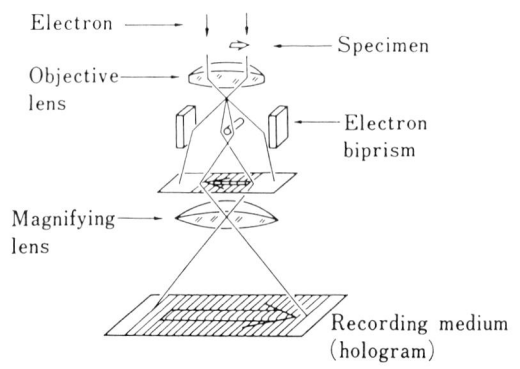

FIGURE 28. Schematic diagram of hologram formation by means of electron microscopy.

Ltd., initiated the study of electron holography to overcome the resolution limit, with the first experiment of "in-line" Fraunhofer-type holography. The pointed cathode developed by Hibi *et al.* was used in the experiment as the electron source, but resolution of the reconstructed image was several tens of angstroms, fairly inferior to that of the normal image because of insufficient coherency of the electron beam.

At that time, a field-emission electron gun providing very high brightness had been successfully used in a scanning electron microscope by Crewe. Tonomura *et al.* also tried to use the field-emission electron gun in electron holography. When the stability of their field-emission gun was improved to a practically useful level, the HU-11C microscope was equipped with a field-emission gun in 1970. As a result, they could obtain 120 Fresnel fringes with 10 s exposure in 1972. More Fresnel fringes (300 with 4 s exposure) were obtained in 1978 by improving the stability of the illumination system of the HU-12A (Tonomura *et al.*, 1979). The brightness of the field-emission gun was as high as 2×10^8 Å/cm^2 · Sr at 100 kV. They lowered the record of the lattice fringe spacing to $d = 0.62$ Å of the half-spacing of Ni{220} using this field-emission gun. They also obtained over 3000 interference fringes when a Möllenstedt-type electron biprism was installed in a position between the objective and the intermediate lenses, which was coherent enough to start experiments in holography.

A hologram was made by an off-axis mode as shown in Fig. 28, where the electron wave through the specimen was superposed on the reference wave passing through the space without the specimen. Figure 29 shows such a hologram and its optically reconstructed image with laser light. Resolution of the reconstructed image was about 5 Å.

Tonomura (1984) began to measure the thickness distribution and magnetic field distribution of the sample from phase information in the

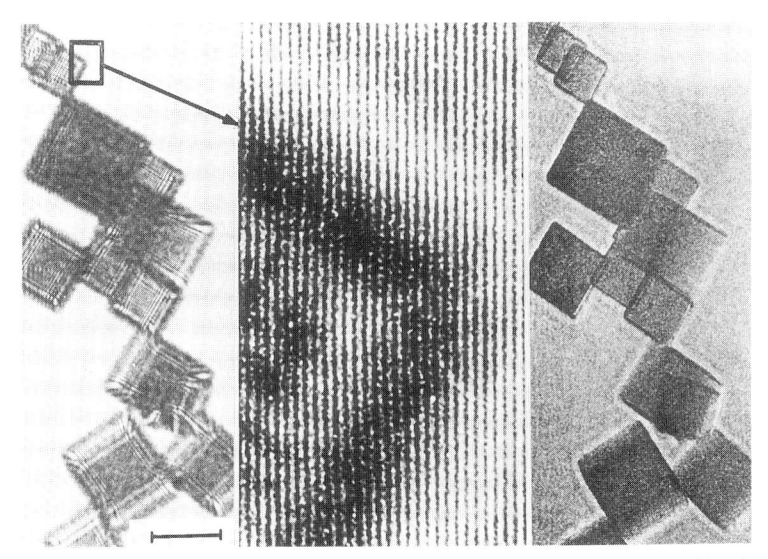

FIGURE 29. Electron hologram (left) and its reconstructed optical image (right).

electrons transmitted through the sample included in the hologram. They could read out the phase information as an interference fringe pattern in the optical reconstraction procedures.

Figure 30 shows a magnetic flux distribution observed in a Co evaporated magnetic tape (Tonomura, 1984), where the fringe pattern shows a flow of magnetic flux, and a leakage flux outside from the specimen edge is also seen in the lower part. Figure 31 shows magnetic flux quanta (fluxon) observed for the first time anywhere (Matsuda *et al.,* 1989), where each band corresponds to a unit flux quantum with the value of $h/2e$.

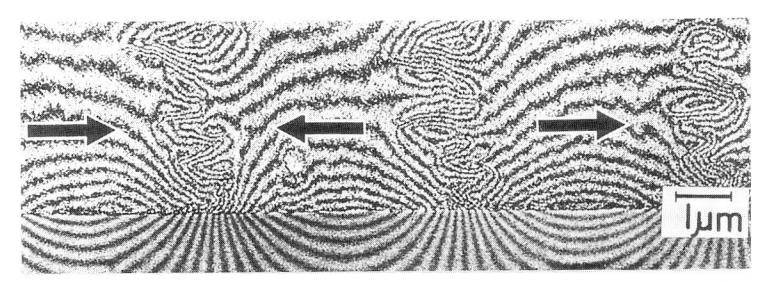

FIGURE 30. Magnetic force lines observed by means of electron holography (Co magnetic tape).

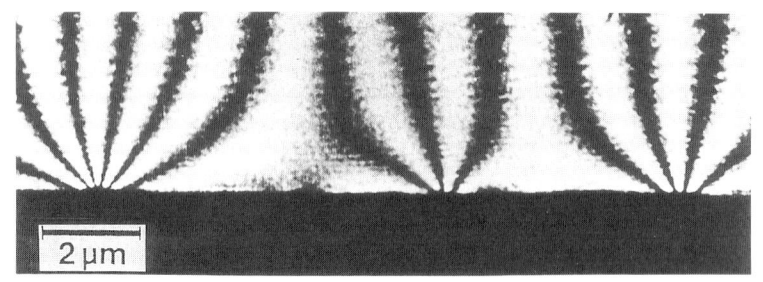

FIGURE 31. Magnetic fluxons observed by electron holography.

A minute strain around a screw dislocation terminated on a GaAs (110) surface was observed with a resolution of 0.1 Å in height by the holographic technique applied to a reflection-type electron microscope (Osakabe *et al.*, 1989). In 1986, there was experimental proof of the Aharanov-Bohm effect, which made a great contribution to basic physics (Tonomura *et al.*, 1986).

Thus, electron holography has grew to be a powerful new measuring method for investigating local physical properties of a sample quantitatively.

D. *Development of Field-Emission-Type Scanning Electron Microscopes*

As described in the preceding section, Hitachi, Ltd., to which Tonomura belonged, had developed a field-emission (FE) gun for electron holography. As its performance reached a practical level around 1970, it was planned to use the FE gun in the scanning electron microscope. In those days, Crewe had succeeded in taking the image of individual atoms using a 5 Å electron beam with his transmission scanning electron microscope equipped with an FE electron gun. In contrast, Hitachi applied the FE gun to a normal scanning electron microscope which used secondary electrons for surface observation. As a result, they obtained remarkablly good resolution of 20 Å in the secondary electron image (Komoda and Saito, 1972). The HFS-2 was produced commercially in the next year as shown in Fig. 32 and has been supplied worldwide. In the 1980s, from a demand for high-level evaluation of materials in industrial fields such as electronics, biotechniques, and new materials, the high-resolution FE-SEM became an indispensible tool for the purpose, and its market expanded rapidly around 1982.

In the field of semiconductors, for instance, it became necessary to use SEM for inspection or quality control of LSI elements, instead of a light microscope, without damage and charge-up. For this purpose, FE-SEM, which could maintain resolution better than 100 Å even at 1 kV or lower accelerating voltage, was examined. The FE-SEM, to which other functions

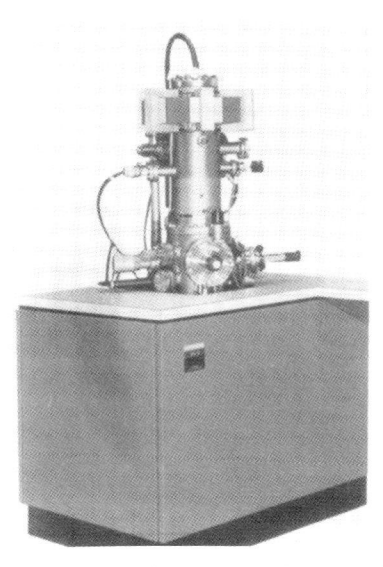

FIGURE 32. Field-emission scanning electron microscope (HFS-2, 1973).

such as size measuring for LSI patterns, LSI testing with potential contrast, and stroboscopic observation were added around 1984, has since been used for in-process purposes on semiconductor plant lines.

So far, it had been thought that the resolving power of SEM in the secondary electron mode would be limited to 10–20 Å, by the secondary electron generation volume. In contrast to this, Tanaka (Tottori University) thought it might be possible to improve the resolution of SEM further by reducing the electron probe size, and he tried, in cooperation with Hitachi, Ltd., to develop a new SEM in which the beam size could be reduced to 5 Å. In parallel with the use of an FE gun, they employed an in-lens system to reduce spherical aberration; in this, the sample was located in a position inside the objective (Nagatani and Saito, 1989). This instrument was completed in 1985. Kuroda *et al.* (1985), collaborators of Hitachi, could observe surface steps of 4.5 Å high on a tungsten crystal in the secondary electron image, which proved the high-resolution capability of the instrument. This ultrahigh-resolution SEM was produced commercially by Hitachi (S-900, 1986), which brought in an era of SEM better than 10 Å.

E. Development of the Nanoprobe Analytical Electron Microscope

Analytical electron microscopes had been realized in the middle of the 1960s, but the analysis area was larger than 0.1 μm and the resolution of the electron image was worse than that of normal electron microscopes. In the 1970s, it was highly desired to make a so-called nanoprobe analytical microscope which would be capable of element analysis of an ultrafine nanometer region, maintaining the high resolution of the electron microscope.

The first trials of the nanoprobe analysis technique began with the development of the high-excitation objective lens by Koike et al. (1970) at JEOL, Ltd. In the objective lens system developed by them, the incident beam was focused by the prefield of the lens into a nanometer-order spot on the specimen and at the same time, an electron diffraction pattern of the sample was formed on the screen by the postfield of the lens as shown in Fig. 33. An image of the sample was observed in the STEM mode by scanning the incident beam, and the electron diffraction pattern of a selected area was obtained by stopping the scanning of the beam. Another capability of nanoprobe element analysis was realized by adding X-ray analysis equipment and an electron energy-loss spectrometer (EELS). In addition to these, a secondary electron image was obtained by adding a secondary electron detector above the specimen. This kind of instrument was produced commercially by JEOL (JEM-100B analytical) and widely accepted for TEM scan.

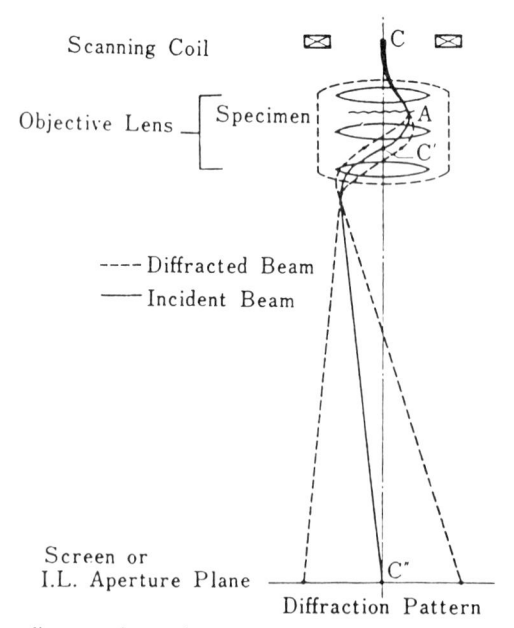

FIGURE 33. Ray diagram of scanning TEM by using a highly excited objective lens.

Later, Koike *et al.* moved to Akashi from JEOL, and developed a new nanobeam analytical microscope (EM-002A, 1983), in which it was possible to switch a high-resolution image over to nanoprobe analysis. As shown in Fig. 34, five lenses were used in the illumination system, including the prefield lens of the objective, for a stable nanoprobe. The objective lens had a symmetrical configuration in both the upper and lower polepieces, and spherical aberration was minimized to the practical limit of the commercial

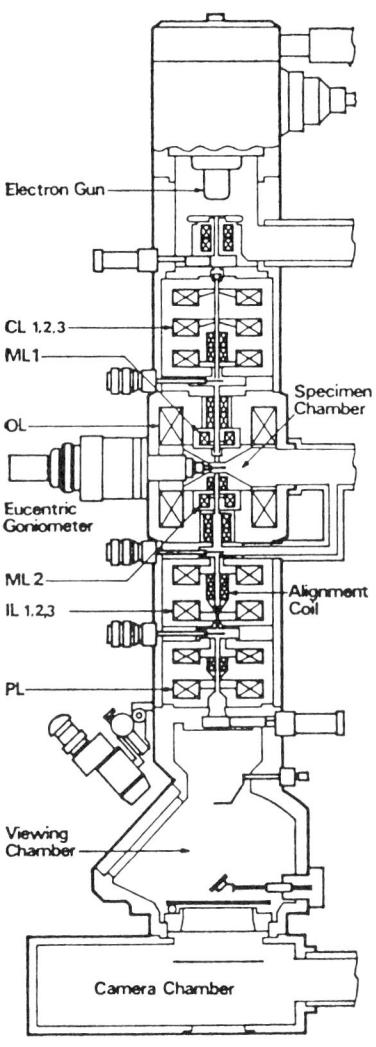

FIGURE 34. Cross-sectional diagram of EM-002A column (1983).

microscope, $C_s = 0.3$ mm, by reducing the lens gap (Yanaka et al., 1983), which made possible a high-resolution and nanoprobe analytical microscope.

Another way to realize the nanoprobe for TEM scan by using an FE electron gun, instead of multiple condenser lens such as the EM-002A, was tried by Koike et al. (1974). They obtained a resolution of 10–20 Å in the SEM mode and 5 Å in the STEM mode.

At Hitachi, Isakozawa et al. (1989) developed a similar FE-TEM and commercialized it (HF-2000), in which the accelerating voltage was 200 kV and the nanoprobe size was 10 A.U. on the sample with a probe current of 1 nA. It was possible to do nanoprobe analysis of about 10 A.U. spatial resolution, as shown in Fig. 35. Moreover, the information limit in the high-resolution image extended to abont 1.6 Å because of the small energy spread of 0.5 eV in the cold field-emission gun. Thus, the nanoprobe high-resolution electron microscope was realized in both name and fact.

F. Recent Technical Trends

1. Ultrahigh-Vacuum Microscope

Specimen contamination had become a troublesome problem as nanoprobe analysis was developed. Therefore, there was a demand to evacuate the specimen chamber to an ultrahigh vacuum better than 10^{-7} Pa. On the other hand, the ultrahigh-vacuum technique was indispensable in the *in situ* observation of the surface reaction process of the sample. In Japan, Yagi et al. (Tokyo Institute of Technology) had studied thin-film formation by deposition in the specimen chamber of an ultrahigh-vacuum microscope developed in cooperation with JEOL in 1978 (Takayanagi et al., 1978). Later, they observed the surface structure transformation of a Si crystal by applying the technique to reflection electron microscopy (REM) (Osakabe et al., 1981).

More recently, Hitachi developed a new ultrahigh-vacuum microscope whose specimen chamber could be evacuated to a 2×10^{-8} Pa (Fig. 36) (Kubozoe et al., 1989). An interesting feature of this instrument was a pretreatment chamber provided in the specimen transfer system, in which it was possible to do various kinds of pretreatments and surface analysis such as LEED, Auger analysis, and mass spectroscopy followed by electron microscope observation without exposure to air. By this trial, instruments for study of surface physics with high-resolution electron microscopy at the atomic level were realized through ultrahigh vacuum.

2. Spin-Polarized SEM

Recently, there occurred a wave of interest in developing a new functional SEM by using new kinds of detectors and measuring techniques. One of

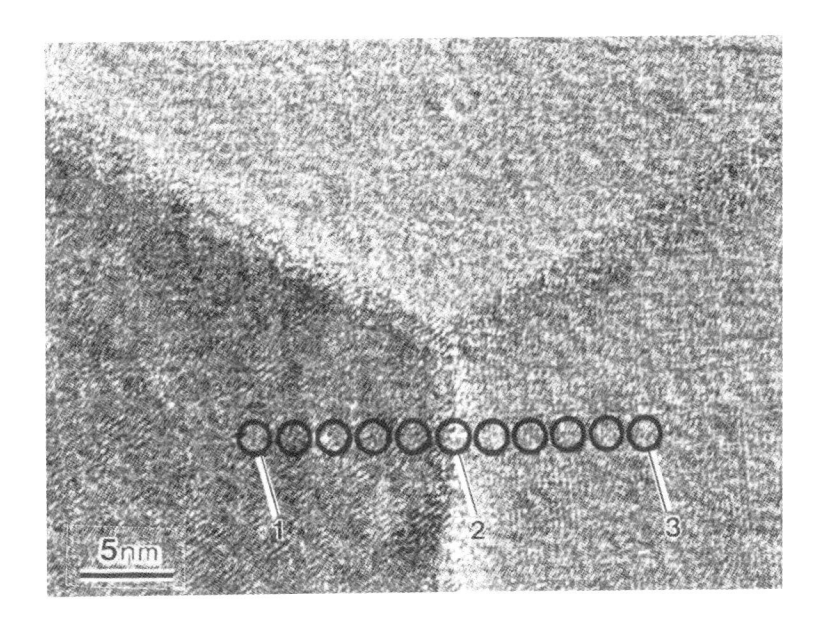

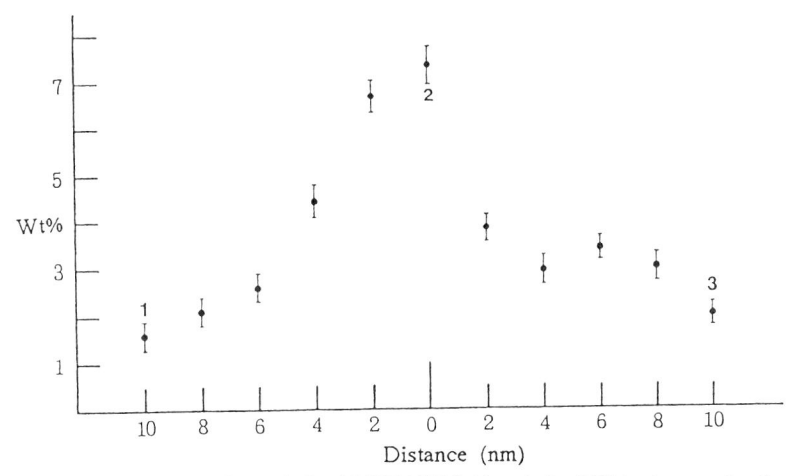

FIGURE 35. Nanoprobe analysis with HF-2000 field-emission TEM, whose probe diameter is 1.5 nm. Illustrated is density distribution of P doped in Si polycrystal.

them was the spin-polarized SEM. In general, an electron beam has spin polarization, but its detection is not easy, so application to electron microscopy had never been tried. At the end of the 1960s, however, a measuring technique was developed, and Chrobok and Hofmann showed that spins of the secondary electron from a ferromagnetic material are aligned in a

718 TSUTOMU KOMODA

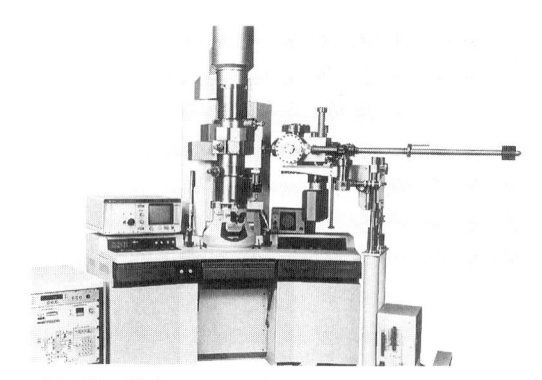

FIGURE 36. Ultrahigh-vacuum electron microscope (H-9000, 1989).

certain direction reflecting the magnetization direction of the magnetic material.

In 1984, Koike and Hayakawa (1984), of Hitachi, Ltd., developed a spin-polarization SEM by which magnetic domains could be observed. The sensitivity of a Mott-type spin detector used in this spin SEM (Fig. 37) was so low as to need two or three orders higher beam current than that of the usual SEM. Therefore, they used an FE electron gun for the purpose and obtained a resolution of 1000 A.U. Figure 38 shows an image of the magnetic domain structure obtained with this instrument, where the local magnetization orientations of the sample were determined from the image contrast.

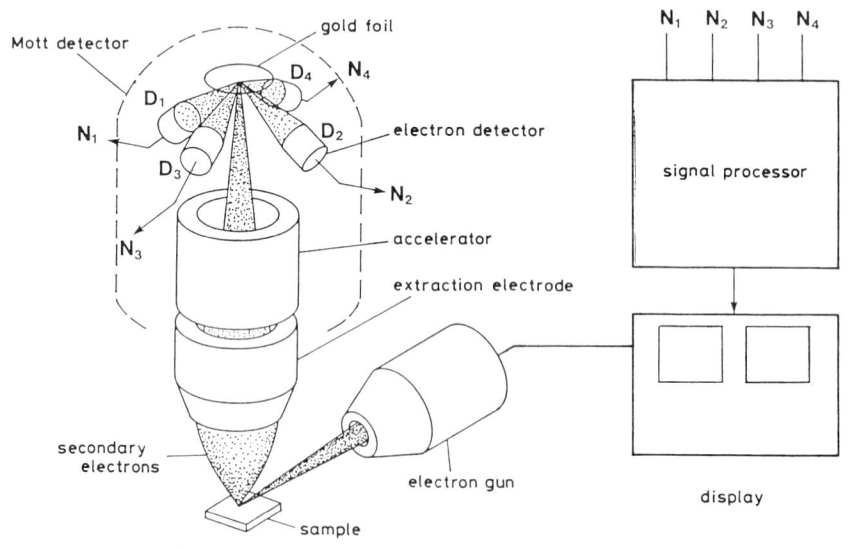

FIGURE 37. Schematic diagram of spin-polarized SEM.

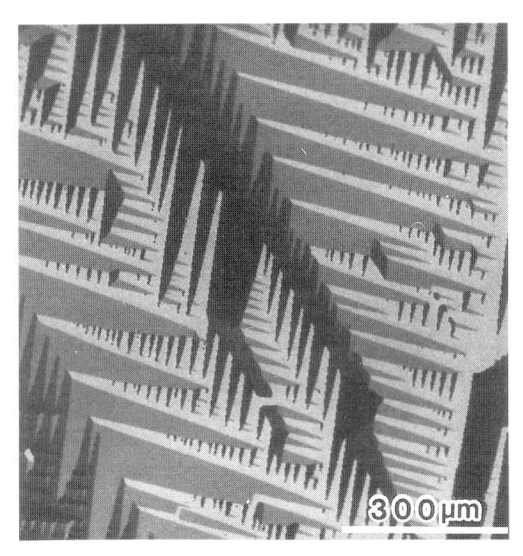

FIGURE 38. Magnetic-domain image of iron (001) surface observed with spin-polarized SEM.

3. *STM/SEM and STM/TEM*

The scanning tunneling microscope was invented by Binnig and Rohrer (IBM Zurich) in 1982 and progressed remarkably thereafter. This microscope has a great advantage of high resolution, better than 1 Å in the direction of height, so that surface structure can be depicted with atomic resolution in real space, but at the same time it has a disadvantage in that the scanned area is limited to a very narrow area by the nature of the piezoelectric element used for driving the scanning probe so that a low-magnification image is not obtainable.

To overcome this difficulty, it was tried to combine an STM with an SEM or a TEM. For instance, Ichinokawa *et al.* (1987), at Waseda University, brought a STEM unit into a commercially available SEM and proved that high-resolution STM images of fine particles and cleaved crystal surface such as Au-Pd deposited particles and MoS_2 can be obtained in addition to normal SEM images. Takata *et al.* (1989), Hitachi, Ltd., installed a STM into the specimen chamber of an ultrahigh-vacuum SEM in 1988. A wide area of the sample was first observed in the SEM mode and a selected area of the sample was then observed by the STM, as shown in Fig. 39, where contamination or abnormality of the tip could be previously checked by the SEM. In future, more comprehensive surface observation or analysis will be achieved by adding a surface analysis function such as RHEED to the STM/SEM.

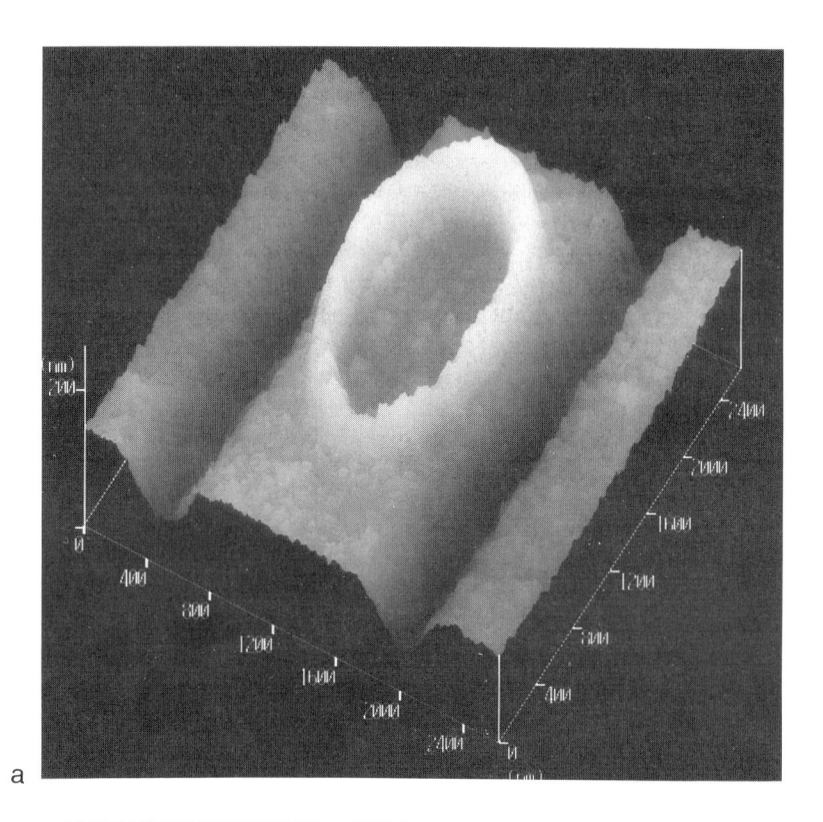

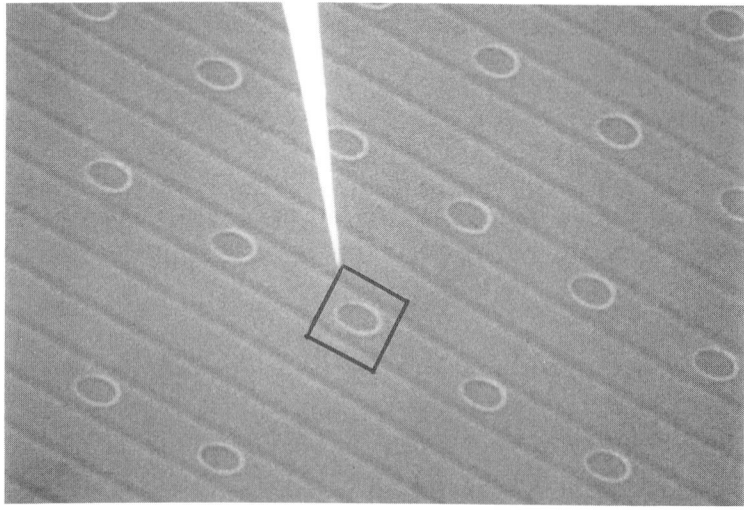

FIGURE 39. STM image (a) and SEM image (b) of optical disk. A white needle in (b) illustrates the STM probe.

On the other hand, Takayanagi et al. (Tokyo Insitute of Technology) developed a STM installed in the side-entry specimen holder of a TEM in cooperation with JEOL (Iwatsuki et al., 1991). With this instrument, a high-resolution transmission image was observed at atomic resolution and at the same time, a STM image of the sample could be observed by moving the probe to the surface. Reflection images could also be observed by the REM mode, namely, a wide area of the specimen surface is dynamically observed and then a selected local structure is observed at atomic resolution in the STM mode. Surface phenomena such as adsorption, chemical reaction, diffusion, and so on, of atoms or molecules on the surface will be analyzed dynamically and microscopically by these new instruments.

III. Conclusion

It may be concluded that recent study and development of electron microscope instruments in Japan have been oriented in a direction of the next generation of instruments, high-resolution observations at the atomic level, and various functions of nanoprobe analysis.

References

Bando, Y., Matui, Y., and Kitani, Y. (1986). *Electron-Microscopy* (in Japanese) **21,** 30.
Hashimoto, H., Endoh, H., Honda, T., and Harada, Y. (1983). *Proc. 7th Int. Conf HVEM,* Berkeley, p. 15.
Hashimoto, H., Tanaka, K., Shimadzu, S., Naiki, T., and Mannami, M. (1958). *Proc. 4th ICEM,* Berlin, vol. 1, p. 477.
Hibi, T. (1956). *J. Electronmicrosc.* **4,** 15.
Hibi, T., and Takahashi, S. (1959). *J. Electronmicrosc.* **7,** 15.
Hibi, T., and Yada, K. (1964). *J. Electronmicrosc.* **13,** 94.
Hibino, M. (1986). *Electron-Microscopy* (in Japanese), **21,** 99.
Horiuchi, S., Matsui, Y., and Kitai, T. (1991). *Proc. Int. Sympo. HVEM,* Osaka, p. 62.
Ichinokawa, T., Miyazaki, Y., Koga, Y. (1987). Ultramicroscopy, *23,* 115.
Isakozawa, S., Kashikura, Y., Sato, Y., Takahashi, T. (1989). *EMSA 47,* p. 112.
Ito, K., Ito, T., and Watanabe, M. (1954). *J. Electronmicrosc.* **2,** 10.
Iwatsuki, M., Murooka, K., and Takayanagi, K. (1991). *J. Electron Microsc.* **40,** 48.
Katagiri, S. (1953). *Electron-Microscopy* (in Japanese) **3,** 21.
Katagiri, S., and Ozasa, S. (1967) *J. Electronmicrosc.* **16,** 120.
Kimura, H. (1959). *J. Electronmicrosc.* **7,** 1.
Kimura, H., Higuchi, H., Maki, M., and Tamura, H. (1966). *J. Electronmicrosc.* **15,** 21.
Kimura, H., and Kikuchi, Y. (1956). *Hitachi Hyoron* (in Japanese) **38,** 1043.
Kobayashi, K., Hashimoto, H., Suito, E., Shimadzu, S., and Iwanaga, M. (1963). *Jpn. J. Appl. Phys.* **2.** 47.

Koike, H., Harada, Y., Goto, T., and Kokubo, Y. (1974). *Proc. 8th ICEM* Canberra, vol. 1, p. 42.
Koike, K., and Hayakawa, K. (1984). *Jpn. J. Appl. Phys.* **23,** L187.
Koike, H., Ueno, K., and Watanabe, M. (1970). *Proc. 7th ICEM,* Grenoble, vol. 1, p. 241.
Komoda, T. (1964). *Jpn. J. Appl. Phys.* **3,** 122.
Komoda, T. (1966). *J. Electronmicrosc.* **15,** 173.
Komoda, T., and Morito, N. (1960). *J. Electronmicrosc.* **9,** 77.
Komoda, T., and Saito, S. (1972). *Proc. 5th. SEM Symp., IIT-RI,* Chicago, p. 129.
Kubozoe, K., Tomita, M., and Matui, I. (1989). *Proc. MRS Symp.* **139,** 259.
Kuroda, K., Hosoki, S., and Komoda, T. (1985). *J. Electron Microsc.* **34,** 179.
Maruse, S. (1958). *Electron Microscopy* (in Japanese) **6,** 148.
Maruse, S. (1959). *Electron Microscopy* (in Japanese) **7,** 158.
Matsuda, T., Hasegawa, S., Igarashi, M., and Kobayashi, (1989). *Phys. Rev. Lett.* **62,** 2519.
Morito, N., Tadano, B., and Katagiri, S. (1959). *J. Electron Microsc.* **7,** 4.
Nagatani, T., and Saito, S. (1989). *Electron Microscopy* (in Japanese) **14,** 107.
Osakabe, N., Endo, J., Matsuda, T., Tonomura, A., and Fukuhara, A. (1989). *Phys. Rev. Lett.* **62,** 1969.
Osakabe, N., Tanishiro, Y., Yagi, K., and Honjo, G. (1981). *Surface Sci.* **109,** 353.
Ozasa, S., Kato, Y., Sugata, E., and Fukai, K. (1972). *J. Electron Microsc.* **21,** 109.
Suzuki, M., Akashi, K., and Tochigi, H. (1968). *Proc. 26th EMSA* **26,** 320.
Tadano, B. (1957). *Electron-Microscopy* (in Japanese) **5,** 143.
Tadano, B., Kimura, H., Uyeda, R., and Sakaki, Y. (1965). *J. Electron Microsc.* **14,** 88.
Tadano, B., Sakaki, Y., Maruse, S., and Morito, N. (1956). *J. Electronmicrosc.* **4,** 5.
Takahashi, N., Ashinuma, K., and Watanabe, M. (1960). *J. Electronmicrosc.* **9,** 104.
Takahashi, N., Takeyama, T., and Ito, K. (1956). *J. Electronmicrosc.* **4,** 16.
Takata, K., Hosoki, S., and Tajima, T. (1989). *Rev. Sci. Instrum.* **60,** 789.
Takayanagi, K., Yagi, K., Kobayashi, K., and Honjo, G. (1978). *J. Phys.* **E11,** 441.
Tanaka, K., and Hashimoto, H. (1953). *Rev. Sci. Instrum.* **24,** 669.
Tonomura, A., Fukuhara, A., Watanabe, H., and Komoda, T. (1968). *Jpn. J. Appl. Phys.* **7,** 295.
Tonomura, A. (1984). *J. Electron Microsc.* **33,** 101.
Tonomura, A., Matsuda, T., Endo, J., Todokoro, H., and Komoda, T. (1979). *J. Electron Microsc.* **28,** 1.
Tonomura, A., Osakabe, T., and Kawasaki, T. (1986). *Phys. Rev. Lett.* **56,** 792.
Uyeda, N., Kobayashi, T., Ishikawa, K., and Fujiyoshi, Y. (1979). *Chem. Scripta* **14,** 47.
Uyeda, R., and Nonoyama, M. (1967). *Jpn. J. Appl. Phys.* **6,** 557.
Watanabe, H. (1954). *J. Phys. Soc. Jpn.* **9,** 920.
Watanabe, H. (1956). *J. Phys. Soc. Jpn.* **11,** 112.
Watanabe, H., Uyeda, R. (1962). *Proc. 5th ICEM,* Philadelphia, vol. 1, p. A-5.
Yada, K., and Hibi, T. (1969). *J. Electron Microsc.* **18,** 226.
Yada, K., and Kawakatsu, H. (1976). *J. Electron Microsc.* **25,** 1.
Yanaka, T., and Shirota, K. (1967). *Proc. 1967 Meeting of Society of EM Japan.*
Yanaka, T., Yonezawa, A., Oosawa, K., and Iwaki, T. (1983). *EMSA, 41* p. 312.

4.5
Electron Microscopes and Microscopy in Japan
4.5G Application of Electron Microscopy to Biological Science

ATSUSHI ICHIKAWA

Yokohama City University
Minamiku
Yokohama 232, Japan

YONOSUKE WATANABE

Department of Pathology
Keio University Medical School
Shinjuku-ku
Tokyo 160, Japan

Application of electron microscopy to biological science started in 1939, when the 37th Subcommittee was established in the 10th Committee of the Japan Society for the Promotion of Science to develop the electron microscope. Dr. K. Sasagawa, who was Professor of Physiology at Osaka Medical College and soon moved to Kyoto University, was the only subcommittee member who specialized in biomedical science at that time. He presented the first electron micrographs of biological specimens in Japan to the subcommittee meeting in 1941, in collaboration with Dr. N. Higashi, who was a bacteriologist at Kyoto University and soon joined the subcommitte. The following remarks will be limited to introducing the contributions made by Japanese bioscientists, not so much in the aspects of biological data themselves but mainly in terms of technical progress in biological electron microscopy.

In the 1940s, electron microscopic studies in the biomedical fields were carried out in only a few research laboratories under an effective cooperation of bioscientists and technical engineers of electron optics. Such cooperative efforts have continued to the present and produced both rapid progress in biological electron microscopy in Japan and the development of Japanese electron microscopes of high quality. In those days, when the ultrathin sectioning technique had not yet been developed, electron microscopic studies of biomedical specimens dealt with direct observation of specimens mounted on a grid coated with a supporting film or shadowed replicas. Thus, the subjects for electron microscopic observation were limited mainly to microorganisms or cell fragments, such as bacteria (Sasagawa and

Higashi, 1941; Terada, 1949), viruses (Sugata, 1941; Higashi, 1943), rickettsia (Kimura and Higashi, 1948), cell fragments of homogenized tissues (Sasagawa and Hosomi, 1949, and/or chromosomes separated from *Drosophila* sperm (Yasuzumi, 1947). In May 1949, the Japanese Society of Electron Microscopy was founded, and several reports were presented at the first meeting. It is unfortunate that we have no documents of the meeting now, though the proceedings of the second meeting, which was held in October of the same year, are preserved. Nine of the 15 reports presented to the second meeting were concerned with biological science. They were electron microscopic observations of cell fragments of homogenized nervous and muscular tissues by Sasagawa and his collaborators (Kyoto University), chromosomes separated from crucian carp erythrocyte by Yasuzumi (Osaka University), liverwort sperm by Sato (University of Tokyo), erythrocyte membrane, bacteria, rickettsia, and viruses by Higashi (Kyoto University), bacterial viruses by Terada (Jikei Medical University in Tokyo) and collodion replicas of the polished surface of teeth by Takuma (Tokyo Dental College), in collaboration with Tsuchikura (Hitachi, Ltd.). Until 1952, society meetings were held twice a year, and thereafter once a year in spring. The number of reports, including special invited lectures, verbal presentations, and posters, presented to the society meeting from 1949 to 1990 is shown in Figure 1. It shows a rapid increase in the number of presentation since 1962. This is due mainly to the increase of electron microscopic studies in biomedical fields. Before the thin sectioning technique was developed, the number of presentations was about 30–60 in total, and one-third or less than one-half of them were concerned with biomedical science. To develop the ultrathin sectioning technique which is indispensable for electron microscopy of biological specimens, a number of attempts to design ultramicrotomes and prepare embedding media were carried out by some bioscientists and mechanical or chemical engineers (cf. Chapter 2.5). Electron micrographs of rat liver cell sections, which show fairly well some details of the intracytoplasmic organelles, were first presented to the 57th meeting of the subcommittee by Y. Watanabe, a pathologist at Keio University in Tokyo, in 1951. He had intensively observed ultrathin sections of various cell types and proposed the concept of the "intracytoplasmic sac" in 1955, which corresponds to the endoplasmic reticulum described by K. R. Porter (1954). Tashiro and Ogura, in Sasagawa's laboratory at Kyoto University, studied rat liver microsomes and revealed that electron-dense particles studded to the microsomal membrane are selectively digested by RNase incubation (1957). A research committee for ultrathin sectioning methods was organized in the society from 1952 to 1953, and workshops about the method were held at Tokyo and Kyoto in 1952 and 1953, respectively. More than 50 bioscientists participated in each workshop. Since the

APPLICATION OF ELECTRON MICROSCOPY 725

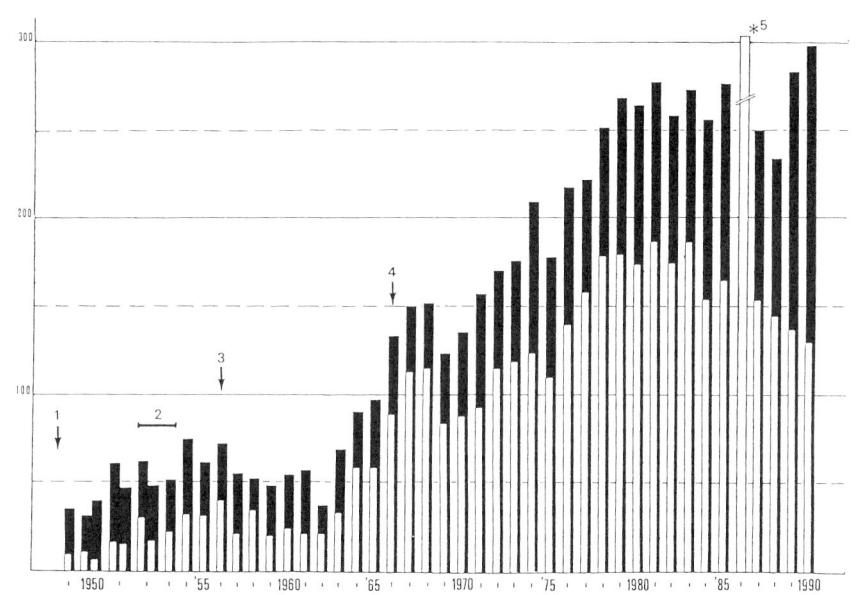

FIGURE 1. Number of papers contributed to annual meetings from 1949 to 1990. Black column, total number; white column, papers in biological science. 1. the JSEM was established. Until 1952, society meetings were held twice a year, and thereafter once a year. 2. A research committee for ultrathin sectioning methods was organized in the society. 3. The First Regional Conference on Electron Microscopy in Asia and Oceania was held in Japan, after the annual meeting. 4. The 6th International Congress on Electron Microscopy was held at Kyoto, besides the annual meeting. 5. The 11th International Congress on Electron Microscopy was held at Kyoto, combined with the annual meeting. The white column shows the number of papers contributed by Japanese participans in the biomedical field (418 of 786 papers).

resin embedding method of Newman, Borysko, and Swerdlow (1949) and the glass knife of Latta and Hartman (1950) were introduced in Japan around 1953, biological electron microscopy has became widely utilized in many laboratories as an indispensible approach to ultrastructural analyses of microorganisms and cells and tissues in both normal and pathological conditions. The number of presentation to the annual meeting increased rapidly after 1963, and more than a half or two-thirds of the total contributions dealt with biological specimens (Fig. 1). Kushida has intensively studied the resin embedding method since 1954 and suggested the advantages of epoxy resin for embedding (1960). The First Regional Conference on Electron Microscopy in Asia and Oceania was held at Tokyo in 1956. Fifty-eight reports from nine countries were presented to the conference, and 11 of 27 contributions from Japan were concerned with biological electron microscopy.

In the 1960s, many of the specimen preparation techniques in conventional histology at the light microscopic level, such as histo- and cytochemistry for sugars, lipids, proteins, and metals, enzyme- and immunocytochemistry, autoradiography, cryosectioning techniques, cell fractionation, etc., were modified for application to biological electron mocroscopy. Development of negative staining methods coupled with progress in electron optics made it possible to achieve high-resolution electron microscopy of cell fine structure at the molecular level. Thus, the biological sciences have made striking progress to an extent never dreamed of in the past. Among the contributions made by Japanese bioscientists, a number of studies are especially noteworthy because of their originality and superb quality. Kanaseki and Kadota (1969) observed the coated vesicles isolated from nerve endings and revealed that each vesicle is contained in a spherical polygonal "basketwork" shown only by the negative staining techniques. The labeling method of actin

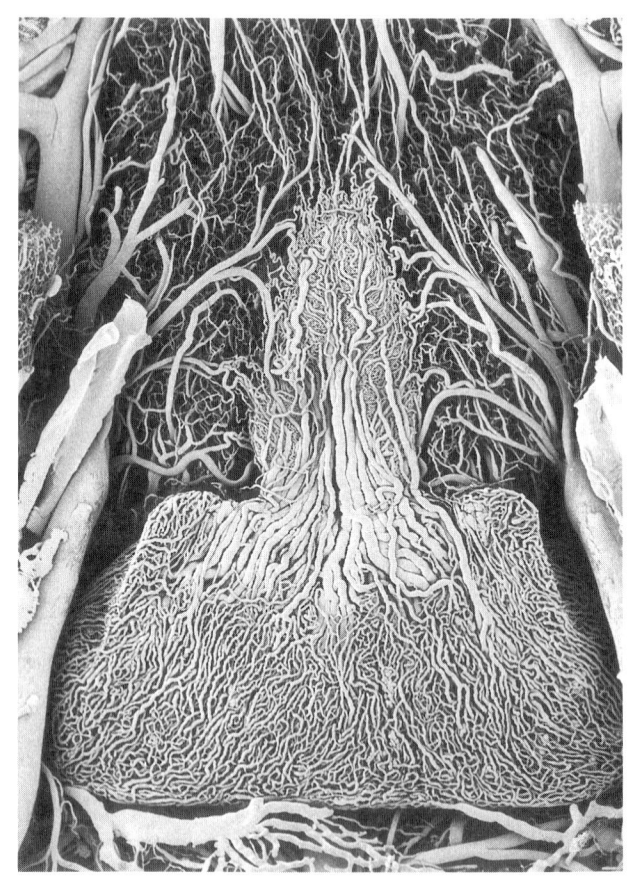

FIGURE 2. Scanning electron micrograph of the vascular system in the anterior pituitary lobe and hyphyseal stalk by Murakami's corrosion casting method. (Courtesy of T. Murakami.)

filaments by arrowhead complex formation with heavy meromyosin designed by Ishikawa *et al.* (1969) greatly promoted the fine structural analysis of the cytoskeletal system in various cell types. Ogawa and his colleagues intensively studied enzyme cytochemical approaches for oxidoreductases, transferases, hydrorases, and others since 1962. See Saito's minireview, 1989. Mizuhira and Kurotaki first succeeded in electron microscopic autoradiography with high resolution using domestic photosensitive emulsion (1964) and contributed to make the approach popular to many investigators in Japan. High-voltage electron micrographs of biological specimens with reasonable resolution were first presented by Hama and his collaborators in 1967. They showed the remarkable merits of high-voltage electron microscopy for biological studies (Hama, 1973). Immunocytochemical methods using enzyme-labeled antibody development by Nakane and Pierce (1967) facilitated immunocytochemical approaches in biological electron microscopy.

In the last two decades, a number of new approaches to biological electron microscopy, such as the freeze-fracture method, X-ray microbeam analysis, high-resolution electron microscopy, frozen sectioning, quick freezing, scanning electron microscopy, and others, have been developed, and remarkable progress in cellular and molecular biology has been made. Among them, scanning electron microscopy has made epochal development in Japan in both aspects of the instrument itself and its application. Since scanning electron microscopy (SEM) was first applied to biological studies in Japan in the late 1960s (Tokunaga, 1967; Fujita *et al.*, 1968), SEM has become widely utilized by biomedical investigators as a useful means for observa-

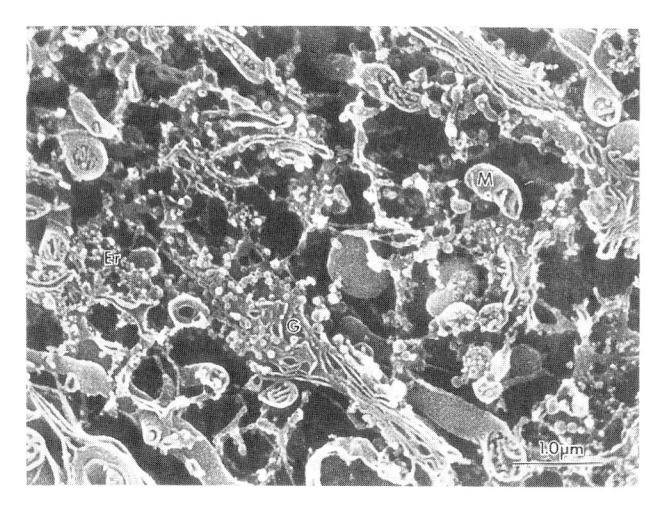

FIGURE 3. Scanning electron micrograph of the intracellular structure of rat motor nerve cell by ultrahigh-resolution SEM. (From Tanaka, 1989.)

tions of surface structure of cells and tissues as well as their three-dimensional organization. Vascular casting methods a using methyl methacrylate mixture developed by Murakami (1971) contributed immensely to microvascular research (Fig. 2). Tanaka and his colleagues developed a number of advantageous preparatory procedures for the SEM study of biological samples, such as the frozen resin cracking method (Tanaka, 1972), the ion-etching method (Tanaka *et al.*, 1976), and the maceration method (Tanaka and Naguro, 1981). These methods, coupled with the development of field-emission SEM, made it possible to observe intracellular structures

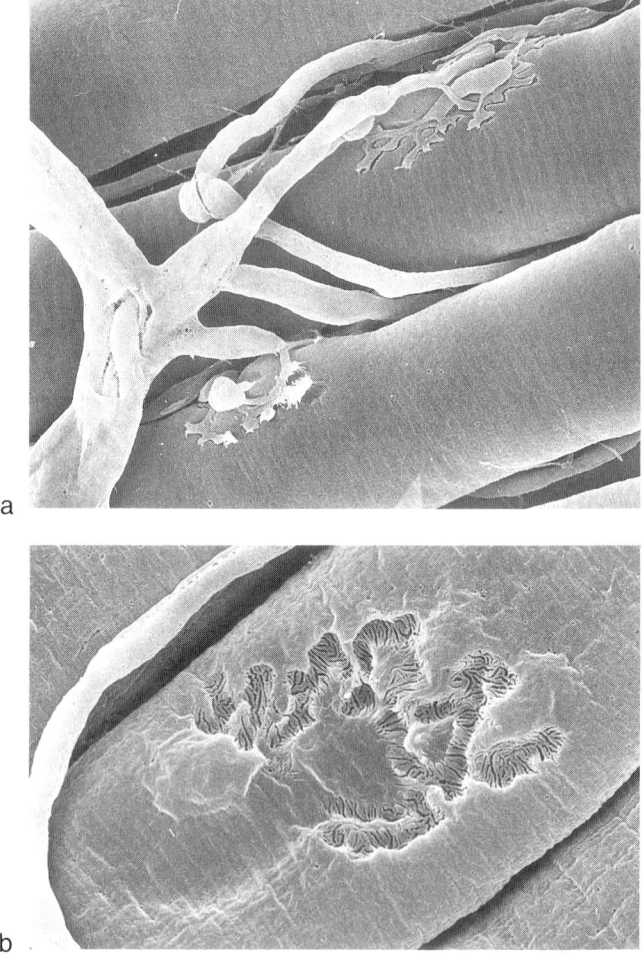

FIGURE 4. Scanning electron micrographs of the motor end plates (a) and the primary and secondary synaptic clefts (b). (From Desaki and Uehara, 1981.)

as well as microorganisms with high resolution (Fig. 3). (Cf. Tanaka's minireview, 1989.) Uehara and his collegues succeeded in showing clearly the three-dimensional organization of the vascular autonomic plexus (Uehara and Suyama, 1978) and neuromuscular junction (Fig. 4) (Desaki and Uehara, 1981) by SEM using a modified method with HCl hydrolysis for removing the extra cellular materials. Among transmission electron microscopic studies of biological specimens in recent years, Mizuhira and Futaesaku (1971) applied tannic acid solution to fixation and block staining for cytochemistry and obtained good results. The metal block freeze-fracture apparatus developed by Nishiura (1977) alerted Japanese investigators to the usefulness of this approach. Tonosaki and Yamamoto (1974) improved the freeze-fracture apparatus for complementary replicas and obtained results of superb quality. A number of high-voltage electron microscopes were provided for general use in several universities and institutes. This facilitated remarkable progress in biological electron microscopy by three-dimensional analysis of the intracellular constituents in thick sections. (Cf. Hama's minireview, 1989.) The quick-freezing method was much improved in the late 1970s and has become the focus of wide attention as the most desirable method to preserve the fine structural features and components of cells and tissues to reflect their living state. Kirino and Hirokawa (1978) were the first to apply this method to thin section study in Japan and obtained good results. Quick freezing followed by freeze-substitution fixation has brought a number of new insights into the cell fine structure, which had been erroneously understood by artefactious modifications caused by conventional chemical fixation (Hirokawa and Kirino, 1980; Ichikawa *et al.*, 1980, 1987; Ichikawa and Ichikawa, 1987; Usukura and Yamada, 1987).

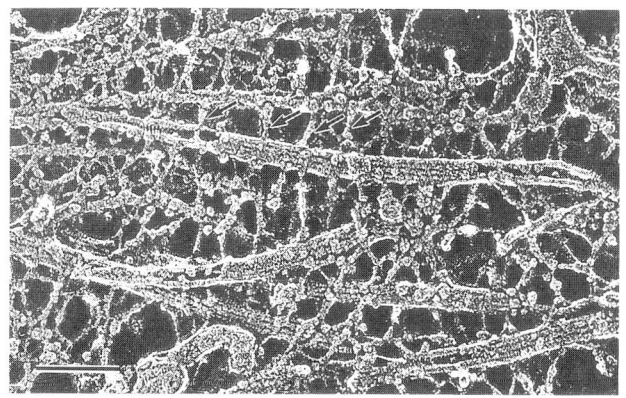

FIGURE 5. Quick-frozen, deep-etched dendrite of a motor neuron in rat spinal cord, showing crossbridges between neurofilaments and microtubules. (From Hirokawa *et al.*, 1988b.)

TABLE I

Main Themes of EM Symposia Held by JSEM, 1954–1990

No.	Year	Main themes
1.	1953	Fine structure of the cytoplasm. EM study on the myofibrils. Electron diffraction and its application. Application of replica method.
2.	1954	Technical problems for improving electron microscope. Dislocation network. Homogeneous slip of crystal. Electron diffraction of metal film. Crystal growth.
3.	1955	EM on iron and ferroalloy. Crystal growth. Microparticles—Nonbiological and biological.
		(EM symposium was not held in 1956 because of the First Asia-Oceania Regional Conference on EM held in Tokyo)
4.	1957	Resolution of electron microscopy. Thin sectioning techniques. Radiation damages. Fixation and electron staining. Electron diffraction.
5.	1958	Periodic structure of substances. Crystallography in metal engineering and biological science.
6.	1959	Image contrast.
7.	1960	Electron illumination system. Contamination caused by electron beam bombardment. Resin embedding method.
8.	1961	Technical problems for improving electron micrograph. Fine structure of the eye. EM study of cancer cells.
9.	1962	Preparation method of foil for EM. EM observations on phenomena in metal engineering. Electron staining. Photographic techniques in EM.
10.	1963	Imaging lens system. EM cytochemistry. Crystallization of microparticles—Metals, micromolecules, macromolecules, and viruses.
11.	1964	High-resolution electron microscopy. Biomembranes.
12.	1965	High-voltage electron microscope. Technical problems for easy operation of highly qualified electron microscope—Alignment and correction of astigmatism. Setoh Prize Lectures.
		(EM symposium in 1966 was not held because of the 6th International Congress on Electron Microscopy held in Kyoto)
13.	1967	Contrast of highly magnified EM images. Dislocation in metal crystal. Biomembranes. Setoh Prize Lectures.
14.	1968	EM observation of atoms and molecules. New applications of electron beams—Electron holography, SEM, and dynamic observation by TEM. Setoh Prize Lectures.
15.	1969	SEM—Fundamentals and application. Protein synthesis and its secretion. Dynamic observation of metals. Setoh Prize Lectures.
16.	1970	Approaches for quantitative electron microscopy. Setoh Prize Lectures.
17.	1971	Stereoscopic electron microscopy. SEM. Freeze-etching method. Replica method. Reconstruction for 3-D structure. Setoh Prize Lectures.
18.	1972	High-resolution TEM. Setoh Prize Lectures. Electron microprobe analysis.
19.	1973	Lattice defect. Radiation damage. Cryotechniques. X-ray microanalysis in biology. Setoh Prize Lectures.
20.	1974	EM of heated or frozen samples. EM of atoms and molecules. Setoh Prize Lectures.
21.	1975	Shadowing and coating. Microcrystals. Modern EM techniques in virology. Resolution limit of EM. Setoh Prize Lectures.

(*continues*)

TABLE I (*Continued*)

No.	Year	Main themes
22.	1976	Beam coherncy and image processing. Element analyses by EM—Present status and future. Cryotechniques. High-voltage EM. Setoh Prize Lectures.
23.	1977	Future images of STEM. Structure and function of biomembranes. Tannic acid in EM. Assessment of EM radioautography. Thin membrane—Its surface structure and physical properties. Setoh Prize Lectures.
24.	1978	Contaminations in EM. Artifacts caused during specimen preparation. EM histochemistry. Physicochemical nature of embedding media. Setoh Prize Lectures.
25.	1979	Recent development in microbeam analyses—General problems, X-ray backscattered and secondary electron images, Auger EM and iron EM. Setoh Prize Lectures.
26.	1980	Accelerating voltage dependence of TEM and STEM images. Rapid freezing method of biological specimens. Setoh Prize Lectures.
27.	1981	Recent progress in electron-probe analyses. Methods and applications of EM to biological hard tissues. Cytoskeleton. Setoh Prize Lectures.
28.	1982	High resolution TEM and SEM—Instrumentations and applications. Setoh Prize Lectures.
29.	1983	200-kV TEM—Fundamentals and applications. Setoh Prize Lectures.
30.	1984	Processing and reconstruction of EM images—Fundamentals (image processing, high-speed digital image processing system) and applications (scanning laser stimulated luminescence system, computerized image processing, 3-D structural analyses, morphometry). Setoh Prize Lectures.
31.	1985	Localization of constituents—Future of high-resolution analyses, development of hardware, atomic specification in metal science and biology, immunoelectron microscopy, EM radioautography, EM molecular biology. Setoh Prize Lectures.
		(EM symposium was not held in 1986 because of the 11th International Congress on Electron Microscopy in Kyoto)
32.	1987	The present status and future view of EM—Application of EM to promotion of new material science, development of ultramodern technology, and life science. Superconducting cryoelectron microscope. Scanning tunneling microscope. Reflection high-energy electron diffraction. Total reflection angle X-ray spectroscopy. Current image processing technology. Accelerating voltage dependence on microanalysis of EDS and EELS. Recent advance of microbeam analysis in biomedical science. HVTEM. HRSEM. Cryotechniques. Immunocytochemistry. Setoh Prize Lectures.
33.	1988	Cytoskeleton. Quantitative EM. Cryotransfer system. Intracellular processing of macromolecules. Setoh Prize Lectures.
34.	1989	Today and tomorrow of analytical techniques. Scanning tunneling microscope. Dynamic, quantitative analyses of biological materials. New expansion of immunelectron microscopic methods. Setoh Prize Lectures.
35.	1990	Atoms, molecules and macromolecules—High-resolution EELS, field emission TEM, STEM, HVTEM. Nonradioactive electron microscopic in situ hybridization. Imaging plate. Histo- and cytochemistry. Metallic superlattice film. Holographic electron interferometry. Crystal in bioscience. 3D EM visualization of actin-myosin system using molecular or heavy metal labeling. Setoh Prize Lectures.

TABLE II

BIOSCIENTISTS AWARDED THE SETOH PRIZE AND TITLES OF THEIR WORKS, 1956–1990

1956	N. Higashi: EM studies in bacteriology and virology.
1957	G. Yasuzumi: EM study on the chromosomes and sperm.
1958	E. Yamada: EM study of the centrioles and retina.
1960	K. Fukai: EM study on influenza virus.
1961	Y. Watanabe: EM study on the fine structures in the cytoplasm.
1962	K. Takeya and S. Koike: EM study on the acid-fast bacteria.
1963	K. Hama: Fine structural study of the synapse
1964	K. Kurosumi: EM study on the morphology of secretion in glands.
1965	T. Nagano: EM study of the testis.
1966	T. Oda: Molecular structure and its biochemical significance of the mitochondrial membranes and the plasma membrane of intestinal epithelial microvilli.
	V. Mizuhira: Application of autoradiography to EM study of biological materials
1967	Y. Hosaka: Fine structural study on the viral globule.
1968	R. Honjin: Highly ordered crystal structure of the vitelline protein molecules.
1969	T. Nei: EM study on freezing and drying of biological specimens.
1970	T. Kanaseki: Fine structural study of the coated vesicle.
1972	T. Fujita, H. Sakaguchi, and J. Tokunaga: SEM study in biomedical science.
1973	K. Ogawa: Electron microscopic enzyme histochemistry.
	K. Tanaka: Development and application of specimen preparation techniques for biological SEM.
1974	E. Shikata: EM study on the plant viruses.
1975	A. Matsumoto and Y. Nonomura: Fine structural analyses of biological specimens with using high-qualified, modern techniques.
1976	M. Nishiura: Development of freeze-fracture method.
1977	J. Tawara, K. Amako, and K. Kumon: Application of high-resolution SEM to microbiology.
1978	A. Ichikawa and M. Ichikawa: EM study on secretion of the salivary glands.
1979	H. Ishikawa: EM study on cell motility.
1980	T. Kawata: EM study of the bacterial cell surface coat.
	T. Suzuki: Development of specimen preparation techniques in biological EM.
1981	H. Fujita: EM study on secretion of the thyroid gland.
	M. Osumi: EM study of the yeast.
1982	H. Yasuda: Fine structural study of the lung and alveolar surfactant.
	A. Tonosaki and H. Washioka: Development of supplemental replication method.
1983	K. Hirosawa: EM radioautographic study on the vitamin A-storing cell system.
1984	Y. Uehara: On the 3D structure of the neuroterminals, with special reference to SEM observation of bared cell surface of nerve endings prepared by HCl hydrolysis.
1985	Y. Shimada: Fine structural study of the cultured myoblastic cells.
	N. Hirokawa: Stereological study on the fine structural relationship of cell membrane to cytoskeleton with using quick-freezing and deep-etching methods.
1986	T. Yamamoto: EM study on absorptive mechanism of the intestinal epithelial cells.
	K. Watanabe and K. Yasuda: Development and application of immunoelectron microscopic techniques.
1987	M. Nakai: Fine structural study of HIV (AIDS virus).
	H. Hirano: Lectin histo- and cytochemistry. An electron microscipic study.
1988	Sh. Tsukita and S. Tsukita: EM study on the molecular architecture of the intercellular junctional apparatus.
1989	T. Y. Yamamoto: EM study on the fine structure of sensory organs.
1990	T. Saito: Electron microscopic enzyme histochemistry of the retina.

Quick freezing followed by freeze-subsitution fixation was applied to sugar histochemistry (Murata et al., 1985), immunocytochemistry (Inoue and Kurosumi, 1985; Ichikawa *et al.,* 1987), and enzyme cytochemistry (Saito and Takizawa, 1987), and superb results were obtained. Hirokawa and his colleagues have been carrying out a number of distinguished cell biological studies by means of quick-freeze, deep-etch replicas (Hirokawa, 1982; Hirokawa *et al.,* 1988a). and combined with immunocytological decoration (Hirokawa *et al.,* 1988b, 1989a, 1989b; Nakata *et al.,* 1990; Harada *et al.,* 1990). Progress in biological electron microscopy in Japan during the past two decades is reviewed in the 1989 supplemental issue of the *Journal of Electron Microscopy.* It contains minireviews on several major topics as follows: high-resolution cryoelectron microscopy of biological macromolecules (Fujiyoshi, 1989), three-dimensional image reconstruction from electron micrographs of biological macromolecules by computer-aided imaging (Wakabayashi, 1989), the freeze-fracture method and its application to cell biology (Fujita, 1989), freeze-substitution for thin-section study of biological specimens (Ichikawa *et al.,* 1989), quick-freeze, deep-etch electron microscopy (Hirokawa, 1989), trends in electron microscopic cytochemistry—34 years of development in Japan (Saito, 1989), state of the art of immunoelectron microscopy (Nakane, 1989), recent advances in biomedical microbeam analysis—from EDX to EELS—imaging analysis of biomedical specimens (Mizuhira, 1989), retrospect and prospect of electron microscopy in biology and virology in Japan (Hosaka, 1989), development of methods for observing of bacteria and fungi by electron microscopy (Osumi, 1989), and biological application of high-voltage electron microscopy (Hama, 1989).

The Electron Microscopy Symposium to discuss updated topics started in 1953 and was thereafter held annually in the autumn as a formal event of the society. The main themes of each symposium, which may reflect the topics at each time from 1954 to 1990, are shown in Table I. The Setoh prize, which is awarded for distinguished work in electron microscopic studies, was found in 1956, and the names of the bioscientists awarded the prize and the titles of their works are listed in Table II.

References

Desaki, J., and Uehara, Y. (1981). *J. Neurocytol.* **10,** 101.
Fujita, H. (1989). *J. Electron Microsc.* **38** (suppl.), S110.
Fujita, T. Inoue, H., and Kodama, T. (1968). *Acta Histol. Jpn.* **29,** 511.
Fujiyoshi, y. (1989). *J. Electron Microsc.* **38** (suppl.), S97.
Hama, K. (1989). *J. Electron Microsc.* **38,** (suppl.), S156.
Hama, K. (1973). *In* "Advanced Technique in Biological Electron Microscopy" (J. K. Koehler, ed.), p. 275. Springer Verlag, Berlin.
Harada, A., Sobue, K., and Hirokawa, N. (1990). *Cell Struct. Function* **15,** 329.

Higashi, N. (1943). *Proc. 24th Meeting 37th Subcommittee* **24**, 2.
Hirokawa, N. (1982). *J. Cell Biol.* **94**, 129.
Hirokawa, N. (1989). *J. Electron Microsc.* **38** (suppl.), S123.
Hirokawa, N., and Kirino, T. (1980). *J. Neurocytol.* **9**, 243.
Hirokawa, N., Shiomura, Y., and Okabe, S. (1988a). *J. Cell Biol.* **107**, 1449.
Hirokawa, N., Hisanaga, S., and Shiomura, Y. (1988b). *J. Neurosci.* **8**, 2769.
Hirokawa, N., Sobue, K., Kanda, K., Harada, A., and Yorifuji, H. (1989a). *J. Cell Biol.* **108**, 111.
Hirokawa, N., Pfister, K. K., Yorifuji, H., Wagner, M. C., Brady, S. T., and Bloom, G. S. (1989b). *Cell* **56**, 867.
Hosaka, Y. (1989). *J. Electron Microsc.* **38** (suppl.), S145.
Ichikawa, A., Ichikawa, M., and Hirokawa, N. (1980). *Am. J. Anat.* **157**, 107.
Ichikawa, A., Yasuda, K., Ichikawa, M., Yamashita, S., Aiso, S., and Shiozawa, M. (1987). *Acta Histochem. Cytochem.* **20**, 601.
Ichikawa, A., Ichikawa, M., and Sasaki, K. (1989). *J. Electron Microsc.* **38**, (suppl.), S118.
Ichikawa, M., and Ichikawa, A. (1987). *Cell Tissue Res.* **250**, 305.
Ichikawa, M., Ichikawa, A., and Kidokoro, S. (1987). *J. Electron Microsc.* **36**, 117.
Inoue, K., and Kurosumi, K. (1982). *J. Electron Microsc.* **31**, 93.
Ishikawa, H. Bischoff, R., and Holzer, H. (1969). *J. Cell Biol.* **43**, 313.
Kanaseki, T., and Kadota, K. (1969). *J. Cell Biol.* **42**, 202.
Kimura, Y., and Higashi, N. (1948). *Shimadzu Rev.* **5**, 36.
Kirino, T., and Hirokawa, N. (1978). *J. Electron Microsc.* **27**, 339.
Kushida, H. (1960). *J. Electronmicrosc.* **9**, 157.
Mizuhira, V. (1989). *J. Electron Microsc.* **38** (suppl.), S142.
Mizuhira, V., and Kurotaki, M. (1964). *Igaku no Ayumi* **49**, 775 (in Japanese).
Mizuhira, V., and Futaesaku, Y. (1971). *In* Arceneaux, C. J. (ed.) *EMSA 29th Ann. Meeting* (C. J. Arceneaux, ed.), p. 494.
Murakami, T. (1971). *Acta Histol. Jpn.* **32**, 445.
Murata, F., Suzuki, S., Tsuyama, S., Suganuma, T., Imada, M., and Furihata, C. (1985). *Histochem. J.* **17**, 967.
Nakane, P. K. (1989). *J. Electron Microsc.* **38** (suppl.), S135.
Nakane, P. K., and Pierce, G. B. (1967). *J. Cell Biol.* **33**, 307.
Nakata, T., Sobue, K., and Hirokawa, N. (1990). *J. Cell Biol.* **110**, 13.
Nishiura, M. (1977). *Denshikenbikyo* **47**, 86 (in Japanese).
Ogawa, K., Shinonaga, Y., and Saito, T. (1962). *Okajima Folia Anat. Jpn.* **38**, 355.
Osumi, M. (1989). *J. Electron Microsc.* **38** (suppl.), S150.
Saito, T. (1989). *J. Electron Microsc.* **38** (suppl.) S129.
Saito, T., and Takizawa, T. (1987). *Acta Histochem. Cytochem.* **20**, 357.
Sasagawa, K., and Higashi, N. (1941). *Proc. 12th Meeting 37th Subcommittee* **14**, 6 (in Japanese).
Sasagawa, K., and Hosomi, T. (1949). *Acta Schol. Med. Univ. Kyoto* **27**, 96.
Sugata, E. (1941). *Proc. Jpn. Soc. Sci.* **17**, 218.
Tanaka, K. (1972). *Naturwiss.* **59**, 77.
Tanaka, K. (1989). *J. Electron Microsc.* **38** (suppl.), S105.
Tanaka, K., Iino, A., and Naguro, T. (1976). *Acta Histol. Jpn.* **39**, 165.
Tanaka, K., and Naguro, T. (1981). *Biomed. Res.* **2** (suppl.), 63.
Tashiro, Y., and Ogura, M. (1957). *Acta Schol. Med. Univ. Kyoto* **34**, 267.
Terada, M. (1949). *Proc. 2nd Meeting JSEM* (in Japanese).
Tokunaga, J. (1967). *Jpn. J. Oral. Biol.* **9**, 23.
Tonosaki, A., and Yamamoto, T. (1974). *J. Ultrastruct. Res.* **47**, 86.
Uehara, Y., and Suyama, K. (1978). *J. Electron Microsc.* **27**, 157.
Usukura, J., and Yamada, E. (1987). *Cell Tissue Res.* **247**, 483.
Wakabayashi, T. (1989). *J. Electron Microsc.* **38** (suppl.), S102.
Watanabe, Y. (1955). *J. Electronmicrosc.* **3**, 43.
Yasuzumi, G. (1947). *Cytologia* **14**, 1.

4.5
Electron Microscopes and Microscopy in Japan
4.5H Application of Electron Microscopy to Biological Science (Microbiology)

YASUHIRO HOSAKA

Department of Virology and Immunology
Osaka University of Pharmaceutical Sciences
Kawai Matsubara
Osaka 580, Japan

TADASHI HIRANO

Jikei University School of Medicine
Miniato-ku
Tokyo 105, Japan

I. Introduction

The study of ultrastructure in the biological world by electron microscopy started with the observation of virus by Dr. E. Ruska and his brother Helmut after the construction of the first electron microscope in the 1930s. In Japan, the electron microscopic study of the world of biology was begun similarly, with the study of viruses by microbiologists. Bacteria are visible in an ordinary microscope and were classified by their morphology before the invention of the electron microscope, but the study of bacterial ultrastructure began later. In the late 1950s, when bacteriologists became interested in the ultrastructure of bacteria and their accessory organs such as flagellae, they began to use electron microscopes.

From the end of World War II to the 1950s, several pioneering microbiologists cooperated closely wtih electron microscope engineers to produce equipment good enough to satisfy their curiosity about the ultrastructure of virus particles and virus-infected cells. They carried out these activities under extremely adverse conditions at research facilities and even in daily life. Most of these microbiologists and engineers took part in the foundation of the Japanese Society of Electron Microscopy in 1949. The researchers who worked actively in the 1950s included the late Dr. Noboru Higashi (Medical School of Kyoto University and later Kawasaki Medical School), the late

Dr. Kenji Takeya (Kyushu University Medical School), Dr. Konosuke Fukai (Research Institute for Microbial Diseases, Osaka University), the late Masanaka Terada (Jikei University School of Medicine), followed by Dr. Isamu Kondo (Jikei University School of Medicine) and Dr. Jutaro Tawara (Medical School of Okayama University, currently Kochi Medical College). All were awarded Seto prizes by the Japanese Society of Electron Microscopy.

In 1966, the 6th International Congress of Electron Microscopy was held in Kyoto, for the first time in Japan (Prof. N. Higashi was President of the IFSEM at that time). This had a profoundly favorable effect on the development of electron microscopic studies in microbiology as well as in other biological fields.

II. Development of Virus Ultrastructure Research

A. Development, 1950–1980s

Fukai and Suzuki (1955) elucidated the relationship of influenza virus physical particle numbers to infectious ones. In the early 1950s, the shadowing technique was used mainly for the observation of free virus particles (virions); this allowed one to see only the gross surface structure of virions. Fukai and the Hitachi group tried to overcome this limitation by constructing a high-voltage electron microscope (HVEM) with a high specimen penetration capability. This plan met with difficulties, but in the meantime a thin section technique had been developed for seeing intracellular structures. This allowed one to distinguish virus growth in cells and to observe a cross section of the virions. Thus, their aim in wanting HVEM was in fact fulfilled with the thin section technique, although not completely, and the plan for HVEM was therefore suspended.

The late Takeya and his colleagues (Takeya *et al.*, 1959a, 1959b; Takeya and Amako, 1968; Amako *et al.*, 1974; Amako and Yasunaka, 1977) studied the morphology of bacteriophages and their interaction with host bacterial cells. They isolated a filamentous bacteriophage from *Pseudomonas aeruginosa,* which has a single-stranded DNA, infects only a limited number of strains of *P. aeruginosa,* and does not use conjugative pili which bave been reported as the attachment organ in filamentous coliphages (Takeya and Amako, 1968; Amako and Yasunaka, 1977). They studied primarily the morphology of coliphage T4 by a high-resolution scanning electron microscope (Amako *et al.*, 1974). They and Kondo *et al.* (1974) (Fig. 1) independently showed by the same method the binding mode of T4 bacteriophages to *Escherichia coli.*

In the early 1960s, the negative staining technique in electron microscopy of Brenner and Horne was introduced into Japan. Using this procedure,

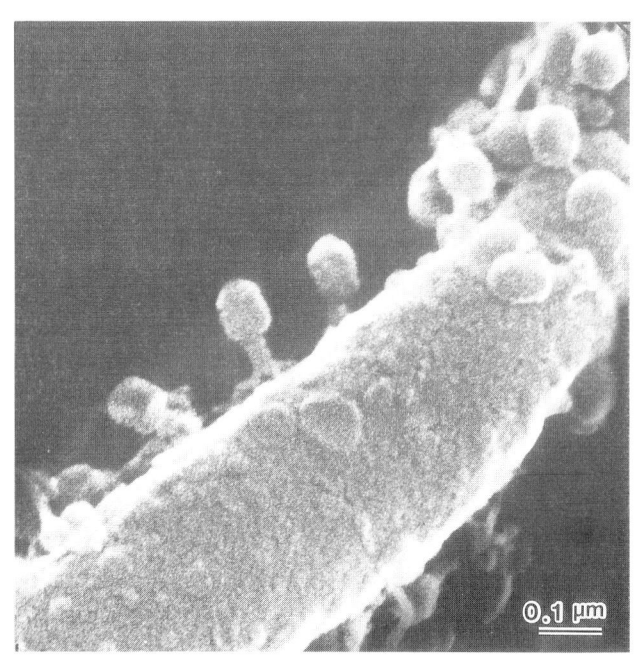

FIGURE 1. High-resolution scanning electron micrograph of T4 phage adsorption onto a host bacteria (*E. coli*). (By Dr. I. Kondo, Emeritus Professor of Jikei Medical School.)

many Japanese virologists elucidated the anatomy and architecture of virions: Hosaka *et al.* (1966) elucidated the organization of paramyxovirus polymorphism, the morphological identification of a unit nucleocapsid (1.1 μm × 18 nm), and the existence of polynucleocapsids. Hosaka (1968) isolated Sendai virus nucleocapsids and demonstrated their flexible nature on a protein monolayer by the shadowing technique (Fig. 2). Nonomura and Kohama (1974) demonstrated that the Sendai virus nucleocapsid is left-handed. Amako and his colleagues (1974) studied phage morphology, likewise by using the negative staining technique.

Ultrathin sectioning and the related immunoelectron microscopy have been useful, particularly for analysis of virus morphogenesis in infected cells. The later Higashi, Matsumoto, and their colleagues (1967) elucidated the virus-RNA synthesis site in chikungunya virus-infected Vero cells using autoradiography in electron microscopy. Nii *et al.* (1968) studied herpes virus morphogenesis by using immunoelectron microscopy, and further, Nii and his colleagues investigated this issue by combined use of high-voltage TEM and SEM (Yoshida *et al.*, 1986), which made it possible to distinguish intra- and extracellular (on the cell surface) mature virions. The crystallike arrangement of the herpes virus capsid in an infected cell nucleus is shown

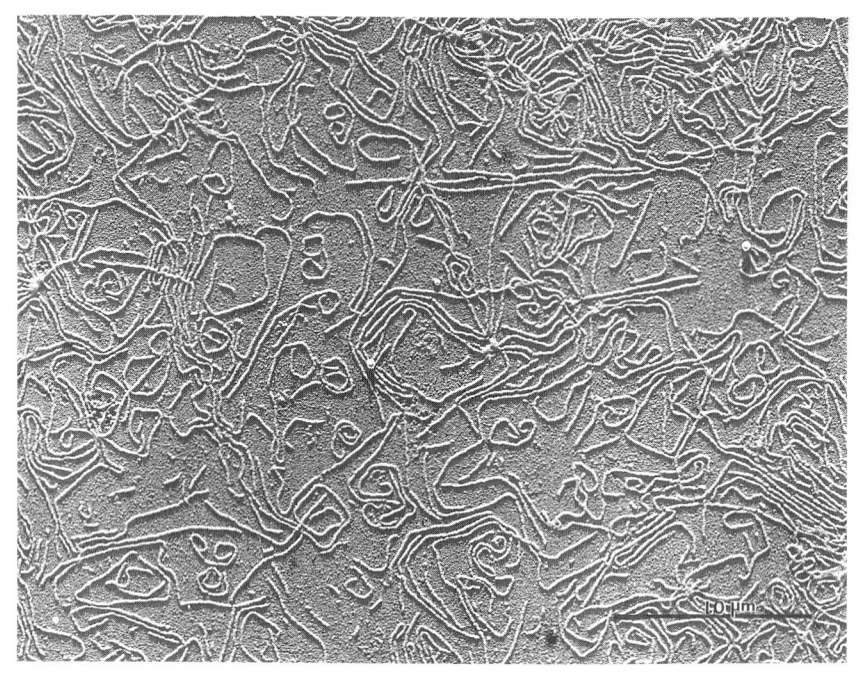

FIGURE 2. Nucleocapsids isolated from parainflueza (Sendai) virus. The alkali-Emasol method (*Virology* **35,** 445, 1968) was used for the NC isolation. Isolated NC were spread on a cytochrome C protein monolayer according to the method of Kleinschmidt and Zahn (1959), and shadowed by platinum-palladium. Flexible rods of unit length of 1.1 µm and its multiplicatives are seen. (By Dr. Y. Hosaka, Osaka University of Pharmaceutical Science.)

in Fig. 3. Kim *et al.* (1979) morphologically distinguished envelope fracture structures of young and aged Sendai virions. Nakai *et al.* (1989) studied AIDS virions and their morphogenesis (Fig. 4). Yasuda *et al.* (1981) studied by immunoelectron microscopy the interaction of noninfectious Sendai virus of uncleaved type and murine cells and demonstrated their degradation in endosomes. This finding suggested that the failure of fusion of the virus envelope with cell membranes gave a signal for activation of endosome proteases. Hirano *et al.* (1962) found a viruslike body in yeast. Scanning electron microscopy was generally used for observation of virus budding and adsorption (Tawara, 1976).

B. The Current Situation

Currently, virologists concerned with electron microscopy are trying to find out what is the native architecture of virions in a medium and to identify

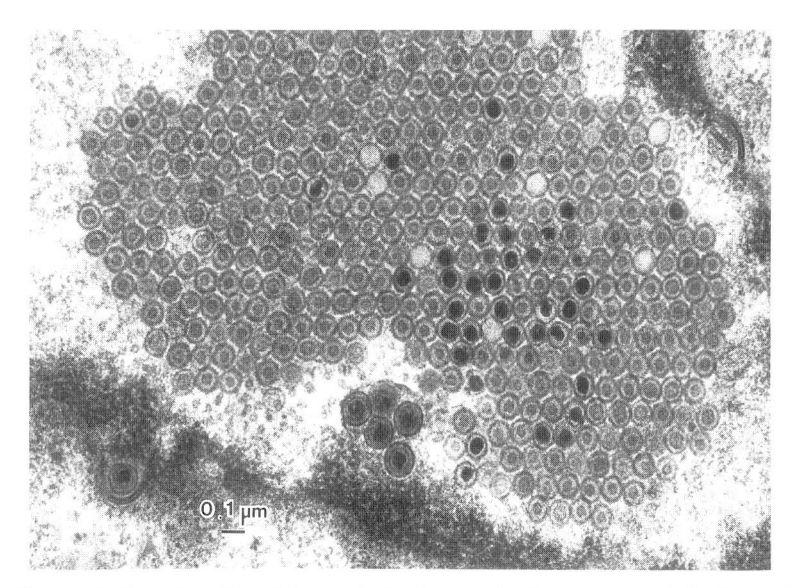

FIGURE 3. Part of an FL cell infected with herpes simplex virus type 2. In the nuclear matrix, a crystallike arrangement of viral capsids is seen. Several enveloped particles in a nuclear vacuole and one in a perinuclear cisterna are also seen. (By D. S. Nii, Okayama University School of Medicine.)

dynamic aspects of virus morphogenesis by using more sophisticated methods. Furthermore, they want to correlate their morphological results with those of gene-oriented molecular biology, and also to provide useful criteria for the diagnosis of virus diseases including AIDS. The following procedures, including the recently developed high-resolution scanning electron microscopy, are increasingly useful in this field. Moreover, when electron microscopy is combined with recent advanced and sophisticated light microscopy, such as video microscopy and confocal microscopy, it will be a great help.

1. *Cryoelectron Microscopy*

The vitrified method developed by Dubochet and his colleagues (1982) in the 1980s became an essential and in fact the only method for observing a native virions in a medium. By this method, Sendai virions were found to be completely spherical in a medium (Hosaka and Watabe, 1988). Yuba *et al.* (1990) found the transfer of virus spikes to the target cell membranes after fusion of influenza virus and liposomes. Yamaguchi *et al.* (1988) successfully applied this method for observations of hepatitis B virus antigens.

Fujiyoshi *et al.* (1988), together with the JEOL group, developed a cryoelectron microscope operating below −200°C using liquid helium, which has a more stable stage and less irradiation damage, and observed DNA and

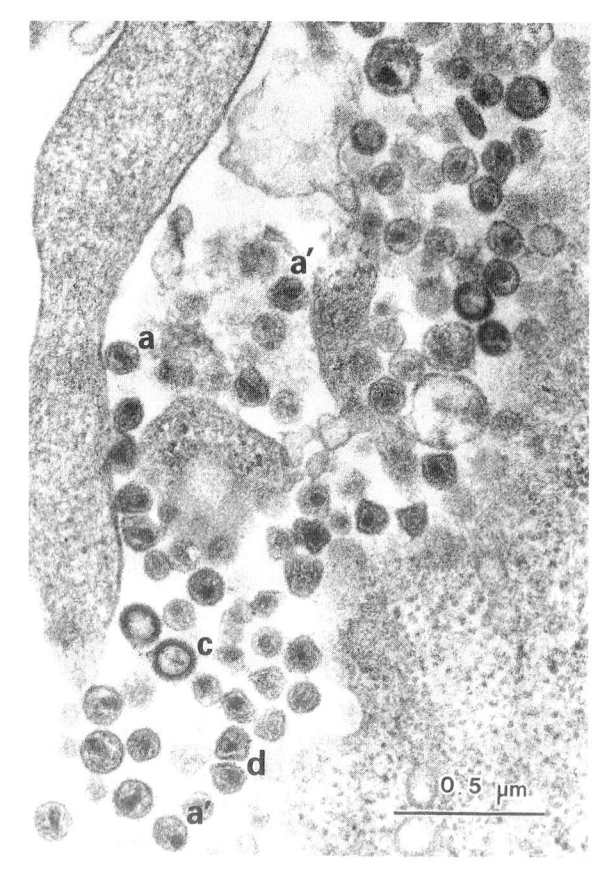

FIGURE 4. HIV particles released from infected human lymphocyte cultures. Various forms of the inner structure are seen: virions with a rodlike nucleoid (a), those with a central or excentral circular nucleoid (a'), that of a doughnut form (c), and that of a teardrop form. (By Dr. M. Nakai, Osaka Medical College.)

influenza virions. However, this type of electron microscope is very expensive, and its domestic distribution is still confined to only two laboratories.

2. *Immunoelectron Microscopy*

As described in the previous section, immunoelectron microscopy has been widely employed and is now being used. Antibodies labeled with various sizes of colloidal golds, and soft embedding materials are commercially available. Furthermore, various monoclonal antibodies are provided commercially. However, frozen sectioning technique is still limited in Japanese virology.

3. Rapid-Freezing Method

The rapid-freezing method is useful for analysis of fast biological reactions between virus envelopes and cellular membranes. Kawasaki *et al.* (1988) found using this method that pit structures on the fracture plane of liposome membranes fused with influenza virus envelopes.

III. Development of Bacterial Ultrastructure Research

A. Development, 1950s–1980s

In the 1950s, the late Higashi and his colleagues studied intracellular development of vaccinia virus, and later switched to the study of larger infectious agents, chlamydia, which, at that time, were regarded as viruses by a minority group. They studied the infectious cycle of chlamydia by electron microscopy, and distinguished ribosomes and nucleoids during intracellular development of trachoma virus and psittacosis virus (Chlamydia) (Higashi *et al.*, 1962; Higashi, 1965), and this work was referred to in a book entitled *Chlamydia and Chlamydia-Induced Diseases,* by J. Storz (Charles C

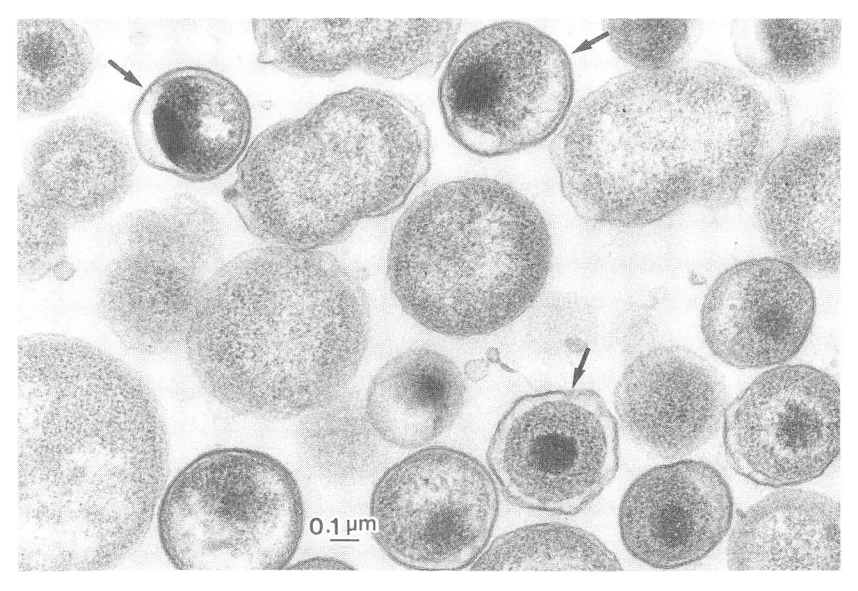

FIGURE 5. Chlamydial bodies of different developmental stages in an intracellular inclusion. The strain used was isolated from a nasopharyngeal swab sepecimen obtained from a psittacosis patient without history of contact with any bird. Arrows indicate a mature form with a nucleoid. Magnification ×60,000. (By Dr. A. Matsumoto.)

Thomas, 1971). Then, in the 1960s, chlamydia was unequivocally classified as Chlamydiales of Bacteriomycota, an agent different from the virus world. Chlamydia makes a unique world in microbiology (Fig. 5).

The late Takeya and his colleagues (1959a, b, 1963) studied the cell wall of *Mycobacterium* and found its unique surface structures, paired fibers, which is now believed to be found only on the cell wall surface of the species of Mycobacteriaceae. Later, Amako and Umeda (1977) employed high-resolution scanning electron microscopy for study of the ultrastructure of bacterial surfaces. Furthermore, Amako's group (Amako *et al.*, 1983; Umeda *et al.*, 1987) studied the bacterial cell morphology, using the rapid freezing and substitution fixation method, to reveal the fine structure of the bacterial capsules (Fig. 6).

Iino's group and Tawara studied the morphology of bacterial flagella. Asakura *et al.* (1968) demonstrated a unidirectional growth of flagellar filaments of *Salmonella* by immunoelectron microscopy, and Suzuki *et al.* (1978) constructed the morphogenetic map of bacterial flagellae by comparative electron microscopy of the incomplete flagellar structures observed

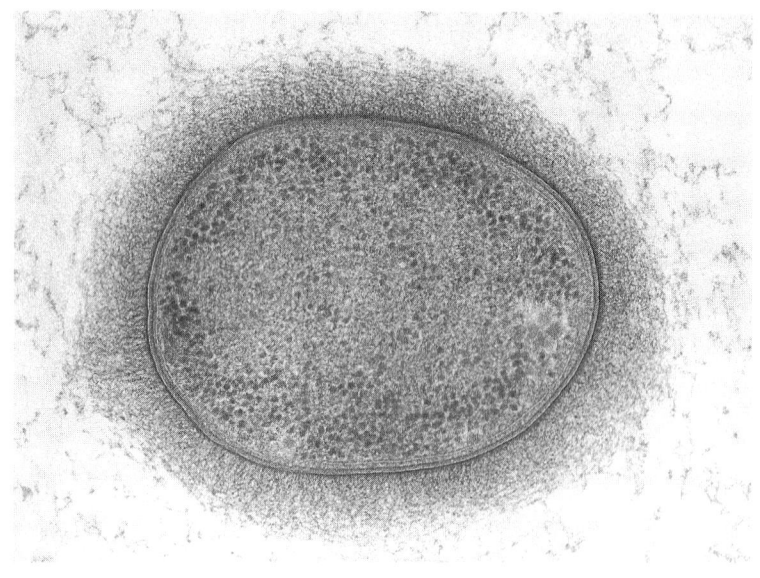

FIGURE 6. Ultrastructure of *Klebsiella pneumoniae* observed by quick-freezing followed by freeze-substitution fixation: network of capsule fibrils. Capsule polysaccharide fibrils of bacteria are destroyed by regular chemical fixation methods, and have been unobservable by transmission electron microscope. Application of quick-freezing followed by freeze-substitution fixation first made individual capsule fibrils visible by electron microscopy, which indicates the superiority of this freezing method. (By Dr. K. Amako, Kyushu University School of Medicine.)

from many nonflagellate mutants of *Salmonella typhimurium.* Tawara (1957) observed that each flagellum of *Vibrio coma* is attached to a granule (blepharoplast) (Fig. 7).

Suganuma and his colleagues studied the plasma membrane and cell wall of *Staphylococcus.* Suganuma (1964, 1966) demonstrated a typical triple-layered unit membrane of the plasma membrane of *Staphyococcus aureus,* and Morioka *et al.* (1986, 1987) found the localization of aminoglyucoside-binding sites in the cell wall and cytoplasm of *E. coli,* and that of carbohydrate bound by wheat germ agglutinin-gold complex in the cross section of *Staphylococcus aureus.* Morioka *et al.* (1976) also studied the ultrastructure of the same bacteria by the freeze-etching technique.

Microtubular and corelike structures in the cytoplasm of stable Staphylococcal L-forms were observed by Eda *et al.* (1976, 1979), and further, filipin sterol complexes in the cytoplasmic membrane of these L-forms were observed by freeze-fracture electron microscopy by Nishiyama and Yamaguchi (1990).

Yamaguchi and his colleagues have mainly studied the morphology of protoplasts of *Proteus* and *E. coli:* They showed the presence of a reversion sequence to bacillary forms of spheroplasts induced by beta-lactam antibiot-

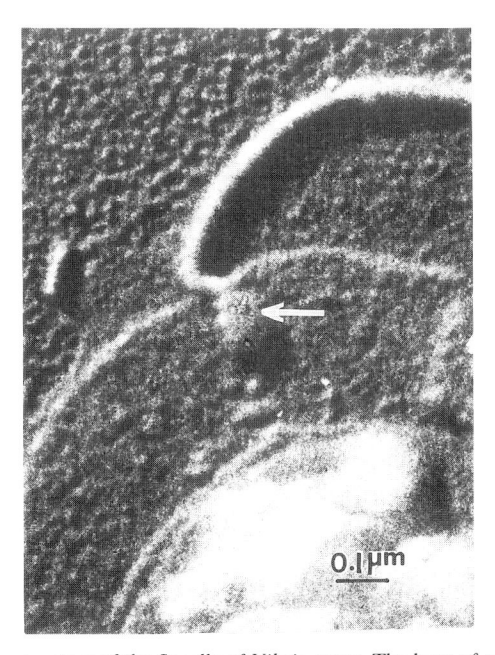

FIGURE 7. Base structure of the flagella of *Vibrio coma.* The base of a flagella is attached to a granule (blepharoplast, indicated by arrow) embedded in the cytoplasm. Granules range in size from 150 to 200 nm. (By Dr. J. Tawara, Emeritus Professor of Kohchi Medical School.)

ics in *Proteus mirabilis* (Tada and Yamaguchi, 1979), and in *Serratia marcescens* (Furutani *et al.*, 1985). Tada and Yamaguchi (1983, 1984) studied characteristics of chromosomal DNA of *E. coli* spheroplasts induced by penicillin G biochemically and electron microscopically.

Fukushi and his colleagues found that the granules of BCG detected by Ziehl-Neelsen's staining method were different from the electron-dense granules of the BCG seen in electron microscopy (Ebina *et al.*, 1957). They also found the intracytoplasmic tubular structures in thin sections of *Mycobacterium avium* (Shinohara *et al.*, 1957).

Kawata and his colleagues studied the ultrastructure of various bacteria. They revealed the presence of the regular arrays in the cell walls of *Lactobacillus fermenti* (Kawata *et al.*, 1974), *L. buchneri* (Masuda and Kawata, 1981), and *Clostridium difficile* (Kawata *et al.*, 1984) by the freeze-etching and negative staining techniques.

B. The Current Situation

Current situations of bacteriology and mycology are described together in a later section.

IV. Development of Yeast Ultrastructure Research

A. Development, 1950s–1980s

In 1955, Yotsuyanagi first studied the mitochondria of yeast by electron microscopy. The mitochondria in yeast cells have been described as spherical, rod-shaped, or threadlike and branched bodies by Hashimoto *et al.* (1958), Hirano and Lindegren (1961), and Yotsuyanagi (1961), independently. A degeneration of mitochondria in yeast has been reported to take place in glucose-repressed cultures (Yotsuyanagi, 1961). Similar phenomena were also observed in respiration-deficient mutants (Yotsuyanagi, 1962), in anaerobically grown cells (Hirano and Lindegren, 1963), and in mitochondriogenesis (Osumi, 1965).

Hirano (1962) found that nuclei of yeast cells are distinct from those of higher organism: The nuclear envelope does not disappear during the nuclear division as in Ascarides. The nucleolus of yeast is observed as an electron-dense region within the nucleus. The plasma membrane in yeast protoplast was found to consist of two electron-dense layers separated by a less dense layer, the overall thickness being approximately 8 nm.

Osumi and her colleagues (Osumi *et al.*, 1985; Yamada *et al.*, 1986; Baba and Osumi, 1987) applied low-voltage, high-resolution SEM to yeasts and observed their nearest native surface ultrastructure.

B. The Current Situation

Current new techniques, previously described, are also available for bacteriology and mycology. Recently, ruthenium tetroxide treatment of yeast cells provides electron-dense and clear images of glucan fibrils on the protoplast surface and intracellular organella, such as filasomes, Golgi apparatus, and secretory vesicles (Fig. 8) (Naito *et al.,* 1991). Immunoelectron microscopy of bacteria is expected to give us intracellular localization and transport of enzymes and other active proteins, particularly ones coded by recombinant plasmid DNA. Vitrified bacterial toxins and other products should be good targets for cryoelectron microscopy. Other cryotechniques such as freeze-fracture and the quick-freezing substitution fixation method will also give more native images of bacteria.

Acknowledgments

The authors would like to thank K. Amako, (Kyushu University School of Medicine), A. Matsumoto (Kawasaki Medical School), T. Iino (Emeritus

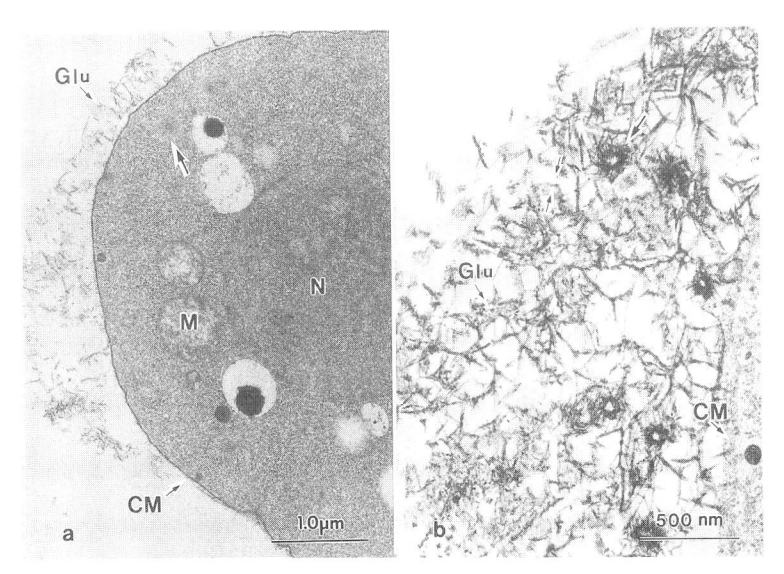

FIGURE 8. Reverting protoplasts *Schizosaccharomyces pombe,* treated with RuO$_4$ after glutaraldehyde-OsO$_4$ fixation (by Dr. M. Osumi). (a) Regeneration of a new cell wall substance, β-glucan (Glu), from the cell surface, and intracellular organella with a higher contrast. Filosomes are indicated by an arrow. CM, cytoplasmic membrane; M, mitochondoria; N, nucleus. (b) Higher-magnification image of fibrillar network of glucan (Glu). The ring-shaped organizing centers (big arrow) and a tube structure (small arrows) of fibrils are seen.

Prof., Tokyo University), T. Kawata (Emeritus Prof., Tokushima University Medical School), M. Nakai (Osaka Medical School), S. Nii, (Okayama University Medical School), M. Osumi, (Japan Women's University), A. Sugawara (Emeritus Prof., Kyoto, Prefecture Medical School), J. Tawara (Kochi Medical School), and J. Yamaguchi (Kinki University School of Medicine) for their cooperation in collecting information and literature references for this review.

REFERENCES

Amako, K., and Yasunaka, K. (1977). *Nature* **267,** 862.
Amako, K., Murata, K., and Umeda, A. (1983). *Microbiol. Immunol.* **27,** 95.
Amako, K., Takeya, K., Nagatani, T., and Saito, M. (1974). *J. Electron Microsc.* **23,** 301.
Amako, K., and Umeda, A. (1977). *J. Ultrastruct. Res.* **58,** 34.
Asakura, S., Eguchi, G., and Iino, T. (1968). *J. Mol. Biol.* **35,** 227.
Baba, M., and Osumi, M. (1987). *J. Electron Microsc. Technique* **5,** 249.
Dubochet, J., Lepault, J., Freeman, R., Berriman, J., and zHomo, J.-C. (1982). *J. Microsc.* (Oxford) **128,** 219.
Ebina, T., Shinohara, C., Saito, H., Fukushi, K., and Suzuki, J. (1957). *Nature* **180,** 42.
Eda, T., Kanda, Y., and Kimura, S. (1976). *J. Bacteriol.* **127,** 1567
Eda, T., Kanda, Y., Mori, C., and Kimura, S. (1979). *Microbiol. Immunol.* **23,** 915.
Fujiyoshi, Y., Yamagishi, H., and Harada, M. (1988). *J. Electron Microsc.* **37,** 294.
Fukai, K., and Suzuki, T. (1955). *Med. J. Osaka Univ.* **6,** 1.
Furutani, A., Tada, Y., and Yamaguchi, J. (1985). *Microbiol. Immunol.* **29,** 901.
Hashimoto, T., Conti, C. F., and Naylor, H. B. (1958). *J. Bacteriol.* **76,** 406.
Higashi, N. (1965). *Exp. Mol. Pathol.* **4,** 24.
Higashi, N., Matsumoto, A., Tabata, K., and Nagatomo, Y. (1967). *Virology* **33,** 5.
Higashi, N., Tamura, A., and Iwanaga, M. (1962). *Ann. N.Y. Acad. Sci.* **98,** 100.
Hirano, T. (1962). *J. Ultrastruct. Res.* **7,** 201.
Hirano, T., and Lindegren, C. C. (1961). *J. Ultrastruct. Res.* **5,** 321.
Hirano, T., and Lindegren, C. C. (1963). *J. Ultrastruct. Res.* **8,** 322.
Hirano, T., Lindegren, C. C., and Bang, Y. N. (1962). *J. Bacteriol.* **83,** 1363.
Hirano, T., Tacreiter, A., Eaves, A., and Kaplan, J. G. (1968). *Cytologia* **33,** 558.
Hosaka, Y., Kitano, H., and Ikeuchi, S. (1966). *Virology* **29,** 205.
Hosaka, Y. (1968). *Virology* **35,** 445.
Hosaka, Y., and Watabe, T. (1988). *J. Virol. Meth.* **22,** 347.
Kawasaki, K., Murata, M., Ohnishi, S., Ikeuchi, Y., and Kanaseki, T. (1988). Abstr. 36th Ann. Meet. Jpn. Virol. (in Japanese), Tokyo.
Kawata, T., Masuda, K., Yoshida, K., and Fujimoto, M. (1974). *Jpn. J. Microbiol.* **118,** 469.
Kawata, T., Takeoka, A., Takumi, T., and Masuda, K. (1984). *FEBS Microbiol. Lett.* **24,** 323.
Kim, J., Hama, K., Miyake, Y., and Okada Y. (1979). *Virology* **95,** 523.
Kondo, I., Hasegawa, N., Aida, S., Takano, Y., and Wababe, T. (1974). *J. Electron Microsc.* **23,** 222.
Masuda, K., and Kawata, T. (1981). *J. Gen. Microbiol.* **124,** 81.
Morioka, H. (1976). *J. Electron Microsc.* **25,** 271.
Morioka, H., Tachibana, M., Amagai, T., and Suganuma, A. (1986). *J. Histochem. Cytochem.* **34,** 909.
Morioka, H., Tachibana, M., and Suganuma, A. (1987). *J. Bacteriol.* **169,** 1358.
Naito, N., Yamada, N., Kobori, H., and Osumi, M. J. (1991). *Electron Microsc.* **40,** 416.

Nakai, M., Goto, T., and Imura, S. (1989). *J. Electron Microsc. Technique* **12,** 95.
Nii, S., Morgan, C., and Rose, H. M. (1968). *J. Virol.* **2,** 517.
Nishiyama, Y., and Yamaguchi, H. (1990). *Microbiol. Immunol.* **34,** 25.
Nonomura, S., and Kohama, K. (1974). *J. Mol. Biol.* **86,** 621.
Osumi, M. (1965). *Bot. Mag.* (Tokyo) **78,** 231.
Osumi, M., Baba, M., Suzuki, T., Watanabe, T., and Nagatani, T. (1985). *Bio-Med. SEM.* **14,** 47.
Shinohara, C., Fukushi, K., and Suzuki, J. (1957). *J. Bacteriol.* **74,** 413.
Suzuki, T., Iino, T., Horiguchi, T., and Yamaguchi, S. (1978). *J. Bacteriol.* **133,** 904.
Suganuma, A. (1964). *J. Cell Biol.* **21,** 290.
Suganuma, A. (1966). *J. Cell Biol.* **30,** 208.
Tada, Y., and Yamaguchi, J. (1979). *J. Electron Microsc.* **28,** 100.
Tada, Y., and Yamaguchi, J. (1983). *Microbiol. Immunol.* **27,** 893.
Tada, Y., and Yamaguchi, J. (1984). *Microbiol. Immunol.* **28,** 853.
Takeya, K., and Amako, K. (1968). *Virology* **28,** 163.
Takeya, K., and Hisatsune, K. (1963). *J. Bacteriol.* **85,** 16.
Takeya, K., Koike, M., Mori, R., Yuda, Y., and Toda, T., (1959a). *J. Bacteriol.* **78,** 311.
Takeya, K., Mori, R., Nakashima, N., Koike, M., and Toda, T. (1959b). *J. Bacteriol.* **78,** 313.
Tawara, J. (1957). *J. Bacteriol.* **73,** 89.
Tawara, J. (1976). *Denshikenbikyou* **11,** 95 (in Japanese).
Umeda A., Ueki, Y., and Amako, K. (1987). *J. Bacteriol.* **169,** 2482.
Yamada, M., Nagatani, T., and Osumi, M. (1986). *J. Electron Microsc.* **15** 35.
Yamaguchi, M., Hirano, T., Sugahara, K., Hirakawa, H., Sugahara, K., Mizokami, H., and Matsubara, K. (1988). *J. Electron Microsc.* **37,** 337.
Yasuda, Y., Hosaka, Y., Fukami, Y., and Fukai, K. (1981). *J. Virol.* **39,** 273.
Yoshida, M., Uno, F., and Nii, S. (1986). *J. Electron Micros.* **35,** 47.
Yotsuyanagi, Y. (1955). *Nature* **176,** 1208.
Yotsuyanagi, Y. (1961). *J. Ultrastruct. Res.* **7,** 121.
Yotsuyanagi, Y. (1962). *J. Ultrastruct. Res.* **7,** 141.
Yuba, D. A., Yoshizawa, C. A., Sato, B. S., and Kume, P. N. (1990). Abstr. 46th Ann. Meet. Jpn. EM (in Japanese), Maebashi, *J. Electron Microsc.* **39,** 296.

4.5
Electron Microscopes and Microscopy in Japan
4.5I Applications to Materials Science

HIROSHI FUJITA[1]

Osaka University, Suita, Osaka 565, Japan

Applications of electron microscopy to materials science have expanded widely since dislocations in thin metal foils themselves could be observed directly. As a result, improvement of the functional features of electron microscopes has been strongly demanded to observe the atomic structures of various materials. In Japan, the functional features of electron microscopes have been very much improved by the challenge of constructing high-voltage electron microscopes in 1963–1965. Later, the use of electron microscopy was further extended into various fields of natural science, especially materials science, and indispensible applications have been carried out to obtain information dynamically by *in situ* experiments with high-voltage electron microscopes, and directly at the atomic scale by high-resolution electron microscopy. In the present report, recent topics in the applications of electron microscopy to materials science in Japan are reviewed.

I. Early Work in Materials Science

Before 1953, the accelerating voltage of electron microscopes in Japan was less than 100 kV, and the thinning technique was also not adequate to make thin foils. Thus, applications of electron microscopy were limited to replicas of materials surfaces and fine materials such as various sols, vapor-deposited films, very fine particles, cleaved thin foils, and very fine materials grown in special atmospheres. Since the surface unevenness of materials is caused by the surface relief resulting from various shear displacements and/or chemical attack, interesting results by replica methods had been obtained on the following subjects: slip bands (Takamura, 1955; Fujita, 1955), surface reliefs of martensites (Nishiyama and Shimizu, 1956, 1958), twinned structures (Nishiyama *et al.*, 1958), fractured surfaces (Nonaka, 1955), etch pits (Nishimura *et al.*, 1951), subboundaries (Taoka and Aoyagi,

---

[1] Present address: Research Institute for Science and Technology, Kinki University, Kowakae, Higashi, Osaka 577, Japan.

1956, 1957), precipitates (Koda and Takeyama, 1953), paints (Terao and Sakata, 1953, 1954) and so on. Furthermore, the extraction replica method was used effectively to determine complex crystal structures of small crystallites such as precipitates and insoluble inclusions (Fukami, 1956). Microstructures of various powders, such as sols (Suito *et al.*, 1954; Suito and Takiyama, 1955), paints (Fukami and Shiyota, 1955, 1960), fine ceramic particles (Fukami and Shiyota, 1955, 1960), and vapor-deposited ones in vacuum and low-pressure gases (Hibi, 1955), were also investigated effectively by electron microscopy. Details of these results were summarized in the book entitled *Theory and Applications of Electron Microscopy, Part III—Applications to Science and Engineering,* edited by K. Kubo (1960).

On the other hand, it is well known that the behavior of bulk materials is very structure sensitive. It was thought that a material's behavior is closely related to the behavior of lattice defects whose density and arrangement are very sensitive to the pretreatments of the materials. In order to make this fact clear, many trials were carried out to observe the materials themselves directly by electron microscopy, instead of by the replica method. Direct observation of thin metal foils was made first by Heidenreich (1949) using foil specimens which were thinned chemically, and such diffraction contrasts as subgrains and bend contours were observed. By using cleaved nonmetallic materials, i.e., mica films, moiré fringes were observed by Mitsuishi *et al.* (1951). Magnified dislocation images were also observed in the moiré fringes of overlapping crystals by Hashimoto *et al.* (1956–1957).

Afterwards, the electropolishing method was used for making thin foils of metals by Bollmann (1956), and the diffraction contrast of subboundaries consisting of dislocations in aluminum was observed with a 100-kV electron microscope. Hirsch *et al.* (1956) also succeeded in observing the movement of dislocations in aluminum with a 100-kV electron microscope. After that, so-called direct observation of metal foils was commonly used. Direct observation of dislocations in metals was first achieved in Japan with a 50-kV electron microscope by Nishiyama and Fujita (1957), using aluminum foil specimens made by a chemical thinning method, and the subgrain grouping, or the subgrain coalescence, in recovery and recrystallization processes of deformed metals was discovered (Fujita, 1961). After that result, the accelerating voltage of Japanese-made conventional electron microscopes was also increased up to 100 kV, and various electropolishing methods were improved for making thin foils of metals and alloys. As a result, electron microscopy was widely used for metallurgical investigations, because much information could be obtained from both the diffraction contrast of microstructures and the corresponding diffraction patterns in metals and alloys. At the same time, fundamental information about diffraction contrast was provided by the kinematic theory, which was developed mainly by Hirsch *et al.* (1960). Detailed calculations of image contrasts by the dynamical theory were made by Kato (1963) and by Howie and Whelan (1961, 1962). The contribution of inelastically scattered electrons to image contrast was also clarified by Kamiya and Uyeda (1961, 1962). Extension of the diffraction theory was made on various electron microscope images by Hashimoto *et al.* (1962, 1977) and by others in Japan by considering the absorption effect.

Besides the electropolishing methods, various thinning methods such as the jet method, the spark cutting and thinning, the acid cutting and thinning, the ion thinning, etc., were improved, so that electron microscopy was widely applied to research fields of nonmetallic as well as metallic ones. Furthermore, precise tilting devices were developed to obtain the necessary contrast of microstructures in foil specimens, and some specimen treatment devices such as heating devices (Takahashi and Mihama, 1957; Nemoto and Koda, 1963), low-pressure environmental cells (Hashimoto *et al.*, 1966), etc., were also improved.

By using direct observation of foil specimens, new information was obtained on the following subjects: Dislocation structures in deformed (Fujita and Nishiyama, 1961, 1962; Takeuchi *et al.*, 1967; Karashima *et al.*, 1968) and annealed specimens (Takeyama and Takahashi, 1970; Furubayashi *et al.*, 1966), boundary structures (Furubayashi, 1971), superlattice structure of ordered crystals (Ogawa *et al.*, 1958), microstructures of alloys (Takahashi and Ashinuma, 1958, 1959) and martensite crystals (Nishiyama and Shimizu, 1959, 1961; Nishiyama *et al.*, 1968, 1968), G.P. zones and various precipitates (Nemoto and Koda, 1965), secondary defects in quenched (Yoshida *et al.*, 1963; Kiritani *et al.*, 1964; Shimomura and Yoshida, 1965) and particle-irradiated specimens (Nemoto *et al.*, 1971), stacking faults (Watanabe, 1966), and so on. These results played an important role for making clear the contributions of both microstructures and lattice defects in various phenomena occurring in metals and other materials.

Furthermore, *in situ* experiments were tried for some research subjects such as the growth process of W-oxide with an environmental cell by using motion picture photography (Hashimoto *et al.*, 1958), the transition from θ' phase to θ phase in Al-Cu alloy with a heating device (Takahashi and Mihama, 1957; Nemoto and Koda, 1963), interaction between moving dislocations and precipitates by thermal stressing (Nemoto and Koda, 1966), the growth process of Mo-oxides (Hashimoto *et al.*, 1966), crystallization of amorphous selenium foils by electron irradiation (Shiojiri, 1967), oxidation of Mo particles and deoxidation of MoO_3 (Ueda, 1959), crystal growth of Hg at $-70°C$ (Honjio *et al.*, 1956). The wall structures of magnetic domains in magnetic materials such as iron, Ni-Fe, and Ni-Co foil specimens were also observed by Lorentz microscopy (Tsukahara and Kawakatsu, 1966, 1972; Tsukahara et al., 1971; Watanabe and Sekiguchi, 1984).

In addition, epitaxial growth process of metallic powders (Ino et al., 1956; Takahashi and Mihama, 1957), structure change of electrodeposited films (Tanabe and Kawasaki, 1972), and microstructures of fine powders deposited in vapor (Uyeda, 1942; Kimoto et al., 1963) and in low-pressure gases (Mihama and Tanaka, 1976) were effectively investigated by using specimens which were treated successively under various conditions outside the microscope.

In these direct observations of foil specimens, the selected-area electron diffraction method was another advantage of electron microscopy. By using this method, crystal structures of very fine crystals and orientation relationships between two different crystals, e.g., martensite crystals and precipitates against the matrix (Shimizu and Okamoto, 1971) and multiply-twined Au particles (Ino et al., 1972), were determined.

As mentioned above, the direct observations of lattice defects and microstructures in foil specimens of materials gave valuable new information to materials science, especially metallurgical objects. Details of those results were summarized in *Interpretation of Electron Microscope Images and Its Applications to Metallurgy*, edited by Z. Nishiyama and S. Koda (1975).

On the other hand, it was clarified that the lattice defects, especially dislocations, are very sensitive to the foil thickness, so that not only the behavior and but also the density are markedly changed in such thin foil specimens. Since material behavior is determined by the behavior of lattice defects, attention was directed to observing the same density and the same behavior, if possible, of lattice defects as those in the bulk material. Most electron microscopists, however, did not expect that the maximum observable thickness of the specimens would increase significantly with increasing accelerating voltage. Furthermore, some investigators worried about blurring of the images due to the inelastic scattering of electrons through crystals, and others about heavy damage or destruction of the specimens during observation due to electron irradiation (Fujita, 1986).

Around 1960, a 300-kV electron microscope operating with a cascade transformer generator was installed in Japan by Kobayashi et al. (1963), and quantitative measurement of the energy dependence of transmissive power of electron waves was carried out by Hashimoto et al. (1964). The relativistic dynamical theory of electron diffraction was also examined by Fujiwara (1962) with a 300-kV electron diffraction instrument. These results, however, gave pessimistic information on the voltage dependence of the maximum observable thickness of the specimens.

In 1963, the first separate-type 0.5-MV transmission electron microscope operating with a Cockcroft-Walton high-voltage generator was installed at the National Research Institute for Metals (NRIM), and was followed by construction of a few 0.5-MV electron microscopes of both separate and symmetry types, operating with the same type of high-voltage generator (Fujita, 1986). The electron microscope at NRIM was used for investigating the following phenomena in 1965–1967 (Fujita et al., 1965, 1967; Fujita, 1966): (1) voltage dependence of the maximum observable thickness for the diffraction contrast of dislocations and other lattice imperfections; and (2) the thickness effect on the behavior of lattice defects. The other instruments were used to study the voltage effect on irradiation damage in polymer specimens (Kobayashi and Ohara, 1966), the critical voltage effect on diffraction (Nagata and Fukuhara, 1967; Watanabe et al., 1968), and other questions. As a result, it was found that the same dislocation behavior as in bulk specimens can be observed at 0.5 MV for light metals whose atomic number is smaller than about 20, and *in situ* experiments were carried out on various phenomena in those metals (Fujita et al., 1965, 1967; Fujita, 1966). Furthermore, TV-VTR systems were also applied to image recording, so that real-time recording of the rapid motion of lattice defects became possible (Imura et al., 1969).

Consequently, electron microscopy became the most effective technique for obtaining topographic information at the atomic scale directly and dynamically, giving the sensational results of both "seeing is believing" and "the materials are alive." In the past 30 years or so, high-voltage electron microscopy and high-resolution electron microscopy have been very much improved in Japan. The challenge of constructing high-voltage electron microscopes has contributed greatly to the improvement of the functional features of electron microscopes, because the technical difficulty of keeping the same overall performance of electron microscopes, i.e., energy fluctuation of electrons, vibration proof, shielding for fluctuation of both electromagnetic field and X-ray, etc., is approximately proportional to the third power of the accelerating voltage.

II. New Trends in Applications to Materials Science

New trends in applications of electron microscopy to materials science in Japan will be reviewed, with reference to the *Journal of Electron Microscopy,* vol. 38 (Supplement), which was a special issue for the 40th anniversary of the Japanese Society of Electron Microscopy (Ura, 1989).

A. Electron Holography

The performance of electron holography has been very much improved by using a field-emission-type electron microscope. By this method, the thickness distribution in the atomic resolution range, the electromagnetic field distribution, the reality of gauge fields [Aharonov-Bohm (AB) effect], etc., have been precisely determined (Tonomura *et al.*, 1986). Figure 1 is electron holographic interferometry showing the existence of the AB effect (Tonomura *et al.*, 1986). A tiny toroidal magnet was selected as the sample [Fig. 1(a)]. The magnetic flux rotates inside the toroid and does not leak outside. Such a sample is cooled down to 5 K and the relative phase shift is measured between two electron beams, one passing through the hole and the other outside the toroid. Measurements were made for various magnetic flux values, but only two kinds of interferograms are observed, as shown in Figs. 1(b) and 1(c). The phase shift is either 0 or π.

This phase-shift quantization implies that the magnetic flux is completely surrounded by the superconductor, and ensures that magnetic fields do not leak outside by the Meissner effect. The observed phase shift of π under ideal condition provides definitive evidence for the existence of the AB effect and also the physical reality of gauge fields.

B. High-Voltage Electron Microscopy

In order to observe the same behavior of lattice defects as in bulk materials, the specimen thickness must be greater than the mean free path of the contributing lattice defects, i.e., about 3 μm in general (Fujita *et al.*, 1965, 1967; Fujita, 1965). Remarkable increase of the maximum observable thickness of specimens with high-voltage electron microscopes (HVEMs) shows a great advantage in the research fields of materials science. By *in situ* experiments which satisfy the above condition, valuable information can be obtained dynamically under various experimental conditions, such as deformation, annealing, electron and ion irradiation, various atmospheres, etc., in a wide temperature range from about 5 K to 2300 K as follows: (1) motion of each individual lattice defect, (2) interaction among the lattice defects, (3) interaction between the lattice defects and microstructures, (4) qualitative determination of physical parameters, and others. Based on the results, mechanisms of the following phenomena occurring in materials have been made clear: (1) mechanical behavior such as yielding, work hardening, fracture, creep, and fatigue deformation (Fujita, 1969; Imura *et al.*, 1969, 1970; Imura and Saka, 1976), (2) mechanical twinning, (3) the martensitic transformation, (4) recovery and recrystallization, (5) precipitation, (6) electron irradiation damage, (7) environmental–material interaction, (8) toughening and strengthening of various ceramic composites, etc. Details of these results are summarized in *In Situ Experiments with HVEM*, edited by H. Fujita (1986), and in *New Directions and Future Aspects of*

APPLICATIONS TO MATERIALS SCIENCE 755

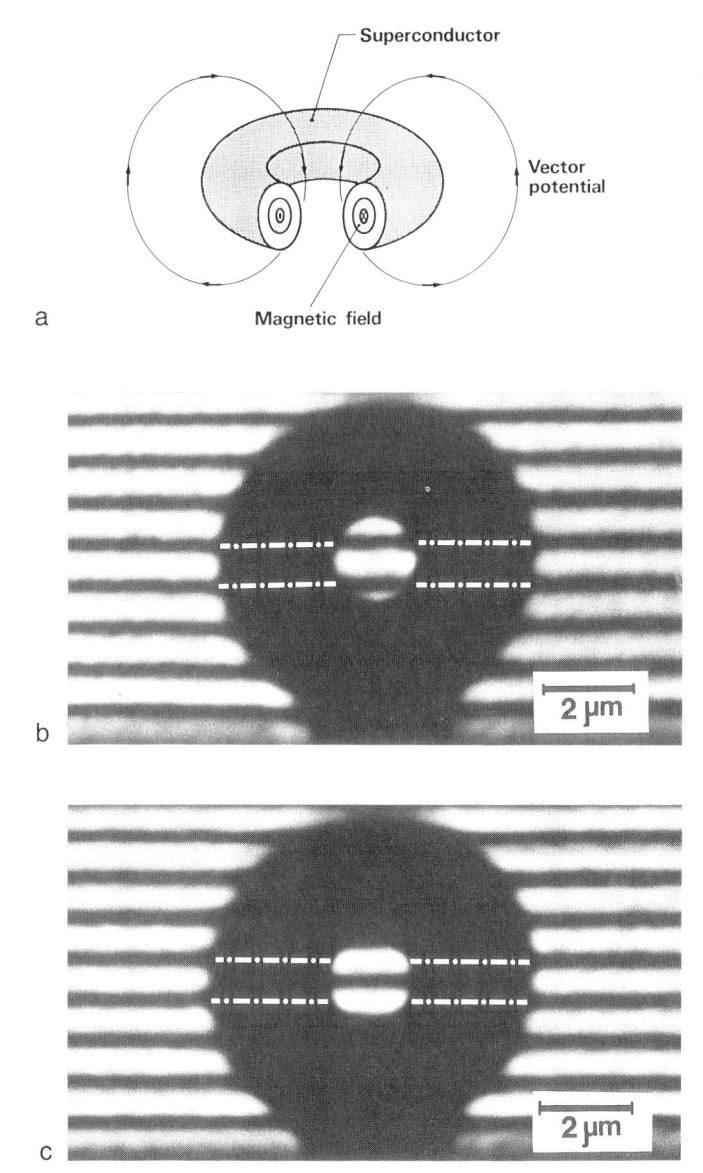

FIGURE 1. Experimental confirmation of the Aharonov-Bohm effect using toroidal magnets completely covered with superconducting layers. A phase shift of π is detected between two electron beams passing inside the hole and outside of the toroid, in the case (c) of an odd number of fluxons trapped within the toroid, although the beams never touch magnetic fields. (a) Schematic of toroidal sample; (b) electron interferogram (phase shift = 0); (c) electron interferogram (phase shift = π).

HVEM, edited by H. Fujita *et al.* (1991). Applications of ultra-HVEM are also summarized by Fujita (1986).

In the items mentioned above, electron irradiation effects are one of the important applications of HVEM, though the rate of irradiation damage is very sensitive to the specimen orientation (Fujita, 1986). The effects have been extensively investigated by Kiritani *et al.* and others in regard to the following subjects: (1) mobility of point defects (Kiritani, 1991; Yoshida and Kiritani, 1975; Fujita *et al.*, 1991); (2) secondary defects in pure metals (Shimomura *et al.*, 1986), (3) irradiation-induced phenomena in alloys and ceramics (Takeyama *et al.*, 1985; Kinoshita, 1985; Kinoshita *et al.*, 1987; Yada *et al.*, 1987).

Figure 2 is an example showing successive stages of superelasticity of a monoclinic $ZrO_2+3.9$ mass% MgO crystal (Fujita, 1986, 1991). That is, the twinned structure observed in Fig. 2(a) disappears in micrograph (b) immediately after the stress is applied, and then appears again in micrograph (c) as soon as the applied stress is removed.

Furthermore, with the remarkable increase of observable thicknesses of specimens, three-dimensional observation of microstructures and quantitative measurements of physical parameters have also been carried out precisely for the following subjects: determinations of (1) formation and migration energies of point defects (Kiritani, 1991), (2) three-dimensional distribution of microstructures (Kiritani, 1991), (3) internal stress, (4) stacking fault energies of crystals (Saka and Imura, 1982), (5) the stress-induced voiding in Al-lines for LSI (Takaoka and Ura, 1991; Okabayashi *et al.*,

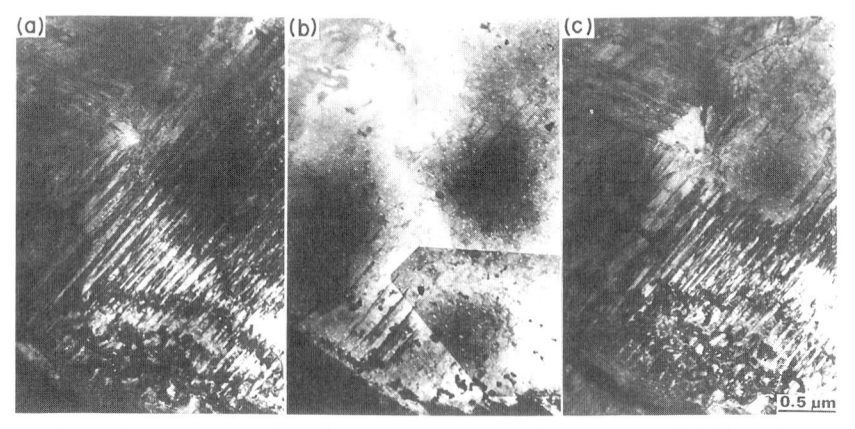

FIGURE 2. Pseudo-elasticity of a $ZrO_2-3.88$ mass% MgO composite. (a) Fine twin structures under no applied stress. (b) When the shock stress is applied at around room temperature, the twin structure disappears immediately. (c) Twin structure appears again as soon as the stress is removed.

1991), etc. Details of these results are summarized in the two books mentioned above on the applications of HVEM.

Besides these applications, the magnetic domain structures of materials have been investigated by using the Lorentz force of magnetized regions by high-voltage electron microscopy (Watanabe *et al.*, 1984). The critical voltage effect has also been used effectively for determination of the structure factor of complex crystal structures (Eguchi *et al.*, 1987; Tomokiyo *et al.*, 1990). Furthermore, a high-voltage STEM with a field-emission gun was constructed in Japan, and new advantages such as usefulness as an analytical microscope, weak damage of biological specimens, etc., have been clarified (Kuroda *et al.*, 1991). New high-resolution HVEMs have also been constructed (Matsui *et al.*, 1991), and the resolving power is expected to approach about 0.1 nm, as mentioned later.

By using the great advantages of high-voltage electron microscopy as mentioned already, a new research field called "microlaboratory" has been introduced. This means that high-voltage electron microscopy is not only a powerful tool for both characterization and identification of materials, but is also an indispensable "microlaboratory" in which various sorts of specimen treatments, including formation of nonequilibrium phases, can be carried out precisely on the atomic scale (Fujita, 1990, 1991). Typical examples are electron irradiation-induced phenomena as follows: (1) crystalline-amorphous solid transition (Mori and Fujita, 1982), and (2) foreign atom implantation in solid materials (Fujita and Mori, 1988). In the amorphization process, the electron irradiation-induced method is superior to earlier methods, such as liquid quenching, ion implantation, etc., in controlling conditions of amorphization. A precise study of amorphization has been carried out by this method (Mori and Fujita, 1982), and the general rules for amorphization were clarified (Fujita, 1990, 1991).

Figure 3 is an example showing electron irradiation-induced foreign atom implantation (Fujita and Mori, 1988), in which a vertical arrow indicates the irradiating direction and a dark particle is a Pb precipitate. Pb atoms were implanted into an Al substrate by 2-MeV electron irradiation, and Fig. 3(b) was taken by rotating the specimen 90° after irradiation at a dosage of 1.08×10^{28} e/m^2. Under the irradiation, the precipitate decreases in size and Pb-implanted regions (↑) are produced within the substrate underneath the precipitate, as shown in Fig. 3(b). The implanted Pb atoms form a supersaturated solid solution of aluminum, but the faint contrast of the implanted region results mainly from the difference in the scattering factor between the two elements. After annealing of the specimen at 573 K for 36 ks, the Pb atoms congregate into small particles, whose size is 2–10 nm in diameter, as seen in Fig. 3(c). The distribution of these small precipitates reflects the concentration profile of lead in the implanted region

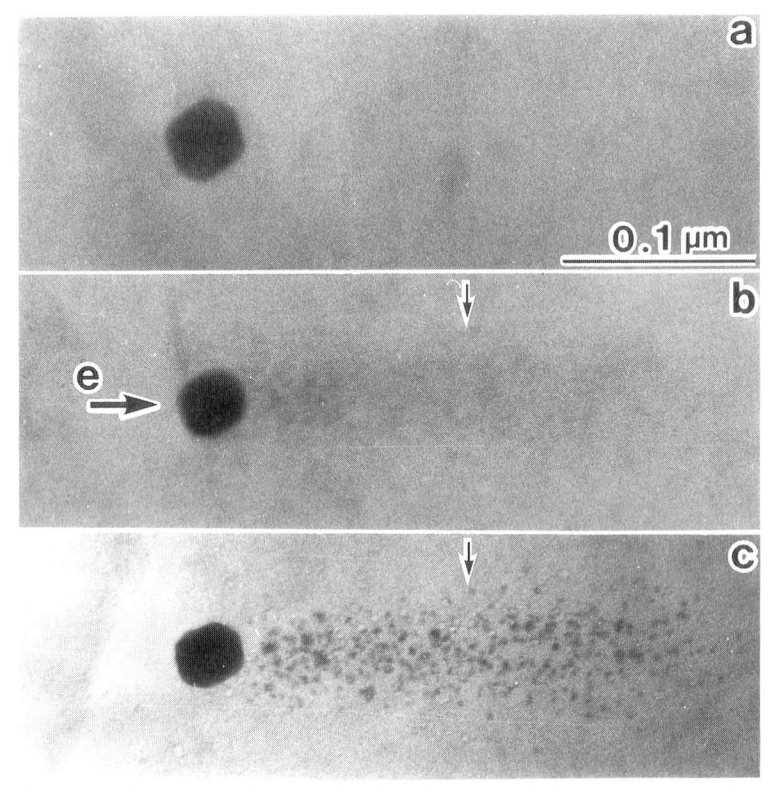

FIGURE 3. Electron irradiation-induced implantation of Pb atoms into an Al crystal. Accelerating voltage, irradiation temperature, and flux are 2 MV, 175 K, and 1.2×10^{24} e/m²s, respectively. Micrograph (a) was taken before irradiation, in which a dark particle is a target Pb precipitate. The same area after irradiation to 1.09×10^{28}e/mm² is depicted in (b). After irradiation, the implanted Pb atoms form a supersaturated solid solution in the Al matrix. These Pb atoms congregate into small particles whose size is 2–10 nm in diameter after annealing at 573 K for 36 ks, as shown in (c).

shown in Fig. 3(b). One point worth noting in Fig. 3(b) is the absence of strain contrast within and around the Pb-implanted region in spite of the large difference in atomic size between Pb and Al atoms (e.g., the former is 22% larger than the latter in diameter). Postirradiation annealing experiments revealed that this is due to the relaxation of lattice distortion by an effective coupling of lead atoms with vacancies (Fujita and Mori, 1988; Fujita, 1990).

In this implantation method, various advantages in the use of a highly accelerated electron beam as the primary irradiation beam can be fully utilized. For example, owing to the high penetration power of electrons, implantation into solids can be carried out without any serious temperature rise or surface damage. Furthermore, by using a microbeam of electrons, the implantation can be confined to a narrow region of the order of 2–3 nm diameter by the electron channeling phenomenon (Fujita, 1986). Therefore, by controlling the irradiation conditions, a variety of nonequilibrium states can be induced in the objects. A systematic study on the stability of artificially produced nonequilibrium solid phases was also made.

C. High-Resolution Electron Microscopy

The use of high-resolution electron microscopy has been extended into various fields of materials science by the development of the resolving power of electron microscopes. In Japan, some HVEMs have been developed for improving image resolution, and 300- and 400-kV electron microscopes are now conventional high-resolution electron microscopes. These results are summarized in *Characterization of Advanced Materials by High Resolution Electron Microscopy and Analytical Electron Microscopy*, edited by S. Nagakura (1990). The resolving power of these electron microscopes approaches about 0.1 nm with HVEMs. The many-wave imaging method has been used widely in these research fields, and the many-wave lattice fringes, i.e., so-called atomic structures, of various organic (Uyeda *et al.*, 1972), inorganic (Cowley and Iijima, 1972; Horiuchi and Matsui, 1974), and metal compounds (Hashimoto *et al.*, 1977; Izui *et al.*, 1978–1979) have been investigated by this method combined with computer-simulated images. Figure 4 shows many-wave lattice fringes of {110} diamond taken with a 300-kV electron microscope (Fujita and Sumida, 1986), and Fig. 5 is a so-called structure image of {110} silicon taken with a 1300-kV electron microscope in comparison with a computer-simulated one (Horiuchi *et al.*, 1991). In these micrographs, the minimum image separation is about 0.1 nm. Figure 6 is an example showing molecular images of chlorinated Cu-phthalocyanine taken with a 500-kV high-resolution electron microscope and computer-simulated images (Kobayashi and Uyeda, 1988). The specimen used was a thin epitaxial film prepared with vacuum deposition on KCl. This kind of work has been carried out by Suito *et al.* (1958), Uyeda *et al.* (1970, 1972), and others in Japan.

High-resolution electron microscopy has been applied to determine atomic structures of the following materials: (1) lattice defects in molecular crystals (Kobayashi and Uyeda, 1988), (2) atomic structures of various grain boundaries and interfaces (Ichinose and Ishida, 1981, 1989; Ichinose *et al.*, 1987), (3) microstructures of iron-based alloys (Nagakura, 1986), (4) ordered structures and intermetallic compounds (Hirabayashi, 1983; Kitano and Komura, 1985), (5) ichosahedral structures of materials (Hiraga

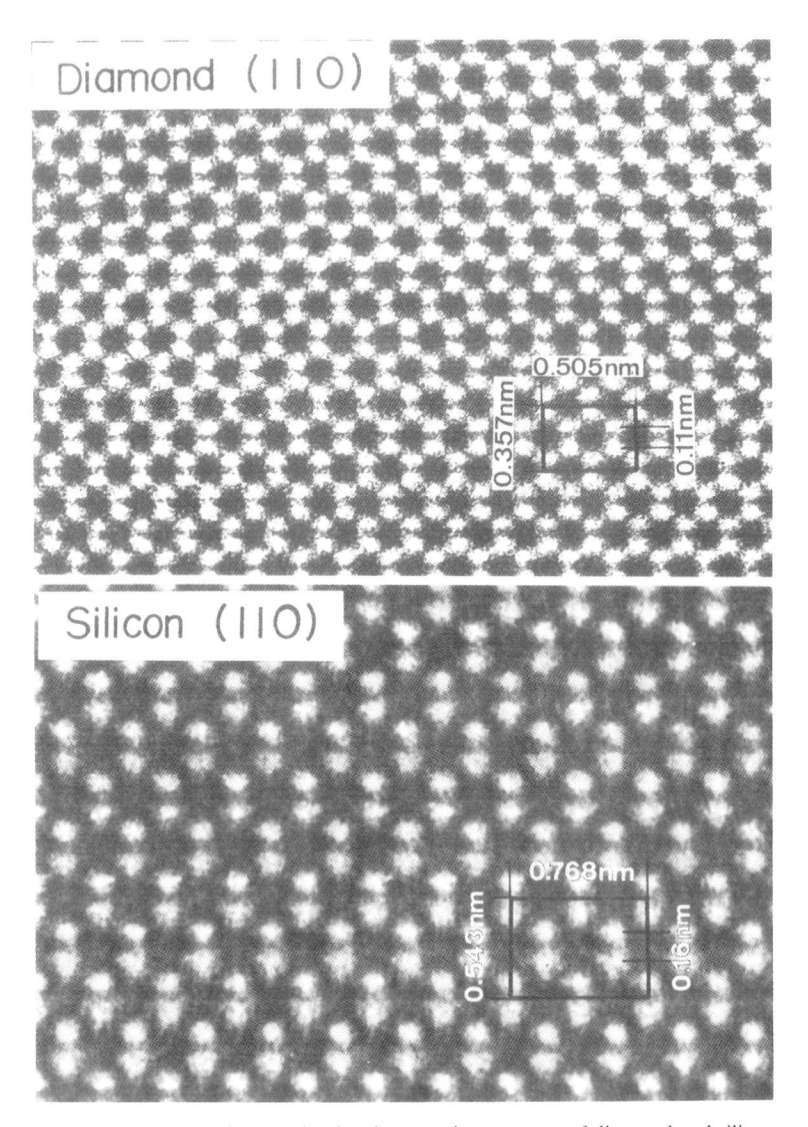

FIGURE 4. Electron micrographs showing atomic structures of diamond and silicon crystals. The micrographs were taken with a 300-kV high-resolution electron microscope.

et al., 1985), (6) porous alumina films (Akahori, 1965), (7) microstructures and crystal growth of inorganic materials (Kakibayashi *et al.*, 1987; Takeda, 1991), (8) amorphous solids (Hirotsu and Akada, 1984; Ishida *et al.*, 1981, 1985; Hamada and Fujita, 1986), (9) industrially important materials (Hiraga *et al.*, 1986, 1987, 1990), (10) atom clusters (Iijima, 1985), (11) carbon nanotubes (Iijima, 1991), etc. In term (4), the cause of incommensu-

APPLICATIONS TO MATERIALS SCIENCE 761

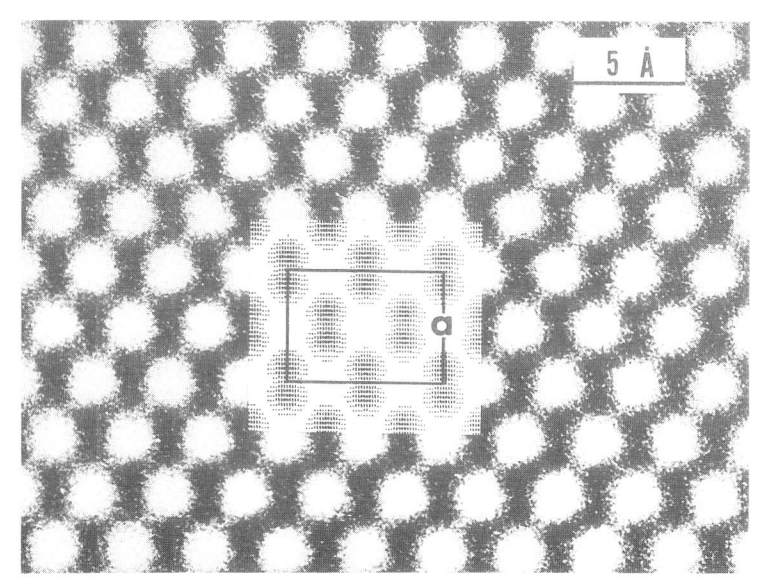

FIGURE 5. A structure image of silicon crystal. The micrograph was taken with a high-resolution 1300-kV HVEM with the electron beam along the [110] axis. Each site of Si atoms is imaged as dark spots. The dark spot images of dumbbell-like pattern correspond to the structure image of silicon [110] incidence.

rate structures has been investigated directly by high-resolution electron microscopy (Nagakura, 1986; Hirabayashi, 1983; Kitano and Komura, 1985), while these studies were carried out only by the electron diffraction method. Relationships between the amorphous structures and microcrystallites in item (8) have been discussed by using various models based on contrast change of observed images (Hirotsu and Akada, 1984; Ishida et al., 1981; Hamada and Fujita, 1986). Furthermore, microstructures of high-T_c oxide ceramics (Hiraga et al., 1987, 1990) are involved in item (7), and microstructures of Sm-Co and Nd-Fe-B permanent magnets are involved in item (9) (Hiraga et al., 1986).

Figure 7 is an example of item (2), and shows the atomic structure of a superlattice of GaAs/AlAs (Tanaka et al., 1987). In the superlattice, the thickness of AlAs is changed from one to 20 molecular layers artificially, as shown in Fig. 7(b). Micrograph (a) is an enlargement of the circled region in micrograph (b) in which each AlAs part consists of only a monomolecular layer. It is recognized in Fig. 7(a) that the interface is flat on the lower side of each monomolecular AlAs layer but rough on the upper side. The difference in the roughness of the AlAs interface coincides with the result of RHEED investigation.

Figure 8 is an example showing the atomic structure of icosahedral quasi-crystal and the corresponding diffraction pattern of an Al-Li-Cu alloy (Hir-

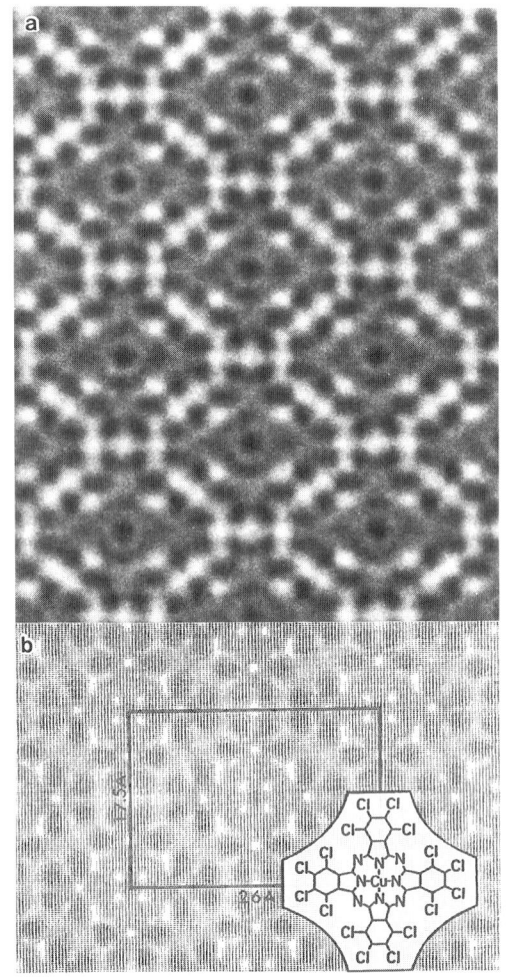

FIGURE 6. Molecular images of chlorinated Cu-phthalocyanine taken with a 500-kV high-resolution electron microscope and computer-simulated images. The specimen was a thin epitaxial film prepared with vacuum deposition on KCl.

aga, 1991). The micrograph was taken with the incident beam parallel to the fivefold symmetry axis. Characteristic image contrasts are composed of a bright ring and 10 bright dots surrounding the ring, and they correspond to icosahedral atom clusters forming the structure of the quasi-crystal.

Surface science is one of the important applications of high-resolution electron microscopy, and an ultrahigh-vacuum, high-resolution electron microscope has been developed for such subjects in Japan (Yagi and Honjio,

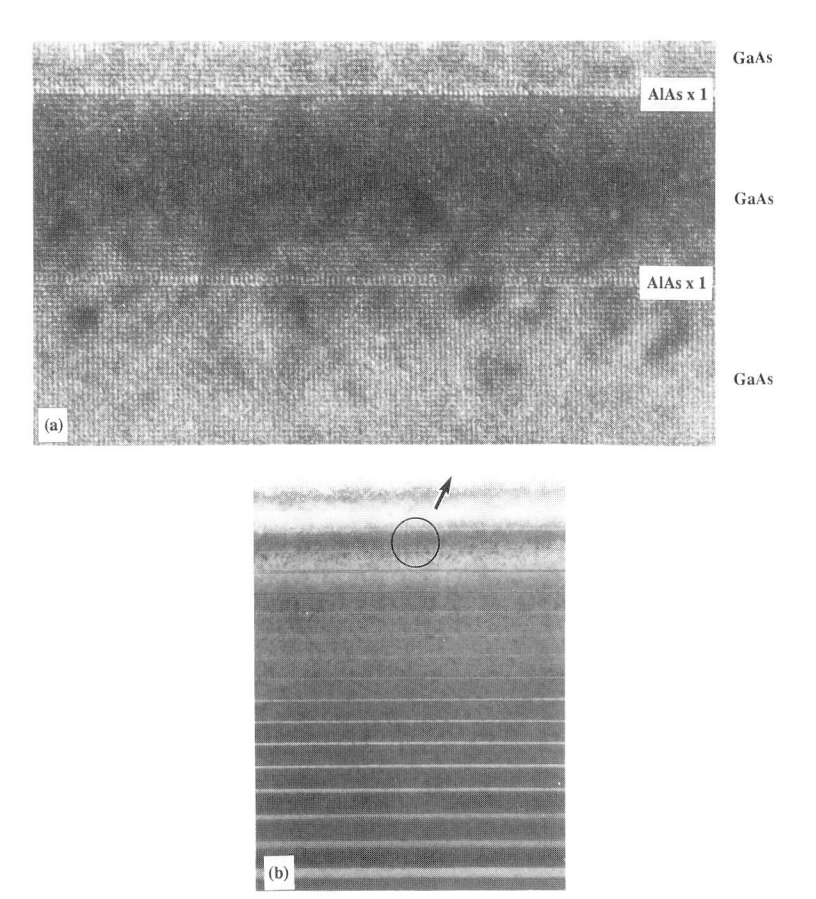

FIGURE 7. High-resolution electron micrograph of monomolecular AlAs layers in GaAs crystals. The thickness of the AlAs layers was changed from 1 to 20 molecular layers in GaAs crystals, as shown in (b). Electron microscope image (a) is an enlargement of a circled region in (b), and shows the atomic structures of the monomolecular (001) plane. The GaAs/AlAs interface is flat at the lower side of each monomolecular AlAs layer, but it is rough at the upper side.

1976). Figure 9 is an example of a profile TEM image of the 5×1 reconstructed structure of a gold (001) surface (Takayanagi et al., 1987). Figure 10 is a high-resolution REM image of the 7×7 reconstructed surface of Si (111) crystal with the corresponding RHEED pattern, in which the 2.3-nm fringes (dark lines) of 7×7 superlattice and the dark lines of the surface steps are clearly resolved (Takayanagi et al., 1987). High-resolution REM has been used effectively for studying adsorption and film growth (Takayanagi et al., 1987). Since REM has an advantage

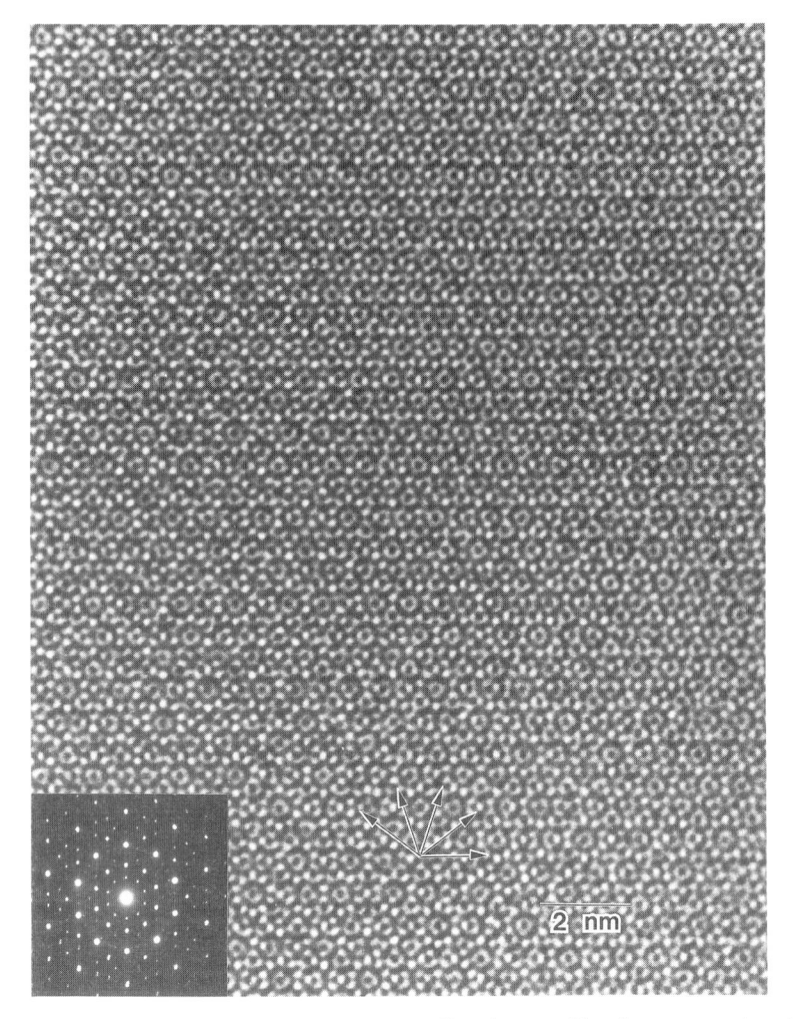

FIGURE 8. Electron micrograph and corresponding electron diffraction pattern of an Al-Li-Cu icosahedral quasi-crystal. Note that the atom clusters are arranged along straight lines parallel to the fivefold directions indicated by arrows.

that bulk specimen surfaces are observable at the atomic scale, this method is applicable to various fields of materials science.

Dynamic information has also been obtained by *in situ* experiments with high-resolution electron microscopes in the following cases: (1) change of local atomic structures during growth of both stacking faults and twins in Au crystals, (2) behavior of ultrafine particles and atom clusters of Au (Iijima, 1985; Iijima and Ichihashi, 1986; Takayanagi *et al.*, 1987), (3) melting processes of small particles embedded in a crystalline matrix (Saka *et al.*, 1985),

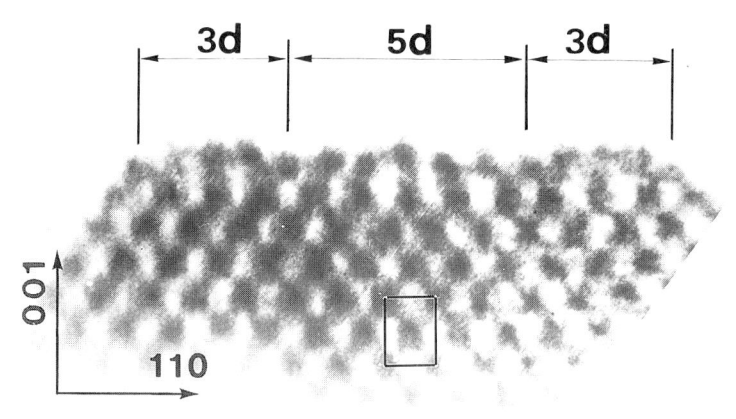

FIGURE 9. Profile TEM image of the 5 × 1 reconstructed structure of gold (001) surface.

(4) mechanisms of phenomena in which the mean free path of contributing lattice defects is extremly small, such as behavior of solute atoms and point defects at low temperature (Fujita and Lu, 1992), and (5) spontaneous alloying of atom clusters (Mori and Yasuda, 1993; Yasuda *et al.*, 1992, 1993).

In these investigations, ultrahigh-vacuum, high-resolution electron microscopes are necessary. Figure 11 shows behavior of a Au-ultrafine particle on a SiO_2 crystal surface, in which both the shape and the atomic arrangement of the particle are frequently changed by multitwinning as in a liquid drop (Iijima, 1986). Figure 12 is a similar *in situ* experiment showing growth of Au-atom clusters on a graphite film (Takayanagi *et al.*, 1987). It is noted in Fig. 12 that the atomic structure also frequently changes during coalescence between different clusters.

The results show that the behavior of atom clusters markedly differs from that of the corresponding bulk materials. Namely, *in situ* experiments give indispensible information about the peculiar behavior of atom clusters.

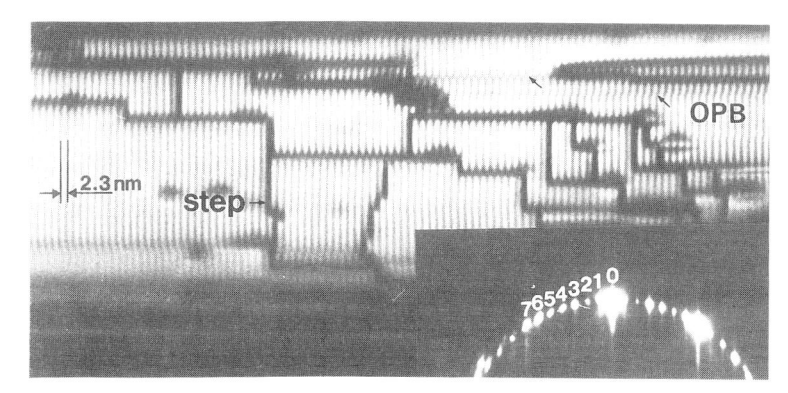

FIGURE 10. High-resolution REM image of the 7 × 7 reconstructed surface of Si (111) crystal.

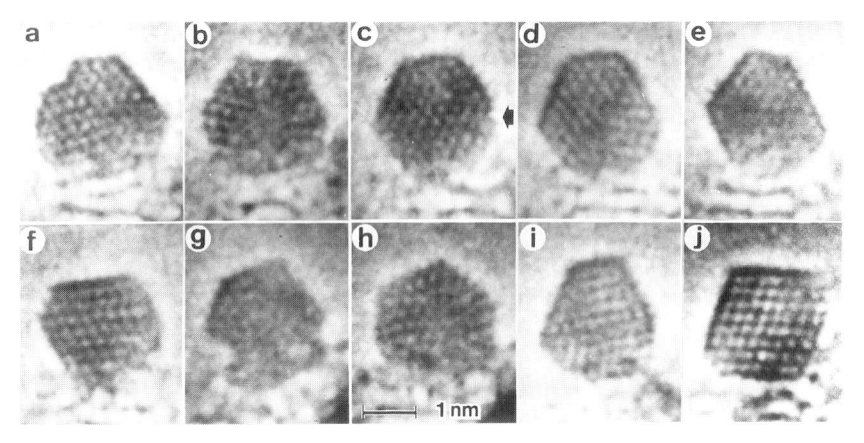

FIGURE 11. Change of atomic structure of an ultrafine Au particle: Electron micrographs showing various shapes of a Au particle 2 nm in diameter reproduced from a VTR tape. The lattice fringes correspond to 0.235 nm of d_{111} spacing. The particle in (a), (d), and (i) is a single twin. A single crystal with a cuboctahedral shape is seen in (e), (f), and (i). It is concluded from the size of cuboctahedron (j) that the particle theoretically contains 459 Au atoms. The particle also transformed into a multiply twinned icosahedral particle, (b) and (h).

III. Other Topics

Convergent-beam electron diffraction has also been used by Tanaka (1986) and Tanaka *et al.* (1987, 1991) and others as a useful analytical tool for studying the space groups and the lattice constants of crystals, characteristics of lattice defects, lattice strain, and composition change in the vicinity of grain boundaries. In order to increase further the reliability of atomic information, electron microscopy has been combined with other methods such as EELS, Auger valence electron microscopy, STM, etc. Microcharacterization is one example.

Microcharacterization is carried out based on the analysis of local chemical composition and atomic structure. Figure 13 is an example showing microcharacterization of a silicon aluminum oxynitride polytype, 15R-$SiAl_4O_2N_4$ (Bando *et al.*, 1986, 1989). This microcharacterization was carried out by the combination of convergent-beam electron diffraction (a), EDS (b), EELS (c), and high-resolution electron microscopy combined with a calculated image contrast (d). All these results were obtained from the observed area in Fig. 13(d). In this crystal, a true space group was not determined uniquely by both X-ray and the convergent selected-area electron diffraction method. From the convergent-beam ED pattern of Fig. 13(*a*), the space group is assigned as R3m. From the EDS spectrum, Fig. 13(*b*), the molar ratio of Si to Al is assigned to be 1:4. The quantitative

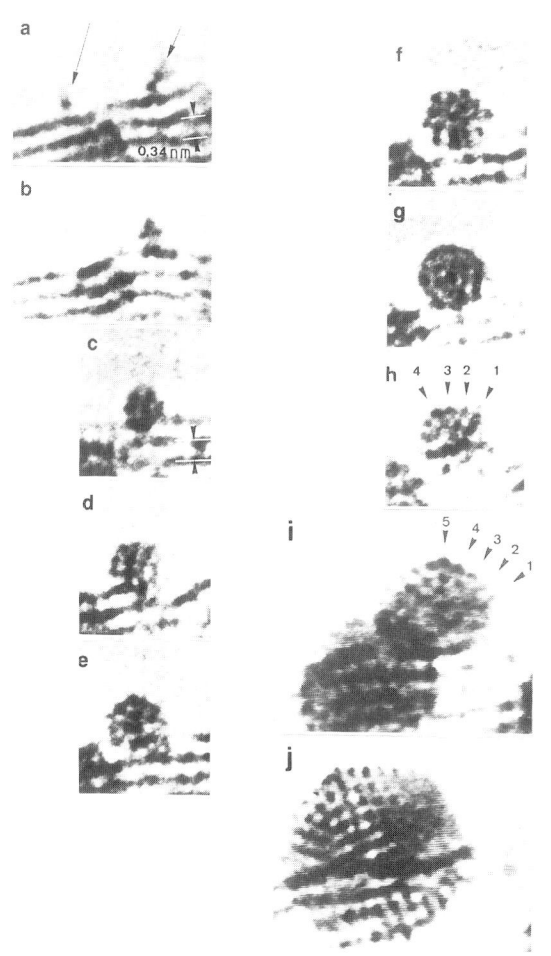

FIGURE 12. (a–j) Electron microscope images of Au atom clusters. An atom cluster is growing on a graphitized carbon surface which is vertical to the plane of the sheet.

analysis of the EELS spectrum, Fig. 13(c), gives the molar ratio of O to N as 2:4. These results show that the chemical composition of the 15R type is determined to be $SiAl_4O_2N_4$. Furthermore, many-wave lattice fringes of the 15R, Fig. 13(d), in which calculated image contrast is inserted, shows that cation sites appear well resolved as dark spots. The possible structure model is derived from the structural analogy of AlN, and then the validity of the structure model obtained is confirmed by image calculation. Thus, it is concluded that the combination of both lattice image and spectroscopic microanalysis leads to accurate determination of crystal structure.

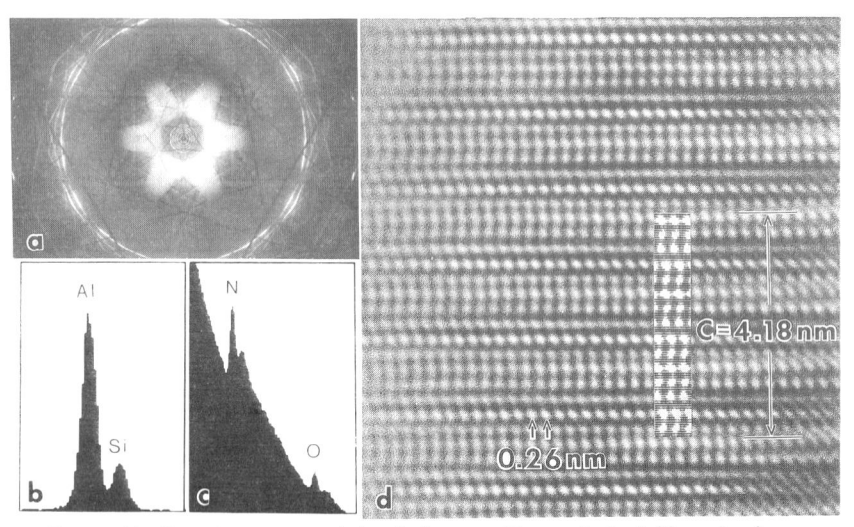

FIGURE 13. Crystal structure and chemical composition analysis of silicon aluminum oxynitride polytype, 15R-SiAl$_4$O$_2$N$_4$, by a combination method. The analysis was carried out by a combination of convergent beam electron diffraction (a), EDS (b), EELS (c), and high-resolution electron microscopy (d). A calculated image contrast is inserted in (d).

Furthermore, determination of atom location by channeling-enhanced microanalysis (ALCHEMI) has been widely used to decide the exact postion of solute atoms in complex crystal structures (Shindo *et al.*, 1990; Nakata *et al.*, 1989), and combination of electron microscopy and other methods such as Auger valence electron spectroscopy (Fujita *et al.*, 1989) and cathodeluminescence (Yamamoto, 1990) has been used to obtain information about valence electrons.

A gigahertz stroboscopic SEM (Hosokawa *et al.*, 1978) and a stroboscopic TEM (Ura and Fujioka, 1989) were developed, and propagation of an electromagnetic field in semiconductor devices was measured by Ura and Fujioka (1989). The instruments are used effectively for studying detailed processes of very rapid phenomena such as phase transformations of various materials, propagation of magnetic domain walls, changes in the atomic structure of atom clusters under various conditions, and others.

REFERENCES

Akahori, H. (1965). *Kinzoku Hyoumen Gijutsu* (in Japanese) **16,** 398.
Bando, Y., Mitomo, M., Kitami, Y., and Izumi, F. (1986). *J. Microsc.* **142,** 235. (1989). *J. Electron Microsc.* **38,** S81.

Bollman, W. (1956). *Phys. Rev.* **103**, 1588.
Cowley, J. M., and Iijima, S. (1972). *Naturforsch.* **27**, 445.
Eguchi, T., Tomokiyo, Y., and Matsuhata, H. (1987). *J. Microsc. Spectrosc. Electron* **12**, 559.
Fujita, E. (1955). *Jpn. Electron Microsc. (in Japanese)* **4**, 63.
Fujita, H. (1961). *Phys. Soc. Jpn.* **16**, 397.
Fujita, H. (1966). *Jpn. J. Appl. Phys.* **5**, 729. (1966). *J. Phys. Soc. Jpn.* **26**, 1605.
Fujita, H. (1986). *J. Electron Microsc. Technique* **3**, 243.
Fujita, H. (1986). *Proc XIth Int. Congress on Electron Microscopy,* Kyoto, vol. 2, p. 1025.
Fujita, H. (ed.-in-chief) (1986). *Proc. Int. Symp. on Behavior of Lattice-Imperfections in Metals—In Situ Experiments with HVEM,* Osaka University.
Fujita, H. (1991). *Ultramicroscopy* **39**, 105. (1990). *Material Trans. JIM* **31**, 523.
Fujita, H., Kawasaki, Y., Furubayashi, E., Kajiwara, S., and Taoka, T. (1967). *Jpn. J. Appl. Phys.* **6**, 214.
Fujita, H. (1969). *J. Phys. Soc. Jpn.* **26**, 331. (1969). *J. Phys. Soc. Jpn.* **26**, 1437.
Fujita, H., and Lu, C. (1992). *Material Trans. JIM* **33**, 897.
Fujita, H., and Mori, H. (1988). *Proc. Int. Symp. on Non-Equilibrium Solid Phases of Metals and Alloys,* Kyoto, Jpn. Inst. Met., p. 37.
Fujita, H., Nakayama, H., and Fuchida, Y. (1989). *Phil. Mag.* **A56**, 873.
Fujita, H., and Nishiyama, Z. (1961). *J. Phys. Soc. Jpn.* **16**, 1893. (1962). *J. Phys. Soc. Jpn.* **17** (suppl. B-II), 200.
Fujita, H., and Sumida, N. (1986). *Hitachi Inst. News,* no. 19, p. 3.
Fujita, H., Taoka, T., and NRIM-500kVEM Group (1965). *J. Electron Microsc.* **14**, 307.
Fujita, H., Ura, K., and Mori, H. (eds) (1991). "New Directions and Future Aspects of HVEM," Special issue of *Ultramicroscopy,* p. 39.
Fujiwara, K. (1962). *J. Phys. Soc. Jpn.* 17 (suppl. B-II), Proc. Int. Conf. on Mag. and Crystal, p. 118.
Fukami, A. (1956). *J. Electron Microsc. Jpn.* **4**, 109.
Fukami, A., and Shiyota, R. (1955). *Shikizai* **28**, 92 (in Japanese). (1960). "Theory and Applications of Electron Microscope Images and Its Applications, Part III," p. 255. Maruzen, Tokyo.
Furubayashi, E. (1971). *Proc. Int. Conf. on Sci. Tech. Iron Steel,* Tokyo Suppl. to *Trans. ISIJ* **119II**, 1245.
Furubayashi, E., Fujita, H., and Taoka, T. (1966). *Proc. 6th Int. Congress on Electron Microscopy,* Kyoto, vol. 1, p. 415.
Hamada, T., and Fujita, F. E. (1986). *Jpn. Appl. Phys.* **25**, 318.
Hashimoto, H., Endo, H., Tanji, T., Ono, A., and Waranabe, E. (1977). *J. Phys. Soc. Jpn.* **42**, 1073.
Hashimoto, H., Howie, A., and Whelan, M. J. (1962). *Phil. Mag.* **5**, 967.
Hashimoto, H., Mannami, M., Nakai, T., Eto, T., and Fujiwara, K. (1960). *J. Electron Microsc.* **9**, 130.
Hashimoto, H., Naiki, Eto, T., and Fujiwara, K. (1966). *Proc. 6th Int. Cong. on Electron Microscopy,* Kyoto, vol. 1, p. 495.
Hashimoto, H., Takai, Y., Yokota, Y., Endoh, H., and Kumao, A. (1979–1980). *Chem. Scripta* **14**, 125; *Jpn. J. Appl. Phys.* **19**, L-1.
Hashimoto, H., Tanaka, K., Kobayashi, K., Shimadzu, S., Naiki, T., and Mannami, M., (1958). *Proc. 4th Int. Conf. on Electron Microscopy,* Berlin, p. 477. (1958). *J. Electron Microsc.* **9**, 130.
Hashimoto, H., Tanaka, K., Kobayashi, K., Suito, E., Shimadzu, S., and Iwanaga, M. (1962). *J. Phys. Soc. Jpn.* **17,** (suppl. B-II), 170. (1964). *J. Appl. Phys.* **35**, 277.
Hashimoto, H., Tanaka, K., and Uyeda, R. (1956). *Proc. 1st Regional Conf. Asia-Oceania on Electron Microscopy,* p 292. (1957). *Acta Crystallog.* **10**, 143.

Heidenreich, R. D. (1949). *J. Appl. Phys.* **20,** 993.
Hibi, T. (1955). *Electron Microsc.* (in Japanese) **4,** 68.
Hirabayashi, M. (1983). *Trans. Inst. Met.* **24,** 317.
Hiraga, K. (1991). *J. Electron Microsc.* **40,** 81.
Hiraga, K., Hirabayashi, M., Inoue, A., and Masumoto, T. (1987). *J. Microsc.* **146,** 245.
Hiraga, K., Hirabayashi, M., and Ishigaki, N. (1986). *J. Microsc.* **142,** 201.
Hiraga, K., Hirabayashi, M., Sagawa, M., and Matsuura, Y. (1985). *Jpn. J. Appl. Phys.* **24,** 130.
Hiraga, K., and Shindo, D. (1990). *Material Trans. JIM* **31,** 567.
Hirotsu, Y., and Akada, R. (1984). *Jpn. J. Appl. Phys.* **23,** L479.
Hirsch, P. B., Horne, R. W., and Whelan, M. J. (1956). *Phil. Mag.* **1,** 677.
Hirsch, P. B., Howie, A., and Whelan, M. J. (1960). *Phil. Trans. R. Soc. A* **252,** 49.
Honjio, G., Kitamura, K., Shimaoka, K., and Mihama, K. (1956). *J. Phys. Soc. Jpn.* **11,** 527.
Howie, A., and Whelan, M. J. (1961). *Proc. R. Soc. A* **263,** 217. (1962). *Proc. R. Soc.* **A267,** 206.
Horiuchi, S., and Matsui, Y. (1974). *Phil. Mag.* **30,**777.
Horiuchi, S., Matsui, Y., Kitami, Y., Yokoyama, M., Suehara, S., Wu, X. J., Matsui, I., and Katsuta T. (1991). *Ultramicroscopy* **30,** 231.
Hosokawa, T., Fujioka, H., and Ura, K. (1978). *Rev. Sci. Instrum.* **49,** 1293.
Ichinose, H., and Ishida, Y. (1981). *Phil. Mag.* **43,** 1253. (1989). *Phil. Mag.* **60,** 555.
Ichinose, H., Ishida, Y., Furuta, T., and Sakaki H. (1987). *J. Electron Microsc.* **36,** 82.
Iijima, S. (1985). *Electron Microsc.* **40,** 246. Iijima, S., and Ichihashi, T. (1986). *Phys. Rev. Lett.* **56,** 616.
Iijima, S. (1991). *Nature* **354,** 56.
Imura, T. (1970). *Electron Microsc.* (in Japanese) **19,** 231.
Imura, T., and Saka, H. (1976). *Mem. Faculty Eng. Nagoya Univ.* **28,** 54.
Imura, T., Saka, H., and Doi, M. (1970). *J. Phys. Soc. Jpn.* **29,** 803.
Imura, T., Saka, H., and Yukawa, N. (1969). *J. Phys. Soc. Jpn.* **26,** 327.
Ino, S., Ogawa, S., and Taoka, T. (1972). *Jpn. J. Appl. Phys.* **11,** 1859.
Ino, S., Watanabe, D., and Ogawa, S. (1956). *J. Phys. Soc. Jpn.* **17,** 1074.
Ishida, Y., Ichinose, H., Shimada, H., and Kojima, H. (1981). *Proc. Int. Conf. on Rapid Quenched Metals,* Sendai, p. 13. Ishida, Y., and Ichinose, H. (1985). *J. Electron Microsc.* **34,** 266.
Iwase, K., and Ogawa, K. (1948). *Powder and Powder Metallurgy* (in Japanese) **2,** 1.
Izui, K., Furuno, S., Nishida, T., and Otsu, H. (1978–1979). *Chem. Scr.* **14,** 99.
Kakibayashi, H., Nagata, F., and Ono, Y. (1987). *J. Appl. Phys.* **26,** 770.
Kamiya, Y., and Uyeda, R. (1961). *J. Phys. Soc. Jpn.* **16,** 136. (1962). *J. Phys. Soc. Jpn.,* Suppl. B-II, p. 191.
Karashima, S., Oikawa, H., and Ogura, T. (1968). *Trans. JIM* **9,** 205.
Kato, N. (1963). *Acta Crystallog.* **16,** 282.
Kimoto, K., Kamiya, Y., Nonoyama, M., and Uyeda, R. (1963). *Jpn. J. Appl. Phys.* **2,** 702.
Kinoshita, C. (1985). *J. Electron Microsc.* **34,** 299.
Kinoshita, C., Nakai, K., and Kitajima, S. (1987). *Material Sci. Forum* **15–18,** 1403.
Kiritani, M. (1991). *Ultramicroscopy* **39,** 135.
Kiritani, M., Shimomura, Y., and Yoshida, S. (1964). *J. Phys. Soc. Jpn.* **19,** 1624.
Kitano, Y., and Komura, Y. (1985). *J. Electron Microsc.* **20,** 101.
Kobayashi, K., Hashimoto, H., Suito, E., Shimadzu, S., and Iwanaga, M. (1963). *Jpn. J. Appl. Phys.* **2,** 47.
Kobayashi, K., and Ohara, M. (1966). *Proc. 6th Int. Cong. on Electron Microscopy,* Kyoto, vol. 1, p. 579.
Kobayashi, T., and Uyeda, N. (1988). *Phil. Mag.* **B57,** 493.
Koda, S., and Takeyama, T. (1953). *J Electron Microsc. Jpn.* **1,** 39.

Kubo, K. (ed.-in-chief) (1960). "Theory and Applications of Electron Microscopy, Part III—Applications to Science and Engineering" (in Japanese), Japanese Society for Electron Microscopy, pp. 2–56. Maruzen, Tokyo.

Kuroda, K., Morita, C., Arai, S., Yokoi, N., and Saka, H. (1991). *Ultramicroscopy* **39**, 58.

Matsui, Y., Horiuchi, S., Bando, Y., Kitami, Y., Yokoyama, M., Suehara, S., Matsui, I., and Katsuta, T. (1991). *Ultramicroscopy* **39**, 8.

Mihama, K., and Tanaka, N. (1976). *J. Electron Microsc.* **25**, 65.

Mitsuishi, T., Nagasaki, N., and Uyeda, R. (1951). *Proc. Jpn. Acad.* **27**, 86.

Mori, H., and Fujita, H. (1982). *Jpn. J. Appl. Phys.* **21**, L494.

Mori, H., and Yasuda, H. (1993). *Intermetallics*, **1**, 30.

Nagakura, S. (1986). *Electron Microsc.* (in Japanese) **21**, 113.

Nagakura, S. (ed.-in-chief) (1990). "Characterization of Advanced Materials by High Resolution Electron Microscopy and Analytical Electron Microscopy," Special Issue of *Material Trans. JIM* **31**.

Nagata, E., and Fukuhara, A. (1967). *Jpn. J. Appl. Phys.* **6**, 1233.

Nakata, Y., Tadaki, T., and Shimizu, K. (1989). *Material Trans. JIM* **7**, 625.

Nemoto, M., and Koda, S. (1963). *J. Inst. Metals Jpn.* **27**, 599.

Nemoto, M., and Koda, S. (1965). *J. Inst. Metals Jpn.* **93**, 164.

Nemoto, M., and Koda, S. (1966). *Trans. JIM* **7**, 235.

Nemoto, M., Koguchi, T., and Sudo, H. (1971). *J. Inst. Metals Jpn.* **35**, 886.

Nishimura, H., Murakami, Y., and Takamura, J. (1951). *Tech. Rep. Eng. Res. Inst. Kyoto Univ.* **1**, 19.

Nishiyama, Z., and Fujita, H. (1957). *Metal Phys. Jpn.* (in Japanese) **3**, 108.

Nishiyama, Z., and Koda, S. (eds.) (1975). "Interpretation of Electron Microscope Images and Its Applications to Metallury" (in Japanese). Maruzen, Tokyo.

Nishiyama, Z., and Shimizu, K. (1956). *J. Electron Microsc. Jpn.* **4**, 51. (1958). *Acta Metall.* **6**, 125.

Nishiyama, Z., and Shimizu, K. (1959). *Acta Metall.* **2**, 432. (1961). *Acta Metall.* **9**, 980.

Nishiyama, Z., Shimizu, K., and Morikawa, H. (1968). *J. Inst. Metals* **32**, 116. (1968). *Trans. JIM* **9**, 307.

Nishiyama, Z., Shimizu, K., and Oka, M. (1958). *Inst. Metals Jpn.* (in Japanese) **22**, 532.

Nonaka, K. (1955). *Electron Microsc.* (in Japanese) **4**, 7.

Ogawa, S., Watanabe, D., Watanabe, H., and Komoda, T. (1958). *Acta Crystallog.* **11**, 872.

Okabayashi, H., Tanikawa, A., Mori, H., and Fujita, H. (1991). *Ultramicroscopy* **39**, 306.

Saka, H., Sakai, A., Kamino, T., and Imura, T. (1985). *Phil. Mag.* **A52**, 29.

Saka, H., Sueki, Y., and Imura, T. (1978). *Phil. Mag.* A37, 273, 291.

Saka, H., and Imura, T. (1982). *Proc. 10th Int. Congress on Electron Microscopy*, Humburg, vol. 2, p. 481.

Shimizu, K., and Okamoto, M. (1971). *Trans. JIM* **12**, 273.

Shimomura, Y., Guinan, M. W., Fukuhara, H., Hahn, P. A., and Kiritani, M. (1986). *J. Nuclear Materials* **141**, 846.

Shimomura, Y., and Yoshida, S. (1965). *J. Phys. Soc. Jpn.* **20**, 1667. (1967). *J. Phys. Soc. Jpn.* **22**, 319.

Shindo, D., Hiraga, K., Takasugi, T., and Tasi, A. P. (1990). *Material Trans. JIM* **7**, 647.

Shiojiri, M. (1967). *Jpn. Appl. Phys.* **6**, 163.

Suito, E., Takiyama, K., and Uyeda, N. (1954). *Bull. Inst. Chem. Res. Kyoto, Univ.* Suppl. issue 18. (1955). Suito, E., and Takiyama, K. *Science* (in Japanese) **25**, 39.

Suito, E., Uyeda, N., Watanabe, H., and Komoda, T. (1958). *Nature* **181**, 332.

Takahashi, N., and Ashinuma, K. (1958). *J. Electron Microsc.* **6**, 29. (1959). *J. Electron Microsc.* **7**, 37.

Takahashi, N., and Mihama, K. (1957). *Acta Metall.* **5,** 159.
Takamura, J. (1955). *Metal Phys. Jpn.* (in Japanese) **1,** 209.
Takaoka, A., and Ura, K. (1991). *Ultramicroscopy* **39,** 299.
Takayanagi, K., Tanishiro, Y., Kobayashi, K., Akiyama, K., and Yagi, K. (1987). *Jpn. J. Appl. Phys.* **26,** L957.
Takeda, S. (1991). *Jpn. J. Appl. Phys.* **30,** L639.
Takeuchi, S., Furibayashi, E., and Taoka, T. (1967). *Acta Metall.* **15,** 1179.
Takeyama, T., Ohnuki, S., and Takahashi, H. (1985). *J. Nuclear Material* **133,** 571.
Takeyama, T., and Takahashi, H. (1970). *J. Inst. Metals Jpn.* **34,** 307.
Tanabe, Y., and Kawasaki, S. (1972). *Kinzoku Hyoumen Gijutsu* (in Japanese) **23,** 521.
Tanaka, M. (1986). *J. Electron Microsc.* **35,** 314. Tanaka, M., Terauchi, M., and Kaneyama, T. (1991). *J. Electron Microsc.* **40,** 211.
Tanaka, M., Ichinose, H., Ishida, Y., and Sakaki, H. (1987). *J. Physique* **48,** C5-105.
Taoka, T., and Aoyagi, S. (1956). *J. Phys. Soc. Jpn.* **11,** 522. Taoka, T., and Sakata, S. (1957). *Acta Metall.* **5,** 236.
Terao, N., and Sakata, S. (1953). *Electron Microsc.* (in Japanese) **3** 80. (1954). *J. Phys. Soc. Jpn.* **9,** 109.
Tomokiyo, Y., Okuyama, T., Matsumura, S., Kuwano, N., and Oki, K. (1990). *Material Trans. JIM* **31,** 641.
Tonomura, A., Osakabe, N., Matsuda, T., Kawasaki, T., Endo, J., Yano, S., and Yamada, H. (1986). *Phys. Rev. Lett.* **56,** 792.
Tsukahara, S., and Kawakatsu, H. (1966). *J. Phys. Soc. Jpn.* **21,** 2551.
Tsukahara, S., Kawakatsu, H., and Taoka T. (1971). *Jernkont. Ann.* **115,** 468 (Proc. 2nd Int. Conf. on HVEM, Stockholm).
Tsukahara, S., and Kawasaki, H. (1972). *J. Phys. Soc. Jpn.* **32,** 72.
Ueda, R. (1959). *J. Inst. Metals Jpn.* **23,** 289, 292, 422, 426.
Uyeda, N., Kobayashi, T., Suito, E., Harada, Y., and Watanabe, M. (1970). *Farad Soc. Francaise Microsc. Electronique* **1,** 23. (1972). *J. Appl. Phys.* **43,** 5181.
Uyeda, R. (1942). *Proc. Phys.-Math. Soc. Jpn.* **24,** 809.
Ura, K. (ed.-in-chief) (1989). "Electron Microscopy of Japan—Present and Future," *J. Electron Microsc.* (suppl.) **38,** S42–S101.
Ura, K., and Fujioka, H. (1989). "Electron Beam Testing," Advances in Electronics and Electron Physics, vol. 73, p. 233.
Watanabe, D., and Sekiguchi, T. (1984). *J. Crystallog Soc. Jpn.* (in Japanese) **19,** 10.
Watanabe, D., Uyeda, R., and Kogiso, M. (1968). *Acta Crystallog.* **A24,** 249.
Watanabe, R. (1966). *J. Inst. Metals Jpn.* **30,** 754.
Yada, K., Tangi, T., and Sunagawa, I. (1987). *Phys. Chem. Miner.* **14,** 197.
Yagi, K., and Honjiyo, G. (1976). *J. Phys. Soc. Jpn.* **40,** 601.
Yamamoto, N. (1990). *Material Trans. JIM* **7,** 659.
Yasuda, H., Mori, H., Komatsu, M., and Takeda, K. (1992). *Appl. Phys. Letters* **18,** 21. *J. Appl. Phys.* **73,** 1100.
Yasuda, H., Mori, H., Takeda, K., and Fujita, H. (1993). "Defect and Diffusion Forum," Trans. Tech. Publications, **95–98,** 697.
Yoshida, S., and Kiritani, M. (1975). *J. Phys. Soc. Jpn.* **38,** 1220.
Yoshida, S., Kiritani, M., and Shimomura, Y. (1963). *J. Phys. Soc. Jpn.* **18,** 175.

4.5
Electron Microscopes and Microscopy in Japan
4.5J Specimen Preparation Techniques

KEIJI YADA[1]

Tohoku University
Katahira
Sendai 980, Japan

In Japan, pioneering work on specimen preparation techniques, one of the most important fields in electron microscopy, was done by groups consisting of the 37th Subcommittee of the Japan Society for the Promotion of Science (JSPS) from the beginning of the 1940s in parallel with the study of instrumentation under the very severe wartime conditions. With the progress of instrumentation, various kinds of specimen techniques were developed, especially in the 1960s and 1970s. These remarkable techniques were compiled into comprehensive volumes by the Japanese Society of Electron Microscopy, Kanto Branch (1970, 1975). As it is difficult to arrange the developments of specimen techniques chronologically, the present article will deal with the relevant subjects.

I. Substrates, Replicas, and Shadowing

A study of the specimen substrate, one of the most fundamental works, was first reported by K. Sasagawa *et al.* (1942) following the method described in *Elektronen Übermikroskopie* by M. von Ardenne, using collodion (cellulose nitrate, or Parllodion in the United States) in amyl acetate solution and specially designed tools. Applications of electron microscope observation to various kinds of samples, including biological ones, were greatly expanded following the development of suitable substrates, though some small particles such as smokes and scales of butterflies had been observed without substrate film. The collodion solution was spread over the surface of warmed water (25–35°C) to prevent the formation of holes in the membrane. This collodion membrane mounted on a specimen holder was somewhat weak against the electron beam inpact as compared with a Formvar membrane, which was introduced to electron microscopy later in the United States, so a careful preillumination of the electron beam was necessary. This was

---

[1] Present address: Aomori Public College, 153-4 Yamazaki, Goshizawa, Aomori 031-01, Japan.

used for a fairly long time, until the introduction of reinforcement or its complete replacement by a carbon film by D. E. Bradley. A small disk with a single hole or a slightly rolled woven mesh was used as the specimen holder in those days. A. Fukami (1952) developed a copper sheet with multiple holes slightly smaller than 0.1 mm in diameter using a photoetching technique; this was called "sheet mesh" and was soon available commercially. B. Tadano and his group studied the surface replica technique which was initially developed by Mahl. They observed the surfaces of metals such as Al and Ni by a plastic one-step method and plastic-metal two-step method, in addition to the metal-oxide method. In the plastic-metal two-step method, they successfully tried a new kind of plastic, methyl methacrylate, besides the ordinarily used plastics, polystyrene and celluloid. Methyl methacrylate yielded very high replication quality when polymerized on the surface to be replicated (Tadano 1947a, 1947b). A satisfactory-quality replica was obtained by Al evaporation onto the replicated plastic, followed by dissolution of the plastic. Figure 1 shows an example of an etched Al surface by the methyl methacrylate-Al two-step method, which was the most prevailing method for a certain period in Japan. Studies of the replica preparation were accelerated at the beginning of the 1950s by metal shadow-casting, introduced by R. C. Williams and R. W. G. Wyckoff. Plastics used for the first-step replication were studied extensively by A. Fukami (1953). He found a much better replicating material, acetyl cellulose, and established a very practical method called "filmy replica" for replica preparation.

FIGURE 1. Replica of etched aluminum prepared by methyl methacrylate-Al method (Tadano, 1947b).

T. Hibi studied the characteristics of various kinds of metallic evaporation films for metallic shadowing with a nozzle system, and discovered a general rule for the degree of the aggregation of deposited metallic films (Hibi, 1952). This work was conducted independently of the work of C. J. Calbick concerning the role of the evaporation material in the resolution of the replica. From Hibi's findings of the granularity of evaporated films, it was pointed out that refractive heavy metals of group A and the transition metal group of the periodic table are generally suitable for shadow casting, tungsten being the most suitable material. This idea was extended to a high-resolution replica (Hibi and Yada, 1954a, 1954b, 1961a) with the development of a new evaporation method called dynamic evaporation. Further efforts to improve the granularity of evaporated films using tungsten and other refractive metals were continued by several workers. S. Matsui *et al.* (1975, 1980) developed an electron-beam point-source evaporator for high-quality shadowing. Adachi *et al.* (1976) applied these ideas and techniques to the high-resolution replica. H. Akahori *et al.* (1986) tried shadowing with high-melting-point metals such as Mo, Ta, and W to freeze replica samples for SEM observation, by electron-beam heating and also by ion-beam sputtering, and proved that these metals are very suitable for high-resolution work, where a coating thickness of 3 nm or less is required, and that W containing 2% carbon can be applied for the conventional freeze replica method.

The selected-area replica initiated by A. W. Agar and R. S. M. Revell in 1952 was not so successful because of its technical difficulty, until the first fruitful results by T. Fourie based on Bradley's method. In Japan, the successive replica method of selected area was development by T. Hibi and K. Yada (1959). They observed the changes of surface structure of the alkali halide, KCl, step by step, during coloring by X-ray irradiation and bleaching of the colored crystal. They also observed the printout and development processes of photographic emulsions (Hibi and Yada, 1961b).

II. Microgrids and Thin Substrates for High-Resolution Work

The microgrid was developed for the first time by S. Sakata utilizing a very interesting method. He breathed onto a spreading collodion solution of amyl alcohol on cold water, to make water droplets for the microgrid formation. A misty film showing a rainbow color to some extent remained on subsequent drying of the solvent and the water droplets. The surface of the dried film showed a beautiful structure of microdroplets of about 1 μm in diameter. However, there were no holes on the film. So the film was mounted on a metal grid and was fire-rinsed for a very short time to

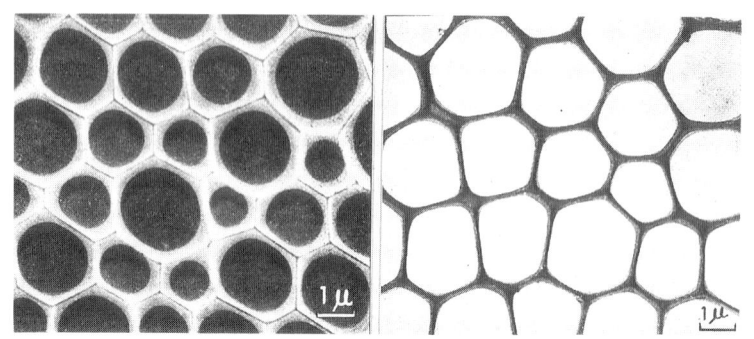

FIGURE 2. Microgrids by Sakata before (left) and after fire rinsing (Sakata, 1958).

make a perforated microgrid as shown in Fig. 2. The microgrid reinforced with carbon was very effective for making samples stable under the intense electron-beam irradiation necessary for high-resolution work. However, Sakata's method, especially the perforation process using fire-rinsing, was very tricky. Therefore, A. Fukami and K. Adachi (1965) developed a new method of self-perforating microgrids using new plastics and procedures. Their method enabled the control of hole sizes from 0.1 μm to 10 μm, as shown in Fig. 3. They also demonstrated the usefulness and importance of a microgrid for high-resoution electron microscopy, especially for the stabilization of samples under intense electron irradiation, which made the microgrid very popular for high-resolution work (Fukami and Adachi, 1965). Another, more convenient way of microgrid preparation, called

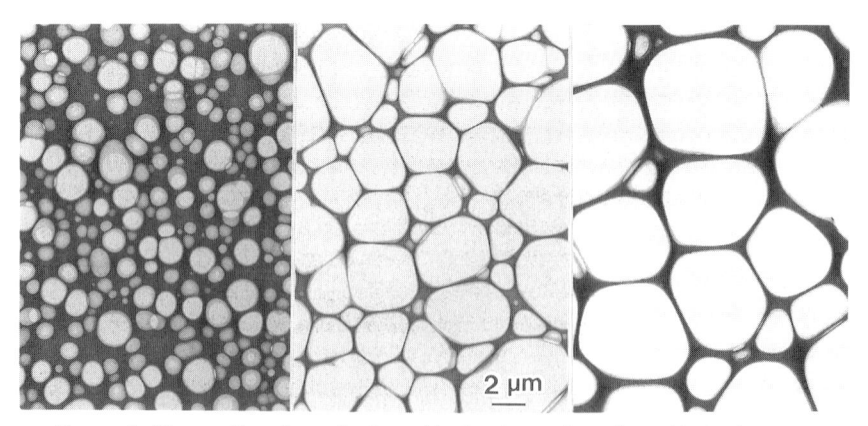

FIGURE 3. Three self-perforated microgrids showing various sizes of holes (Fukami and Adachi, 1965).

"instant microgrid," was developed by K. Saito et al. (1990). Trials of microgrids having much smaller holes were made by several workers, an "ultrafine grid" with 10–100 nm holes by Tanaka et al. (1975), an "alumina supermicrogrid" with 20–100 nm holes by Fujiyoshi and Uyeda (1978), and a "nanogrid" with 5–50 nm holes by Koreeda (1980). Ultrafine grids, were made from tropomyosin, one of the muscle proteins in the form of fibrous polymer, placed over the conventional microgrid for observation of double-stranded structure of DNA filament bridging a small hole. The supermicrogrid, made of aluminum anodic oxide film, was also used to observe the double-stranded structure of DNA fiber prepared by the freeze-drying method, and giving a much clearer image (Fujiyoshi and Uyeda, 1981). The nanogrid was made by ion etching of a double film of Al 0.5 nm/carbon 20 nm in thickness under a suitable etching condition such as 0.2 torr, 0.2 mA/cm^2, 500 V, 2 min. Figure 4 shows the supermicrogrid and nanogrid.

With the development of these microgrids, it became possible to observe many kinds of samples such as tiny particles, fibers, and thin sections, so as to give better-quality images without disturbance due to the substrate. On the other hand, experiments were made to make an extremely thin substrate by combination with a microgrid. A. Fukami et al. (1972) tried to prepare a thin carbon film, 5 nm or less in thickness, mounted on a microgrid. S. Sakata et al. (1991) developed a new way to prepare a much thinner carbon substrate, 1 nm, in thickness, called "limit film."

A. Tanaka et al. (1978, 1986) developed a new method of preparing plasma-polymerized hydrocarbon film using methane, ethylene, naphthalene, etc. The plasma-polymerized film is structureless and is very strong both chemically and mechanically, so that it can be used not only as a

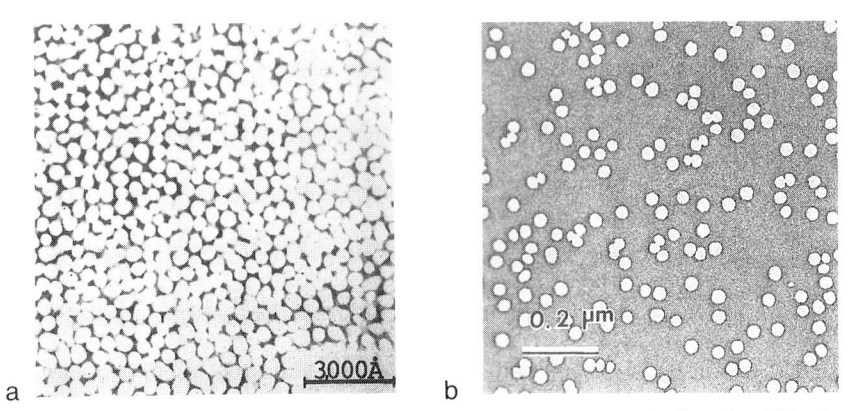

FIGURE 4. (a) Supermicrogrid made of anodic aluminum oxide film (Fujiyoshi and Uyeda, 1978). (b) Nanogrid by ion etching of Al/carbon double films (Koreeda, 1980).

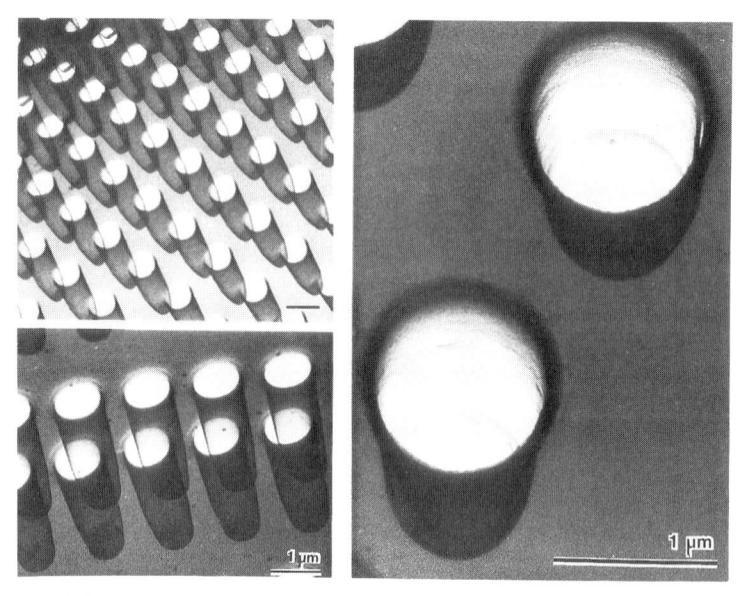

FIGURE 5. (counterclockwise from top left) Plasma-polymerization film replicas of trenches in Si wafer at low and high magnifications (Tanaka et al., 1978).

surface replica, but also as a substrate. Figure 5 shows a replica of trenches in a silicon wafer using the plasma-polymerization method.

Substrates with less granularity than carbon were tried by several workers. R. Uyeda (1967) proposed the use of thin Be platelet (Be smokes formed in Ar atmosphere) prepared by K. Kimoto and I. Nishida (1967). Hashimoto et al. (1971) used thin crystalline graphite substrates for the observation of Th atoms. K. Mihama and N. Tanaka (1976) made thin beryllium oxide (BeO) films, 5 nm in thickness, by vacuum deposition on a cleavage surface of sodium chloride. The film is fairly structureless, though it is polycrystalline and over about 100 A.U. across. An effective procedure for making the specimen or substrate membrane hydrophilic by a glow-discharge. (Fleischer et al., 1967) became popular in Japan after a report containing practical tips and notes by K. Fukai et al. (1972).

III. Ultramicrotomy Techniques

In Japan, the first experiments with ultrathin sectioning were conducted by K. Kobayashi around 1942 for the observation of wood tissue, using a modified microtome for light microscopy. About 10 years later, several groups, including K. Sasagawa *et al.* (1952), K. Sasaki *et al.* (1952), K. Kobayashi and S. Shimadzu (1952), I. Kuroha and T. Sakurai (1952), and H. Eto and S. Sakata (1953) started to work on the instrumentation of ultramicrotomy, stimulated by the successful work of D. C. Peace and R. F. Baker (1948). The feed systems of their instruments were all mechanical. A thermal feed system was employed in 1955 by T. Fujiwara, in cooperation with JEOL. A little later, other groups, Japan Microtome (1954), Shimadzu (1955), JEOL (1955), and Hitachi (1957), successfully developed microtomes with their own feed mechanisms. Their performance, however, was not good enough to surpass that of the existing instruments, such as the LKB (I, III) and Porter Blum (MT-1, 2), and the production of these Japanese ultramicrotomes gradually ceased. T. Sakai (1985) gives a detailed review.

On the other hand, theoretical consideration of ultrathin cutting was proposed by T. Fujiwara as early as 1958. The peripheral aspects of ultramicrotomy, such as measurement of the section thickness, glass knife making, embedding, and so on, progressed. The glass knife-making technique developed by H. Latta and J. F. Hartmann in 1950 was modified so as to produce a better knife more easily by T. Sakai and was reported by Z. Takahashi (1959). Studies of embedding plastics were carried out by H. Kushida and his coworker (Kushida, 1961, 1963, 1964, 1965; Kushida and Fujita, 1968) over a long period. The diamond knife, which was originally developed by H. Fernandez-Moran in 1953, was introduced into Japan in 1957 by T. Sakai and was gradually distributed to users. Meantime, some domestic diamond knives were developed and are now available commercially. The

FIGURE 6. Thin section of multistructure silver halide grain (Inoue, 1989).

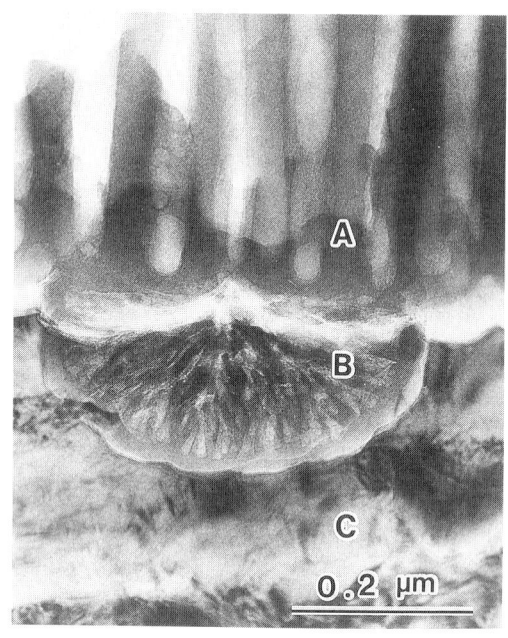

FIGURE 7. Thin section of anodic aluminum film (Takahashi et al., 1973).

sapphire knife, first developed for light microscopy, was refined for electron microscopy by Sakura Seiki, Ltd. I. Yamamoto *et al.* tested the characteristics of the knives called Saphatom with various knife angles (30°, 45°, 60°) and proved their suitability for biological samples, especially for serial sectioning (Yamamoto and Maruyama, 1985; Yamamoto *et al.*, 1986). By the end of the 1960s, ultramicrotomy had become very popular as a powerful and indispensable technique in the biological field, as glass knife making and the diamond knife became popular and widespread. In parallel with this, the ultramicrotomy of industrial materials was tried and many interesting results were obtained, such as cross sections of rubber (Sakai, 1985; Kato, 1968), metals (Ogura, 1965), audio tape (Sakai and Arakawa, 1960), fibrous clay minerals (Yada, 1967, 1971), photographic emulsions (Genda and Sakaguchi, 1965; Kurosaki *et al.*, 1972; Inoue, 1989), oxides, ceramics, and semiconductors such as GaAs (Sakai, 1986). Figure 6 shows a cross section of a multistructural silver halide grain which was embedded in Epon, sectioned 40 nm in thickness, and observed with a liquid-nitrogen specimen holder (Inoue, 1989). Figure 7 shows an example of the fine structure of aluminum anodic oxide film (Takahashi *et al.*, 1973), where A is a steady current stage, B is a steeply increasing current stage followed by an abrupt current drop, and C is matrix metal.

IV. Preparation of Thin Specimens of Metallic and Nonmetallic Materials

Studies of electrolytic polishing were accelerated for various kinds of metals and alloys for years in Japan by a group led by Z. Nishiyama (1962). The existing window method and a newly developed jet stream method by T. Takeyama *et al.* (1966) became very popular, and the devices for these methods are now available commercially.

Thin-film preparation by epitaxy was studied for various kinds of metals and metallic compounds by several groups. The groups of S. Shirai (1961) and S. Ogawa (Ino *et al.*, 1964) clarified the critical temperatures of epitaxy of metals for various cases of substrates including a cleavage method in vacuum, vacuum degree, etc.

Epitaxy was also extended to nonmetallic materials. One example is the preparation of epitaxial film of organic materials, Cu-phthalocyanine, by E. Suito and his group (Suito *et al.*, 1962; Ashida, 1966). They found that three kinds of epitaxy of Cu-phthalocyanine take place depending on the substrate temperature of the mica substrate. This work became the basic background for the high-resolution investigations of chlorinated Cuphthalocyanine in which each atom constituent of the phthalocyanine molecule was resolved (Uyeda *et al.*, 1979). The *in situ* observation of epitaxial growth of thin films was greatly improved by the first realization of an ultrahigh-vacuum electron microscope (Takayanagi *et al.*, 1974). This worked at an ambient vacuum of 10^{-7} Pa around the specimen combined with new preparation methods of thin substrates with clean surfaces (Takayanagi *et al.*, 1975; Yagi and Honjo, 1976), up to recent development of atomic-resolution surface microscopy in both TEM and REM work (Takayanagi, 1989). Preparation of thin crystals of nonmetallic materials was studied from the end of the 1950s by several groups. S. Sakata (1960) succeeded in making, for the first time, thin films of water-soluble ionic crystals, such as K_2PtCl_4 and NaCl, by blowing up an aqueous salt solution containing 0.1% gelation using a nebulizer at elevated temperature. Upon drying the droplets, a concentric growth of thin crystallites remained, exhibiting the initial state of crystallization. Thin crystallites of K_2PtCl_4 were used as a test sample for lattice fringe observation.

T. Hibi and K. Yada (1960) made thin crystallites of alkali halides by a rapid vacuum-drying method or a rapid freeze-drying method from aquaous solutions using a vacuum pump. This method can be used to produce other water-soluble ionic crystals, such as $MgCl_2$, $CaCl_2$, K_2PtCl_4, etc. Alkali halide samples made in this way were used for the study of lattice defects induced by electron-beam irradiation (Hibi and Yada, 1962). Y. Kawamata and T. Hibi (1965) made thin alkali halide specimens by cleavage and chemical polishing from bulk single crystals using water as an etchant. Electron beam-induced defects in the alkali halides were studied by Y. Kawamata (1973, 1976; Kawamata and Hibi, 1965), and by K. Tanaka *et al.* (1963). Figure 8 shows dislocation loops induced in a KBr crystal by two-step electron irradiation at 20°C, 1 mA/cm^2, and at −110°C, 0.1 mA/cm^2 (Kawamata, 1976). G. Honjo and his group tried an "electron-beam flash thinning method" on ionic crystals such as alkali halides and magnesium oxide in the microscope column (Yagi and Honjo, 1964; Honjo *et al.*, 1966). This method was fairly effective with MgO crystal, cleaving thin enough under the intense beam flashing when it was previously thinned down to the order of 2–10 μm. Ion thinning, which is applicable to either metallic or insulating materials, was tried as a specimen preparation for electron microscopy at the beginning of the 1960s (Tomoda *et al.*, 1963) by a commercial apparatus with a Penning-type Ar ion gun. Nowadays, ion etching techniques have become very popular as an indispensable means for the study of various kinds of materials after the accumulation of much

FIGURE 8. Complex dislocation loops introduced in thin KBr by two-step electron irradiations: I, interstitial type; V, vacancy type (Kawamata, 1976).

technical knowhow, especially in the field of property estimation of semiconductor materials using TEM and SEM. Domestic ion-etching devices are available in addition to foreign ones.

V. Critical-Point Drying Method

The critical-point drying method initiated by T. F. Anderson using CO_2 gas was modified, using Dry Ice, by K. Tanaka and A. Iino (1974) and became popular in Japan. This alternative method has many merits, and well-preserved samples were demonstrated in high-resolution SEM. Figure 9 shows a SEM imag of the AIDS virus prepared by critical-point drying (Tanaka, 1989).

VI. Rapid Freezing Techniques and Related Applications

After the first work on freeze-fracture replicas by Steere, many workers in Japan tried to realize rapid freezing for well-preserved biological samples without crystalline ice formation, followed by cryosectioning, freeze-etching, freeze-fracture replication, or freeze-substitution using liquid propane, nitrogen, or helium. In particular, an improvement of the metal block method under controlled temperature and degree of vacuum by Nishiura and Okada (1977) created a driving force to establish the metal block-type freeze-replica devices now commercially available. J. Usukura and

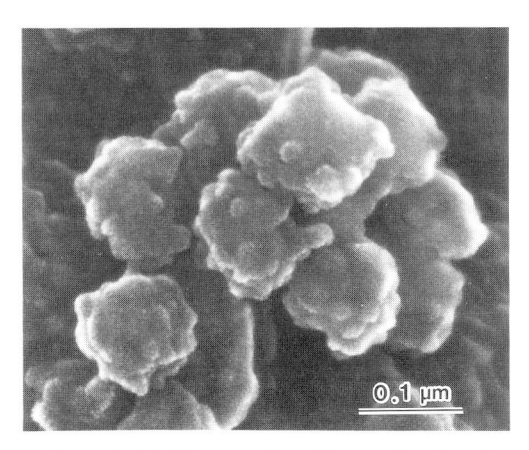

FIGURE 9. AIDS viruses budded from a cultured human lymphocyte prepared by critical-point drying method (Tanaka, 1989).

E. Yamada (1980) developed a rapid-freeze device using liquid nitrogen as coolant.

H. Akahori *et al.* (1980) developed a modified apparatus based on Heuser's device for freeze-fracture replication using liquid helium (He-I, He-II), expecting much rapid freezing and better preservation. Their device was further improved by J. Usukura *et al.* (1982). H. Akahori *et al.* (1989) made a portable type of rapid-freezing device which permitted both metal contact freezing and immersion freezing with a handy-type nitrogen container.

Complementary freeze-fracture replication, initiated by E. Wehrli *et al.* (1970), was improved by A. Tonosaki *et al.* (1974, 1980). They successfully designed a new specimen holder incorporating the procedures of fracture, rotation, and landing of the specimen in vacuum. They also developed an improved vacuum evaporator with a high-speed vacuum pump and liquid nitrogen traps to produce good-quality twin replicas reliably in a short time. Figure 10 shows a typical example of the complementary freeze-fracture replica of epithelial cells of mouse renal tubule and a schematic drawing of the procedures.

The freeze-substitution method initiated by W. L. Simpson in 1941 and modified by N. Feder and R. L. Sidman in 1958 was tested by A. Ichikawa *et al.* (1980) to study tissues, followed by sectioning, and good results, due to better preservation, were obtained. Freeze-substitution fixation followed by sectioning was applied to the study of microorganisms such as *Escherichia coli* and *Bucillus subtilis* and yielded new findings (Amako *et al.*, 1983, 1985).

A freeze-drying method using tert-butyl alcohol was successfully developed by T. Inoue and H. Osatake (1988), and by H. Akahori *et al.* (1988). This method should be mentioned as an epoch-making one to replace the critical-point drying method.

VII. Ion Etching and Ion Sputter Coating

It was found, as mentioned earlier, that a glow discharge is effective in making the specimen hydrophilic (Fukai *et al.*, 1972), in parallel with the well-known ion sputtering. T. Fujita *et al.* (1974) found that the ac glow discharge is suitable for etching biological specimens under certain discharge conditions. A series of extensive studies on ion sputtering of biological samples for SEM observation (Akahori and Fukuoka, 1975; Akahori *et al.*, 1989, 1990) led to the production of a universal type of ion coating device by numerous manufacturers, with functions such as hydrophilization, etching, sputter coating, and ionic coating abilities.

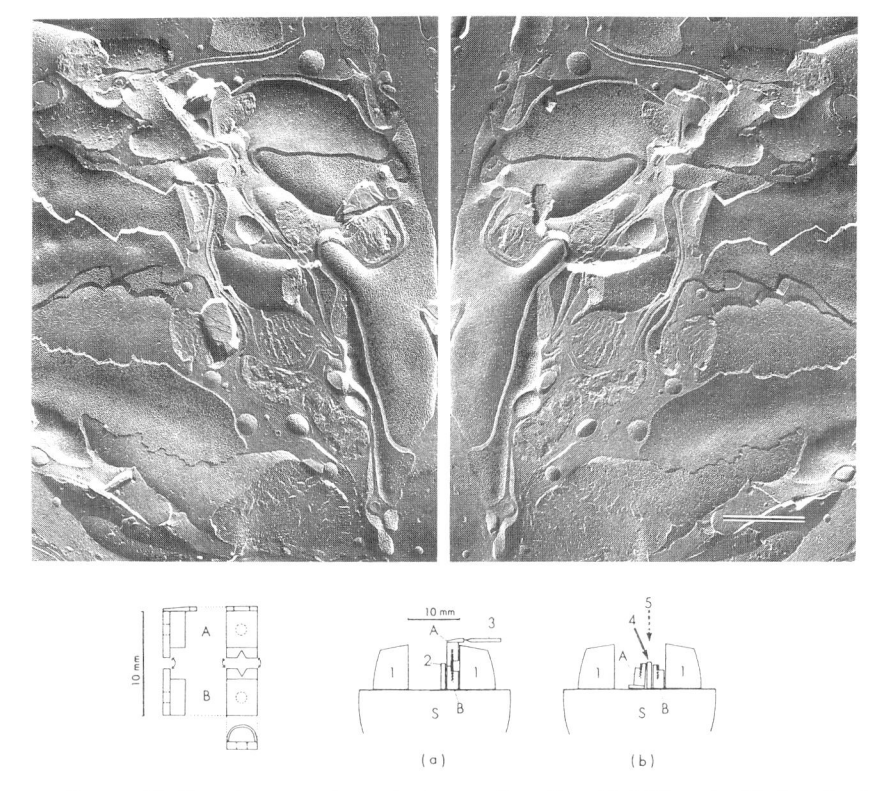

FIGURE 10. Complementary freeze-fracture replica of basal infolding of epithelia cells of mouse renal tubule (Tonosaki *et al.*, 1980), and schematic drawing of the procedure (Tonosaki and Yamamoto, 1974).

VIII. Gas Evaporation Method

A unique method for making fine metal particles was developed in Japan by R. Uyeda and his group. The origin of this method, called the gas evaporation method, is rather old, dating to the time of World War II, when zinc was evaporated in the air as an infrared light sensor of guided bombs. About 15 years later, they began the gas evaporation experiment again (Kimoto *et al.*, 1963) under the stimulus of a theoretical anticipation of interesting properties of fine particles by R. Kubo (1962) and established a new experimental method by which a series of fine metal and semimetal particles were tried extensively. The samples were analyzed crystallographically by electron microscopy and electron diffraction. Figure 11 shows the evaporation methods using an inert gas in the early and in the refined

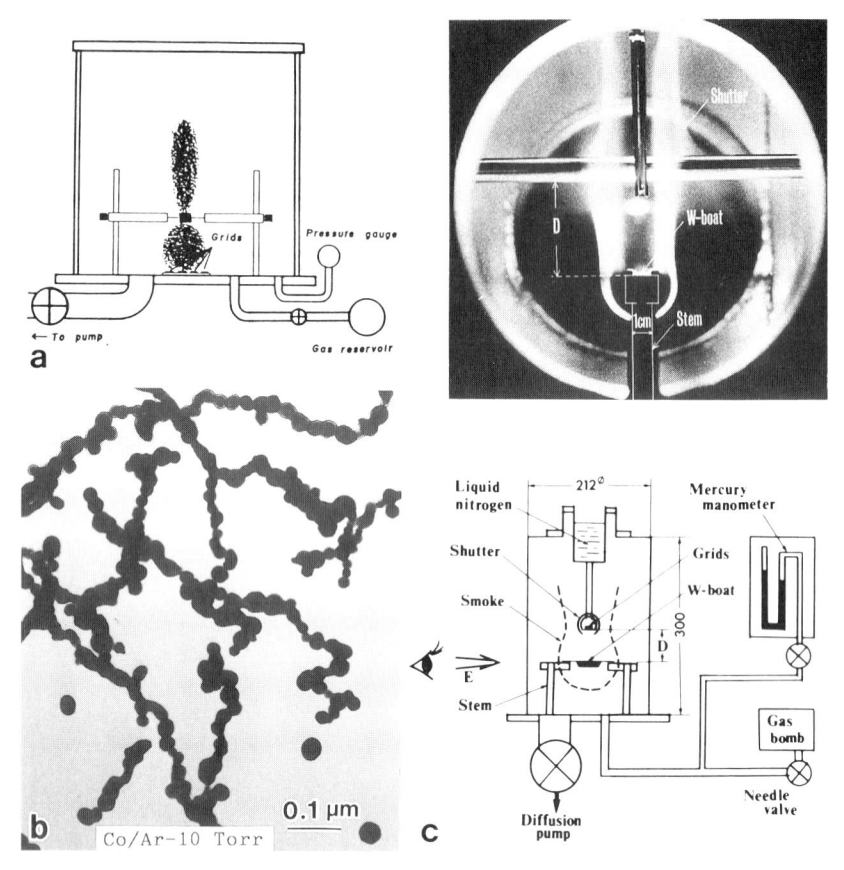

FIGURE 11. (a) Early evaporation chamber used for the gas evaporation method, (b) Co particles made with the chamber (Kimoto et al., 1963), and (c) refined device for collecting specimens (Yatsuya et al., 1973).

experiments (Kimoto et al., 1963; Yatsuya et al., 1973), whose results have been summarized by Kimoto and Nishida (1975) and by Uyeda (1987).

IX. Wet-Cell Microscopy

The idea of an environmental cell by which living specimens could be observed electron microscopically was proposed by E. Ruska in 1942. In Japan, Fukami and his group have studied wet-cell microscopy for a long time by developing a membrane-enclosed type of environmental cell and established wet-cell microscopy, which enabled them to observe dynamic reaction processes by injecting a liquid reagent into the specimen, controlled at a certain environmental gas condition, such as wet muscle and photographic silver halides under electron microscope observation (Fukami and Murakami, 1974; Fukushima *et al.,* 1985, 1986; Fukami *et al.,* 1987).

References

Adachi, K., Hojou, K., Kato, M., and Kanaya, K. (1976). *Ultramicroscopy* **2,** 17.
Akahori, H., and Fukuoka, T. (1975). *J. Electron Microsc.* **24,** 44.
Akahori, H., Ishii, H., Nonaka, I., and Yoshida, H. (1988). *Biomed. SEM* **17,** 50. (in Japanese).
Akahori, H., Ishii, H., Nonaka, I., and Yoshida, H. (1989). *J. Electron Microsc.* **38,** 159.
Akahori, H., Nakajima, Y., Terasawa, K., and Ishii, H. (1986). *J. Electron Microsc.* **35,** 202.
Akahori, H., Yamada, E., Usukura, J., and Takahashi, H. (1980). *J. Clin. Electron Microsc.* **13,** 576.
Akahori, H., Yoshida, T., Ishii, H., and Nonaka, I. (1990). *Biomed. SEM* **19,** 32 (in Japanese).
Akahori, H., Yoshida, T., Nonaka, I., and Ishii, H. (1989). *Biomed. SEM* **18,** 33 (in Japanese).
Amako, K., Murata, K., and Umeda, A. (1983). *Microbiol. Immunol.* **27,** 95.
Amako, K., and Takade, A. (1985). *J. Electron Microsc.,* **34,** 13.
Ashida, M. (1966). *Bull. Chem. Soc. Jpn.* **39,** 2632.
Fleischer, S., Fleischer, B., and Stockenius, W. (1967). *J. Cell Biol.* **32,** 193.
Fujita, T., Nagatani, T., and Hattori, A. (1974). *Arch. Histol. J.* **36,** 195.
Fujiwara, T. (1958). *J. Electronmicros.* **17,** 1 (in Japanese).
Fujiyoshi, Y., and Uyeda, N. (1978). *J. Electron Microsc.* **27,** 75.
Fujiyoshi, Y., and Uyeda, N. (1981). *Ultramicroscopy* **7,** 189.
Fukai, K., Hosaka, Y., and Nishimura, A. (1972). *J. Electron Microsc.* **21,** 331.
Fukami, A. (1952). *Electron Microsc.* **21,** 126 (in Japanese).
Fukami, A. (1953). *J. Electron Micros.* **1,** 28.
Fukami, A. (1954). *J. Electron Micros.* **2,** 20.
Fukami, A. (1955). *J. Electron Micros.* **4,** 36 (in Japanese).
Fukami, A., and Adachi, K. (1965). *J. Electron Microsc.* **14,** 112.
Fukami, A., Adachi, K., and Katoh, M. (1972). *J. Electron Microsc.* **21,** 99.
Fukami, A., Fukushima, K., Ishikawa, A., and Ohi, K. (1987). *Proc. 45th EMSA Meeting,* Baltimore, p. 142.
Fukami, A., and Murakami, S. (1974). *J. Electron Microsc.* **9,** 4 (in Japanese).
Fukushima, K., Ishikawa A., and Fukami, A. (1985). *J. Electron Microsc.* **34,** 47.
Fukushima, K., Sugi, H., and Murakami, S. (1986). *Proc. 11th Int. Congress on Electron Microscopy,* Kyoto, vol. 1, p. 329.
Genda, S., and Sakaguchi, T. (1965). *J. Soc. Photogr. Sci. technol. Jpn.* **28,** 131 (in Japanese).
Hashimoto, H., Kumao, A., Yotsumoto, H., and Ono, A. (1971). *Jpn. J. Appl. Phys.* **10,** 1115.
Hibi, T. (1952). *J. Appl. Phys.* **23,** 957.
Hibi, T., and Yada, K. (1954a). *J. Appl. Phys.* **25,** 712.
Hibi, T., and Yada, K. (1954b). *Proc. Int. Conf. on Electron Microscopy,* London, 460.
Hibi, T., and Yada, K. (1959). *J. Phys. Soc. Jpn.* **14,** 445.
Hibi, T., and Yada, K. (1960). *J. Electron Microsc.* **9,** 101.
Hibi, T., and Yada, K. (1961a). *Sci. Rep. RITU,* **A13,** 105.
Hibi, T., and Yada, K. (1961b). *J. Electronmicrosc.* **10,** 51.
Hibi, T., and Yada, K. (1962). *J. Appl. Phys.* **33,** 3530.

Honjo, G., Shinozaki, S., and Sato, H. (1966). *Appl. Phys. Lett.* **9,** 25.
Ichikawa, A., Ichikawa, M., and Hirokawa, N. (1980). *Am. J. Anat.* **157,** 107.
Ino, S., Watanabe, D., and Ogawa, S. (1964). *J. Phys. Soc. Jpn.* **19,** 881.
Inoue, M. (1989). *Konica Tech. Rep.* **2,** 116.
Inoue, T., and Osatake, H. (1988). *Arch. Histol. Cytol.* **51,** 53.
Japan Society of Electron Microscopy, Kanto-Branch (1970). "Specimen Preparation Techniques for Electron Microscopy." Seibundo-Shinkosha, Tokyo (in Japanese).
Japan Society of Electron Microscopy, Kanto Branch (1975). "Preparation Techniques of Biological Specimen for Electron Microscopy." Maruzen, Tokyo (in Japanese).
Kato, K. (1968). *J. Electron Microsc.* **17,** 29 (in Japanese).
Kawamata, Y. (1973). *Phys. Stat. Sol. (a)* **19,** 331.
Kawamata, Y. (1976). *J. Physique,* Colloque C7-502, supplement to No. 12, vol. 37.
Kawamata, Y., and Hibi, T. (1965). *J. Phys. Soc. Jpn.* **20,** 242.
Kimoto, K., Kamiya, Y., Nonoyama, M., and Uyeda, R. (1963). *Jpn. J. Appl. Phys.* **2,** 702.
Kimoto, K., and Nishida, I. (1967). *Jpn. J. Appl. Phys.* **6,** 1047.
Kimoto, K., and Nishida I. (1975). *Electron Microsc.* **10,** 22 (in Japanese).
Koreeda, A. (1980). *J. Electron Microsc.* **29,** 61.
Koreeda, A. (1986). *Proc. XIth Int. Congress on Electron Microscopy,* Kyoto, p. 345.
Kubo, R. (1962). *J. Phys. Soc. Jpn.* **17,** 975.
Kurosaki, K., Kagawa, Y., and Hase, H. (1972). *Proc. 33th Meeting of Applied Physics Society,* Sapporo, p. 376.
Kushida, H. (1961). *J. Electron Microsc.* **10,** 16.
Kushida, H. (1963). *J. Electron Microsc.* **12,** 72.
Kushida, H. (1964). *J. Electron Microsc.* **13,** 107; **13,** 139; **13,** 200.
Kushida, H. (1965). *J. Electron Microsc.* **14,** 275.
Kushida, H., and Fujita, K. (1968). *J. Electron Microsc.* **17,** 349.
Matsui, S., Kawakatsu, H., and Adachi, K. (1975). *J. Appl. Phys.* **24,** 227; *Bull. Electrotech. Lab.* **39,** 231 (in Japanese).
Matsui, S., Kawakatsu, H., and Adachi, K. (1980). *Bull. Electrotech. Lab.* **44,** 595 (in Japanese).
Mihama, K., and Tanaka, N. (1976). *J. Electron Microsc.* **25,** 65.
Nishiura, M., and Okada, S. (1977). *Electron Microsc.* **12,** 36 (in Japanese).
Nishiyama, Z. (1962). Report of specimen preparation method for electron microscopy, supported by grants-in-aid for general research projects from the Ministry of Education, Science and Culture (in Japanese).
Ogura, I. (1965). *J. 2nd Coll. Eng. Nihon Univ.* **B6,** 43; *J. Electron Microsc.* **16,** 216.
Saito, K., Iwahashi, M., and Futaesaku, Y. (1990). *Proc. 46th Annu. Meeting Electron Microscopy Society of Japan,* Maebashi, p. 311.
Sakai, T. (1985). Ultramicrotomy. *In* "Ultra Precision Working Manual" (A Kobayashi, ed.), pp. 378–401. New Techniques Development Center, Tokyo (in Japanese).
Sakai, T. (1986). *Proc. XIth Int. Congress on Electron Microscopy,* Kyoto, p. 1487.
Sakai, T., and Arakawa, S. (1980). *J. Electronmicrosc.* **9,** 61.
Sakata, S. (1958). *J. Electron Microsc.* **6,** 75.
Sakata, S. (1960). *J. Electron Microsc.* **8,** 46.
Sakata, S., Hosumi, S., and Watanabe, H. (1991). *J. Electron Microsc.* **40,** 67.
Sasagawa, S., Watanabe, S., and Higashi, N. (1942). Report of the 37th Subcommittee of JSPS, No. 20-3 (in Japanese).
Shirai, S., Fukuda, Y., and Nomura, M. (1961). *J. Phys. Soc. Jpn.* **16,** 1989.
Suito, E., Uyeda, N., and Ashida, M. (1962). *Nature* **194,** 273.

Tadano, B. (1947a). Report of the 37th Subcommittee of JSPS, No. 45-6, No. 45-8 (in Japanese).
Tadano, B. (1947b). *Oyo Butsuri* **16,** 113 (in Japanese).
Takahashi, H., Nagayama, M., Akahori, H., and Kitahara, A. (1973). *J. Electron Microsc.* **22,** 149.
Takahashi, Z. (1959). *J. Electron Microsc.* **8,** 76. (in Japanese).
Takayanagi, K. (1989). *J. Electron Microsc.* **38** (suppl.), S58.
Takayanagi, K., Yagi, K., Kobayashi, K., and Honjo, G. (1974). *Jpn. J. Appl. Phys.* **suppl. 2, part 1,** 533.
Takayanagi, K., Yagi, K., Kobayashi, K., and Honjo, G. (1975). *J. Cryst. Growth* **28,** 168.
Takeyama, T., Hachinohe, M., and Sato, Y. (1966). *J. Electron Microsc.* **15,** 269 (in Japanese).
Tanaka, A., and Sekiguchi, Y. (1986). *Proc. 11th Int. Congress on Electron Microscopy,* Kyoto, p. 221.
Tanaka, A., Sekiguchi, Y., Kurita, T., and Kuroda, S. (1978). *J. Electron Microsc.* **27,** 378.
Tanaka, K. (1989). *J. Electron Microsc.* **38,** (suppl.), S105.
Tanaka, K., and Iino, A. (1974). *Stain Technol.* **49,** 203.
Tanaka, K., Mannami, M., and Izumi, K. (1963). *J. Phys. Soc. Jpn.* **18,** (suppl. 3), 350.
Tanaka, M., Higashi-Fujime, S., and Uyeda, R. (1975). *Ultramicroscopy* **1,** 7.
Tomoda, Y., Nagata, S., and Soejima, Y. (1963). *Tech. Rep. Osaka Univ.* **13,** 63.
Tonosaki, A., and Yamamoto, T. Y. (1974). *J. Ultrastruct. Res.* **47,** 86.
Tonosaki, A., Yamasaki, M., Washioka, H., and Mizoguchi, J. (1980). *Arch. Histol. Jpn.* **43,** 115.
Usukura, J., Akahori, A., Takahashi, H., and Yamada, E. (1982). *J. Electron Microsc.* **32,** 180.
Usukura, J., and Yamada, E. (1980). *J. Electron Microsc.* **29,** 376.
Uyeda, N., Kobayashi, T., Ishizuka, K., and Fujiyoshi, Y. (1979). *Chem. Scr.* **14,** 47.
Uyeda, R. (1967). *Proc. Jpn. Appl. Phys. Soc. Meeting* (in Japanese).
Uyeda, R. (1987). In "Morophology of Crystals" (I. Sunagawa, ed.), pp. 367–508. TERRAPUB, Tokyo.
Wehrli, E., Muhlethaler, K., and Moor, H. (1970). *Exp. Cell Res.* **59,** 336.
Yada, K. (1967). *Acta Cryst.* **23,** 704.
Yada, K. (1971). *Acta Cryst.* **A27,** 659.
Yagi, K., and Honjo, G. (1964). *J. Phys. Soc. Jpn.* **19,** 1892.
Yagi, K., and Honjo, G. (1976). *J. Phys. Soc. Jpn.* **40,** 601.
Yamamoto, I., and Maruyama, H. (1985). *J. Electron Microsc.* **34,** 442.
Yamamoto, I., Tachibana, T., and Maruyama, H. (1986). *Proc. XIth Congress on Electron Microscopy,* Kyoto, p. 2191.
Yatsuya, S., Kasukabe, S., and Uyeda, R. (1973). *Jpn. J. Appl. Phys.* **12,** 1675.

4.6
Towards Atomic Resolution

FRIEDRICH LENZ

Institute of Applied Physics
University of Tübingen
D-72076 Tübingen, Germany

In this account I shall not attempt to give a review of the progress in biological and medical applications which refer to structural details at least 10 times coarser than the interatomic distances in solids, because the relatively strong specimen damage in biological specimens does not allow one to use electron densities high enough to reduce noise to the level required for atomic resolution. Further, other experts would be much more competent to report on progress in this field.

I am well aware that a review article of limited length can never be complete. Further, it is an inevitable drawback of any attempt to identify the contributions of one country to a common effort of scientists from many nations, that progress in science is always an international effort and never restricted to the scientists of one nation. This report may appear biased with respect to contributions to this progress which have been made in institutes or teams or by colleagues with whom I had the privilege of cooperating closely, as in Düsseldorf, Aachen, and since 1960 in Tübingen. Thus I should like to apologize to all colleagues at home and abroad whose names I have failed to mention even though they have made important contributions to progress in electron microscopy.

In postwar Germany, research on electron microscope instrumentation and application was concentrated at a few companies, universities, and other governmental or semigovernmental research laboratories. Siemens and Halske AG in Berlin continued the development of electron microscopes with Ernst Ruska and his co-workers, among them Otto Wolff, Siegfried Leisegang, and Käthe Müller, until 1955. Siemens stopped the production of electron microscopes in 1979, evidently under the increasing pressure of Japanese competition. The latest version of a high-performance Siemens electron microscope was the ST 100F, of which only eight instruments were produced. This instrument turned out to be too expensive to find a large market. The Carl Zeiss company, which had been reestablished in Oberkochen in southwestern Germany after the dispossession and destruction of the old Carl Zeiss Foundation in Jena by the Communists in Eastern Germany, developed simpler but efficient types of electron microscopes, initially in cooperation with the Süddeutsche Laboratorien (SDL) in Mosbach under Ernst Brüche, who continued the electron optical tradition of the Allgemeine Elektrizitäts-Gesellschaft (AEG). At SDL a great number of electrostatic electron microscopes was manufactured based on design work by Boersch, Brüche, Mahl, and Scherzer at the AEG research laboratory in Berlin. At the SDL some new interesting methods of electron microscope investigations of crystals were developed, such as selected-area diffraction and imaging using only electrons which had been diffracted into one defined diffraction peak. From 1954 until 1973, Hans Mahl was the scientific director of the Department of Electron Microscopy and Electron Optics of Carl Zeiss, Oberkochen. Mahl, who died in 1988, was one of the early pioneers of electron microscopy, who contributed greatly to the realization of the first electrostatic electron microscope, and to the improvement of instrumentation and specimen preparation, especially by his invention of the oxide replica method. One specific feature of the Zeiss microscopes is the incorporation of an energy filter which not only allows one to remove inelastically scattered electrons from the image but also to produce images using only electrons which have suffered a characteristic energy loss. Originally a Castaing-Henry filter was used, but recently it has been replaced by a magnetic "Omega" energy filter designed in cooperation with Rose in Darmstadt, and Krahl in Berlin-Dahlem, which has improved imaging and filtering properties.

Among the SDL co-workers were Gottfried Möllenstedt, Otto Rang, Otto Scherzer, and R. Seeliger. Before and during the war, Möllenstedt had worked with W. Kossel at the Technical University at Danzig. During that time, Kossel and Möllenstedt invented convergent-beam electron diffraction (CBED), which is now an important method in solid-state research. In 1953, Möllenstedt took up the Chair for Applied Physics at Tübingen University, where, over the following decades, he succeeded in establishing another important center of electron physical research which is presently headed by his successor in office, Karl Heinz Herrmann. Before his call to Tübingen, Herrmann had been an active member, first of the Laboratory for Electron Optics of Siemens and Halske AG in Berlin, where electron microscopes, at that time the Elmiskop 101, were designed, tested, and produced, and then at the Fritz Haber Institute at Berlin-Dahlem with Ernst Ruska. During that time, Herrmann studied in detail the physical basis of image intensifiers and other recording systems. Together with D. Krahl, he made us aware that the detection quantum efficiency of such systems is an important figure of merit. Presently, he is continuing research in this field and is much involved in the progress of electron holography.

In 1954, Scherzer, who had already been an Associate Professor of Theoretical Physics at Darmstadt since 1935, accepted a call to a full Professorship for Theoretical and Applied Physics at Darmstadt. After his death in 1982, his successor in office was not interested primarily in electron optics or electron microscopy, but the tradition of theoretical electron optics is continued by Harald Rose, who, even without the advantage of a full chair with its associated funding, is leading a very efficient team of young, highly motivated physicists, who have made significant contributions especially in the field of imaging theory, correction of aberrations, nonaxisymmetric multipole systems, energy filters with a curved axis, and the effect of Coulomb interactions within the electron beam. Hans Boersch, who had worked at the AEG Research Laboratories in Berlin from 1935 to 1940, and then at a University Institute for Chemistry in Vienna from 1941 to 1946, took up a position at the Physikalisch-Technische Bundesanstalt in Braunschweig, where he worked until 1954, when he received a call to a Chair for Physics at the Technical University in Berlin. Boersch made and initiated many interesting experiments in the field of electron wave optics. The broadening of the energy distribution in an electron beam due to Coulomb interaction between the electrons which he was the first to observe became known as "Boersch effect." After Boersch's retirement in 1974 (†1986), his large Institute was incorporated into an even larger Optical Institute, in which Heinz Niedrig and Oleg Bostanjonglo continued work in the field of electron optics, e.g., some important experimental and theoretical work on electron backscattering and on ultrashort-pulse electron microscopy. Boersch's successor in office in Braunschweig was K. J. Hanszen who, together with a team of co-workers, made many contributions, mainly to wave-optical electron microscopic and electron holographic imaging. Another very active center of electron microscopy was formed at the University of Münster, where we owe much progress in scanning microscopy to Gerhard Pfefferkorn (†1989), Ludwig Reimer, and others. Reimer has also made many valuable experiments on scattering and contrast, and has written books on transmission electron microscopy (1984) and scanning electron microscopy (1985), which have become indispensible sources of information in many laboratories.

Ernst Ruska, who had, together with Bodo von Borries, managed the development and production of electron microscopes and a laboratory for applications in Berlin, stayed with Siemens until 1955, continuing the improvement and production of magnetic high-performance electron microscopes (ÜM100, Elmiskop I). During that time, E. Ruska and his co-workers achieved important improvements of the electron microscope, such as stigmators, the double condenser, provisions for illuminating small specimen areas, dark-field imaging with defined reflections, cooled specimen stages, and electronic control of voltage and current stability. Afterwards, E. Ruska became the director of the Institute for Electron Microscopy at the Fritz Haber Institute of the Max Planck Society in Berlin-Dahlem, where he continued research on electron microscope instrumentation. Among such instruments was an electron microscope for high resolution with a single-field condensor-objective lens designed in cooperation with W. D. Riecke. One of the advantages of this lens was that it had low values for its coefficients of spherical (0.5 mm at 100 kV) and chromatic (0.7 mm) aberration. Improved designs for cooling the environment were developed in order to reduce the unwanted effects of specimen contamination, and methods to reduce the effects of mechanical vibrations were studied. In 1986, Ernst Ruska was awarded the Nobel Prize for Physics for his fundamental electron optical papers and the design of the first electron microscope.

Bodo von Borries left Berlin after the war and established the Rheinisch Westfälisches Institut für Übermikroskopie in Düsseldorf, where he designed and tested a new type of magnetostatic electron microscope, and organized a Guest Institute for applications. In 1953, von Borries was given the Chair for Elektronenoptik und Feinmechanik at the Technical University at Aachen. After his death in 1956, the Chair in Aachen was renamed Applied Physics, and his successor worked in the field of semiconductors. The Institute at Düsseldorf, now renamed the Institute for Biophysics and Electron Microscopy, became part of the Medical Academy, now the University of Düsseldorf, and Helmut Ruska (†1973) became its director. Helmut Ruska, brother of Ernst Ruska, was an early pioneer in the application of electron microscopy to medicine and biology. He had begun studying viruses and bacteria at the Siemens laboratory in Berlin-Spandau as early as 1938. After the war he established an Institute for Micromorphology in Berlin, and continued working at medical applications, especially the design and application of microtomes for ultrathin sections, at Albany, New York, from 1952 to 1958.

In his book, *Supermicroscopy* (1949), B. von Borries dedicated one chapter to the "Limits of Supermicroscopy." Among other limits, he discussed resolution as an optical, and contrast as a specimen-related limit. In 1938, R. Rebsch, a co-worker of O. Scherzer, had derived an expression for the limit of resolution, $\delta = C(C_s\lambda^3)^{1/4}$, which is still frequently used. In this formula, C_s is the coefficient of spherical aberration, λ is the electron wavelength, and C is a dimensionless factor close to unity. Assuming realistic values for the accelerating voltage and the maximum field strength in the magnetic objective lens, a theoretical resolution limit of 0.3 nm would follow, i.e., a value of the order of the interatomic distances in solids. In practice, however, around 1950 even the best instruments yielded resolutions not much better than 2 nm. By 1942, W. Glaser had already given an explanation for this discrepancy between theoretical and practical resolution, and attributed the poor performance of the instruments to deviations of the lens field from axial symmetry, which resulted in axial astigmatism, an additional aberration which had been disregarded in the above formula for the theoretical resolution limit. Based on Hans Mahl's observation that axial astigmatism could be corrected using a cylinder lens rotatable around the optical axis, O. Rang, another co-worker of O. Scherzer, designed in 1949 at the SDL in Mosbach an element for the correction of axial astigmatism of electrostatic lenses for which Scherzer coined the term "Stigmator," and other types of electric and magnetic stigmators were shortly after designed and tested in other institutes and companies. I remember how happy Walter Glaser, the theoretical electron optician and author of the standard textbook *Elektronenoptik,* was when, at the meeting of the German and Austrian Societies for Electron Microscopy at Innsbruck in 1953, it became evident that after correction of axial astigmatism the discrepancy between theoretical and practical resolution had been greatly reduced.

The German electron microscopists have, since the early 1930s, always had the support of strong theoretical groups. Walter Glaser's habilitation thesis (Prague, 1933) treated geometric-optical imaging using electron rays. He introduced the concept of an electron-optical index of refraction and derived a theory of third-order aberrations based on the Eikonal principle. Otto Scherzer, who also engaged in theoretical electron optics as early as 1933, derived independently another shape of an electron-optical aberration theory, preferring the trajectory method and complex coordinates, but the results of the two authors were the same. Whereas Glaser cooperated mainly with von Borries and E. Ruska in the Siemens group, which used magnetic lenses, Scherzer worked mainly with Brüche in the AEG group, where electrostatic systems were used, and later with the SDL in Mosbach. The theoretical groups in Germany are still strong, e.g., H. Rose in Darmstadt, and E. Kasper in Tübingen, who made significant contributions to the improvement of methods in numerical electron optics and, together with P. Hawkes in Toulouse, France, has written three volumes of a comprehensive work on principles of electron optics.

The discussion of contrast in von Borries' book was based on the concept of the mean free path of a primary electron within the specimen: The bright-field contrast of a thin specimen is practically equal to the probability for an electron to be elastically scattered, i.e., to the ratio of the specimen thickness over the mean free path. For thicker specimens and for dark-field contrast, similar relations were used. Since the mean free path is inversely proportional to the density of the specimen and to the elastic scattering cross section of the atoms forming the specimen, the contrast was given as a function of "mass thickness," i.e., of the product of specimen density and thickness. Such relations are still used for "area contrast" to describe the relative image current density of areas whose lateral dimensions are large compared to the resolution limit. They cannot, however, explain the contrast of single atoms or other object details with dimensions comparable to the resolution limit.

In the early years of electron microscopy, electrons were considered to be particles rather than waves. Even early diagrams explaining the superior resolving power of electrons versus visibile light used the particle concept, showing how the image of a human hand spread on a table plate could be formed by pouring some fine-grained sand over the hand, whereas, on the other hand, the same effect could not be achieved using apples. It was, however, evident that some effects in electron optics could only be understood using the wave concept. This was evidently true for electron diffraction by crystals and electron microscope images of crystals, whose contrast was found to depend not only on mass thickness but also on the direction of the incident beam. Reliable values for the atomic scattering cross sections, which were needed to explain area contrast, could only be obtained using quantum mechanical concepts. In 1939, H. Boersch had shown that in the shadow image of an edge, Fresnel fringes could be observed. In 1949, O. Scherzer published a theoretical paper on the contrast of a single atom in which he computed the wave-optical image of a phase disk taking into account spherical aberration and defocusing. One important result of this paper was that optimum contrast is obtained if the image is slightly underfocused. In modern terminology we would say that this optimum defocus yields not maximum "contrast" but maximum width of the spatial frequency transmission band, but at that time such terminology was not used by electron microscopists. This optimum focus for phase contrast is now generally called Scherzer focus. It took several years before most electron microscopists in Germany became fully aware of the importance of Scherzer's paper, which had appeared at a time when the *Journal of Applied Physics* was hard to get in Germany. In the early 1950s it had been observed that micrographs even of noncrystalline objects showed some granularity, and that the apparent grain size seemed to depend on defocus. An explanation for this effect was attempted in 1956 by B. von Borries and F. Lenz, who studied defocused images of periodic objects and used Huygens' principle to derive a relation between the defocus needed to obtain maximum contrast and the spatial frequency of the periodic structure. In 1964, F. Thon included the effect of spherical aberration and made more detailed investigations into the dependence of contrast transfer on spatial frequency. Following a suggestion by W. Hoppe, he analyzed experimentally the transmission of the different spatial frequencies using an optical diffractometer. In the following years, the new concepts of wave aberration and transfer functions began to play an increasing role in electron optics, gradually replacing the old concepts of resolution and contrast as figures of merit for an instrument. Walter Hoppe (†1986) and his school in München, who made great efforts to realize three-dimensional electron microscopy, contributed considerably to the propagation of these new concepts. I think that the most important point in the new way of looking at electron microscopic imaging is that wave aberration and transfer functions are properties of the instrument only, and not of the object. With this view, the object appears as the source of information at the input of an information channel that transmits the different (spatial) frequencies of which the input signal is composed to the image which plays the role of the receiver. As in other communication channels, a frequency-dependent transfer factor determines how well a certain frequency in the input signal is transmitted to the output. The old concept of "resolution" is closely related to the "bandwidth" of the channel, and the old concept of "area contrast" is simply the value of the contrast transfer function for zero frequency. If a user knows the transfer function of an instrument he or she knows much more than the information contained in two numerical values such as "limit of resolution" and "area contrast." The progress which is being made in observing and evaluating transfer functions on-line is the basis of the development of automatic or semiautomatic procedures for the the on-line achievement of optimum alignment, stigmatization, and focus.

The new concepts, however, did not change the fact that resolution was still given by Scherzer's limit, $\delta = C(C_s\lambda^3)^{1/4}$, i.e., by wavelength and spherical aberration. Thus, most attempts to improve the resolving power of electron microscopes were directed to either reducing the wave length, i.e., increasing the beam voltage, or to correcting spherical aberration. In Germany, prototype electron microscopes for 200 kV were designed by Manfred von Ardenne and at the Siemens laboratories, but later no serious attempts were made to use electron energies of the order of 1 MeV or beyond. On the other hand, the correction of spherical aberration remained a fascinating project. Scherzer had shown as early as 1936 that the coefficient C_s of third-order spherical aberration of an electron lens cannot be made negative if the lens field is axially symmetric, time-independent, and free of space charges; his co-workers, R. Rebsch and W. Tretner, derived expressions for a lower limit for C_s values. Many attemps have been made to design lenses with zero spherical aberration, either by placing a charged foil across the electron beam, by using high-frequency ac fields, or by using systems deviating from axial symmetry, such as multipoles. The design of lenses using charged foils or high-frequency ac fields has not yet led to practicable systems, but some encouraging results have been obtained with multipole systems. In 1953, R. Seeliger described a purely electrostatic system based on an idea of Scherzer's consisting of two round lenses, two cylindrical lenses, and three octopoles, which was realized and tested to show that third-order spherical aberration could in principle be corrected, but it failed to be applicable to practical electron microscopy because the number of mechanically adjustable elements was too great to permit stable and sufficiently accurate alignment. Another, still more sophisticated system was designed by Scherzer and his co-workers in Darmstadt in the years from 1972 to 1982. It consisted of five electric and magnetic multipoles and could be shown to correct third-order spherical as well as axial chromatic aberration. Again, however, the electrical and mechanical stabilization of the many multipole elements was too complicated to incorporate the system into an electron microscope for routine operation. In 1990 Rose suggested a new, less complicated, and purely magnetic multipole design, which is less sensitive to misalignment. It consists of the objective lens, two telescopic round lens doublets, and two sextupoles in the node planes of the second doublet. Since Rose's new design does not contain quadrupoles, variations of multipole strengths do not affect the paraxial properties of the system. Thus, the alignment and stabilization of the sextupoles is less critical than that of quadrupoles. For the same reason, axial chromatic aberration cannot be corrected together with spherical aberration, because it is a paraxial property. If the system can be realized as designed, it can be used to give C_s a negative value instead of making it vanish. An analysis of the transfer function taking into account fifth-order spherical aberration shows that now the width of the transfer band depends on defocus and C_s, i.e., it can be maximized by selecting suitable values of defocus *and* C_s, and it is hoped that using Rose's corrector the resolution limit can be reduced to values below one angström unit. A possible field of application is the physics of solids, which has traditionally been studied at the Max Planck-Institut für Metallforschung at Stuttgart (A. Seeger, M. Rühle), and in recent years increasingly at the Institut für Metallphysik at Göttingen University (P. Haasen) and the Institut für Festkörperforschung, Forschungszentrum Jülich (K. Urban). At all three institutes the transmission electron microscope is one of the most important tools of research.

Other ideas for improving resolution, such as hollow-cone illumination or the use of zone plates consisting of a system of concentric opaque and/or phase-shifting rings as objective apertures, have been suggested and tested experimentally, but they have not led to practical improvements of instruments for routine operation.

Every new concept in high-resolution electron microscopy that allows the use of a wider transfer band must, for its practical realization, use an electron source of sufficient brightness, because the value of the transfer functions, especially at higher spatial frequencies, is depressed by the limited lateral and/or longitudinal coherence of the primary electron beam illuminating the object. The important concept of "brightness" in electron optics was first introduced in 1939 by B. von Borries and E. Ruska who coined the German word "Richtstrahlwert," defined as current density per unit solid angle. In 1956, Sakaki and Möllenstedt in Tübingen showed that pointed cathodes using field-emission and/or the Schottky effect were capable of producing significantly higher brightness values than the thermal cathodes which were then mostly used. Further studies of field-emission guns were continued in Tübingen, mainly by R. Speidel (†1987). The high degree of coherence of such guns has not only enabled Möllenstedt and his co-workers to perform a number of excellent electron interference experiments, it is also the basis of present and future development of high-resolution transmission electron microscopes and of high-resolution electron holography.

Another, completely different method using field emission, applicable to the study of solid conducting surfaces, was found by Gerd Binnig and Heinrich Rohrer, who were awarded the Nobel Prize for Physics in 1986 together with Ernst Ruska. They invented a scanning tunneling electron microscope (STM) in which a fine-pointed field emitter positively charged at several tens of millivolts is mechanically moved at a close distance, of the order of 0.5 nm, over a solid surface using piezoelectric scanners. During the scanning the distance is controlled, keeping constant the emitted tunnel current of a few nanoamperes. Since the tip of the emitter has a radius of curvature of atomic dimensions and can even consist of a single atom, the instrument is capable of determining atomic positions on the surface. The STM is extremely sensitive to the surface depth profile and can detect differences in height of the order of picometers; it is not even necessary to keep the specimen surface under vacuum. A recent variation of the STM, named the atomic force microscope, can be applied to insulating surfaces. Its tip is similar to that of the STM but is mounted on a cantilever.

As mentioned above, wave optical concepts have become more and more important in conventional transmission electron microscopy. Textbooks on quantum mechanics often begin explaining the most important quantum mechanical concepts by describing the analog of Young's double-slit experiment with electrons. For example, in the Feynman Lectures of 1969, the double-slit experiment with electrons is still described as a "thought experiment" and used to explain the wave aspect of quantum particles, and a warning is added: "We should say right away that you should not try to set up this experiment . . . . This experiment has never been done in just this way. The trouble is that the apparatus would have to be made on an impossibly small scale to show the effects we are interested in. We are doing a thought experiment which we have chosen because it is easy to think about." Actually, the experiment had already been realized by Claus Jönsson, a co-worker of Möllenstedt, in 1960. It was known from light optics that for a successful interference experiment it is not necessary that the dimensions of the diffracting structure be of the order of the wavelength. Thus it is possible to observe the diffraction pattern of a football using visible light, even though the radius of the ball exceeds the wavelength by a factor of more than 10,000. The important point is that the coherence condition must be satisfied. Another, even earlier experiment in Tübingen by Möllenstedt and Düker (1956) was still more suited to realize two-beam interferences with electrons. It is an analog of the light optical biprism of Young and Fresnel. The electron beam coming from a very fine real electron source is split up by a charged thin filament, the electrostatic field of which deflects in opposite directions the coherent parts of the beam passing on both sides. Thus, after passing the biprism filament, there are two coherent partial beams which seem to originate from two separate coherent virtual electron sources instead of one real source. If the biprism filament carries a positive charge, the two coherent beams overlap, and a large number of interference fringes can be observed in the overlapping region. Möllenstedt's biprism forms the basis of many subsequent electron interference experiments in analogy with optical interferometric experiments, where the primary beam is split up into two coherent parts, which, after passing through different media, are superimposed again to form an interference pattern which allows one to measure properties of the media through which the partial beams have passed. One such electron interferometric experiment (W. Bayh, Tübingen, 1962), which has stirred much controversy although it confirmed the predictions of quantum mechanical theory, was the phase shift of electron waves by the magnetic vector potential even in the absence of the Lorentz force acting directly on the electrons, a typical quantum mechanical effect.

In the context of our subject, it is noteworthy that Möllenstedt's biprism has made possible high-resolution electron holography (H. Lichte, 1986)—strictly speaking, image-plane off-axis holography. A biprism is incorporated in an electron microscope equipped with a field-emission electron gun. The two coherent electron beams are guided symmetrically toward each other through the electron microscope, but only one of them, the carrier beam, passes through the object, while the other serves as a reference beam. The interference pattern ("hologram") obtained after superposition shows a magnified image of the object with superimposed interference fringes. Roughly speaking, one may say that the phase shift which the carrier wave has suffered when passing through the object gives rise to a local fringe shift in the hologram, whereas a loss of amplitude of the carrier wave in the object leads to a loss in fringe visibility. The hologram is recorded and stored. The evaluation of the stored hologram is done numerically. Important features of electron holography are that the phase and amplitude parts of an object function describing the interaction of the object with the carrier beam can be determined separately and independently, and it is possible to correct the aberrations in the numerical evaluation process. Presently, a fringe width of down to 0.3 Å and a resolution limit of 1.5 Å at 100 kV has been achieved.

Many valuable contributions have been made by German colleagues to the direct imaging of solid surfaces by electron emission microscopy and by scanning microscopy, and the performance of the instruments and their detectors has been considerably improved. Early pioneering work in scanning electron microscopy had been done by Max Knoll and Manfred von Ardenne, but the production of commercial scanning microscopes started only lately by German companies such as Leitz and Zeiss. Although the scanning electron microscope as applied to surfaces has become a valuable tool in many fields of science and industry, it is not included in this report since its resolution is far from atomic, whereas, on the other hand, the scanning transmission electron microscope (STEM) is approaching atomic resolution.

The foundation of the DGE (Deutsche Gesellschaft für Elektronenmikroskopie) in 1949 is treated in G. Schimmel's contribution. The DGE meetings were initially held annually, but after the foundation of the International Federation of Societies for Electron Microscopy (IFSEM), it was decided that International EM Conferences should be held once in four years (1954, 1958, 1962, etc.), Regional Conferences in between (1956, 1960, 1964, etc.), and national or other local conferences in the odd years (1955, 1957, 1959, etc.).

APPENDIX I

MEETINGS OF THE DEUTSCHE GESELLSCHAFT FÜR
ELEKTRONENMIKROSKOPIE (DGE)

1949	Mosbach	Foundation of the DGE
1950	Bad Soden	
1951	Hamburg	
1952	Tübingen	
1953	Innsbruck	With Austria
1955	Münster	

(continues)

APPENDIX I (*Continued*)

1957	Darmstadt	
1958	Berlin	International Conference
1959	Freiburg	
1961	Kiel	
1963	Zürich	With Switzerland
1965	Aachen	
1967	Marburg	
1969	Wien	With Austria
1971	Karlsruhe	
1973	Lüttich	With Belgium and the Netherlands
1975	Berlin	
1977	Münster	With the Royal Microscopical Society, Oxford
1979	Tübingen	
1981	Innsbruck	With Austria
1982	Hamburg	International Conference
1983	Antwerpen	With Belgium
1985	Konstanz	With Austria and Switzerland
1987	Bremen	
1989	Salzburg	With Austria and Switzerland
1991	Darmstadt	
1993	Zürich	With Australia and Switzerland
1995	Leipzig	

PRESIDENTS AND SECRETARIES OF THE DGE

1949	E. Ruska	B. von Borries
1950	E. Brüche	B. von Borries
1951	H. Ruska	B. von Borries
1952	U. Hofmann	B. von Borries
1953	W. Glaser	B. von Borries
1954/55	H. Mahl	B. von Borries
1956/57	H. Kehler	B. von Borries
1958	G. Möllenstedt	H. Kehler
1959	W. Bargmann	H. Kehler
1960/61	D. Peters	H. Kehler
1962/63	E. Ruska	H. Kehler
1964/65	H. Boersch	H. Kehler
1966/67	K. E. Wohlfahrt-Bottermann	H. Kehler
1968/69	L. Wegmann	H. Kehler
1970/71	P. Giesbrecht	G. Schimmel
1972/73	L. Reimer	G. Schimmel
1974/75	P. Sitte	G. Schimmel
1976/77	G. Pfefferkorn	G. Schimmel
1978/79	H. Themann	G. Schimmel
1980/81	F. Lenz	G. Schimmel
1982/83	E. Zeitler	G. Schimmel

(*continues*)

Appendix I (Continued)

1984/85	H. Plattner	G. Schimmel
1986/87	W. Baumeister	B. Tesche
1988/89	K. H. Herrman	B. Tesche
1990/91	H. Niedrig	B. Tesche
1992/93	K. Zierold	B. Tesche
1994/95	M. Rühle	W. Hert

Honorary Members of the DGE

Hans Busch	1884–1973
Max von Laue	1879–1960
Ernst Ruska	1906–1988
Ernst Brüche	1900–1985
Max Knoll	1897–1985
Hans Mahl	1909–1988
Otto Scherzer	1909–1982
Hans Boersch	1909–1986
Helmut Ruska	1908–1973
Walter Glaser	1906–1960
Bodo von Borries	1905–1956
Gottfried Möllenstedt	1912–
Gerhard Pfefferkorn	1913–1989
Helmut Kehler	1908–1994
Heinz Bethge	1919–
Friedrich Lenz	1922–

References

Reimer, L. (1984). "Transmission Electron Microscopy," Physics and Image Formation and Microanalysis, Vol. 36, Springer Series in Optical Sciences, Berlin, Heidelberg, New York, Tokyo.
Reimer, L. (1985). "Scanning Electron Microscopy," Physics and Image Formation and Microanalysis, Vol. 45, Springer Series in Optical Sciences, Berlin, Heidelberg, New York, Tokyo.
Rang, O. (1949). *Physikalische Blätter* **5,** 78–80; *Optik* **5,** 518–530.
von Borries, B., and Lenz, F. (1956). Über die Entstehung des Kontrastes im elektronenmikroskopischen Bild. In "Electron Microscopy, Proceedings of the Stockholm Conference." September 1956 (F. S. Sjöstrand and J. Rhodin, eds.), pp. 60–64. Almqvist & Wiksell, Stockholm.
von Ardenne, M. (1941). Über ein 200kV-Universal-Elektronenmikroskop mit Objektabschattungsvorrichtung, *Z. Phys.* **117,** 657–688.
Müller, H. O., and Ruska, E. (1941). Ein Übermikroskop für 220 kV Strahlspannung, *Koll. Z.* **95,** 21–25.

4.7
The Construction of Commercial Electron Microscopes in China

LAN YOU HUANG

KYKY Scientific Instrument Research and Development Center
Chinese Academy of Sciences
Beijing, China

I. Introduction

Not long after the overthrow of the Nationalist government in China in 1949, the new government asked the Chinese Academy of Sciences to identify a mysterious piece of equipment found during the clearing of a warehouse that had belonged to the old Bureau of Broadcasting. It turned out to be a transmission electron microscope (TEM) made by the Metropolitan Vickers Company in Manchester. Who had ordered it and for what purpose remains a mystery, but it was with this instrument that electron microscopy was begun in China under the supervision of Prof. Qian Ning Zhao, later to become the first President of the Chinese Electron Microscopy Society.

II. Early Days[1]

In 1955, when Prof. Qian was involved in drawing up the first Five-Year Plan, he was told by the Soviet specialist responsible for the plan that China would not be capable of manufacturing a TEM of its own for at least 10 years. In 1958, apart from some early Soviet type EM3 TEMs, the most modern electron miscroscope in operation in China was an East German Zeiss electrostatic model, a birthday present to Chairman Mao from President Pieck of East Germany. It could not be operated above 30 kV. In March 1958, the "Great Leap Forward" was in the air. Everyone was looking for some impressive project to work on. I had just returned to China after 13 years of education in the United States and also in Germany,

---

[1] A brief account of the first attempts to build an electron microscope in China was published in *Optik* **92**, 23–26 (1992).

where I obtained my Ph.D. under Prof. Möllenstedt at Tübingen University. At the Institute of Electronics of the Chinese Academy of Sciences, where I began my career, I did my best to impress the director with numerous topics that I thought might be useful. The expression on his face, however, told me that none of them measured up to the "Great Leap Forward" concept! In desperation, I proposed building a transmission electron miscroscope. That certainly made an instant impression, but I was told that the Institute of Optics and Fine Mechanics (IOFM) already had plans to build one. In April, I visited this institute, some 1000 km away, in northeastern China, and met the decision makers, headed by Prof. Wang Da Heng, the father of optics in China. They listened patiently to my boasting about having studied electron optics in Germany, then they told me that they had no immediate plans for making a TEM. Something must have happened, though, during our discussions, for that very afternoon, they told me to move my things into the institute right away and start work on the TEM! In spite of my protests, they wanted it finished, not in a couple of years, but in early September of that very year! Their motivation was very simple: They wanted it as a "first fruit" of the opening year of the "Great Leap Forward." Moreover, they wanted it presented in Beijing as a "9th birthday" gift to the People's Republic of China on 1 October! My suggestion was that the only hope of meeting such a tight deadline was to copy an existing industrial model. This brought a quick response. The very next day the president of the Academy of Sciences agreed over the phone to lend the IOFM their newly purchased Japanese JEM-T4 TEM. Work started in June. The institute had, in fact, set up eight major projects for the "birthday." The TEM was the newest member of the group.

Professor Wang Da Heng, a man of extraordinary insight, had surrounded himself with a talented group of co-workers, whose qualities I learned to appreciate more and more as the years went by. The young physicists Lin Taiji and Zhu Huanwen and a mechanical engineer named Wang Hongyi were assigned to work with me. They, and the other engineers on the support team, together with the machinists in the workshop, were excellent. They had to work under a severe handicap, however; up-to-date equipment and component parts were simply not available. To make matters worse, there were no suppliers of parts and accessories, no catalogs or "Yellow Pages." A planned economy, it was said, did not need such things. With eight "unplanned" major projects demanding parts immediately, the problem seemed insurmountable. Among the talented people in the institute that I marveled at was one, Miss Yu, a young purchasing agent, who obtained a supply of vacuum bellows for me, going just on my vague description over a long-distance telephone. Nobody in IOFM had ever seen a vacuum bellows before; I myself did not even know the proper name for it in Chinese! Miss Yu managed to buy the correct items in an automobile depot!

Among the talented people around the place, perhaps the most important was a group of dedicated young Party members with university backgrounds, the junior executives who coordinated the programs. One of them, a young cadré named Sun Gongyu, was assigned to the electron microscope project. In spite of a complex maze of problems, the junior executives ensured that everything fell into place at the proper time. It was a masterpiece of organization! Thus, at 2.45 a.m. on 19 August 1958, we obtained our first image! Figure 1 shows a newspaper cutting recording the event. The resolution had reached 10 nm by September, when the president of the Academy of Sciences and the directors of all the institutes under the umbrella of the academy gathered at the IOFM to celebrate the successfully concluded set of the academy's "birthday gifts," of which the IOFM ones were the most successful and the TEM was the star. It was a great day for all of us who had labored for months without much sleep. The "birthday presents" were then shipped to Beijing for a full-scale public exhibition. Throughout the entire project, I was always uneasy that something might come up for which I did not know the answer. I was therefore grateful that while I was in Tübingen I had picked up not only some sophisticated skills, but also some useful practical tricks and ideas for makeshift construction. I remembered the words of Mr. Speidel, the well-known workshop master (*Werkstattmeister*) at Tübingen: "If you are going back to China, postwar Germany is the right place to come to. You will learn how to make things

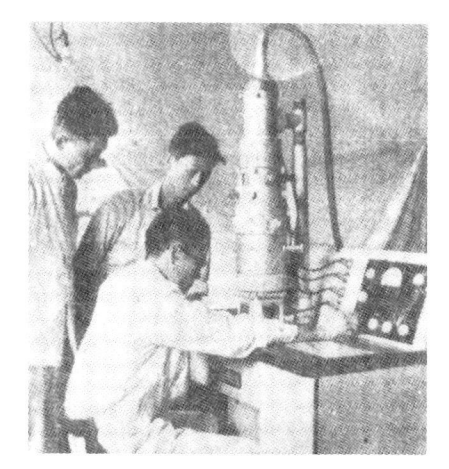

FIGURE 1. Newspaper cutting of the prototype of the first commercial TEM, constructed in China. Resolution 10 nm. Operational on 19 August, 1958. The author is at the controls.

from scratch, with your own hands." So tricks such as vacuum-leak hunting with alcohol and a Tesla coil, or testing the sensitivity of the column to stray ac fields, using the stray field generated by a soldering iron, always made a big impression on bystanders. There were also other makeshift methods that I am not quite so proud of. The ceramic insulator for the electron gun was delivered very late. In the meantime I had one made in Perspex. It would soften after 15 min of use, but it worked splendidly while it was cool!

At the end of 1958 we started to design a 100-kV TEM. This time we were allowed until the next "birthday" to build it. I had never designed a complete TEM before, and my colleagues had only their experience in copying the JEOL T-4. All I had to go on was a brochure of the Siemens Elmiskop, which I had brought from Germany. Although Tübingen had the tradition of using electrostatic lenses exclusively, a legacy from the Zeiss Mosbach laboratory via Möllenstedt, I had learned how to design magnetic lenses from an Englishman, Dr. D. Jones, who had gained his Ph.D. under Dr. V. E. Cosslett at the Cavendish Laboratory. He gave me reprints by Liebmann, Mulvey, Haine, etc. With the help of Jones, I had actually built the first magnetic lens to appear in Möllenstedt's laboratory. This was one of the many benefits which students enjoyed from the various international exchange visits that Möllenstedt encouraged.

At first, we had fancy ideas about the design of the TEM, but when the time actually came to put them onto the drawing board, we found ourselves subconsciously falling back on the Siemens design. No one wanted to risk delaying the "gift." The physicists Zhu and Lin left the EM group. In their place a new physicist, Yao Jun En, was assigned to assist me in the electron optical design. Administratively, he was also the leader of the IOFM TEM group. I went over with him very thoroughly all I knew about the electron optical design of a TEM and passed onto him all the literature that I had brought from Germany. These included articles by Ruska, von Borries, Liebmann, Haine, Mulvey, etc., together with two very useful handbooks by von Ardenne, *Handbuch der Elektronenphysik, Ionenphysik und Uebermikroskopie,* and a volume of Flugge's *Handbuch der Physik* with a section on electron miscroscopy by Leisegang. Yao made a big impression on me by his ability to read not only English but also some German, which saved me a lot of effort. Not only that, he also knew some Russian, which all Chinese scientists were required to learn at that time, although it did not, in fact, help us. Yao was a good student and a diligent worker. For one who had never even operated a commercial TEM before, he caught on pretty quickly. With Yao and Wang Hongyi (the mechanical engineer who worked with me in 1958) to assist me, the work proceeded much more quickly than I had expected. I think Wang Hongyi had a more difficult time than I, since there was no detailed information about the design of the mechanical stage or of the camera chamber.

In the late spring of that year, the column was ready for assembly, but the electronics and the console, which had been subcontracted to the Shanghai Medical Factory, had not arrived. We had to make a wooden bench to support the column and the vacuum system. We used an old high-voltage supply from 1958 and batteries to power the lenses and the gun filament. This enabled us to make a start on our experiments in electron microscopy. A raw university graduate, Zen Zhaowei, was assigned to help me at the microscope. Yao, who was not very skillful at experimental work, was kept very busy doing the liason work with the supporting departments of the IOFM. A typical night's work went as follows: Zen would get a good vacuum in the column, find the electron beam, align the column as best he could, and then wake me up in the dormitory. I would then continue the experiments on the column to find the cause of various malfunctionings. On locating a fault, I would remove the suspect section of the column and take it along to the workshop, where it would be modified there and then according to my verbal instructions. I then reassembled the column and looked for the next problem. In those days the machine shop, even in the dead of night, was lit up like a ballroom, with all machines running. Modifications were made on the spot. There was no time to modify the drawings or enter them into a log book. This practice, not uncommon among impatient researchers, was one of the reasons why, later on, the Nanjing and Shanghai groups had such difficulty in reproducing our results on a commercial instrument. When the electronics and the microscope console eventually arrived from Shanghai, I was astonished to see the console looking exactly like that of the Siemens instrument. In order to avoid confusion, we had indeed agreed to adopt the general Siemens layout, but to make it look exactly like that of Siemens must have involved an extraordinary amount of unnecessary extra effort!

The debugging of the electronics took more time than we expected. Most of it was due to confusion, with too many hands trying to do too many things in a hurry, but we also ran into the inevitable problems of earth loops and component damage by high-voltage flashovers. By the time it was ready for the resolution test, we were only a few days away from the "birthday"! After working all night, and feeling pretty groggy, I succeeded, on the morning 26 September, in taking the resolution test micrographs. My train was due to depart at 9 a.m. Comrade Sun begged me to stay for just another day, but I had to think of my department at the Institute of Electronics, which had also prepared something for the "birthday." I took the micrographs out of the fixing solution and examined them while they were still wet. My excited colleagues surrounded me, anxiously waiting for the result. I quickly found some details with 5 nm separation and rushed off to the station. Later that day, when the film had dried, my colleagues found separations of 2.5 nm. This became the official resolution of the 100-kV TEM type XD-100.

On the same day, the XD-100 was quickly taken to pieces, crated, and put on a specially reserved section of the train to Beijing, to be exhibited on 1 October. Yao and Wang rode in the same freight compartment and kept a continuous watch over it all the way to Beijing. At Beijing, Yao told me, they had only 2 days to get it up and running. The first setback was that the coil of the objective was shorted out by a leak in the cooling-water system. They had to repair the leak and then take the coil to a sympathetic institute in Beijing to get it baked out. After all this, it took them a day to get a good vacuum in the column. They only managed to get an image at the last minute before the opening of the exhibition. Figure 2 shows Zen Zhaowei with the XD-100 at the exhibition, probably showing the awed visitors the magnified image of a mosquito wing, one of the few specimens at the time that needed no preparation!

An important source of help in our development program was the Institute of Metals Research, which manufactured iron polepiece material for

FIGURE 2. Zen Zhaowei at the controls of the XD-100 at the October Exhibition in Beijing (1959).

us. There I met Prof. Kuo Kexin, a very helpful scientist who had studied in Sweden. He promised me that he would have some Permendur specially made for us. Later, Prof. Kuo became increasingly interested in electron microscopy. He introduced diffraction contrast to Chinese materials scientists. His work on quasi-crystals won worldwide recognition. From 1984 to 1993, as president of the Chinese Electron Microscope Society, he did much toward building up the society to the respected status it now enjoys.

The year 1959 was the year of the "anti-rightist-tendency" movement. Therefore the "gift for the birthday" took on a double political significance. A huge model of the TEM headed the procession of the Chinese Academy of Sciences within the gigantic parade, which marched past Chairman Mao, waving his hand from the top of the Tiananmen (Heavenly Peace Gate) to the huge crowd below. I was among the invited spectators below Tiananmen. It was an impressive sight.

I did not publish anything on the 1958 and 1959 TEMs, because I thought the work was somewhat lacking in originality. Professor Gong, the vice director of the IOFM, who organized the project along with Prof. Wang, thought, however, that something ought to be written. When I showed no interest, he wrote an article himself, with my agreement, for *Science News* (Gong, 1959), a journal for general readers in science. Later, Yao surprised me by publishing, without my knowledge, an article on the XD-100 (Yao, 1965), under his name only, with the consequence that many electron microscopists thought that he was the first to make an electron microscope in China, especially since everybody knew that the IOFM was the cradle of EM in China and that Yao was from the IOFM, but I was not. I never said anything in public about it, but this unfortunate event separated us for many years. After the Cultural Revolution I got over it when I realized that both of us had suffered much more than we deserved, he probably more than I. After 1959, EM activity at the IOFM came virtually to a stop. The drawings of the 1958 and 1959 TEMs were given to the Nanjing and the Shanghai Medical Equipment Factories, respectively. In fact, they were given to many other factories as well. The number of factories, universities, and institutes that attempted to build a TEM at that time amounted to about 14. Some were not even associated with us. Only Nanjing and Shanghai succeeded. This squandering of effort represented only an insignificant fraction of the waste of resources that occurred during the "Great Leap," which, together with atrocious weather, brought about a three-year famine in China.

What took us a few months to accomplish at the institute took the two factories several years. Many reasons could be given for this. Much later, I began to understand the fundamental cause. Although the drive in those days was, in truth, a sense of patriotism, of one's "duty to serve the people," the only way to realize this in practice, however, was through the government hierarchy. Then "to serve the people" degenerated into "to please your superior" at every level of the hierarchy. The call for "selfless aid to your brother organization" becomes an empty slogan if the brother happens to be in another branch of the hierarchy. In fact, if the superior of both "brothers" does not himself direct the "aid," no technology transfer is possible. Throughout these years, I observed this to be as true as any law in physics.

It was in 1964 that the Shanghai group led by Zhu Zhufu brought our design into production. As a result of this event, a postage stamp was issued of this TEM, as shown in Fig. 3, and put into a set showing new industrial products made in China. However, this TEM looked so much like the Siemens Elmiskop that when the American stamp-collecting electron microscopist Simine Short wrote an article (Short, 1990) in the *EMSA Bulletin* on postage stamps showing EMs, he wrote: "2000 Elmiskops sold worldwide. The People's Republic of China bought one of these modern, highly sophisticated instruments and included . . . the Siemens Elmiskop in a set of set of eight stamps in 1966, showing new industrial products. This stamp, by the way, seems to be the earliest stamp issued to show a TEM." (N.B.:

FIGURE 3. Postage stamp of the Shanghai Factory version of the XD-100 TEM.

The year 1966 should read 1964.) The eight "new products" were, in fact, all made in China.

I first met Zhu in 1959, when he came to Changchun to collaborate with us. He was a pleasant, good-looking young man with a bright smile. He made an impression on me with his winning ways and his eagerness to learn. Although he was directly responsible solely for the electronics, he was curious about everything. He would sit and watch me prepare the resolution-calibration specimens and ask questions. He said he did not have a college degree because his family could not afford it, so he was taking courses at night school. Things were particularly bad during the Great Leap Forward in 1958, when he worked long hours along with everyone else, but still went to his classes. He told me how hard everyone worked then. At one time a girl disappeared after having worked around the clock for some days. Three days later she was found sleeping in a corner. She had slept 36 hours, nonstop, and woke up thinking it was still the same day!

Zhu not only became a good electronics engineer, he also learned everything about designing an electron optical column. At the height of his career he led the construction of a 100-kV TEM capable of 0.2 nm line resolution regularly and 0.14 occasionally. This is the present (1994) record in China, and nothing better is in sight.

For this TEM he put the high-voltage supply into a gas tank, to avoid the instability due to oil convection. Although this proved impractical for manufacture, he did achieve a stability of one part in a million, which helped him to get the record resolution. He also became a "People's Representative," a very high honor in China. But let us go back to the 1960s.

Around 1965, the Shanghai and Nanjing engineers improved the resolution of the instrument to 0.7 nm and 2 nm, respectively. With these and later models of their own, they each produced about 300 EMs. Especially in the days of China's comparative isolation from the rest of the world, these EMs greatly helped the country. A summary of their achievements is listed in Appendix I. However, the reliability of these EMs and the service facilities available were very poor, so that each user had to employ a full-time university graduate to look after the instrument. In spite of repeated complaints, there was no way built into the hierarchical system to remedy this situation.

III. The Scientific Instrument Factory (KYKY)

More than in any other country, the Chinese history of electron microscopy was strongly influenced by the pattern of political events in the country.

Although the Shanghai and Nanjing groups would have very different tales to tell, the general course of their history is similar. The complete list of all electron microscopes produced in China together with the names of the key people on the design teams, at the institute or factory concerned, is set out in detail in Appendix I. Appendix II lists the names of leading Chinese electron opticians and their specialist areas. I will concentrate on KYKY, which I know from first hand. What started out as the Beijing Scientific Instrument Factory (SIF) is now the Beijing Scientific Instrument R&D Center. In the late 1970s it got an abbreviated name, KYKY, from its Chinese title (Ke Yi Ke Yuan), meaning literally "Science Instruments Science Academy," or more specifically, "Scientific Instruments of the Science Academy." To avoid confusion I will mainly use the name KYKY, since this is how we are better known outside China, but I may occasionally use the alternative names, where appropriate.

A favorite saying during political rallies at KYKY, in order to show the wise leadership of the Party, or that of the Party Secretary of the factory, was its success story "from engineering vises to electron microscopes." The factory had started by making crude bench vises for machine shops of the institutes. It sounded like a "rags-to-riches" story, but in fact it was more complicated than that. Before the reform of the 1980s, and to a lesser extent even now, Chinese society used to be governed by a vertically structured hierarchy, whose spine was the Communist Party. Each organization could interact successfully only with its immediate superior or its subordinate. So, if an institute wanted machining done, it could not buy such service horizontally from society. The only way forward was to have its own machine shop. Every institute had to have equipment for every process—for cutting, casting, heat treatment, plating, painting, the lot. Theoretically, they could get help from their "relatives" within the same clan, but in fact every director knew that this did not work. With so many institutes asking for expensive equipment, the Chinese Academy of Sciences (CAS) decided to set up a factory and equip it with the best machines, with the main purpose of serving all institutes. So was the Scientific Instrument Factory (KYKY) conceived and born. Little did the officials of the CAS realize that this idea violated the "principle of hierarchy." In any hierarchical system, one's success does not come from recognition by society, as, for example in a market-oriented system, but from recognition by one's superior, and therefore, in such a system, pleasing one's peers is never a road to success. So, sooner or later, KYKY had to go its own way and leave the institutes back with their own machine shops. KYKY's Party Secretary and Director Tian, like all good cadrés, was an expert in cultivating contacts and connections. His backer was none less than the vice president of the Chinese Academy of Science, Comrade Pei. Director Tian, an intelligent man, knew he would never get any recognition just by serving other institutes. So, with the help of his protector, he obtained permission to make instruments of his own. He recruited competent and talented people in both management and engineering. An important recruit was Mr. Xiang Pengju, who became the vice director of KYKY, and, after his retirement, the secretary of the Chinese Electron Microscope Society, a position he still holds today. The first instrument that the factory took on was a mass spectrometer, a copy of a Soviet model, which they built in record time. With it, the position of KYKY was firmly established. Although other institute directors complained, they had to be satisfied with their own machine shops. No one within the Chinese Academy of Sciences management raised any further objections.

The next major goal would be the electron microscope, the most prestigious one being the TEM. The construction of a TEM was, however, in the hands of the Institute of Optics and Fine Mechanics (IOFM). Party protocol did not allow the taking over of the "rice bowl" of a sister organization. To get a foot in the door, KYKY asked me, while I was still at the Institute of Electronics, to assist them in building an electron emission microscope, something the IOFM would not mind; it was actually needed to study cathodes at the Institute of Electronics. I found the KYKY group headed by Hu Mingxiang very quick on the uptake and easy to liase with. Fortunately, I still had my notes for building an electron emission microscope from my Ph.D. days in Tübingen. The emission microscope (Fig. 4) was built with surprising speed. Vice Director Xiang, a smart

FIGURE 4. KYKY electron emission microscope (1964).

operator, was then ready for the next move. He came to me and said that he wanted to manufacture the electron trajectory tracer we had developed. This Gabor type of "tricycle" trajectory tracer had been designed by an extremely talented young engineer of mine, Jiang Jungji, who came with me to the IOFM in 1958 to build the first TEM. I explained to Xiang that we had built the trajectory tracer for the design of microwave tube guns, in tandem with a space charge-simulating electrolytic tank, a very specialized application (Fig. 5). I thought that such an instrument would not be much use for other purposes. But Xiang insisted. Seven trajectory tracers were built and sold to institutes that dealt with vacuum tubes. So far as I know, none of them was put to good use. Xiang, however, now had his foot in the door of electron optics. After these successes, the Scientific Instruments Factory was poised to attack the manufacture of electron microscopes. Xiang said, "Let the Institute of Optics and Fine Mechanics do research on ordinary electron microscopes. The Scientific Instrument Factory should develop and manufacture high-resolution electron microscopes."

Around 1963, when the Chinese Academy of Sciences was undergoing a process of reorganization, the IOFM's party secretary and Comrade Sun approached me and asked me if I would be prepared to head their Electron Optical Department, which they planned to move to Shanghai, where they hoped to set up a branch institute specializing in "electro-optics". Somehow, they associated electron optics with this. I thought about it but decided that I did not want to leave Beijing. This was a very fortunate decision, because three years later, during the Cultural Revolution, factionalism plus politics made the Shanghai IOFM a deadly place to be in. Two scientists that I knew very well from the Institute of Electronics, who were specialists in laser optics, committed suicide there. Poor Comrade Sun, who gave me such wonderful support during the building of the first TEMs in 1958 and

FIGURE 5. KYKY electrolytic tank and trajectory tracer, with space-charge simulating electrolytic tank (1963). The current-injecting needles are at the bottom of the tank.

1959, also committed suicide. He was an underground party member in his university days, before the "liberation." Many of his activities were done in secret, some not directly under the eyes of the Party, and therefore there were periods in his records that could not be clarified. People took advantage of that and persecuted him.

KYKY, with all this background experience, was now ready, with the support of Comrade Pei, to take up electron microscopy during the reorganization of the Chinese Academy of Sciences. The electron microscope department of the IOFM, headed by Yao, was transferred to the KYKY in 1964. Secretary Tian and Director Xiang thus got their way. At the same time, they recruited me and my group to work on electron-beam lithography. KYKY was a better place to work than the Institute of Electronics, because it had wonderful equipment, very good machinists, and competent management.

IV. The DX-2 Microscope

The honeymoon period of KYKY's original electron microscope group (the Beijing group) and Yao's group (the Changchun group) did not last long. A dispute began as to which design should be adopted for the next transmission electron microscope (TEM). This dispute surfaced again as a bitter political struggle during the Cultural Revolution two years later. At the time, however, it was settled amicably. Since I had been a consultant for both sides, I was made the judge to decide which design to accept. I chose Yao-Wang's design. Wang had been an electron microscope designer since 1958. I thought the KYKY design was too close a copy of the Hitachi HU-11. However, I discovered a mistake in the design of Yao's illuminating system. He had designed it as a probe-forming lens of an electron microprobe, striving for as small a spherical aberration as possible, so that it could handle a large "optimum aperture angle" to give as large a probe current as possible. The concept was of course not suitable for the TEM illuminating system, where the illuminating aperture and the current in the illuminating spot are limited by other factors than those of a microprobe. When I pointed out this mistake, I was shown a copy of the letter which I had written to Prof. Wang, the director of the IOFM, when I was asked to examine the design previously, consenting to the design of the whole electron optical column. It was a rather embarrassing oversight on my part! So, the SIF design for the illuminating system was then adopted. Once this had been settled, the work proceeded with remarkable speed, not as fast as at IOFM in 1958 and 1959, but at a speed we were never to experience again in China. Never again would we see such optimistic patriotism, such selfless enthusiasm among the masses. True, the enthusiasm of the masses was misused at the time. All that activity came to practically nothing. By comparison, the present Party policy seems at last to be leading to a better future, but those of us who experienced the spirit of the period between 1958 and 1965 still miss it. A year and a half later, in 1965, the TEM model DX-2, shown in Fig. 6, was completed, achieving a resolution of 0.5 nm, a record at the time. Credit must be given to Party Secretary Tian for his ability to mobilize the masses and for getting support from the Chinese Academy of Sciences. Likewise, Director Xiang must be mentioned for his great organizing skills and Yao for his pinpointing and concentration on crucial technical aspects, such as the precision manufacture of the polepieces and the development of high-stability accelerating voltage supplies. Through Vice President Pei, assignments were given to the Institute of Metals Research and to the Institute of Metallurgy to develop iron-cobalt alloys of high purity and homogeneity for the polepieces. A special team of experts, each concentrating on different aspects of polepiece fabrication and testing, was set up. The final test was the measurement of astigmatism under operational conditions. A Hitachi HU-11 was imported for the purpose. As a result, polepieces with a residual astigmatism below 1 μm were obtained regularly. Model DX-2 won great acclaim from the Chinese Academy of Science and from the press, to the great satisfaction of all who had taken part in the project. Because of the Cultural Revolution, there was no publication on the DX-2, only a collectively written internal report. During a "summing-up meeting," however, in which all the participants looked back with hindsight to see what had been done correctly and what could have been done better, the discussion tended to be critical of Yao.

FIGURE 6. KYKY DX-2 TEM, resolution 0.5 nm, 100 kV (1965).

It was alleged that he did not solve any of the practical problems that arose in the experimental stage of model DX-2. To be sure, most of the young scientists in China at the time were pushed into the leadership of groups in entirely new fields, without having much practical experience behind them. Moreover, Yao was busy with many other aspects of the program and had little time to spare for experimental work. I also found myself in a similar situation at the Institute of Electronics. The tension between the "Beijing group" and Yao and Wang erupted into a political "affair" a few months later during the Cultural Revolution.

V. The Cultural Revolution

The Cultural Revolution at the Scientific Instrument Factory started with a "Big Character Poster" against Yao and Wang. As in the case of the Shanghai IOFM, differences of opinion and ill-feeling among fellow scientists was made use of by Party cadrés. Although Chairman Mao had intended to use the movement to purge the Party, the Party cadrés diverted the fire away from themselves and toward "plausible enemies." Yao, with landlord and "rightist" elements among his family, and Wang, who had worked for the Japanese during the occupation of Manchuria, seemed plausible targets. The Party cadrés, headed by Secretary Tian, relied on the Beijing group as their masses for the movement. I was protected by the Beijing group because of the help I had given them. In particular, they appreciated my help in locating a major source of trouble during the testing of the DX-2, which got things stuck for a long time. I traced the fault to the effect on the column of a strong stray ac magnetic field caused by an earth fault on an electric cable outside the window. My troubles in the Cultural Revolution were to come later, when another set of cadrés came to power and another set of masses was used.

For Yao, this was a difficult time. Later he told me that it was the worst time in his life and career. It was the fear and the loneliness that was the worst. If he was made "an enemy of the people," his life and the lives of his family would be finished. Loneliness added to the despair. Nobody talked to him. He felt grateful when an acquaintance gave him a surreptitious nod. In the building where he lived, there were several scientists who got similar treatment. Yao's wife told me that their son Yao Nan would pretend to play outside the building with the children of the other scientists, about the time the parents were due to come home. If they did not see their parents they would cautiously ask the other grown-ups for news about them. (Yao Nan, by the way, later gained his Ph.D. under Cowley at

Arizona and is now doing electron microscopy at Princeton.) Fortunately for Yao, his persecution did not last long. The offensive soon turned against the party cadrés themselves, and Secretary Tian became the main target. He, I must say, suffered more than Yao when the Red Guards took control of the Revolution. During its next phase, in which the Army took control, the Scientific Factory was taken over by the Academy of Military Science. A hundred and eighty scientists, engineers, and other employees of "bad" family origin were left in the Chinese Academy of Science, which was in a state of paralysis. The Academy of Military Science was particularly interested in the electron-beam lithography machine, which at that time had been assembled and was ready for testing. They took most of the engineers, leaving those who were politically unreliable. I, fortunately, was considered one such. Later on, to try to prove once more that science could be done much better and more quickly through mass movements, the Beijing Municipality organized a "69/6" project, in which our lithography machine was included. It was claimed that "international standards" were achieved; in fact, I know that the thing never worked. I was told that almost all the components in the computer broke down one after the other, and the wiring would come off "like loose strands of hair."

A. Science in the Cultural Revolution

"69/6" in Chinese stands for 1969/June. In that year there was a series of movements that had to do with science. The movement to discredit Einstein, the biggest bourgeois scientist, was already over. There was the movement "to walk out of the Ivory Tower" of the Academy of Sciences or other such places so as to "join the workers and peasants, so as to learn from their wisdom and outlook." It was said that academic knowledge led to conservatism. In the wake of the "walk out" movement was the "electronic centralism" movement. It was claimed that electronics was the center of all sciences and hence was all set to lead to the next industrial revolution. Such a revolution was to be led not by the bourgeois intellectuals but by the working masses. Although all these ideas came from Mao, the animator of the movements themselves was Chen Boda, the Party's ideological specialist. In June 1969, Wen Yucheng, Commander-in-Chief of Beijing Military District, organized this "69/6" event, which included countless subprojects, all having to do with microelectronics. A fantastic amount of money must have been spent on these projects, of which only a few trickled down to KYKY, the lithography machine being one of them, and later, two ion-beam and two electron-beam projects. Judging from inside information that leaked out from these projects and the general mess that was made of them, almost nothing useful came out of them. In fact, "69/6" was a knee-jerk response to the rush for microminiaturization that was taking place in the West. In China it took on a mainly political character. It was said, for example, that in the next people's war, battles would be fought in the mountains, inaccessible to tanks and planes, so instead of conventional radar stations, miniradars had to be made, so that both antenna and station could be carried on the backs of soldiers or horses.

B. The 5–7 School

"69/6" did not come to the CAS until early 1970. In the autumn of 1969, the Chinese Academy of Science, with nothing better to do, sent all its scientists and other employees to the countryside for "reeducation" in a "5–7 Cadrés School." For those outside China, this might sound like the worst thing in the Cultural Revolution. In fact, for most of us, it was the easiest ordeal we had to bear during the Revolution. The hard manual labor was stimulating. Sleeping with 50 others under the same roof was at least companionable. Political indoctrination did not have to be taken too seriously. Above all, we were not singled out as targets of the class struggle. Peasants were invited to "reeducate" us. These honest peasants would condemn the "Great Leap Forward," which Chairman Mao allowed no one to criticize. They told us how their crops were ruined by it and how people had died of starvation, something the Party was trying to hide. They praised the cadrés that Mao had purged. The Army cadré who was responsible for us looked very embarrassed on these occasions, and we, the bourgois intellectuals, would listen and smile wryly.

The worst part of the 5–7 school was the food. It was a monotonous diet of rough grains and boiled cabbage, day in and day out. After a couple of months of this, the bourgois intellectuals began to crave protein. Comrade Lo, a portly engineer, found things worse than it was for the rest of us. He collected all the sparrows that the others brought down for him with sling shots, salted them, and hung them out on a line to dry. He told everybody that he would hold a "hundred chicken (sparrows) feast" one day. I had left the school before that feast took place. One day, Comrade Lo learned that someone had killed a snake and buried it in the field. The Chinese have a superstition against all snakes, poisonous or not. Lo took a spade and dug it out, skinned it, cleaned it and chopped it into small sections, and cooked it with spices over an open fire. A crowd gathered, with great interest, to watch him at it. They commented on the appetizing aroma from the pot, but no one dared to try snake meat; the round shape of the snake was far too obvious. I was the first to volunteer. It was delicious. After that, everybody rushed to their bedsides to fetch their bowls and in no time there was no snake left, not even the soup.

On another occasion the Army people, who kept pigs, buried a piglet that had suffocated under its mother's weight. A small group of us got the news and went and dug it up under the cover of darkness, fearing that the soldiers might report us; after all, hardship was part of our training! We roasted it. It tasted so good that someone suggested that perhaps some wine would go well with it. Someone said that he had seen a bottle of wine under the bed next to his and went back to fetch it. The label on the bottle was that of a famous rice wine. One of our gourmets tasted it and said that the taste was indeed worthy of its name. The next day the owner of the bottle told us that it was the cheapest wine available and that he had bought it to soak his Chinese herbal medicine in to make a concoction which he applied to his septic foot! Just to nauseate us, he said he regularly wiped his foot with the liquid, using a cotton wad, which he dipped repeatedly in the bottle!

In 1970, new cadrés were assigned to the Chinese Academy of Science to restore the Scientific Instrument Factory.

The 5–7 school was a terrible time for Yao. He was not used to physical labor, and he was lonely. To make matters worse, he lost both his father and brother while he was at the school.

C. The "69/6" Project

Early in 1970, new cadrés were assigned by the CAS to rebuild KYKY, which at that time had only 180 people, plus a few newcomers. Work started right away on the 69/6 project. I can only describe the ion-beam project that I actually took part in. The idea was what later became known as the focused ion-beam (FIB) method, in which an ion beam is focused onto an IC chip. By changing the ion source, one could either deposit thin metal lines onto the circuit or alternatively sputter away unwanted lines. Sophisticated commercial apparatus of this sort, with high-brightness liquid metal sources, is now available. In 1969 it must have been in the scientific literature, but we had stopped reading that for fear of being ridiculed or criticized. The project was a collaborative effort among KYKY, Qinghua University, and Beijing University. The Qinghua team was headed by a junior researcher called Chen. He must have read about it in the literature and proposed this project, but he always made it sound like his own idea. He constantly reminded people, especially in front of the cadrés, that the idea was "an original Chinese idea," quoting from Mao, who encouraged such action. This crazy guy even signed a contract with an IC factory promising this wonder machine within a year. To show his fellow feeling for members of the working class, Chen used every occasion to repeat the story of how he saw this old worker in the IC factory who had to look at ICs through the light microscope for hours on end, trying to repair them with a needle. He would then say that we must work twice as hard and deliver the machine ahead of schedule to prevent our working-class brothers from ruining their eyes. I tried to convince Chen that the brightness of known ion sources was insufficient to meet the requirements of a practical machine and that doing research on an idea was a long way from making a usable product. He was evasive. I doubt if he understood what I was saying. I discussed the problem with other colleagues from KYKY and from Beijing University. They all agreed with me but advised me to keep my mouth shut. The cadré in charge of our project was a veteran who had formerly been a bodyguard for an Army officer. His body was riddled with bullet wounds (as I saw later with my own eyes). He was a straightforward man, but he had very little education. When he asked me whether this machine would work or not, I think he really did want to know, but he wanted an answer that he could understand: a simple yes or no—would it work or would it not? I used an analogy to help him: I said, "Would a bicycle work without tires?", suggesting that it might work, but its performance might be seriously impaired. When, later, military officers took over KYKY's leadership, and a new wave of "class struggle" called for new targets, they made an example of me. I was charged with "maliciously defiling the Proletariat Headquarters," ridiculing a project they had formally examined and approved, by describing it as "a bicycle without tires." I pictured Commander Wen panting along on a bicycle without tires, but I took their point.

For this 69/6 project, I built two ion sources, a Duoplasmatron source, to work with metaloorganic gases and a Penning sputter ion source with an aluminum target. With the latter, I succeeded in getting an aluminum line deposited onto a flat substrate, but before we went any further, Comrade Chen and the people from Qinghua and Beijing University were called back to their respective universities to take part in the next phase of the Cultural Revolution. The whole project slowly fizzled out.

D. The Military Representatives

At KYKY this new phase was led by three "Military Representatives," as they were called. These officers mistrusted me from the start. They could not understand how a bourgeois like me had escaped punishment. Although the leader of the three was relatively reasonable, the second-in-command always looked at me with a hostile attitude. He made it very obvious to me that my time was yet to come. I was not given any more projects except odds and ends, and when those were finished, I did pure manual labor, such as cleaning the yards and bricklaying for a garbage dump. Although in the end nothing serious happened to me, I had a taste of that hopeless, lonesome feeling of an outcast. There were others at KYKY who were tormented to breaking point. One disappeared after leaving what looked like a suicide note; he was never seen again. The other was Comrade Wang, a competent and highly respected cadré with a university education. He joined the Party as an underground member before the "liberation," when he was a university student. Like Comrade Sun of the IOFM, he was not able to account for all his actions in those days. He hanged himself in his home and was discovered there by his 12-year-old son. Wang had been continually tormented by one of the military representatives, a low-ranking, uneducated soldier who worked overtime, probably trying to extort a confession out of him, while the other representatives were away.

Besides the ion-beam project, there was also the ion-implantation machine and two ion-beam reticle writing machines, one of them made outside KYKY, by Liu Xuping (later co-constructor of the first SEM in China) as part of the "walk out" program, and the other by Yao and Wu Minjun inside KYKY (but not part of the "69/6"). Yao and Wu did some effective work on the design of deflector coils with reduced distortion, and got better than 10-μm accuracy in a field of 32 mm $\times$ 32 mm. Their experiments were facilitated by the development at the Institute of Chemistry of a plastic film that changed color upon electron bombardment. The normally transparent film changed to an orange color where it was exposed to electrons. The resolution was good and the sensitivity was sufficient for the experiments. Later, Yao and Wu converted their instrument into an optical projection monitor, using this film for an Air Force Research Institute that was trying to develop something for tracing out the flight paths of aircraft.

If only Comrade Wang could have hung on for just a few months more! The whole farce of the Cultural Revolution was about to be exposed by the attempted assassination of Mao by his designated successor Lin Biao, whose escape ended in his death in an airplane crash in Mongolia on 13 September 1971. After that, persecution of ordinary citizens, which was serving the purpose of a front, eased up and died out with partial regaining of power by Deng Xiaoping in 1972. At the time, the whole country was wasting away. Production had been at a standstill for many years. There was a shortage of everything. Unable to offer employment to youngsters just out of school, Mao sent them to the countryside to be "reeducated." There they saw how the peasants had been impoverished by the People's Commune. The youngsters were bullied by the ignorant, greedy, but all-powerful local rural cadrés. Parents bribed the cadrés to get their children better treatment. Many children escaped and went back to the city. There they had no legal status, no income, no food ration, but lots of time. The inevitable consequence was the ruin of a generation of youngsters and a sudden rise in crime. Society was falling apart. It was in this mess that Deng became Vice Premier in 1972.

VI. The DX-3 and the Ion Microprobe

Deng started by putting everything back into its former condition. All the ministries made plans for reconstruction. So did the CAS, under its new director, Hu Yaobang, a respected cadré, who later became premier of China and still later, chairman of the Communist Party before he was dethroned. It was the commemoration of his death by students that triggered the Tianmen incident of 1989. At KYKY, Xiang was called back from the 5-7 school to make a plan in which the electron microscope once more took its proper place. I felt the relaxation of attitude toward me when Representative Wang, the leader of the military representatives, who, although he had never directly maltreated me, had hardly ever talked to me before, came to me for my opinion about SEM. Apparently, Ge Zhousheng, originally from the IOFM, had made a proposal to build one. I told Wang that already in 1956, my supervisor, Prof. Möllenstedt, after a visit to Cambridge University, predicted that SEM would be the microscope of the future. At the 1964 Prague Conference on Electron Microscopy I had heard favorable comments on SEM. In 1965, when I went with KYKY Party Secretary Tian to Japan on a fact-finding tour, we saw well-advanced prototypes of SEMs at both JEOL and Hitachi. In our report to the National Science Commission, we recommended that we should start making SEMs. There were two people from the Shanghai group in our delegation. Soon after our return, they bought a Japanese SEM with the intention of copying it. But for the Cultural Revolution, they would have finished it quickly. Some time after the talk with Representative Wang, Comrade Wu Zuoli came and asked me if I would be interested in either one of two projects, an SEM or an ion microprobe project. Wu was a cadré sent to KYKY by the CAS in 1970. He was to be the director of KYKY from 1981 to 1986. I picked the ion microprobe because I thought it might be more challenging.

During the 1960s, the CAS thought it had a better bunch of research staff than was usual in the universities. Since that time, a system has been established whereby research institutes can take on and train graduate students. Basic graduate courses were given centrally by the University of Science and Technology, which belongs to the CAS. Around 1964, two students from Beijing University, Liu Xuping and Ge Zhaoshen, decided to go into electron optics. Liu came to the Institute of Electronics to study under me, and Ge went to the IOFM to study under Yao. Both ended up at KYKY. Liu was the thoughtful, steady, and studious type, Ge the daring, active type and a good organizer. It was no wonder that Ge was the one to propose the SEM project at the right time, and became the leader of the group. However, the two worked together marvellously and made equal contributions to the project. At the beginning they acquired a good mechanical engineer, Huang Tongjie, to begin on the design of the specimen stage, while they did preliminary experiments using the illuminating system of the DX-2 TEM as the probe-forming system. The Everhart-Thornley detector and the deflector were placed in the specimen chamber and the specimen was put into the normal TEM specimen stage. With this lashup they gained enough experience and data to complete the design of the SEM DX-3. A team of 14 people was assigned to take part in the project. The electronics, the first all-transistor design for an EM in China, was carried out by Rong Denian and Wang Zhaoyou. Much time was spent in protecting the semiconducting elements against destruction by high-voltage discharges. The DX-3, the first Chinese SEM, shown in Fig. 7, was announced in 1975 with a resolution of 10 nm.

At that time, Chinese scientists had just resumed their work. They had learned from the literature about SEM, which had already become a widely used tool everywhere except in China. The demand for SEM was high, but because of a shortage of hard currency, very few were imported. Customers used to beg us to make one for them. The situation was what a Cambridge Instruments salesman once called a "salesman's paradise." The DX-3 be-

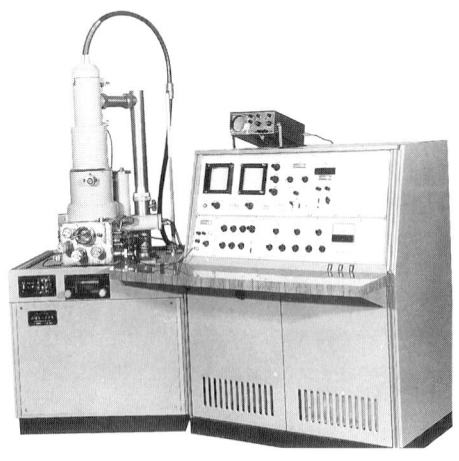

FIGURE 7. KYKY DX-3, resolution 10 nm, 30 kV (1975). The first Chinese scanning electron microscope.

came the first instrument to be produced in quantity by KYKY. Together with the DX3-A, which had a wavelength-dispersion X-ray spectrometer (WDS) and was designed by Yao Junen and Yu Jianji, a total of 65 units were sold.

While the DX-3 was being built in the EM Department, I led a group in the Mass Spectrometer Department in the design of an ion-microprobe mass analyser. We picked the Liebl instead of the Castaing-Slodzian type of analyzer because of its versatility. The design of the ion source, a cold-cathode duoplasmatron source, and the probe-forming system, with a Septier long-working-distance objective, was quite straightforward. So was the mass spectrometer, a second-order aberration-free system proposed by Matsuda, and a Dali-type detector. The secondary ion-extraction system for extracting ions from the specimen, a two-dimensional system, was designed using the classical graphical method and with the electron trajectory tracer of the Institute of Electronics. In order to match the exit conditions of the extractor to the entrance condition of the mass spectrometer, I designed a special lens, which we called the "beta lens" (Huang and Hu, 1981). It was an Einzel lens, with the central round electrode cut into four equal segments. When equal voltages are applied on each segment, it acts as a round lens. When the voltages on the two X electrodes and the two Y electrodes are equal and opposite, the lens acts essentially as a quadrupole lens. In an intermediate state, it becomes a focusing lens superimposed on a quadrupole. By properly adjusting the voltages on the electrodes, one can adjust, independently, the focusing power in the x and y directions. Similarly, a deflection field can be produced. With this compact device placed immediately behind the extractor, it was easy to focus and align the spectrometer in both the x and y directions and obtain maximum resolution and intensity.

After the design was finished and the column tested, I left it to the project group to complete the instrument, which is shown in Fig. 8. A spatial

FIGURE 8. KYKY prototype ion-microprobe mass analyzer (1978): spatial resolution 2 m at 20 kV, mass resolution 2000, sensitivity, 2 parts per billion for boron in silicon.

resolution of 2 μm at 20 keV, a mass resolution of 2000, and a sensitivity of 2 parts per billion (boron in silicon) were obtained. I was told that the beta lens performed just as expected and was an indispensible component of the design. A brief description of the ion microprobe was published in the proceedings of 1980 Oslo Mass Spectrometry Conference, under the name of the Ion Microprobe Research Group (1979). Creativity was still considered at the time to exist only in the collectives.

VII. The DX-4 and DX-5

In 1973 I went back to the EM Department and started to build a new TEM, the DX-4, the last TEM to be built at KYKY. Unlike the management team under Comrade Tian, who emphasized results, the new leaders of KYKY concentrated totally on politics. Since paying too much attention to nonpolitical work was politically risky at the time, they just let things go along at their own pace. Discipline and coordination within KYKY was poor. However, this had its good side. For the first time, I was able to sit down and think through problems. The ion microprobe and the DX-4 TEM were the two projects that I really enjoyed, at least during the design stage.

Comparing the lens data by Liebmann, Dugas, El Kareh, and Mulvey, I got together a reliable set of lens data, from which I derived a series of empirical formulas, which were easy to manipulate and to interpret (Huang, 1977; Huang and Liu, 1991). These included a simple, user-friendly formula for minimum focal length at a given field strength in the gap of both the condenser and projector lenses. The design of the magnetic circuits was also difficult in the absence of today's computers. Liu Xuping and I used model calculations for our respective DX-4 and DX-5 objectives and obtained excellent results. The chat about magnetic circuit design that I had had with Tom Mulvey during the 1964 Prague EM Conference gave us important guidance, especially with regard to the choice and treatment of soft iron and the allowable flux density within the magnetic circuit. The scheme of Kynaston and Mulvey for compensating distortion in a three-lens TEM column at low magnification was fine, but to get even lower magnification, an extra lens is needed immediately below the objective. Inspired by Le Poole's presentation at the 1964 Prague EM Conference, I designed a minilens that could be placed immediately below the objective. This solved the low-magnification problem, but it was not easy to manufacture such lenses commercially.

Misalignment problems had always troubled us in our earlier designs. My calculations showed that even a few micrometers of polepiece misalignment would cause noticeable image displacement. With an adjustable upper polepiece, I was able to remove gross misalignment, but even at its best, residual misalignment caused the image to move in a peculiar pattern when the objective current was varied over a relatively wide range. By assuming the existence of a deflecting field just below the objective polepiece, I was able to show theoretically that the image would move in the same manner observed experimentally. Nonuniform gaps between the polepiece and yoke could cause nonuniform leakage fields and hence a deflecting field. By using thin, nonmagnetic spacer cylinders between the polepiece and lens yoke, to make the leakage field more axisymmetrical, a trick I learned from JEOL, I was able to reduce the displacements by an order of magnitude, to an acceptable level.

After the design stage, further progress depended on support from other departments. This was not forthcoming in KYKY at the time. For example, we found a large stray ac field coming from a power line outside our room. Our urgent request for another room was turned down, but we were given permission to build ourselves another laboratory! Many of us were quite good bricklayers after our training in the 5–7 school and from building air-raid shelters in 1971, when Mao thought that the third world war was coming! But when the laboratory was built, the Factory Medical Service took it over! In the end we could only do high-resolution work at night. I even made a 2-m Helmholz coil to try to cancel the field, but without success. The project survived many political and physical upheavals, such as the ousting of Deng from his position as vice premier, which meant that paying too much attention to work was once more considered "bourgeois." Another factor was the death of Premier Zhou En Lai, which later touched off the first Tiananmen incident, a spontaneous mass demonstration against Madame Mao, which was brutally suppressed. To add to the turmoil, there was the great earthquake of Tangshan, reaching more than 8 on the Richter scale, in which 800,000 people (unofficial figure) were said to have perished. Beijing, some 200 km away, felt the tremor at 6.7 on the Richter scale. The ensuing panic stopped all work for weeks. Then there was the death of Mao and the ensuing fall of Madame Mao and the "Gang of Four." All these political events were accompanied by much indoctrination and "discussions" and "criticism meetings." However, the parade around Tiananmen Square after the downfall of the Gang of Four, although orchestrated, showed genuine joy among the masses.

The death of Mao helped the DX-4 in an unexpected manner. The DX-4 got into the Number One Project, the project to preserve Mao's corpse. It was an honor, and anyone who got into that project would get some priority within his own organization. Some hospital that did not have a TEM but wanted to cash in on this project dragged us into it by claiming to the authorities that they would use our DX-4 to examine tissue taken from Mao's body. For a while things went well in KYKY, but it didn't last. Many groups were involved in the prestigious Project Number One, but in fact there was no money in it for anyone! After much effort, the DX-4 attained a resolution of 0.2 nm. Figure 9 shows the 1980 version of the DX-4. It was the first microprocessor-controlled TEM in China.

In parallel with the construction of the DX-4, Liu Xuping built the second SEM, the DX-5, shown in Fig. 10. The column was more compact than that of the DX-3 and had a better finish. It had a resolution of 6 nm and all the standard features of contemporary SEMs, including split-screen, dual-magnification displays, as well as image-processing facilities. We had thus proved to ourselves that we were capable of building experimental SEMs on a par with the best commercial ones worldwide. Only later did we come to realize that we lacked the engineering knowhow to transform our DX-3, -4, and -5 into viable commercial products.

In 1983 the CAS commissioned the Institute of Electrical Engineering (IEE), KYKY, and the Institute of Electro-Optics (IE-O) to build a variable-shape beam electron lithography machine. At that time the import-

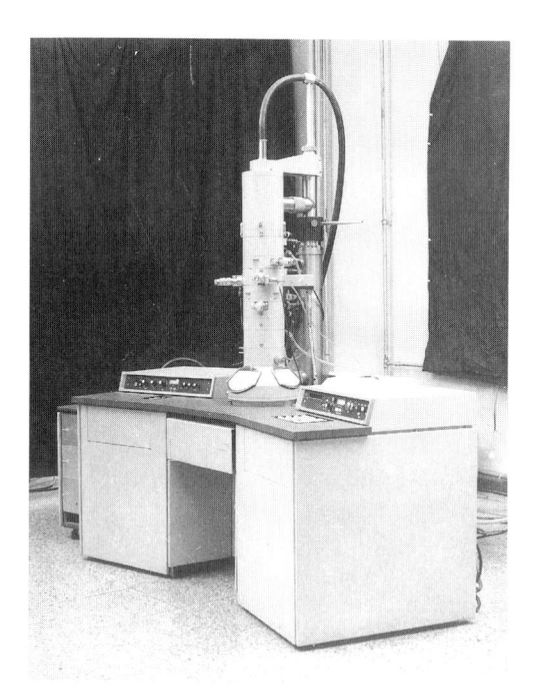

FIGURE 9. KYKY DX-4 (1980), resolution 0.2 nm, 100 kV: the first microprocessor-controlled TEM in China.

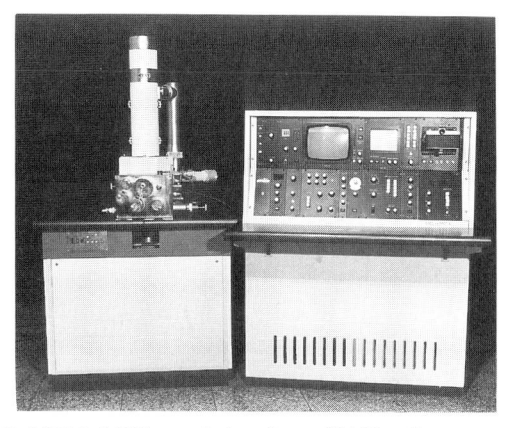

FIGURE 10. DX-5 SEM (1980): resolution 6 nm, 30 kV: split-screen, dual-magnification display, image processing.

ing of such equipment was banned by COCOM. The IEE was the coordinator and also responsible for some of the high-speed electronics and the computer control system. KYKY was responsible for the electron optical column and IE-O for the laser interferometer stage. The project was completed in 1991 with a minimum spot size of 1 μm^2 over a field of 2 mm × 2 mm. The current density in the probe was 0.4 A/cm^2, overlay and stitching accuracy was 0.1 μm, and wafer size was 110 mm × 110 mm. Of all the lithography machines built in China, this one was the best. However, in this type of equipment, it is notoriously difficult to maintain dust-free and contamination-free conditions in the vacuum chamber and to ensure the mechanical and electrical stability needed for writing a circuit that is absolutely free from fabrication errors. This machine was unfortunately not able to meet these stringent requirements for production purposes.

VIII. KYKY-AMRAY

KYKY was famous for its "firsts" and "bests," but when it came down to the finished commercial product, there were problems for users. All the DX- and WDX-series instruments had serious reliability problems. These were more serious than those arising from microscopes made by the Shanghai and Nanjing groups, whose technical staff were more familiar with proper production procedures and quality control. It took us a long time at KYKY to realize that it was our problem and not our customers' problem. The changeover from the role of the "hero" hailed by users to that of their humble servant was a painful and difficult one. By the end of the 1970s, however, sweeping reforms had started in China. The Chinese Academy of Science began to ask KYKY to rely more and more on its own sales revenue. At the same time, we were subject to Western-style business management—much earlier, in fact, than the Shanghai and Nanjing groups, which were in industry. Recognition of a problem is one thing, solving it is another. When I became the deputy director of KYKY, in charge of R&D, I made the move to emphasize R&D in engineering and quality improvement. This idea won praise all around, but when I submitted the annual R&D plan for putting the idea into practice, nobody wanted to know. The director of KYKY as well as our superior within the CAS, both Party cadrés, disliked my plan because, instead of seeing impressive projects proposed, they now had a daunting list of apparently trivial items with technical descriptions they could not understand and whose significance they were unable to appreciate. Our superior, a lady, said "Why not put everything under one heading and call it 'Improvement of reliability,'" to which she allocated only a small amount of funding. She wanted to see more "meaningful" projects. The decisive resistance, however, came from the engineers themselves. They thought that the "trivial" problems put forward by me were just dull and boring, although they were, in fact, very difficult to solve. They surmised that after tedious, time-consuming efforts on their part, the problem might be found to boil down to something very mundane! They pointed out to me that under CAS rules, the engineers were supposed to concentrate on sophisticated scientific engineering work, which alone earned them professional credit, et cetera, et cetera.

At the beginning of the 1980s, as the reform bit more deeply, the CAS cut off most of the funding to KYKY. We had to rely on sales revenue from customers, who, at that time, had no faith in the reliability of our products. Just then, however, the government was actively encouraging the importation of foreign technology. Several of us saw this as the only quick way to overcome our quality control problem. The first company I approached was ETEC, when I found out that the Chinese-American who played tennis with me had a father on ETEC's board of directors, but it did not work out. ETEC wanted $2 million for the technology. I talked to Zeiss, Philips, Cambridge Instruments, and two Japanese companies. In the end, I contacted AMRAY, a small company whose name I happened to see in an advertisement.

The president of AMRAY, Mr. Cameron, is a unique capitalist, at least not the stereotype the Communists made capitalists out to be, namely, people who are after every penny you have. Mr. Cameron made it clear that he wanted no part of the Chinese market. What went on in his mind, I cannot tell. Was it fascination for China, the mysterious land that had just opened up? Sympathy for KYKY? For me, who gave up U.S. citizenship to go back to China? Or was it a dislike for his business adversaries, the Japanese, who were dominating the Chinese market? Could it also have been the Catholic in him wanting to help the helpless? Whatever it was, it was not just business. After thinking about it for a year, he rang me back and said he would give us the technology for free. Not only that—he said he would also pay for the living expenses of our engineers during their training at AMRAY. I was often praised for my skill in these negotiations, getting all that from a "capitalist." I had a hard time convincing them that I did just the opposite. I tried to tell Mr. Cameron that he did not have to do all this. Whatever the reason, Mr. Cameron did a good deed and won the warmest feelings of gratitude from many Chinese people, including some Party officials.

During my search for an industrial partner, I had come to know a business consultant, Mr. Paterson, who helped me in making the initial contact with AMRAY. I benefited tremendously from the long talks I had with him. He is not only a business management expert, but also a philosopher, well versed in Chinese philosophy, especially in the writings of Sun Zi, the military philosopher, and their application to modern business. If my success with AMRAY and later with other companies had anything to do with my meagre understanding of American business, it was due largely to his teaching.

The AMRAY 1000B project was straightforward at the technical level, but opposition within the EM Department was overwhelming. Most of the engineers were not, in fact, involved in the project. At the time, sales revenues still depended on them. They could not see the necessity for outside technology. They still had the erroneous idea that electron optics was the primary technology in an EM. It was probably a fact that, at that time, we knew more about electron optics than they did at AMRAY. It was at this point that Jin He Ming, head of the EM Department, proved himself. (He is now director of KYKY.) He pacified the opposition and steered the 1000B project to success. In 1985, the first batch of AMRAY-KYKY 1000Bs came off the assembly line with a resolution of 6 nm and an accelerating voltage of 30 kV.

We soon won the confidence of our customers with our new product, so from a business standpoint we were successful, but all the romance had gone out of our work! The large staff of physicists who specialized in electron optics stopped waiting their turn to design a new instrument, and gradually left us. Among them were Ge Zhao Sheng and Liu Xu Ping. The electrical engineers then rose from second- to first-class citizens. New development depended primarily on them. The hero of the next generation of SEM, the KYKY 2000 (1989), shown in Fig. 11, was Wang Ke Ding (Wang et al., 1990). The model after that, the KYKY-3000 (1994), had a

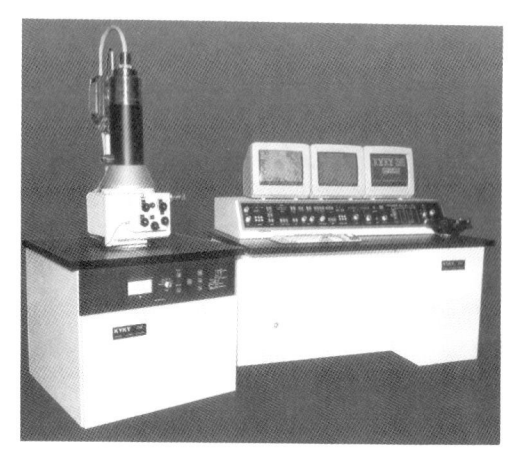

FIGURE 11. KYKY 2000 (1989): digital SEM with 512 × 512 frame grabber and image processing software.

more powerful frame grabber. It was designed by Wang's student, Zhang Yong Ming, a young computer engineer. Electron optically, we improved the resolution of the 3000 model from 6 nm to 4.5 nm by decreasing the upper bore of the objective and the thickness of the lower pole plate. This decreased the chromatic aberration of the lens significantly without appreciably increasing the spherical aberration. Classical designs of objectives paid far too much attention to spherical aberration! Over the years, a total of 200 SEMs were sold by KYKY.

IX. Electron-Probe X-Ray Microanalysis

The first proposal to build an electron X-ray microanalyzer in China was that of Wang Daheng, Director of the Institute of Optics and Fine Mechanics (IOFM), in 1958, after he saw in an exhibition in the Soviet Union, the one designed by I. Borovskii at the Baikov Institute in Moscow. This instrument was unusual inasmuch as the wavelength-dispersive crystal was fixed in position on the Rowland circle, while the heavy electron optical column was moved physically around the Rowland circle in order to select the appropriate Bragg angle for a chosen chemical element to be analyzed. The crystal itself was thin and bent to twice the radius of curvature of the Rowland circle. The X rays from the specimen were not reflected by the crystal in the conventional manner, but passed through it into the X-ray detector. Borovskii's design worked well in demonstrating the principle of electron-probe X-ray microanalysis in the laboratory but was unsuitable for commercial production. Nevertheless, work on this type of microprobe analyzer started in China in 1959 with Sun Zongyuan in charge of the project. By the end of 1959, an experimental model had been constructed and shown at an exhibition. After that, the work came to a halt until Sun Zongyuan joined the EM group and came to KYKY. I remember discussing Mulvey's article with him, when, in 1965, KYKY bought an AEI microprobe analyzer to help Sun with his experiments. In 1966, Sun was already able to analyze elements from uranium to magnesium on his own machine. Before he finished the prototype, the Cultural Revolution had begun, but he still did some work in redesigning the spectrometer. Sun, a mechanical engineer by training, was keen to design a spectrometer in which the X-ray take-off angle from the specimen would remain constant for different elements. When the Scientific Instruments Factory was taken over by the military, he obtained permission to join the Nanjing Optical Factory and continue his work. Upon his arrival at the Nanjing Factory, Sun Zongyuan finished his new and improved X-ray spectrometer and built an electron-probe X-ray microanalyzer with a resolution of 1 μm and a wavelength resolution of 1–5% for elements between sodium and uranium; the peak/background ratio was 300 at Fe with a maximum count rate of 70,000 counts/s/μA. After this, Sun left the Nanjing Factory and went to the Nanjing Observatory to work on X-ray astronomy. At the Nanjing Factory, Wang Qimei took over the work and in 1976 the new X-ray microprobe, X-D1, shown in Fig. 12, was completed. This was the first electron-probe X-ray microanalyzer produced in China. It had a spatial resolution of 200 nm, and a wavelength resolution of 0.1–0.5% (sodium to uranium) and 1–5% (carbon to fluorine). Maximum intensity was 200,000 counts/s/μA.

FIGURE 12. The X-01 (1976), the first electron-probe X-ray microanalyzer produced commercially in China: probe size 200 nm, accelerating voltage 50 kV, 4-channel, fully focusing X-ray spectrometer (Nanjing Factory).

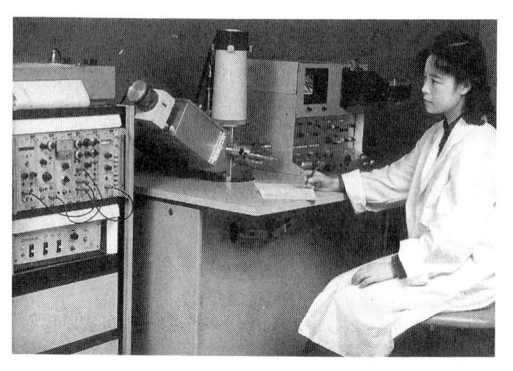

FIGURE 13. Mini-SEM DXS-X2 (1983): 15 nm, 25 kV (Nanjing Factory).

The peak/background ratio was around 300, except for the element carbon, where it was only 20. Four fully focusing spectrometers were incorporated. Two of them had Rowland circles of 140-mm radius; the other two had radii of 250 mm. In 1983, Wang designed a small-wavelength dispersive spectrometer to go with his digital mini-SEM DXS-X2. This is shown in Fig. 13, on the left of the column. The minimum probe size was 15 nm and the accelerating voltage was 25 kV. In 1989, these ideas were developed further. Figure 14 shows the DXS-3 SEM, with a resolution of 6 nm at 35 kV, with a wavelength-dispersive spectrometer as in the DXS-2. This combination is still being sold to Chinese customers who cannot afford to buy energy-dispersive X-ray (EDX) systems or who do not want to be bothered with the inconvenient liquid-nitrogen cooling systems required by EDX detectors. Wang's design is the only Chinese wavelength-dispersive system (WDS) that is still being sold today.

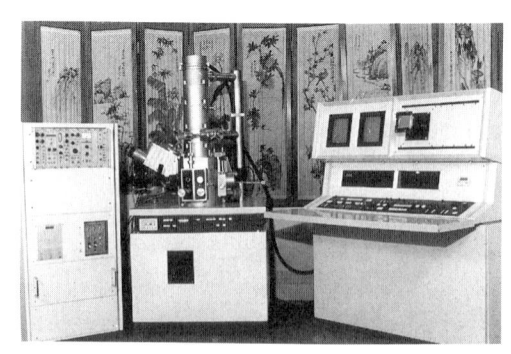

FIGURE 14. DXS-3 SEM (1989): resolution 6 nm, 35 kV, with mini wavelength-dispersive spectrometer (WDS) (Nanjing Factory).

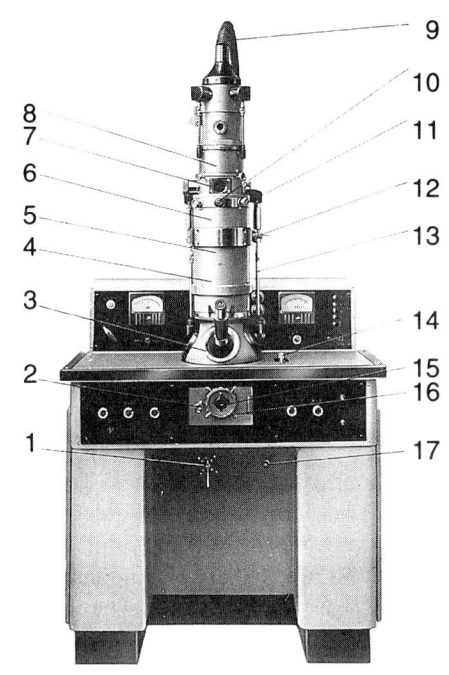

COMMERCIAL ELECTRON MICROSCOPE CONSTRUCTION 837

FIGURE 15. XD-301 TEM (1961): resolution 10 nm, 50 kV (Nanjing Factory). Key: 1, camera valve; 2, camera door latch; 3, viewing screen; 4, projector; 6, objective; 7, specimen chamber; 8, condenser; 9, EHT cable; 10, objective aperture; 11, stigmator; 12, selected area aperture; 13, specimen stage movement; 14, camera operating button; 15, camera isolation valve; 16, film wind-on control; 17, air inlet valve.

At KYKY, Yao Junen and Yu Jianye copied the wavelength-dispersive (WDS) system of the Japanese U3 in 1977; this could readily be fitted to the DX-3, converting it into the DX-3A. Yu Jianye and other engineers in the group had to solve many technical problems before achieving success; in the end they achieved a specification close to that of the Japanese model. Quite a few were sold in the early days in China, when energy-dispersive X-ray (EDX) analysis was not known in China. However, the demand for it dropped because of maintenance problems and the growing popularity of EDX.

X. TEMs FROM THE NANJING FACTORY

The Nanjing Factory started their TEM activity in 1958. It is interesting to see how their designs developed over the years. Figure 15 shows their earliest (1960) model, a copy of the IOFM's 1958 TEM, which they named XD-301,

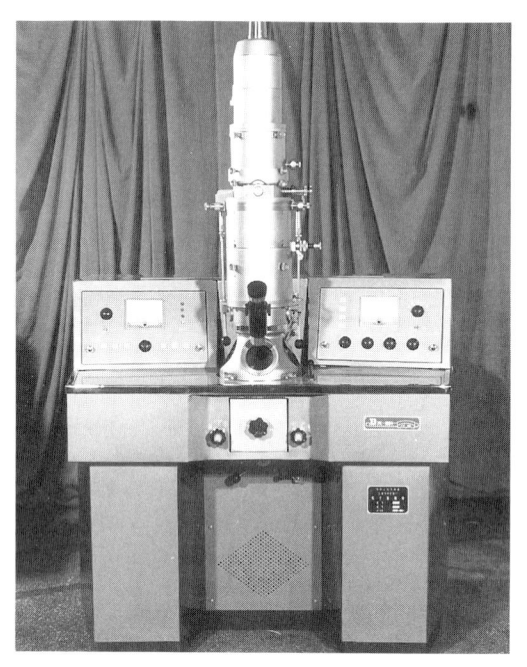

FIGURE 16. XD-201 TEM (1967): resolution 2 nm, 80 kV (Nanjing Factory).

with a resolution of 10 nm and an acceleration voltage of 50 kV. It had a single condenser lens, an objective, and intermediate and final projector lens. Three viewing windows were provided, and a simple telescope was fitted for viewing the final screen to assist in focusing. This was followed in 1967 by the XD-201, shown in Fig. 16, an 80-kV TEM with an improved resolution of 2 nm. The column is more robust, but the positioning of the lens and other controls are not very user-friendly. By 1976, however, the design of the new DXT-10 (Fig. 17) showed the increasing expertise of the Nanjing designers and the influence of international exhibitions of electron microscopes, such as those of the IFSEM, which set a high standard for all manufacturers. The accelerating voltage was now 100 kV, an essential requirement for metallurgical electron microscopy. The resolution was now improved to 1 nm, a useful value for imaging dislocations in metals, for example. In addition, the more comfortable binocular viewing of the fluorescent screen was provided as standard. Other activities can be seen in Appendix I.

XI. THE SHANGHAI FACTORY

The Shanghai Factory's first production model, the DXA2-8, a copy of the IOFM's 1959 TEM, came out in 1964. Their most successful and best-selling

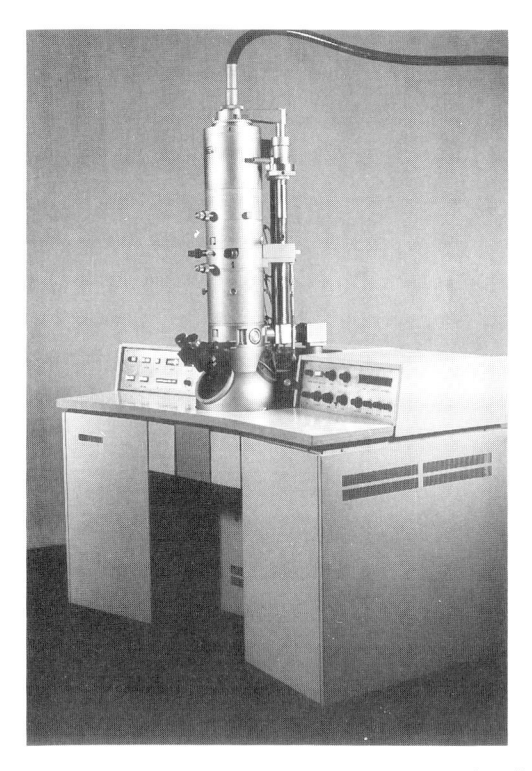

FIGURE 17. DXT-10 TEM (1976): resolution 1 nm, 100 kV (Nanjing Factory).

models were the (1968) DXA4-10 TEM (63 produced) and the (1977) DXB2-12 (23 produced).

Figure 18 shows three DXA4-10s on test. The resolution was 0.7 nm at 100 kV. The DXA4-10 was well engineered, but it relied heavily on mechanical adjustment of the column by the operator; the Siemens influence is still apparent. The DXB2-12 (1977), shown in Fig. 19, obviously broke new ground. Operating at 120 kV and with a resolution of 0.2 nm, this was an ideal specification for advanced electron microscopy in materials science. The column has cleaner lines, with preset mechanical adjustments and electronic beam alignment. A single viewing window is provided, helping to reduce external magnetic fields compared with the previous three-window arrangement. Binocular viewing of the fluorescent screen was also a great improvement in operating convenience. Finally, the DXT-5 TEM (1991), shown in Fig. 20, should be mentioned. This was designed as a teaching microscope. It operated at 50 kV, with a resolving power of 5 nm. In 1992, the DXS-10A mini-SEM, shown in Fig. 21, was produced. This was a user-friendly SEM, intended for routine work. It had a resolution of 10 nm and several automatic features to make operation easy.

FIGURE 18. Three DXA4-10 TEMs on test (1968): resolution 0.7 nm, 100 kV (Shanghai Factory).

XII. THE FUTURE

China, therefore, now has the technical expertise to produce the whole range of electron microscopes currently needed. The population of China now stands at around 12 billion, with a multitude of universities and factories. In spite of these facts, China at present appears to have only a small market for electron microscopes. Nevertheless, there are, in fact, three manufacturers, a consequence of the "planned economy," which meant in practice that nearly everything was decided by a whim on the part of some influential bureaucrat. KYKY was the brainchild of Comrade Tian and his protector, CAS Vice President Pei; the Shanghai group arose from a well-meaning gesture on the part of the mayor of Shanghai, who thought, quite rightly of course, that electron microscopes were "good things to have around the place." Furthermore, the Minister at the Ministry of Machinery would never give up making EMs on the grounds that there were already two other manufacturers in this restricted field. Making electron microscopes was, after all, their legitimate domain. With the recent establishment

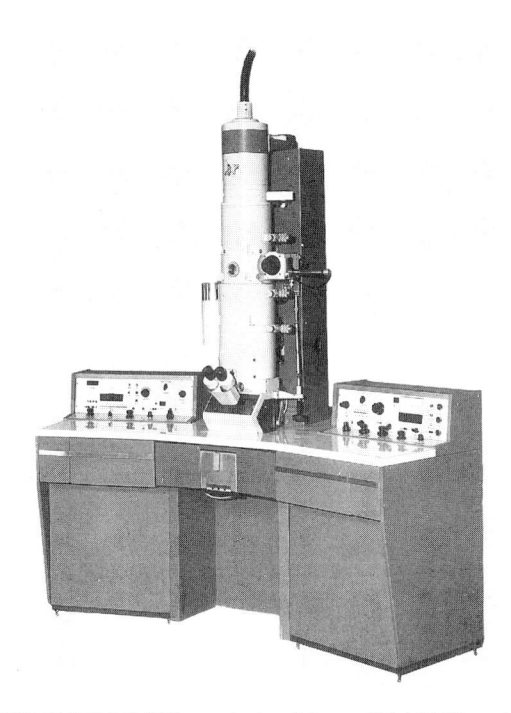

FIGURE 19. DXB2-12 TEM (1977): resolution 0.2 nm, 120 kV (Shanghai Factory).

of a "market economy," only one, or perhaps even no manufacturer of EMs is needed in China. Already, subsidies have been cut for KYKY and the Nanjing Factory. Only the Shanghai group is still (partially) subsidized by the Local Authority of Shanghai. KYKY's average sales over the years have been around 20 units a year. Sales by Nanjing and Shanghai are probably somewhat less. If subsidy ends, the EM business could end with it.

KYKY can probably survive with sales of 15 annually. In that case, the major problem would be that of attracting gifted young recruits. Given a talented workforce, we can develop not only better EMs, but also other profitable commodities. Our present key people are not getting any younger, so we urgently need new young recruits. KYKY, being a state-owned organization, is handicapped by many government restrictions, wage restriction being one of them. Today, the most gifted tend to go abroad. This has applied, for example, to all my own graduate students. The less talented tend to turn to foreign companies or private companies, where they can get many times the salaries KYKY are allowed to offer. Another basic problem is that of the market itself. The domestic market for EMs should lie in the basic industries, both heavy and light. However, the outlook for these is not good at present. The growth sector of the Chinese economy

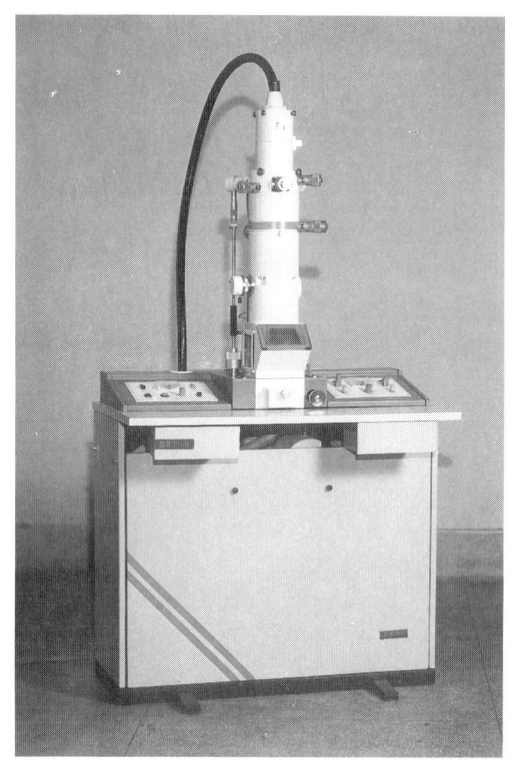

FIGURE 20. DXT-5 TEM (1991): a teaching microscope, resolution 5 nm, 50 kV (Shanghai Factory).

FIGURE 21. DXS-10A (1992): resolution 10 nm, 30 kV; some automatic features (Shanghai Factory).

is largely in the consumer market, rather than in the basic industries, which are almost totally state-owned and largely unprofitable. There is, of course, no quick solution to the problem, and it will take time for China to make a satisfactory transition to the new economy. There is no immediate growth in sight for the EM market. It could be that the manufacture of EMs in China, as in some other countries, was a transient phenomenon, although for many of us, it has lasted a whole professional lifetime. Looking back on my career, I can honestly say that it was great fun while it lasted!

APPENDIX I: SUMMARY OF EM MANUFACTURING ACTIVITIES IN CHINA

The Chinese Academy of Science

Institute of Optics and Fine Mechanics, Changchun

Year	EM production	Units made
1958	First TEM in China, 50 kV, 10 nm	1
1959	First Chinese-designed TEM	1
1959	Experimental X-ray microprobe	1

Beijing Scientific Instruments Factory, KYKY

Name Changes

    1958   Scientific Instrument Factory
    1982   KYKY Scientific Instrument Factory
    1991   KYKY Scientific Instrument R&D Center

Year	EM production	Units made
1962	Electron trajectory tracer	9
1964	Emission EM FD-1	1
1965	TEM DX-2, 0.5 nm, 100 kV	6
1966	X-ray microprobe, unfinished, moved to Nanjing	
1966	Electron-beam lithography machine, unfinished, moved to Changsha	
1966	Cultural Revolution	
1971	Ion-beam circuit lithograph; stopped	
1973	Ion-beam implantation machine, 60 kV, 10 μA	1
1974	E-beam reticle maker	1
1975	SEM DX-3, 10 nm, 30 kV (1st SEM in China)	65

(*continues*)

APPENDIX I (*Continued*)

Year	EM production	Units made
1977	WDX with Rowland circle 140 mm	1
1980	TEM DX-4, 0.2 nm, 100 kV	5
	SEM DX-5, 6 nm, 30 kV	8
	FEG SEM with Auger and ESCA; development stopped	
1985	AMRAY-KYKY 1000B SEM, 6 nm, 30 kV	131
1987	KYKY 1000G, SEM, 6 nm, 30 kV; domestic electronics	2
1989	KYKY 2000 digital SEM, 512 × 512 frame grabber	9
1990	Electron optical column of EB variable-shape lithography machine, 1-μm line width; with China Institute of Electrical Engineering	1
1993	KYKY 3000, prototype digital SEM, 4.5 nm, 1024 × 1024 frame grabber	1

Key People (Family Name First)

Huang Lanyou	Wang Keding	Wang Hongyi
Yao Junen	Wang Zhaoyou	Huang Tongjie
Ge Zhaosheng	Ma Xiuzhen	Qian Tianyu
Liu Xuping	Rong Denian	Fan Sirong
Zen Zhaowei	Liu Zenfu	Zhu Nentong

The Nanjing Group

Name Changes

    1958–1965   Nanjing Factory of Educational Instruments
    1965–   Nanjing Jiangnan Optical Factory

Year	EM production	Units made
1958	Received training and drawings from the IOFM of the Chinese Academy of Sciences	—
1961	TEM XD-301, 10 nm, 50 kV	10
1964	TEM XD-302, 4 nm, 50 kV	38
1967	TEM XD-201, 2 nm, 80 kV	72
1966	Cultural Revolution	
1972	First X-ray microprobe in China (Na to U)	1
1976	X-01 microprobe, spatial revolution 200 nm, 50 kV, 4 channels	3
1976	TEM DX-10, 1 nm, 100 kV	42
1979	Mini-SEM DXS-1, 30 nm, 20 kV	7
1983	Mini-SEM DXS-X2, 15 nm, 25 kV, mini-WDX	52
1984	TEM DXT-10C, 60° goniometer	6
1989	SEM DXS-3, 6 nm, 35 kV, mini-WDX, digital SEM	2

(*continues*)

APPENDIX I (*Continued*)

Year	EM production	Units made
1993	TEM H-600A, 0.2 nm, 100 kV, analytical TEM, licensed from Hitachi	32
1993	Low-cost TEM DXT-100G, 1 nm, 100 kV	3

Key People (Family Name First)

Wei Jinjun	Chen Xuiyian	Yuan Liying
Wang Qimei	Guan Guoshu	Gao Chenxin
Jiang Jiaying	Zhu wenxin	

The Shanghai EM Group

Name Changes

1958–1961	Shanghai Precision Medical Equipment Factory
1961–1964	Shanghai Instrument Factory
1964–1975	Shanghai Electron Optical Institute
1975–1986	Xinyao Instrument Factory
1986–	Shanghai Electron Optical Institute

Date	EM production	Units made
1959	Jointly with the IOFM, built the first 100 kV TEM DXA-1 in China	1
1961	Electron diffraction camera, cold cathode; copy of Trüb Täuber model	1
1963	Trial production of the TEM DXA-1, 5 nm, 100 kV	3
1964	TEM DXA2-8, 2 nm, 80 kV	2
1965	TEM DXA3-8, 0.7 nm, 80 kV	5
1968	TEM DXA4-10, 0.7 nm, 100 kV	63
1973	TEM DXB1, 0.5 nm, 100 kV	2
1977	TEM DXB2-12, 0.2 nm, 120 kV	23
1986	TEM DXC1, 0.7 nm, 75 kV	2
1989	TEM DXT-5, 5 nm, teaching TEM for universities	2
1990	TEM DXT-200, 0.2 nm, 200 kV (200-CX, licensed from JEOL)	1
1991	TEM DXT-5A, 3 nm, 50 kV, cost $6000 only	5
1991	TEM DXT-100, 0.45 nm, 100 kV	2

Scanning Electron Microscopes

Date	EM production	Units made
1977	SMDX-1P, 20 nm, 50 kV	2
1978	WDX	1

(*continues*)

Appendix I (*Continued*)

Year	EM production	Units made
1978	TSM-1, 30 nm, 17 kV, desk-top	74
1980	DXS-1P, 15 nm, 20 kV	2
1981	TSM-2, 20 nm, 20 kV, desk-top	160
1985	DXS-10, 10 nm, 30 kV, mini-SEM	22
1992	DXS-10A, 10 nm, 30 kV, some auto-facilities	5
1993	DXS-4, 6 nm, 30 kV (Hitachi S2500)	1

Key People (Family Name First)

Zhu Zhufu	Wang Huifang	Hu Jinshen
Zhan Jiaxiong	Hong Peitai	
Xu Rongtang	Shen Jinde	

APPENDIX II: CHINESE ELECTRON OPTICIANS

The Chinese manufacturers of electron microscopes, for social, material, or political reasons, were not able to concentrate on originality. However, Chinese electron opticians, mostly theorists, who could pick their own subjects and needed only paper and pencil to do creative work of a high standard, worked in various institutes and universities. A representative, but not exhaustive, list, together with workplace and speciality, is given below. The family name is listed first.

Ximen, Jiye. Beijing University. Doctorate in China under Prof. O. I. Seman (Soviet Union), who was in China from 1956 to 1958. High-order aberration theory, combined field systems, multipole systems, wide-beam systems, ion optics.

Zhou, Liwei. Beijing Institute of Technology. Doctorate 1966, Liningrad Institute of Electrical Engineering, USSR. Broad-beam electron optics, image tubes.

Tong, Linshu. Southeast University (Nanjing). TV monitors, CRT, etc., multipole systems, aberration correction.

Tang, Tiantong. Xiang Transportation University. Multipole systems, e-beam interactions.

Ding, Shouqian. Nankai University. Third- and fifth-order deflection aberrations.

Chen, Ergang. Yunnan University. Emitters, aberration correction.

Sun, Borao. Qinghua University. FIB, CRT, CAD.

Li, Yu. Shanghai Academy of Mechanical Engineering. Combined focusing-deflection systems, group theory applied to electron optics.
Gu, Changxin. Fudan University. System optimization.
Li, Shengpei. Chinese Academy of Sciences Institute of Electronics. High-density electron beams.

Acknowledgments

The author wishes to thank KYKY, and the Jiangnan and Xinyao factories, for supplying photographs of the electron microscopes shown in this article and for permission to publish them. I also wish to thank the editor, Tom Mulvey, for his encouragement in writing this personal account and his help in its editing.

References

Gong Zhutong (1959). The first electron microscope made in China. *Ke yue tongbao* (*Science News*) (in Chinese).
Huang, L. Y. (1977). Formulas for the optical properties of magnetic electron lenses (in Chinese). *Wuli Xuebao* (*Acta Phys.*) **26,** 256.
Huang, L. Y., and Hu, Z. H. (1981). Int. A lens with independent properties in the x and y direction. *J. Mass Spectrom. Ion Phys.* **37,** 309.
Huang, L. Y., and Liu, X. P. (1991). In "Dianzi xianweijing he dianzi guangxue," ("Electron Microscope and Electron Optics), chap. 3 (in Chinese). Science Publisher, Beijing.
Ion Microprobe Research Group (1979). *Proc. 8th. Int. Conf. Mass Spectrometry,* Oslo, p. 1667 (reported by Ji Tongding).
Short, S. (1992). A stamp collector's view of electron microscopy, *EMSA Bull.* **20,** 132.
Wang, K. D., Ma, Y., Zhang, Y. M., Li, M. Q., Qian, T. Y., and Guo, L. Y. (1990). A digital SEM (in Chinese). *Chinese J. Electron Microscopy* **9,** 291.
Yao Junen (1965). The design of a 100kV TEM. *Ke yue yiqi* (*Scientific Instruments*) 1 suppl. 1) (in Chinese).

APPENDIX
Conference Proceedings and Conference Abstracts

Compiled by Peter W. Hawkes

The following material began as a list of all the International Congresses and European Regional and Asia-Pacific Regional Conferences on Electron Microscopy, with full publication details of their proceedings. This was felt to be valuable, both as a general reference and to permit these proceedings to be cited in the preceding chapters in condensed form. Several other series of published conference proceedings or abstracts were also included. It then occurred to me that much more information about national meetings and small group meetings is in fact available in print than is commonly realized, and that a fuller list than I originally intended might well be a useful tool for future students of our subject. The list now includes conferences on charged particle optics, commonly known as CPO; the intermittent Balkan congresses on electron microscopy; the British EMAG and earlier Institute of Physics meetings (EMG), for which bound proceedings have been issued since 1971; the (Electron) Microscopy Society of America meetings, for which bound proceedings volumes have been issued since 1967; the biennial meetings of the Italian Society of Electron Microscopy (Società Italiana di Microscopia Elettronica, SIME), available in book form since 1987; the meetings of the German Society (Deutsche Gesellschaft für Elektronenmikroskopie, DGEM) and the French Society (Société Française de Microscopie Electronique, SFME, now Société Française des Microscopies, SFμ); the annual meetings of the Microscopical Society of Canada/ Société de Microscopie du Canada, for which Proceedings/Résumés des Communications have been published since the first meeting in 1974; the annual meetings of the Electron Microscopy Society of Southern Africa/ Elektronmikroskopievereniging van Suidelike Afrika, for which Proceedings/Verrigtinge have been published since 1971; the series of All-Union meetings held in the former Soviet Union since 1950, the proceedings of all of which are published in the physics series of *Izvestiya Akademii Nauk SSSR/Bull. Acad. Sci USSR* (now without SSSR/USSR); and the annual meetings of the Japanese Society of Electron Microscopy, the abstracts of which began to appear in *Journal of Electron Microscopy* in 1963. I am well aware that there are many other candidates for inclusion and shall perhaps prepare a more complete list one day, but I hope that even this imperfect version will save future students some effort. Meetings abstracts of a number of other societies used to be published in *Ultramicroscopy;* these are easily found and are usually not included here. The abstracts of the annual meetings of the Irish Society form a supplement to the *Proceedings of the Royal Microscopical Society,* and they too are not listed. For Chinese meetings, see the *Journal of the Chinese Electron Microscope Society,* and for Latin American congresses, see *Revista de Microscopía Electrónica,* which changed its title in 1983 to *Microscopía Electrónica y Biología Celular.* Such details as I have of forthcoming congresses are also given, the dates of these should not be regarded as immutable.

The information in this list has come from many sources, and I wish to thank most sincerely all who have sent me details or provided proceedings volumes; Prof. Hatsujiro Hashimoto and Prof. Koichi Kanaya have been particularly generous with their time, and Mr. Joe Britton, Dr. Walter Hert, Mr. Clive Jones, Mlle. Dorothée Lotthé, Mrs. Celia Snyman, Prof. E. Sukedai, Dr. Bernd Tesche, Prof. Ugo Valdrè, Dr. Michael Witcomb, and Prof. Keiji Yada have likewise been very helpful. As so often in the past, I am grateful to successive librarians of the Cambridge Scientific Periodicals Library for maintaining so excellent a collection, and on this occasion I must add my appreciation of the holdings of the London Patent Office Library, notably in Micawber Street, where almost any title may "turn up."

Conference Publications

1942

EMSA 1 [First National Conference on the Electron Microscope]: Hotel Sherman, Chicago, IL, 27–28 November 1942; transcript of proceedings reproduced in "EMSA and Its People, the First Fifty Years," by S. P. Newberry (Electron Microscopy Society of America, 1992)

1943

[EMG] 1943: Royal Society, London, 29 November 1943

1944

EMSA 2: La Salle Hotel, Chicago, IL, 16–18 November 1944; *J. Appl. Phys.* **16** (1945), 263–266
[EMG] 1944: National Physical Laboratory, June 1944

1945

EMSA 3: Frick Chemical Laboratory, Princeton University, Princeton, NJ, 30 November–1 December 1945; *J. Appl. Phys.* **17** (1946), 66–68
[EMG] 1945: National Institute of Medical Research, Hampstead, June 1945

1946

EMSA 4: Mellon Institute, Pittsburgh, PA, 5–7 December 1946; *J. Appl. Phys.* **18** (1947), 269–273
[EMG] 1946/1: Midland Hotel, Manchester, 16–17 January 1946
EMG 1946/2: Clarendon Laboratory, Oxford, 17–18 September 1946; *J. Sci. Instrum.* **24** (1947), 113–119 (V. E. Cosslett)

1947

EMSA 5: Franklin Institute, Philadelphia, PA, 11-13 December 1947; *J. Appl. Phys.* **19** (1948), 118-126

EMG 1947/1: British Medical Association, Tavistock Square, London, 20-21 March 1947; *J. Sci. Instrum.* **25** (1948), 23-27 (J. Sayer, E. F. Brown, and J. B. Todd)

EMG 1947/2: Physics Department, University of Leeds, 16-17 September 1947; *J. Sci. Instrum.* **25** (1948), 167-170 (V. E. Cosslett)

1948

EMSA 6: E. F. Burton Memorial Meeting, University of Toronto, 9-11 September 1948; *J. Appl. Phys.* **19** (1948), 1186-1192

EMG 1948/1: Anatomy Theatre and Physics Department, King's College London, 7-8 April 1948; *J. Sci. Instrum.* **25** (1948), 328-331 (V. E. Cosslett)

EMG 1948/2: Cavendish Laboratory, Cambridge, 20-23 September 1948; *J. Sci. Instrum.* **26** (1949), 163-169, and *Nature* **163** (1949), 32-34 (V. E. Cosslett)

1949

Delft, 1949: "Proceedings of the Conference on Electron Microscopy," Delft, 4-8 July 1949 (Houwink, A. L., Le Poole, J. B., and Le Rütte, W. A., eds.; Hoogland, Delft, 1950)

EMSA 7: National Bureau of Standards, Washington, DC, 6-8 October 1949; *J. Appl. Phys.* **21** (1950), 66-72

EMG 1949/1: The Investigation of Biological Systems by the Electron Microscope and by X-Ray Analysis, Buxton, May 1949; *Br. J. Appl. Phys.* **1** (1950), 57-59 (I. M. Dawson and M. F. Perutz)

EMG 1949/2: Metallurgical Applications of the Electron Microscope, Royal Institution, London, 16 November 1949 (organized by the Institute of Metals and seven other Learned Societies); *IoM Monograph and Report Series,* no. 8 (Institute of Metals, London, 1950); *Nature* **165** (1950), 390-393 (C. J. B. Clews)

DGEM 1: Mosbach, 23-24 April 1949; *Optik* **5** (1949), 457-575, and *Phys. Blätt.* **5** (1949), 237 (E. Brüche)

JSEM 1: Engineering Department, University of Tokyo, 13 May 1949 [no Proceedings]

JSEM 2: School of Medicine, Kyoto University, 5 October 1949 [Preprints distributed to participants at these and subsequent JSEM meetings]

1950

Paris, 1950: "Comptes Rendus du Premier Congrès International de Microscopie Electronique," Paris, 14-22 September 1950 (Editions de la Revue d'Optique Théorique et Instrumentale, Paris, 1953), 2 vols.

EMSA 8: Statler Hotel, Detroit, MI, 14-16 September 1950; *J. Appl. Phys.* **22** (1951), 110-116
EMG 1950: Joint Convention on Modern Microscopy (with RMS), Newcastle, 18-21 April 1950
DGEM 2: Bad Soden, 14-16 April 1950; *Optik* **7** (1950), 185-335, and *Phys. Blätt.* **6** (1950), 327 (B. von Borries)
JSEM 3: Engineering Department, University of Tokyo, 23 March 1950
JSEM 4: School of Medicine, Osaka University, 5 November 1950
Moscow 1950: "Proceedings of the 1st Soviet All-Union Conference on Electron Microscopy," Moscow, 15-19 December 1950; *Izv. Akad. Nauk SSSR (Ser. Fiz.)* **15** (1951), nos. 3 and 4 (no English translation)

1951

Washington DC, 1951: "Electron Physics. Proceedings of the NBS Semicentennial Symposium on Electron Physics," Washington, DC, 5-7 November, 1951; issued as *National Bureau of Standards Circular* **527** (1954)
EMSA 9: Franklin Institute, Philadelphia, PA, 8-10 October 1951; *J. Appl. Phys.* **23** (1952), 156-164
EMG 1951: St. Andrews, 19-20 June 1951; *Br. J. Appl. Phys.* **3** (1952), 25-29, and *Nature* **168** (1951), 819-821 (D. G. Drummond and G. Liebmann)
DGEM 3: Hamburg, 18-20 May 1951; *Phys. Verhandl.* **2** (1951), 63-92
JSEM 5: School of Medicine, University of Tokyo, 8 July 1951
JSEM 6: Engineering Department, Kyoto University, 8-9 December 1951

1952

EMSA 10: Hotel Statler, Cleveland, OH, 6-8 November 1952; *J. Appl. Phys.* **24** (1953), 111-118
EMG 1952: H. H. Wills Physical Laboratory, University of Bristol, 16-18 September 1952; *Br. J. Appl. Phys.* **4** (1953), 1-5 (V. E. Cosslett, J. B. Nutting, and R. Reed); *Nature* **170** (1952), 861-863 (V. E. Cosslett)
DGEM 4: Tübingen, 6-9 June 1952; *Optik* **9** (1952), 189-191, and **10** (1953), 1-205; *Phys. Verhandl.* **3** (1952), 97-124
JSEM 7: School of Medicine, Keio University, 24-25 May 1952
JSEM 8: School of Medicine, University of Nagoya, 16-17 October 1952

1953

EMSA 11: Pocono Manor Inn, Pocono Manor, PA, 5-7 November 1953; *J. Appl. Phys.* **24** (1953), 1414-1426
EMG 1953/1: Recent Research in Electron Optics, Imperial College, London, 15-16 May 1953; *Nature* **172** (1953), 61-62 (O. Klemperer)
EMG 1953/2: Birkbeck College, London, 10-11 November 1953; *Br. J. Appl. Phys.* **5** (1954), 165-170, and *Nature* **173** (1954), 340-341 (C. E. Challice)

DGEM 5: Innsbruck, 16–19 September 1953; *Optik* **11** (1954), 97ff, and *Phys. Verhandl.* **4** (1953), 83–124
JSEM 9: Engineering High School, Institute of Tokyo, 23–24 May 1953

1954

London, 1954: "The Proceedings of the Third International Conference on Electron Microscopy," London, 15–21 July 1954 (Ross, R., ed.; Royal Microscopical Society, London, 1956)
Gent, 1954: "Rapport Europees Congrès Toegepaste Electronenmicroscopie," Gent, 7–10 April 1954; edited and published by G. Vandermeersche (Uccle-Bruxelles, 1954)
EMSA 12: Moraine-on-the-Lake Hotel, Highland Park, IL, 14–16 October 1954; *J. Appl. Phys.* **25** (1954), 1453–1468
JSEM 10: School of Medicine, Jikei University, Tokyo, 29–30 April 1954

1955

EMSA 13: Pennsylvania State University, University Park, PA, 27–29 October 1955; *J. Appl. Phys.* **26** (1955), 1391–1398
EMG 1955: Department of Chemistry, University of Glasgow, 5–7 July 1955; *Br. J. Appl. Phys.* **7** (1956), 89–93 (C. E. Challice)
DGEM 6: Münster, 28–31 March 1955; *Phys. Verhandl.* **6** (1955), 9–42
JSEM 11: Yokohama Medical University, 7–8 May 1955
Toulouse, 1955: "Les Techniques Récentes en Microscopie Electronique et Corpusculaire," Toulouse, 4–8 April 1955 (CNRS, Paris, 1956)

1956

Stockholm, 1956: "Electron Microscopy. Proceedings of the Stockholm Conference," 17–20 September 1956 (Sjöstrand, F. J., and Rhodin, J., eds.; Almqvist and Wiksells, Stockholm, 1957)
Tokyo, 1956: "Electron Microscopy. Proceedings of the First Regional Conference in Asia and Oceania," Tokyo, 1956 (Electrotechnical Laboratory, Tokyo, 1957)
EMSA 14: University of Wisconsin, Madison, WI, 10–12 September 1956; *J. Appl. Phys.* **27** (1956), 1389–1398
EMG 1956/1: Electron Microscopy of Fibres, Department of Textile Industries, University of Leeds, 3–4 January 1956; *Br. J. Appl. Phys.* **8** (1957), 1–8, and erratum, 218 (C. E. Challice and J. Sikorski)
EMG 1956/2: Departments of Chemistry and Physics, University of Reading, 24–26 July 1956; *Br. J. Appl. Phys.* **8** (1957), 259–269 (C. E. Challice)
JSEM 12: Kyotofuritsu School of Medicine, Kyoto, 13–14 June 1956

1957

EMSA 15: Massachusetts Institute of Technology, Cambridge, MA, 9–11 September 1957; *J. Appl. Phys.* **28** (1957), 1368–1386

EMG 1957: Department of Botany, University College of North Wales, Bangor, 10–12 September 1957; *Br. J. Appl. Phys.* **9** (1958), 306–312 (H. W. Emerton)
DGEM 7: Darmstadt, 23–25 September 1957; *Phys. Verhandl.* **8** (1957), 211–240
JSEM 13: Jikei Medical University, Tokyo, 14–15 May 1957
SIME, 1957: Rome, 7 May 1957

1958

Berlin, 1958: "Vierter Internationaler Kongress für Elektronenmikroskopie," Berlin, 10–17 September 1958, "Verhandlungen" (Bargmann, W., Möllenstedt, G., Niehrs, H., Peters, D., Ruska, E., and Wolpers, C., eds.; Springer, Berlin, 1960), 2 vols.
EMSA 16: Santa Monica Civic Auditorium, Santa Monica, CA, 7–9 August 1958; *J. Appl. Phys.* **29** (1958), 1615–1626
EMG 1958: On Precipitation in Alloys and on Metal Structures (with X-Ray Analysis Group), London, 28 November 1958; *Br. J. Appl. Phys.* **10** (1959), 438–444 (A. Franks and R. S. M. Revell)
JSEM 14: Kyoto University Medical Department, 24–26 June 1958
Moscow 1958: "Proceedings of the 2nd Soviet All-Union Conference on Electron Microscopy," Moscow 9–13 May 1958; *Izv. Akad. Nauk SSSR (Ser. Fiz.)* or *Bull. Acad. Sci. USSR (Phys. Ser.)* **23** (1959), nos. 4 and 6

1959

EMSA 17: Ohio State University, Columbus, OH, 9–12 September 1959; *J. Appl. Phys.* **30** (1959), 2024–2042
EMG 1959: Washington Singer Laboratory, University of Exeter, 7–10 July 1959; *Br. J. Appl. Phys.* **11** (1960), 22–32 (J. A. Chapman and M. J. Whelan)
DGEM 9: Freiburg i. Br., 18–21 October 1959; *Phys. Verhandl.* **10** (1959), 189–207
JSEM 15: Faculty of General Education, Tohoku University, Sendai, 13–15 May 1959
SIME, 1959: Atti del II Congresso Italiano di Microscopia Electronica, Fondazione Carlo Erba, Palazzo Visconti, Milan, May 1959 (Società Italiana di Microscopia Elettronica)
SFME 1959: Paris, 18–19 December 1959

1960

Delft, 1960: "The Proceedings of the European Regional Conference on Electron Microscopy," Delft, 29 August–3 September 1960 (Houwink, A. L., and Spit, B. J., eds.; Nederlandse Vereniging voor Elektronenmicroscopie, Delft, n.d.), 2 vols.
EMSA 18: Marquette University, Milwaukee, WI, 29–31 August 1960; *J. Appl. Phys.* **31** (1960), 1831–1848
EMG 1960: Co-ordination of Light and Electron Microscopy (with RMS), Leeds, 31 March–1 April 1960; *J. R. Microsc. Soc.* **79** (1959–1960), part 3, 179–274; *Nature* **186** (1960), 672–673 (V. E. Cosslett)

JSEM 16: Department of Agriculture, Tokyo University, 19–20 May 1960
Leningrad 1960: "Proceedings of the 3rd Soviet All-Union Conference on Electron Microscopy," Leningrad, 24–29 October 1960; *Izv. Akad. Nauk SSSR (Ser. Fiz.)* or *Bull. Acad. Sci. USSR (Phys. Ser.)* **25** (1961), no. 6

1961

EMSA 19: Pittsburgh Hilton Hotel, Pittsburgh, PA, 23–26 August 1961; *J. Appl. Phys.* **32** (1961), 1626–1646
EMG 1961: Department of Chemistry, University of Nottingham, 10–14 July; *Br. J. Appl. Phys.* **12** (1961), 585–591 (P. M. Kelly and R. Reed)
DGEM 10: Kiel, 24–27 September 1961; *Mikroskopie* **17** (1962), 25–56, and *Phys. Verhandl.* **12** (1961), 133–154
JSEM 17: Jikei Medical University, Tokyo, 18–19 May 1961
SIME 1961: Atti del III Congresso Italiano di Microscopia Elettronica, Fondazione Carlo Erba, Milan, 3–4 March 1961 (Società Italiana di Microscopia Elettronica)
SFME 1961: Lyon–Villeurbanne, 16–17 February 1961

1962

Philadelphia, 1962: "Electron Microscopy. Fifth International Congress for Electron Microscopy," Philadelphia, PA, 29 August–5 September 1962 (Breese, S. S., ed.; Academic Press, New York, 1962), 2 vols.
JSEM 18: Engineering Faculty, Nagoya University, 19–20 May 1962
SFME 1962: Toulouse, 15–16 February 1962
EMSSA 1: University of the Witwatersrand, Johannesburg, December 1962

1963

EMSA 21: Denver CO, 28–31 August 1963; *J. Appl. Phys.* **34** (1963), 2502–2534
EMAG 1963/1: Cavendish Laboratory, Cambridge, 2–6 July 1963; *Br. J. Appl. Phys.* **14** (1963), 733–740 (R. B. Nicholson, W. C. Nixon, and D. H. Warrington)
EMAG 1963/2: Electron Probe Microanalysis, Department of Physics, University of Reading, 26–27 September 1963; *Br. J. Appl. Phys.* **15** (1964), 113–120, and *J. Sci. Instrum.* **41** (1964), 61–65 (P. Duncumb, J. V. P. Long, and D. A. Melford)
DGEM 11: Zürich, 22–25 September 1963 (with the Swiss Society); *Mikroskopie* **19** (1964), 1–73
JSEM 19: 19th Scientific Meeting of the Society of Electron-Microscopy, Japan, Hiroshima University, 18–19 May 1963; *J. Electron Microsc.* **12** (1963), 110–126
Sumy 1963: "Proceedings of the 4th Soviet All-Union Conference on Electron Microscopy," Sumy, 12–14 March 1963; *Izv. Akad. Nauk SSSR (Ser. Fiz.)* or *Bull. Acad. Sci. USSR (Phys. Ser.)* **27** (1963), no. 9

SIME 1963: Atti del IV Congresso Italiano di Microscopia Elettronica, Padova, 25–26 November 1963 (Società Italiana di Microscopia Elettronica)
SFME 1963: Orsay, 14–16 February 1963; *J. Microscopie* **2** (1963), 1–43
EMSSA 2: University of the Witwatersrand, Johannesburg, December 1963

1964

Prague, 1964: "Electron Microscopy 1964. Proceedings of the Third European Regional Conference," Prague, 26 August–3 September 1964 (Titlbach, M., ed.; Publishing House of the Czechoslovak Academy of Sciences, Prague, 1964), 2 vols.
EMSA 22: Detroit, MI, 13–16 October 1964; *J. Appl. Phys.* **35** (1964), 3074–3102
EMAG 1964: Use of Electron Microscopy, Diffraction and Probe Analysis in the Identification of Precipitates in Solids, National Physical Laboratory, Teddington, 16–17 April 1964; *Br. J. Appl. Phys.* **15** (1964), 867–870 (K. F. Hale and R. F. Braybrook)
JSEM 20: 20th Scientific Meeting of the Japanese Society of Electron-Microscopy, Tokushima University, 16–17 May 1964; *J. Electron Microsc.* **13** (1964), 29–54
SFME 1964: Strasbourg, 10–12 February 1964; *J. Microscopie* **3** (1964), 1–60
EMSSA 3: Council for Scientific and Industrial Research, Pretoria, 1 December 1964

1965

Calcutta, 1965: "Proceedings of the Second Regional Conference on Electron Microscopy in Far East and Oceania," Calcutta, 2–6 February 1965 (Electron Microscopy Society of India, Calcutta)
EMSA 23: New York, NY, 25–28 August 1965; *J. Appl. Phys.* **36** (1965), 2603–2632
EMAG 1965: Non-conventional Electron Microscopy, Engineering Department, Cambridge, 31 March–2 April 1965
DGEM 12: Aachen, 26–30 September 1965 (with the Dutch Society); *Mikroskopie* **21** (1966), 105–174
JSEM 21: 21st Scientific Meeting of the Japanese Society of Electron Microscopy, Kagoshima University, 15–16 May 1965; *J. Electron Microsc.* **14** (1965), 127–161
Sumy 1965: "Proceedings of the 5th Soviet All-Union Conference on Electron Microscopy," Sumy, 6–8 July 1965; *Izv. Akad. Nauk SSSR (Ser. Fiz.)* or *Bull. Acad. Sci. USSR (Phys. Ser.)* **30** (1966), no. 5
SIME 1965: Atti del V Congresso Italiano di Microscopia Elettronica, Bologna, 5–7 October 1965 (Società Italiana di Microscopia Elettronica)
SFME 1965: Marseille, 1–3 March 1965; *J. Microscopie* **4** (1965), 99–175
EMSSA 4: University of the Witwatersrand, Johannesburg, 7 December 1965

1966

Kyoto, 1966: "Electron Microscopy 1966. Sixth International Congress for Electron Microscopy," Kyoto, 28 August–4 September 1966 (Uyeda, R., ed.; Maruzen, Tokyo, 1966), 2 vols.

EMSA 24: San Francisco Hilton, San Francisco, CA, 22–25 August 1966; *J. Appl. Phys.* **37** (1966), 3919–3952
EMAG 1966: Electron Microscope Study of Chemically-Grown Surface Films, Glasgow, 7–8 July 1966
JSEM 22: 22nd Scientific Meeting of the Japanese Society of Electron Microscopy, Tokyo University, 7–9 April 1966; *J. Electron Microsc.* **15** (1966), 30–66
SFME 1966: Bordeaux, 23–25 May 1966; *J. Microscopie* **5** (1966), no. 3, 1a–91a
EMSSA 5: University of Pretoria, Pretoria, 6 December 1966; *S. Afr. J. Sci.* **63** (1967), 184–187

1967

EMSA 25: "Proceedings of the 25th Anniversary Meeting Electron Microscopy Society of America," Chicago, IL, 29 August–1 September 1967 (Arceneaux, C. J., ed.; Claitor, Baton Rouge, LA, 1967)
EMAG 1967: Electron Optics, Instrumentation and Quantitative Electron Microscopy, Buchanan Arts Theatre, University of St. Andrews, 19–21 September 1967
DGEM 13: Marburg, 17–21 September 1967; *Optik* **27** (1968), 61, and *Mikroskopie* **23** (1968), 61–129
JSEM 23: 23rd Scientific Meeting of the Japanese Society of Electron Microscopy, Kyushu University, 5–6 May 1967; *J. Electron Microsc.* **16** (1967), 179–220
Novosibirsk 1967: "Proceedings of the 6th Soviet All-Union Conference on Electron Microscopy," Novosibirsk, 11–16 July 1967; *Izv. Akad. Nauk SSSR (Ser. Fiz.)* or *Bull. Acad. Sci. USSR (Phys. Ser.)* **32** (1968), nos. 6 and 7
SIME 1967: Atti del VI Congresso Italiano di Microscopia Elettronica, Siena, 29–31 October 1967 (Società Italiana di Microscopia Elettronica)
SFME 1967: Brussels (with the Belgian Society), 22–24 May 1967; *J. Microscopie* **6** (1967), no. 4, 1a–93a
EMSSA 6: South African Institute for Medical Research, Johannesburg, 5 December 1967; *S. Afr. J. Sci.* **64** (1968), 413–419

1968

Rome, 1968: "Electron Microscopy 1968. Pre-Congress Abstracts of Papers Presented at the Fourth Regional Conference," Rome, 1–7 September 1968 (Bocciarelli, D. S., ed.; Tipographia Poliglotta Vaticana, Rome, 1968), 2 vols.
EMSA 26: "Proceedings of the 26th Annual Meeting Electron Microscopy Society of America," New Orleans, LA, 16–19 September 1968 (Arceneaux, C. J., ed.; Claitor, Baton Rouge, LA, 1968)
EMAG 1968: Scanning Electron Microscopy, Cambridge, 8–10 July; *Phys. Bull.* **19** (1968), 423–424 (K. C. A. Smith)
JSEM 24: 24th Scientific Meeting of the Japanese Society of Electron Microscopy, Yamanashi University, 11–12 May 1968; *J. Electron Microsc.* **17** (1968), 237–282
SFME 1968: Lille, 17–20 May 1968; *J. Microscopie* **7** 1968), no. 4, 1a–65a
EMSSA 7: University of Pretoria, 3 December 1968; *S. Afr. J. Sci.* **65** (1969), 296–300

1969

HVEM Monroeville, 1969: "Current Developments in High Voltage Electron Microscopy (First National Conference)," Monroeville, 17–19 June 1969 [Proceedings not published, but *Micron* **1** (1969), 220–307, contains official reports of the meeting based on the session chairmen's notes]

EMSA 27: "Proceedings of the 27th Annual Meeting Electron Microscopy Society of America," St. Paul, MN, 26–29 August 1969 (Arceneaux, C. J., ed.; Claitor, Baton Rouge, LA, 1969)

EMAG 1969/1: Dynamic Experimentation in Electron Optical Instruments, Imperial College, London, 14–15 April 1969

EMAG 1969/2: Non-conventional Electron Microscopy, St. Catherine's College, Oxford, 14–16 July 1969

DGEM 14: Wien, 22–25 September 1969 (with the Austrian Society); *Optik* **31** (1970), 111–112, and *Mikroskopie* **26** (1970), 81–144

JSEM 25: 25th Scientific Meeting of the Japanese Society of Electron Microscopy, Jikei University School of Medicine, 10–11 May 1969; *J. Electron Microsc.* **18** (1969), 198–233

Kiev 1969: "Proceedings of the 7th Soviet All-Union Conference on Electron Microscopy," Kiev, 14–21 July 1969; *Izv. Akad. Nauk SSSR (Ser. Fiz.)* or *Bull. Acad. Sci. USSR (Phys. Ser.)* **34** (1970), no. 7

SIME 1969: Atti del VII Congresso Italiano di Microscopia Elettronica, Modena, 28–30 September 1969 (Società Italiana di Microscopia Elettronica)

SFME 1969: Lausanne (with the Swiss Society), 19–21 May; *J. Microscopie* **8** (1969), no. 4, 1a–99a

EMSSA 8: University of the Witwatersrand, Johannesburg, 3 December 1969, *S. Afr. J. Sci.* **66** (1970), 324–326

1970

Grenoble, 1970: "Microscopie Electronique 1970. Résumés des Communications Présentées au Septième Congrès International," Grenoble, 30 August–5 September 1970 (Favard, P., ed.; Société Française de Microscopie Electronique, Paris, 1970), 3 vols.

HVEM Stockholm, 1970: "The Proceedings of the Second International Conference on High-Voltage Electron Microscopy," Stockholm, 14–16 April 1971; published as *Jernkontorets Annaler* **155** (1971), no. 8

EMSA 28: "Proceedings of the 28th Annual Meeting Electron Microscopy Society of America," Houston, TX, 5–8 October 1970 (Arceneaux, C. J., ed.; Claitor, Baton Rouge, LA, 1970)

EMAG 1970/1: Application of High-Voltage Electron Microscopy, Harwell, 2–3 April 1970; *Proc. R. Microsc. Soc.* **5** (1970), 96–124

EMAG 1970/2: Scanning Electron Microscopy in Materials Science, Newcastle-upon-Tyne, 7–9 July 1970

JSEM 26: 26th Scientific Meeting of the Japanese Society of Electron Microscopy, Science Museum, Tokyo, 20–22 May 1970; *J. Electron Microsc.* **19** (1970), 284–324

EMSSA 9: University of Natal, Durban, 4 December 1970 [no published record]

1971

EMSA 29: "Proceedings of the 29th Annual Meeting Electron Microscopy Society of America," Boston, MA, 9–13 August 1971 (Arceneaux, C. J., ed.; Claitor, Baton Rouge, LA, 1971)

EMAG, 1971: "Electron Microscopy and Analysis. Proceedings of the 25th Anniversary Meeting of the Electron Microscopy and Analysis Group of the Institute of Physics," Cambridge, 29 June–1 1971 (Nixon, W. C., ed.; Institute of Physics, London, 1971), *Conference Series* **10**

DGEM 15: Karlsruhe, 19–23 September 1971; *Optik* **35** (1972), 257–345, and *Mikroskopie* **28** (1972), 321–367

JSEM 27: 27th Scientific Meeting of the Japanese Society of Electron Microscopy, Kyoto 18–20 May 1971; *J. Electron Microsc.* **20** (1971), 215–265

Moscow 1971: "Proceedings of the 8th Soviet All-Union Conference on Electron Microscopy," Moscow, 15–20 November 1971; *Izv. Akad. Nauk SSSR (Ser. Fiz.)* or *Bull. Acad. Sci. USSR (Phys. Ser.)* **36** (1972), nos. 6 and 9

SIME 1971: Abstracts of the VIII Italian Congress for Electron Microscopy, Milano, 23–25 September 1971; *J. Submicrosc. Cytol.* **4** (1972), 101–134

SFME 1971: Caen, 7–10 May 1971; *J. Microscopie* **11** (1971), 1–106

EMSSA 10: Council for Scientific and Industrial Research, Pretoria, 3 December 1971; *Electron Microscopy Society of Southern Africa Proceedings/Elektronenmikroskopievereniging van Suidelike Afrika Verrigtinge* **1** (1971)

1972

Manchester, 1972: "Electron Microscopy 1972. Proceedings of the Fifth European Congress on Electron Microscopy," Manchester, 5–12 September 1972 (Institute of Physics, London, 1972)

EMSA 30: "Proceedings of the 30th Annual Meeting Electron Microscopy Society of America and First Pacific Regional Conference on Electron Microscopy," Los Angeles, CA, 14–17 August 1972 (Arceneaux, C. J., ed.; Claitor, Baton Rouge, LA, 1972)

JSEM 28: 28th Scientific Meeting of the Japanese Society of Electron Microscopy, Okayama, 23–25 May 1972; *J. Electron Microsc.* **21** (1972), 203–257

SFME 1972: Nantes, 29–31 May 1972; *J. Microscopie* **14** (1972), no. 2, 1a–108a

EMSSA 11: University of the Witwatersrand, Johannesburg, 30 November–1 December 1972; *Electron Microscopy Society of Southern Africa Proceedings/Elektronenmikroskopievereniging van Suidelike Afrika Verrigtinge* **2** (1972)

1973

HVEM Oxford, 1973: "High Voltage Electron Microscopy. Proceedings of the Third International Conference," Oxford, August 1973 (Swann, P. R., Humphreys, C. J., and Goringe, M. J., eds.; Academic Press, London and New York, 1974)

EMSA 31: "Proceedings of the 31st Annual Meeting Electron Microscopy Society of America," New Orleans, LA, 14–17 August 1973 (Arceneaux, C. J., ed.; Claitor, Baton Rouge, LA, 1973)

EMAG, 1973: "Scanning Electron Microscopy: Systems and Applications," Newcastle-upon-Tyne, 3–5 July 1973 (Nixon, W. C., ed.; Institute of Physics, London, 1973), *Conference Series* **18**

DGEM 16: Liège, 3–6 September 1973 (with the Belgian and Dutch Societies); *Optik* **40** (1974), 233–283, and *Beiträge zur Elektronenmikroskopische Direktabbildung von Oberflächen* **6** (1974)

JSEM 29: 29th Scientific Meeting of the Japanese Society of Electron Microscopy, Tokyo, 18–20 May 1973; *J. Electron Microsc.* **22** (1973), 281–320

Tbilisi 1973: "Proceedings of the 9th Soviet All-Union Conference on Electron Microscopy," Tbilisi, 28 October–2 November 1973; *Izv. Akad. Nauk SSSR (Ser. Fiz.)* or *Bull. Acad. Sci. USSR (Phys. Ser.)* **38** (1974), no. 7

SIME 1973: Abstracts of the IX Italian Congress for Electron Microscopy, Saint-Vincent, 8–10 October 1973; *J. Submicrosc. Cytol.* **6** (1974), 103–141

SFME 1973: Dijon, 4–7 June 1973; *J. Microscopie* **17** (1973), no. 3, 1a–116a

EMSSA 12: University of Pretoria, Pretoria, 29–30 November 1973; *Electron Microscopy Society of Southern Africa Proceedings/Elektronenmikroskopievereniging van Suidelike Afrika Verrigtinge* **3** (1973)

1974

Canberra, 1974: "Electron Microscopy 1974. Abstracts of Papers Presented to the Eighth International Congress on Electron Microscopy," Canberra, 25–31 August 1974 (Sanders, J. V., and Goodchild, D. J., eds.; Australian Academy of Science, Canberra, 1974), 2 vols.

Sarajevo, 1974: "Electron Microscopy 1974. Pre-Congress Abstracts of Papers Presented at the First Balkan Congress on Electron Microscopy," Sarajevo, 22–26 May 1974 (Devidé, Z., Dobardžić, R., Jerković, L., Marinković, V., Pantić, V., Pejowski, S., and Pipan, N., eds.)

EMSA 32: "Proceedings of the 32nd Annual Meeting Electron Microscopy Society of America," St. Louis, MO, 13–15 August 1974 (Arceneaux, C. J., and G. W. Bailey, eds.; Claitor, Baton Rouge, LA, 1974)

JSEM 30: 30th Scientific Meeting of the Japanese Society of Electron Microscopy, Osaka, 22–24 May 1974; *J. Electron Microsc.* **23** (1974), 201–245

SIME 1974: Atti del Convegno del Gruppo Strumentazione e Tecniche non Biologiche della Società Italiana di Microscopia Elettronica, Laboratorio di Chimica e Tecnologia dei Materiali e dei Componenti per l'Elettronica [LAMEL], Bologna, 28 June 1974 (Cooperativa Libreria Universitaria, Bologna, 1975)

SFME 1974: Rennes, 27–30 May 1974; *J. Microscopie* **20** (1974), no. 1, 1a–104a

EMSSA 13: University of the Witwatersrand, Johannesburg, 28–29 November 1974. *Electron Microscopy Society of Southern Africa Proceedings/Elektronenmikroskopievereniging van Suidelike Afrika Verrigtinge* **4** (1974)

MSC 1: Ontario Science Centre, Toronto, 24–25 June 1974

1975

HVEM Toulouse: "Microscopie Electronique à Haute Tension. Textes des Communications Présentées au 4è Congrès International," Toulouse, 1–4 Septembre 1975 (Jouffrey, B., and Favard, P., eds.; SFME, Paris, 1976)

EMSA 33: "Proceedings of the 33rd Annual Meeting Electron Microscopy Society of America," Las Vegas, NV, 11–15 August 1975 (Bailey, G. W., and Arceneaux, C. J., eds.; Claitor, Baton Rouge, LA, 1975)
EMAG, 1975: "Developments in Electron Microscopy and Analysis. Proceedings of EMAG 75," Bristol, 8–11 September 1975 (Venables, J. A., ed.; Academic Press, London and New York, 1976)
DGEM 17: Berlin, 21–26 September 1975; *Optik* **45** (1976), 105–109, and *Mikroskopie,* **32** (1976), 145–190, and 204–255
JSEM 31: 31st Scientific Meeting of the Japanese Society of Electron Microscopy, Tokyo, 22–24 May 1975; *J. Electron Microsc.* **24** (1975), 179–218
SIME 1975: Abstracts of the X Italian Congress for Electron Microscopy, Rosa Marina di Ostuni (Brindisi), 2–4 October 1975; *J. Submicrosc. Cytol.* **8** (1976), 243–268
SFME 1975: Montreal (with the Canadian Society), 26–30 May; *J. Microsc. Biol. Cell.* **23** (1975), no. 2, 1a–88a
EMSSA 14: University of Pretoria, Pretoria, 27–28 November 1975; *Electron Microscopy Society of Southern Africa Proceedings/Elektronenmikroskopievereniging van Suidelike Afrika Verrigtinge* **5** (1975)
MSC 2: Pavillon Comtois, Université Laval, Quebec City, 15–17 June 1975

1976

Jerusalem 1976: "Electron Microscopy 1976. Proceedings of the Sixth European Congress on Electron Microscopy," Jerusalem, 14–20 September 1976 (Brandon, D. G. (vol. I), and Ben-Shaul, Y. (vol. II), eds.; Tal International, Jerusalem, 1976), 2 vols.
EMSA 34: "Proceedings of the 34th Annual Meeting Electron Microscopy Society of America," Miami Beach, FL, 9–13 August 1976 (Bailey, G. W., ed.; Claitor, Baton Rouge, LA, 1976)
JSEM 32: 32nd Scientific Meeting of the Japanese Society of Electron Microscopy, Nagoya, 20–22 May 1976; *J. Electron Microsc.* **25** (1976), 175–224
Tashkent 1976: "Proceedings of the 10th Soviet All-Union Conference on Electron Microscopy," Tashkent, 5–8 October 1976; *Izv. Akad. Nauk SSSR (Ser. Fiz.)* or *Bull. Acad. Sci. USSR* (Phys. Ser.), **41** (1977), nos. 5 and 11.
SFME 1976: Clermont-Ferrand, 9–11 June; *J. Microsc. Biol. Cell.* **26** (1976), nos. 2–3, 1a–35a
EMSSA 15: Rand Afrikaans University, Johannesburg, 2–3 December 1976; *Electron Microscopy Society of Southern Africa Proceedings/Elektronenmikroskopievereniging van Suidelike Afrika Verrigtinge* **6** (1976)
MSC 3: University of Ottawa, Ottawa, 20–23 June 1976

1977

HVEM, Kyoto, 1977: "High Voltage Electron Microscopy 1977. Proceedings of the Fifth International Conference on High Voltage Electron Microscopy," Kyoto, 29 August–1 September 1977 (Imura, T., and Hashimoto, H., eds.; Japanese Society of Electron Microscopy, Tokyo, 1977); published as a supplement to *Journal of Electron Microscopy* **26** (1977)

Istanbul, 1977: "Abstracts of Communications, Second Balkan Congress on Electron Microscopy," Istanbul, 25-30 September 1977 (Erbengi, T., Chairman Sci. Prog. Comm.; Istanbul Faculty of Medecine and Turkish Society of Electron Microscopy, Istanbul)

EMSA 35: "Proceedings of the 35th Annual Meeting Electron Microscopy Society of America," Boston, MA, 22-26 August 1977 (Bailey, G. W., ed.; Claitor, Baton Rouge, LA, 1977)

EMAG 1977: "Developments in Electron Microscopy and Analysis. Proceedings of EMAG 77," Glasgow, 12-14 September 1977 (Misell, D. L., ed.; Institute of Physics, Bristol, 1977); *Conference Series* **36**

DGEM 18: Münster, 4-9 September 1977; *Optik* **52** (1978/9), 261-262; *Mikroskopie* **34** (1978), 79-112 and 134-193

JSEM 33: 33rd Scientific Meeting of the Japanese Society of Electron Microscopy, Fukuoka, 12-14 May 1977; *J. Electron Microsc.* **26** (1977), 225-276

SIME 1977: Cosenza, 10-12 October 1977; *J. Microsc. Spectrosc. Electron.* **3** (1978), no. 1, 1ab-14ab

SFME 1977: Nice, 31 May-2 June 1977; *J. Microsc. Spectrosc. Electron.* **2** (1977), no. 3, 1a-18a

EMSSA 16: Newlands Hotel, Cape Town, 1-2 December 1977; *Electron Microscopy Society of Southern Africa Proceedings/Elektronenmikroskopievereniging van Suidelike Afrika Verrigtinge* **7** (1977)

MSC 4: University of Western Ontario, London, 13-15 June 1977

1978

Toronto, 1978: "Electron Microscopy 1978. Papers Presented at the Ninth International Congress on Electron Microscopy," Toronto, 1-9 August 1978 (Sturgess, J. M., ed.; Microscopical Society of Canada, Toronto, 1978), 3 vols.

JSEM 34: 34th Scientific Meeting of the Japanese Society of Electron Microscopy, Sapporo, 20-22 June 1978; *J. Electron Microsc.* **27** (1978), 333-390

SFME 1978: Nancy, 23-25 May 1978; *J. Microsc. Spectrosc. Electron.* **3** (1978), no. 3, 1a-16a, 271-386, and 551-631

EMSSA 17: Council for Scientific and Industrial Research, Pretoria, 4-5 December 1978; *Electron Microscopy Society of Southern Africa Proceedings/Elektronenmikroskopievereniging van Suidelike Afrika Verrigtinge* **8** (1978)

1979

EMSA 37: "Proceedings of the 37th Annual Meeting Electron Microscopy Society of America," San Antonio, TX, 13-17 August 1979 (Bailey, G. W., ed.; Claitor, Baton Rouge, LA, 1979)

EMAG 1979: "Electron Microscopy and Analysis, 1979. Proceedings of EMAG 79," Brighton, 3-6 September 1979 (Mulvey, T., ed.; Institute of Physics, Bristol, 1980); *Conference Series* **52**

DGEM 19: Tübingen, 9-11 September 1979; *Optik* **55** (1980), 217-218; *Mikroskopie* **36** (1980), part 1, 274-282 and 282-318, part 2, 336-355

JSEM 35: 35th Scientific Meeting of the Japanese Society of Electron Microscopy, Takarazuka, 23-25 May 1979; *J. Electron Microsc.* **28** (1979), 201-262; "Development of Electron Microscopy and Its Future," Proceedings 30th Anniversary of Japanese Society of Electron Microscopy, 22 May 1979; *J. Electron Microsc.* **28** (1979), suppl.

Tallin, 1979: "Proceedings of the 11th Soviet All-Union Conference on Electron Microscopy," Tallin, October 1979; *Izv. Akad. Nauk SSSR (Ser. Fiz.)* or *Bull. Acad. Sci. USSR (Phys. Ser.)* **44** (1980), nos. 6 and 10

SIME 1979: Abstracts of the Papers Presented at the 12th Congress of the Italian Society of Electron Microscopy, Ancona, 20–22 September 1979; *Ultramicroscopy* **5** (1980), 363–428

SFME 1979: Lyon-Villeurbanne, 21–23 May 1979; *J. Microsc. Spectrosc. Electron.* **4** (1979), no. 3, 1a–26a, 269–519, and 581–612

EMSSA 18: University of Port Elizabeth, Port Elizabeth, 3–5 December 1979; *Electron Microscopy Society of Southern Africa Proceedings/Elektronenmikroskopievereniging van Suidelike Afrika Verrigtinge* **9** (1979)

MSC 6: University of British Columbia, Vancouver, 15–17 June 1979

1980

The Hague, 1980: "Electron Microscopy 1980. Proceedings of the Seventh European Congress on Electron Microscopy," The Hague, 24–29 August 1980 [Brederoo, P., and Boom, G. (Vol. I), Brederoo, P., and de Priester, W. (Vol. II), Brederoo, P., and Cosslett, V. E. (Vol. III), and Brederoo, P., and van Landuyt, J. (Vol. IV), eds.]; Vols. I and II contain the proceedings of the Seventh European Congress on Electron Microscopy, Vol. III those of the Ninth International Conference on X-Ray Optics and Microanalysis, and Vol. IV those of the Sixth International Conference on High Voltage Electron Microscopy, Antwerp, 1–3 September 1980 (Seventh European Congress on Electron Microscopy Foundation, Leiden, 1980)

Giessen, 1980: "Charged Particle Optics. Proceedings of the First Conference on Charged Particle Optics," Giessen, 8–11 September 1980 (Wollnik, H., ed.), *Nuclear Instrum. Meth.* **187** (1981), 1–314

EMSA 38: "Proceedings of the 38th Annual Meeting Electron Microscopy Society of America," San Francisco, CA, 4–8 August 1980 (Bailey, G. W., ed.; Claitor, Baton Rouge, LA, 1980)

JSEM 36: 36th Scientific Meeting of the Japanese Society of Electron Microscopy, Yokohama, 27–29 May 1980; *J. Electron Microsc.* **29** (1980), 274–339

SFME 1980: Poitiers, 4–6 June 1980; *J. Microsc. Spectrosc. Electron.* **5** (1980), no. 3, 1a–18a, 451–527, and 539–727

EMSSA 19: University of the Witwatersrand, Johannesburg, 1–3 December 1980; *Electron Microscopy Society of Southern Africa Proceedings/Elektronenmikroskopievereniging van Suidelike Afrika Verrigtinge* **10** (1980).

MSC 7: Memorial University, St. John's, Newfoundland, 5–7 June 1980

1981

EMSA 39: "Proceedings of the 39th Annual Meeting Electron Microscopy Society of America," Atlanta, GA, 10–14 August 1981 (Bailey, G. W., ed.; Claitor, Baton Rouge, LA, 1981)

EMAG, 1981: "Electron Microscopy and Analysis, 1981. Proceedings of EMAG 81," Cambridge, 7–10 September 1981 (Goringe, M. J., ed.; Institute of Physics, Bristol, 1982); *Conference Series* **61**

DGEM 20: Innsbruck, 23–27 August 1981 (with the Austrian Society); *Optik* **62** (1982), 329–331; *Beiträge zur Elektronenmikroskopische Direktabbildung von Oberflächen* **14** (1981)

JSEM 37: 37th Scientific Meeting of the Japanese Society of Electron Microscopy, Kyoto, 20–22 May 1981; *J. Electron Microsc.* **30** (1981), 213–280

SIME 1981: Florence, 30 September–3 October, 1981; *J. Microsc. Spectrosc. Electron.* **7** (1982), no. 3, 29a–32a

SFME 1981: Besançon, 25–27 May 1981; *J. Microsc. Spectrosc. Electron.* **6** (1981), no. 3, 1a–10a and 345–462

EMSSA 20: University of Durban Westville, Durban Westville, 2–4 December 1981; *Electron Microscopy Society of Southern Africa Proceedings/Elektronenmikroskopievereniging van Suidelike Afrika Verrigtinge* **11** (1981)

MSC 8: McGill University, Montreal, 13–15 June 1981

"New Trends of Electron Microscopy in Atom Resolution Materials Science and Biology," Proceedings of the First Chinese-Japanese Electron Microscopy Seminar, Dalian, 27–31 July 1981 (Hashimoto, H., Kuo, K. H., and Ko, T., eds.; Science Press, Beijing, 1982)

1982

Hamburg, 1982: "Electron Microscopy, 1982. Papers Presented at the Tenth International Congress on Electron Microscopy," Hamburg, 17–24 August 1982 (Deutsche Gesellschaft für Elektronenmikroskopie, Frankfurt, 1982), 3 vols.

EMSA 40: "Proceedings of the 40th Annual Meeting Electron Microscopy Society of America," Washington, DC, 9–13 August 1982 (Bailey, G. W., ed.; Claitor, Baton Rouge, LA, 1982)

JSEM 38: 38th Scientific Meeting of the Japanese Society of Electron Microscopy, Komaba Eminence, Tokyo, 26–28 May 1982; *J. Electron Microsc.* **31** (1982), 285–344

Sumy 1982: "Proceedings of the 12th Soviet All-Union Conference on Electron Microscopy," Sumy 1982; *Izv. Akad. Nauk SSSR* (*Ser. Fiz.*) or *Bull. Acad. Sci. USSR* (*Phys. Ser.*) **48** (1984), no. 2

SFME 1982: Reims, 25–27 May 1982; *J. Microsc. Spectrosc. Electron.* **7** (1982), no. 2, 1a–28a, 315–423, and 487–553

EMSSA 21: Rhodes University, Grahamstown, 1–3 December 1982; *Electron Microscopy Society of Southern Africa Proceedings/Elektronenmikroskopievereniging van Suidelike Afrika Verrigtinge* **12** (1982)

MSC 9: University of Alberta, Edmonton, 11–13 June 1982

1983

HVEM Berkeley, 1983: "Proceedings of the Seventh International Conference on High Voltage Electron Microscopy," Berkeley, 16–19 August 1983 (Fisher, R. M., Gronsky, R., and Westmacott, K. H., eds.); published as a Lawrence Berkeley Laboratory Report, LBL-16031, UC-25, CONF-830819

EMSA 41: "Proceedings 41st Annual Meeting Electron Microscopy Society of America," Phoenix, AZ, 8–12 August 1983 (Bailey, G. W., ed.; San Francisco Press, San Francisco, 1983)

EMAG, 1983: "Electron Microscopy and Analysis, 1983. Proceedings of EMAG 83," Guildford, 30 August–2 September 1983 (Doig, P., ed.; Institute of Physics, Bristol, 1984); *Conference Series* **68**

DGEM 21: Antwerp, 11–16 September 1983 (with the Belgian Society); *Beiträge zur Elektronenmikroskopische Direktabbildung von Oberflächen* **16** (1983)

JSEM 39: 39th Scientific Meeting of the Japanese Society of Electron Microscopy, Aichi Sangyo Boeki-kan, Nagoya, 31 May–2 June 1983; *J. Electron Microsc.* **32** (1983), 219–287

SIME 1983: Abstracts of the Papers Presented at the 14th Congress of the Italian Society of Electron Microscopy, Ferrara, 21–24 September 1983; *Ultramicroscopy* **12** (1983/4), 87–166

SFME 1983: Liège (with the Belgian Society), 16–19 May; *J. Microsc. Spectrosc. Electron.* **8** (1983), no. 2, 1a–48a, 111–278, and 341–488

EMSSA 22: University of the Witwatersrand, Johannesburg, 30 November–2 December 1983; *Electron Microscopy Society of Southern Africa Proceedings/Elektronenmikroskopievereniging van Suidelike Afrika Verrigtinge* **13** (1983)

MSC 10: Chalk River Nuclear Laboratories, Chalk River, Ontario, 17–19 May 1983

"Recent Developments of Electron Microscopy," Proceedings of the Second Chinese-Japanese Electron Microscopy Seminar, Beijing, 17–19 October 1983 (Hashimoto, H., Ko, T., Kuo, K. H., and Ogawa, K., eds.; Japanese Society of Electron Microscopy, Tokyo)

1984

Budapest, 1984: "Electron Microscopy 1984. Proceedings of the Eighth European Congress on Electron Microscopy," Budapest, 13–18 August 1984 (Csanády, Á., Röhlich, P., and Szabó, D., eds.; Programme Committee of the Eighth European Congress on Electron Microscopy, Budapest, 1984), 3 vols.

Singapore, 1984: "Conference Proceedings 3rd Asia Pacific Conference on Electron Microscopy," Singapore, 29 August–3 September 1984 (Chung Mui Fatt, ed.; Applied Research Corporation, Singapore)

EMSA 42: "Proceedings 42nd Annual Meeting Electron Microscopy Society of America Jointly with Microscopical Society of Canada, Eleventh Annual Meeting," Detroit, MI, 13–17 August 1984 (Bailey, G. W., ed.; San Francisco Press, San Francisco, 1984)

JSEM 40: 40th Scientific Meeting of the Japanese Society of Electron Microscopy, Sendai Shimin Kaikan, Sendai, 27–29 June 1984; *J. Electron Microsc.* **33** (1984), 261–322

SFME 1984: Montpellier, 21–24 May; *J. Microsc. Spectrosc. Electron.* **9** (1984), no. 1, 1a–51a and 147–340

EMSSA 23: University of Stellenbosch, Stellenbosch, 5–7 December 1984; *Electron Microscopy Society of Southern Africa Proceedings/Elektronenmikroskopievereniging van Suidelike Afrika Verrigtinge* **14** (1984)

1985

EMSA 43: "Proceedings 43rd Annual Meeting Electron Microscopy Society of America," Louisville, KY, 5–9 August 1985 (Bailey, G. W., ed.; San Francisco Press, San Francisco, 1985)

EMAG, 1985: "Electron Microscopy and Analysis, 1985. Proceedings of EMAG 85." Newcastle-upon-Tyne, 2–5 September 1985 (Tatlock, G. J., ed.; Institute of Physics, Bristol, 1986); *Conference Series* **78**

DGEM 22: Konstanz, 15–21 September 1985 (with the Swiss and Austrian Societies); *Optik* (1985), suppl. 1, or *Eur. J. Cell Biol.* (1985), suppl. 10; see also *Beiträge zur Elektronenmikroskopische Direktabbildung von Oberflächen* **18** (1985)

JSEM 41: 41st Scientific Meeting of the Japanese Society of Electron Microscopy, Hokkaido University, Sapporo, 25–27 June 1985; *J. Electron Microsc.* **34** (1985), 183–247

SIME 1985: Abstracts of the Papers Presented at the XV Congress of the Italian Society of Electron Microscopy, Rome, 28–31 May 1985; *Ultramicroscopy* **17** (1985), 399–409

SFME 1985: Strasbourg, 28–31 May 1985; *J. Microsc. Spectrosc. Electron.* **10** (1985), no. 2, 1a–55a, 149–249, and 311–514

EMSSA 24: University of Natal, Pietermaritzburg, 4–6 December 1985; *Electron Microscopy Society of Southern Africa Proceedings/Elektronenmikroskopievereniging van Suidelike Afrika Verrigtinge* **15** (1985)

MSC 12: University of New Brunswick, Fredericton, 18–20 May 1985

"Recent Developments of Electron Microscopy," Proceedings of the Third Chinese-Japanese Electron Microscopy Seminar, Hanzhou, 4–7 November 1985 (Hashimoto, H., Kuo, K. H., Lee, K., and Ogawa, K., eds.; Japanese Society of Electron Microscopy, Tokyo)

1986

Kyoto, 1986: "Electron Microscopy 1986. Proceedings of the XIth International Congress on Electron Microscopy," Kyoto, 31 August–7 September 1986 (Imura, T., Maruse, S., and Suzuki, T., eds.; Japanese Society of Electron Microscopy, Tokyo), 4 vols.; published as a supplement to *J. Electron Microsc.* **35** (1986).

Beijing, 1986: Proceedings of the International Symposium on Electron Optics, Beijing, 9–13 September 1986 (Ximen, J.-Y., ed.; Institute of Electronics, Academia Sinica, 1987)

Albuquerque, 1986: "Charged Particle Optics. Proceedings of the Second International Conference on Charged Particle Optics," Albuquerque, NM, 19–23 May 1986 (Schriber, S. O. and Taylor, L. S., eds.) *Nuclear Instrum. Meth. Phys. Res.* **A258** (1987), 289–598

EMSA 44: "Proceedings 44th Annual Meeting Electron Microscopy Society of America," Albuquerque, NM, 10–15 August 1986 (Bailey, G. W., ed.; San Francisco Press, San Francisco, 1986)

SFME 1986: Nantes, 9–11 June 1986; *J. Microsc. Spectrosc. Electron.* **11** (1986), no. 2, 1a–55a, 129–214, and 359–396

EMSSA 25: Potchefstroom University for Christian Education, Potchefstroom, 3–5 December 1986; *Electron Microscopy Society of Southern Africa Proceedings/Elektronenmikroskopievereniging van Suidelike Afrika Verrigtinge* **16** (1986)

MSC 13: McMaster University, Hamilton, Ontario, 13–15 June 1986

1987

EMSA 45: "Proceedings 45th Annual Meeting Electron Microscopy Society of America," Baltimore, MD, 2–7 August 1987 (Bailey, G. W., ed.; San Francisco Press, San Francisco, 1987)

EMAG, 1987: "Electron Microscopy and Analysis, 1987. Proceedings of EMAG 87," Manchester, 8–9 September 1987 (Brown, L. M., ed.; Institute of Physics, Bristol and Philadelphia, 1987); *Conference Series* **90**

DGEM 23: Bremen, 13–19 September 1987; *Optik* **77** (1987), suppl. 3, or *Eur. J. Cell Biol.* **44** (1987), suppl. 19

JSEM 43: 43rd Scientific Meeting of the Japanese Society of Electron Microscopy, Kanagawa-kenritsu Kenmin Hall, Yokohama, 27–29 May 1987; *J. Electron Microsc.* **36** (1987), 294–352

Sumy 1987: "Proceedings of the 13th Soviet All-Union Conference on Electron Microscopy," Sumy, October 1987; *Izv. Akad. Nauk SSSR (Ser. Fiz.)* or *Bull. Acad. Sci. USSR (Phys. Ser.)* **52** (1988), no. 7.

SIME, 1987: "Atti del XVI Congresso di Microscopia Elettronica," Bologna, 14–17 October 1987; *Microscopia Elettronica* **8** (2) (suppl.)

SFME 1987: Talence, 20–22 May; *J. Microsc. Spectrosc. Electron.* **12** (1987), no. 3, 1a–39a and 353–421

EMSSA 26: University of the Witwatersrand, Johannesburg, 2–4 December 1987; *Electron Microscopy Society of Southern Africa Proceedings/Elektronenmikroskopievereniging van Suidelike Afrika Verrigtinge* **17** (1987)

MSC 14: University of Manitoba, Winnipeg, 17–19 June 1987

"Recent Developments of Electron Microscopy," Proceedings of the Fourth Chinese-Japanese Electron Microscopy Seminar, Kunming, 8–12 November 1987 (Hashimoto, H., Kuo, K. H., Lee, K., and Ogawa, K., eds.; Japanese Society of Electron Microscopy, Tokyo)

1988

York, 1988: "Proceedings of the Ninth European Congress on Electron Microscopy," York, 4–9 September 1988 (Goodhew, P. J., and Dickinson, H. G., eds.; Institute of Physics, Bristol and Philadelphia, 1988), *Conference Series* **93**, 3 vols.

Bangkok, 1988: "Electron Microscopy 1988. Proceedings of the IVth Asia-Pacific Conference and Workshop on Electron Microscopy," Bangkok, 26 July–4 August 1988 (Mangclaviraj, V., Banchorndhevakul, W., and Ingkaninun, P., eds.; Electron Microscopy Society of Thailand, Bangkok)

EMSA 46: "Proceedings 46th Annual Meeting Electron Microscopy Society of America Jointly with Microscopical Society of Canada, Fifteenth Annual Meeting," Milwaukee, WI, 7–12 August 1988 (Bailey, G. W., ed.; San Francisco Press, San Francisco, 1988)

JSEM 44: 44th Scientific Meeting of the Japanese Society of Electron Microscopy, Sendai Shimin Kaikan, Sendai, 1–3 June 1988; *J. Electron Microsc.* **37** (1988), 232–286.

SFME 1988: Villeneuve d'Ascq-Lille (with the Belgian Society), 17–20 May 1988; *J. Microsc. Spectrosc. Electron.* **13** (1988), no. 3, 1a–64a, 65a–72a, 257–312, and 451–510

EMSSA 27: University of Natal, Durban, 6–8 December 1988; *Electron Microscopy Society of Southern Africa Proceedings/Elektronenmikroskopievereniging van Suidelike Afrika Verrigtinge* **18** (1988)

1989

Athens, 1989: "Proceedings III Balkan Congress on Electron Microscopy," Athens, 18–22 September 1989 (Margaritis, L. H., ed.)

EMSA 47: "Proceedings 47th Annual Meeting Electron Microscopy Society of America," San Antonio, TX, 6–11 August 1989 (Bailey, G. W., ed.; San Francisco Press, San Francisco, 1989)

EMAG, 1989: "EMAG-MICRO 89. Proceedings of the Institute of Physics Electron Microscopy and Analysis Group and Royal Microscopical Society Conference," London, 13–15 September 1989 (Goodhew, P. J., and Elder, H. Y., eds.; Institute of Physics, Bristol and New York, 1990); *Conference Series* **98,** 2 vols.

DGEM 24: Salzburg, 10–16 September 1989 (with the Austrian and Swiss Societies); *Optik* **83** (1989), suppl. 4, or *Eur. J. Cell Biol.* **49** (1989), suppl. 27

JSEM 45: 45th Scientific Meeting of the Japanese Society of Electron Microscopy, Osaka Shoko Kaigishu, Osaka, 31 May–2 June 1989, Abstracts of the Presentation in Commemoration of the 40th Anniversary of the Japanese Society of Electron Microscopy; *J. Electron Microsc.* **38** (1989), 250–320

SIME, 1989: "Atti del XVII Congresso di Microscopia Elettronica," Lecce, 4–7 October 1989; *Microscopia Elettronia* **10** (2) (suppl.)

SFME 1989: Grenoble-Saint Martin d'Hères, 8–12 July 1989; *J. Microsc. Spectrosc. Electron.* **14** (1989), no. 2, 1a–80a, 315–384, and 387–414

EMSSA 28: University of Pretoria, Onderstepoort, 5–7 December 1989; *Electron Microscopy Society of Southern Africa Proceedings/Elektronenmikroskopievereniging van Suidelike Afrika Verrigtinge* **19** (1989)

MSC 16: University of Guelph, Guelph, Ontario, 30 May–2 June 1989

1990

Seattle, 1990: "Electron Microscopy 1990. Proceedings of the XIIth International Congress for Electron Microscopy," Seattle, WA, 12–18 August 1990 (Peachey, L. D., and Williams, D. B., eds.; San Francisco Press, San Francisco 1990), 4 vols.

Toulouse, 1990: "Charged Particle Optics. Proceedings of the Third International Conference on Charged Particle Optics," Toulouse, 24–27 April 1990 (Hawkes, P. W., ed.); *Nuclear Instrum. Meth. Phys. Res.* **A298** (1990), 1–508

JSEM 46: 46th Scientific Meeting of the Japanese Society of Electron Microscopy, Gunma University, Maebashi, 17–19 May 1990; *J. Electron Microsc.* **39** (1990), 275–349

Suzdal 1990: "Proceedings of the 14th Soviet All-Union Conference on Electron Microscopy," Suzdal, October and November 1990; *Izv. Akad. Nauk SSSR (Ser. Fiz.)* or *Bull. Acad. Sci. USSR (Phys. Ser.)* **55** (1991), no. 8

SFME 1990: Toulouse, 19–22 June 1990

EMSSA 29: Rhodes University, Grahamstown, 5–7 December 1990; *Electron Microscopy Society of Southern Africa Proceedings/Elektronenmikroskopievereniging van Suidelike Afrika Verrigtinge* **20** (1990)

MSC 17: Dalhousie University, Halifax, Nova Scotia, 10–12 June 1990

"Recent Developments of Electron Microscopy," Proceedings of the Fifth Chinese-Japanese Electron Microscopy Seminar, Urumqi, 5–7 September 1990 (Hashimoto, H., Kuo, K. H., Lee, K., and Ogawa, K, eds.; Japanese Society of Electron Microscopy, Tokyo)

1991

EMSA 49: "Proceedings 49th Annual Meeting Electron Microscopy Society of America," San Jose, CA, 4–9 August 1991 (Bailey, G. W., and Hall, E. L., eds.; San Francisco Press, San Francisco, 1991)

EMAG, 1991: "Electron Microscopy and Analysis 1991. Proceedings of EMAG 91," Bristol, 10–13 September 1991 (Humphreys, F. J., ed.; Institute of Physics, Bristol, Philadelphia, and New York, 1991); *Conference Series* **119**

DGEM 25: Darmstadt, 1–7 September 1991; *Optik* **88** (1991), suppl. 4 [sic], or *Eur. J. Cell Biol.* **55** (1991), supplement 33

JSEM 47: 47th Scientific Meeting of the Japanese Society of Electron Microscopy, Osaka Sun Palace, Suita, Osaka, 22–24 May 1991; *J. Electron Microsc.* **40** (1991), 234–300

SIME, 1991: "Atti del XVIII Congresso di Microscopia Elettronica," Padova, 24–28 September 1991; *Microscopia Elettronica* **12** (2) (suppl.)

SFME 1991: Barcelona (with the Spanish Society), 2–5 July 1991

EMSSA 30: University of Cape Town, Cape Town, 4–6 December 1991; *Electron Microscopy Society of Southern Africa Proceedings/Elektronenmikroskopievereniging van Suidelike Afrika Verrigtinge* **21** (1991)

MSC 18: Health Sciences Centre, University of Calgary, Calgary, Alberta, 23–26 June 1991

"Recent Developments of Electron Microscopy," Proceedings of the Sixth Chinese-Japanese Electron Microscopy Seminar, Okayama, 5–9 November 1991 (Hashimoto, H., Kuo, K. H., Lee, K., and Ogawa, K, eds.; Japanese Society of Electron Microscopy, Tokyo)

1992

Granada, 1992: "Electron Microscopy 92. Proceedings of the 10th European Congress on Electron Microscopy," Granada, 7–11 September 1992 (Ríos, A., Arias, J. M., Megías-Megías, L., and López-Galindo, A. (Vol. I), López-Galindo, A., and Rodríguez-García, M. I. (Vol. II), and Megías-Megías, L., Rodríguez-García, M. I., Ríos, A., and Arias, J. M. (Vol. III), eds.; Secretariado de Publicaciones de la Universidad de Granada, Granada), 3 vols.

Beijing, 1992: "Electron Microscopy I and II. 5th Asia-Pacific Electron Microscopy Conference," Beijing, 2–6 August 1992 (Kuo, K. H., and Zhai, Z. H., eds.; World Scientific, Singapore, River Edge, NJ, London, and Hong Kong), 2 vols.

EMSA 50: "Proceedings 50th Annual Meeting Electron Microscopy Society of America, 27th Annual Meeting Microbeam Analysis Society, Nineteenth Annual Meeting Microscopical Society of Canada/Société de Microscopie du Canada," Boston, MA, 16–21 August 1992 (Bailey, G. W., Bentley, J., and Small, J. A., eds.; San Francisco Press, San Francisco, 1992), 2 vols.

JSEM 48: 48th Scientific Meeting of the Japanese Society of Electron Microscopy, Makuhari Messe, Chiba, 2–4 June 1992; *J. Electron Microsc.* **41** (1992), 277–320

SFME 1992: Rouen–Mont Saint-Aignan, 30 June–3 July 1992

EMSSA 31: University of Natal, Pietermaritzburg, 2–4 December 1992; *Electron Microscopy Society of Southern Africa Proceedings/Elektronenmikroskopievereniging van Suidelike Afrika Verrigtinge* **22** (1992)

1993

EMSA 51: "Proceedings 51st Annual Meeting Electron Microscopy Society of America," Cincinnati, OH, 1–6 August 1993 (Bailey, G. W., and Rieder, C. L., eds.; San Francisco Press, San Francisco, 1993)

EMAG, 1993: "Electron Microscopy and Analysis 1993. Proceedings of EMAG 93," Liverpool, 15–17 September 1993 (Craven, A. J., ed.; Institute of Physics, Bristol, Philadelphia, and New York, 1994); *Conference Series* **138**

DGEM 26: Zürich, 5–11 September 1993 (with the Swiss and Austrian Societies); *Optik* **94** (1993), suppl. 5, or *Eur. J. Cell Biol.* **61** (1993), suppl. 39

JSEM 49: 49th Scientific Meeting of the Japanese Society of Electron Microscopy, International Conference Center, Kobe, 26–28 May 1993; *J. Electron Microsc.* **42** (1993), 244–283

SIME, 1993: "Proceedings Multinational Congress on Electron Microscopy," Parma, 13–17 September 1993; *Microscopia Elettronica* **14** (2) (suppl.)

SFME 1993: Villeurbanne, 29 June–3 July 1993

EMSSA 32: Berg-en-Dal, Kruger National Park, 1–3 December, 1993; *Electron Microscopy Society of Southern Africa Proceedings/Elektronenmikroskopievereniging van Suidelike Afrika Verrigtinge* **23** (1993).

MSC 20: University of Toronto, 3–5 June 1993

"Recent Developments of Electron Microscopy," Proceedings of the Seventh Chinese-Japanese Electron Microscopy Seminar, Zhang Jia Jie, 1–4 November 1993 (Hashimoto, H., Kuo, K. H., Lee, K., and Suzuki, T., eds.; Japanese Society of Electron Microscopy, Tokyo)

1994

Paris 1994: "Electron Microscopy 1994. Proceedings of the 13th International Congress on Electron Microscopy", Paris, 17–22 July 1994 [Jouffrey, B., Colliex, C. Chevalier, J. P., Glas, F., Hawkes, P. W., Hernandez–Verdun, D., Schrevel, J., and Thomas, D. (Vol. 1), Jouffrey, B., Colliex, C. Chevalier, J. P., Glas, F., and Hawkes, P. W. (Vols. 2A and 2B), and Jouffrey, B., Colliex, Hernandez-Verdun, D., Schrevel, J., and Thomas, D. (Vols. 3A and 3B), eds.; Editions de Physique, Les Ulis, 1994]

Tsukuba 1994: "Charged Particle Optics. Proceedings of the Fourth International Conference on Charged Particle Optics," Tsukuba, 3–6 October 1994 (Ura, K., Hibino, M., Komuro, M., Kurashige, M., Kurokawa, S., Matsuo, T., Okayama, S., Shimoyama, H., and Tsuno, K., eds.), *Nuclear Instrum. Meth. Phys. Res.* **A363** (1995), 1–496

MSA 52: "Proceedings 52nd Annual Meeting Microscopy Society of America, 29th Annual Meeting Microbeam Analysis Society," New Orleans, LA, 31 July–5 August 1994 (Bailey, G. W., and Garratt-Reed, A. J., eds.; San Francisco Press, San Francisco, 1994)

JSEM 50: 50th Scientific Meeting of the Japanese Society of Electron Microscopy, Hokutopia, Kita-ku, Tokyo, 25–27 May 1994; *J. Electron Microsc.* **43** (1994), 213–253

Chernogolovka 1994: Proceedings of the 15th Conference on Electron Microscopy, Chernogolovka, May 1994; *Izv. Akad. Nauk (Ser. Fiz.)* or *Bull. Russ. Acad. Sci. (Phys.)* **59** (1995), No. 2.

EMSSA 33: University of Port Elizabeth, Port Elizabeth, 29 November–2 December 1994; *Electron Microscopy Society of Southern Africa Proceedings/Elektronenmikroskopievereniging van Suidelike Afrika Verrigtinge* **24** (1994)

MSC 21: University of Montreal, 12–15 June 1994

1995

MSA 53: "53rd Annual Meeting Microscopy Society of America," Kansas City, KS, 13–16 August 1995; *J. Microsc. Soc. Am. Proceedings, Microscopy and Microanalysis 1995* (Bailey,

G. W., Ellisman, M. H., Henniger, R. A., and Zaluzek, N. J., eds.; Jones & Begell, New York and Boston, 1995).

EMAG 1995: "Electron Microscopy and Analysis 1995. Proceedings of EMAG 95." Birmingham, 12–15 September 1995 (Cherns, D., ed.; Institute of Physics, Bristol and Philadelphia, 1995) Conference Series **147**

DGEM 27: Leipzig 10–15 September 1995; *Optik* **100** (1995) Supplement 6 or *Eur. J. Cell Biol.* **67** (1995) (Suppl. 41)

JSEM 51: "51st Annual Meeting of the Japanese Society of Electron Microscopy," Riihga Royal Hotel, Sakai, Osaka, 24–26 May 1995; *J. Electron Microsc.* **44** (1995), 231–265

SIME 1995: "Atti XX Congresso di Microscopia Elettronica," Rimini, 11–14 September 1995; *Microscopia Elettronica* **16**(2) (suppl.)

SFME 1995: Lausanne (with the Swiss and Belgian Societies), 26–30 June 1995

EMSSA 34: Aventura Resort, Warmbaths, 30 November–1 December 1995; *Electron Microscopy Society of Southern Africa Proceedings/Elektronenmikroskopievereniging van Suidelike Afrika Verrigtinge* **25** (1995)

MSC 22: University of Ottawa, 4–7 June 1995

"Recent Developments of Electron Microscopy," Proceedings of the Eighth Chinese-Japanese Electron Microscopy Seminar, Wuishan 1–3 May 1995 (Hashimoto, H., Kuo, K. H., Lee, K., and Suzuki, T, eds.; Japanese Society of Electron Microscopy, Tokyo)

1996

Dublin 1996: "Electron Microscopy 1996. Proceedings of the 11th European Conference on Electron Microscopy," Dublin, 26–30 August 1996

Hong Kong 1996: "Electron Microscopy 1996. Proceedings of the 6th Asia-Pacific Congress on Electron Microscopy," Hong Kong, 1–5 July, 1996

MSA 54: "54th Annual Meeting Microscopy Society of America, Twenty-third Annual Meeting Microscopical Society of Canada/Société de Microscopie du Canada, 31st Annual Meeting Microbeam Analysis Society," Minneapolis, MN, 11–15 August, 1996

SFμ 1996: Rennes, 24–28 June

1997

MSA 55: "55th Annual Meeting Microscopy Society of America," Cleveland, OH

EMAG 1997: "Electron Microscopy and Analysis 1997. Proceedings of EMAG 97," Cambridge, 1–5 September 1995 (Institute of Physics, Bristol, Philadelphia and New York, 1997/8)

DGEM 28: Regensburg, 7–12 September 1997 (with the Swiss and Austrian Societies)

1998

Cancún 1998: "Electron Microscopy 1998. Proceedings of the 14th International Congress on Electron Microscopy," Cancún, 26 September–2 October 1998

MSA 56: "56th Annual Meeting Microscopy Society of America," Atlanta, GA

1999

MSA 57: "57th Annual Meeting Microscopy Society of America," Portland, OR

2000

MSA 58: "58th Annual Meeting Microscopy Society of America," Philadelphia, PA

2001

MSA 59: "59th Annual Meeting Microscopy Society of America," Long Beach, CA

INDEX

A

AEG, *see* Allgemeine Electrizitäts-Gesellschaft
AEI Research Laboratories
  commercial microscope production, 453–457, 463–464
  Corinth, 275, 498
  electron diffraction camera production, 490
  electron microprobe production, 464–465
  EM 6B, 475–477
  EM 6G, 480
  EM 7, 492–493, 531
  EM 801, 480–481
  EMMA-4, 490–491
  scanning electron microscope research, 468–469
  withdrawal from microscope production, 469, 530–532
Africa, *see* Southern Africa
Allgemeine Electrizitäts-Gesellschaft (AEG)
  EM 5 development, 426–427
  EM 6, 428
  EM 7, 433, 438
  EM 8-I, 438–439
  EM 8-II, 445, 447
  World War II impact, 428
  Zeiss collaboration, 428, 433
Analytical electron microscope, commercial development in Europe, 490–492
Anticontamination devices, Japanese contributions, 697–698
Asian-Pacific electron microscopy conference
  first meeting, 605–606
  history of meetings, 607, 629
Atomic imaging
  history at international meetings, 402–403, 408
  Japanese contributions, 615–618, 623, 625–626
  processing by fast Fourier transform, 622–623
Australia
  Australian Society for Electron Microscopy
    awards, 46–47
    origin, 39–40, 46
    responsibilities, 46
  contributions to electron microscopy
    early history, 40–42, 52
    instrumentation, 50–51
    theory and techniques, 51–52
  electron microscopy centers, 47–50
  National Committee for Electron Microscopy
    initiatives, 44, 46
    national conferences, 42–44
    newsletter, 46
    origin, 42
  research funding sources, 39

Austria
  Austrian Society for Electron Microscopy
    meetings and symposia, 55–56
    origin, 55
    publications, 57
  contributions to electron microscopy, see
    Glaser, Walter

B

Bacteria
  flagella studies, 280, 742
  Japanese contribution to ultrastructural
    studies, 735, 741–744
Belgium
  Claude, Nobel Prize, 76–77
  contributions to electron microscopy, see
    Marton, Ladislaus
  instrumentation
    Antwerp facilities, 77–78
    distribution, 76–77
    history, 72–74
  Society for Electron Microscopy
    meetings and conferences, 75
    origin, 74–75
Boersch ray path
  Boersch effect, 793
  development, 139
Burton, Eli
  Burton Society of Electron
    Microscopy, 87
  contributions to electron microscopy,
    79–82
  Kohl's influence, 80–81

C

Cambridge Instruments
  computer control of microscopes,
    510–511
  electron microprobe production, 464–466
  scanning electron microscope production,
    469–470, 505–506, 509
Cameca
  Camebax instrument, 504
  electron microprobe production,
    464–465, 519
  MEB 07, 471
  niche marketing, 532–533
Camscan
  electron microscope development,
    470–471, 503–504
  niche marketing, 532–533

Canada
  Burton Society of Electron
    Microscopy, 87
  contributions to electron microscopy, see
    also Burton, Eli; Hall, Cecil
  commercial contributions, 85–86
  transmission electron microscope
    prototype
      design, 82–84
      resolution, 84–85
  Microscopical Society of Canada
    founding members, 91
    meetings, 88, 90
    membership, 90
    origin, 88
    publications, 88
CAPSEM see Committee of Asia-Pacific
  Societies for Electron Microscopy
CESEM see Committee of European
  Societies for Electron Microscopy
China, commercial production of electron
  microscopes
  cultural revolution impact on science,
    819–823
  early history, 805–813
  electron opticians, listing, 846–847
  first prototype, 805–808
  market, 840–841, 843
  Nanjing Jiangnan Optical Factory
    DXS-3 SEM 836
    DXT-10, 838
    XD-201, 838
    XD-301, 838
    X-ray microanalysis, 834–837
  Scientific Instrument Factory (KYKY)
    AMRAY union, 831–833
    1000B SEM, 833
    DX-2, 817–819
    DX-3, 825–828
    DX-4, 828–830
    DX-5, 828, 830–831
    electrolytic tank production, 816
    first electron emission microscope,
      815–816
    impact of politics, 814–817, 819–825
    KYKY 2000, 833
    KYKY 3000, 833–834
    origins, 814
    sales, 841
  Shanghai Electron Optical Institute,
    transmission electron microscopes,
    838–839

INDEX 877

summary of manufacturing activities, 843–846
XD-100, 810–812
Committee of Asia-Pacific Societies for Electron Microscopy (CAPSEM)
conferences, 31–32, 597, 629
international federation, 629–631
origin, 31
Committee of European Societies for Electron Microscopy (CESEM)
objectives, 29, 31
origin, 28–29
Convergent-beam electron diffraction
apparatus, 587–588
Gottfried's research, 585–591, 593–594
intensity minima, 593–594
Japanese contributions, 766
Kikuchi lines, 585, 591, 593
Kossel effect, 585
specimen preparation, 590
Critical-point drying method, Japanese contributions, 783
Cryoelectron microscopy, Japanese contributions, 739
Crystal growth, Japanese contributions, 603–604, 608–610, 698–700, 751–752
CSF
electron microscope development, 430, 437, 504
M IX, 437, 446

D

Diffraction contrast, Japanese contributions, 601–602

E

Electron diffraction, *see* Convergent-beam electron diffraction
Electron diffraction camera
commercial development in Europe, 427, 441, 489–490, 500–501, 519–520
Japanese contributions, 599–600
Electron microprobe
commercial development in Europe, 460–461, 464–466
Southern Africa research, 341–342
Electron microscope, *see specific microscopes*

Electron Microscope Society of America (EMSA)
formation, 4, 357
growth, 358–360
international relationships, 358–359, 364, 368
meetings, 357–358, 374–375
presidents, 357, 374–375
publications, 359–361, 851–875
Electron optics, history at international congresses, 405–412
Electrostatic electron microscope
development at Japanese universities, 247–248, 251–256, 263–267
invention and development, 137–141
transmission electron microscope, commercial production, 141
Electrostatic lens
Einzel lens, 139
Japanese contributions, 685–688, 705–707
patent, 137
Electrotechnical Laboratory, *see* Japan
Embedding resin, French research contributions, 98–99
Emission electron microscope, commercial development in Europe, 486, 489–490
EMSA, *see* Electron Microscope Society of America
Environmental scanning electron microscope (ESEM), development, 510
ESEM *see* Environmental scanning electron microscope
Europe, electron microscope production
chronology of microscope manufacture, 538–571
manufacturers and model numbers, 535–538
production and use trends, 571–573
European Regional Meetings
biological research history, 387–391
timetable, 386

F

Field-emission gun
French development, 118
holography application, 709–712
Japanese contributions, 710–713
scanning electron microscopes, 143–144, 712–713

878 INDEX

France
  electron microscopy
    commercial development, 484
    early pioneers, 93–95, 97–98, 101–102, 104–106
    facilities, 96
    first application for financial support, 102–104
    historical contributions
      biology, 93–100
      electron guns, 118
      electron probe microanalysis, 120–127
      high-voltage electron microscopy, 112–115, 117–120
      metallurgy, 106–110
      radiation defects, 118–119
      specimen holders, 117–118
      thin films, 110–112, 120
  French Society for Electron Microscopy
    meetings, 128
    origin, 128
    publications, 128–129, 861–876
Freezing, rapid technique, Japanese contributions, 740–741, 783–784

G

Gabor, Denis, holography invention, 139, 398
Gas evaporation method, Japanese contributions, 785–786
German Democratic Republic
  contributions to electron microscopy
    biology, 175
    materials science, 175, 178
    metallurgy, 175
    microscope production, 171–172, 175, 448–449
  international activity, 179–180
  monographs, 178–179
  national conferences, 178, 180
  resignation from German Society for Electron Microscopy, 157
  societies, 178
German Society for Electron Microscopy
  board members
    original members, 151, 153
    table, 165–168, 802–803
  constitution, 153, 155, 161–165

dues, 158–159
Ernst Ruska Prize Foundation, 158
honorary members, 170, 803
meetings
  first conference, topics, 151–152
  international conferences, 155–157
  locations, 802
  minutes of first meeting, 151, 159–160
origins, 150–153, 155–159, 801
publications, 157–158, 852–876
working groups, 157
Germany
  contributions to electron microscopy, *see also* Knoll, Max; Ruska, Ernst
    electrostatic electron microscope, 137–141
    field emission microscope, 143–144
    magnetic electron microscope, 132–137
    scanning electron microscope, 142–143
  electron microscopy growth, 149–150
Glaser, Walter
  contributions to electron microscopy, 59–66, 795
  Grundlagen der electronenoptik, 59–64, 66
Golgi apparatus, discovery, 97

H

Hall, Cecil, contributions to electron microscopy, 81–82
Hashimoto, Hatsujiro
  career
    Cambridge, 607–611
    Hiroshima, 598–599
    Kyoto, 599–604, 612–618
    Osaka, 619–620, 622–623
  reflections of historic events in electron microscopy, 597, 605–607, 626–631
High-resolution electron microscopy
  Japanese contributions, 759–765
  United States contributions, 367, 369
High-voltage electron microscopy
  commercial production in Europe, 440–441, 492–494
  French contributions, 112–115, 117–120
  history of use, 394–395
  Japanese contributions, 612–615, 700–704, 736–737, 754, 756–759
  United States contributions, 366, 369

INDEX 879

Hiroshima University
  electron diffraction research, 598–599
  nuclear attack, 598–599
Hitatchi
  HF-2000, 716
  HFS-2, 712
  HS series, 688
  HU-1, 653–654
  HU-2, 654–655
  research
    anticontamination devices, 697–698
    electron energy loss, 691–692
    pointed cathode, 690–691
  S-900, 713
  World War II impact, 655
Holography
  electron microscopy application, 140–141, 181, 183, 398, 410, 709–712, 753–754, 800–801
  invention, 139, 398
  Japanese contributions, 709–712, 753–754
Hungary
  contributions to electron microscopy, *see also* Marton, Ladislaus
    holography, 181, 183
    pioneers, 183
    history of electron microscopy, 184
  Hungarian Group for Electron Microscopy
    future developments, 191
    history, 181, 184
    international congresses, 184–186, 189

I

ICEM *see* International Congresses for Electron Microscopy
ICSU, *see* International Council of Scientific Unions
IFSEM *see* International Federation of Societies for Electron Microscopy
Image contrast, discussion at international meetings, 397–398
Immunoelectron microscopy, Japanese contributions, 740
International Assembly for Electron Microscopy, formation, 6
International Committee for Electron Microscopy, record of meetings
  establishing the committee, 6, 12–16

International Congresses for Electron Microscopy (ICEM)
  biological research history, 387–391
  content of meetings, 33
  electron optics research history, 405–412
  history, 386–387
  instrument exhibitions, 34, 394–395
  materials science research history
    applications sessions, 399–400
    atomic resolution, 402–403, 408
    commercial developments, 394–395
    holography, 398
    home-built projects, 395–397
    image contrast, 397–398
    workshops, 401
  oversight, 32–33
  proceedings, 403, 406, 408, 411
  value of attendance, 385
International Council of Scientific Unions (ICSU), association with IFSEM 3, 5–7, 9–10
International Federation of Societies for Electron Microscopy (IFSEM)
  congresses, *see* International Congresses for Electron Microscopy
  constitution, 8–10, 18–19, 21, 32
  Executive Committee, 22–23
  General Assembly, 22, 33
  General Secretary, 23, 25
  growth, 9, 25, 27, 34–35
  industry relations, 33–34
  international congresses, 32–33
  International Council of Scientific Unions, association, 3, 5–7, 9–10
  Joint Commission, rise and fall, 5–7
  membership, 30
  member societies, 25, 27, 30, 34
  objectives, 21
  origin, 3–5, 155–156
  presidents and secretaries, 9, 20, 22, 24
  proposal by International Committee, 7, 16–18
  regional committees, 28–29, 31–32
  responsibilities, 10–11
Ion etching, Japanese contributions, 784
Italy
  contributions to electron microscopy
    biomedical sciences, 202–204
    fixation and staining, 203

880   INDEX

Italy (*continued*)
   specimen stages, 205–206, 208, 213
   superconductors, 213
   electron microscopy laboratories
      Ispra, 199
      Rome, 193–194, 202, 212
      University of Bologna laboratory
         Cambridge connection, 204–206, 209
         diploma for technical experts, 211–212)
         founding, 195–197
         internal collaborative projects, 201–202
         LAMEL materials laboratory, 209–210
         microscope acquisition, 196–198, 200
         objectives, 194
         publications, 210
         recent organization of departments, 210–211
         solid-state physics group, 199–200
         support service for external users, 200–202
         teaching activity, 210, 213
   industry-related research, 202
   Italian Society of Electron Microscopy (SIME)
      founding, 198–199
      international congresses, 208–209, 213
      publications, 861–876

J

Japan
   commercial production of microscopes
      companies, *see specific companies*
      history, 217, 226, 243, 257–258, 662–664
   contributions to electron microscopy, *see also specific universities*
      accelerating voltage elevation, 707–709
      anticontamination devices, 697–698
      atomic imaging, 615–618, 623, 625–626
      bacterial ultrastructure, 735, 741–744
      biological specimen preparation, 725, 727–729, 733
      convergent-beam electron diffraction, 766
      critical-point drying method, 783
      cryoelectron microscopy, 739
      crystal growth experiments, 603–604, 608–610, 698–700, 751–752
      cytology, 682
      diffraction contrast experiments, 601–602
      electron diffraction, 599–600, 693–694
      electron lens aberrations, 685–688, 705–707
      electron microscope development, 599–601, 679–680
      electropolishing methods, 750–751
      field-emission gun, 710–713
      gas evaporation method, 785–786
      high-resolution electron microscopy, 759–765
      high-voltage electron microscopy, 612–615, 700–704, 736–737, 754, 756–759
      holography, 709–712, 753–754
      immunoelectron microscopy, 740
      ion etching, 784
      materials science, 619–620, 622, 749–768
      microbiology, 681–682
      microcharacterization, 766–768
      microgrid development, 775–778
      nanoprobe analytical electron microscope, 713–716
      rapid freezing techniques, 740–741, 783–784
      replicas, 774–775
      scanning electron microscopy, 704–705, 712–713
      scanning tunneling microscopy, 719, 721
      specimen manipulation, 694–696, 773–774
      spin-polarized SEM 716–718
      thin films, 611, 614
      thinning methods, 750–751, 781–783
      ultrahigh-vacuum microscope, 716
      ultramicrotomy, 778–781
      ultrathin sectioning, 725
      virus ultrastructure, 682–683, 735–741
      wet-cell microscopy, 786–787
      X-ray microanalysis, 696
      yeast ultrastructure, 744–745
   Electrotechnical Laboratory
      electron lens contributions, 258

INDEX 881

instrumentation, 257–258
specimen temperature research, 258, 260–261
history of early research
   cradle period, 232–234
   improvement and reformation period, 235–236, 239–241
   period of application and newcomers in manufacturing, 241–243
Japanese Society of Electron Microscopy (JESM)
   founding, 217, 723
   historical activities in electron microscopy, 223, 724
   honorary members, 225
   membership, 225
   papers submitted to meetings, growth, 724–725
   presidents, 224, 227
   publications, 225–226, 627, 852–877
   Ronbun Prize, 222, 226
   Seto Prize, 218–221, 226, 732
   symposia themes, history, 730–731
Japanese Society for the Promotion of Science, 37th Subcommittee
   activities reported in presented papers
     aberration calculations in electron optics, 232, 235, 243
     microscope construction, 233–236, 238–239
     stabilization of high tension, 233, 235
   establishment
     background, 217, 227–228
     committee members, 229
     first meeting, 229–231, 723
     prospectus, 229–230
     research subjects, 230
   industrial research, 240–241, 243
   open-invitation to member's laboratories, 236, 239
   policies, 231–232
   training of young researchers, 240
   translation of von Ardenne's book, 240
   World War II impact, 234, 241–242
JEOL
   DA-1, 660
   founding, 659–660
   JE-100B, 714
   JEM-5 reflection electron microscope, 694
   JEM-T1, 661–662

JESM, *see* Japanese Society of Electron Microscopy
Jugoslav Iskra, microscope production, 484–485

K

Knoll, Max, electron microscope invention, 132–135, 142
Kyoto University
   atomic imaging
     dark-field images, 616–618, 623
     magnification and contrast, 615–616
   crystal growth experiments, 603–604, 608–610
   cytology, 682
   diffraction contrast experiments, 601–602
   electron diffraction camera development, 599–600
   electron microscope development, 599–601, 679–680
   high-voltage electron microscopy, 612–615
   microbiology, 681–682
   thin films, 611, 614
   virology, 682–683

L

Le Poole, Jan
   commercial collaboration, 278–279, 287, 289
   contributions to electron microscopy
     metallic and semiconductor materials science, 289–292
     solid catalysts, 294
     vapor-deposited metal films, 293
   microscope development, 273–275, 284, 288, 395, 433–435

M

Magnetic electron microscope
   commercial production, 136–137
   development at Japanese universities, 245–247, 251–252
   invention, 132–135
   resolution, 134–135
Marton, Ladislaus
   electron microscope development, 67–70, 102

882 INDEX

Marton, Ladislaus (*continued*)
 heavy metal stain development, 69, 136, 183
 improvements in electron microscopy, 70–71, 183
Metropolitan-Vickers Electrical Company
 EM 2, 431–433
 EM 3, 435, 441
 EM 3A, 450–452
 EM 4, 442–443
 EM 5, 440–441
 microscope production history, 430–433
Microgrid, Japanese contributions, 775–778
Microtome, *see* Ultramicrotomy
Mills sample holder, development, 50
Möllenstedt, Gottfried
 biprism, 800
 convergent-beam electron diffraction research, 585–591, 593–594
Multislice theory, development, 51

N

Netherlands
 contributions to electron microscopy research
  biological specimens, 294–297
  cryofixation, 296
  lenses, 288–289
  Philips innovations, 289
  software, 297
 Delft University of Technology
  bacterial flagella studies, 280
  Le Poole microscope, 273–275, 288
  microscope construction, 275, 288
  organization of electron microscopy research, 273–274
  research staff, 279
  specimen preparation, 279–281
  yeast research, 271–273, 276, 281, 283–284
 Dutch Society for Electron Microscopy, 298–299
 growth of electron microscopy, 281–284, 295–296
 World War II impact on research, 276–278

O

Okayama University, atomic imaging, 625–626

OPL, commercial microscope production, 460
Osaka University
 atomic image processing by fast Fourier transform, 622–623
 high-voltage electron microscopy, 703–704
 Mark series microscope development, 263–267
 materials science research
  stacking faults, 619–620
  vacancies, 620, 622

P

Patent
 electron microscope, 134
 electrostatic lens, 137
Philips
 computer control of microscopes, 511–512, 514
 EM 75, 443–444
 EM 100, 433–435, 439
 EM 200, 457–458
 EM 201, 485
 EM 300, 480–481, 481, 485
 EM 400, 496–499
 PSEM 500, 502
 SEM 505, 505
 TWIN lens, 498–499, 514

R

Rapid freezing technique, Japanese contributions, 740–741, 783–784
RCA, commercial microscope development, 479
Replica technique
 development, 150, 281, 338, 440, 462
 Japanese contributions, 774–775
 Southern Africa contributions, 338
Resolution
 atomic, *see* Atomic imaging
 improvement over time
  history, 791–801
  scanning electron microscopy, 525, 528
  transmission electron microscopy, 524, 526
 Scherzer's limit, 797
Robinson detector, development, 51

Ruska, Ernst
  career, 132–137, 416–418, 639, 792–793
  electron microscope invention and development, 11–12, 65, 132–135, 416, 462–463
  Ernst Ruska Prize, 158
  magnetic lens development, 134
  Nobel Prize, 131, 145, 159, 794
  scanning electron microscope, early thoughts, 635–638
Russia
  commercial microscope development at Sumy, 428–429, 437–438, 454, 460, 470, 472–473, 482–484, 492, 494–496, 499–500, 502–503, 510, 516–517, 532
  electron diffraction camera production, 489–490, 500–501
  electron microprobe production, 465
  pioneers, 523, 574

S

Scandinavia
  contributions to electron microscopy
    cell ultrastructure, 307–308
    enzyme cytochemistry, 308–309
    materials science, 310–311
    methacrylate embedding method, 306
    ultramicrotome, 306, 310
  growth of microscopy, 306–312, 314
  Scandinavian Society for Electron Microscopy
    board, 316–317
    courses, 319
    formation, 315
    growth, 315–316
    meetings, 317–319
    scientific impact, 319–320
    topics of research, 319
  Siegbahn electron microscope
    characteristics, 303–304
    commercial production, 304–306
    development, 301–302
Scanning electron microscope
  commercial development in Europe, 466–471, 501–506, 509–513
  development, 142–143, 466–469, 638–639, 643–644
  Japanese contributions, 704–705, 712–713
  resolution improvement over time, 525, 528

Scanning transmission electron microscope (STEM), commercial development in Europe, 506–508, 519–520, 644–645
Scanning tunneling electron microscopy
  invention, 799
  Japanese contributions, 719, 721
Shimadzu Corporation
  early research interests
    bacteriology, 669
    diatom taxonomy, 670
    gene imaging, 669
    metallurgy, 669
    photoemulsion, 668–669
  SM-1, 665–666, 670
  SM-1A, 666–667
  SM-1B, 667
Sickle cell anemia, French research contributions, 95–96
Siegbahn, Manne, see Scandinavia
Siemens
  electrostatic astigmatism corrector, 446
  Elmiskop 1, 454–455, 462–463
  Elmiskop 2, 443
  Elmiskop 51, 482
  Elmiskop 101, 480
  Elmiskop 102, 495
  Elmiskop CT, 150, 495–496, 529
  prototype electron microscope development, 418–422
  ST 100F, 507, 531, 791
  Übermikroskop, 423–429, 436
  ÜM 100, 436, 446
  withdrawal from microscope production, 530–531, 791
  World War II impact, 427–428
Southern Africa
  contributions to electron microscopy
    andrology, 333
    botany, 331–332
    ceramics, 341
    dentistry, 332
    earth science, 341–342
    electron microprobes, 341–342
    embryology, 332–333, 335
    instrumentation, 330–331
    medicine, 332–336
    physical sciences, 336–341
    plastic deformation, 338–341
    replica technique, 338
    semiconductors, 337

Southern Africa (*continued*)
  specimen preparation, 331
  thin films, 336–337
  virology, 334–335
electron microscopy facilities, 323–325, 343
Electron Microscopy Society of Southern Africa
  conferences, 325–327
  delegates to international meetings, 329
  formation, 323
  international federation, 325
  presidents, 330
  prizes, 327–328
  publications, 326–327
  workshops, 328–329
  future of electron microscopy, 343–344
  language barrier, 325
Spin-polarized scanning electron microscopy, Japanese contributions, 716–718
STEM *see* Scanning transmission electron microscope
Sumy, *see* Russia

T

Temperature, rise in specimen due to electron bombardment, 258, 260–261
Tesla, commercial microscope production, 452, 475, 495, 503, 506, 512, 517–518, 532–534
Thin films
  French contributions, 110–112, 120
  Japanese contributions, 611, 614
  Southern Africa contributions, 336–337
Tohoku University
  electrostatic electron microscope development, 247–248
  magnetic-type emission microscope development, 245–247
  pointed cathode studies, 689–690
Toshiba Corporation
  EUL series, 676–677
  Toshiba No. 1, 673, 675
  Toshiba No. 2, 675–676

Transmission electron microscope
  commercial development in Europe, 472–486, 494–499, 513–519, 639–641, 645–647
  field emission gun instrument, 402
  invention, 11–12, 65, 132–135
  resolution improvement over time, 144, 402, 524, 526
Trüb Täuber Company, commercial microscope production, 453–454, 486, 488

U

Ultrahigh-vacuum electron microscope
  Japanese contributions, 716
  Vacuum Generators Company
    niche marketing, 532–533
    scanning electron microscope development, 508, 519
Ultramicrotomy
  instrument development, 386, 461–462
  Japanese contributions, 725, 778–781
  Scandinavian contributions, 306, 310
United States
  Electron Microscopy Society of America, *see* Electron Microscope Society of America
  history of electron microscopy
    commercial microscopes, 354–356, 360–362, 365–366
    computer technology, 368–369
    electron optics, 347–348
    high-resolution electron microscopy, 367, 369
    high-voltage electron microscopy, 366, 369
    information sources, 348–352
    materials science, 363–364
    meetings, 348, 357
    microscope construction, 353–354
    research sites, 352–353, 356, 367
    significant events, 375–381
    specimen preparation, 362–363
University of Tokyo, microscope development
  electrostatic electron microscope, 251–256
  magnetic electron microscope, 251–252

INDEX

V

Vacuum Generators Company
　niche marketing, 532–533
　scanning electron microscope
　　development, 508, 519
Virus
　cryoelectron microscopy, 739
　French contribution to ultrastructural
　　studies, 94–95, 99–100
　immunoelectron microscopy, 740
　Japanese contribution to ultrastructural
　　studies, 682–683, 735–741
　rapid-freeze method, 740–741
von Ardenne, Manfred
　impact of World War II, 649–650
　relationship with Ruska, 639–641
　scanning electron microscope
　　development with Siemens, 638–639,
　　643–644
　scanning transmission electron
　　microscope development, 644–645
　translation of works into Japanese, 240,
　　649
　transmission electron microscopy
　　contributions, 639–641, 645–647
von Borries, Budo, contributions to
　electron microscopy, 794–796

W

Wet-cell microscopy, Japanese
　contributions, 786–787

WF Company, commercial microscope
　production, 429, 452, 458, 477–479,
　482–483

X

X-ray microanalysis
　development in China, 834–837
　development in Japan, 696, 704
X-ray projection microscope, invention, 143

Y

Yeast
　Dutch electron microscopy research,
　　271–273, 276, 281, 283–284
　Japanese contribution to ultrastructure,
　　744–745

Z

Zeiss
　collaboration with AEG, 428, 433
　computer control of microscopes, 516
　EF4, 472–473
　EF5, 473
　EF6, 473–474, 488–489
　electron microscope development, 448,
　　458
　EM 9, 473–474, 486
　EM 10, 494–495
　EM 109, 499
　EM 902, 515–516
　EM 912, 516
　energy filter incorporation, 792